ATOMIC AND MOLECULAR PROCESSES

PURE AND APPLIED PHYSICS

A SERIES OF MONOGRAPHS AND TEXTBOOKS

CONSULTING EDITOR

H. S. W. MASSEY

University College, London, England

Volume 1. F. H. Field and J. L. Franklin, Electron Impact Phenomena and the Properties of Gaseous Ions. 1957

Volume 2. H. Koppermann, Nuclear Moments. English Version Prepared from the Second German Edition by E. E. Schneider. 1958

Volume 3. Walter E. Thirring, Principles of Quantum Electrodynamics. Translated from the German by J. Bernstein. With Corrections and Additions by Walter E. Thirring. 1958

Volume 4. U. Fano and G. Racah, Irreducible Tensorial Sets. 1959

Volume 5. E. P. Wigner, Group Theory and Its Application to the Quantum Mechanics of Atomic Spectra. Expanded and Improved Edition. Translated from the German by J. J. Griffin. 1959

Volume 6. J. Irving and N. Mullineux, Mathematics in Physics and Engineering. 1959

Volume 7. Karl F. Herzfeld and Theodore A. Litovitz, Absorption and Dispersion of Ultrasonic Waves. 1959

Volume 8. Leon Brillouin, Wave Propagation and Group Velocity. 1960

Volume 9. Fay Ajzenberg-Selove (ed.), Nuclear Spectroscopy. Parts A and B. 1960

Volume 10. D. R. Bates (ed.), Quantum Theory. In three volumes. 1961

Volume 11. D. J. Thouless, The Quantum Mechanics of Many-Body Systems. 1961

Volume 12. W. S. C. Williams, An Introduction to Elementary Particles. 1961

Volume 13. D. R. Bates (ed.), Atomic and Molecular Processes. 1962

Volume 14. Amos de-Shalit and Igal Talmi, Nuclear Shell Theory. In press

In preparation

Walter H. Barkas. Nuclear Research Emulsions. In two volumes.

ATOMIC and MOLECULAR PROCESSES

Edited by

D. R. BATES

Department of Applied Mathematics
The Queen's University of Belfast
Belfast, Northern Ireland

1962

Academic Press New York and London

ACADEMIC PRESS INC.
111 FIFTH AVENUE
NEW YORK 3, N. Y.

United Kingdom Edition
Published by
ACADEMIC PRESS INC. (LONDON) LTD.
Berkeley Square House, London, W. 1

Library of Congress Catalog Card Number 62-13122

PRINTED IN THE UNITED STATES OF AMERICA

Contributors

S. K. Allison, *Enrico Fermi Institute for Nuclear Studies*, University of Chicago, Chicago, Illinois

M. Baranger, *Service de Physique Mathématique*, Centre d'Études Nucléaires de Saclay, Gif-sur-Yvette (Seine-et-Oise), France

D. R. Bates, *Department of Applied Mathematics*, The Queen's University of Belfast, Belfast, Northern Ireland

L. M. Branscomb, *The National Bureau of Standards*, Washington, D. C.

J. D. Craggs, *Department of Electrical Engineering*, University of Liverpool, Liverpool, England

R. W. Crompton, *C.S.I.R.O.*, Adelaide, South Australia

A. Dalgarno, *Department of Applied Mathematics*, The Queen's University of Belfast, Belfast, Northern Ireland

R. W. Ditchburn, *Physics Department*, University of Reading, Reading, England

W. L. Fite, *John Jay Laboratory for Pure and Applied Science*, General Dynamics Corp., San Diego, California

M. Garcia-Munoz, *Enrico Fermi Institute for Nuclear Studies*, University of Chicago, Chicago, Illinois

R. H. Garstang, *University of London Observatory*, Mill Hill Park, London, England

H. R. Griem, *U. S. Naval Research Laboratory*, Washington, D. C.

J. B. Hasted, *Department of Physics*, University College, London, England.

L. G. H. Huxley, *The National University*, Canberra, A. C. T., Australia

A. C. Kolb, *U. S. Naval Research Laboratory*, Washington, D. C.

J. D. Lambert, *Physical Chemistry Laboratories*, University of Oxford, Oxford, England

E. A. Mason, *Institute for Molecular Physics*, University of Maryland, College Park, Maryland

B. L. Moiseiwitsch, *Department of Applied Mathematics*, The Queen's University of Belfast, Belfast, Northern Ireland

R. W. Nicholls, *Department of Physics*, University of Western Ontario, London, Ontario, Canada

U. Öpik, *Physics Department*, University of Reading, Reading, England

J. C. Polanyi, *Department of Chemistry*, University of Toronto, Toronto, Canada

A. N. Prasad, *Department of Electrical Engineering*, The University of Liverpool, Liverpool, England

J. Sayers, *Department of Electronic Physics*, University of Birmingham, Birmingham, England

M. J. Seaton, *Department of Physics*, University College, London, England

A. L. Stewart, *Department of Applied Mathematics*, The Queen's University of Belfast, Belfast, Northern Ireland

J. T. Vanderslice, *Institute for Molecular Physics*, University of Maryland, College Park, Maryland

Preface

This compilation, which is designed primarily as a reference book for research scientists, is concerned with radiative and collisional processes involving atoms or molecules. It provides surveys covering the following topics: forbidden and allowed lines and bands, photoionization, photodetachment; recombination, attachment; elastic and inelastic scattering of electrons, energy loss by slow electrons; collision broadening of spectral features; encounters between atomic systems including range, energy loss, excitation, ionization, detachment, charge transfer, elastic scattering, mobility, diffusion, relaxation in gases, and chemical reactions. A chapter is devoted to the use of high temperature shock waves and accounts are given of the other main experimental methods. The relevant theoretical work is also described, detailed mathematics being avoided as far as possible.

The main emphasis is placed on the developments which have taken place in the past decade, that is, since the publication of the first edition of the great treatise by Massey and Burhop *Electronic and Ionic Impact Phenomena*. These developments were stimulated by the growth of interest in such fields as space science, astrophysics, and plasma physics. They were rendered possible by remarkable technical advances which have benefited directly not only experimentalists but also (through fast digital computing) theorists.

Thanks must be given to the staff of the Academic Press for their determined efforts to ensure that a thick volume reviewing work done up to almost the end of 1961 should appear early in 1962.

D.R.B.

Department of Applied Mathematics
The Queen's University of Belfast
Belfast, Northern Ireland

February 1962

Contents

4. Photodetachment

L. M. Branscomb

5. High-Temperature Shock Waves

A. C. Kolb and H. R. Griem

6. Attachment and Ionization Coefficients

A. N. Prasad and J. D. Craggs

7. Electronic Recombination

D. R. Bates and A. Dalgarno

1.

Forbidden Transitions

R. H. Garstang

1 *Introduction*

1.1 Discovery and Importance of Forbidden Transitions

Early in the history of spectroscopy empirical rules were developed to enable the prediction of spectral lines to be accomplished from the energy levels of the atoms. These rules became known as selection rules; they enabled one to select, from all the possible transitions between pairs of energy levels, those which might be expected to be observable. These selection rules were subsequently justified by quantum mechanics. As the subject progressed some lines were discovered which violated the selection rules, and such lines became known as forbidden lines. The first forbidden transitions to be recognized as such were the $^2D - {}^2S$ transitions in the alkali metals, observed by Datta in 1922. Other lines observed in the laboratory were the $6^3P_2 - 6^1S_0$ line of mercury (Rayleigh, 1927), the mercury line $6^3P_0 - 6^1S_0$ by Fukuda (1926), and the auroral line $2p^4\,{}^1S_0 - 2p^4\,{}^1D_2$ of oxygen (McLennan and Shrum, 1925). The study of forbidden lines received its greatest stimulus when Bowen (1928) identified many of the strongest lines in the spectra of gaseous nebulae as being due to forbidden transitions in O II, O III, and N II. Many more forbidden lines were discovered subsequently in celestial objects, and a few were produced in laboratory sources. The appearance of the forbidden lines in celestial objects indicates the presence of unusual physical conditions, particularly low densities, when the frequency of collisional de-excitation of atomic levels is much reduced, and radiative de-excitation by forbidden transitions becomes important. Observations of forbidden lines are thus of importance in astrophysics because of the information which they can yield on the conditions in their source. A number of forbidden transitions were found in molecular spectra from about 1930 onwards. Van Vleck (1934) established the nature of the atmospheric absorption bands of oxygen, Vegard and Kaplan studied forbidden bands in N_2, and others were identified. The occurrence of such bands in the telluric spectrum has led to much of the interest in their study.

1.2 Terminology

A number of definitions of *forbidden* transitions have been proposed. The traditional definition divides spectrum lines into two groups, those which satisfy all the selection rules are termed *permitted lines*, all the others are called *forbidden lines*. This definition is not entirely adequate, for many of the selection rules are only approximate, and the strengths of the forbidden lines vary greatly with atomic number for atoms of the same electronic structure. An alternative definition calls lines forbidden if the probability of their occurrence is very small compared with the probability of the strongest transitions between levels of similar total quantum numbers (Mrozowski, 1944). Other authors refer to those lines which are due to magnetic dipole or electric quadrupole radiation as multipole radiation (Rubinowicz, 1949). Notwithstanding these definitions, a practical terminology has arisen which is described below and used in this chapter.

In atomic spectroscopy, all transitions which violate the rigorous selection rules for electric dipole radiation in free atoms are termed *forbidden transitions*. This category includes all magnetic dipole and electric quadrupole transitions, two-quantum processes, electric dipole radiation enforced by perturbations external to the atom, and electric dipole radiation caused by the atomic nucleus. Electric dipole transitions which violate only certain approximate selection rules (e.g., $4s^2\,{}^1S_0 - 4s4p\,{}^3P_1$ in Ca I, which violates the rule $\Delta S = 0$) are not called forbidden transitions.

In molecular spectroscopy all transitions which violate any selection rules, whether rigorous or not, are called forbidden. Thus, intercombinations (e.g., ${}^3\Pi - {}^1\Sigma$) are included among forbidden molecular transitions. In polyatomic molecules transitions made possible by vibronic interactions are also included among forbidden transitions.

In atomic spectroscopy forbidden lines are denoted by square brackets, e.g., the auroral line is described as occurring in the spectrum of [O I].

2 *Forbidden Lines in Atomic Spectra*

2.1 Introduction

In accordance with the terminology discussed earlier, all transitions which violate the rigorous selection rules for electric dipole radiation in free atoms are termed forbidden transitions. The selection rules for electric dipole, magnetic dipole, and electric quadrupole radiation are listed in Table I. The notation used is the standard one: L, S, and J are, respectively, the orbital, spin, and total angular momenta of the atomic electrons, M is the magnetic quantum number (component of J) and n is the principal quantum number. The parity is $(-1)^{\Sigma l_i}$ where l_i is the azimuthal quantum number of the ith electron. The selection rules (1), (2), and (3) are rigorous in the absence of nuclear perturbations and two-quantum processes. Rule (4) holds only when configuration interaction is negligible, and rules (5) and (6) hold only for LS-coupling. Forbidden lines may arise from several causes.

(a) The rigorous selection rules, (1)-(3), may be violated for electric dipole radiation, but allowed for magnetic dipole or electric quadrupole radiation.

TABLE I

SELECTION RULES IN ATOMIC SPECTRA

Electric dipole	Magnetic dipole	Electric quadrupole
(1) $\Delta J = 0, \pm 1$	$\Delta J = 0, \pm 1$	$\Delta J = 0, \pm 1, \pm 2$
$(0 \nleftrightarrow 0)$	$(0 \nleftrightarrow 0)$	$(0 \nleftrightarrow 0, \frac{1}{2} \nleftrightarrow \frac{1}{2}, 0 \nleftrightarrow 1)$
(2) $\Delta M = 0, \pm 1$	$\Delta M = 0, \pm 1$	$\Delta M = 0, \pm 1, \pm 2$
(3) Parity change	No parity change	No parity change
(4) One electron jump	No electron jump	One or no electron jump
$\Delta l = \pm 1$	$\Delta l = 0$	$\Delta l = 0, \pm 2$
	$\Delta n = 0$	
(5) $\Delta S = 0$	$\Delta S = 0$	$\Delta S = 0$
(6) $\Delta L = 0, \pm 1$	$\Delta L = 0$	$\Delta L = 0, \pm 1, \pm 2$
$(0 \nleftrightarrow 0)$		$(0 \nleftrightarrow 0, 0 \nleftrightarrow 1)$

(b) The approximate selection rules, (4)-(6), may be violated.
(c) The atoms may be subject to external perturbations.
(d) Nuclear perturbations may be appreciable.
(e) A two-quantum process may take place.

Lines produced by (b) above [without (a), (c), (d), or (e)] are not usually termed "forbidden" (see § 1.2).

We shall discuss first the general theory of magnetic dipole and electric quadrupole radiation, then consider calculations and observations on individual atoms, and finally discuss the remaining types of forbidden transitions. Review articles on these subjects have been published by Borisoglebskii (1958), Rubinowicz (1949), and Mrozowski (1944).

2.2 THEORY OF MAGNETIC DIPOLE AND ELECTRIC QUADRUPOLE RADIATION

The basic theory of magnetic dipole and electric quadrupole radiation was given by Condon and Shortley (1951). They gave the formulae for transition probabilities in terms of the matrix elements of the magnetic dipole and electric quadrupole moments, and quoted the formulae of Rubinowicz for the relative strengths of the Zeeman components of a line and of the lines of a multiplet in quadrupole radiation. The theory was extended by Shortley (1940), who showed how many of the general methods used for electric dipole intensity calculations could be extended to the electric quadrupole case. In particular, Shortley showed how to perform calculations for the intermediate coupling conditions which are often encountered in practical problems, and how to obtain absolute line strengths. He prepared the ground for detailed computations on the astrophysically important p^n ions later performed by Shortley and associates (1941). Important work was performed independently by Pasternack (1940). Some of the methods used by Shortley could not readily be applied to more complicated atoms, and for this purpose Garstang (1957a, 1958a) reformulated the quadrupole intensity theory by means of the methods of Racah (1942, 1943).

The theory of magnetic dipole radiation can be worked out explicitly in LS-coupling. Pasternack (1940) and Shortley (1940) gave the necessary formulae. The magnetic dipole moment ($\mathbf{M}$) is given by

$$\mathbf{M} = -\frac{e}{2mc}(\mathbf{L} + 2\mathbf{S}). \tag{1}$$

The spontaneous emission transition probability A_m between upper level αJ and lower level $\alpha' J'$ is given by

$$A_m(\alpha J, \alpha' J') = \frac{1}{2J+1} \frac{64\pi^4\nu^3}{3hc^3} S_m(\alpha J, \alpha' J') \tag{2}$$

where

$$S_m(\alpha J, \alpha' J') = S_m(\alpha' J', \alpha J) = \sum_{M,M'} |(\alpha JM \mid \mathbf{M} \mid \alpha' J'M')|^2. \tag{3}$$

This expression for S_m can be evaluated explicitly, and the final results are:

$$S_m(SLJ, SLJ+1)$$
$$= [+]^2 \frac{(J-S+L+1)(J+S-L+1)(J+S+L+2)(S+L-J)}{4(J+1)} \tag{4}$$

$$S_m(SLJ, SLJ) = \frac{(2J+1)}{4J(J+1)}[S(S+1) - L(L+1) + 3J(J+1)]^2. \tag{5}$$

These are in units of $(-eh/4\pi mc)^2$; the sign of $S_m^{1/2}$ is that of the quantity in square brackets. An important result which emerges from the calculations leading to these formulae is that in LS-coupling, magnetic dipole radiation takes place only between two levels of the same term. Intermediate coupling calculations can be carried out in the standard manner (see Pasternack, 1940; Shortley *et al.*, 1941) starting from the above LS-coupling line strengths.

The theory of quadrupole radiation in LS-coupling is more complicated, and the electron configuration of the atoms must be taken into consideration. The quadrupole moment is a dyadic $\mathcal{N}$, where according to Shortley (1940)

$$\mathcal{N} = -e \sum_s (\mathbf{r}_s \mathbf{r}_s - \tfrac{1}{3} r_s^2 \mathcal{I}) \tag{6}$$

$$\mathcal{I} = \mathbf{i}\,\mathbf{i} + \mathbf{j}\,\mathbf{j} + \mathbf{k}\,\mathbf{k} \tag{7}$$

and the summation s runs over the electrons. The spontaneous emission transition probability is given by (αJ is the upper level)

$$A_q(\alpha J, \alpha' J') = \frac{1}{2J+1} \cdot \frac{32\pi^6 \nu^5}{5hc^5} S_q(\alpha J, \alpha' J') \tag{8}$$

and

$$S_q(\alpha J, \alpha' J') = S_q(\alpha' J', \alpha J) = \sum_{M,M'} |(\alpha JM | \mathcal{N} | \alpha' J'M')|^2. \tag{9}$$

The absolute square of a dyadic is the sum of the absolute squares of its nine elements. $\mathcal{N}$ is expressed in terms of x, y, and z; it can be expressed in terms of five second order spherical harmonics, $T_q^{(2)}$ ($q = -2, -1, 0, +1, +2$). The matrix elements of these were evaluated by Racah (1942) techniques, and the summation over M and M' and over the nine elements of $\mathcal{N}$ carried out. The result, obtained by Garstang (1957a), was that

$$S_q(\alpha J, \alpha' J') = \tfrac{2}{3} |[\alpha J || T^{(2)} || \alpha' J']|^2, \tag{10}$$

where $(\alpha J || T^{(2)} || \alpha' J')$ is a typical matrix element in Racah's theory. This is of course equivalent to the formulae of Shortley (1940, Eq. 12). Rubinowicz' formulae can also be obtained quite easily by Racah's methods.

The further development of the theory is also by Racah methods. The matrix element has a J dependence in LS-coupling given by (Racah, 1942, formula 44b)

$$\frac{[\alpha SLJ || T^{(2)} || \alpha' SL' J']}{[\alpha SL || T^{(2)} || \alpha' SL']} = (-1)^{S-L-J'} [(2J+1)(2J'+1)]^{1/2} W(LJL'J'; S2) \tag{11}$$

$W(LJL'J'; S2)$ is an algebraical function of the quantum numbers, which can be evaluated in any particular case. This formula is equivalent to those of Rubinowicz for relative line strengths in a multiplet. The matrix element $(\alpha SL || T^{(2)} || \alpha' SL')$ can be reduced by means of coefficients of fractional parentage (Racah, 1943) to elements of the form $(S_1 L_1 l_n SL || T_n^{(2)} || S_1 L_1 l'_n SL')$, where $T_n^{(2)}$ is a one-electron operator acting on the electron n. These elements may be reduced to the form $(l_n \,||\, T_n^{(2)} \,||\, l'_n)$, and these one-electron elements can be evaluated explicitly in terms of radial integrals of the form

$$s_q(nl, n'l') = e \int_0^\infty r^2 P(nl) P(n'l') dr, \tag{12}$$

where $P(nl)$ is the radial wave function of an (nl) electron. Intermediate coupling calculations are performed for quadrupole radiation in the same way as for dipole radiation, starting from LS-coupling line strengths.

The general theory of magnetic dipole and electric quadrupole radiation for p^2, p^3, and p^4 configurations for any arbitrary coupling was considered by Shortley, Aller, Baker, and Menzel. For complicated configurations it is more profitable to tabulate only the LS-coupling line strengths (for quadrupole radiation), and to proceed with intermediate coupling line strengths for both magnetic dipole and electric quadrupole radiation for the atom under consideration. This was done for the p^2, p^3, d^2, and d^3 configurations by Pasternack (1940), (see also Garstang, 1957a) and for p^2 and p^3 by Shortley (1940). Great care must be taken to ensure a consistent choice of phases in all such calculations. The configurations d^4 and d^5 were studied by Garstang (1957a) and transitions $d^n - d^{n-1}s$, $d^{n-1}s - d^{n-1}s$, and $d^{n-1}s - d^{n-2}s^2$ were studied in general and numerical results tabulated for $n = 7$ and $n = 9$ by Garstang (1958a).

In the course of performing detailed calculations on various atoms, it is often necessary to proceed by empirical methods. Certain parameters occur in the intermediate coupling part of the calculation, and numerical values of these parameters are required. They can usually be determined by fitting to experimental data on energy levels. Sometimes there are theoretical relationships between some of the parameters, and the empirically determined parameters may not satisfy these relationships. It is therefore pertinent to ask whether the use of empirical parameters, rather than purely theoretical parameters, will give satisfactory transition probabilities. The differences between theoretical and empirical parameters are usually ascribed to configuration interaction. Layzer (1954) has shown that for highly ionized atoms only configurations with the same set of principal quantum numbers and the same parity need be considered. Thus, for [Ca XV], an especially interesting case, the basic configuration is $1s^2 2s^2 2p^2$ and only the interaction with $1s^2 2p^4$ need be taken into account. Garstang (1956) studied the effect of interaction between these two configurations on the strengths of forbidden lines in the $1s^2 2s^2 2p^2$ configuration. He showed that if observed energy levels are used instead of theoretical energies in line-strength formulae which neglect configuration interaction the major part of the effect of configuration interaction is nevertheless taken into account. Thus the use of empirical energies in the usual formulae may be expected to yield more accurate transition probabilities than the use of theoretical energies. This result has been proved also for the $1s^2 2s^2 2p^4$ configuration. It is dangerous to generalize from simple cases to complex atoms with many interacting configurations, but the work does at least give some support to Garstang's conjecture that for this type of calculation the observed energies should be used. This conclusion does not necessarily hold for other types of calculation. Garstang (1955) discussed the effect of configuration interaction in [O I] and showed that it makes probably only a few per cent difference in the transition probability of $\lambda 5577$. The uncertainty due to the quadrupole radial integral is probably greater.

2.3 Laboratory Observations and Identification of Forbidden Lines

Some forbidden lines have been observed in absorption. The $^2D - {}^2S$ series in potassium was observed (Datta, 1922) in this way. More observations have been made in emission in discharges. Those whose intensity increases as the square of the electric current may be attributed to electric dipole radiation forced by collisions and interatomic fields. Those whose intensity is a maximum for low currents are spontaneous emissions, for which the metastable upper states are destroyed for high currents (high collision frequencies). Long discharge tubes were employed by McLennan and Shrum (1925) and others, in their work on [O I]. The tube contained a rare gas and a very small amount of oxygen. The rare gas decreases the probability of de-excitation of metastable oxygen atoms by collision with other atoms or with the walls of the tube, and reduces the energy of free electrons, preferentially exciting the lower oxygen levels. Another method used a quartz tube with external electrodes, to which an alternating high voltage was applied. Hollow cathode discharge tubes and other methods have been used. With one or two exceptions, only the forbidden lines of neutral atoms have been produced in the laboratory. For ionized atoms, identifications are only possible by means of wavelengths calculated from independently known energy levels. For transitions between fine and hyperfine structure levels microwave techniques have been employed. Reviews of two principal methods have been written by Kellogg and Millman (1946) and Gordy (1948), and reference may also be made to Ramsey (1956) and to Series (1959).

The nature of the auroral line $\lambda 5577$ was investigated by McLennan and Shrum (1925). They succeeded in producing it from oxygen, its wavelength agreed extremely closely with the interferometric wavelength of the auroral line, and both lines had very small widths. Paschen (1930a, b) showed that the red auroral lines at $\lambda\lambda 6300$, 6364 were due to oxygen. Conclusive proof that the $\lambda 5577$ line was an electric quadrupole transition was obtained by Frerichs and Campbell (1930), who showed that the observed Zeeman effect for the line agreed with theoretical predictions for electric quadrupole radiation but not with those for electric or magnetic dipole radiation. Segrè and Bakker (1931) showed by the Zeeman effect that the transitions $^2S - {^2D}$ in sodium and potassium were electric quadrupole radiation, and confirmed the theoretical work of Rubinowicz. Niewodniczanski (1934) studied the line $6p^2$ $^3P_1 - {^1S_0}$ in [Pb I] at $\lambda 4618$ and by the Zeeman effect gave the first proof of the occurrence of magnetic dipole radiation. Since these early researches a number of other forbidden lines have been produced in the laboratory, and in some cases observations of the Zeeman effect have established the nature of the lines. Observation and theory are in excellent agreement, and no real doubt remains about the interpretation of the observations. A recent example of such work is that of Martin and Corliss (1960) on the transitions $\lambda 4460$ $^3P_1 - {^1S_0}$ and $\lambda 7283$ $^3P_2 - {^1D_2}$ in the $5p^4$ configuration of [I II].

Most magnetic dipole transitions in the optical region of the spectrum are intercombinations. One case recently studied for which $\Delta S = 0$ is the $^3P_1 - {^3P_2}$ transition in [Bi II] $6p^2$ (Cole and Mrozowski, 1954; Cole, 1960). Some transitions of this type have been studied by microwave techniques (Lamb, 1957). For example, Lamb and Maiman (1957) have measured the $3^3P_1 - 3^3P_2$ interval in He I.

Observations of the hyperfine structure of a number of lines have also provided evidence as to their nature. References to work of this type up to 1949 may be found in the review of Rubinowicz (1949). It was shown that the intensities of hyperfine components could be obtained from the usual formulae for LS-coupling multiplets (for the same type of radiation) by replacing the quantum numbers S, L, and J by J, I, and F. The nature of various transitions in [Hg II], [Pb I], and [Bi I] were ascertained from the relative intensities of their hyperfine components. The observations are in good agreement with theoretical predictions for the transitions concerned. Other atoms which have been studied are [I II] (Martin and Corliss, 1960), [As I] and [Sb I] (Hults and Mrozowski, 1952), [Bi II] (Cole, 1960), and [Pb II] (Cole, 1961).

Other lines which have been produced in the laboratory include

[Po I] (Mrozowski, 1956), [P I] (Mrozowski, 1954), [N I] (Kaplan, 1939a, b), [Se I] (Ruedy and Gibbs, 1934), [Te I] (Niewodniczanski and Lipinski, 1938), [Xe II], [Xe III], and [Rn II] (Edlén, 1944), and [Br II] (Martin and Tech, 1961).

Two other types of magnetic dipole radiation have been observed. If an atom is placed in a weak external magnetic field, each fine structure level (given S, L, and J) is split up into components labeled by the magnetic quantum number M. It is then possible to have magnetic dipole transitions between two Zeeman components of the same level, with $\Delta M = \pm 1$. Here $\Delta L = 0$, $\Delta S = 0$, $\Delta J = 0$, $\Delta n = 0$ and all the selection rules for magnetic dipole radiation are satisfied. The transitions are in the microwave region, and take place by absorption or induced emission; spontaneous emission is very weak. The energy difference is measured directly by the frequency of the microwave radiation. This method has been used in determining magnetic fields. The $2s\,^2S_{1/2}$ $M = \frac{1}{2} \rightarrow M = -\frac{1}{2}$ transition in hydrogen was used in this way (Lamb, 1951).

An atom whose nucleus has a magnetic moment $\boldsymbol{\mu}$ and spin momentum $\mathbf{I}$ shows hyperfine structure. The magnetic moment of the nucleus is given by

$$\boldsymbol{\mu} = \mu_0 \gamma \mathbf{I} \tag{13}$$

where γ is the Landé nuclear factor and μ_0 $(= e\hbar/2Mc)$ is the Bohr nuclear magneton. The magnetic moment interacts with the magnetic field produced by the electrons to give hyperfine splitting of the energy levels. The total angular momentum of the electrons $\mathbf{J}$ is coupled with $\mathbf{I}$ to give a resultant angular momentum $\mathbf{F}$. In atoms for which $I \neq 0$, the level J is split into components with differing F values; the quantum numbers J, I, and F satisfy the relation $|J - I| \leqslant F \leqslant |J + I|$, and successive F's differ by integers. Transitions between hyperfine components of a level have $\Delta L = \Delta S = \Delta J = 0$, and are forbidden for electric dipole radiation because the transitions are between states of the same parity, but the transitions are allowed for magnetic dipole radiation. The selection rule is found to be $\Delta F = \pm 1$. The transitions are in the microwave region, and measures of their frequency give the hyperfine structure with great accuracy. The outstanding example of a transition of this kind occurs in the ground state of hydrogen. This is $1s\,^2S_{1/2}$. The proton has spin $\frac{1}{2}$, and consequently the ground state is in reality a very close doublet with $F = 0$ and $F = 1$ for the two levels. The frequency of the transition between these levels has been calculated theoretically and measured experimentally, and is 1420.40573 ± 0.00005 Mc/sec (Kusch, 1955). This is equivalent to a wavelength of approximately 21 cm. This line is of quite exceptional importance in astronomy (see § 6). Similar lines between hyperfine structure levels are known for numerous other atoms, and have been used for determining nuclear spins, magnetic moments, and electric quadrupole moments. The spin of the cesium nucleus of mass 133 has $I = \frac{7}{2}$, so that the ground state $6s\ ^2S_{1/2}$ has two hyperfine levels $F = 3$ and $F = 4$, the separation being 9192 Mc/sec. In a magnetic field the $F = 3 \leftrightarrow F = 4$ transition breaks up into a large number of components, 14 of which have been observed by Roberts *et al.* (1946). The transition $m_F = 0 \leftrightarrow m_F = 0$ is used as the basis of the atomic clock (Essen and Parry, 1957). Extensive references on hyperfine structure and fine structure in atoms studied by microwave techniques will be found in Ramsey (1956).

In addition to the observation of forbidden lines in the laboratory and their subsequent identification, mention must be made of the identification of lines in astrophysical sources. Bowen (1928) identified various lines in gaseous nebulae with [O II], [O III], and [N II]. Soon afterwards Merrill (1928) identified lines of [Fe II] in the peculiar star η Carinae. During the following years many more forbidden lines were identified in gaseous nebulae and peculiar stars. The following are certainly present: [N I], [N II], [O I], [O II], [O III], [F IV], [Ne III], [Ne IV], [Ne V], [S I], [S II], [S III], [Cl III], [Cl IV], [A III], [A IV], [A V], [K IV], [K V], [Ca V], [Mn V], [Mn VI], [Fe II], [Fe III], [Fe IV], [Fe V], [Fe VI], [Fe VII], [Ni II], [Ni III], and [Cu II]. For a detailed study of the spectrum of a planetary nebula with identifications see Aller *et al.* (1955). Garstang (1958b) discussed the identification of [Fe IV] in the star RR Telescopii. Another major advance was made in 1942 when Edlén (1945) identified certain emission lines in the solar corona as due to [Fe X], [Fe XI], [Fe XIII], [Fe XIV], [Fe XV], [Ni XII], [Ni XIII], [Ni XV], [Ni XVI], [Ca XII], [Ca XIII], [Ca XV], [A X], and [A XIV]. The identifications were based on extrapolating along isoelectronic sequences. The identification of [Ca XV], though now reasonably certain, has been controversial, and a good discussion of the problem was given by Rohrlich (1956). Some additional identifications have been proposed by Pecker and Rohrlich (1961), including [Cr XI], [Mn XII], and [Mn XIII], and various ions of potassium, vanadium, and cobalt.

One interesting point which has emerged in work by Bowen (1955, 1960) is that the wavelengths of forbidden lines measured on high-dispersion plates of gaseous nebulae are more accurate than those calculated from laboratory ultraviolet spectroscopic data. The astronomical observations serve to refine our knowledge of the low energy levels of the atoms concerned.

2.4 Magnetic Dipole and Electric Quadrupole Transition Probabilities for Individual Atoms

Detailed calculations of the transition probabilities for individual atoms have been made by a number of authors. The most important paper in this field is undoubtedly that of Pasternack (1940), and quite a few of the more recent papers represent small improvements on his work or applications to other atoms. The calculations which he made were based on the theory of intermediate coupling taking as the perturbation the spin interaction of an electron with its own orbit. In general the results so obtained gave intensity ratios of various lines in good agreement with astronomical observations. In one important case, however, the theory failed to agree with observation. In [O II] the ratio of the intensities of the pair of nebular lines at $\lambda 3727$ due to the transitions $^4S - {}^2D$ was widely different from the calculated intensity ratio. The discrepancy was explained by Aller and associates (1949). In the p^3 configuration the lines $^4S - {}^2D$ are forbidden for both magnetic dipole and electric quadrupole radiation to the first order in the spin-orbit interaction, but have nonvanishing intensities when the second order spin-orbit effects are included. Aller, Ufford, and Van Vleck showed that for this particular transition the effect of spin-other-orbit and spin-spin interactions between pairs of electrons is large and comparable with the second order effect of spin-orbit interaction. Only the latter effect had been included in the earlier calculations of Pasternack and others. A number of other studies of ions with p^3 configurations have been made. The corrections introduced by spin-other-orbit and spin-spin interactions are large for $2p^3$ configurations, and appreciable for $3p^3$ configurations; they are probably negligible for higher principal quantum numbers. Garstang (1951) has examined the effects in the p^2 and p^4 configurations of a number of ions. It has been shown that while the energy levels are appreciably affected for $2p^2$ and $2p^4$ configurations there is little change in the transition probabilities. The effects are negligible for np^2 and np^4 configurations with $n \geqslant 3$. References to the results of detailed calculations on transition probabilities of forbidden lines are included in Table II. Authors whose contributions have not been mentioned in Table II because their results have been superseded, or published independently by other authors, include Condon, Ufford and Gilmour, Naqvi, Obi, Yamanouchi and Horie, K. Huang, Yilmaz, and Nikitin.

The transition probability of the 21-cm magnetic dipole line of hydrogen was calculated by Wild (1952). An interesting point which arises in this calculation is that the nuclear magnetic moment may be

TABLE II

BIBLIOGRAPHY[a] OF CALCULATED TRANSITION PROBABILITIES FOR MAGNETIC DIPOLE AND ELECTRIC QUADRUPOLE RADIATION

Ion	References	Ion	References	Ion	References
A III	1, 2	Fe III, V	9	Ne IV	14
A IV	1, 3	Fe IV	10	Ni I	8
A V	1	Fe VI, VII	1	Ni II, III	15
A X, XIV	4	Fe X, XIII	4	Ni XII, XIII, XV, XVI	4
A XI	2	Fe XI	1, 4	O I	5, 13
Al VI, VII, VIII	1	Fe XIV	4, 11	O II	16
Br II	23	Fe XV	2, 12	O III	5
C I	5	H I	19	P I, II	1
Ca II	2	I I	8	Pb I	21
Ca V, VI	1	I II	20	Po I	8
Ca VII	1, 2	K I	22	Rb I	22
Ca XII	4	K IV, VI	1, 2	Rn II	18
Ca XIII	6	K V	1	S I	1, 2
Ca XV	7	Kr III	2	S II	1, 3, 17
Cl II	1, 2	Mg V, VI, VII	1	S III	1
Cl III, IV	1	Mn V, X	1	Sc VI, VII	1
Cr IV, IX	1	Mn VI	8	Si I, VII, VIII	1
Cs I	22	N I	13	Si X	8
Cu II	8	N II	5	Ti VII	1
F II, IV	5	Na I	22	V VIII	1
F III	1	Na IV, V, VI	1	Xe II	18
Fe II	8	Ne III, V	5	Xe III	2, 18

REFERENCES

1. Pasternack (1940)
2. Osterbrock (1951)
3. Naqvi and Talwar (1957)
4. Edlén (1945)
5. Garstang (1951)
6. Garstang (1952b)
7. Garstang (1956)
8. Garstang (unpublished)
9. Garstang (1957b)
10. Garstang (1958c)
11. Froese (1957)
12. Blaha (1957)
13. Garstang (1955)
14. Garstang (1960)
15. Garstang (1958b)
16. Seaton and Osterbrock (1957)
17. Garstang (1952a)
18. Edlén (1944)
19. Wild (1952)
20. Martin and Corliss (1960)
21. Gerjuoy (1941)
22. Condon and Shortley (1951, p. 256)
23. Martin and Tech (1961)

[a] Where several authors have worked on the same lines in an ion, only one reference has been quoted, generally that which in the writer's opinion contains the most accurate results.

neglected. In computing energies and relative electric dipole line strengths of components in hyperfine structure the usual LS-coupling formulae may be used provided that the quantum numbers S, L, and J are replaced by J, I, and F. In the calculation of magnetic dipole line strengths one requires elements of the form

$$\left\langle SL(J)IF \,\middle|\, \frac{\gamma e\hbar}{2Mc}\,\mathbf{I} + \frac{e\hbar}{2mc}\,(\mathbf{L} + 2\mathbf{S}) \,\middle|\, SL(J)IF' \right\rangle \tag{14}$$

($\gamma I = 2.6$ for the proton). Owing to the size of the proton mass M compared with the electron mass m, we can neglect the term involving I. The usual methods are then applied to reduce the matrix to the form

$$\left\langle SLJ \,\middle|\, \frac{e\hbar}{2mc}\,(\mathbf{L} + 2\mathbf{S}) \,\middle|\, SLJ \right\rangle \tag{15}$$

and finally this is evaluated. The line strength of the 21-cm line of hydrogen obtained in this way is 3 atomic units. The spontaneous emission transition probability is 2.85×10^{-15} sec^{-1}.

There is a need for laboratory observations of selected forbidden line intensities. Gerjuoy (1941) and Mrozowski (1944) have discussed the intensities in Pb I and shown how the observations can be used to derive a value of s_q (the quadrupole radial integral). In p^2 and p^4 configurations, relative intensity measurements of transitions $^1S_0 - {}^1D_2$, $^1S_0 - {}^3P_2$, and $^1S_0 - {}^3P_1$, when made, are independent of the method of exciting the lines, and give valuable checks on the theory. The same applies to the lines $^1D_2 - {}^3P_2$, $^1D_2 - {}^3P_1$, and $^1D_2 - {}^3P_0$. Measurements have been attempted for [O I] by Liszka and Niewodniczanski (1958) who obtained for the ratio of the lines $\lambda 2958/\lambda 2972$ a value 1/45, compared with a theoretical value of 1/210 (Garstang, 1951). A determination of the ratio $\lambda 2972/\lambda 5577$ would be valuable. Similar measurements might be attempted in other atoms (e.g., Hg I, Bi I, and Te I). Omholt (1959) has studied the variation of the ratio of [O I] $\lambda 5577$ to permitted N_2^+ bands in rapidly changing aurorae, and deduces a lifetime of at least 0.7 ($\pm$ 0.1) sec for the O I ^1S state, in accord with the theoretical results of Garstang (1951, 1955). Stoffregen and Derblom (1960) and Omholt (1960) have studied the red auroral lines in the same way, obtaining (Omholt) the life of the O I ^1D state as at least 190 sec, compared with the value 110 sec computed by Garstang (1951).

used a cadmium sample in which the odd isotopes had been enriched, and showed that the intensity of the forbidden lines was proportional to the amount of odd isotope present. The lines were strongest for low discharge currents, showing that electric dipole radiation forced by external effects was not the cause of the lines. Kessler (1950) performed some similar experiments on mercury, using various isotopic mixtures, and proving that the intensities of the lines were proportional to the amount of odd isotopes present.

2.7 Transitions Due to Two-Quantum Processes

The transition $2s\ ^2S_{1/2} - 1s\ ^2S_{1/2}$ in hydrogen is forbidden for all types of radiation in the usual approximation, and the $2s\ ^2S_{1/2}$ level is therefore strictly metastable. There are, however, three processes which lead to a finite lifetime for the state. Owing to the Lamb shift the $2s\ ^2S_{1/2}$ level in hydrogen is displaced upwards from the $2p\ ^2P_{1/2}$ level by 0.035 cm^{-1} (Lamb, 1951), and the spontaneous emission probability of this line is 8×10^{-10} sec^{-1}. Thus, the $2s\ ^2S_{1/2}$ level is no longer strictly metastable even for electric dipole radiation. A second process which is possible is magnetic dipole radiation, in a higher approximation than the usual one used in deriving the selection rules. When the Pauli treatment of electron spin is used, $\Delta n = 0$ for magnetic dipole radiation. If relativistic wave functions are used, weak magnetic dipole radiation becomes possible, with a transition probability of order 10^{-5} sec^{-1} (Breit and Teller, 1940). There is a third process, which leads to a still higher transition probability, and which appears to be the most probable radiative process for the $2s - 1s$ transition. This is two-quantum emission. The quantum theory of radiation usually assumes that only one quantum is absorbed or emitted at once, and of course there is no reason why two or more quanta should not be involved in a process, provided in a case such as the $2s - 1s$ transition in hydrogen that the sum of the energies of the two quanta emitted simultaneously is equal to the energy of the transition. The process was studied by Spitzer and Greenstein (1951) and by Kipper (1952). A more accurate investigation was published by Shapiro and Breit (1959). The transition probability for the $2s - 1s$ transition in hydrogen is 8.2 sec^{-1}. This is electric dipole radiation, and because of the arbitrary division of energy between the two quanta (i.e., the arbitrary energy of the virtual intermediate state) it forms a continuum. Its main interest is that it contributes to the observed continuum in planetary nebulae (Seaton, 1955). No other investigations of two-quantum emission in optical spectra

2.5 Lines Induced by External Fields

Some lines are known whose intensity in a discharge increases as the square of the electric current. These lines are forced electric dipole radiation, produced by interionic fields. Let us consider the case of singly-excited helium. If the interionic Stark fields are weak, so that the Stark splitting of the levels is small compared with the separation of levels with the same n and different l, then the selection rules for levels which can interact because of the Stark perturbations are that the parity must be opposite and $\Delta l = \pm 1$. If the perturbing level combines with a third level by an allowed transition ($\Delta l = \pm 1$), then the perturbed level can combine with the third level and we get a line which satisfies $\Delta l = 0$ or ± 2. If the interionic Stark fields are strong, there is no restriction on l (Bethe and Salpeter, 1957, p. 332). The latter situation applies to certain lines in He I. Among these, one at $\lambda 4143.23$ has been ascribed to a transition $2p\ ^1P_1 - 6g\ ^1G_4$. This line was studied by Jacquinot and Brochard (1943) and by Brochard and Jacquinot (1945) and was shown by means of its Zeeman effect to be due to forced electric dipole radiation. Many other similar lines are known in He I, including $\lambda 4469.95\ 2p\ ^3P - 4f\ ^3F$ which has been found in certain B-type stars. Jenkins and Segrè (1939) showed that in potassium the higher members of the series $4S - nS$, $4S - nD$ were due to enforced dipole radiation because of their Zeeman effect. Only the lowest member of the $4S - nD$ series is due predominantly to electric quadrupole radiation.

Another interesting example of forced electric dipole radiation was given by Shenstone (1953). In the Pd I spectrum transitions from both $4d^9\ 5d\ ^3P_1$ and $4d^9\ 6p\ ^3P_1^0$ to the odd terms in $4d^9\ 5p$ are known, with the odd-odd transitions about one-tenth of the intensity of the odd-even transitions. The forbidden transitions arise because of the interaction of the two $^3P_1$ levels (which are 3 cm^{-1} apart) in the ionic fields. There are several more similar transitions in Pd I, and some are known in other atomic spectra. For an outline of the relevant theory, see Condon and Shortley (1951, Chapter XVII).

2.6 Transitions Due to Nuclear Perturbation

Among the early discoveries of forbidden transitions were two lines of Hg I. Rayleigh (1927) observed the line $\lambda 2270\ 6s^2\ ^1S_0 - 6s6p\ ^3P_2$ and Fukuda (1926) observed $\lambda 2656\ 6s^2\ ^1S_0 - 6s6p\ ^3P_0$. The former line is rigorously forbidden for electric dipole and magnetic dipole radiation because $\Delta J = 2$, and for electric quadrupole radiation by the parity rule. The second line is forbidden for all types of radiation by the rule $J = 0 \nleftrightarrow J = 0$. Some other origin must be sought for this radiation. Bowen suggested that nuclear interaction with the outer electrons was responsible. Many nuclei have magnetic moments, and these interact with the magnetic field produced by the orbital motion of the electrons. This leads to hyperfine structure—a splitting of the energy levels—and it may lead to the appearance of certain forbidden lines.

The theory of the intensity of these lines was studied by Opechowski (1938). Using the notation introduced in § 2.3, the interaction energy of $\boldsymbol{\mu}$ with the magnetic field $\mathbf{H}$ produced by the electrons is

$$W = -\boldsymbol{\mu} \cdot \mathbf{H}. \tag{16}$$

If the wave functions of the complete system are $| JIF\rangle$, then the interaction between neighboring fine structure levels is given by matrix elements of the form $\langle JIF \mid W \mid J'IF'\rangle$. $\mathbf{I}$ depends only on the nuclear coordinates, $\mathbf{H}$ on the electron coordinates, and hence, using standard formulae (Condon and Shortley, 1951, p. 71) we find that $\langle JIF \mid W \mid J'IF'\rangle = 0$ unless

$$F' = F$$
$$J' = J, J \pm 1\ (0 \nleftrightarrow 0). \tag{17}$$

When considering the levels of a term we see that only neighboring levels ($\Delta J = \pm 1$) perturb each other, and within each such perturbing pairs of levels only the hyperfine levels with $\Delta F = 0$ interact. The diagonal elements ($\Delta J = 0$) give the contribution of the nuclear moment to the hyperfine structure. (Another contribution arises from the nuclear quadrupole moment.) From Condon and Shortley's formulae it is found that

$$\langle JIF \mid W \mid JIF\rangle = C(J, J)\,[F(F+1) - J(J+1) - I(I+1)] \tag{18}$$

$$\langle JIF \mid W \mid J-1\,IF\rangle = C(J, J-1)$$
$$\times \sqrt{(F+J-I)(F-J+I+1)(F+J+I+1)(J+I-F)} \tag{19}$$

where the C's can be determined from hyperfine structure observations or computed from theory.

Let us consider mercury as an example. For the Hg^{199} isotope, $I = \frac{1}{2}$. For the $6s6p\ {}^3P_0$ level $J = 0$ $F = \frac{1}{2}$. For the ${}^3P_1$ level $J = 1$, $F = \frac{3}{2}$ and $\frac{1}{2}$. Thus, the ${}^3P_0$ and ${}^3P_1$ levels interact through their $F = \frac{1}{2}$ components. Treating this interaction as a small perturbation, the wave

function of the ${}^3P_0$ state is expressed in terms of the unperturbed wave functions by the equation

$$| {}^3P_0\rangle = | {}^3P_0\rangle_0 + \frac{\langle {}^3P_0, \frac{1}{2}, \frac{1}{2} \mid W \mid {}^3P_1, \frac{1}{2}, \frac{1}{2}\rangle}{E({}^3P_0) - E({}^3P_1)} \,| {}^3P_1, F = \tfrac{1}{2}\rangle. \tag{20}$$

The electric dipole matrix element is given by

$$\langle {}^3P_0 \mid P \mid {}^1S_0\rangle = \frac{\langle {}^3P_0, \frac{1}{2}, \frac{1}{2} \mid W \mid {}^3P_1, \frac{1}{2}, \frac{1}{2}\rangle}{E({}^3P_0) - E({}^3P_1)} \langle {}^3P_1, F = \tfrac{1}{2} \mid P \mid {}^1S_0\rangle \tag{21}$$

since by the selection rules $\langle {}^3P_0 \mid P \mid {}^1S_0\rangle_0 = 0$ for the unperturbed wave functions. The elements $\langle {}^3P_1 \mid P \mid {}^1S_0\rangle_0$ for $F = \frac{1}{2}$ and $\frac{3}{2}$ are nonzero because of spin-orbit interaction, and are responsible for the observed strength of the $\lambda 2537\ {}^1S_0 - {}^3P_1$ line. The intensity ratio of the lines is thus

$$\frac{\text{Intensity of } \lambda 2656\ ({}^1S_0 - {}^3P_0)}{\text{Intensity of } \lambda 2537\ ({}^1S_0 - {}^3P_1)} = \left| \frac{\langle {}^3P_0, \frac{1}{2}, \frac{1}{2} \mid W \mid {}^3P_1, \frac{1}{2}, \frac{1}{2}\rangle}{E({}^3P_0) - E({}^3P_1)} \right|^2 f_1 \tag{22}$$

where the factor $f_1\ (= \frac{1}{3})$ is inserted because the $F = \frac{1}{2} - F = \frac{1}{2}$ transition has only one-third of the intensity of the whole ${}^1S_0 - {}^3P_1$ line. This calculation is repeated for the Hg^{201} isotope, and the mean intensity ratio computed allowing for the abundance ratios of the isotopes. For even isotopes $I = 0$; such isotopes make no contribution to the forbidden line intensities. For mercury, Opechowski found the intensity ratio $\lambda 2656/\lambda 2537$ to be 5×10^{-9}. This is in satisfactory agreement with th intensity ratio observed by Gaviola (1929). The ratio of the intensiti of the ${}^1S_0 - {}^3P_0$ line from the Hg^{199} and Hg^{201} isotopes separately w in reasonable agreement with observations made by Mrozowski (19 The nuclear quadrupole moment does not contribute to the inter in this case, for such a moment can lead to $J = 0$ levels being pertu by $J = 2$, and a small admixture of ${}^3P_2$ characteristics into the ${}^3P_0$ function has no effect on the line strengths for electric dipole ra because $\langle {}^3P_2 \mid P \mid {}^3P_0\rangle = 0$.

The same type of explanation will suffice to explain the app of the $\lambda 2270\ {}^1S_0 - {}^3P_2$ line, although here several pairs of h levels with $\Delta F = 0$ contribute to the intensity, and the nuclea pole moment may also give a contribution. The correctn explanation was demonstrated by Mrozowski (1945).

Further experimental proof of the above interpretatio forbidden lines has been obtained. Holmes and Deloume (a number of lines in Cd I, including $5s^2\ {}^1S_0 - 5s5p\ {}^3P_0$ a

have come to the writer's notice. Cases are known of multiple quantum transitions in microwave spectra. Kusch (1956) observed, in K^{39} atoms, transitions with $\Delta F = 0$, $\Delta m = \pm 1, \pm 2, \pm 3$, and ± 4, and the theory of these was examined by Salwen (1955, 1956a, b; see also Series, 1959).

3 *Forbidden Transitions in Diatomic Molecular Spectra*

3.1 Introduction

During recent years there has been an increasing interest in the study of forbidden transitions in diatomic molecules. As mentioned earlier the definition of forbidden transitions used here is wider than that customary in dealing with atomic spectra. In diatomic molecules by far the strongest transitions are those which correspond to an alternating electric dipole moment of the molecule, and usually Hund's cases *a* or *b* are good approximations to the coupling conditions. Transitions violating the selection rules holding under these conditions are designated as forbidden transitions. They are weaker than permitted transitions, and usually require long absorbing paths for observation as absorption bands. Bernstein and Herzberg (1948) have described an apparatus using multiple reflections in a tube 20 meters long, with up to 250 transits, and a pressure of up to 10 atm giving a maximum absorption length of 50 km-atm.

The general selection rules for electric dipole radiation in diatomic molecules are:

(1) If J is the total angular momentum

$$\Delta J = 0, \pm 1 \qquad \text{(but not } 0 \leftrightarrow 0\text{)}. \tag{23}$$

This is a standard rule of general validity for any atomic system.

(2) Positive (+) terms combine only with negative (−) terms.

A rotational level is called positive or negative according to whether the total eigenfunction remains unchanged or changes sign for reflection at the origin. The selection rule follows by considering integrals of the form $\int \psi_m^* \mathbf{M} \psi_n \, d\tau$. If these integrals are not to vanish, the integrand must not change sign on reflection at the origin, and since the dipole moment $\mathbf{M}$ changes sign, one of ψ_m, ψ_n must change sign and the other must not.

(3) For identical nuclei, symmetric (*s*) terms combine only with symmetric terms, and antisymmetric (*a*) with antisymmetric.

A term is symmetric or antisymmetric according to whether the total eigenfunction remains unchanged or changes sign for an exchange of the nuclei. This rule is also proved by considering integrals of the form $\int \psi_m^* \mathbf{M} \psi_n \, d\tau$. $\mathbf{M}$ is unchanged on interchanging nuclei, and hence, for nonvanishing integrals, either both ψ_m, ψ_n change sign, or neither change sign.

(4) For nuclei of equal charge (but not necessarily of equal mass), even (g) electronic states combine only with odd (u) electronic states.

An electronic state is even or odd according to whether the electronic eigenfunction remains unchanged or changes sign for a reflection at the center of symmetry. The proof is similar to that of rule (2) except that the electronic transition moment alone is considered. For homonuclear molecules (4) can be proved from (2) and (3). For heteronuclear molecules the s, a property does not exist, and (4) is no longer rigorous, although it is a very good approximation if the interaction between electronic and rotational motion is small.

In addition to these general selection rules, there are others which are valid only for certain cases of coupling. For case (a), the electronic orbital and spin angular momenta are strongly coupled to the internuclear axis, with components Λ and Σ, respectively, and the resultant of these has component $\Omega = |\Lambda + \Sigma|$. Ω is loosely coupled with the rotation N to give the total angular momentum J. For case (b) the electronic orbital angular momentum (component Λ) is strongly coupled to the axis, Λ is coupled with N to give a resultant K; K is coupled with the spin S to give J.

In both cases (a) *and* (b)

(5) $$\Delta S = 0. \qquad (24)$$

This is true for molecules, as it is for atoms, only if spin-orbit interaction is negligible.

(6) $$\Delta\Lambda = 0, \pm 1. \qquad (25)$$

This rule corresponds to the selection rule for M_L for atoms in a strong field, and is rigorous only if rotation-electronic and spin-orbit interactions are negligible.

If *both* initial and final electronic states belong to *Hund's case* (a) we have, for small spin-orbit interaction,

(7) $$\Delta\Sigma = 0, \qquad (26)$$

which is analogous to the rule for M_S for atoms in a strong field. If spin-orbit interaction is large, rules (5), (6), and (7) no longer hold, but, analogous to the rule for M_J for atoms, we have

(8) $$\Delta\Omega = 0, \pm 1. \quad (27)$$

Furthermore, if $\Omega = 0$ for both electronic states

(9) $$\Delta J \neq 0 \quad \text{for} \quad \Omega = 0 \leftrightarrow \Omega = 0. \quad (28)$$

If *Hund's case* (*b*) holds for *both* initial and final states, then

(10) $$\Delta K = 0, \pm 1, \quad (29)$$

(11) $$\Delta K \neq 0 \quad \text{for} \quad \Sigma \leftrightarrow \Sigma \text{ transitions.} \quad (30)$$

These rules are analogous to (1) and (9) and are valid provided that the interaction of K and S is negligible.

A Σ state is called Σ^+ or Σ^- according to whether its electronic eigenfunction remains unchanged or changes sign upon reflection in any plane passing through the internuclear axis. In a similar way to the proof of (2) it can be shown that

(12) $$\Sigma^+ \not\leftrightarrow \Sigma^-. \quad (31)$$

This rule is valid only if the spin-orbit interaction and the interaction of rotation and electronic motion are negligible. For $\Sigma^+ - \Sigma^-$ transitions $\Delta K = \pm 1$ is rigorously forbidden by rule (2), $\Delta K = 0$ becomes allowed by the interaction of rotation and electronic motion, and $\Delta K = 0$ and $\Delta K = \pm 2$ become allowed by spin-orbit interaction (Present, 1935).

In considering the application of these rules it is essential to note that for all singlet ($S = 0$) states the distinction between coupling cases (a) and (b) disappears, and singlet states can be considered as belonging to either. Furthermore, all Σ states ($\Lambda = 0$) belong to case (b). For Σ^+ states, rotational levels with even K (for all J) have "+" symmetry and those with odd K have "—" symmetry; for Σ^- states even K levels have "—" symmetry and odd K levels have "+" symmetry. For Π and Δ states the levels for each J are doubly degenerate (Λ-type doubling), one component has "+" symmetry and the other has "—" symmetry. For the case of identical nuclei, the g states have "+" levels with symmetry s and "—" levels with symmetry a; for u states the "+" levels have symmetry a and the "—" levels have symmetry s. For homonuclear molecules nuclear spin leads to intensity alternation of the lines in a band, and if the nuclei have zero spin (e.g., O_2) alternate lines are missing, and antisymmetric (a) levels are absent.

Violations of the selection rules are caused by perturbations or other higher order effects which are neglected in the simple theory. If the rigorous selection rules are violated, magnetic dipole, electric quadrupole, or forced electric dipole radiation are present. The breakdown of other selection rules may be caused by spin-orbit interactions (which can cause the mixing of $^1\Sigma$ and $^3\Pi_0$ states, and other similar cases with $\Delta S \neq 0$) and there are other perturbations possible (analogous to configuration interaction in atoms) which can cause the mixing of states such as $^1\Sigma^+$ and $^1\Pi$. The perturbed state takes on some of the properties of the perturbing state, and weak transitions become possible which are forbidden in the absence of perturbations.

The distribution of the intensity of intercombination transitions ($\Delta S \neq 0$) among the various rotational branches which occur has been studied by several authors. Schlapp (1932, 1937) studied the transitions $^1\Sigma^+ - {}^3\Sigma^+$, $^1\Sigma^- - {}^3\Sigma^-$, $^1\Sigma^- - {}^3\Sigma^+$, $^1\Sigma^+ - {}^3\Sigma^-$, $^1\Sigma - {}^3\Pi$, and $^1\Sigma - {}^3\Delta$. His most important qualitative result was that $^1\Sigma^+ - {}^3\Sigma^-$ and $^1\Sigma^- - {}^3\Sigma^+$ transitions have intensities comparable with those of $^1\Sigma^+ - {}^3\Sigma^+$ and $^1\Sigma^- - {}^3\Sigma^-$, in spite of the rule (12) being violated for the former transitions and not for the latter. Intensity distributions for $^1\Sigma - {}^3\Pi$ were given by Budó (1937), for $^4\Sigma^\pm - {}^2\Sigma^\pm$, $^4\Sigma - {}^2\Pi$, and $^4\Sigma^\pm - {}^2\Sigma^\mp$ by Budó and Kovács (1940), for $^4\Pi - {}^2\Sigma$ and $^4\Pi - {}^2\Pi$ by Kovács and Budó (1941), for $^3\Sigma - {}^1\Delta$ by Van Vleck (1934), for $^1\Sigma - {}^3\Delta$, $^3\Sigma - {}^3\Delta$, and $^1\Pi - {}^3\Delta$ by Kovács (1960), and for $^3\Sigma_g^+ - {}^3\Sigma_u^-$ by Present (1935).

There are three types of forbidden transitions in molecules: (*a*) those which are rigorously forbidden for electric dipole radiation, (*b*) those which violate approximate selection rules, and (*c*) those rigorously forbidden for free molecules but which become allowed by external perturbing influences. Examples of these kinds of transitions will be discussed in the following subsections.

3.2 Transitions Which Are Rigorously Forbidden for Electric Dipole Radiation

If electric dipole is rigorously forbidden, the only radiation possible arises from higher multipoles. Such radiation is much weaker than allowed electric dipole radiation, and in practice it is found that only magnetic dipole and electric quadrupole radiation need be considered. The selection rules for these may be determined in a similar way to those for electric dipole radiation; the rules are given in Table III, numbered corresponding to the electric dipole rules and distinguished by primes. It may be noted that the band structure of magnetic dipole transitions

TABLE III

SELECTION RULES FOR MAGNETIC DIPOLE AND ELECTRIC QUADRUPOLE RADIATION

	Coupling	Magnetic dipole	Electric quadrupole
(1)′	General	$\Delta J = 0, \pm 1$ $(0 \nleftrightarrow 0)$	$\Delta J = 0, \pm 1, \pm 2$ $(0 \nleftrightarrow 0, 0 \nleftrightarrow 1, \frac{1}{2} \nleftrightarrow \frac{1}{2})$
(2)′	General	$+ \leftrightarrow +, - \leftrightarrow -, + \nleftrightarrow -$	$+ \leftrightarrow +, - \leftrightarrow -, + \nleftrightarrow -$
(3)′	General	$s \leftrightarrow s, a \leftrightarrow a, s \nleftrightarrow a$	$s \leftrightarrow s, a \leftrightarrow a, s \nleftrightarrow a$
(4)′	General	$g \leftrightarrow g, u \leftrightarrow u, g \nleftrightarrow u$	$g \leftrightarrow g, u \leftrightarrow u, g \nleftrightarrow u$
(5)′	(a) and (b)	$\Delta S = 0$	$\Delta S = 0$
(6)′	(a)	$\Delta \Lambda = \pm 1$ if $\Delta \Sigma = 0$; $\Delta \Lambda = 0$ if $\Delta \Sigma = \pm 1$	$\Delta \Lambda = 0, \pm 1, \pm 2$
	(b)	$\Delta \Lambda = 0, \pm 1$	
(7)′	(a)	See (6)′	$\Delta \Sigma = 0$
(8)′	(a)	$\Delta \Omega = \pm 1$	$\Delta \Omega = 0, \pm 1, \pm 2$
(9)′	(a)	—	$\Delta J \neq 1$ for $\Omega = 0 \leftrightarrow \Omega = 0$
(10)′	(b)	$\Delta K = 0, \pm 1$	$\Delta K = 0, \pm 1, \pm 2$
(11)′	(b)	$\Delta K = 0$ for $\Sigma - \Sigma$ transitions	$\Delta K = 0, \pm 2$ for $\Sigma - \Sigma$ transitions
(12)′	(b)	$\Sigma^+ \nleftrightarrow \Sigma^-$	$\Sigma^+ \nleftrightarrow \Sigma^-$

is the same as that of similar electric dipole transitions. The two types may be distinguished by intensity considerations, or by knowledge of the electronic states involved. Van Vleck (1934) showed that the rotational intensity distribution for magnetic dipole transitions is the same as for electric dipole radiation in similar transitions. Thus, for example, ${}^1\Pi_g - {}^1\Sigma_g^+$ has the same rotational intensity distribution as ${}^1\Pi_u - {}^1\Sigma_g^+$.

The Lyman-Birge-Hopfield (LBH) bands of N_2 are an interesting example of a magnetic dipole transition. They occur in the ultraviolet, and are due to the transition $a^1\Pi_g \leftrightarrow X^1\Sigma_g^+$. They have been observed both in emission and in absorption. Herzberg (1946) verified the assignment to a magnetic dipole transition. The bands are very weak, molecular orbital theory predicts that the lowest ${}^1\Pi$ state of N_2 should be ${}^1\Pi_g$, and the ground state is undoubtedly ${}^1\Sigma_g^+$. Furthermore, the transitions $q'\,{}^1\Pi_u - X\,{}^1\Sigma_g^+$ and $q'\,{}^1\Pi_u - a\,{}^1\Pi_g$ are known; so are $t'\,{}^1\Sigma_u^+ - X\,{}^1\Sigma_g^+$ and $t'\,{}^1\Sigma_u^+ - a'\Pi_g$, and there are other similar cases. One cannot have three electronic states connected by three electric dipole transitions, by rule (4), which is rigorous for N_2. Hence one of the transitions must be magnetic dipole or electric quadrupole. It is therefore very likely that the LBH bands are $a^1\Pi_g - X\,{}^1\Sigma_g^+$. The identification of electric quadrupole branches by Wilkinson and Mulliken (1957) finally confirmed the assignment. The LBH bands violate the rigorous rule (4), and of course (2), but they satisfy all the other relevant selection rules. Jarmain *et al.* (1955) have given relative vibrational transition probabilities. Wilkinson (1957) has carried out a high dispersion study of the LBH bands, and Douglas and Rao (1958) have identified the same transition in the spectrum of P_2. Lichten (1957) measured the mean lifetime of the $a^1\Pi_g$ state in N_2, for magnetic dipole radiation only, as 1.7×10^{-4} sec, i.e., a spontaneous emission transition probability of 6000 sec^{-1}.

Other examples of magnetic dipole transitions occur in O_2. The red region of the telluric spectrum shows several absorption bands which have been assigned to a transition $b\,{}^1\Sigma_g^+ \leftarrow X\,{}^3\Sigma_g^-$ of O_2. The nature of the upper state was a puzzle—rotational intensity studies had suggested ${}^1\Sigma_u^-$—until Van Vleck (1934) showed that if the transition was due to magnetic dipole radiation the upper state could be ${}^1\Sigma_g^+$, the intensity relations would still agree with observation, and the type of state would agree with predictions based on molecular orbital theory. The transition is very weak, absolutely, because not only is it a magnetic dipole transition, violating rule (4), but also it is on intercombination transition, violating rule (5)′ and as has been shown by Schlapp (1932) this leads to violations also of rules (11)′ and (12)′. Other relevant magnetic dipole selection rules are obeyed. The intensity is derived from a permitted transition by spin-orbit perturbation, probably of the ${}^1\Sigma^+$ state by a ${}^3\Pi_0$ perturbing term. A comprehensive study of the structure of the transition has been published by Babcock and Herzberg (1948). Four branches, of types ${}^RR$, ${}^PP$, ${}^RQ$, and ${}^PQ$ are observed, as expected. The intensity of the transition has been studied by Childs and Mecke (1931), Allen (1937), van de Hulst (1945), and de Jager (1956). Converting their absorption coefficients into spontaneous emission transition probabilities for whole bands, we find the values in Table IV. If one adopts

TABLE IV

SPONTANEOUS EMISSION TRANSITION PROBABILITIES (SEC^{-1}) FOR INDIVIDUAL BANDS IN $b^1\Sigma_g^+ - X^3\Sigma_g^-$ OF OXYGEN

Author	$A(0, 0)$ band	$B(1, 0)$ band	$\alpha(2, 0)$ band
Childs and Mecke (1931)	0.13	—	—
Allen (1937)	0.060	0.0055	0.00027
van de Hulst (1945)	0.17	0.010	0.00031
de Jager (1956)	0.10	—	—

the relative vibrational transition probability given by Fraser and associates (1954) for the (0, 0) band, 0.93, the spontaneous emission probability for the whole electronic transition is 0.14, 0.064, 0.18, and 0.11 as determined by the four authors, respectively.

Mulliken predicted the existence of a low-lying $^1\Delta_g$ state of O_2 on the basis of the electron configuration of the molecule. This state was first observed in the spectrum of liquid oxygen, and subsequently the transition $a\,^1\Delta_g - X\,^3\Sigma_g^-$ was identified in the telluric spectrum at 1.26μ. A detailed analysis was made by Herzberg and Herzberg (1947). The assignment of the transition to magnetic dipole radiation was fully confirmed. *O*-, *P*-, *Q*-, *R*-, and *S*-form branches ($\Delta K = -2, -1, 0, +1, +2$, respectively) were observed, as expected, and there is no $P(1)$ line observed, which is also as expected since the $^1\Delta$ state has no $J = 0$ and 1 levels. The (1, 0) band is much weaker than the (0, 0) band. Furthermore, the intensity of the $^1\Delta_g - {}^3\Sigma_g^-$ system is much less (perhaps 0.05 times) that of the $^1\Sigma_g^+ - {}^3\Sigma_g^-$ system. Possibly this is because the former system violates the selection rule $\Delta\Lambda = 0, \pm 1$, [rule (6)′] as well as rule (4) and rule (5)′. Rule (10)′ is also broken, since $\Delta K = \pm 2$ is observed, but, as Present (1935) showed, this is a consequence of spin-orbit interaction, which also leads to the breakdown of rule (5)′. The band probably arises because of mixing between the $^3\Sigma_1^-$ and perturbing $^1\Pi$ states and between $^1\Delta$ and perturbing $^3\Pi_2$ states. Both sets of oxygen bands are also allowed as electric quadrupole radiation, and branches with $\Delta J = \pm 2$ would then occur; such branches have not yet been observed. Relative vibrational transition probabilities have been given by Nicholls *et al.* (1960), while Vallance Jones and Harrison (1958) obtained a spontaneous emission transition probability of 1.9×10^{-4} sec^{-1} for the (0, 0) band.

A recent study of a metastable level in hydrogen deserves mentioning. Lichten (1960) has shown by a molecular beam magnetic resonance technique that H_2 has a metastable $1s\sigma 2p\pi\, c^3\Pi_u$ state with $v = 0$. This state (the $K = 0$ level is at 94794 cm^{-1} above the hydrogen molecular ground state) can only decay to the (unstable) $1s\sigma 2p\sigma\, b^3\Sigma_u^+$ state by magnetic dipole transitions. States of $c^3\Pi_u$ with $v \geqslant 1$ can decay to $1s\sigma 2s\sigma\, a^3\Sigma_g^+$ by electric dipole radiation.

One other type of magnetic dipole transition has been observed in O_2. The electronic ground state $X^3\Sigma_g^-$ is made up of a series of very close triplets, with $J = K - 1$, K and $K + 1$, and K taking odd values only. Transitions $\Delta J = \pm 1$ within each triplet are possible by magnetic dipole radiation. The transitions violate rule (4) for electric dipole radiation. The triplet intervals were studied theoretically by Schlapp (1937) and observationally by Babcock and Herzberg (1948) from optical spectra. Beringer (1946) found a microwave absorption at 0.5-cm wavelength which was shown by Van Vleck (1947) to be due to a blend of the $J = K \leftrightarrow J = K + 1$ transitions for many values of K. This blend was later resolved, by Burkhalter *et al.* (1950), and the observations agreed reasonably well with the positions of the levels predicted from the formulae of Schlapp. Anderson *et al.* (1951) observed a transition at about 2.5-mm wavelength which they identified as the $J = 0 \leftrightarrow J = 1$ transition for $K = 1$, and Richardson (1958) observed the $J = 7 \leftrightarrow J = 8$ transition for $K = 7$. Simple formulae for the magnetic dipole line strengths can be obtained (Van Vleck, 1947). They are, in atomic units,

$$S_m(K; J, J+1) = \frac{4K(2K+3)}{K+1} \tag{32}$$

$$S_m(K; J, J-1) = \frac{4(K+1)(2K-1)}{K}. \tag{33}$$

The spontaneous emission transition probabilities are easily calculated from these line strengths. For O_2, $K = 1$, we find for $J = 1 \rightarrow J = 2$, $A = 5.9 \times 10^{-10}$ sec^{-1} and for $J = 1 \rightarrow J = 0$, $A = 4.4 \times 10^{-9}$ sec^{-1}.

Mention should be made of the observation by Kusch (1954) of multiple quantum transitions between Zeeman levels in O_2. In the $K = 1$, $J = 2$ level he found transitions $\Delta M = \pm 1$, ± 2, and ± 3. The theory of these was considered by Salwen (1955, 1956a).

Electric quadrupole transitions in molecular spectra are very rare; only two cases of quadrupole electronic transitions are known, and there is one observed example of a quadrupole rotation-vibration spectrum.

The first case of a quadrupole electronic transition was found by Wilkinson and Mulliken (1957). During a high dispersion study of the absorption spectrum of N_2 in the vacuum ultraviolet, some 20 weak lines were found which could not be attributed to the P, Q, and R branches of the magnetic dipole LBH bands (mentioned earlier in this section). These weak lines were shown to be members of S and O branches, with $\Delta J = +2$ and $\Delta J = -2$, respectively. The selection rules show that such transitions can only be due to electric quadrupole radiation. Rules (1), (2), and (4) for electric dipole radiation are violated, all the rules for electric quadrupole radiation are satisfied. The quadrupole transitions are weaker than the magnetic dipole transitions by a factor of about 7. The absorption coefficients for the quadrupole lines are of the order of 10^7 times those for the quadrupole rotation-vibration spectrum of hydrogen (discussed in the sequel): the nitrogen quadrupole lines have Einstein spontaneous transition probabilities (Wilkinson and

Mulliken, 1959) of 9×10^2 sec^{-1}. The discovery of weak quadrupole branches of the LBH bands removes all doubt about Herzberg's (1946) identification of them as predominantly magnetic dipole transitions.

The second example of an electric quadrupole transition was found by Noxon (1961) in O_2. He observed the Q branch of the (0, 0) band of the $b^1\Sigma_g^+ - a^1\Delta_g$ transition and estimated for the whole (0, 0) band a spontaneous emission probability of 2.5×10^{-3} sec^{-1}.

In addition to forbidden electronic transitions, it is possible to have rotation-vibration spectra which are rigorously forbidden for electric dipole radiation but which are allowed for electric quadrupole radiation. The most important example is in H_2. The transitions all occur within the ground electronic state $^1\Sigma_g^+$. Levels with even J ($\equiv K$) have "+" and s symmetries, those with odd J have "−" and a symmetries. For electric dipole radiation transitions $\Delta J = \pm 1$ violate the rule $s \nleftrightarrow a$, transitions $\Delta J = 0$ violate the rule $+ \nleftrightarrow -$, all violate the rule $g \leftrightarrow u$, and there is in any case no permanent dipole moment so that electric dipole radiation is rigorously forbidden. $^1\Sigma_g^+$ states have no magnetic dipole moment, so that magnetic dipole radiation does not occur. Thus quadrupole radiation is the only possibility. Transitions $\Delta J = \pm 1$ violate the rigorous quadrupole rules (2)′ and (3)′, but $\Delta J = 0, \pm 2$ are allowed (except $0 \nleftrightarrow 0$). $\Delta J = +2, 0, -2$ give rise to S, Q, and O branches, with first lines $S(0)$, $Q(1)$, $O(2)$ having lower $J'' = 0, 1, 2$, respectively. On using his multiple transit apparatus Herzberg (1949, 1950a) succeeded in observing this quadrupole spectrum, the first quadrupole radiation observed in molecular spectroscopy. On using a path of up to 50 km-atm, eight lines were observed: $Q(1)$, $S(0)$, $S(1)$, and $S(2)$ in each of the bands (2-0) and (3-0). The intensities of these lines seemed to be proportional to the pressure, so that the lines were not due to enforced dipole radiation. The linewidths were very small, < 0.05 cm^{-1} even at 10-atm pressure, as is expected for quadrupole radiation. (See § 3.4 for this same spectrum appearing due to enforced dipole radiation.) Approximate wavelengths are given in Table V.

TABLE V

WAVELENGTHS (A) IN THE HYDROGEN ROTATION-VIBRATION SPECTRUM

Band	Line			
	$S(0)$	$S(1)$	$S(2)$	$Q(1)$
(1-0)	22232	21217	20336	24065
(2-0)	11892	11619	11379	12380
(3-0)	8273	8151	8046	8497
(4-0)	6435	6367	6313	6566

Intensity calculations for quadrupole transitions have been performed by James and Coolidge (1938) and by Bates and Poots (1953). It was shown that the line strength is given by

$$S(\alpha J, \alpha' J') = \left| \int \Psi_{\alpha J}^{*} (\mathcal{N} - \tfrac{1}{3} \mathcal{N}_s \mathcal{I}) \Psi_{\alpha' J'} d\tau \right|^2 \quad (34)$$

where $\Psi_{\alpha J}$, $\Psi_{\alpha' J'}$ are the normalized wave functions of the initial and final states, $\mathcal{N}$ is the quadrupole moment operator, $\mathcal{N}_s$ is its scalar, $\mathcal{I}$ is the idemfactor, and the square of the modulus denotes the double dot product of the tensor and its conjugate, i.e., the sum of the squares of its nine components. The wave functions are assumed to be products of electronic, vibrational and rotational functions, $\psi_e(\mathbf{r} \mid R) \psi_{vJ}(R) \psi_{JM}$ where R is the internuclear distance and $\mathbf{r}$ the electron coordinates, z being taken along the internuclear axis. Then the matrix element above can be reduced to the form

$$S(\alpha J, \alpha' J') = | \bar{N}(v', J'; v'', J'') |^2 S(J', J'') \quad (35)$$

where $S(J', J'')$ is obtained by integrating over the rotational wave functions and summing over the magnetic components. James and Coolidge found that

$$S(J, J+2) = \frac{(J+1)(J+2)}{2J+3} \qquad (S \text{ branch}) \quad (36)$$

$$S(J, J) = \frac{2J(J+1)(2J+1)}{3(2J-1)(2J+3)} \qquad (Q \text{ branch}) \quad (37)$$

$$S(J, J-2) = \frac{J(J-1)}{2J-1} \qquad (O \text{ branch}) \quad (38)$$

The quantity $\bar{N}$ is given by

$$\bar{N} = \int \psi_{v', J'}^{*}(R) \, N(R) \, \psi_{v'', J''}(R) \, R^2 dR \quad (39)$$

and if we define, for the coordinates of an electron

$$\overline{x^2} = \int \psi_{e'}^{*}(\mathbf{r}|R) x^2 \, \psi_{e''}(\mathbf{r}|R) d\mathbf{r} \quad (40)$$

and $\overline{z^2}$ similarly, we find that

$$N(R) = \tfrac{1}{2} R^2 + 2(\overline{x^2} - \overline{z^2}) \quad (41)$$

for H_2, and

$$N(R) = \tfrac{1}{2} R^2 + \overline{x^2} - \overline{z^2} \quad (42)$$

for H_2^+.

James and Coolidge studied transitions within the ground state $^1\Sigma_g^+$ of H_2. The theory above may be applied to this case by assuming the same initial and final electronic wave functions. Their final results for the (1-0), (2-0), and (3-0) bands were that the $S(0)$ lines had absorption coefficients 8.1×10^{-9}, 7.3×10^{-9}, and 1.5×10^{-9} times that of the first line of the R branch of the fundamental band of HCl. The Einstein spontaneous emission transition probabilities of the three $S(0)$ lines calculated from James and Coolidge's data are 8.0×10^{-8}, 2.6×10^{-7}, and 1.1×10^{-7} sec^{-1}, respectively. The intensity decrease for the overtone bands is slower than for most molecular spectra, partly because of a (frequency)^3 factor, instead of a first power as for dipole radiation, and partly because of the large anharmonicity of H_2.

Bates and Poots (1953) used exact wave functions for H_2^+ to study its quadrupole rotation-vibration spectrum. The Einstein spontaneous emission transition probabilities for the $S(0)$ lines in the (1-0), (2-0), and (3-0) bands are 1.6×10^{-7}, 3.1×10^{-8}, and 5×10^{-9} sec^{-1}, respectively. Many of the other lines in the S, Q, and O branches have comparable transition probabilities.

3.3 Transitions Violating Approximate Selection Rules

The selection rule (5), $\Delta S = 0$, is the one most commonly broken, due to the effect of spin-orbit interaction. An example is the Cameron system, $a^3\Pi - X^1\Sigma^+$ in CO, studied in detail by Rao (1949), and for which Jarmain *et al.* (1955) gave relative vibrational transition probabilities. Similar transitions have been observed in several molecules, including gallium, indium, and thallium halides and the halogen molecules; recent papers include that of Barrow and associates (1957) on GaF. The system $d^3\Pi - X^1\Sigma^+$ in CO was found by Tanaka *et al.* (1957), and the system $C^3\Pi - X^1\Sigma_g^+$ in N_2 by Tanaka (1955). A band $^3\Delta_u - X^1\Sigma_g^+$ was found in N_2 by Wilkinson (1959). The only intercombination bands of higher multiplicity so far observed with certainty are a $^4\Sigma^- - {}^2\Pi$ transition in GeH (Klemen and Werhagen, 1953).

Intercombination transitions between Σ states are known. One type is exemplified by some of the Hopfield-Birge transitions, $a^1\,{}^3\Sigma^+ - X^1\Sigma^+$ in CO. This transition is made possible by spin-orbit interactions. The Vegard-Kaplan bands in N_2 are of this type, $A^3\Sigma_u^+ - X^1\Sigma_g^+$. Relative vibrational transition probabilities were given by Jarmain *et al.* (1953). The Vegard-Kaplan bands have recently been observed in absorption by Wilkinson (1959) in 13 meter-atm of N_2; they have previously been observed only in emission and there may well have been many unsuccessful attempts to get absorption spectra. Their spontaneous emission transition probability was found by Wilkinson and Mulliken (1959) to be 40 $\sec^{-1}$. Herzberg (1953) has observed a system $^1\Sigma_u^- - X^3\Sigma_g^-$ in O_2. The structure of these is similar to that of the well-known red atmospheric bands, and the nature of the upper state was decided on the basis of theoretical considerations on the terms expected from the known electronic configurations of oxygen. One of the surprises is that the $^1\Sigma_u^- - X^3\Sigma_g^-$ is so weak. The calculations also predicted a state $^3\Delta_u$, and Herzberg found some very weak bands which he ascribed to the $^3\Delta_u - X^3\Sigma_g^-$ transition. This transition is only forbidden by the rule $\Delta\Lambda = 0, \pm 1$ [rule (6)] and, again, its weakness is surprising.

An excellent example of a transition violating the $\Sigma^+ \leftrightarrow \Sigma^-$ rule (12) was found by Herzberg in O_2. The transition was shown to be $A^3\Sigma_u^+ - X^3\Sigma_g^-$. These bands are of very low absolute intensity, and were studied in detail by Herzberg (1952b) using his multiple reflection technique (giving a path length of 800 meter-atm). *S* and *O* branches were observed, showing that spin-orbit interactions are responsible for the observed intensity. The Herzberg bands have been observed in afterglows in oxygen-nitrogen mixtures (Barth and Kaplan, 1957). Jarmain and associates (1955) gave relative vibrational transition probabilities for the Herzberg bands. A band system $a'\ ^1\Sigma_u^- - X^1\Sigma_g^+$ in N_2 was observed by Wilkinson (1959), by Ogawa and Tanaka (1959, 1960), and by Wilkinson and Mulliken (1959), who determined the spontaneous emission transition probability to be 25 $\sec^{-1}$. The intensity is probably due to the mixing of $a'\ ^1\Sigma_u^-$ with nearby $^1\Pi_u$ states, but Wilkinson and Mulliken showed that there are several possible $^1\Pi_u$ states, so that detailed intensity calculations would be difficult. The two transitions just discussed do not involve a change of multiplicity. The transitions $^3\Sigma^+ - {}^1\Sigma^-$ and $^1\Sigma^+ - {}^3\Sigma^-$ were shown by Schlapp (1932) to occur with an intensity comparable with that of $^3\Sigma^+ - {}^1\Sigma^+$ or $^3\Sigma^- - {}^1\Sigma^-$ transitions, and they are nothing like so weak as $\Sigma^+ - \Sigma^-$ transitions with no change of multiplicity. Spin-orbit perturbations are responsible for the intensity. An example of this type of transition is $e^3\Sigma^- - X^1\Sigma^+$ of CO, discussed by Herzberg and Hugo (1955), and the similar $Y^3\Sigma_u^- - X^1\Sigma_g^+$ in N_2 has been discussed by Ogawa and Tanaka (1960) and by Wilkinson (1960).

As was seen in the previous subsection (§ 3.2), the rotation-vibration spectrum is very highly forbidden for homonuclear diatomic molecules. In the case of heteronuclear molecules, for example, HD, there is a permanent dipole moment and there is no distinction between symmetric and antisymmetric rotational levels. The rule that electronic states of the same parity do not combine is now valid only to a first approximation, and it may be violated by interaction of electronic motion with rotation and vibration, giving a weak electric dipole rotation-vibration spectrum. The simplest example is HD, which was first observed by Herzberg (1950b) and later studied in detail by Durie and Herzberg (1960). The transitions are within the electronic state $^1\Sigma_g^+$, violating rule (4), transitions $\Delta J = 0$ violate the rigorous rule (2), and only $\Delta J = \pm 1$ are allowed. Durie and Herzberg confirmed this, finding six or seven lines in each of the (1-0), (2-0), (3-0), and (4-0) bands and identifying P and R branches with a zero gap in between. No pressure broadening was detectable, and the paths used were not long enough to observe quadrupole radiation. The theory of the intensities of these transitions was studied by Wick (1935) and, quantitatively, by Wu (1952). The overtone band intensities fall off very slowly; this is qualitatively because the dipole moment is produced by the vibration and increases with it. Wu's formulae were applied to HD^+ by Bates and Poots (1953), who found a transition probability 18 sec^{-1} for the (1-0) band. They found a value of 0.02 sec^{-1} for the (1-0) band in $^{14}N^{15}N^+$.

3.4 Transitions Induced by External Fields

There are several cases known where a selection rule, which holds exactly for free molecules, may be violated in the presence of external fields, collisions with other molecules, and so on. Such radiation is of the electric dipole type, and is known as enforced dipole radiation. Infrared forbidden rotation-vibration spectra of O_2, N_2, and H_2 were observed by Crawford *et al.* (1949) and Welsh *et al.* (1949). They used absorbing paths of 85 cm and pressures up to 100 atm. The intensities of the spectra varied as the square of the gas pressure, which shows that collisions are the cause of the spectra. The selection rule which operates was shown to be $\Delta J = 0, \pm 2$; this is in agreement with the work of Condon (1932), who studied the effect on a molecule of external electric fields. Later work by Crawford *et al.* (1950) and Welsh *et al.* (1951) confirmed the interpretation of these spectra, and the latter paper reported observations in H_2, O_2, and N_2 at twice the fundamental frequencies. In the (1-0) band of hydrogen, a maximum at 4155 cm^{-1} was identified as the Q branch, and the lines $S(0)$ and $S(1)$ could be seen. These features were also seen in the (2-0) branch. The line widths are large, ~ 250 cm^{-1}, as they should be for enforced dipole radiation (cf. § 3.2, quadrupole radiation).

There are examples of electronic transitions becoming allowed because of enforced dipole radiation. Thus, in oxygen at high pressure the $^1\Delta_g - {}^3\Sigma_g^-$ system is greatly enhanced relative to the $^1\Sigma_g^+ - {}^3\Sigma_g^-$ system. The $^1\Delta_g - {}^3\Sigma_g^-$ system has been studied in the liquid oxygen spectrum by Herzberg and Herzberg (1947), Cho *et al.* (1956), and by others. Such observations provided the first evidence for the existence of the $^1\Delta_g$ state. The intensities of the bands in high-pressure gaseous oxygen increase proportional to the square of the density, so in this case, and probably also for liquid oxygen, induced electric dipole radiation is present much more strongly than magnetic dipole radiation.

4 Forbidden Transitions in Polyatomic Molecular Spectra

4.1 Introduction

The intensity of an electronic band in the spectrum of a polyatomic molecule is proportional to the square of the transition dipole moment. To a first approximation this moment may be calculated with the nuclei in their equilibrium position. For many of the more symmetrical molecules, however, symmetry considerations impose a zero value for the equilibrium dipole moment. In such cases weak transitions may still be observed. These vibronic transitions are due to the mixing of the wave functions of the symmetrical state with nontotally symmetric wave functions of certain vibrational states. During these vibrations the dipole moment will have nonzero instantaneous values, although the average value will still be zero. These considerations need not be restricted to cases when the equilibrium transition moment is exactly zero; for weakly allowed transitions the vibronic interactions may give appreciable contributions to the band intensities. Another class of weak transitions arises from spin-orbit interaction, which allows finite transition probabilities between states of different multiplicities. Still more highly forbidden transitions are possible because of the combined effects of vibronic interactions and spin-orbit interactions.

4.2 General Theory

The theory of vibronic transitions in polyatomic molecules was first discussed by Herzberg and Teller (1933), who outlined the principles involved. No detailed computations were made at that time. Only from about 1955 were attempts made at carrying out numerical calculations. Papers by Murrell and Pople (1956), Albrecht (1960a), and

Liehr (1960) contain useful accounts of the theory and additional references.

The Born-Oppenheimer approximation (that owing to the rapid motions of the electrons compared to the nuclei the electrons at any time can be regarded as being in a steady state in the field of the instantaneous position of the nuclei) is used to express the vibronic wave function Ψ_{kj} of the molecule in the form

$$\Psi_{kj} = \Theta_k(x, Q)\, \Phi_{kj}(Q), \tag{43}$$

where $\Theta_k(x, Q)$ is the electronic wave function of the kth electronic state for fixed Q, $\Phi_{kj}(Q)$ is the vibrational wave function of the jth vibrational state of electronic state k, and x and Q refer to the complete set of coordinates required to locate all the electrons and all the nuclei, respectively. The coordinates Q are taken as zero in equilibrium. The transition moment $\mathbf{M}_{gi,kj}$ between vibronic states described by quantum numbers gi, kj is given by

$$\mathbf{M}_{gi,kj} = \int \Psi_{gi}^* \{\mathbf{m}_e(x) + \mathbf{m}_n(Q)\}\, \Psi_{kj}\, dx\, dQ, \tag{44}$$

where $\mathbf{m}_e(x)$ and $\mathbf{m}_n(Q)$ are, respectively, the electronic and nuclear terms of the electric dipole moment operator; g will denote the ground state. Upon substituting the expression (43) for Ψ into (44), we find that the term involving $\mathbf{m}_n(Q)$ vanishes because of the orthogonality of the electronic wave functions Θ_k and $\Theta_g (g \neq k)$, and the remaining term may be written

$$\mathbf{M}_{gi,kj} = \int \Phi_{gi}^*(Q)\, \mathbf{M}_{gk}(Q)\, \Phi_{kj}(Q)\, dQ \tag{45}$$

where

$$\mathbf{M}_{gk}(Q) = \int \Theta_g^*\, (x, Q)\, \mathbf{m}_e(x)\, \Theta_k(x, Q)\, dx \tag{46}$$

is the variable electronic transition moment.

The transition probability from all vibrational levels of state k to the state g, i is proportional to

$$\sum_j |\, \mathbf{M}_{gi,kj}\, |^2 = \int \Phi_{gi}^*(Q)\, \mathbf{M}_{gk}^2(Q)\, \Phi_{gi}(Q)\, dQ = \langle \mathbf{M}_{gk}^2(Q) \rangle_{ii} \tag{47}$$

where the rule for a matrix product has been used. $\langle\, \rangle_{ii}$ denotes the average. If B_i is the Boltzmann weighting factor for vibrational state i

($\Sigma B_i = 1$), then the total transition probability from state g to state k is proportional to

$$\sum_i B_i \left(\sum_j | \mathbf{M}_{gi,kj} |^2\right) = \sum_i B_i \langle \mathbf{M}^2_{gk}(Q) \rangle_{ii}. \tag{48}$$

We now assume that we can expand the electronic wave function in the form

$$\Theta_k(x, Q) = \Theta^0_k(x) + \sum_s \lambda_{ks}(Q)\, \Theta^0_s(x). \tag{49}$$

This is the basis of the Herzberg-Teller theory, and effectively assumes that we may treat nuclear motion as a perturbation on the electronic Schrödinger equation, taking for the zero order functions those for the vibrationless state. If the Hamiltonian of the molecule is written

$$H = H_0 + H_1(Q) \tag{50}$$

where H_0 is the symmetrical part (for no vibration) and $H_1(Q)$ the change in H induced by the vibration, then first order perturbation theory shows that

$$\lambda_{ks}(Q) = \frac{1}{E^0_s - E^0_k} \int \Theta^0_k(x)\, H_1(Q)\, \Theta^0_s(x)\, dx, \tag{51}$$

with E^0_s and E^0_k the unperturbed energies of the electronic states s and k, respectively. Then

$$\mathbf{M}_{gk}(Q) = \mathbf{M}^0_{gk} + \sum_s \lambda_{ks}(Q)\, \mathbf{M}^0_{gs} + \sum_t \lambda_{gt}(Q)\, \mathbf{M}^0_{tk} \tag{52}$$

where

$$\mathbf{M}^0_{gk} = \int \Theta^{0*}_g(x)\, \mathbf{m}_e(x)\, \Theta^0_k(x)\, dx. \tag{53}$$

The only part of the Hamiltonian which contains Q is the Coulomb interaction between the electrons and the nuclei, and for small displacements $H_1(Q)$ involves the displacements linearly. Equation (51) may then be expressed in the form

$$\lambda_{ks}(Q) - \sum_a \Lambda_{ksa} Q_a \tag{54}$$

where Q_a are the individual coordinates of the nuclei. Substituting (54) in (52), and the result in (48), we find the total transition probability is given to the second order in the Λ's by

$$(\mathbf{M}^0_{gk})^2 + \sum_{i,a} B_i \langle Q^2_a \rangle_{ii} \left(\sum_s \Lambda_{ksa} \mathbf{M}^0_{gs} + \sum_t \Lambda_{g\,ta} \mathbf{M}^0_{kt} \right)^2. \tag{55}$$

The g-k transition contains a forbidden component if for at least one s neither Λ_{ksa} nor $\mathbf{M}^0_{gs}$ vanishes or for at least one t neither $\Lambda^0_{g\,ta}$ nor $\mathbf{M}^0_{kt}$ vanishes. If one $\mathbf{M}^0_{gs}$ is nonvanishing, s must be the upper state of a transition from g which is allowed. The symmetry of such states may be determined by group theory for any particular molecule. For Λ_{ksa} not to vanish, the product $\Theta^{0*}_k H_1 \Theta^0_s$ must contain at least once the totally symmetric representation of the group of the molecule, and since the symmetries of Θ^0_k and Θ^0_s are known the possible symmetries of H_1 may be determined.

If $\mathbf{M}^0_{gk} = 0$ the transition is forbidden in the absence of vibration. If, in addition, one $\mathbf{M}^0_{gs} \neq 0$ Θ^0_k and Θ^0_s are of different symmetries. Elementary group theory shows that $\mathbf{M}^0_{gs} \neq 0$ only if H_1 is not totally symmetrical, and hence that the vibrations are not totally symmetrical. If $\mathbf{M}^0_{gk} \neq 0$ totally symmetric vibrations may contribute to the intensity.

The general conclusion from the theory sketched here is that if a transition is forbidden in the absence of vibration, it may acquire a finite strength by interactions with nontotally symmetric vibrations.

We shall not discuss theoretically the effects of spin-orbit interaction, but in the next section we shall refer briefly to some of the work which has been done on this type of perturbation. The intercombination transitions to which it gives rise are much weaker than those produced by vibronic interactions.

4.3 Numerical Results

The most important examples of electronic bands which are made allowed by vibronic interactions are in benzene. There are three absorption bands of benzene in the ultraviolet region of the spectrum. One at $\lambda 1850$ A is ascribed to a transition from the ground state $({}^1A_{1g})$ to an excited state ${}^1E_{1u}$, and this transition is allowed even in the absence of vibration. It has an oscillator strength $f = 0.6$ (Potts, 1955b). A band at $\lambda 2150$ A is ascribed to ${}^1A_{1g} - {}^1B_{1u}$, with $f = 0.1$ (Potts, 1955b). A third band at $\lambda 2650$ A is due to the transition ${}^1A_{1g} - {}^1B_{2u}$ with $f = 0.0014$ (Klevens and Platt, quoted by Murrell and Pople, 1956).

There has been some controversy about the identifications of these upper states, but they now seem settled. Detailed examination shows that the $^1B_{1u}$ and $^1B_{2u}$ states are made accessible by e_{2g} vibrations. Murrell and Pople (1956) applied the Herzberg-Teller theory to the calculation of the intensities of the $\lambda 2650$ and $\lambda 2150$ bands. They found for the $\lambda 2650$ band a value $f \simeq 0.01$, and for $\lambda 2150$ $f \simeq 0.4$, both values rather higher than the experimental values. Calculations were made by Liehr (1957, 1958) who obtained for $\lambda 2650$ $f = 0.0003$ to 0.006, and for $\lambda 2150$ $f = 0.01$ to 0.06, according to the assumptions made. Albrecht (1960b) has made extensive calculations on benzene, obtaining for $\lambda 2650$ $f = 0.0062$ and for $\lambda 2150$ $f = 0.053$. The agreement between theory and experiment is not very good, but the discrepancies can probably be accounted for by the crudeness of the wave functions assumed.

A second molecule which has proved of particular interest in connection with forbidden bands is formaldehyde (H_2CO). This has an absorption system in the ultraviolet near $\lambda 3500$ A. In the electronic ground state ($^1A_1$) the molecule is planar. The relevant excited state is of type $^1A_2$, and group theory shows that the transition $^1A_2 \leftarrow {}^1A_1$ is forbidden if the excited state is planar. There is good evidence that the excited state is nonplanar (Robinson and DiGiorgio, 1958), and it seems probable that vibronic interactions are responsible for the observed intensities of the bands. This has been confirmed by Pople and Sidman (1957), who have shown that there is a nontotally symmetric out-of-plane vibration which leads to intensity mixing chiefly from the $^1B_2 \leftarrow {}^1A_1$ transition. Their calculations gave $f = 3 \times 10^{-4}$, compared with an observed value 2.4×10^{-4}.

A number of other organic compounds have been studied in this way, including various substituted benzenes (Albrecht, 1960a, b; Potts, 1955b), olefins (Potts, 1955a), and coronene (Sidman, 1955).

The second type of forbidden transition—those connecting states of different multiplicities and caused by spin-orbit interaction—has been investigated for several substances. Sometimes these intercombination bands show in normal absorption spectra. In others they are observed by phosphorescence, a strong afterglow emitted after excitation by ultraviolet light. These bands arise because of a low-lying metastable triplet state. Spin-orbit interaction allows this to mix with an excited singlet state, and the transition from the singlet ground state to the excited triplet state becomes weakly allowed. The observed lifetimes (McClure, 1949 and other references) vary from 10^{-4} to 100 sec or more. Aromatic compounds have the longest lifetimes, and presumably the smallest spin-orbit interactions. Among the evidence supporting the attribution of these finite lifetimes to spin-orbit interactions may be mentioned the increase in transition probabilities when (say) bromine replaces chlorine in halogen substituents, the transition probability increasing with the square of the spin-orbit parameter for the atom, as one would expect on theoretical grounds.

A weak transition in formaldehyde near λ3900 A has been identified as a $^3A_2 \leftarrow {}^1A_1$ transition (DiGiorgio and Robinson, 1959), Sidman (1958) has calculated by allowing for spin-orbit interaction that $f = 1.5 \times 10^{-7}$, and DiGiorgio and Robinson have confirmed this interpretation by polarization observations and by estimating an experimental oscillator strength of about 10^{-6}.

In benzene, a weak band at λ3400 A has been ascribed to a transition $^3B_{2u} \leftarrow {}^1A_{1g}$. Evans (1956) has shown that dissolved oxygen enhances these bands and their intensity as normally observed is not due mainly to spin-orbit interaction within the benzene molecule, but to spin-orbit perturbations produced by the oxygen. Craig and associates (1958) have shown that in oxygen-free benzene, the triplet-singlet transition has $f \leqslant 7 \times 10^{-12}$, or the triplet lifetime $\geqslant$ 300 sec; the bands were not visible through 22.5 meters liquid path length. Hameka and Oosterhoff (1958) studied the theoretical oscillator strengths, ascribing the $^3B_{2u} \leftarrow {}^1A_{1g}$ intensity to mixing of $^3B_{2u}$ and $^1B_{1u}$. As mentioned earlier the $^1B_{1u} \leftarrow {}^1A_{1g}$ transition derives its intensity from vibronic interactions, so that $^3B_{2u} \leftarrow {}^1A_{1g}$ has a finite strength only because of the combined effects of vibronic and spin-orbit interactions. Hameka and Oosterhoff obtained $f = 8.8 \times 10^{-12}$, corresponding to a lifetime of 190 sec. There is a possibility that the upper state should be $^3B_{1u}$ (instead of $^3B_{2u}$) and if that were so the lifetime would be increased to 800 sec, or $f = 2.1 \times 10^{-12}$.

Other molecules which have been studied in connection with intercombination transitions include acetone (Hameka and Oosterhoff, 1958) and thiophosgene (Burnell, 1956).

5 *Forbidden Transitions in Crystals*

In crystalline salts many atoms make up closed shells by losing valency electrons or by completing shells, and the ions so formed, being in ^1S states, are of little interest. In the case of the transition elements in the long periods and in the case of the rare earth elements, the ions which remain have incomplete d^n or f^n shells. These have ground states which are degenerate. It is a good approximation to regard the ion of the transition element or rare earth as being situated in an electric field formed by the remaining ions of the crystalline lattice. This crystalline field, together with the spin-orbit interaction of the electrons in the incomplete shell of the ion, causes the degenerate levels of the ion to be split into many levels of distinct energies. Considerable interest attaches to the spectroscopic transitions which take place between all the various energy levels.

The transitions in the optical region of the spectrum take place between levels separated by intervals comparable with those in the free d^n or f^n ion, and closely correlated with them. The transitions are weak, many of the lines having oscillator strengths of order 10^{-6}; they range from 10^{-5} to 10^{-10}. This suggests that they are forbidden at least to some degree of approximation. The lines are sharp, and occur in groups of nearly the same wavelengths. The lines are interpreted as being due to transitions within the d^n or f^n shells, and are forbidden (by Laporte's rule) for free ions since they connect states of the same parity. They must be due either to forced electric dipole radiation or to magnetic dipole or electric quadrupole radiation. Study of the occurrence of the lines in groups has shown that each group usually corresponds to one possible transition in the free ion, and the various lines of the group arise from transitions between the levels into which the ion states are split by the crystalline field. It might therefore be expected that the transitions are themselves produced by the perturbing effects of the crystal field. Detailed calculations have confirmed this. Estimates of the intensity of transitions produced by crystal field perturbations have been made by Broer and associates (1944) and they find values of 10^{-6} to 10^{-8} for the oscillator strengths of typical transitions. In the case of magnetic dipole and electric quadrupole radiation transitions are permitted between states of the same parity, so that the crystal field perturbations can be ignored. Intermediate coupling theory is required for the magnetic dipole transitions, taking the spin-orbit interaction as the perturbation. Estimates of the oscillator strengths for magnetic dipole and electric quadrupole transitions by Broer, Gorter, and Hoogschagen were 2×10^{-8} and 2×10^{-9}, respectively. On grounds of predicted intensities it seems probable that most of the stronger transitions are due to forced electric dipole radiation.

Further evidence for the nature of the transitions has been obtained by the study of the behavior of polarized light passing through large single crystals of the salt. Magnetic dipole and electric dipole radiation behave in a different manner. In this way it has been shown that a very large number of transitions are due to electric dipole radiation. For example, Sayre *et al.* (1955) have proved that all lines from $PrCl_3$

crystals are due to electric dipole transitions, and Stout (1959) has shown that in manganous fluoride the transitions from the $^6A_{1g}$ ground state to the excited quartet states have oscillator strengths of order 10^{-7} and are all electric dipole radiation.

Although most lines in crystal spectra are due to enforced electric dipole radiation, there are two examples of magnetic dipole radiation. One occurs in crystals containing the Eu^{3+} ion; the free ion has a $4f^6$ configuration, and the transitions observed occur between levels (in the free ion) ^7F and ^5D. Polarization studies by Sayre and Freed (1956, and earlier papers by Freed) have shown that in the crystals the transitions $^7F_0 - {}^5D_1$, $^7F_2 - {}^5D_1$ and $^7F_1 - {}^5D_0$ are due to magnetic dipole radiation. The transition $^7F_0 - {}^5D_1$ has an oscillator strength of 10^{-8}. Several other transitions have been observed, and are due to electric dipole radiation. Among these is the transition $^7F_0 - {}^5D_0$, with an oscillator strength about 10^{-10}. This is of a still higher degree of forbiddenness, because the selection rule $J = 0 \nleftrightarrow J = 0$ is violated; this rule holds even in intermediate coupling and for all types of radiation, and its breakdown is due to the crystal field perturbations. The other example of magnetic dipole radiation was found by Rosa (1943) in Dy^{3+} ions.

C. M. Herzfeld (1957) has discussed the appearance of [N I] lines in solids, with half-lifetimes of 15 sec, instead of 20 hours for the free atoms. The transitions appear to be made possible because the terms $2p^23s\ {}^2P$ and $2p^33s\ {}^4P$ interact by spin-orbit interaction, and the terms $2p^23s\ {}^2P$ and $2p^3\ {}^2D$ interact by crystal field perturbations. Thus the $2p^3\ {}^2D$ term acquires some of the characteristics of $2p^33s\ {}^4P$, and because the transition $2p^23s\ {}^4P - 2p^3\ {}^4S$ is allowed, the transition $2p^3\ {}^4S - 2p^3\ {}^2D$ becomes weakly allowed.

Space does not permit of any further discussion of the spectra of crystals. Reference may be made to articles by Runciman (1958), Tanabe and Sugano (1954), Kotani *et al.* (1960), Hellwege (1948), Moffitt and Ballhausen (1956), and McClure (1959). There does not appear to be any known example of electric quadrupole radiation being observed in crystals. Many of the considerations discussed above for crystals also apply to the spectra of solutions, but owing to the variable perturbations we get a broad absorption band corresponding to each group of lines in crystal spectra. Magnetic dipole and electric quadrupole intensities are the same in solutions as in crystals.

6 *Forbidden Transitions in Astrophysics*

Space does not permit a detailed discussion of the occurrence of forbidden transitions in celestial objects and of the information which has been obtained from them, and we shall only give references to the more important fields of work. An important general reference is Merrill (1956).

Forbidden lines were first identified in gaseous nebulae, and they are the origin of many prominent spectrum lines. Many objects show lines from several stages of ionization of the same atom. The temperatures, electron densities and ionic abundances in the nebulae can be deduced from the intensities of the forbidden lines. The loss of energy in the forbidden lines is the chief mechanism controlling the electron temperature of the nebular gas. Excellent reviews on gaseous nebulae have been written by Aller (1956) and by Seaton (1960).

Forbidden lines were discovered by Edlén in the solar corona. Their high degree of ionization provided strong evidence for an electron temperature of roughly 10^6 °K in the corona, and this in turn accounted for the absence of absorption and ordinary emission lines of hydrogen and the metals, for the breadth of the coronal lines, for the extended size of the corona, and for observations of solar radio emission. Useful reviews on the subject have been given by Edlén (1945), Allen (1954), and van de Hulst (1953).

Forbidden transitions occur in the terrestrial atmosphere. The O_2 $b^1\Sigma_g^+ - X^3\Sigma_g^-$ bands are strong in the telluric spectrum, occurring near λ7594 [(0-0) band], λ6867 [(1-0) band] and λ6277 [(2-0) band]. The $a^1\Delta_g - X^3\Sigma_g^-$ transition in O_2 is also present (Babcock and Moore, 1947). It was during the study of the λ7594 band that weak bands due to the isotopic molecules $O^{16}O^{18}$ and later $O^{16}O^{17}$ were discovered, these leading to the determination of the relative isotope abundances in oxygen. Forbidden emission lines [O I], [N I], [O II], [N II], the Vegard-Kaplan bands of N_2, and the atmospheric and Herzberg bands of O_2 occur in either or both the auroral and airglow spectra. For reviews on these subjects see Goldberg (1954), Chamberlain and Meinel (1954), and Bates (1960).

The presence of methane and ammonia in the atmospheres of the giant planets and the low mean densities of the planets made it seem probable that much of their atmospheres might be composed of H_2 and N_2. Spectroscopic detection seemed impossible because neither H_2 nor N_2 has any permitted transitions in the accessible spectrum.

However, Herzberg (1938) suggested that the quadrupole rotation-vibration spectrum of H_2 might be observable in their atmospheres. He predicted the wavelengths which would be observed. The presence of the $S(0)$ line of the (3-0) band in Uranus and Neptune was established by Herzberg (1952a); its origin was probably pressure-induced electric dipole radiation rather than quadrupole radiation. Kiess *et al.* (1960) have identified the lines Q(1), S(0), S(1), and S(2) of the (3-0) band of H_2 in Jupiter; these are probably quadrupole lines.

Forbidden lines occur in numerous peculiar stellar spectra, including novae. A comprehensive discussion of the occurrence of forbidden lines may be found in Merrill (1956), who also gives an extensive bibliography.

The λ5577 line of [O I] is present in the solar absorption spectrum, and was studied in detail by Bowen (1948) and by Cabannes and Dufay (1948a), who also found the lines $\lambda\lambda$6300, 6364 of [O I]. These lines are of interest because permitted absorption lines in oxygen in the accessible spectrum arise from high energy levels, and analysis of these transitions involves uncertainties in the solar atmospheric structure. Bowen showed that the forbidden lines can give an improved solar abundance of oxygen, and Osterbrock and Rogerson (1961) have rediscussed this. Cabannes and Dufay (1948b) identified a number of absorption lines of [Fe II], Babcock and Moore (1947) identified a line of [Si I] at λ10991 and [N I] may possibly be present (Dufay, 1952).

In § 2.3 we discussed the magnetic dipole transition at frequency 1420 Mc/sec between the two hyperfine levels of the hydrogen ground state. This transition has proved to be of exceptional importance in radioastronomical studies of our galaxy. It was discovered as an emission line in 1951 by three independent groups in Australia, Holland, and the United States. In spite of the low transition probability of the line it has been extensively observed. In the spiral arms of our galaxy there are clouds of hydrogen gas, with a mean density of about 1 hydrogen atom per cubic centimeter. The absorption coefficient computed from the transition probability is about 1 per kiloparsec (3.08×10^{18} cm). It is therefore of the right order to be important for problems of large scale structure. Many observations have been made and much information obtained. For a general account of this work see Oort (1959).

References

Albrecht, A. C. (1960a) *J. chem. Phys.* **33**, 156.
Albrecht, A. C. (1960b) *J. chem. Phys.* **33**, 169.
Allen, C. W. (1937) *Astrophys. J.* **85**, 156.

Allen, C. W. (1954) *Rep. Progr. Phys.* **17**, 135.
Aller, L. H. (1956) "Gaseous Nebulae," Chapman and Hall, London.
Aller, L. H., Ufford, C. W., and Van Vleck, J. H. (1949) *Astrophys. J.* **109**, 42.
Aller, L. H., Bowen, I. S., and Minkowski, R. (1955) *Astrophys. J.* **122**, 62.
Anderson, R. S., Johnson, C. M., and Gordy, W. (1951) *Phys. Rev.* **83**, 1061.
Babcock, H. D., and Herzberg, L. (1948) *Astrophys. J.* **108**, 167.
Babcock, H. D., and Moore, C. E. (1947) Carnegie Inst. Washington Publ. **579**, p. 12.
Barrow, R. F., Dodsworth, P. G., and Zeeman, P. B. (1957) *Proc. Phys. Soc.* **A70**, 34.
Barth, C. A., and Kaplan, J. (1957) *J. chem. Phys.* **26**, 506.
Bates, D. R. (1960) *In* "Physics of the Upper Atmosphere" (J. A. Ratcliffe, ed.), Chapts. 5 and 7. Academic Press, New York and London.
Bates, D. R., and Poots, G. (1953) *Proc. Phys. Soc.* **A66**, 784.
Beringer, R. (1946) *Phys. Rev.* **70**, 53.
Bernstein, H. J., and Herzberg, G. (1948) *J. chem. Phys.* **16**, 30.
Bethe, H. A., and Salpeter, E. E. (1957) *In* "Handbuch der Physik" (S. Flügge, ed.), Vol. 35, p. 88, Springer, Berlin.
Blaha, M. (1957) *Bull. Astron. Inst. Czech.* **8**, 34.
Borisoglebskii, L. A. (1958) *Uspekhi fiz. Nauk.* **66**, 603; *Soviet Phys.—Uspekhi* **1**, 211.
Bowen, I. S. (1928) *Astrophys. J.* **67**, 1.
Bowen, I. S. (1948) *Rev. mod. Phys.* **20**, 109.
Bowen, I. S. (1955) *Astrophys. J.* **121**, 306.
Bowen, I. S. (1960) *Astrophys. J.* **132**, 1.
Breit, G., and Teller, E. (1940) *Astrophys. J.* **91**, 215.
Brochard, J., and Jacquinot, P. (1945) *Ann. Phys. (Paris)* **20**, 508.
Broer, L. J. F., Gorter, C. J., and Hoogschagen, J. (1944) *Physica* **11**, 231.
Budó, L. (1937) *Z. Phys.* **105**, 579.
Budó, L., and Kovács, I. (1940) *Z. Phys.* **116**, 393.
Burkhalter, J. H., Anderson, R. S., Smith, W. V., and Gordy, W. (1950) *Phys. Rev.* **79**, 651.
Burnell, L. (1956) *J. chem. Phys.* **24**, 620.
Cabannes, J., and Dufay, J. (1948a) *C. R. Acad. Sci. (Paris)* **226**, 1569.
Cabannes, J., and Dufay, J. (1948b) *C. R. Acad. Sci. (Paris)* **226**, 2032.
Chamberlain, J. W., and Meinel, A. B. (1954) *In* "The Earth As a Planet" (G. P. Kuiper, ed.), Chapt. 11. Univ. Chicago Press, Chicago, Illinois.
Childs, W. H. J., and Mecke, R. (1931) *Z. Phys.* **68**, 344.
Cho, C. W., Allin, E. J., and Welsh, H. L. (1956) *J. chem. Phys.* **25**, 371.
Cole, C. D. (1960) *Bull. Amer. Phys. Soc.* **5**, 412.
Cole, C. D. (1961) *Bull. Amer. Phys. Soc.* **6**, 74.
Cole, C. D., and Mrozowski, S. (1954) *Phys. Rev.* **93**, 933.
Condon, E. U. (1932) *Phys. Rev.* **41**, 759.
Condon, E. U., and Shortley, G. H. (1951) "Theory of Atomic Spectra." Cambridge Univ. Press, London and New York.
Craig, D. P., Hollas, J. M., and King, G. W. (1958) *J. chem. Phys.* **29**, 974.
Crawford, M. F., Welsh, H. L., and Locke, J. L. (1949) *Phys. Rev.* **75**, 1607.
Crawford, M. F., Welsh, H. L., McDonald, J. C. F., and Locke, J. L. (1950) *Phys. Rev.* **80**, 469.
Datta, S. (1922) *Proc. Roy. Soc.* **A101**, 539.
de Jager, C. (1956) *Bull. Astron. Inst. Neth.* **13**, 9.
DiGiorgio, V. E., and Robinson, G. W. (1959) *J. chem. Phys.* **31**, 1678.
Douglas, A. E., and Rao, K. S. (1958) *Canad. J. Phys.* **36**, 565.

Dufay, J. (1952) *Ann. Astrophys.* **15**, 359.
Durie, R. A., and Herzberg, G. (1960) *Canad. J. Phys.* **38**, 806.
Edlén, B. (1944) *Phys. Rev.* **65**, 248.
Edlén, B. (1945) *Monthly Not. Roy. Astron. Soc.* **105**, 323.
Essen, L., and Parry, J. V. L. (1957) *Phil. Trans.* **A250**, 45.
Evans, D. F. (1956) *Nature* **178**, 534.
Fraser, P. A., Jarmain, W. R., and Nicholls, R. W. (1954) *Astrophys. J.* **119**, 286.
Frerichs, R., and Campbell, J. S. (1930) *Phys. Rev.* **36**, 1460.
Froese, C. (1957) *Monthly Not. Roy. Astron. Soc.* **117**, 615.
Fukuda, M. (1926) *Sci. Pap. Inst. Phys. Chem. Res. (Tokyo)* **4**, 171.
Garstang, R. H. (1951) *Monthly Not. Roy. Astron. Soc.* **111**, 115.
Garstang, R. H. (1952a) *Astrophys. J.* **115**, 506.
Garstang, R. H. (1952b) *Astrophys. J.* **115**, 569.
Garstang, R. H. (1955) *In* "The Airglow and the Aurorae" (E. B. Armstrong and A. Dalgarno, eds.), p. 324. Pergamon, New York.
Garstang, R. H. (1956) *Proc. Cambridge Phil. Soc.* **52**, 107.
Garstang, R. H. (1957a) *Proc. Cambridge Phil. Soc.* **53**, 214.
Garstang, R. H. (1957b) *Monthly Not. Roy. Astron. Soc.* **117**, 393.
Garstang, R. H. (1958a) *Proc. Cambridge Phil. Soc.* **54**, 383.
Garstang, R. H. (1958b) *Monthly Not. Roy. Astron. Soc.* **118**, 234.
Garstang, R. H. (1958c) *Monthly Not. Roy. Astron. Soc.* **118**, 572.
Garstang, R. H. (1960) *Monthly Not. Roy. Astron. Soc.* **120**, 201.
Gaviola, E. (1929) *Contrib. estud. Cienc. fís. y mat., Ser. mat. fis. (La Plata)* **5**, 65.
Gerjuoy, E. (1941) *Phys. Rev.* **60**, 233.
Goldberg, L. (1954) *In* "The Earth As a Planet" (G. P. Kuiper, ed), Chapt. 9. Univ. Chicago Press, Chicago, Illinois.
Gordy, W. (1948) *Rev. mod. Phys.* **21**, 668.
Hameka, H. F., and Oosterhoff, L. J. (1958) *Molecular Phys.* **1**, 358.
Hellwege, K. H. (1948) *Ann. Phys. (Leipzig)* [6] **4**, 97, 127, 136, 143, 150, 357.
Herzberg, G. (1938) *Astrophys. J.* **87**, 428.
Herzberg, G. (1946) *Phys. Rev.* **69**, 362.
Herzberg, G. (1949) *Nature* **163**, 170.
Herzberg, G. (1950a) *Canad. J. Res.* **A28**, 144.
Herzberg, G. (1950b) *Nature* **166**, 563.
Herzberg, G. (1952a) *Astrophys. J.* **115**, 337.
Herzberg, G. (1952b) *Canad. J. Phys.* **30**, 185.
Herzberg, G. (1953) *Canad. J. Phys.* **31**, 657.
Herzberg, G., and Hugo, T. J. (1955) *Canad. J. Phys.* **33**, 757.
Herzberg, G., and Teller, E. (1933) *Z. phys. Chem.* **B21**, 410.
Herzberg, L., and Herzberg, G. (1947) *Astrophys. J.* **105**, 353.
Herzfeld, C. M. (1957) *Phys. Rev.* **107**, 1239.
Holmes, J. R., and Deloume, F. (1952) *J. Opt. Soc. Amer.* **42**, 77.
Hults, M., and Mrozowski, S. (1952) *Phys. Rev.* **86**, 587.
Jacquinot, P., and Brochard, J. (1943) *C. R. Acad. Sci. (Paris)* **216**, 581.
James, H. M., and Coolidge, A. S. (1938) *Astrophys. J.* **87**, 438.
Jarmain, W. R., Fraser, P. A., and Nicholls, R. W. (1953) *Astrophys. J.* **118**, 228.
Jarmain, W. R., Fraser, P. A., and Nicholls, R. W. (1955) *Astrophys. J.* **122**, 55.
Jenkins, F. A., and Segrè, E. (1939) *Phys. Rev.* **55**, 545.
Kaplan, J. (1939a) *Phys. Rev.* **55**, 598.
Kaplan, J. (1939b) *Phys. Rev.* **55**, 858.

Kellogg, J. B. M., and Millman, S. (1946) *Rev. mod. Phys,* **18**, 323.
Kessler, K. G. (1950) *Phys. Rev.* **77**, 559.
Kiess, C. C., Corliss, C. H., and Kiess, H. K. (1960) *Astrophys. J.* **132**, 221.
Kipper, A. Y. (1952) *Pub. Astron. Obs. Tartu* **32**, 63.
Kleman, B., and Werhagen, E. (1953) *Ark. Fys.* **6**, 399.
Kotani, M., Tanabe, Y., and Sugano, S. (1960) *Progr. theor. Phys. Suppl. No.* **14**.
Kovács, I. (1960) *Canad. J. Phys.* **38**, 955.
Kovács, I., and Budó, L. (1941) *Z. Phys.* **117**, 612.
Kusch, P. (1954) *Phys. Rev.* **93**, 1022.
Kusch, P. (1955) *Phys. Rev.* **100**, 1188.
Kusch, P. (1956) *Phys. Rev.* **101**, 627.
Lamb, W. E. (1951) *Rep. Progr. Phys.* **14**, 19.
Lamb, W. E. (1957) *Phys. Rev.* **105**, 559.
Lamb, W. E., and Maiman, T. (1957) *Phys. Rev.* **105**, 573.
Layzer, D. (1954) *Monthly Not. Roy. Astron. Soc.* **114**, 692.
Lichten, W. (1957) *J. chem. Phys.* **26**, 306.
Lichten, W. (1960) *Phys. Rev.* **120**, 848.
Liehr, A. D. (1957) *Canad. J. Phys.* **35**, 1123.
Liehr, A. D. (1958) *Canad. J. Phys.* **36**, 1588.
Liehr, A. D. (1960) *Rev. mod. Phys.* **32**, 436.
Liszka, L., and Niewodniczanski, H. (1958) *Acta phys. Polon.* **17**, 345.
McClure, D. S. (1949) *J. chem. Phys.* **17**, 905.
McClure, D. S. (1959) *Solid. State Phys.* **9**, 399.
McLennan, J. C., and Shrum, G. M. (1925) *Proc. Roy. Soc.* **A108**, 501.
Martin, W. C., and Corliss, C. H. (1960) *J. Res. Nat. Bur. Stand.* **A64**, 443.
Martin, W. C., and Tech, J. L. (1961) *J. Opt. Soc. Amer.* **51**, 591.
Merrill, P. W. (1928) *Astrophys. J.* **67**, 391.
Merrill, P. W. (1956) "Lines of the Chemical Elements in Astronomical Spectra," *Carnegie Inst. Wash. Publ.* **610**.
Moffitt, W., and Ballhausen, C. J. (1956) *Ann. Rev. phys. Chem.* **7**, 107.
Mrozowski, S. (1938) *Z. Phys.* **108**, 204.
Mrozowski, S. (1944) *Rev. mod. Phys.* **16**, 153.
Mrozowski, S. (1945) *Phys. Rev.* **67**, 161.
Mrozowski, S. (1954) *Phys. Rev.* **93**, 933.
Mrozowski, S. (1956) *J. Opt. Soc. Amer.* **46**, 663.
Murrell, J. N., and Pople, J. A. (1956) *Proc. Phys. Soc.* **A69**, 245.
Naqvi, A. M., and Talwar, S. P. (1957) *Monthly Not. Roy. Astron. Soc.* **117**, 463.
Nicholls, R. W., Fraser, P. A., Jarmain, W. R., and McEachran, R. P. (1960) *Astrophys. J.* **131**, 399.
Niewodniczanski, H. (1934) *Acta phys. Polon.* **3**, 285.
Niewodniczanski, H., and Lipinski, F. (1938) *Nature* **142**, 1160.
Noxon, J. F. (1961) *Canad. J. Phys.* **39**, 1110.
Ogawa, M., and Tanaka, Y. (1959) *J. chem. Phys,* **30**, 1354.
Ogawa, M., and Tanaka, Y. (1960) *J. chem. Phys.* **32**, 754.
Omholt, A. (1959) *Geofys. Publ.* **21**, 1.
Omholt, A. (1960) *Planet. Space Sci.* **2**, 246.
Oort, J. H. (1959) *In* "Handbuch der Physik" (S. Flügge, ed.), Vol. 53, p. 100. Springer, Berlin.
Opechowski, W. (1938) *Z. Phys.* **109**, 485.
Osterbrock, D. E. (1951) *Astrophys. J.* **114**, 469.

Osterbrock, D. E., and Rogerson, J. B. (1961) *Pubs. Astron. Soc. Pacific* **73**, 129.
Paschen, F. (1930a) *Naturwiss.* **18**, 752.
Paschen, F. (1930b) *Z. Phys.* **65**, 1.
Pasternack, S. (1940) *Astrophys. J.* **92**, 129.
Pecker, C., and Rohrlich, F. (1961) *Mém. Soc. roy. Sci. Liège* [5] **4**, 190.
Pople, J. A., and Sidman, J. W. (1957) *J. chem. Phys.* **27**, 1270.
Potts, W. J. (1955a) *J. chem. Phys.* **23**, 65.
Potts, W. J. (1955b) *J. chem. Phys.* **23**, 73.
Present, R. D. (1935) *Phys. Rev.* **48**, 140.
Racah, G. (1942) *Phys. Rev.* **62**, 438.
Racah, G. (1943) *Phys. Rev.* **63**, 367.
Ramsey, N. F. (1956) "Molecular Beams." Oxford Univ. Press, London and New York.
Rao, K. N. (1949) *Astrophys. J.* **110**, 304.
Rayleigh, Lord (1927) *Proc. Roy. Soc.* **A117**, 294.
Richardson, J. M. (1958) *J. appl. Phys.* **29**, 137.
Roberts, A., Beers, Y., and Hill, A. G. (1946) *Phys. Rev.* **70**, 112.
Robinson, G. W., and DiGiorgio, V. E. (1958) *Canad. J. Chem.* **36**, 31.
Rohrlich, F. (1956) *Astrophys. J.* **123**, 521.
Rosa, A. M. (1943) *Ann. Phys. (Leipzig)* **43**, 161.
Rubinowicz, A. (1949) *Rep. Progr. Phys.* **12**, 233.
Ruedy, J. E., and Gibbs, R. C. (1934) *Phys. Rev.* **46**, 880.
Runciman, W. A. (1958) *Rep. Progr. Phys.* **21**, 30.
Salwen, H. (1955) *Phys. Rev.* **99**, 1274.
Salwen, H. (1956a) *Phys. Rev.* **101**, 621.
Salwen, H. (1956b) *Phys. Rev.* **101**, 623.
Sayre, E. V., and Freed, S. (1956) *J. chem. Phys.* **24**, 1213.
Sayre, E. V., Sancier, K. M., and Freed, S. (1955) *J. chem. Phys.* **23**, 2060.
Schlapp, R. (1932) *Phys. Rev.* **39**, 806.
Schlapp, R. (1937) *Phys. Rev.* **51**, 342.
Seaton, M. J. (1955) *Monthly Not. Roy. Astron. Soc.* **115**, 279.
Seaton, M. J. (1960) *Rep. Progr. Phys.* **23**, 313.
Seaton, M. J., and Osterbrock, D. E. (1957) *Astrophys. J.* **125**, 66.
Segrè, E., and Bakker, C. J. (1931) *Z. Phys.* **72**, 724.
Series, G. W. (1959) *Rep. Progr. Phys.* **22**, 280.
Shapiro, J., and Breit, G. (1959) *Phys. Rev.* **113**, 179.
Shenstone, A. G. (1953) *Proc. Roy. Soc.* **A219**, 419.
Shortley, G. H. (1940) *Phys. Rev.* **57**, 225.
Shortley, G.H., Aller, L.H., Baker, J.G., and Menzel, D.H. (1941) *Astrophys. J.* **93**, 178.
Sidman, J. W. (1955) *J. chem. Phys.* **23**, 1365.
Sidman, J. W. (1958) *J. chem. Phys.* **29**, 644.
Spitzer, L., and Greenstein, J. L. (1951) *Astrophys. J.* **114**, 407.
Stoffregen, W., and Derblom, H. (1960) *Nature* **185**, 28.
Stout, J. W. (1959) *J. chem. Phys.* **31**, 709.
Tanabe, Y., and Sugano, S. (1954) *J. Phys. Soc. Japan* **9**, 753, 766.
Tanaka, Y. (1955) *J. Opt. Soc. Amer.* **45**, 663.
Tanaka, Y., Jursa, A. S., and Le Blanc, F. (1957) *J. chem. Phys.* **26**, 862.
Vallance Jones, A., and Harrison, A. W. (1958) *J. atmos. terrest. Phys.* **13**, 45.
van de Hulst, H. C. (1945) *Ann. Astrophys.* **8**, 12.
van de Hulst, H. C. (1953) *In* "The Sun" (G. P. Kuiper, ed.), p. 267ff. Univ. Chicago Press, Chicago, Illinois.

Van Vleck, J. H. (1934) *Astrophys. J.* **80**, 161.
Van Vleck, J. H. (1947) *Phys. Rev.* **71**, 413.
Welsh, H. L., Crawford, M. F., and Locke, J. L. (1949) *Phys. Rev.* **76**, 580.
Welsh, H. L., Crawford, M. F., McDonald, J. C. F., and Chisholm, D. A. (1951) *Phys. Rev.* **83**, 1264.
Wick, G. C. (1935) *Atti real. accad. Lincei* **21**, 708.
Wild, J. P. (1952) *Astrophys. J.* **115**, 206.
Wilkinson, P. G. (1957) *Astrophys. J.* **126**, 1.
Wilkinson, P. G. (1959) *J. chem. Phys.* **30**, 773.
Wilkinson, P. G. (1960) *J. chem. Phys.* **32**, 1061.
Wilkinson, P. G., and Mulliken, R. S. (1957) *Astrophys. J.* **126**, 10.
Wilkinson, P. G., and Mulliken, R. S. (1959) *J. chem. Phys.* **31**, 674.
Wu, T.-Y. (1952) *Canad. J. Phys.* **30**, 291.

2.

Allowed Transitions

R. W. Nicholls and A. L. Stewart

1 Introduction

In this chapter we are concerned with single quantum electric dipole radiation from atoms and diatomic molecules. The early work on atoms is fully described in "Resonance Radiation and Excited Atoms" (Mitchell and Zemansky, 1934) and extensive tables of results are contained in "Astrophysical Quantities" (Allen, 1955). Comparable reviews of the work on diatomic molecules are not available.

For detailed descriptions of spectroscopic terminology and for the basic theory of dipole radiation the reader is referred to the standard treatises by Pauling and Goudsmit (1930), White (1934), Condon and Shortley (1935), and Herzberg (1944, 1950).

2 Basic Concepts and Formulae

Consider an excited atomic system with ω_m states i and ω_n states j

belonging to the degenerate levels m and n with eigenenergies E_m and E_n, respectively, and suppose that $E_m < E_n$. Then neglecting stimulated emission the rate per unit volume at which photons are radiated in transitions from j to i is $N_j A_{ji}$, where N_j is the number density in the state j and A_{ji} is the Einstein coefficient for spontaneous emission of the component line. Accordingly, the total (energy) intensity of the emitted line of frequency

$$\nu_{nm} = E_{nm}/h = (E_n - E_m)/h \tag{1}$$

is

$$I_{nm} = E_{nm} \sum_{i,j} N_j A_{ji}. \tag{2}$$

if we assume that the number densities in the different states of the same level are equal, which will be the case if the system has been excited isotropically, then the number density of atoms in the level n is given by

$$N_n = \omega_n N_j \tag{3}$$

and we can write

$$I_{nm} = N_n E_{nm} A_{nm} \tag{4}$$

where

$$A_{nm} - \sum_{i,j} A_{ji}/\omega_n \tag{5}$$

is the Einstein coefficient for spontaneous transitions from level n to level m. The related quantity

$$\tau_n = 1\Big/\sum_m A_{nm}, \tag{6}$$

the summation being over all lower levels, is the radiative lifetime of the upper level.

It is customary† to describe the inverse process of absorption by means of the oscillator strength f_{mn}, which is related to the Einstein A coefficient through

$$f_{mn} = \frac{mc^3\hbar^2\omega_n}{2e^2E_{nm}^2\omega_m} A_{nm}. \tag{7}$$

Thus, the absorption coefficient K_ν (in cm^{-1}), which measures the attenuation of a uniform beam of photons of frequency ν, on integration

† Condon and Shortley (1935) define an oscillator strength $f(n, m)$ which is related to f_{mn} through $f(n, m) \quad f_{mn}$.

over the frequency spread of the absorption line centered at ν_{nm} is

$$\int K_\nu \, d\nu = \frac{\pi e^2}{mc} N_m f_{mn}. \tag{8}$$

This is an important result since it shows that the integrated absorption coefficient is constant if N_m is constant, irrespective of the physical process responsible for the line broadening. A similar relation holds for the integrated absorption cross section (in cm^2), which is obtained from (8) on division by the number density N_m.

The quantum theory of radiation yields for the case of single quantum electric dipole radiation

$$A_{ji} = \frac{4E_{nm}^3}{3\hbar^4 c^3} S_{ji} \tag{9}$$

where

$$S_{ji} = |\mathbf{R}_{ij}|^2 = |(i \mid \mathbf{P}^e \mid j)|^2 \tag{10}$$

and

$$\mathbf{P}^e = -\sum_s e\mathbf{r}_s', \tag{11}$$

the position of the sth electron being specified with respect to the nucleus by the vector $\mathbf{r}_s'$, is the dipole moment of the atom; S_{ji} is a measure of the strength of the component line in a specific state of polarization and propagated in a particular direction. However, the total intensity of the line from level n to level m depends instead on

$$S_{nm} = \sum_{i,j} S_{ji} \tag{12}$$

as can be seen by substituting (9) into (5) and (4). The so-called line strength S_{nm} is symmetrical with respect to interchange of the levels n and m; it can be shown to weight equally all directions of propagation and all directions of polarization, and so is a convenient theoretical measure of line intensities from isotropically excited atoms.

The selection rules for electric dipole radiation from atoms following from (10) are listed in Table I of Chapter 1. The oscillator strengths and spontaneous transition probabilities (second^{-1}) associated with a line of wavelength λ (angstroms) and strength S_{nm} are given by the formulae

$$f_{mn} = \frac{3.04 \times 10^2 \, S_{nm}}{\omega_m \lambda}, \tag{13}$$

$$A_{nm} = \frac{2.02 \times 10^{18} \, S_{nm}}{\omega_n \lambda^3}. \tag{14}$$

By appeal to the Schrödinger equation satisfied by the electronic wave functions ψ_i and ψ_j, the dipole length matrix element (Bethe and Salpeter, 1957)

$$\mathbf{M}^l_{ij} = \int \psi_i^* \sum_s \mathbf{r}'_s \psi_j \, d\tau' \tag{15}$$

may be expressed as the dipole velocity integral

$$\mathbf{M}^v_{ij} = \frac{\hbar^2}{mE_{nm}} \int \psi_i^* \sum_s \nabla'_s \psi_j \, d\tau' \tag{16}$$

or in the dipole acceleration form

$$\mathbf{M}^a_{ij} = \frac{\hbar^2}{mE^2_{nm}} \int \psi_i^* \left(\sum_s \nabla'_s V\right) \psi_j \, d\tau' \tag{17}$$

where for an atom of nuclear charge Z

$$V = -\sum_s \frac{Ze^2}{r'_s} + \frac{1}{2}\sum_{s,t} \frac{e^2}{r'_{st}} \tag{18}$$

and

$$\sum_s \nabla'_s V = \sum_s \frac{Ze^2 \mathbf{r}'_s}{r_s^3}\,. \tag{19}$$

3 *Calculations of Atomic Line Strengths*

The calculation of atomic line strengths requires a knowledge of the wave functions of the states involved in the transition. References to the wave functions presently available are provided in the books by Hartree (1957) and Slater (1960) and in the article by Knox (1957). The line strength (12) involves the dipole length integral and is almost always evaluated in this form, rarely in the dipole velocity or dipole acceleration equivalent forms. The three formulae agree only if ψ_i and ψ_j are known exactly, or if self-consistent field representations are used together with the self-consistent values for E_{nm} and the total electron potential energy when it can be uniquely specified.

3.1 Russell-Saunders Coupling

In Russell-Saunders coupling (Condon and Shortley, 1935) we can write

$$S_{nm} = \mathscr{S}(\mathscr{M})\mathscr{S}(\mathscr{L})\,\sigma_{nm}^2 \tag{20}$$

where $\mathscr{S}(\mathscr{M})$ depends on the particular multiplet of the transition array, $\mathscr{S}(\mathscr{L})$ depends on the particular line of the multiplet, and

$$\sigma_{nm}^2 = (4l^2 - 1)^{-1}\left\{\int_0^\infty P_n P_m r'\,dr'\right\}^2 \tag{21}$$

where P_n/r' and P_m/r', normalized in atomic units, are the radial wave functions for the optical electron in configurations with azimuthal quantum numbers l and $l-1$. The tables of relative multiplet strengths, $\mathscr{S}(\mathscr{M})$, compiled by Goldberg (1935) and Menzel and Goldberg (1936) and of relative line strengths $\mathscr{S}(\mathscr{L})$ compiled by White and Eliasen (1933) and Russell (1936) are given in the book by Allen (1955) together with the sum rules for the total strength of the transition array derived by Goldberg (1936) and Menzel (1947). As a check on these tables and to extend them to cover all cases of astrophysical interest, Rohrlich (1959a) has derived expressions for $\mathscr{S}(\mathscr{L})$ and $\mathscr{S}(\mathscr{M})$ which can be evaluated by means of tables of the coefficients of fractional parentage and tables of Racah coefficients (Racah, 1942, 1943; Edmonds, 1957; Kelly and Armstrong, 1959). A variety of useful sum rule expressions has also been derived by Rohrlich (1959b).

The work of Bates and Damgaard (1949), which employs the Coulomb approximation to P_n and P_m (discussed further by Seaton, 1958), must receive special mention because of the wide applicability of the tables published. The method gives results of high accuracy for all transitions in the lighter, simple systems (cf. Trefftz *et al.*, 1957; Mastrup and Wiese, 1958; Richter, 1961) and meets with considerable success even in complex systems. The self-consistent field approximation to σ_{nm}^2 appears to be less satisfactory than the Coulomb approximation, even when exchange is included as in the Hartree-Fock approximation. Attempts to improve on the self-consistent field approximation by addition of a polarization potential to the central field have been made by Biermann and Lübeck (1948,† 1949) and in the case of the Hartree-Fock approximation by Biermann and Trefftz (1949), Trefftz (1949),

† In this paper Biermann and Lübeck calculate the oscillator strength for the 3082 line of Al I assuming it to arise from the transition $3s^23p\ ^2P - 3s^23d\ ^2D$ and obtain 0.71. Burgess *et al.* (1960) however, confirm the belief of A. G. Shenstone that the ^2D term very probably belongs to the configuration $3s3p^2$ and it follows that this value is incorrect.

and Trefftz and Biermann (1952). The effects of departure from Russell-Saunders coupling due to configuration interaction (Trefftz, 1950, 1951) and spin-orbit interaction (Shortley, 1935; Gottschalk, 1948) have been investigated in a few cases. The calculations of Trefftz (1951) allowed for polarization as well as configuration interaction. For the $4\,^1S - 4\,^1P$ resonance transition in calcium the oscillator strengths 1.63 and 1.31 were obtained using the dipole length and dipole velocity forms, respectively, which are to be compared with the self-consistent field value, 2.37, the Hartree-Fock value, 1.17, and the value 1.86 obtained from the Coulomb approximation. All the values quoted use the experimental eigenenergies. Such a diversity of calculated values is not unusual and illustrates the lack of precision which still exists in calculations of oscillator strengths of complex atoms. When account is taken of the ease of calculation using the Coulomb approximation, the relative success in predicting the resonance oscillator strength in calcium is gratifying.

A method of calculating line strengths which is especially suited to ionized systems is being investigated by Layzer (1961). The wave function of the system is expanded in inverse powers of the nuclear charge and subsequently modified by the introduction of a screening parameter determined from the associated energy expansion to first order. The transition integrals are then evaluated in terms of the resulting screened hydrogenic radial functions. The method has been shown by Varsavsky (1958, 1961) to yield reliable results in many cases.

3.2 Oscillator Strength Sum Rules

Apart from the astrophysical interest in its spectrum (cf. Athay and Johnson, 1960) helium provides a simple system which can be used to test the various approximations which are made in the calculation of oscillator strengths, and so has received considerable attention from theorists (Trefftz *et al.*, 1957, and references therein). The large number and high accuracy of the values calculated for helium, particularly for transitions from the ground state, makes it possible to test a number of theoretical relationships involving oscillator strengths, for example, the continuity condition at the spectral head (Hargreaves, 1929; Vinti, 1933) and the oscillator strength sum rules (Dalgarno and Lynn, 1957). With energy expressed in rydbergs and all other quantities in atomic units, these may be written

$$S(-4) = \sum_n \frac{f_{mn}}{E_{nm}^4} - \beta \tag{22}$$

where β can be determined from experimental data on the refractive index and the Verdet constant of the system in the level m and the summation includes integration over the continuum.

$$S(-2) \equiv \sum_n \frac{f_{mn}}{E_{nm}^2} = \frac{\alpha}{4} \tag{23}$$

where α is the polarizability of the system in the level m.

$$S(-1) \equiv \sum_n \frac{f_{mn}}{E_{nm}} = \frac{1}{3}\left(m \left| \left\{ \sum_{s=0}^{\mathcal{N}} \mathbf{r}'_s \right\}^2 \right| m\right) \tag{24}$$

where $\mathcal{N}$ is the number of electrons in the system.

$$S(0) \equiv \sum_n f_{mn} = \mathcal{N} \tag{25}$$

$$S(1) \equiv \sum_n f_{mn} E_{nm} = \frac{4}{3}\left[- E_m + \left(m \left| \sum_{s,t} \mathbf{p}'_s \cdot \mathbf{p}'_t \right| m\right)\right] \tag{26}$$

where $\mathbf{p}'_s$ is the momentum of the sth electron.

$$S(2) \equiv \sum_n f_{mn} E_{nm}^2 = \frac{16\pi Z}{3}\left(m \left| \sum_s \delta(\mathbf{r}'_s) \right| m\right) \tag{27}$$

where $\delta(\mathbf{r})$ is the three-dimensional Dirac δ-function. Properly normalized one can regard these sums as giving mean values of powers of the excitation energy E_{nm}, weighted according to the oscillator strength of the associated transition. Thus,

$$\frac{d}{dk} \log S(k) = \sum_n f_{mn} E_{nm}^k \log E_{nm} \Big/ \sum_n f_{mn} E_{nm}^k \tag{28}$$

is the expression for the logarithm of the mean excitation energy, $\log I$, which controls the stopping powers of atoms towards fast charged particles when $k = 0$ (Livingston and Bethe, 1937), and, when $k = 2$, is the expression for the logarithm of the average excitation energy, $\log K$, defined by Bethe (1947), which enters the expression for the Lamb shift.

Comparisons of the values of $S(k)$ obtained by summation of the best available oscillator strengths with those calculated directly or given by experiment have been made by Dalgarno and Lynn (1957), Kabir and Salpeter (1957), and Dalgarno and Stewart (1960) for the transitions from the ground state of helium. Similar comparisons have been made by Dalgarno and Kingston (1958) for the $2\,^1S$ and $2\,^3S$ metastable levels of helium. By assigning values to the unknown oscillator strengths and by modifying those which were uncertain, it was possible for Dalgarno and his collaborators to satisfy exactly the relations with $k = -1$, 0, 1, and 2. Finally, the amended oscillator strengths for helium were used in the remaining sum rule expressions and in (28) to calculate a wide range of atomic properties with high accuracy. Pekeris (1959) and Dalgarno (1960) have shown that a reliable analytic fit to $S(k)$ can be obtained using only a few of the sum rule expressions. This method has been used with success to calculate refractive indices and Verdet constants for the inert gases (Dalgarno and Kingston, 1960) and log I for H, He, Li, and Be (Dalgarno, 1960), but is not so successful when applied to the calculation of log K.

3.3 Results

Lists of references to calculations of line strengths are provided in the review by Garstang (1955) and in the Reports of the I.A.U. Commission, 14a (1955, 1958, 1961). Since these lists were compiled, the only calculations of line strengths which seem to have been made are: hydrogenlike transitions (Herdan and Hughes, 1961; Karzas and Latter, 1961); $3s^2 3p\ ^2P - 3s3p^2\ ^2S$, Al I (Eddy *et al.*, 1961).

4 Theory of Molecular Line Strengths

If the nuclei of a diatomic molecule could be held stationary, separated by a distance r, then the emission or absorption of radiation could occur only through a change in state of the electrons and, just as for atoms, line spectra would be observed. The intensity of such lines would be related to a line strength

$$S_{nm}(r) \equiv \sum_{i,j} |\, \mathfrak{R}^e_{ij}(r) \,|^2 \tag{29}$$

where the summation is over the degenerate upper and lower levels and

$$\mathfrak{R}^e_{ij}(r) = \int \psi_i^*(\mathbf{r}_s, r) \left\{ - \sum_s e\mathbf{r}_s \right\} \psi_j(\mathbf{r}_s, r)\, d\tau, \tag{30}$$

ψ_i and ψ_j being the wave functions for the electrons with coordinate vectors $\mathbf{r}_s$ fixed with respect to the nuclear axis. An oscillator strength could similarly be defined

$$f_{mn}(r) = \frac{2m}{3\hbar^2 e^2 \omega_m} \{U^n(r) - U^m(r)\} \sum_{i,j} | \mathfrak{R}^e_j(r) |^2 \tag{31}$$

where $U^n(r)$ and $U^m(r)$ are the potentials provided by the charged nuclei and the electrons and are a measure of the energies of the upper and lower electron levels, respectively.

A slow vibrational or rotational motion of the molecules should have little effect on the electron state of the molecule but rather distribute the light intensity associated with the ideal electronic lines over many lines or over a broad continuum. According to the approximation of Born and Oppenheimer (1927), the form of this distribution is decided by the Hönl-London and Franck-Condon factors, as we shall see later. The total intensity, it seems reasonable to suppose, is little changed and for this reason (29) and (31) are sometimes referred to as "absolute" line and oscillator strengths for the molecule, even though their relationship with the total intensity is not the simple one which can be written down for atoms.

The basic theory of electric dipole radiation from diatomic molecules is identical with that for atoms, the line strength being given by (10), where now i and j specify states of the molecule, particular modes of nuclear rotation and vibration being associated with each state of electron motion, and $\mathbf{P}$ is the total dipole moment of the molecule. The selection rules are discussed in Chapter 1, § 3.

4.1 The Born-Oppenheimer Approximation

In the Born-Oppenheimer (1927) approximation the diatomic molecule wave function is written

$$\psi_{ivJ\Lambda M} = \psi_i(\mathbf{r}_s, r) \frac{1}{r} \psi_v(r)\, \psi_{J\Lambda M}(\theta, \chi, \varphi) \tag{32}$$

where ψ_i is the wave function for the electrons with coordinate vectors $\mathbf{r}_s$ specified with respect to the molecular axis, and with angular momentum $\Lambda\hbar$ about the molecular axis; ψ_v and $\psi_{J\Lambda M}$ specify the states with vibrational quantum number v and rotational quantum numbers J and M and θ, χ, and φ are the Euler angles relating the coordinate system on the molecule to the fixed coordinate system (Goldstein, 1953). Thus,

$$\mathbf{R}^{jv'J'\Lambda'M'}_{iv''J''\Lambda''M''} = (jv'J'\Lambda'M' \mid \mathbf{P}^e \mid iv''J''\Lambda''M'') + (j \mid i)\,(v'J'M' \mid \mathbf{P}^n \mid v''J''M'') \tag{33}$$

where $\mathbf{P}^e$ and $\mathbf{P}^n$ are the electronic and nuclear electric dipole moments, respectively, the second term vanishing if, as we shall assume, an electronic transition takes place. The first matrix element is evaluated by transforming the electric dipole moment to coordinates $\mathbf{r}_s$ fixed with respect to the molecular axis through

$$\mathbf{P}^e = -\sum_s e\mathbf{r}'_s = \left\{-\sum_s e\mathbf{r}_s\right\} \cdot \mathbf{D}(\theta, \chi, \varphi) \tag{34}$$

where $\mathbf{D}(\theta, \chi, \varphi)$ is the dyadic whose elements are the direction cosines relating the coordinate axes fixed in the molecule to the fixed coordinate system. This gives

$$\mathbf{R}^{jv'J'\Lambda'M'}_{iv''J''\Lambda''M''} = \left\langle jv' \middle| \left\{-\sum_s e\mathbf{r}_s\right\} \middle| iv'' \right\rangle \cdot (J'\Lambda'M' \mid \mathbf{D}(\theta, \chi, \varphi) \mid J''\Lambda''M'') \tag{35}$$

The matrix elements $(J'\Lambda'M' \mid \mathbf{D} \mid J''\Lambda''M'')$, which determine the selection rules in the approximation here considered, have been evaluated by Dennison (1926), Kronig and Rabi (1927), and Rademacher and Reiche (1927). The squares of the transition moments $\mathbf{R}^{jv'J'\Lambda'M'}_{iv''J''\Lambda''M''}$ on summation over the degenerate quantum numbers M' and M'' and over the degenerate electron states give

$$S^{nv'J'\Lambda'}_{mv''J''\Lambda''} = S^{J'\Lambda'}_{J''\Lambda''}\, p_{v'v''} \tag{36}$$

where $S^{J'\Lambda'}_{J''\Lambda''}$ is the Hönl-London (1925) factor which is well tabulated (Rademacher and Reiche, 1927; Herzberg, 1950) and where

$$p_{v'v''} = \sum_{i,j} \left| \int \psi_{v'}\, \Re^e_{ij} \psi_{v''}\, dr \right|^2. \tag{37}$$

The dimensionally incorrect description *vibrational transition probability* is sometimes given to $p_{v'v''}$, but the term *band strength* suggested by Chamberlain (1961) is preferable.

From the line strength $S^{nv'J'\Lambda'}_{mv''J''\Lambda''}$, expressions for the associated transition probabilities, oscillator strengths, lifetimes, and intensities can be written down, the intensity of the $nv'J'\Lambda' - mv''J''\Lambda''$ line, for example, being given by

$$I_{v'J',v''J''} = N_{J'} \frac{4E^4_{v'J',v''J''}\, S^{nv'J'\Lambda'}_{mv''J''\Lambda''}}{3\hbar^4 c^3 \omega_n (2J' + 1)} \tag{38}$$

where ω_n is the weight of the upper electron level or potential energy function and $N_{J'}$ is the number density in the rotational state J'.

Since

$$\sum_{J''} S^{J'\Lambda'}_{J''\Lambda''} = (2J' + 1) \tag{39}$$

the total intensity of all rotational lines in the $v' - v''$ band is

$$I_{v'v''} = \sum_{J',J''} I_{v'J',v''J''} = N_{v'} \frac{4E^4_{v'v''}\, p_{v'v''}}{3\hbar^4\, c^3\, \omega_n} \tag{40}$$

where $N_{v'} = \Sigma_{J'} N_{J'}$, if the band is sufficiently sharp so that an average energy jump $E_{v'v''}$ for the whole band can be defined. $E_{v'v''}$ can usually be taken without serious error to be $E_{v'0},\, _{v''0}$ (Robinson and Nicholls, 1961). A mean Einstein A coefficient and oscillator strength $f_{v'v''}$ can, in these circumstances, be derived using (4) and (5) and a mean lifetime associated with the vibrational level v' using (6).

The total intensity of all bands in the band system

$$I_{nm} = \sum_{v',v''} I_{v'v''} \tag{41}$$

is of little interest because of the difficulty of defining an average jump for the band system. However, if the quantity

$$p_{v'v'} = \sum_{v''} p_{v'v''} = \sum_{i,j} (v' \mid \{\Re^e_{ij}\}^2 \mid v') \tag{42}$$

which enters the sum

$$\sum_{v''} \frac{I_{v'v''}}{E^4_{v'v''}} = N_{v'} \frac{4p_{v'v'}}{3\hbar^4\, c^3\, \omega_n} \tag{43}$$

is approximately independent of v', then such sums may be used to estimate approximately the ratios of the population densities in the upper vibrational levels and hence vibrational temperatures (cf. Herzberg, 1950). In such circumstances, one can also sum over the upper vibrational quantum number giving

$$\sum_{v',v''} \frac{I_{v'v''}}{E^4_{v'v''}} = N_n \frac{4\bar{p}_{v'v'}}{3\hbar^4\, c^3\, \omega_n} \tag{44}$$

where $N_n = \Sigma_{v'} N_{v'}$ and $\bar{p}_{v'v'}$ is a suitably averaged value of $p_{v'v'}$ for the upper vibrational states. To associate a mean Einstein A coefficient for the whole band system using (4), we need to define an average energy jump for the band system $\overline{E^4_{v'v''}}$ so that

$$I_{nm} = \overline{E^4_{v'v''}} \sum_{v'v''} \frac{I_{v'v''}}{E^4_{v'v''}} = N_n \frac{4\overline{E^4_{v'v''}}\, \bar{p}_{v'v'}}{3\hbar^4\, c^3\, \omega_n}\,. \tag{45}$$

Usually $p_{v'v''}$ and $\overline{E^4_{v'v''}}$ vary considerably with v' and v'' so that suitable averages are difficult to specify and electronic oscillator strengths or transition probabilities determined from the experimental data on the basis of (45) can claim order of magnitude accuracy only.

Similar considerations apply to the theory of absorption.

4.2 Franck-Condon Factors and r-Centroids

If $\Re^e_{ij}(r)$ is replaced by some mean value $\overline{\Re^e_{ij}}$ and taken outside the integral sign (37) becomes

$$p_{v'v''} = q_{v'v''} \sum_{i,j} | \overline{\Re^e_{ij}} |^2 \tag{46}$$

where

$$q_{v'v''} = | \int \psi_{v'} \psi_{v''} \, dr |^2. \tag{47}$$

Following a suggestion of Bates (1952) $q_{v'v''}$ is called the Franck-Condon factor because of its obvious association with the Franck-Condon principle.

Alternatively, account can be taken of the variation of the electron transition moments with internuclear distance by the method of r-centroids. This method was developed by Jarmain (1953) and by Fraser (1954a), who noticed that, using Morse-type vibrational wave functions, the wave function product in (37) behaved effectively on integration like a delta function. Thus, we can write approximately

$$\psi_{v'} \psi_{v''} / \int \psi_{v'} \psi_{v''} \, dr = \delta(r - \bar{r}_{v'v''}) \tag{48}$$

where

$$\bar{r}_{v'v''} = \int \psi_{v'} \, r \, \psi_{v''} \, dr / \int \psi_{v'} \psi_{v''} \, dr \tag{49}$$

is the r-centroid for the transition. On substitution in (37) this gives

$$p_{v'v''} = q_{v'v''} \sum_{i,j} | \Re^e_{ij}(\bar{r}_{v'v''}) |^2. \tag{50}$$

The range of validity of the r-centroid approximation has been discussed by Fraser (1954a).

Assuming its validity and the availability of arrays of values of $q_{v'v''}$ and $\bar{r}_{v'v''}$, it is possible to use (50) together with experimental band intensities to trace out the variation of $\Sigma_{i,j} | \Re^e_{ij}(\bar{r}_{v'v''}) |^2$ as $\bar{r}_{v'v''}$ varies, that is, to plot the electronic transition moment as a function of internuclear distance (Fraser, 1954a). For substitution of (50) into (40) gives

$$S_{nm}(\bar{r}_{v'v''}) = \sum_{i,j} | \Re^e_{ij} (\bar{r}_{v'v''}) |^2 = \frac{1}{N_{v'}} \left[\frac{3\hbar^4 c^3 \omega_n I_{v'v''}}{4 q_{v'v''} E^4_{v'v''}} \right] \tag{51}$$

where all quantities are known except $N_{v'}$, which, however, is constant along a v'' progression in which v' is constant. A different plot is obtained for each value of v', the ordinate scales of each being proportional to $N_{v'}$. By judicious adjustments of the scale, a continuous curve can be obtained. Such curves have been obtained in the following cases: N_2, first positive (Turner and Nicholls, 1954a, b); N_2, second positive, and N_2^+, first negative (Wallace and Nicholls, 1955); CN, red (Dixon and Nicholls, 1958); CN, violet (Ferguson and Nicholls, 1961); O_2^+, second negative, CO, Ångstrom, and CO, third positive (Robinson and Nicholls, 1958a); NO, β, and NO, γ (Robinson and Nicholls, 1958b); CO^+, comet tail, BO, α, and BO, β (Robinson and Nicholls, 1960); O_2, Schumann-Runge (Hébert and Nicholls, 1961; Treanor and Wurster, 1960); SiN, violet (Ferguson *et al.*, 1961); BN, violet (Stevens and Nicholls, 1961); BeO, $^1\Sigma - {}^1\Sigma$ (Tawde and Murthy, 1960a). The method has also been used to interpret intensity measurements previously reported. For example, Nicholls (1956) has discussed OH violet, CN violet, C_2 Swan, and O_2 Schumann-Runge band systems; Biberman *et al.* (1959) have discussed O_2; and Ortenberg (1960) has discussed TiO, α.

4.3 Calculations of Vibrational Intensities

It is clear from (46) and (50) that the Franck-Condon factor plays a very important role in the determination of intensity distributions in band systems. The calculation of arrays of Franck-Condon factors requires a knowledge of the solutions of the Schrödinger equation for molecular vibration,

$$\frac{d^2\psi_v}{dr^2} + \frac{2\mu}{\hbar^2}\{E_v - U(r)\}\psi_v = 0, \tag{52}$$

μ being the reduced mass of the pair of constituent nuclei, $U(r)$ the potential provided by the electrons and the charged nuclei, and E_v being the vibrational energy belonging to the state ψ_v. Theoretical estimates of $U(r)$ are never so accurate as those obtained from spectroscopic data. The usual procedure is to determine $U(r)$ as a numerical function by the methods developed by Rydberg (1931), Klein (1932), and Rees (1947) and improved upon by Jarmain (1956, 1959, 1960a, b) and by Vanderslice and his collaborators (cf. Tobias *et al.*, 1960 and references therein). To determine a wide range of solutions of (52) and to use them to calculate Franck-Condon factors is a formidable computational task. Until recently it has indeed been necessary to resort to various simplifying approximations. Although now redundant, these are still instructive.

If $U(r)$ is represented by a simple harmonic potential, analytic expressions for $q_{v'v''}$ can be derived (Hutchisson, 1930). In the work of Hutchisson (1931), Gaydon and Pearse (1939), Pillow (1949, 1950, 1951, 1952, 1953a, b, 1954a, b, 1955), Pillow and Rowlatt (1960), Nicholls (1950), Montgomery and Nicholls (1951), and Turner and Nicholls (1951), this approximation has been refined to allow for the distortion of $U(r)$ from simple harmonic form. Bates (1952) has included the distortion effect by a perturbation procedure and published a useful double entry table from which Franck-Condon factors for v', $v'' = 0, 1, 2$ can be read.

Morse (1929) has shown that molecular potential energy curves can be fitted by

$$U(r) = D_e \{1 - e^{-\beta(r-r_e)}\}^2 \tag{53}$$

where D_e is the dissociation energy and r_e the equilibrium internuclear distance, and that the resulting Schrödinger equation can be rigorously solved. It is not possible, however, to evaluate $q_{v'v''}$ analytically with such Morse wave functions and recourse must be made to numerical integration (Bates, 1949a; Jarmain and Nicholls, 1954). Analytic integration is possible if β_n and β_m in the Morse wave functions for the upper and lower potential energy curves are replaced by an average β (Fraser and Jarmain, 1953). Methods of correcting the results obtained have been developed (Jarmain and Fraser, 1953; Fraser, 1954b) and used (Jarmain *et al.*, 1953, 1955, 1960; Fraser 1954b; Fraser *et al.*, 1954; Nicholls *et al.*, 1959, 1960; Ortenberg, 1960). The most recent analytic integration procedure, based on the WKB method (Biberman and Yakubov, 1960; Yakubov, 1960) has the virtue of accuracy for high vibrational quantum numbers where, however, the realism of the Morse potential is in doubt. Wu (1952) in a paper critical of the Pillow distortion procedure has also considered the WKB method, and it has been used by Wyller (1953, 1958).

To calculate r-centroids without the labor of direct numerical integration, Fraser (1954a), following the suggestion of Jarmain (1953), has used the relation

$$E_{v'v''}(v' \mid v'') = (v' \mid U^n - U^m \mid v'') \tag{54}$$

which is readily derived from (52). Substitution of (48) in this yields

$$E_{v'v''} = U^n(\bar{r}_{v'v''}) - U^m(\bar{r}_{v'v''}) \tag{55}$$

which can be solved for $\bar{r}_{v'v''}$ (Nicholls and Jarmain, 1956). Arrays of $\bar{r}_{v'v''}$ have been computed for a large number of systems by this method (Nicholls and Jarmain, 1956; Nicholls *et al.*, 1956; Nicholls, 1958; Jarmain *et al.*, 1960; Tawde and Murthy, 1960b; Ortenberg, 1960).

Many of the calculations reported in this section are being repeated using fast computers, the accuracy being limited only by the accuracy of the empirical potentials. (Nicholls, 1961).

4.4 Calculations of Absolute Molecular Oscillator Strengths

If the electronic wave functions for fixed nuclei are known over a range of internuclear distances, then the electron transition moments $\mathfrak{R}^e_{ij}(r)$ can be calculated and the strength of the transition expressed in terms of the absolute oscillator strength $f_{mn}(r)$. The paper by Allen and Karo (1960) is a compilation of the molecular wave functions presently available.

Since the early work of Mulliken (1939a, b, c, 1940a, b, c) and Lyddane *et al.* (1941), reviewed by Mulliken and Rieke (1941), only a few calculations of absolute oscillator strengths have been made, mostly for a single value of the internuclear distance. The calculation of Bates (1949b) on the N_2^+ first negative band system employed LCAO wave functions based on Slater and Hartree-Fock separated atoms wave functions and led to the conclusion that the oscillator strength lay between 0.025 and 0.095. The astrophysically important transition $X^2\Pi - A^2\Delta$ of the CH radical has likewise been treated using LCAO wave functions (Stephenson, 1951; see also Bates, 1951a). With the use of LCAO wave functions based on hybridized separated-atom wave functions, Shull (1950, 1951) carried out calculations on the N_2^+ first negative, C_2 Swan, and C_2 Délandres-d'Azambuja systems using both dipole length and dipole velocity formulations. By varying the degree of hybridization, the dipole length and dipole velocity values were made to agree giving the values 0.18, 0.23, and 0.18, respectively, which are about 40% larger than the dipole length values. The recent measurement of the oscillator strength of the N_2^+ first negative system by Bennett and Dalby (1959) gives the value 0.0348, which lies within the limits estimated by Bates (1949b), and makes questionable the high degree of hybridization favoured by Shull (see Table IV).

Fraser (1954c), using LCAO wave functions with hybridized Slater atomic orbitals, calculated the electronic transition moment of the N_2 first positive and N_2 second positive systems at a number of internuclear separations. His work agrees with the measurements of Turner and Nicholls (1954b) for the first positive system, but the agreement with the measurements of Wallace and Nicholls (1955) for the second positive system is not so good. The computations of Mulliken and Rieke (1941) using the LCAO, MO, and AO methods for the oscillator strength of the Lyman and Werner transitions in H_2 have been repeated by Shull (1952) using the dipole velocity formula. The two sets of results are in close agreement if the charge parameter for the σ_g bonding orbital takes on the value 1.2 obtained by the variation method.

Recently an exhaustive investigation of the Lyman transition in H_2 has been made by Ehrenson and Phillipson (1961). By successively improving the wave functions they showed that the f values calculated using (15)-(17) converged towards a unique value.

To examine the extent of the inaccuracy inherent in the calculation of molecular oscillator strengths using approximate molecular orbitals, Bates and his collaborators have carried out an extensive series of calculations on H_2^+ and HeH^{2+} for which exact fixed nuclei wave functions are available (Bates *et al.*, 1953b; Bates and Carson, 1956) over a wide range of internuclear separations. The transitions studied, together

TABLE I

Molecule	Transitions	Range of r (atomic units)	Reference
H_2^+	$1s\sigma - 2p\sigma$	0–9	Bates (1951b)
H_2^+	$1s\sigma - 2p\pi$, $2p\sigma - 3d\pi$, $2p\pi - 3d\pi$	0–5	Bates *et al.* (1953a)
H_2^+	$1s\sigma - 3p\sigma$, $1s\sigma - 4p\sigma$, $1s\sigma - 4f\sigma$, $2p\sigma - 2s\sigma$, $2p\sigma - 3s\sigma$, $2p\sigma - 3d\sigma$	0–5	Bates *et al.* (1954)
H_2^+	$2p\pi - 2s\sigma$, $2p\pi - 3s\sigma$, $2p\pi - 3d\sigma$, $3d\pi - 3p\sigma$, $3d\pi - 4p\sigma$, $3d\pi - 4f\sigma$	0–5	Lewis *et al.* (1955)
HeH^{2+}	$1s\sigma - 2s\sigma$, $1s\sigma - 3d\sigma$, $2s\sigma - 3d\sigma$	0–5	Dalgarno *et al.* (1956)
HeH^{2+}	$1s\sigma - 2p\pi$, $2p\pi - 2s\sigma$, $2p\pi - 3d\sigma$	0–5	Arthurs and Hyslop (1957a)
HeH^{2+}	$1s\sigma - 2p\sigma$,[a] $2p\sigma - 2s\sigma$, $2p\sigma - 3d\sigma$, $2p\sigma - 2p\pi$	0–5	Arthurs *et al.* (1957)

[a] Extended from $r = 5$ to $r = 10$ by Arthurs and Hyslop (1957b).

with the references are presented in Table I. The opportunity was taken to compare the exact oscillator strengths with those obtained in the length and velocity formulations using approximate orbitals. The main conclusions have been summarized by Dalgarno *et al.* (1957). The comparison shows that the LCAO approximation often yields results which are seriously in error and that the dipole velocity formula is not always to be preferred to the dipole length matrix element. In Figs. 1

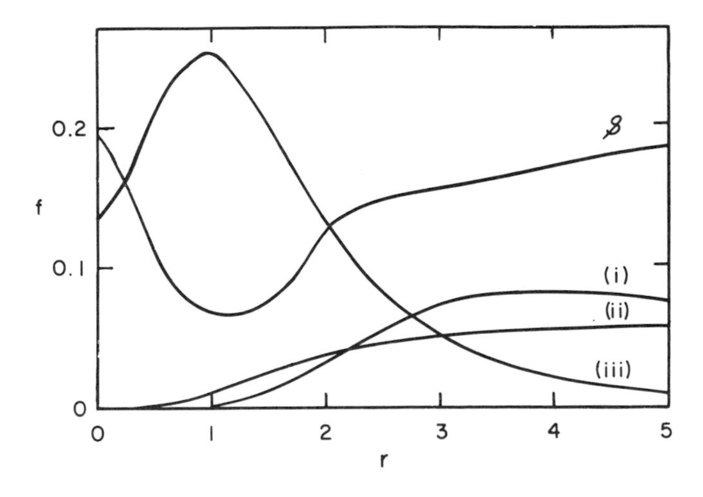

FIG. 1. Oscillator strengths of HeH^{2+}. Curve (i), $1s\sigma - 3d\sigma$; curve (ii), $1s\sigma - 2s\sigma$; curve (iii), $1s\sigma - 2p\sigma$. $\mathscr{S}$, residual sum (see text).

and 2 a representative set of oscillator strength curves for HeH^{2+} are displayed, demonstrating the rapid variation of oscillator strength with internuclear distance obtained. Since the oscillator strength sum rule (25) applies, and since for the present problem (ter Haar, 1952)

$$\sum_{n''l''} f(n'l'\sigma - n''l''\sigma \mid r) = \tfrac{1}{3} \tag{56}$$

$$\sum_{n''l''} f(n'l'\sigma - n''l''\pi \mid r) = \tfrac{2}{3}, \tag{57}$$

it is possible to estimate the sum, $\mathscr{S}(n'l'\sigma \mid r)$, of the oscillator strengths for all the transitions left out of consideration. This curve is also presented in Figs. 1 and 2 for transitions arising from the $1s\sigma$ and $2p\pi$ levels, respectively.

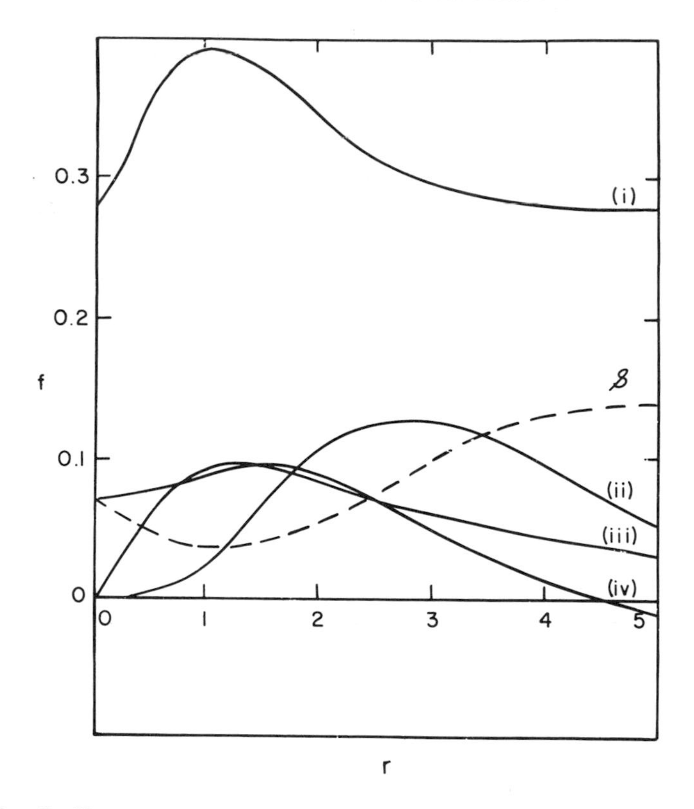

FIG. 2. Oscillator strengths of HeH^{2+}. Curve (i), $1s\sigma - 2p\pi$; curve (ii), $2p\sigma - 2p\pi$; curve (iii), $2p\pi - 2s\sigma$; curve (iv), $2p\pi - 3s\sigma$. $\mathscr{S}$, residual sum (see text).

5 *Measurements of Atomic Transition Probabilities*

The direct relationship (8) between the integrated absorption coefficient of a spectrum line and the oscillator strength f of the transition has enabled measurement of atomic oscillator strengths from absorption profiles of atomic lines provided that the number of absorbing atoms in the light path is known accurately. In cases where the material concerned is easily kept in the vapor phase, a furnace is often used to maintain a thermal distribution of energy levels in the absorbing column of gas. The number of molecules in the absorbing path can then be determined from a knowledge of gas pressure, gas temperature, and the assumption of a Boltzmann distribution. This direct method has been extensively used for lines of astrophysical interest by R. B. and A. S.

King (1935, 1938; R. B. King, 1942, 1947, 1948) who designed a special furnace for their researches. R. B. King and his colleagues (Bell *et al.*, 1958, 1959) have refined the method by making optical absorption measurements through atomic beams of such metals as Mn, Cu, and Fe. The concentration of atoms in the absorbing beam of light is measured from the kinematics of a very light pan (which is suspended by a fine weak quartz helix) on which the atomic beam impinges. Absolute measurements of this kind are difficult and slow to make.

It is inconvenient to use absorption techniques and furnaces for high melting point materials. Recourse has then to be made to measurements of intensities of spectral lines in emission. These intensities are controlled through (2) by the Einstein A coefficient which is given by (5) and (9). The intensity of the line in emission is often defined as the energy radiated by 1 cc of gas in all directions through an optically thin gas.

Allen and Asaad (1955, 1957) have measured, by photographic photometry, oscillator strengths in emission for astrophysically important atoms of the iron group. They photographed the spectrum of an arc between electrodes of copper containing very dilute amounts of the metallic elements studied. The relative oscillator strengths of the lines were inferred from photographic intensity measurements which could be made on a relatively routine basis for a large number of lines. These relative oscillator strengths were placed upon an absolute scale by comparison with lines of known absolute oscillator strength of Cr, Mn, and Fe. Allen (1960) has critically discussed the available oscillator strengths of neutral atoms of the iron group.

Lochte-Holtgreven (1958) and his collaborators have developed high-temperature ($\sim$30,000°K) constricted, water-cooled arcs in which the spectra of atoms and ions are thermally excited. Absolute oscillator strengths and arc temperatures have been inferred from intensity measurements in emission (Mastrup and Wiese, 1958; Hey, 1959; Wiese and Shumaker, 1961).

A number of other methods which do not involve direct spectroscopic observations have been developed. Many of the classical methods reviewed by Mitchell and Zemansky are optical in nature (e.g., magnetic rotation, dispersion). In the magnetic rotation method of Weingeroff (1931), the rotation of the plane of polarization of light of frequency in the neighborhood of an absorption line is used to measure the oscillator strength of the line. The method was applied by Stephenson (1951) to the resonance lines of the alkaline metals Na, K, Li, Rb, and Tl. The method does not require a knowledge of vapor pressure of the metal.

The hook method of Rogestwensky (Rogestwensky, 1912; Rogestwensky and Penkin, 1960), which is much used in the Soviet Union, is extremely simple in principle, as demonstrated by Ladenburg and Wolfsohn (1930). It is very powerful but requires that the material concerned be easily maintained in the vapor state. A Jamin interferometer is crossed with a stigmatic spectrograph and the system illuminated with a continuous source of radiation. When both the tubes of the interferometer are evacuated and the compensating plate removed, horizontal interference fringes appear in the continuous spectrum. On replacing the compensating plate they become oblique. If a gas with an absorption line at an accessible wavelength is now introduced into the tube, the oblique interference fringes become hook shaped symmetrically on both sides of the absorption line. The wavelength separation of two hooks, symmetrically placed with regard to the absorption line, is proportional to the square root of the oscillator strength. Ostrovskii and Penkin (1960) recently have measured oscillator strengths for 65 lines of barium between 3889 and 7911 A by the hook method.

The postwar development of millimicrosecond resolving time electronics and the accompanying method of delayed coincidences (Dunworth, 1940; Bell *et al.*, 1952) for nuclear lifetime measurements has made direct measurement of atomic lifetimes possible.

Early attempts to infer atomic lifetimes from decay of spectral features in pulsed discharges, or from the distance over which the brightness of the spectral feature in a gas flowing at a known velocity decays to $1/e$ of its brightness were always suspect because the measured lifetime might be influenced by processes in the plasma.

The method of delayed coincidences involves two independent channels of detection in which counts can be registered. The two channels also together supply the inputs to a coincidence circuit which registers pulses which arrive at it in coincidence to within the resolving time of the circuit. Variable time delays (cables of known equivalent length) are used to delay the arrival at the input of the coincidence circuit of pulses from one channel relative to pulses from the other by a controlled time.

Consider the two cascade transitions $n \to p$, $p \to m$ in which the lifetime τ_p of the intermediate state p is sought. Photons from $n \to p$ activate a photomultiplier which energizes the upper channel of the detector, and photons from $p \to m$ similarly are counted in the lower channel of detection. A variable delay T is introduced in the upper channel and the coincidence counting rate C_{coin} is recorded as a function of T. The $\log C_{\text{coin}}$ vs. T curve is linear, and of slope $-1/\tau_p$ at values of T greater than the resolving time of the circuit. The requirements of low source strength to prevent overloading of the electronics and to maximize the true coincidence counting rate, and of high optical collection efficiency of photons have been emphasized by Brannen *et al.* (1955a, b).

When an attempt is made to apply the method directly to optical cascade transitions, it is often found that one of the pair of transitions lies in the visible region and the other lies in the infrared or vacuum ultraviolet, for both of which simple detectors are not available. Consequently, effective single channel modifications of the method have been developed.

Heron *et al.* (1954, 1956) excited, in a series of experiments, the 3^1P, 4^3S, 3^3P, 4^3P, and 3^1D levels of helium with a 30-volt electron beam chopped at a pulse repetition rate of 10 kc/second into triangular pulses of 4×10^{-8} second width. The upper channel of their coincidence circuit was activated by the pulses from the electron beam and the lower channel was activated, in various respective experiments by photons from the 3889 A ($3^3P - 2^3S$), 4713 A ($4^3S - 2^3P$), 3188 A ($4^3P - 2^3S$), 5875 A ($3^3D - 2^3P$), and 5016 A ($3^1P - 2^1S$) transitions, each selected by a suitable series of filters. Their measured lifetimes, compared with theoretical predictions are given in Table II.

TABLE II

LIFETIMES OF HELIUM LEVELS

Helium level	Lifetime	
	Measured (units of 10^{-8} second)	Calculated[a]
3^1P	7.4 $\pm$ 0.1	7.4
4^3S	6.75 $\pm$ 0.10	6.4
3^3P	11.5 $\pm$ 0.5	9.7
4^3P	15.3 $\pm$ 0.2	13.8
3^3D	1.0 $\pm$ 0.5	1.39

[a] Bates and Damgaard (1949).

The method has recently been improved by Bennett *et al.* (1960). They have a grating spectrometer in place of filters for isolation of lines between 1800 and 12,000 A, and use a 256-channel pulse-height analyzer with pulse-height time conversion in their detecting electronics which has a 10^{-9} second resolving time. Preliminary observations on neon have been made.

Bradley (1956) and Sagalyn (1956) proposed an extension to the method of Brannen *et al.* (1955a,b) for the study of lifetimes of hyperfine levels. Geometrical and source limitations would probably render this unworkable.

A bibliography of recent measurements is contained in the Reports of the I.A.U. Commission, 14a (1955, 1958, 1961).

6 Measurements of Molecular Transition Probabilities

A number of workers have measured emission band intensities to obtain band strength or $p_{v'v''}$ arrays. Systems of interest in astrophysics or aeronomy have received the most attention. Many of the investigations depend on photographic photometry with all of its attendant uncertainties of calibration. Some photoelectric measurements have also been made. The relative error is far greater for weak than for strong bands. A knowledge of the population in the upper level concerned is required and thus thermal excitation (e.g., furnaces, arcs) was often used. Examples of such measurements of intensities in emission of common molecular band systems are seen in the work of Ornstein and Brinkman (1931) on CN violet, Elliott (1931, 1933) on BO α and β, R. B. King (1948) on C_2 Swan, Phillips (1954) on TiO α and Floyd and King (1959) on CN violet.

It has of course been traditional for years to record subjective estimates of relative intensity in terms of plate blackenings. Such estimates are mainly used as an aid in identification although in some cases (Nicholls, 1954, 1960b) they can be used semiquantitatively.

Limitations in reproducibility of response of the photographic emulsion due to such phenomena as reciprocity failure and the intermittency effect make absolute photographic photometry a relatively inaccurate technique compared to photoelectric photometry which has been increasingly used during the past few years. There are, however, some nonsteady light sources which do not adapt to photoelectric scanning methods and force the continued use of photographic procedures. Under such circumstances the null method of Young (1954) which uses the photographic plate somewhat like a null galvanometer at zero deflection is to be recommended.

A further photometric problem arises when spectral features exhibit moderate to severe overlap in wavelength. Although apparently objective methods can be used to divide the area under the profile of two overlapped bands, the unique division of the area which minimizes error is hard to find. The rotational line intercept method (Robinson and

Nicholls, 1958a,b, 1960, 1961; Hébert, 1960) has been developed to treat cases of severe overlap between bands under conditions of high enough dispersion that a few individual lines per band are sufficiently resolved to be used.

Following the introduction of the r-centroid concept, Fraser (1954b) suggested that *all* the bands of the system could be used in the determination of a *smooth* array of $p_{v'v''}$ values by substitution of the continuous curve $S_{nm}(r)$, obtained with the aid of (51), into (50).

By analogy with the well-established concept of oscillator strengths in the case of atomic transitions, the oscillator strength $f_{v'v''}$ which is associated with an individual band may be written

$$\frac{2m}{3\hbar^2 e^2 \omega_m} E_{v'v''} \, p_{v'v''} \tag{58}$$

and in recent years there has been much activity to measure, from absorption spectra, and in some cases from emission spectra, absolute and relative values of oscillator strengths per band for a number of systems of atmospheric molecules. The band oscillator strength is an alternative to $p_{v'v''}$, and, as can be seen in the above definition, is proportional to $E_{v'v''} \, p_{v'v''}$.

A certain amount of confusion, however, has arisen by the attempt on the part of some authors to assign quantitatively an oscillator strength f_{mn} to the complete molecular transition by equations such as

$$f_{mn} = f_{v'v''} \, E_{00} / q_{v'v''} \, E_{v'v''} \tag{59}$$

which may be formally derived from (58) by asserting that $f_{mn} = \Sigma_{v'} f_{v'0}$ and assuming that $S_{nm}(r)$ is independent of r and that E_{00} is an *average* energy quantum for the whole band system. Neither assumption is really valid and thus although the $f_{v'v''}$ measurements described below are of undoubted value, the usefulness of assigning an f_{mn} to a complete transition is open to question. Any claim that f_{mn} is meaningful to better than an order of magnitude should be viewed with extreme scepticism. In some practical applications, such as estimation of radiant heat transfer in missile re-entry, order of magnitude estimates of f_{mn} are of engineering use.

Weber and Penner (1957) measured the absolute absorption coefficients of the (0, 0), (0, 1), and (2, 0) NO γ bands at a pressure ($\sim$ 100 psia) high enough for the rotational structure in each band to be smeared out. They assigned oscillator strengths to the individual bands from their measurements.

Erkovich (1959) by incorrect reanalysis of low-pressure absorption measurements of Marmo (1953) and Mayence (1952) claimed that the work of Weber and Penner was open to question. Penner (1960), however, has indicated the error in Erkovich's procedure and given a useful view of the current status of NO γ oscillator strengths.

Weber and Penner (1957) also assigned an effective oscillator strength to the whole system. In view of the foregoing discussion, it is doubtful whether this is very meaningful.

Bethke (1959a) recently measured photoelectrically band by band absorption coefficients of the pressure-broadened NO spectrum between 1700 and 2300 A. Bands of the β, γ, δ, and ϵ systems were involved. Oscillator strengths of each of a significant number of bands of each system were determined. Bethke used a similar summing and averaging procedure to that described previously to assign effective oscillator strengths f_{mn} to the complete systems. In view of his assumptions of the constancy of $S_{nm}(r)$ and the uncertainties in the $q_{v'v''}$ values he used, it is doubtful that the ± 7 and $\pm 5\%$ accuracy which he claims for f_{mn} is meaningful. His individual $f_{v'v''}$ are extremely valuable nevertheless. Erkovich and Pisarevskii (1960a, b) have re-examined the β, γ, and δ bands of NO using the work of Marmo and Mayence and computing the effect of oscillator strength for each system. In view of Penner's (1960) valid criticisms of their method where they assumed constancy of $S_{nm}(r)$, their values are not quantitatively of great use.

Keck *et al.* (1959) have made a survey of the spectral intensity distribution in emission of shock heated air in which the luminosity from band systems of NO, N_2, O_2, and CN (from impurities) were reported. After photographic survey they measure photoelectrically, in a series of presumably identical repetitive shocks, the luminous intensity from 2000 to 10,000 A. They infer oscillator strengths for each of the band systems (including NO β and γ) from a combination of their measurements and a highly idealized theory of excitation conditions. In view of the strong CN impurity features, the idealized theory (no detailed account is given of some of the steps in it) the consecutive nature of the experiments and the inherent technical difficulties of transient emission spectral radiometry, their oscillator strengths must be treated as preliminary. In the case of NO γ they differ from those of Weber and Penner and of Bethke by a factor of 24.

The O_2 Schumann-Runge band system has received attention from a number of experimenters, some of whom have assigned effective oscillator strengths to it. For this system the change Δr_e in internuclear separation involved is very large (~ 0.4 A). Consequently, the band system is spread over a wide spectral range (1700-4500 A). Ditchburn and Heddle (1954) working at low pressure measured the absorption coefficients of bands of the $v'' = 0$ progressions by photographic photometry. They interpreted their measurements in terms of oscillator strengths for each band and assuming $S_{nm}(r)$ to be constant, assigned an effective oscillator strength to the bands which they added to a previously measured (Ditchburn and Heddle, 1953) oscillator strength for the photodissociation continuum.

Bethke (1959b) made similar measurements in absorption on pressure broadened bands using photoelectric detection. He derived oscillator strengths for each band and by using the same summation procedure as that adopted for NO (Bethke, 1959a) derived an effective oscillator strength f_{mn} for the whole system. His $f_{v'v''}$ values are extremely valuable data for the $v'' = 0$ progression; but his value of f_{mn} is open to the same objections as those discussed for NO.

Treanor and Wurster (1960) in studying the O_2 Schumann-Runge system made photographic flash absorption measurements through shock heated oxygen for bands the $v' = 0$, 1, 2 progressions. The bands are significantly rotationally extended in the hot oxygen and are thereby subject to severe overlap. The rotational line intensity method (Robinson and Nicholls, 1958a, 1961) was therefore adopted to infer, from the intensity of a few unoverlapped lines in each band, the absorption intensity of the whole band. Their $f_{v'v''}$ values were thought of as a measure of $p_{v'v''}$ and from a rough comparison in v'' progressions between the variation of $f_{v'v''}$ and $q_{v'v''}$ (shifted 1.5 to lower values of v''), Treanor and Wurster concluded that $S_{nm}(r)$ varied little over the progression and that its average value could be adopted for the whole system. They then infer an effective oscillator strength f_{mn} for the whole band system using an effective average E. Again, their band-by-band oscillator strengths $f_{v'v''}$ are extremely valuable, but the assumption of constancy of $S_{nm}(r)$ is not substantiated in the work of Hébert and Nicholls (1960). Neither is it easy in such extensive band systems as the Schumann-Runge to assign an unambiguous average E. Thus, the significance (to better than an order of magnitude) of the oscillator strength of the whole system is doubtful.

Hébert and Nicholls (1961) have measured photoelectrically in emission relative intensities of 30 Schumann-Runge bands, $v' = 0$, $v'' = 9$-19; $v' = 1$, $v'' = 8$-12, 16-20; $v' = 2$, $v'' = 7$-9, 15, 16, 19, and 21, using the rotational line intercept method. The variation of $S_{nm}(r)$ with r was derived, as was an array of smooth band strengths $p_{v'v''}$ using calculated extensive arrays of r-centroid and Franck-Condon factors (Nicholls, 1960b). In each progression a plot of $p_{v'v''}$ vs. v'' is smooth and undulating. The $p_{0v''}$ values of Hébert and Nicholls and the $f_{0v''}$ values of Treanor and Wurster agree well in general trend of

TABLE III

WAVELENGTHS, FRANCK-CONDON FACTORS,[a] AND SMOOTHED BAND OSCILLATOR STRENGTHS FOR THE O_2 SCHUMANN-RUNGE SYSTEM[b, c]

v' \ v''	7	8	9	10	11	12	13	14	15	16	17	18	19	20	21	
	2566.6	2661.8	2763.4	2870.6	2984.3	3105.1	3233.8	3371.6	3517.6	3674.3	3842.2	4022.3	4215.9	4469.5	—	λ(A)
0	4.72–3	1.25–2	2.8–2	5.36–2	8.66–2	1.21–1	1.46–1	1.528–1	1.379–1	1.0739–1	7.21–2	4.16–2	2.05–2	8.60–3	—	q (power of 10 follows entry)
	1.9–3	3.36–3	5.7–3	7.95–3	8.3–3	8–3	6.7–3	4.9–3	3.4–3	2.2–3	1.4–3	0.8–3	0.47–3	0.24–3	—	f (power of 10 follows entry)
	2521.9	2614.7	2712	2815.1	2924.2	3040.2	3162.5	3293.7	3433.4	3584	3743.3	3913.9	4097	4294	4476	
1	2.049–2	4.26–2	7.08–2	9.27–2	9.19–2	6.29–2	2.18–2	1.03–4	1.83–2	6.7–2	1.137–1	1.3–1	1.12–1	7.59–2	4.137–2	
	10–3	14.5–3	23–3	16–3	10.9–3	5.2–3	1.4–3	4.7–6	0.6–3	1.5–3	2.6–3	2.6–3	2.4–3	1.9–3	1.5–3	
	2481	2570	2664	2762	2868	2979	3093	3223	3358	3501	3651	3837	3988	4174	4374	
2	4.42–2	6.89–2	7.82–2	5.8–2	2–2	7.35–7	2.14–2	6.07–2	6.96–2	3.56–2	1.92–3	1.54–2	7–2	1.18–1	1.25–1	
	25.4–3	28.7–3	23.7–3	12–3	2.86–3	0	1.44–3	2.84–3	2.36–3	0.9–3	0.046–3	0.3–3	1.41–3	2.65–3	3.4–3	

[a] Nicholls (1960b).

[b] Hébert (1960), Hébert and Nicholls (1961).

[c] Treanor and Wurster (1960).

variation with v''. However, the $v' = 1, 2$ progressions have $f_{v'v''}$ values below the smooth curve for $v'' = 2, 3$ (wavelengths less than 3000 A). This could be an artefact of photographic photometry. The relative smoothed $p_{v'v''}$ values of Hébert and Nicholls have therefore been made absolute by scaling onto the $v' = 0$ data of Treanor and Wurster. Table III gives the resulting smoothed $f_{v'v''}$ values.

The shock tube measurements of Keck *et al.* (1958, 1959) of effective oscillator strengths of O_2 Schumann-Runge must be treated with some reserve for the same reasons as those cited previously for observations on the NO bands. Further, the reanalysis of their 1958 data by Biberman *et al.* (1959) must have been treated with the same reserve. Heddle (1960) discusses the effect upon linewidth and oscillator strength of the predissociation in the $B^3\Sigma_u^-$ state of the system previously discovered by Wilkinson and Mulliken (1957) and Carroll (1959).

Lifetime measurements have recently been made for the upper levels of a number of molecular transitions using the delayed coincidence method discussed previously. Bennett and Dalby (1959, 1960a, b) have measured radiative lifetimes of levels of the $B^2\Sigma$ state of N_2^+, $A^1\Pi$ state of CO^+, $A^2\Delta$ and $B^2\Sigma$ states of CH^+, and the $A^2\Delta$ state of NH. Dayton *et al.* (1960) have measured the lifetime of levels of the $B^2\Sigma^+$ state of NO_2^+. Their results are summarized briefly in Table IV.

TABLE IV

LIFETIMES OF MOLECULAR LEVELS

Molecule	Electronic state	Vibrational level	Bands measured		Mean radiative lifetime (seconds)
N_2^+	$B^2\Sigma^+$	0	(0, 0)		$6.58 \pm 0.35 \times 10^{-8}$
CO^+	$A^2\Pi$	1	(1, 0) (1, 1)	av	$2.78 \pm 0.2 \times 10^{-6}$
		2	(2, 0) (2, 1)	av	$2.61 \pm 0.2 \times 10^{-6}$
		3	(3, 0)		$2.36 \pm 0.15 \times 10^{-6}$
		4	(4, 0)		$2.22 \pm 0.13 \times 10^{-6}$
		5	(5, 0)		$2.11 \pm 0.13 \times 10^{-6}$
CH	$A^2\Delta$	0	(0, 0)		$5.6 \pm 0.6 \times 10^{-7}$
	$B^2\Sigma$	0	(0, 0)		$1.0 \pm 0.4 \times 10^{-6}$
NH	$A^2\Pi$	0, 1	(0, 0) (1, 1)	av	$4.25 \pm 0.6 \times 10^{-7}$
NO_2^+	$B^2\Sigma^+$	0	(0, 0)		$2.6 \pm 0.2 \times 10^{-7}$

References

Allen, C. W. (1955) "Astrophysical Quantities." Univ. London Press (Athlone), London.

Allen, C. W. (1960) *Monthly Not. Roy. Astron. Soc.* **121**, 299.

Allen, C. W., and Asaad A. S. (1955) *Monthly Not. Roy. Astron. Soc.* **45**, 521.

Allen, C. W., and Asaad A. S. (1957) *Monthly Not. Roy. Astron. Soc.* **117**, 36.

Allen, L. C., and Karo, A. M. (1960) *Rev. mod. Phys.* **32**, 275.

Arthurs, A. M., and Hyslop, J. (1957a) *Proc. Phys. Soc.* **A70**, 489.

Arthurs, A. M., and Hyslop, J. (1957b) *Proc. Phys. Soc.* **A70**, 849.

Arthurs, A. M., Bond, R. A. B., and Hyslop, J. (1957) *Proc. Phys. Soc.* **A70**, 617.

Athay, R. G., and Johnson, H. R. (1960) *Astrophys. J.* **131**, 413.

Bates, D. R. (1949a) *Proc. Roy. Soc.* **A196**, 217.

Bates, D. R. (1949b) *Proc. Roy. Soc.* **A196**, 586.

Bates, D. R. (1951a) *Proc. Phys. Soc.* **A64**, 936.

Bates, D. R. (1951b) *J. chem. Phys.* **14**, 1122.

Bates, D. R. (1952) *Monthly Not. Roy. Astron. Soc.* **112**, 614.

Bates, D. R., and Carson, T. R. (1956) *Proc. Roy. Soc.* **A234**, 207.

Bates, D. R., and Damgaard, A. (1949) *Phil. Trans.* **A242**, 101.

Bates, D. R., Darling, R. T. S., Hawe, S. C., and Stewart, A. L. (1953a) *Proc. Phys. Soc.* **A66**, 1124.

Bates, D. R., Ledsham, K., and Stewart, A. L. (1953b) *Phil. Trans.* **A246**, 215.

Bates, D. R., Darling, R. T. S., Hawe, S. C., and Stewart, A. L. (1954) *Proc. Phys. Soc.* **A67**, 533.

Bell, R. E., Graham, R. L., and Petch, R. E. (1952) *Canad. J. Phys.* **30**, 35.

Bell, G. D., Davis, M. H., King, R. B., and Routly, P. M. (1958) *Astrophys. J.* **127**, 775.

Bell, G. D., Davis, M. H., King, R. B., and Routly, P. M. (1959) *Astrophys. J.* **129**, 437.

Bennett, R. G., and Dalby, F. W. (1959) *J. chem. Phys.* **31**, 434.

Bennett, R. G., and Dalby, F. W. (1960a) *J. chem. Phys.* **32**, 1111.

Bennett, R. G., and Dalby, F. W. (1960b) *J. chem. Phys.* **32**, 1716.

Bennett, W. R., Javan, A., and Ballick, E. A. (1960) *Bull. Amer. Phys. Soc.* [2]**5**, 496.

Bethe, H. A. (1947) *Phys. Rev.* **72**, 339.

Bethe, H. A., and Salpeter, E. E. (1957) *In* "Handbuch der Physik" (S. Flügge, ed.), Vol. 35, p. 88. Springer, Berlin.

Bethke, G. W. (1959a) *J. chem. Phys.* **31**, 662.

Bethke, G. W. (1959b) *J. chem. Phys.* **31**, 669.

Biberman, L. M., and Yakubov, I. T. (1960) *Optics and Spectrosc.* **8**, 155.

Biberman, L. M., Erkovich, S. P., and Soshnikov, V. N. (1959) *Optics and Spectrosc.* **7**, 346.

Biermann, L., and Lübeck, K. (1948) *Z. Astrophys.* **25**, 325.

Biermann, L., and Lübeck, K. (1949) *Z. Astrophys.* **26**, 43.

Biermann, L., and Trefftz, E. (1949) *Z. Astrophys.* **26**, 213.

Born, M., and Oppenheimer, J. R. (1927) *Ann. Phys.* **84**, 457.

Bradley, L. C. (1956) *Phys. Rev.* **102**, 293.

Brannen, E., Hunt, F. R., Adlington, R. H., and Nicholls, R. W. (1955a) *Nature* **175**, 810.

Brannen, E., Hunt, F. R., Adlington, R. H., and Nicholls, R. W. (1955b) *Phys. Rev.* **99**, 1658.

Burgess, A., Field, G. B., and Michie, R. W. (1960) *Astrophys. J.* **131**, 529.

Carroll, P. K. (1959) *Astrophys. J.* **129**, 799.

Chamberlain, J. W. (1961) Conversation with R. W. Nicholls.

Condon, E. U., and Shortley, G. H. (1935) "The Theory of Atomic Spectra." Cambridge Univ. Press, London and New York.
Dalgarno, A. (1960) *Proc. Phys. Soc.* **76**, 422.
Dalgarno, A., and Kingston, A. E. (1958) *Proc. Phys. Soc.* **72**, 1053.
Dalgarno, A., and Kingston, A. E. (1960) *Proc. Roy. Soc.* **A259**, 424.
Dalgarno, A., and Lynn, N. (1957) *Proc. Phys. Soc.* **A70**, 802.
Dalgarno, A., and Stewart, A. L. (1960) *Proc. Phys. Soc.* **76**, 49.
Dalgarno, A., Lynn, N., and Williams, E. J. A. (1956) *Proc. Phys. Soc.* **A69**, 610.
Dalgarno, A., Moiseiwitsch, B. L., and Stewart, A. L. (1957) *J. chem. Phys.* **26**, 965.
Dayton, I. E., Dalby, F. W., and Bennett, R. G. (1960) *J. chem. Phys.* **33**, 179.
Dennison, D. M. (1926) *Phys. Rev.* **28**, 318.
Ditchburn, R. W., and Heddle, D. W. O. (1953) *Proc. Roy. Soc.* **A220**, 61.
Ditchburn, R. W., and Heddle, D. W. O. (1954) *Proc. Roy. Soc.* **A226**, 509.
Dixon, R. N., and Nicholls, R. W. (1958) *Canad. J. Phys.* **36**, 127.
Dunworth, J. V. (1940) *Rev. sci. Instrum.* **11**, 167.
Eddy, J. A., House, L. L., and Zirin, H. (1961) *Astrophys. J.* **133**, 299.
Edmonds, A. R. (1957) "Angular Momentum in Quantum Mechanics." Princeton Univ. Press, Princeton, New Jersey.
Ehrenson, S., and Phillipson, P. E. (1961) *J. chem. Phys.* **34**, 1224.
Elliott, A. (1931) *Z. Phys.* **67**, 75.
Elliott, A. (1933) *Proc. Phys. Soc.* **45**, 627.
Erkovich, S. P. (1959) *Optics and Spectrosc.* **6**, 193.
Erkovich, S. P., and Pisarevski, Y. V. (1960a) *Optics and Spectrosc.* **8**, 160.
Erkovich, S. P., and Pisarevski, Y. V. (1960b) *Optics and Spectrosc.* **9**, 141.
Ferguson, H. I. S., and Nicholls, R. W. (1961) Unpublished work.
Ferguson, H. I. S., Stevens, A. E., and Nicholls, R. W. (1961) Unpublished work.
Floyd, A. L., and King, R. B. (1959) *J. Opt. Soc. Amer.* **45**, 249.
Fraser, P. A. (1954a) *Canad. J. Phys.* **32**, 515.
Fraser, P. A. (1954b) *Proc. Phys. Soc.* **A67**, 939.
Fraser, P. A. (1954c) Transition Probabilities of Molecular Band Systems, XII. Approximate Calculations of the Electronic Transition Moment. Rept No. 17, Contract AF 19(122)-470. University of Western Ontario, London, Ontario.
Fraser, P. A., and Jarmain, W. R. (1953) *Proc. Phys. Soc.* **A66**, 1145.
Fraser, P. A., Jarmain, W. R., and Nicholls, R. W. (1954) *Astrophys. J.* **119**, 286.
Garstang, R. H. (1955) *Vistas in Astron.* **1**, 268.
Gaydon, A. G., and Pearse, R. W. B. (1939) *Proc. Roy. Soc.* **A173**, 37.
Goldberg, L. (1935) *Astrophys. J.* **82**, 1.
Goldberg, L. (1936) *Astrophys. J.* **84**, 11.
Goldstein, H. (1953) "Classical Mechanics." Addison-Wesley, Reading, Massachusetts.
Gottschalk, W. M. (1948) *Astrophys. J.* **108**, 326.
Hargreaves, J. (1929) *Proc. Cambridge Phil. Soc.* **25**, 75.
Hartree, D. R. (1957) "The Calculations of Atomic Structures." Wiley, New York.
Hébert, G. R. (1960) Ph. D. thesis, University of Western Ontario, London, Ontario.
Hébert, G. R., and Nicholls, R. W. (1960) Unpublished work.
Hébert, G. R., and Nicholls, R. W. (1961) *Proc. Phys. Soc.* **78**, 1024.
Heddle, D. W. O. (1960) *J. chem. Phys.* **32**, 1889.
Herdan, R., and Hughes, T. P. (1961) *Astrophys. J.* **133**, 294.
Heron, S., McWhirter, R. W. P., and Rhoderick, E. H. (1954) *Nature* **174**, 564.
Heron, S., McWhirter, R. W. P., and Rhoderick, E. H. (1956) *Proc. Roy. Soc.* **A234**, 565.
Herzberg, G. (1944) "Atomic Spectra and Atomic Structure," 2nd ed. Dover, New York.

Herzberg, G. (1950) "Spectra of Diatomic Molecules," 2nd ed. Van Nostrand, Princeton, New Jersey.
Hey, P. (1959) *Z. Phys.* **157**, 79.
Hönl, H., and London, F. (1925) *Z. Phys.* **33**, 803.
Hutchisson, E. (1930) *Phys. Rev.* **36**, 410.
Hutchisson, E. (1931) *Phys. Rev.* **37**, 45.
I.A.U. Commission, 14a (1955) *Trans. int. astr. Un.* **9**, 214.
I.A.U. Commission, 14a (1958) *Trans. int. astr. Un.* **10**, 220.
I.A.U. Commission, 14a (1961) *Trans. int. astr. Un.* **11** (In Press).
Jarmain, W. R. (1953) Unpublished.
Jarmain, W. R. (1956) Unpublished.
Jarmain, W. R. (1959) *J. chem. Phys.* **31**, 1137.
Jarmain, W. R. (1960a) *Canad. J. Phys.* **38**, 217.
Jarmain, W. R. (1960b) Transition Probabilities of Molecular Band Systems, XVII. Tabulated Klein-Dunham Potential Energy Functions for Fifteen States of—. Sci. Rept. No. 5, Contract AF 19 (604) 4560.
Jarmain, W. R., and Fraser, P. A. (1953) *Proc. Phys. Soc.* **A66**, 1153.
Jarmain, W. R., and Nicholls, R. W. (1954) *Canad. J. Phys.* **32**, 201.
Jarmain, W. R., Fraser, P. A., and Nicholls, R. W. (1953) *Astrophys. J.* **118**, 228.
Jarmain, W. R., Fraser, P. A., and Nicholls, R. W. (1955) *Astrophys. J.* **122**, 55.
Jarmain, W. R., Ebisuzaki, R., and Nicholls, R. W. (1960) *Canad. J. Phys.* **38**, 510.
Kabir, P. K., and Salpeter, E. E. (1957) *Phys. Rev.* **108**, 1256.
Karzas, W. J., and Latter, R. (1961) *Astrophys. J. Suppl.* **6**, 167.
Keck, J. C., Camm, J. C., and Kivel, B. (1958) *J. chem. Phys.* **28**, 724.
Keck, J. C., Camm, J. C., Kivel, B., and Wentink, T. (1959) *Ann. Phys. (New York)* **7**, 1.
Kelly, H. P., and Armstrong, B. H. (1959) *Astrophys. J.* **129**, 876.
King, A. S. (1948) *Astrophys. J.* **108**, 429.
King, R. B. (1942) *Astrophys. J.* **95**, 78.
King, R. B. (1947) *Astrophys. J.* **105**, 376.
King, R. B. (1948) *Astrophys. J.* **108**, 87.
King, R. B., and King, A. S. (1935) *Astrophys. J.* **82**, 377.
King, R. B., and King, A. S. (1938) *Astrophys. J.* **87**, 24.
Klein, O. (1932) *Z. Phys.* **76**, 226.
Knox, R. S. (1957) *Solid State Phys.* **4**, 413.
Kronig, R., and Rabi, I. (1927) *Phys. Rev.* **29**, 262.
Ladenburg, R., and Wolfsohn, G. (1930) *Z. Phys.* **63**, 616.
Layzer, D. (1961) 10th Intern. Conf. on Astrophys., Liège. *Mém. Soc. Roy. Sci. Liège* (5) **4**.
Lewis, J. T., McCarroll, M R. C, and Moiseiwitsch, B. L. (1955) *Proc. Phys. Soc.* **A68**, 565.
Livingston, M. S., and Bethe, H. A. (1937) *Rev. mod. Phys.* **9**, 245.
Lochte-Holtgreven, W. (1958) *Rep. Progr. Phys.* **12**, 312.
Lyddane, R. H., Rogers, F. T., and Roach, F. E. (1941) *Phys. Rev.* **60**, 281.
Marmo, F. F. (1953) *J. Opt. Soc. Amer.* **43**, 1186.
Mastrup, F., and Wiese, W. (1958) *Z. Astrophys.* **44**, 259.
Mayence, J. (1952) *Ann. Phys. (Paris)* **7**, 453.
Menzel, D. H. (1947) *Astrophys. J.* **105**, 126.
Menzel, D. H., and Goldberg, L. (1936) *Astrophys. J.* **84**, 1.
Mitchell, A. C. G., and Zemansky, M. W. (1934) "Resonance Radiation and Excited Atoms." Cambridge Univ. Press, London and New York.

Montgomery, C. E., and Nicholls, R. W. (1951) *Phys. Rev.* **82**, 565.
Morse, P. M. (1929) *Phys. Rev.* **34**, 57.
Mulliken, R. S. (1939a) *J. chem. Phys.* **7**, 14.
Mulliken, R. S. (1939b) *J. chem. Phys.* **7**, 20.
Mulliken, R. S. (1939c) *Astrophys. J.* **89**, 283.
Mulliken, R. S. (1940a) *Phys. Rev.* **57**, 500.
Mulliken, R. S. (1940b) *J. chem. Phys.* **8**, 234.
Mulliken, R. S. (1940c) *J. chem. Phys.* **8**, 382.
Mulliken, R. S., and Rieke, C. A. (1941) *Rep. Progr. Phys.* **8**, 231.
Nicholls, R. W. (1950) *Phys. Rev.* **77**, 421.
Nicholls, R. W. (1954) *Canad. J. Phys.* **32**, 722.
Nicholls, R. W. (1956) *Proc. Phys. Soc.* **A69**, 741.
Nicholls, R. W. (1958) *J. atmos. terrest. Phys.* **12**, 211.
Nicholls, R. W. (1960a) *Nature* **186**, 715.
Nicholls, R. W. (1960b) *Canad. J. Phys.* **38**, 705.
Nicholls, R. W. (1961) *J. Res. Nat. Bur. Stds.* **65A**, 451.
Nicholls, R. W., and Jarmain, W. R. (1956) *Proc. Phys. Soc.* **A69**, 253.
Nicholls, R. W., Robinson, D., Parkinson, W. H., and Jarmain, W. R. (1956) *Proc. Phys. Soc.* **A69**, 713.
Nicholls, R. W., Fraser, P. A., and Jarmain, W. R. (1959) *Combustion and Flame* **3**, 13.
Nicholls, R. W., Fraser, P. A., Jarmain, W. R., and McEachran, R. P. (1960) *Astrophys. J.* **131**, 399.
Ornstein, L. S., and Brinkman, H. C. (1931) *Proc. Roy. Acad. Amsterdam* **34**, 33.
Ortenberg, F. S. (1960) *Optics and Spectrosc.* **9**, 82.
Ostrovskii, Y. I., and Penkin, N. P. (1960) *Optics and Spectrosc.* **9**, 371.
Pauling, L., and Goudsmit, S. (1930) "The Structure of Line Spectra." McGraw-Hill, New York.
Pekeris, C. L. (1959) *Phys. Rev.* **115**, 1216.
Penner, S. S. (1960) *J. Opt. Soc. Amer.* **50**, 627.
Phillips, J. G. (1954) *Astrophys. J.* **119**, 274.
Pillow, M. E. (1949) *Proc. Phys. Soc.* **A62**, 237.
Pillow, M. E. (1950) *Proc. Phys. Soc.* **A63**, 940.
Pillow, M. E. (1951) *Proc. Phys. Soc.* **A64**, 772.
Pillow, M. E. (1952) *Proc. Phys. Soc.* **A65**, 858.
Pillow, M. E. (1953a) *Proc. Phys. Soc.* **A66**, 1064.
Pillow, M. E. (1953b) *Mem. Soc. Roy. Sci. Liège* **13**, 145.
Pillow, M. E. (1954a) *Proc. Phys. Soc.* **A67**, 780.
Pillow, M. E. (1954b) *Proc. Phys. Soc.* **A67**, 847.
Pillow, M. E. (1955) *Proc. Phys. Soc.* **A68**, 547.
Pillow, M. E., and Rowlatt, A. L. (1960) *Proc. Phys. Soc.* **75**, 162.
Racah, G. (1942) *Phys. Rev.* **62**, 438.
Racah, G. (1943) *Phys. Rev.* **63**, 367.
Rademacher, H., and Reiche, F. (1927) *Z. Phys.* **41**, 453.
Rees, A. L. G. (1947) *Proc. Phys. Soc.* **A59**, 998.
Richter, J. (1961) *Z. Astrophys.* **51**, 177.
Robinson, D., and Nicholls, R. W. (1958a) Intensity Measurements of Molecular Spectra, IV. Intensity Measurements on Overlapped Bands. Sci. Rep. 5, Contract AF 19(604) 1718. University of Western Ontario, London, Ontario.
Robinson, D., and Nicholls, R. W. (1958b) *Proc. Phys. Soc.* **71**, 957.
Robinson, D., and Nicholls, R. W. (1960) *Proc. Phys. Soc.* **75**, 817.

Robinson, D., and Nicholls, R. W. (1961) *J. Quant. Spectrosc. and Rad. Trans.* **1**, 76.
Rogestwensky, D. (1912) *Ann. Phys. (Paris)* **39**, 307.
Rogestwensky, D., and Penkin, N. P. (1960) *J. Phys. U.S.S.R.* **5**, 1941.
Rohrlich, F. (1959a) *Astrophys. J.* **129**, 441.
Rohrlich, F. (1959b) *Astrophys. J.* **129**, 449.
Russell, H. N. (1936) *Astrophys. J.* **83**, 129.
Rydberg, R. (1931) *Z. Phys.* **73**, 376.
Sagalyn, P. L. (1956) *Phys. Rev.* **102**, 293.
Seaton, M. J. (1958) *Monthly Not. Roy. Astron. Soc.* **118**, 504.
Shortley, G. H. (1935) *Phys. Rev.* **47**, 295.
Shull, H. (1950) *Astrophys. J.* **113**, 352.
Shull, H. (1951) *Astrophys. J.* **114**, 546.
Shull, H. (1952) *J. chem. Phys.* **20**, 18.
Slater, J. C. (1960) "Quantum Theory of Atomic Structure," Vols. I and II. McGraw-Hill, New York.
Stephenson, G. (1951) *Proc. Phys. Soc.* **A64**, 666.
Stevens, A. E., and Nicholls, R. W. (1961) Unpublished work.
Tawde, N. R., and Murthy, N. S. (1960a) *Proc. Indian Acad. Sci.* **51**, 219.
Tawde, N. R., and Murthy, N. S. (1960b) *Bull. Soc. Roy. Sci. Liège* **29**, 325.
ter Haar, D. (1952) *Proc. Roy. Soc. Edinburgh* **A63**, 381.
Tobias, I., Fallon, R. J., and Vanderslice, J. T. (1960) *J. chem. Phys.* **33**, 1638.
Treanor, C. E., and Wurster, W. H. (1960) *J. chem. Phys.* **32**, 758.
Trefftz, E. (1949) *Z. Astrophys.* **26**, 240.
Trefftz, E. (1950) *Z. Astrophys.* **28**, 67.
Trefftz, E. (1951) *Z. Astrophys.* **29**, 287.
Trefftz, E., and Biermann, L. (1952) *Z. Astrophys.* **30**, 275.
Trefftz, E., Schlüter, A., Dettmar, K. H., and Jörgens, K. (1957) *Z. Astrophys.* **44**, 1.
Turner, R. G., and Nicholls, R. W. (1951) *Phys. Rev.* **82**, 290.
Turner, R. G., and Nicholls, R. W. (1954a) *Canad. J. Phys.* **32**, 468.
Turner, R. G., and Nicholls, R. W. (1954b) *Canad. J. Phys.* **32**, 475.
Varsavsky, C. M. (1958) Ph. D. thesis, Harvard University, Cambridge, Massachusetts (unpublished).
Varsavsky, C. M. (1961) *Astrophys. J. Suppl.* **6**, 75.
Vinti, J. P. (1933) *Phys. Rev.* **44**, 524.
Wallace, L. V., and Nicholls, R. W. (1955) *J. atmos. terrest. Phys.* **7**, 101.
Weber, D., and Penner, S. S. (1957) *J. chem. Phys.* **26**, 860.
Weingeroff, M. (1931) *Z. Phys.* **67**, 699.
White, H. E. (1934) "Introduction to Atomic Spectra." McGraw-Hill, New York.
White, H. E., and Eliasen, A. V. (1933) *Phys. Rev.* **44**, 753.
Wiese, W. L., and Shumaker, J. B. (1961) *J. Opt. Soc. Amer.* **51**, 937.
Wilkinson, P. G., and Mulliken, R. S. (1957) *Astrophys. J.* **125**, 594.
Wu, T.-Y. (1952) *Proc. Phys. Soc.* **A65**, 965.
Wyller, A. (1953) *Mem. Soc. Roy. Sci. Liège* **13**, 917.
Wyller, A. (1958) *Astrophys. J.* **127**, 763.
Yakubov, I. T. (1960) *Optics and Spectrosc.* **9**, 212.
Young, B. G. (1954) M. S. thesis, University of Western Ontario, London, Ontario.

3.

Photoionization Processes

R. W. Ditchburn and U. Öpik

1 *Introduction*

This chapter is concerned with the theoretical calculation and experimental determination of transition probabilities corresponding to photoionization of atoms, molecules, and radicles. It is mainly concerned with "optical processes," i.e., with photoionization associated with the absorption of light of wavelengths 7000-300 A ($\sim$ 2-$\sim$40 e.v.). Photons in this energy range are capable of removing from the atom the valence electrons and, sometimes, electrons which are somewhat more strongly bound.

There are, in principle, three problems: (a) to determine the threshold energy for photoionization, (b) to determine the variation of the transition probability with the energy of the photon when this is above the threshold, and (c) to determine the products of the photoionization of a molecule, e.g., whether the absorption of a photon of given energy produces O_2^+ or O^+, and, if either is possible, of the relative probabilities.

It would be desirable to have a general method of calculation which could be tested by experiments on a fairly wide range of atoms and molecules. When agreement had been found, such a method could be applied with confidence to excited states of atoms and to radicles which are difficult to produce in the laboratory but which may be important in relation to astrophysics. The present situation is not very favorable. An exact calculation for the hydrogen atom is available and may be applied, *mutatis mutandis* to systems in which the electron moves in a nearly coulombian field, i.e., to the deep X-ray levels and to highly excited levels of the valence electron, corresponding to "nonpenetrating" orbits. In calculations on the removal of valence electrons from the ground state, general formulae based on approximate wave functions do not always give even the correct order of magnitude. Reasonably good agreement is obtained by special assumptions applicable to a given atom, but calculations of this type cannot be unambiguously extended to other systems.

2 Experimental

The transition probability may be obtained by measuring:

(a) the absorption coefficient for a gas of known concentration;

(b) the number of ions produced per cubic centimeter when radiation of known intensity is incident upon a gas of known concentration;

(c) the recombination radiation due to capture of electrons by positive ions.

Methods (a) and (b) determine the atomic or molecular cross section (σ_λ) for photoionization by radiation as a function of the wavelength.† Method (c) determines the capture cross section (σ_v) for electrons of velocity v when they are captured directly into the appropriate discrete state of an atom or molecule. We shall require the use of the following relations:

$$h\nu = \frac{hc}{\lambda} = (V_\mathrm{i} + V_\mathrm{e})\frac{e}{300} = V_\mathrm{p}\frac{e}{300} \tag{1}$$

where ν and λ are the frequency and wavelength of the incident radiation; V_p and V_e are the potentials corresponding to the energies of the incident photon and of the emitted electron (in volts) and V_i is the ionization potential.

By the principle of detailed balance, or directly by quantum mechanics, one can show that

$$\frac{\sigma_v}{\sigma_\lambda} = \frac{\omega_i e}{2\omega_f mc^2}\frac{V_\mathrm{p}^2}{300 V_\mathrm{e}} \tag{2}$$

† σ_λ is defined by Eq. (4).

where ω_i and ω_f are the statistical weights of the initial state and the residual system in the final state, respectively. The spin degeneracy of the free electron has already been taken into account in the derivation of the formula, and must *not* be included in ω_f.

It is convenient to express σ_λ and σ_ν in megabarns (1 Mb $= 10^{-18}$ cm^2).

When a parallel beam of radiation passes through a gas at pressure p and temperature T, the absorption in a thin layer dl is given by

$$\frac{dI}{I} = \alpha \left(\frac{273}{T} \frac{p}{760}\right) dl \tag{3}$$

where I is the radiation incident on the layer and dI is the fraction absorbed in it, α is the absorption coefficient (in cm^{-1}) for the gas at standard temperature and pressure. The conditions for the validity of Eq. (3) are:

(a) The radiation must be so nearly monochromatic that the effective absorption does not change as the radiation advances through the gas (Lambert's law), i.e., α independent of l.

(b) α must be independent of p and T so that dI is proportional to p/T, i.e., to the concentration (Beer's law). This is so if there is no effective change in the composition of the gas (e.g., by dissociation of molecules) when p and T vary and if the absorption coefficient is not significantly affected by interatomic or intermolecular collisions.

When α has been measured, σ_λ is obtained from the relation

$$N\sigma_\lambda = 10^{18}\alpha \tag{4}$$

where N is Avogadro's number (2.69×10^{19} atoms/cc). The factor 10^{18} is required because α is measured in cm^{-1} and σ_λ in megabarns.

2.1 Measurement of Absorption Coefficient

Nearly all measurements of absorption due to photoionization have been made by single-beam spectrophotometry. Typical arrangements are shown in Figs. 1 and 2. In the method shown in Fig. 1, light from

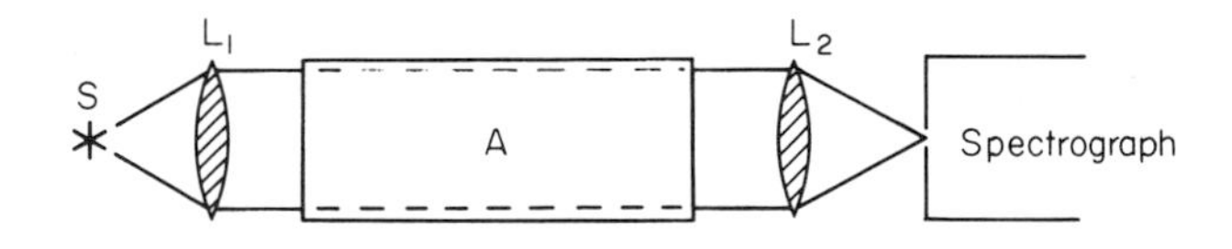

FIG. 1. Measurement of absorption—single-beam photographic method.

the source S is collimated by L_1 and focused on the slit of the spectrograph by L_2. Spectra are photographed when the vessel A is (a) evacuated and (b) filled with the absorbing gas at known temperature and pressure. Standard methods of photographic photometry are used to measure the ratio I_0/I, where I_0 is the intensity without and I is that with the absorbing gas present. The absorption coefficient is then given by

$$\alpha = \left(\frac{I}{273} \frac{760}{pl}\right) \log_e \left(\frac{I_0}{I}\right). \tag{5}$$

In this equation, which is obtained by integrating (3), l is the length of the absorbing column.

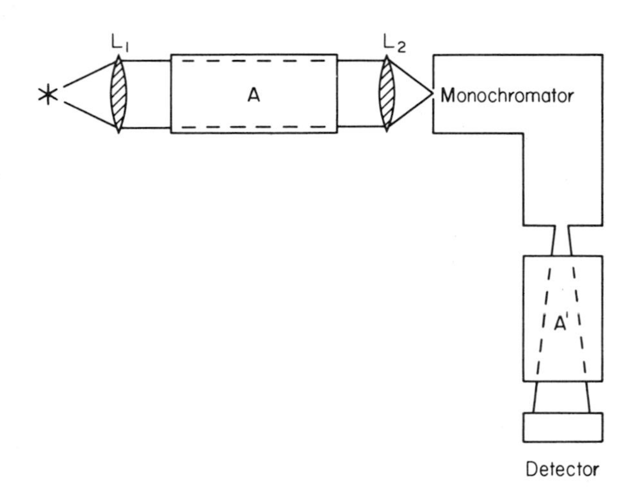

FIG. 2. Measurement of absorption—single-beam method with photoelectric detector. Absorption vessel may be either at A or at A′.

In the method shown in Fig. 2, the absorption vessel may be placed either at A or at A′. The latter position exposes the gas to a much weaker beam of radiation and is desirable if there is any likelihood of the gas being changed chemically by the radiation (e.g., the formation of O_3 when O_2 is irradiated). In either method, for wavelengths shorter than 2000 A, air should be excluded from the path. If the gas is not chemically active, the vacuum spectrograph or monochromator may then be used as the absorption vessel. For wavelengths below 1200 A, no lenses are available and mirrors are inefficient. The appropriate modifications of the system have been described by several workers (Weissler, 1956; Ditchburn, 1955; Platt and Klevens, 1944).

Many of the substances whose photoionization cross sections are of interest are chemically active (e.g., the alkali metals) so that they attack quartz and other substances which may be used as transparent windows. The absorption vessel shown in Fig. 3 was developed by Ditchburn

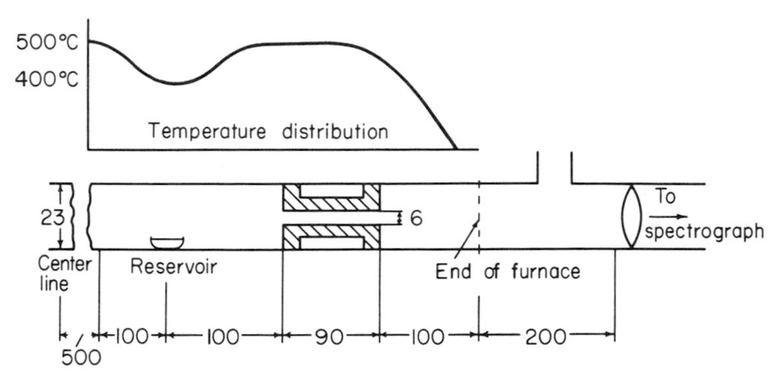

FIG. 3. Absorption vessel for chemically active vapors (right-hand half shown).

and associates (1943) for active vapors such as sodium vapor. The solid is contained in a metal tube with constrictions near the ends. The tube is heated to a little beyond the constrictions and is filled with an inert gas. If the pressure of the inert gas is a little greater than the vapor pressure of the alkali metal, the space between the constrictions is filled with the vapor at a pressure corresponding to the temperature of the solid. The vapor pressure falls sharply through the constrictions to near zero outside. The value of $\int p\,dl$ can be calculated with sufficient accuracy. The inert gas acts as a transparent and nonreactive plug which holds the vapor in the central part of the tube.

In the photographic method, one has the advantage that a great deal of information is recorded on one plate. This is important in experiments on the absorption of vapors where the adjustment of exact temperatures is difficult. The detectors used in the method shown in Fig. 2 are more nearly linear than the photographic plate and, for nonreactive gases used at room temperature, this method is capable of giving accurate results more quickly. When high resolution is required (for investigation of the shape of the curve very near the series limit or of autoionization), the spectrographic method is preferable.

In experiments on photoionization, it is sometimes found that Lambert's law is obeyed but Beer's law is not. The coefficient α, calculated from (5), is not independent of p. If, for example, weakly bound molecules are present, then if p varies and T is constant, we have

$$\alpha(p) = \alpha(0) + \beta p \tag{6}$$

since under these conditions the number of atoms is proportional to p and the number of molecules to p^2. The absorption of the atoms may then be obtained by plotting $\alpha(p)$ against p and extrapolating to zero pressure.

2.2 Measurement of Ionization Current

If all the absorption is associated with photoionization, the number of ion pairs formed is equal to the number of photons absorbed. If n (the number of incident photons) and n_i (the number of ion pairs) are measured for known p, T, and l, then α may be calculated by inserting $n_i/n = dI/I$ in (3) or $n/(n - n_i) = I_0/I$ in (5). In this method the ion current is measured by inserting electrodes in the vessel placed at A′. At the time when some experiments of this type were carried out, measurement of the small ion currents in the presence of surface leaks was very difficult. This problem has now been solved, and the main difficulty is the calibration of detectors to measure the absolute value of the number of photons in a rather weak beam (note that the absorption method requires only the *ratio* of two intensities neither of which is known in absolute units). For this reason, the ion-current method is, in general, less accurate than the absorption method, but it has the advantage that only the absorption due to photoionization is measured. This is very important in experiments on molecular gases where continua due to photodissociation may overlap continua due to ionization.

In some of the early experiments, the weak photoionization currents were increased by space-charge amplification. It now appears probable that the amplification ratio is not independent of the energy of the ions produced, so that the method gives only a general indication of the variation of σ_λ with λ.

2.3 Observations of Recombination Spectra

σ_v may be obtained by measurements on the recombination radiation emitted by a plasma if the concentrations of electrons and positive ions (and the distribution of the relative velocities) are known. Probes are used to measure an electron temperature. The electrical measurements required are difficult and this method is not susceptible of high accuracy. It is, however, the only available method of measuring cross sections associated with highly excited states of atoms or molecules.

2.4 Analysis of Products

In some situations, more than one photoionization process is possible for a given wavelength. The relative probabilities of these processes are obtained by analyzing the positive ions in a mass spectrometer. Ditchburn and Arnot (1929) showed that the ionization of potassium vapor produces K^+ and not K_2^+. More accurate and extensive measurements have recently been made by Schönheit (1957) and by Weissler (1956). See also Herzog and Marmo (1957), Hurzeler *et al.* (1957, 1958), and Weissler *et al.* (1959).

3 *Theoretical*

Consider an atomic or molecular system in an initial (not necessarily pure) state i, having an energy E_i, measured with respect to the lowest state in which one of its electrons is at rest at an infinite distance. Consider processes in which the system absorbs a photon making a transition to a state f, in which one of its electrons is free. The set of all states of the system in which one electron is free and the residual system is in a state of a definite energy is called a continuum of the system. Light of frequency ν may cause transitions to those continua whose lowest energy does not exceed $E_i + h\nu$.

The quantum-mechanical formula for the cross section for the absorption of a light quantum of frequency ν, accompanied by such a transition, is

$$\sigma(\nu) = \frac{8\pi^3\nu}{3c}\frac{1}{\omega_i}\sum_i\sum_f\left|\int\psi_f^*\sum_\mu e_\mu \mathbf{r}_\mu \psi_i d\tau\right|^2. \tag{7}$$

Here the functions ψ_i are wave functions of the ω_i-fold degenerate initial state of the atomic or molecular system, Σ_i denoting the sum over all ω_i of these states. The functions ψ_f are continuum eigenfunctions of the energy, belonging to the eigenvalue $E_i + h\nu$; these functions are in energy normalization, i.e., normalized so that, if the system is in a state represented by

$$\Phi = \int a(E)\,\psi_f(E)\,dE \tag{8}$$

where Φ is normalized to unity and $\psi_f(E)$ belongs to the energy E, then the probability of finding the energy of the system between E and $E + dE$ is $|a(E)|^2 dE$. Σ_f denotes summation over the states in a continuum if the cross section for transitions into that continuum is required, or over the states of all those continua to which transitions are energetically possible if the total cross section is required. e_μ is the charge, and $\mathbf{r}_\mu$ the position vector with respect to the center of mass, of the μth particle of the system. The integration is over all coordinates of all particles and includes summation over the spin coordinates.

Other kinds of normalization for ψ_f than that previously described are often used, with corresponding modifications in the multiplying factor in (7); see, e.g., Bates (1946b). Equation (7) may be considered exact for our purposes.

For the photoionization of a hydrogen-like system, consisting of one electron in the field of a nucleus of charge number Z, (7) has been exactly evaluated by Sugiura (1927, 1929), Gaunt (1930), and more completely, by Menzel and Pekeris (1935); see also Burgess (1958) for a correction to one of Menzel and Pekeris' formulae. The total cross section for the absorption of light of frequency ν is found to be

$$\sigma(\nu, n) = g(32\pi^2 e^6 R Z^4)/(3^{3/2}\, h^3 \nu^3 n^5) \tag{9}$$

where $R = 2\pi^2 e^4 m/(h^3 c)$ is the Rydberg constant, n is the principal quantum number of the initial state, and g is a complicated factor, which has been evaluated for several values of n and over a range of energies of the ionized state by Menzel and Pekeris (1935). For additional, more recent work, see Armstrong and Kelly (1959) and Olsson (1959).

If we set $g = 1$ in Eq. (9) the formula reduces to that derived by Kramers (1923). Since g does not differ from unity by more than about 10 or 20 % (except for values of $h\nu$ only slightly in excess of the ionization threshold) and tends to unity as $n \to \infty$, it is, for many practical purposes, sufficiently accurate to calculate the cross sections of hydrogen-like systems by Kramers' formula. However, Page (1939) and Bates (1939) have pointed out that it is incorrect to use this formula for systems which have more than one electron by replacing Z by some effective value—the results are liable to be incorrect even as regards the order of magnitude. An improved generalization of this method, intended to be applied to atoms or ions having not very many electrons, has been given by Bates (1946a).

The most frequently used approximation for many-electron atoms is the central-field approximation, in which wave functions are taken to be linear combinations of determinants whose elements are one-electron wave functions (Condon and Shortley, 1935, Chapter VI). Each one-electron wave function $u\ (n\ l\ m_l\ m_s)$, called a spin-orbital, is specified by the principal, azimuthal, orbital magnetic, and spin quantum numbers, n, l, m_l, and m_s, respectively. We shall here mean by a shell the set of all $2(2l + 1)$ spin-orbitals which belong to the same values of n and l. We take the distribution of the electrons between the shells to be the same in all those determinants which we combine to obtain an approximate wave function, i.e., we neglect configuration interaction. It is then sufficient to consider only those transitions in which one electron, called the active electron, is removed from one of the shells into the continuum, while the distribution of the remaining electrons, called the passive electrons, between the shells remains unchanged. Then the cross section for transition into any one continuum of the complete system is (see Bates, 1946b)

$$A\mathscr{C}\left\{C_{l-1}\left|\int_0^\infty R^i(n,l;r)\,R^f(\epsilon,l-1;r)r^3dr\right|^2 + C_{l+1}\left|\int_0^\infty R^i(n,l;r)\,R^f(\epsilon,l+1;r)r^3dr\right|^2\right\}. \tag{10}$$

Here $R^i(n, l; r)$ is the radial wave function of the active electron in the initial state, $R^f(\epsilon, l'; r)$ is that in the final state, n is the principal quantum number, ϵ is the kinetic energy of the free electron at $r = \infty$, l or l' is the azimuthal quantum number, and r is the distance from the nucleus. The factor A depends on the normalization of the continuum wave function and is in general a function of ν and ϵ. C_{l-1} and C_{l+1} are numerical coefficients which depend on the spectroscopic type of the initial and the final states and have been tabulated by Bates (1946b); a formula for these coefficients is given by Armstrong (1959).

The coefficient $\mathscr{C}$ is a correcting factor to allow for the fact that the spin-orbitals $u^f(n\,l\,m_l\,m_s)$ in the residual system are slightly different from the corresponding spin-orbitals $u^i(n\,l\,m_l\,m_s)$ in the initial state. Provided that $u^f(n'\,l\,m_l\,m_s)$ is as nearly as possible orthogonal to $u^i(n\,l\,m_l\,m_s)$ whenever $n' \neq n$, it is a sufficiently good approximation to write

$$\mathscr{C} = \prod_{\substack{\text{passive}\\ \text{electrons}}} \left|\int_0^\infty R^i(n,l;r)\,R^f(n,l;r)r^2dr\right|^2. \tag{11}$$

Since in a determinantal wave function a set of spin-orbitals may always be replaced by an equal number of linearly independent linear combinations of the same spin-orbitals, and since the u^i's differ only slightly from the u^f's, it can be shown that it is always possible to satisfy the orthogonality requirement closely enough to make formula (11) a valid approximation.

The chief difficulties in using formula (10) are connected with the evaluation of the integrals of the form

$$\int_0^\infty R^i(n, l; r)\, R^f(\epsilon, l'; r) r^3 dr \tag{12}$$

either (i) because the positive contributions to the integral are nearly equal in magnitude to the negative contributions, so that the integral is the small difference of two large quantities; or (ii) because the magnitude of the continuum orbital near the origin, for a given asymptotic amplitude (which alone determines the normalization), is very sensitive to the exact form of the effective potential field. The difficulty (i) is serious in the case of many neutral atoms, especially sodium and potassium (see Bates, 1946b, for a review); the difficulty (ii) occurs with some negative ions (Bates and Massey, 1943).

Because of these difficulties, wave functions of the active electron calculated by the Hartree or Hartree-Fock methods are usually not accurate enough. One way of improving the accuracy is to allow for the correlation between the active electron and the other electrons by adding a term

$$-\tfrac{1}{2} p/(r_p^2 + r^2)^2$$

to the self-consistent potential energy, to represent the attraction between the active electron and the dipole moment induced by the active electron on the residual system. Here r_p is nearly equal to the mean radius of the state polarized, and p is the polarizability, and can be found by theory (Buckingham, 1937; Kirkwood, 1932), as was done in the case of the O^- ion by Bates and Massey (1943), who used this method to improve the accuracy of the continuum wave function; or the polarizability may be so determined as to make one of the calculated energies agree with the observed value, as was done for O^- by Klein and Brueckner (1958), who used the polarizability correction in obtaining both the bound and the free wave function, and obtained excellent agreement with the experimental results of Branscomb *et al.* (1958).

The main contribution to an integral of the form (12) comes from rather large values of r, where the effective field in which the electron moves may in many cases be taken to be Coulombic. This, together with the observed energy and the requirement that the bound wave function shall tend to zero as $r \to \infty$, is sufficient to determine the bound wave function for these large values of r apart from the normalizing factor, which factor can, however, be sufficiently accurately estimated.

The method of Bates and Damgaard (1949; see Chapter 2) for the intensities of discrete lines is based on these considerations. The difficulty in applying the method to photoionization is that although we can find the *general* solution of the radial Schrödinger equation for the continuum wave function at the relevant large values of r, we do not know what particular solution to choose unless we integrate outwards from the nucleus, which is too inaccurate because we do not know the effective potential field near the nucleus well enough.

This difficulty was overcome by Seaton (1958; see also Seaton, 1955, for a simpler but less complete method). Following Ham (1955), Seaton writes the general solution of the one-electron radial Schrödinger equation for a *pure* Coulomb field in an analytic form (in effect):

$$F(\epsilon, l; r) = f(\epsilon, l; r) + B(\epsilon, l)\, g(\epsilon, l; r) \tag{13}$$

where f and g are two linearly independent solutions of that equation, f being the one that remains finite as $r \to 0$ [it must be remembered that $F(\epsilon, l; r)$ can be approximately equal to the actual wave function only for sufficiently large values of r, well outside the cloud of the other electrons]; ϵ and l have the same significance as before. The analytic form of f and g is such that these functions satisfy the radial equation for *all* finite values of ϵ, and, *regarded as functions of the energy* ϵ at constant l and (finite nonzero) r, are analytic over the whole of the complex ϵ-plane except at $|\epsilon| = \infty$. This can be shown to be true of $F(\epsilon, l; r)$ also, provided that $F(\epsilon, l; r)$ is that particular solution which at large values of r is almost equal to $R(\epsilon, l; r)$, where $R(\epsilon, l; r)$ is the solution, finite at the origin (but not necessarily at $r = \infty$), of the radial equation of the *actual* problem, i.e., for the effective field of the residual system. It follows that $B(\epsilon, l)$ is also an analytic function of ϵ, and can therefore be expanded as a power series in ϵ, which is valid for *all* finite values of ϵ, and may, in particular, be used to find $B(\epsilon, l)$ for real positive values of ϵ, i.e., the continuum. Now $B(\epsilon, l)$ is known, for all those negative values of ϵ which are eigenvalues of the Schrödinger equation of the actual problem, from a relationship between $B(\epsilon_{nl}, l)$ and the quantum defect μ_{nl}, the discrete eigenvalue ϵ_{nl} being expressed as

$$\epsilon_{nl} = -hcRz^2/(n - \mu_{nl})^2 \tag{14}$$

where ze is the excess of positive over negative charge on the residual system. So the coefficients in the power series for B can be obtained from the resulting infinite set of equations, the quantities μ_{nl} being determined from the observed energies by (14).

Burgess and Seaton (1960) have applied this method to a large number of cases. They find that the method gives good results except when there is a strong cancellation between the positive and negative contributions to the transition integral.

Very little theoretical work has been done on the photoionization of molecules and molecular ions because for these it is even more difficult to obtain accurate electronic wave functions than for atomic systems. Exact electronic wave functions can be obtained only for the hydrogen molecular ion, H_2^+, for which the cross section was calculated by Bates *et al.* (1953). Dalgarno (1952) has calculated the photoionization cross section of methane; he used for the active electron wave functions appropriate to a spherically symmetrical field and obtained results in a reasonable agreement with experiments. Geltman (1958) has found the energy dependence of the cross section of diatomic molecules near the threshold.

Lastly, it should be mentioned that in order to obtain the total effective cross section for the removal of an electron from an atomic or molecular system through the action of light, it is usually (except in the case of some light atoms) necessary to consider also, in addition to direct transitions into the continuum, transitions to discrete states in which the energy exceeds the lowest energy of a continuum. From such a state the system may make a radiationless transition into the continuum (autoionization), and if such a transition is allowed, it usually has a very much higher probability than a radiative transition to a lower state. This effect has, for example, been found important for thallium (Marr, 1954b). Detailed information about the occurrence and the term values of states which have sufficiently high energies for autoionization to take place will be found in the work of Beutler and collaborators (1933; Beutler and Demeter, 1934; Beutler *et al.*, 1936), Garton (1950, 1952), and Garton and Codling (1960). Some quantitative measurements of the intensities and widths of the corresponding absorption lines are given for cadmium by Garton and Rajaratnam (1955), xenon by Pery-Thorne and Garton (1960), and for calcium by Ditchburn and Hudson (1960) and Kaiser (1960).

For additional information, especially on points not discussed here, the reader is referred to the following publications: Bates (1946a, b), Biberman and Norman (1960), Burgess and Seaton (1958, 1960), Finkelnburg (1938), Geltman (1958), Unsöld (1955), Weissler (1956), Wigner (1948), Ditchburn (1954, 1956), and Marr (1955).

TABLE I — RESULTS FOR ATOMS AND ATOMIC IONS[a]

System	State	V_i	λ_T	σ_T	λ_m	σ_m	References
H		13.6	912	6.3			1-4, 31
He		24.6	504	7.4			5-10
Li		5.39	2300	2.5			11-15, 10
Be		9.32	1330	(8.2)			16
B		8.30	1490	19			17, 18
C		11.3	1100	(11)			17, 18, 10, 20
N		14.6	852	(9)			19, 20, 17, 18, 21, 10
O		13.6	910	(2.6)			20, 17, 22-24, 10
F		17.4	713	(6)			17, 18
Ne	L_2/L_3	21.6	575	4.0	400	8.0	25-27, 17, 18, 53
	L_1	48.5	256	2.5			
Na		5.14	2412	0.12	1900	0.013	27 28, 30, 32, 10
Mg		7.64	1620	1.18			33, 10
Al		5.98	2070				34
Ar	M_3	15.8	787	35			
	M_2	15.9	778				7, 35-37
	M_1	32	420	(3)			
K		4.34	2860	0.012	2700	0.006	38-42, 27, 10, 51
Ca		6.11	2028	0.45	1930	<0.1	
			1589	(0.45)			43-45, 29
Ga		6.00	2070	(0.2)			46, 34
Kr		14.0	885				
		14.7	845	35	800	35	47, 37
Rb		4.18	2970	0.11		<0.004	41, 40, 27
In		5.79	2140	(0.3)			48
Xe		12.1	1020				37
Cs		3.89	3185	0.22	2750	0.08	49, 41, 40, 27, 52
Tl		6.11	2030	4.5	1730	<0.1	50
He^+		54.4	228	1.6			1-4
C^+		24.4	508	3.7			16, 17
N^+		29.6	419	(6.4)			16, 17
O^+		35.2	353	(8.1)			16, 17, 10
F^+		35.0	354	2.5			16, 17
Ne^+		41.0	302	4.5			16, 17, 10
Na^+		47.3	262	7.1			16, 17, 10
Mg^+		15.0	825				54, 10
Si^+		16.3	760				54, 10
Ca^+		11.9	1040	0.17			10

[a] In the second column the initial state of the atom or ion is given only if it is not the ground state. V_i = ionization potential (ev), λ_T = threshold wavelength (A), σ_T = cross section at the spectral head, λ_m, σ_m are the wavelength and cross section of either the minimum or the maximum (whichever occurs for the system under consideration) absorption associated with the ionization process for which V_i is given. Figures in parentheses are either less accurately known or less certainly associated with the process listed.

Key to references in Table I:

1. Kramers (1923).
2. Menzel and Pekeris (1935).
3. Olsson (1959).
4. Armstrong and Kelly (1959).
5. Vinti (1933).
6. Wheeler (1933).
7. Lee and Weissler (1955).
8. Axelrod and Givens (1959).
9. Stewart and Wilkinson (1960).
10. Burgess and Seaton (1960).
11. Hargreaves (1929).
12. Trumpy (1931).
13. Tunstead (1953).
14. Stewart (1954).
15. Propin (1960).
16. Bates (1946a).
17. Bates (1946b).
18. Bates (1939).
19. Ehler and Weissler (1955).
20. Bates and Seaton (1949).
21. Boldt (1959b).
22. Bates *et al.* (1939).
23. Yamanouchi and Kotani (1940).
24. Boldt (1959a).
25. Lee and Weissler (1953).
26. Ditchburn (1960).
27. Seaton (1951).
28. Ditchburn *et al.* (1953).
29. Jutsum (1954).
30. Rudkjøbing (1940).
31. Suguira (1927).
32. Trumpy (1928).
33. Ditchburn and Marr (1953).
34. Vainshtein and Norman (1960).
35. Wainfan *et al.* (1955).
36. Dalgarno (1952).
37. Beutler (1935).
38. Ditchburn *et al.* (1943).
39. Ditchburn and Arnot (1929).
40. Lawrence and Edlefsen (1929).
41. Mohler and Boeckner (1929).
42. Bates (1947).
43. Ditchburn and Hudson (1960).
44. Kaiser (1960).
45. Bates and Massey (1941).
46. Garton (1952).
47. Pery-Thorne and Garton (1960).
48. Marr (1954a).
49. Braddick and Ditchburn (1934).
50. Marr (1954b).
51. Ditchburn (1928).
52. Ditchburn (1937).
53. Seaton (1954).
54. Biermann and Lübeck (1949).

4 *Results*

The results are summarized in Tables I and II and in Figs. 3-5. For hydrogen-like systems (H, He^+, Li^{++}, etc.), there are no experimental results, but (9) is effectively an exact calculation. This equation may be expected to yield useful results when applied to ionization of those highly excited states of other atoms in which one valence electron has a Bohr orbit which does not penetrate the core or approach it closely. As stated in § 3, this equation should not be applied to calculations on valence electrons in their normal states. For Na, improved calculations by methods described in § 3, yield results in agreement with experiment. It is necessary to choose the exact value of the polarizability to fit the experiments, but the value required is found to agree with less accurate estimates from other sources. The calculations agree with experimental results for photon energies up to about 2 volts above the

TABLE II

RESULTS FOR MOLECULES[a]

System	V_i	λ_T A	λ_m	σ_m	References
H_2	15.4	804	780	7.4	1-4
	12.6	1019			
O_2	16.1	770	550	22	5, 7, 4, 24, 25
N_2	15.6	792	750	26	6, 7, 4, 8, 22, 24
CH_4	12.8	967	960	(56)	9, 4, 10, 13
H_2O	12.5	985	346	(100)	11, 4, 12
CO	14.21	868	600	16.5	23
	(16.5)	(750)		~1	
	(20)	(620)		~1	
CO_2	14.4	860	(800)	18	23, 4
	(17.7)	(700)	(600)	(15)	
NH_3	10.3	1210	(900)	10	13-15
	(16.5)	(750)	(700)	(25)	
N_2O	12.9	960	700	35	16, 17
		?	(357)	(30)	
		?	(225)	(30)	
NO	9.25	1340	880	20	15, 18-20
	(14)	(880)		?	
	(16)	(730)		?	
	(18.7)	(660)	(600)	(20)	
H_2^+	16.3	763	400	0.67	21

[a] V_i = ionization potential (ev), λ_T = threshold wavelength (A), λ_m, σ_m are wavelength and cross section of maximum absorption associated with the ionization process for which V_i is given. Figures in parentheses are either less accurately known or less certainly associated with the process listed. Weissler (1956) gives further data on 0_3, C_2H_2, and C_2H_4, Lowrey and Watanabe (1958) on C_2H_4O, Walker and Weissler (1955b) on C_2H_4 and C_2H_2, and Morrison *et al.* (1960) on Br_2, I_2, HI, and CH_3I.

1. Beutler *et al.* (1936).
2. Beutler and Jünger (1936).
3. Lee and Weissler (1952).
4. Wainfan *et al.* (1955).
5. Lee (1955).
6. Weissler *et al.* (1952).
7. Wainfan *et al.* (1953).
8. Maunsell (1955).
9. Ditchburn (1955).
10. Dalgarno (1952).
11. Astoin *et al.* (1953).
12. Astoin (1956).
13. Sun and Weissler (1955a).
14. Walker and Weissler (1955a).
15. Watanabe (1954).
16. Walker and Weissler (1955c).
17. Astoin and Granier (1955).
18. Sun and Weissler (1955b).
19. Granier and Astoin (1956).
20. Watanabe *et al.* (1953a).
21. Bates *et al.* (1953).
22. Astoin and Granier (1957).
23. Sun and Weissler (1955c).
24. Watanabe and Marmo (1956).
25. Weissler and Lee (1952).

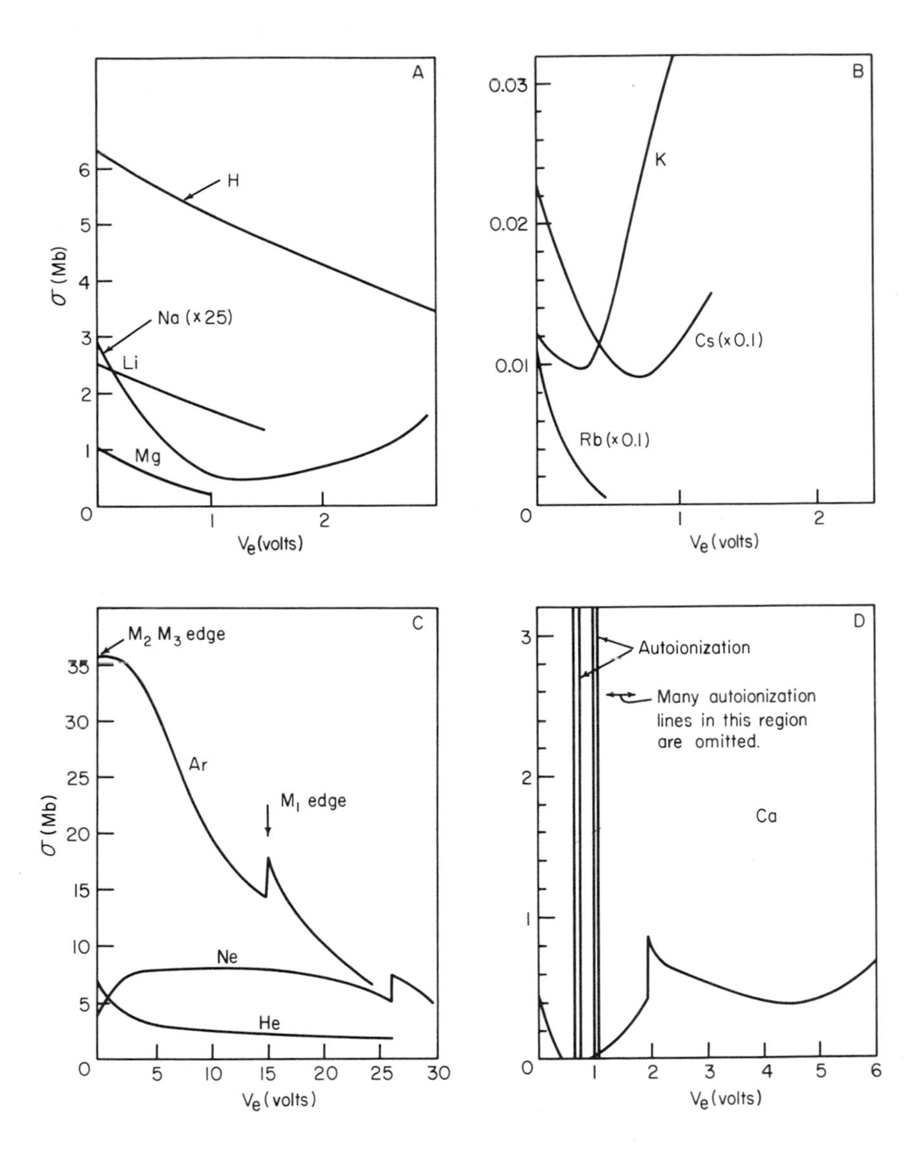

FIG. 4. Absorption cross sections (σ_λ) for atoms. Abscissa are voltages of the emitted electron. Note that ordinate scales for *A*, *B*, *C*, and *D* are not the same and that ordinates for Na are × 25.

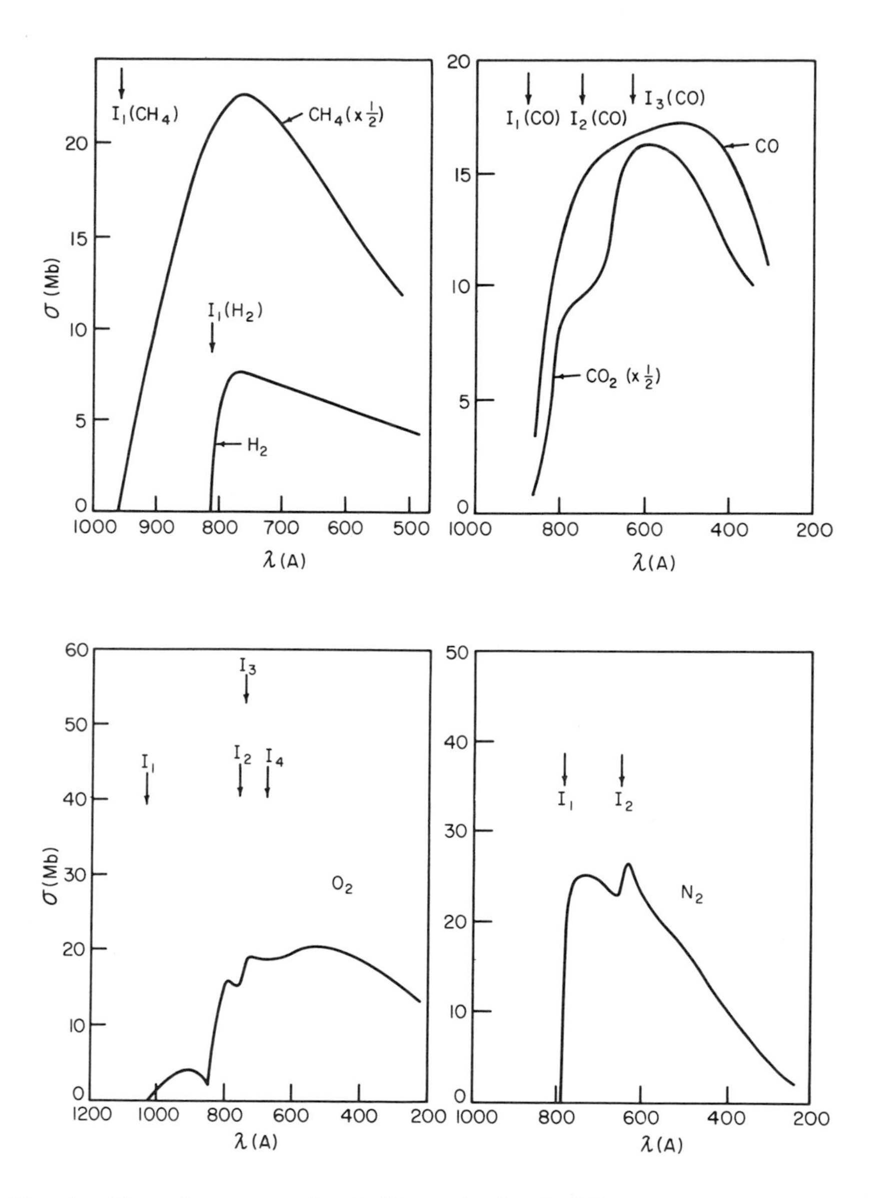

FIG. 5. Absorption cross sections (σ_λ) for molecules. I_1 (CO) represents the threshold wavelength corresponding to the first ionization potential of CO and other ionization potentials are similarly indicated.

threshold energy but not at higher energies. For K, Rb, and Cs, the balancing of positive and negative contributions to the integral (see § 3) is so close that calculation becomes impossible, though it has been shown that a reasonable value of the polarizability is consistent with the general shape of the curve of σ_λ versus λ. It is a little unsatisfactory that the agreement between theory and experiment for lithium is not better than it is. For He, Ne, and Ar, this method yields results in reasonably good agreement with experiment. For calcium, calculation without the polarizability correction gives a value for σ_T which is 50 times larger than the observed value, but the improved calculation gives results which agree with experiment in the region near the series limit but deviate from it rapidly at shorter wavelengths.

Autoionization. For many elements, of which calcium (see Fig. 4D) is typical, autoionization (see § 3) is very important. The f-value associated with autoionization is very often greater (and sometimes greater by a factor of 10 or more) than the f-value associated with removal of the valence electron. Autoionization is known to be strong for Ca, Ba, In, Ga, Cu, Tl, and many other elements. In calculating cross sections for photoionization, autoionization must always be considered and can seldom be regarded as a correction. Unfortunately, few theoretical calculations are available.

There are few calculations of ionization probabilities for molecules. Preionization is probably important. The chief experimental difficulty is the separation of overlapping continua corresponding to different processes of dissociation, ionization, preionization, and predissociation.

References

Armstrong, B. H. (1959) *Proc. Phys. Soc.* **74**, 136.
Armstrong, B. H., and Kelly, H. P. (1959) *J. Opt. Soc. Amer.* **49**, 949.
Astoin, N. (1956) *C. R. Acad. Sci. (Paris)* **242**, 2327.
Astoin, N., and Granier, J. (1955) *C. R. Acad. Sci. (Paris)* **241**, 1736.
Astoin, N., and Granier, J. (1957) *C. R. Acad. Sci. (Paris)* **244**, 1350.
Astoin, N., Johannin-Gilles, A., and Vodar, B. (1953) *C. R. Acad. Sci. (Paris)* **237**, 558.
Axelrod, N. N., and Givens, M. P. (1959) *Phys. Rev.* **115**, 97.
Bates, D. R. (1939) *Monthly Not. Roy. Astron. Soc.* **100**, 25.
Bates, D. R. (1946a) *Monthly Not. Roy. Astron. Soc.* **106**, 423.
Bates, D. R. (1946b) *Monthly Not. Roy. Astron. Soc.* **106**, 432.
Bates, D. R. (1947) *Proc. Roy. Soc.* **A188**, 350.
Bates, D. R., and Damgaard, A. (1949) *Phil. Trans.* **A242**, 101.
Bates, D. R., and Massey, H. S. W. (1941) *Proc. Roy. Soc.* **A177**, 329.
Bates, D. R., and Massey, H. S. W. (1943) *Phil. Trans.* **A239**, 269.
Bates, D. R., and Seaton, M. J. (1949) *Monthly Not. Roy. Astron. Soc.* **109**, 698.

Bates, D. R., Buckingham, R. A., Massey, H. S. W., and Unwin, J. J. (1939) *Proc. Roy. Soc.* **A170**, 322.
Bates, D. R., Öpik, U., and Poots, G. (1953) *Proc. Phys. Soc.* **A66**, 1113.
Beutler, H. (1933) *Z. Phys.* **86**, 495, 710.
Beutler, H. (1935) *Z. Phys.* **93**, 177.
Beutler, H., and Demeter, W. (1934) *Z. Phys.* **91**, 131, 143, 202, 218.
Beutler, H., and Guggenheimer, K. (1934) *Z. Phys.* **87**, 19, 176, 188; **88**, 25.
Beutler, H., and Jünger, H. O. (1936) *Z. Phys.* **100**, 80; **101**, 285.
Beutler, H., Deubner, A., and Jünger, H. O. (1936) *Z. Phys.* **98**, 181.
Biberman, L. M., and Norman, G. É. (1960) *Optika i Spektrosk.* **8**, 433.
Biermann, L., and Lübeck, K. (1949) *Z. Astrophys.* **26**, 43.
Boldt, G. (1959a) *Z. Phys.* **154**, 319.
Boldt, G. (1959b) *Z. Phys.* **154**, 330.
Braddick, H. J. J., and Ditchburn, R. W. (1934) *Proc. Roy. Soc.* **A143**, 472.
Branscomb, L. M., Burch, D. S., Smith, S. J., and Geltman, S. (1958) *Phys. Rev.* **111**, 504.
Buckingham, R. A. (1937) *Proc. Roy. Soc.* **A160**, 94.
Burgess, A., (1958) *Monthly Not. Roy. Astron. Soc.* **118**, 477.
Burgess, A., and Seaton, M. J. (1958) *Rev. mod. Phys.* **30**, 992.
Burgess, A., and Seaton, M. J. (1960) *Monthly Not. Roy. Astron. Soc.* **120**, 121.
Condon, E. U., and Shortley, G. H. (1935) "The Theory of Atomic Spectra," Cambridge Univ. Press, London and New York.
Dalgarno, A. (1952) *Proc. Phys. Soc.* **A65**, 663.
Ditchburn, R. W. (1928) *Proc. Roy. Soc.* **A117**, 486.
Ditchburn, R. W. (1937) *Z. Phys.* **107**, 719.
Ditchburn, R. W. (1954) *In* "Rocket Exploration of the Upper Atmosphere" (R. L. F. Boyd and M. J. Seaton, eds.), p. 313. Pergamon, New York.
Ditchburn, R. W. (1955) *Proc. Roy. Soc.* **A229**, 44.
Ditchburn, R. W. (1956) *Proc. Roy. Soc.* **A236**, 216.
Ditchburn, R. W. (1960) *Proc. Phys. Soc.* **75**, 461.
Ditchburn, R. W., and Arnot, F. L. (1929) *Proc. Roy. Soc.* **A123**, 516.
Ditchburn, R. W., and Hudson, R. D. (1960) *Proc. Roy. Soc.* **A256**, 53.
Ditchburn, R. W., and Marr, G. V. (1953) *Proc. Phys. Soc.* **A66**, 655.
Ditchburn, R. W., Tunstead, J., and Yates, J. G. (1943) *Proc. Roy. Soc.* **A181**, 386.
Ditchburn, R. W., Jutsum, P. J., and Marr, G. V. (1953) *Proc. Roy. Soc.* **A219**, 89.
Ehler, A. W., and Weissler, G. L. (1955) *J. Opt. Soc. Amer.* **45**, 1035.
Finkelnburg, W. (1938) "Kontinuierliche Spektren," Springer, Berlin.
Garton, W. R. S. (1950) *Nature* **166**, 150.
Garton, W. R. S. (1952) *Proc. Phys. Soc.* **A65**, 268.
Garton, W. R. S., and Codling, K. (1960) *Proc. Phys. Soc.* **75**, 87.
Garton, W. R. S., and Rajaratnam, A. (1955) *Proc. Phys. Soc.* **A68**, 1107.
Gaunt, J. A. (1930) *Phil. Trans.* **A229**, 163.
Geltman, S. (1958) *Phys. Rev.* **112**, 176.
Granier, J., and Astoin, N. (1956) *C. R. Acad. Sci. (Paris)* **242**, 1431.
Ham, F. S. (1955) *Solid State Phys.* **1**, 127.
Hargreaves, J. (1929) *Proc. Cambridge Phil. Soc.* **25**, 75.
Herzog, R. F., and Marmo, F. F. (1957) *J. chem. Phys.* **27**, 1202.
Hurzeler, H., Inghram, M. G., and Morrison, J. D. (1957) *J. chem. Phys.* **27**, 313.
Hurzeler, H., Inghram, M. G., and Morrison, J. D. (1958) *J. chem. Phys.* **28**, 76.
Jutsum, P. J. (1954) *Proc. Phys. Soc.* **A67**, 190.

Kaiser, T. R. (1960) *Proc. Phys. Soc.* **75**, 152.
Kirkwood, J. G. (1932) *Phys. Z.* **33**, 57.
Klein, M. M., and Brueckner, K. A. (1958) *Phys. Rev.* **111**, 1115.
Kramers, H. A. (1923) *Phil. Mag.* **46**, 836.
Lawrence, E. O., and Edlefsen, N. E. (1929) *Phys. Rev.* **34**, 233, 1056.
Lee, P., and Weissler, G. L. (1952) *Astrophys. J.* **115**, 570.
Lee, P., and Weissler, G. L. (1953) *Proc. Roy. Soc.* **A220**, 71.
Lee, P., and Weissler, G. L. (1955) *Phys. Rev.* **99**, 540.
Lee, P. (1955) *J. Opt. Soc. Amer.* **45**, 703.
Lowrey, A., III, and Watanabe, K. (1958) *J. chem. Phys.* **28**, 208.
Marr, G. V. (1954a) *Proc. Phys. Soc.* **A67**, 196.
Marr, G. V. (1954b) *Proc. Roy. Soc.* **A224**, 83.
Marr, G. V. (1955) *Proc. Phys. Soc.* **A68**, 544.
Maunsell, C. D. (1955) *Phys. Rev.* **98**, 1831.
Menzel, D. H., and Pekeris, C. L. (1935) *Monthly Not. Roy. Astron. Soc.* **96**, 77.
Mohler, F. L., and Boeckner, C. (1929) *J. Res. Nat. Bur. Stand.* **3**, 303.
Morrison, J. D., Hurzeler, H., Inghram, M. G., and Stanton, H. E. (1960) *J. chem. Phys.* **33**, 821.
Olsson, P. O. M. (1959) *Ark. Fys.* **15**, 131, 159, 289.
Page, T. L. (1939) *Monthly Not. Roy. Astron. Soc.* **99**, 385.
Pery-Thorne, A., and Garton, W. R. S. (1960) *Proc. Phys. Soc.* **76**, 833.
Platt, J. R., and Klevens, H. B. (1944) *Rev. mod. Phys.* **16**, 182.
Propin, R. K. (1960) *Optika i Spektrosk.* **8**, 300.
Rudkjøbing, M. (1940) *Publ. Kjøbenhavns Obs.* **18**, 1.
Schönheit, E. (1957) *Z. Phys.* **149**, 153.
Seaton, M. J. (1951) *Proc. Roy. Soc.* **A208**, 408, 418.
Seaton, M. J. (1954) *Proc. Phys. Soc.* **A67**, 927.
Seaton, M. J. (1955) *C. R. Acad. Sci. (Paris)* **240**, 1317.
Seaton, M. J. (1958) *Monthly Not. Roy. Astron. Soc.* **118**, 504.
Stewart, A. L. (1954) *Proc. Phys. Soc.* **A67**, 917.
Stewart, A. L., and Wilkinson, W. J. (1960) *Proc. Phys. Soc.* **75**, 796.
Sugiura, Y. (1927) *J. Phys. Radium* **8**, 113.
Sugiura, Y. (1929) *Bull. Inst. Phys. Chem. Res. Tokyo, Sci. Papers* **11**, 1.
Sun, H., and Weissler, G. L. (1955a) *J. chem. Phys.* **23**, 1160.
Sun, H., and Weissler, G. L. (1955b) *J. chem. Phys.* **23**, 1372.
Sun, H., and Weissler, G. L. (1955c) *J. chem. Phys.* **23**, 1625.
Trumpy, B. (1928) *Z. Phys.* **47**, 804.
Trumpy, B. (1931) *Z. Phys.* **71**, 720.
Tunstead, J. (1953) *Proc. Phys. Soc.* **A66**, 304.
Unsöld, A. (1955) "Physik der Sternatmosphären." Springer, Berlin.
Vainshtein, L. A., and Norman, G. É. (1960) *Optika i Spektrosk.* **8**, 149.
Vinti, J. P. (1933) *Phys. Rev.* **44**, 524.
Wainfan, N., Walker, W. C., and Weissler, G. L. (1953) *J. appl. Phys.* **24**, 1318.
Wainfan, N., Walker, W. C., and Weissler, G. L. (1955) *Phys. Rev.* **99**, 542.
Walker, W. C., and Weissler, G. L. (1955a) *J. chem. Phys.* **23**, 1540.
Walker, W. C., and Weissler, G. L. (1955b) *J. chem. Phys.* **23**, 1547.
Walker, W. C., and Weissler, G. L. (1955c) *J. chem. Phys.* **23**, 1962.
Watanabe, K. (1954) *J. chem. Phys.* **22**, 1564.
Watanabe, K., and Marmo, F. F. (1956) *J. chem. Phys.* **25**, 965.
Watanabe, K., Marmo, F. F., and Inn, E. C. Y. (1953a) *Phys. Rev.* **91**, 1155.

Weissler, G. L. (1956) *In* "Handbuch der Physik" (S. Flügge, ed.), Vol. 21, p. 304. Springer, Berlin.
Weissler, G. L., and Lee, P. (1952) *J. Opt. Soc. Amer.* **42**, 200.
Weissler, G. L., Lee, P., and Mohr, E. I. (1952) *J. Opt. Soc. Amer.* **42**, 84.
Weissler, G. L., Samson, J. A. R., Ogawa, M., and Cook, G. R. (1959) *J. Opt. Soc. Amer.* **49**, 338.
Wheeler, J. A. (1933) *Phys. Rev.* **43**, 258.
Wigner, E. P. (1948) *Phys. Rev.* **73**, 1002.
Yamanouchi, T., and Kotani, M. (1940) *Proc. Phys. Math. Soc. Japan* **22**, 60.

4.

Photodetachment

L. M. Branscomb

1 Introduction

Photodetachment is a special case of photoionization, referring to the photoionization of a negative ion (Massey, 1950; Branscomb, 1957a). The initial state is a negative ion in a radiation field, and the final state is a neutral atom or molecule and a free electron. Thus, in photodetachment an ion is destroyed, while in the photoionization of a neutral atom an ion is created.

There are four principal sources of scientific interest in photodetachment cross sections. First, photodetachment provides a uniquely precise and unequivocal method for determining binding energies of atomic negative ions (or the electron affinities of the corresponding neutral atoms). In some cases it is equally appropriate for molecular ion binding energies. Second, we can calculate from the photodetachment cross section between specific initial and final states the cross section for the reverse process between the same states: the radiative attachment of an electron to a neutral atom. This form of attachment is important in monatomic gases at low pressures. The resulting radiation is sometimes a significant part of the emission continuum from arcs or shock-heated gases. Third, negative ion photodetachment provides a source of continuous opacity in the visible and infrared spectra of hot gases and stellar atmospheres (Woolley and Stibbs, 1953). In particular, the H^- continuum provides the dominant source of continuum at wavelengths longer than the Balmer limit in the sun and many other stars. Fourth, negative ion structure and photodetachment cross sections provide a particularly sensitive test for the approximations of atomic theory. The distortion of the core electrons of an atom by the valence electron is a dominant phenomenon in negative ion binding energy calculations. The distortion makes a smaller percentage effect in the corresponding neutral atom case, where it can be treated as a perturbation on the static central field.

This chapter will summarize the available results from negative ion spectroscopy and the relevant physical principles which may be inferred from the observations and from theoretical considerations. Spectra of atomic ions will be emphasized. At the close of the chapter a table will be provided summarizing the available information on atomic negative ion binding energies (electron affinities). No attempt will be made to provide a full discussion of data given in this table, for this would lead us into the discussion of subject matter far outside the scope of this volume.

2 *Negative Ion Energy States*

Certain properties of the energy levels of negative atomic ions are important in the photodetachment process. The binding energies of negative ions range from nearly zero (boron, perhaps nitrogen) to nearly 4 ev (chlorine) (Table III). Thus, the absorption thresholds are found primarily in the infrared and visible regions of the spectrum. This is a matter of importance to astronomers, to whom the ultraviolet spectrum is accessible only through rocket and satellite borne instruments. Hence, a photodetachment continuum may dominate the opacity in the visible and near infrared.

It appears that atomic negative ions normally have only one bound state (Branscomb, 1957a). If excited states are to be found, it is likely on theoretical grounds (Massey, 1950, p. 3 ff) that the number of such states is finite, owing to the short range nature of the forces binding the extra electron to the neutral atomic core. [The rigor of the proof has, however, been questioned by Wu (1953).] In addition, such excited states as may be found in light ions are likely to be metastable and to lie very close to the continuum. The stability of a negative ion depends heavily on the extent to which the extra electron shares the attractive field of the nucleus with the other electrons. If, in the light elements, the negative ion is excited into a state of higher principle quantum number than that describing the ion's ground state, one may expect the electron to be so shielded from the Coulomb field of the nucleus that a stable state will not result. It is possible, however, that several of the light paramagnetic ions which have two or three low-lying terms of the ground state configuration, may find an excited term lying in the region of stability (Bates and Moiseiwitsch, 1955; Johnson and Rohrlich, 1959). The most favorable case is thought to be Si^-, whose ground $3p^3\ {}^4S$ state may have a binding energy of about 1.4 ev (Table III) with the ${}^2D$ excited term of this configuration less than a volt above the ground ${}^4S$ term. Boron, aluminium, carbon, and phosphorus may also have bound metastable excited terms, based on these estimates.

The absorption spectrum of a negative atomic ion is therefore expected to consist of perhaps one or two forbidden lines in the visible, and a pure continuum starting from a frequency corresponding to the binding energy of the ion ground state. This continuum may be structured by the presence of autoionization transitions involving absorption to unstable, doubly excited states. It is also possible that a group of allowed lines may be found immediately to the longwave side of the continuum limit. Such lines might result from very weakly bound higher configurations. Even if they exist, they may be inherently unobservable in plasmas because of Stark and collision broadening.† Of the atomic ions which might have bound metastable terms, only C^- has been studied by photodetachment experiments (Seman and Branscomb, 1961). In this case a continuum does appear at longer wavelengths than the threshold for absorption from the ground state. This data will be described in more detail later.

Excited states in atomic ions have been searched for theoretically in the simplest atoms, hydrogen and helium. Configuration interaction calculations with up to 55 configuration wave functions by Weiss (1960)

† Negative ions in such states would be enormous.

for atoms and ions with 1-4 electrons strongly suggest that neither the ^2S state of He^- nor the ^3S metastable state of H^- are bound. (Weiss' lithium affinity is given in Table III.)

Certain doubly excited states of H^- and He^- have been shown to be metastable. Most of these states decay very rapidly through autoionization. A few of them do not. The most interesting example of the metastable doubly excited negative ion state is the $^4P_{5/2}$ state of He^- which was shown to be stable theoretically by Holøien and Midtdal (1955).

The experimentally demonstrated (Hiby, 1939; Windham *et al.*, 1958; Dukel'skii *et al.*, 1956) appearance of He^- ions in beams of high-energy helium positive ions passing through low pressures of rare gases is presumed to result from the formation of ions in this $^4P_{5/2}$ state by double charge capture. A similar situation prevails in nitrogen, whose $N^-(^1D)$ state may lie below $N(^2D) + e$, and whose spontaneous decay to $N(^4S) + e$ is inhibited by angular momentum selection rules. Negative ions of both helium and nitrogen are now being used in tandem accelerators.

Doubly excited states of H^- have been extensively investigated by Hylleraas, Holøien, and others with the conclusion that no metastable long-lived state can exist in H^-, but that the $(2p^2)^3P$ state is probably bound (Holøien, 1960a, b).

3 Theoretical Considerations

3.1 Photodetachment Cross Section

The usual theoretical description of photodetachment considers a negative ion of dipole moment **p** interacting with an oscillating electric field of infinite extent. Time-dependent perturbation theory gives for the probability of a transition per unit time

$$P \propto \nu \, | \langle \Psi_d \, | \, \mathbf{p} \, | \, \Psi_c \rangle |^2 \, \rho(E_k) \tag{1}$$

where $\rho(E_k)$ is the density of states in the continuum per unit energy range corresponding to the energy E_k of the ejected electron. (If the electron affinity of the atom is A, then $E_k = h\nu - A$.) The subscripts d and c refer to wave functions for the discrete (initial) and continuum (final) states. The density of states is proportional to the square root of the final electron energy so that the photodetachment cross section (or absorption coefficient k_ν) is

$$\sigma \propto \nu \, E_k^{1/2} \, | \langle \Psi_d \, | \, \mathbf{p} \, | \, \Psi_c \rangle |^2. \tag{2}$$

The matrix element for the dipole moment may be transformed. If the wave functions Ψ_d and Ψ_c are exact solutions of the n-electron Schrödinger equation, three expressions for the matrix element are then formally equivalent and are referred to as dipole length, velocity, and acceleration formulae:

$$M = \int \Psi_d(r_i \dots r_n) \sum_i z_i \Psi_c(r_i \dots r_n)\, dt_i \dots dt_n \qquad \text{(dipole length)} \qquad (3)$$

$$= \frac{1}{h\nu} \int \Psi_d(r_i \dots r_n) \sum_i \frac{\partial}{\partial z_i} \Psi_c(r_i \dots r_n)\, dt_i \dots dt_n \qquad \text{(dipole velocity)} \qquad (4)$$

$$= \left(\frac{1}{h\nu}\right)^2 \int \Psi_d(r_i \dots r_n) \sum_i \frac{z_i}{r_i^3} \Psi_c(r_i \dots r_n)\, dt_i \dots dt_n \quad \text{(dipole acceleration).} \; (5)$$

Here z_i is the component in the oscillating electric field in the direction of the radius vector length r_i, of the ith electron.

Since all calculations of photodetachment cross sections are made with approximate initial and final state wave functions, the extent to which using approximate Ψ_d and Ψ_c these formulae give identical answers affords a test of the accuracy of the functions used. Further, the different formulae test different regions of electron coordinate space. The dipole length formula emphasizes regions of large r, the velocity formula intermediate regions, and the acceleration formulae small values of r. The greatest contribution to the total energy using variational Hylleraas wave functions comes from intermediate values of r. For this reason the dipole velocity form is generally preferred when using these functions (Chandrasekhar, 1945a). A function which in its entirety gives an excellent minimization of the energy may nonetheless be in error in regions of space which contributed heavily to the photodetachment cross section. As we shall see later, the best calculations now available on the H^- photodetachment cross section give agreement within about 20 % between experiment, dipole length, and dipole velocity formulae. The wave functions are not yet sufficiently accurate to give reasonable cross sections with the acceleration formula.

3.2 Photodetachment Threshold Laws

Without specifying the particular atomic system under study, some generalizations can be made about the limiting energy dependence of the dipole length matrix element, from which the limiting form of (3) will give us the shape of the cross section at threshold. The separation of variables in the Schrödinger equation leads to the familiar $l(l+1)/r^2$ contribution to the potential in the radial wave equation. This centrifugal barrier term falls off more rapidly with distance than the Coulomb potential ($1/r$) which occurs in the ordinary photoionization problem. In photodetachment, however, the angular momentum repulsion is dominant at large distances over other terms in the short-range potential of the neutral atom. Hence, we find that in photoionization the threshold energy dependence of the cross section is the same for all atoms (constant and finite at threshold); the photodetachment cross section varies with energy at threshold in a manner dependent on the angular momentum of the electron in the final state, and hence on the dipole selection rules and the angular momentum of the valence electron in the initial bound state (Massey, 1950; Branscomb, 1957a). The result in the approximation that the continuum function can be represented by a plane wave or a short-range static central field is (Massey, 1950; Wigner, 1948; Branscomb *et al.*, 1958) that

$$\sigma \propto \nu k^{2l+1} (a_0 + a_1 k^2 + a_2 k^4 \ldots) \tag{6}$$

where l is the lowest angular momentum component of the continuum state to which an electric dipole transition is allowed from the discrete initial state. In terms of the ejected electron energy, we have for H^-:

$$\sigma_{1s \to kp} \propto \nu E_k^{3/2} (a_0 + a_1 E_k + a_2 E_k^2 + \ldots). \tag{7}$$

For p electron shells such as C^-, O^-, etc.,

$$\sigma_{np \to ks} \propto \nu E_k^{1/2} (a_0 + a_1 E_k + a_2 E_k^2 + \ldots). \tag{8}$$

The effect of inclusion of exchange and polarization effects on the threshold laws has not been thoroughly investigated. The inclusion of a polarization potential ($-\alpha r^{-4}$) may have the effect of introducing logarithmic terms and odd as well as even powers of k in the expansion of the right-hand side of (6), but the lead term appears to be unaffected (Geltman, 1960).

We therefore expect the photodetachment cross section for H^- and alkali metal ions to start from threshold with zero slope while detachment from O^-, C^-, and other atomic ions with valence p shells rises with infinite slope at threshold. Although we have no *a priori* knowledge of the energy range over which this behavior will hold, it is qualitatively obvious that the more abrupt threshold associated with p electron detachment into the s continuum state facilitates the experimental determination of the binding energy using the photodetachment threshold. The form of the expansion of the cross section in powers of the ejected electron energy [Eqs. (7) and (8)] is also used in the interpretation of low-resolution experimental data near the threshold for the purpose of determining the electron affinity.

Similar expansions have been made for negative ions of diatomic molecules in the Born-Oppenheimer approximation (Geltman, 1958). In this case the cross section takes the threshold form, similar to (6):

$$\sigma \propto \nu k^m (a_0 + a_1 k^2 + a_2 k^4 + ...) \quad (9)$$

where the exponent m, which is given in Table I, depends on the symmetry of the molecular ion and on the projected angular momentum

TABLE I[a]

EXPONENT m IN EQ. (9) FOR DIATOMIC MOLECULE PHOTODETACHMENT

		$\vert \lambda_0 \vert = 0$	$\vert \lambda_0 \vert$ even	$\vert \lambda_0 \vert$ odd
Heteronuclear molecule		1	$2 \vert \lambda_0 \vert - 1$	$2 \vert \lambda_0 \vert - 1$
Homonuclear	Gerade symmetry	3	$2 \vert \lambda_0 \vert - 1$	$2 \vert \lambda_0 \vert + 1$
	Ungerade symmetry	1	$2 \vert \lambda_0 \vert + 1$	$2 \vert \lambda_0 \vert - 1$

[a] See Geltman (1958).

along the molecular axis λ_0 associated with the electron orbital from which the electron is ejected. From this table we can see, for example, that for a homonuclear ion like O_2^- in a $(\pi_g)^3$ configuration, the value of m is 3, and we expect $\sigma \sim \nu k^3$. A heteronuclear molecule ion like OH^- with a valence π-electron would give a cross section rising from threshold as νk.

4 *Experimental Method for Photodetachment Studies*

Photodetachment cross sections have been studied using modulated crossed-beam methods.† One collects and measures the small current of free electrons which are produced in high vacuum by photon absorption at the intersection of two mutually perpendicular beams, one of mass-

† *Note added in proof:* Berry *et al.* (1961) have achieved excellent observations of absorption spectra of atomic halogen negative ions created in a shock wave. This method provides the first observations of photodetachment spectra in the ultraviolet and provides much higher optical resolution than does the crossed beam method. Preliminary electron affinity values obtained by this method are given in Table III. However, it is difficult to determine the absorption cross section by this method.

analyzed negative ions, the other of filtered, intense visible light. The principles of the apparatus (Smith and Branscomb, 1960) used at the National Bureau of Standards for this purpose are illustrated in Fig. 1.

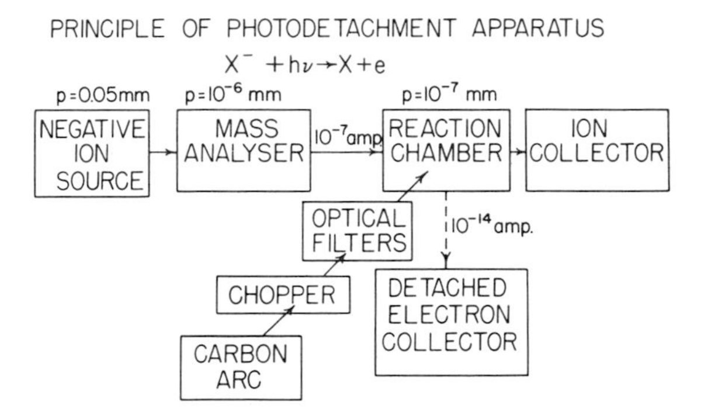

FIG. 1. This block diagram indicates the principle of operation of an apparatus for the study of negative ion photodetachment.

If monochromatic light of sufficient intensity were available, the photodetached electron signal would be related to the other observed quantities and the cross section by the relation

$$j_e = kj_i \frac{I}{h\nu} \sigma \frac{1}{v_i} \tag{10}$$

where k is a geometrical constant; j_e and j_i are electron and ion currents, respectively; I is the intensity of the light (and $I/h\nu$ the photon flux); σ the photodetachment cross section, and v_i the ion speed in the reaction chamber. Thus, measurements of two currents, the ion mass, kinetic energy, and the photon flux would suffice to determine the relative photodetachment cross section. In fact, rather broad bands of continuum have been used in the photon beam, with the consequence that a somewhat more refined analysis of the wavelength dependent parameters must be made to permit the cross section to be extracted from the observed data.

Four essential requirements must be met in a photodetachment apparatus. First, one requires a suitable ion source and mass selector which are capable of delivering to the reaction chamber a collimated beam of one species of identified ions of low noise characteristics. Glow or hot cathode arc discharges of special design have been used. The mass selector has been an astigmatic, 90° sector field instrument with post-deceleration so that the ions pass through the sector field at 2500-volts energy, but through the reaction chamber at only about 300 ev.

Beams of more than 10^{-7} amp of ions of H^-, O^-, and other abundant species are readily obtained. The mass analyzer, glow discharge source, and reaction chamber are shown in Fig. 2.

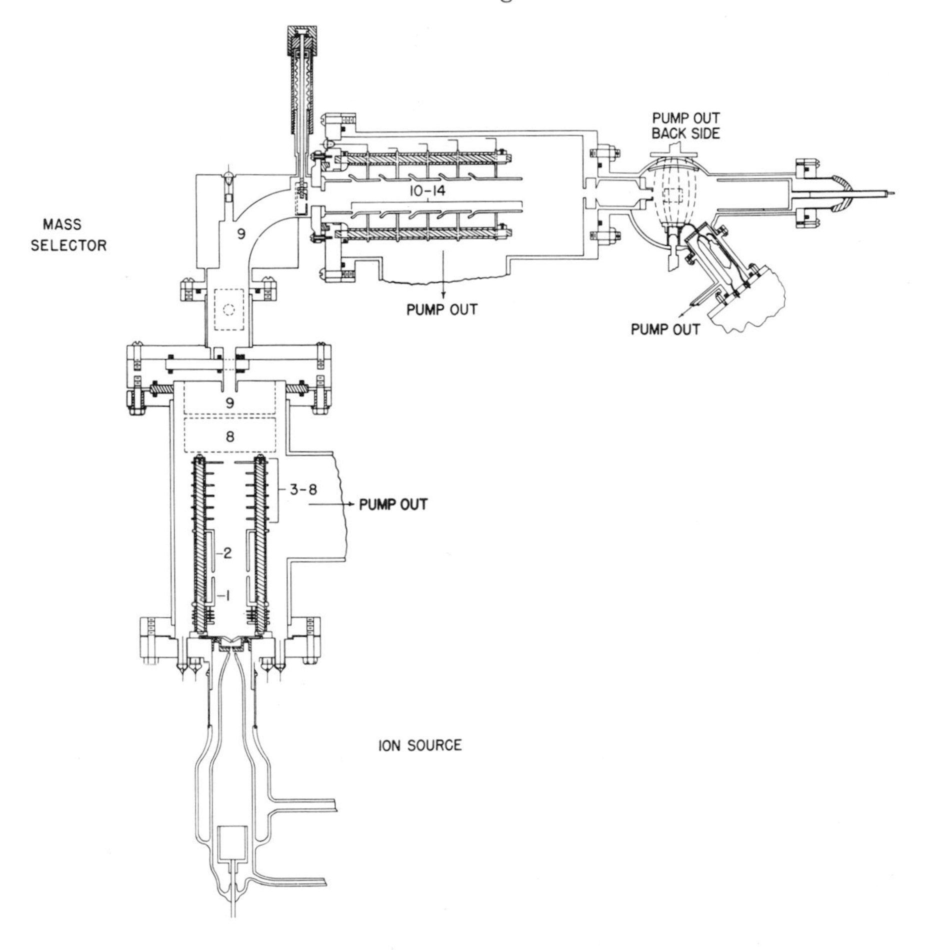

FIG. 2. Negative ion photodetachment apparatus, showing the ion source, accelerating electrodes, 90° sector field mass selector, deceleration chamber, and reaction chamber. The dashed lines in the reaction chamber suggest the inhomogeneous magnetic field used to trap the free electrons. The modulated radiation enters this chamber normal to the drawing through a quartz window not illustrated.

One requires a suitable optical system and light source providing of the order of 1 watt of quasi-monochromatic radiation at the point of intersection of photon and ion beams and in a geometry permitting accurate radiometric monitoring of the photon beam. These requirements were met with a high-intensity carbon arc lamp and the $f/1.5$ optical system shown in Fig. 3. The light beam was filtered by a set of 26 band-pass combinations ranging in wavelength from 0.4 to 1.7 μ. Each is formed from a selection of glass absorption filters, multilayer

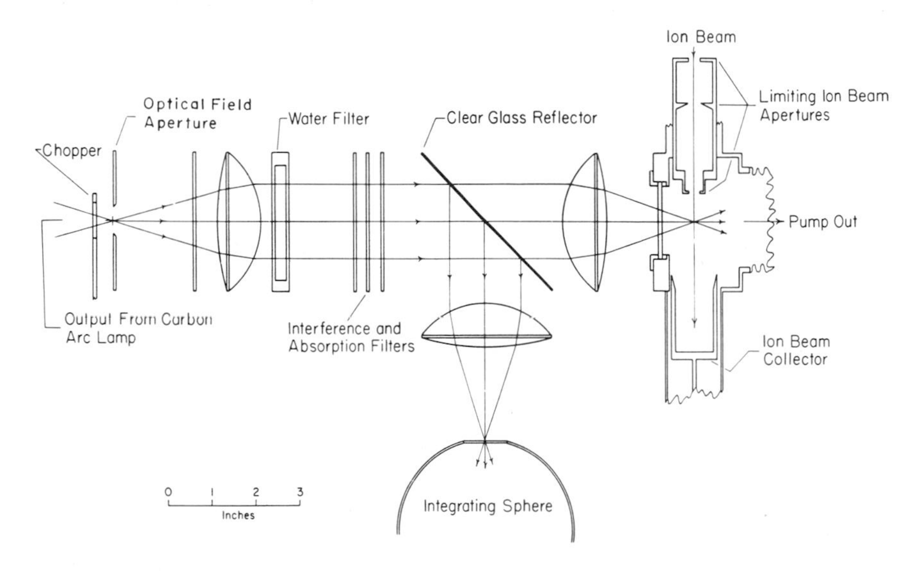

FIG. 3. Optical system used in measurements of relative photodetachment cross sections. The radiation from the carbon arc is imaged on the field aperture at the left by means of an elliptical mirror. The radiation is monitored in the integrating sphere by a black bolometer inserted into this sphere. The fragment of the reaction chamber illustrated is a horizontal section through the chamber illustrated in the previous figure.

interference reflection filters, and circulating-liquid cells. Typical filter combinations have transmittance of about 50 % at the peak and widths of about 500 A at half-maximum transmission. For example, filter No. 7 is compounded of the elements whose transmissions are shown in Fig. 4.

The third requirement is a suitable arrangement in the reaction chamber for detecting all of the photodetached electrons with a minimum interference from scattered ions, slow secondary ions, and electrons photoelectrically ejected from metals or stripped from the negative ions in background gas collisions. The superposition of weak electric and magnetic fields perpendicular to the ion and photon beams suffices to collect slow electrons without affecting the ion beam. The largest source of spurious signal arises from negative ion collisional detachment in the background vacuum. To minimize this noise contribution to the signal, the photon beam is modulated at an audio frequency, and narrow-band phase-sensitive detection is used.

In consequence of the noise current arising from electrons stripped from the ion beam, the signal-to-noise ratio in photodetachment experiments is proportional to the ratio of photon flux to background gas

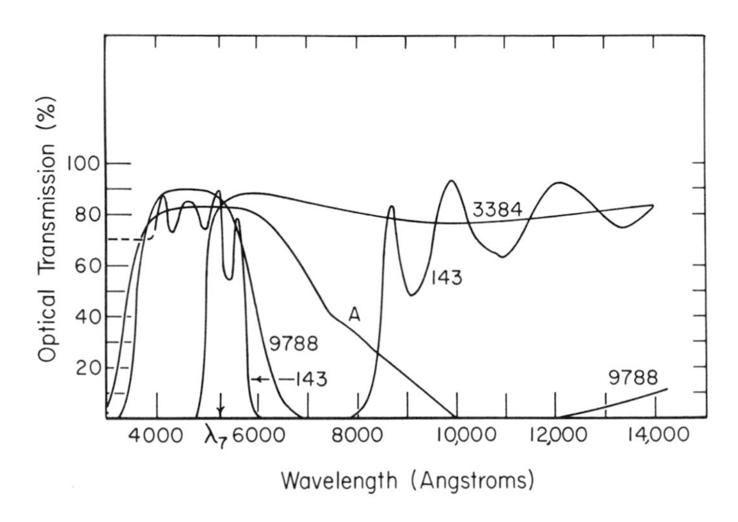

FIG. 4. A typical band-pass filter combination is made up of the elements whose transmissions are illustrated. Filter 143 is a multilayer interference filter. Numbers 3384 and 9788 are glass absorption filters, and curve A is a flowing water absorption cell with colored-glass windows. This combination, called No. 7, was used as the reference wavelength for most photodetachment experiments.

density. It is independent of the ion beam current if this current exceeds a minimum value set by amplifier input noise. The use of an alternating current electrometer amplifier to detect the detached electron beam has required ion beams to exceed 2×10^{-8} amp for useful measurement, yielding about 2×10^{-14} amp of alternating current signal. Recent work with better vacuum and with electron multiplier detection permits photodetachment measurements with beams of about 5×10^{-10} amp, thus making possible the study of a wider range of negative ion species (Seman and Branscomb, 1961).

Because of the difficulties involved in the radiometric calibrations and in proving that all of the photodetached electrons are collected, absolute cross sections are much more difficult to obtain than relative values. In 1955 an experiment was performed to measure the absolute integrated cross section for H^- (Branscomb and Smith, 1955a). The apparatus (Smith and Branscomb, 1955a), shown in Fig. 5, incorporated a radio-

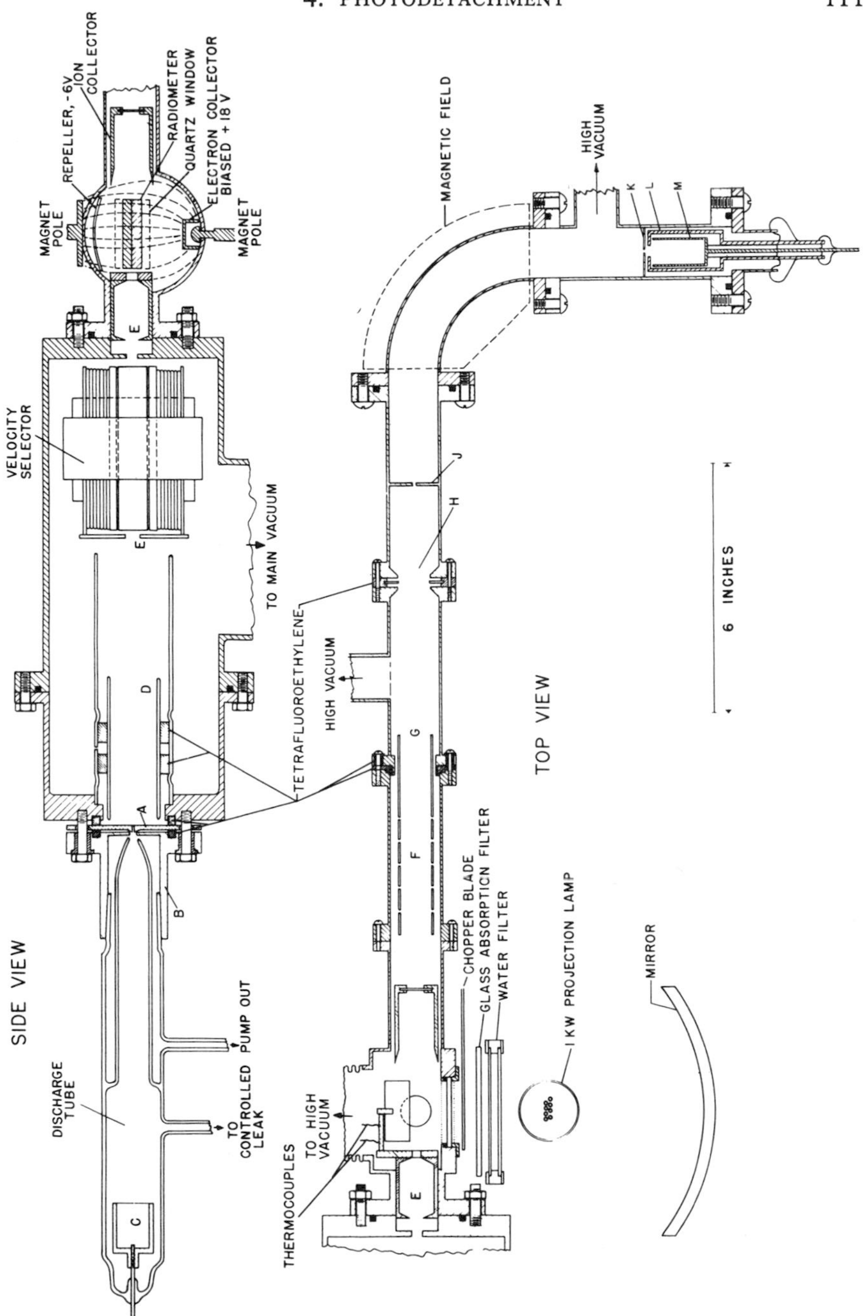

FIG. 5. Apparatus used to determine the integrated absolute cross section for photodetachment from H^-. Here the velocity selector (crossed electric and magnetic fields) is used as a crude mass analyzer, with the 90° sector field instrument after the reaction chamber providing verification that the resolution was adequate. A tungsten projection lamp provided the light source.

metric detector behind the ion beam to monitor the radiation from a 1-kw tungsten projection lamp. The radiometer was calibrated in the radiation of this lamp in the absence of the mirror and light filters, the lamp having previously been compared with radiation standards. The experiment involves the observation of the probability that an ion suffers photodetachment while passing through the beam of absolutely calibrated radiation:

$$P_{\text{exp}} = j_e/j_i. \tag{11}$$

This probability is compared with the prediction of any theoretical cross section which one wishes to test from the relationship

$$P_{\text{theo}} = k/hcv_i \int I(\lambda)\, \sigma(\lambda)\, \lambda\, d\lambda. \tag{12}$$

This equation is derived from (10) by integration over wavelength. For example, when an early calculation of Chandrasekhar (1945b) of the H^- absorption coefficient was integrated over the experimental photon flux distribution, Branscomb and Smith (1955a) found $P_{\text{exp}}/P_{\text{theo}} = 1.01$ ($\pm$ 0.02) ($\pm$ 0.10), where the figures in parenthesis describe, respectively, the reproducibility of the result and an estimate of the possible systematic errors.

It has subsequently developed that virtually all of the modern calculations of the H^- cross section (in the dipole velocity formulation) when compared with this experiment through the foregoing integration give consistency with the integrated, absolute experiment within the 10 % limitation. However, the shapes of the theoretical cross sections vary significantly and it appeared that an accurate relative cross section was required. This was accomplished by Smith and Burch (1959), who measured the H^- cross section from 4000A to 13,000 A with statistical errors of about 1 % and stated probable errors from all causes of about 2% relative to a control filter at 5280 A. The resulting cross section is shown in fig. 6.

This experimental relative cross section can now be normalized to absolute value by insertion in (12) with an undetermined normalizing factor and equating this expression to (11). One need rely on theory only for interpolation of the experimental cross section from 13,000 A to the threshold at 16,400 A, which makes a very small contribution to the uncertainty. Subsequent measurements of photodetachment cross section of ions other than H^- have been put on an absolute basis by comparison with this cross section of H^-, or by comparison with O^- whose absolute integrated cross section was determined at the same, time the absolute H^- absorption was measured (Smith and Branscomb, 1955a).

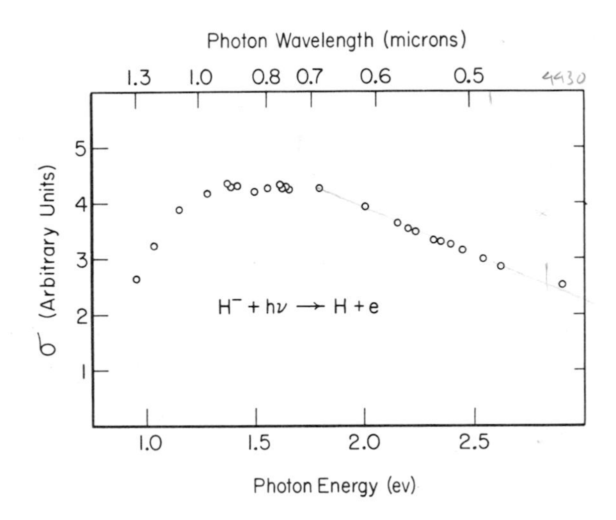

FIG. 6. Experimental values of the relative photodetachment cross section for H^-, as determined by Smith and Burch (1959).

5 The H^- Photodetachment Cross Section

The experimental result shown in Fig. 6 can be compared with a number of recent theoretical values. From this comparison we may learn something about the accuracy of the calculated bound and free wave functions for an electron in the field of a hydrogen atom.

5.1 THEORETICAL CROSS SECTIONS FOR H^-

Since the original calculation of Bates and Massey, much effort has been expended on improvement of both the bound and continuum wave functions. The early calculations (Chandrasekhar, 1945a; Bates and Massey, 1940) used a simple plane wave for the ejected electron, multiplied by the unperturbed hydrogen atom ground state function, suitably symmetrized in the electron coordinates. The next improvement on this part of the problem was the use of the same unperturbed hydrogen atom function multiplied by the wave function of an electron moving in the static Hartree field of the hydrogen atom (Chandrasekhar, 1945b; Chandrasekhar and Elbert, 1958).

Each of these two types of continuum function was used with a series of increasingly accurate Hylleraas-type H^- bound state functions determined by the Ritz variational principle of energy minimization. Most often used have been the 11-parameter functions of Henrich (1944) and the 20-parameter function of Hart and Herzberg (1957). In most of these calculations the accuracy of the wave functions was tested by comparison of the photodetachment cross sections calculated with dipole length and momentum matrix elements. The conclusion from these comparisons is that for a given choice of plane wave or static central-field continuum function, the improvement in the cross section resulting from use of a 20-parameter bound state function rather than an 11-parameter one is only a few per cent (Chandrasekhar, 1958). Further, cross sections calculated with the same Hart-Herzberg H^- function but with either plane-wave or static central-field continuum functions are virtually

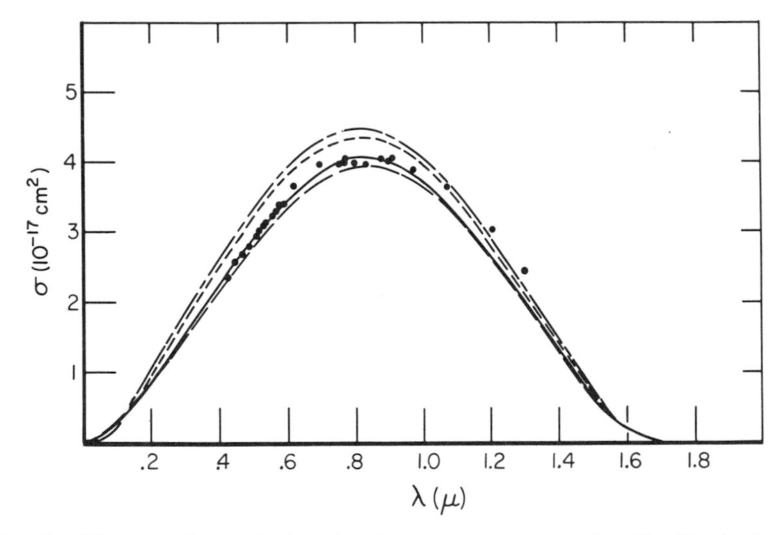

FIG. 7. The experimental photodetachment cross section for H^- (black dots) is compared with four theoretical calculations. All of these calculations used dipole velocity matrix elements with the Hart-Herzberg 20-parameter bound state wave function. The continuum functions used were: (— - — -), static central field (Chandrasekhar and Elbert, 1958); (- - -), plane wave (Chandrasekhar, 1958); (——) exchange central field (John, 1960); (— —) variational (Geltman and Krauss, 1960). Here the experimental points are normalized to the exchange central field calculation at 5280 A.

identical in shape and differ by only about 3 % in magnitude (Fig. 7). Yet we cannot assume from this insensitivity of the dipole velocity cross section to change in the initial and final state wave functions that we are closely approaching the correct solution, for there remains in each case a marked difference between the length, velocity and acceleration matrix elements (Fig. 8).

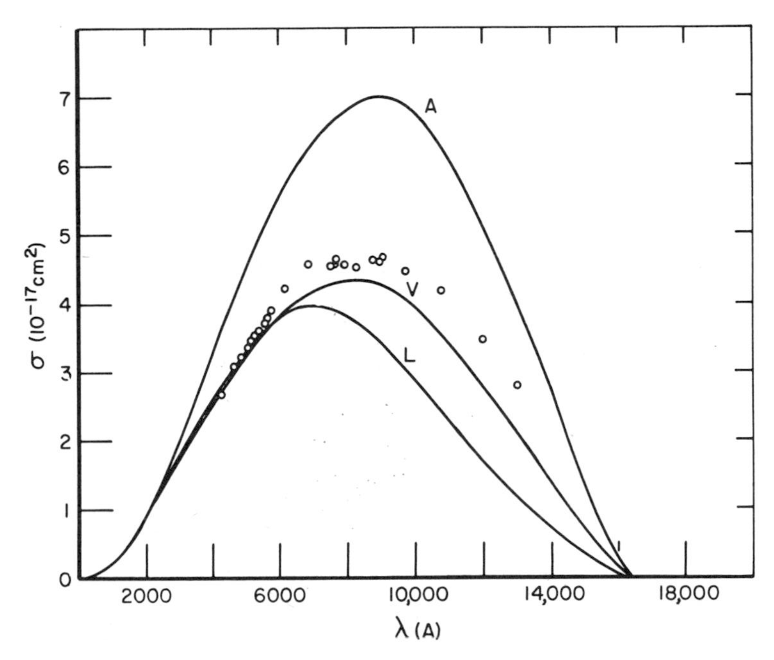

FIG. 8. H^- photodetachment cross sections with plane wave free state calculated using dipole acceleration (A), velocity (V), and length (L) matrix elements (Chandrasekhar, 1958). Experimental points are normalized to the velocity curve and to 5280 A. The calculation is due to Chandrasekhar with the acceleration curve provided by Geltman and Krauss (1960).

Barring any deficiencies in the Hylleraas function for the bound state, further improvement in the H^- cross section is to be found in the inclusion of polarization effects and the correct accounting for electron exchange. John (1960) has solved numerically the Hartree-Fock exchange equations for the electron in the atomic hydrogen field and computed the dipole length and velocity curves of which the former is shown in Fig. 7. A significant improvement in agreement with the experimental data is found, the velocity curve agreeing with the relative experimental curve within experimental error from 4000 to about 9000 A, and falling about 10 % low at longer wavelengths.

In the calculation by John, the effects of distortion of the atom by the free electron, other than those included in the Hartree-Fock functions, are still not included. It is expected that these effects, particularly the atomic polarization potential, will have their largest effect near the threshold energy where the behavior of the potential at large distances is important. This may explain the divergence of the experiment from theory in the region 1.0-1.4 μ.

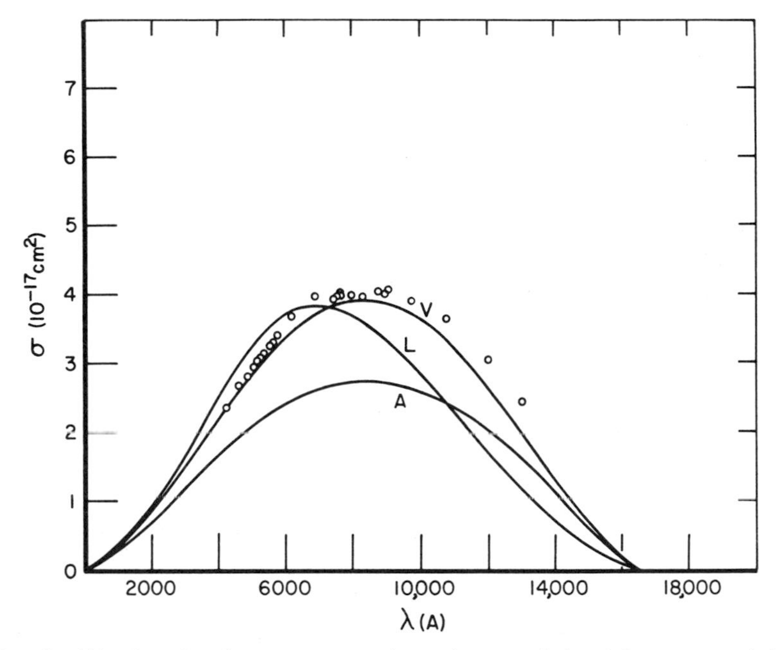

FIG. 9. H^- photodetachment cross section using a variational free state and dipole acceleration, velocity, and length matrix elements, due to Geltman and Krauss (1960). Again the experiment is normalized to the velocity curve.

An alternative method of treating the free state is by means of the Hulthen-Kohn stationary expressions for the scattering phase shift. This has been done by Geltman and Krauss (1960) using a trial function containing 4 linear parameters and a screening parameter. At short distances, the function is made up of a linear combination of $1s$, $2s$, and $3s$ hydrogenic functions. The bound state function used in this calculation was again the Hart-Herzberg 20-parameter Hylleraas function. The greatest limitation of the continuum function used by Geltman and Krauss is the failure to account for angular correlations between the two electrons. This could be accomplished through the inclusion of the $2p$

electron configuration in the trial function.† The result of Geltman's calculation is nearly identical to that of John, and is shown in Fig. 7. A comparison of Figs. 7-9 shows that the dipole velocity calculation is indeed the most accurate one with the initial and final wave functions so far used. Further, refinements which bring the velocity curve closer and closer to the experimental values still leave large differences between the length, velocity, and acceleration cross sections, though they are beginning to converge.

5.2 Oscillator Strength Sum Rules

One final test of the accuracy of these curves must be made. Are any of them inconsistent with the oscillator strength sum rules? As applied to H^-, the most useful formulae (Chandrasekhar and Krogdahl, 1943; Dalgarno and Kingston, 1959) are:

$$\frac{mc}{\pi e^2} \int_{\nu_0}^{\infty} \sigma_\nu^{\mathrm{tot}} \, d\nu = N, \tag{13}$$

and

$$\frac{mc}{\pi h a_0} \int_{\nu_0}^{\infty} \nu^{-1} \sigma_\nu^{\mathrm{tot}} \, d\nu = \frac{2}{3\, a_0^2} \int \Psi_0^* \left(\sum \mathbf{r}_i\right)^2 \Psi_0 \, d\tau_1 \dots d\tau_n \tag{14}$$

when $\sigma_\nu^{\mathrm{tot}}$ is the total photoionization cross section from the H^- ground state, N is the number of electrons on the ion ($N = 2$), and Ψ_0 is the H^- ground state wave function.

In application to a negative atomic ion with only one bound state, the sum rules are particularly easy to apply. One need not know an infinite set of transition probabilities. The integration over the continuum must include all possible final states of the neutral atom, but the contribution of photodetachment into highly excited electronic states or double photoionization will contribute rather little to the total integral. By using Henrich's 11-parameter H^- ground state function, Dalgarno and Kingston (1959) evaluated the right-hand side of (14) as 14.4. Geltman and Martin (1960) confirmed this value, using the Hart-Herzberg 20-parameter function to obtain 14.22. On neglecting electron-

† The recent advances in calculations of electron scattering from atomic hydrogen portend the availability in the near future of *p*-wave continuum functions which will give a very accurate theoretical curve for H^- photodetachment. See, for example, Abstracts of the Conference on Electronic and Atomic Collision Phenomena, Boulder, Colorado, June, 1961, W. A. Benjamin, Inc., New York, 1961.

ically excited or ionized final states, we are then left with two inequalities which must be satisfied:

$$\frac{mc}{\pi e^2} \int_{\nu_0}^{\infty} \sigma_\nu \, d\nu < 2, \tag{15}$$

and

$$\frac{mc}{\pi h a_0} \int_{\nu_0}^{\infty} \nu^{-1} \sigma_\nu \, d\nu < 14.22. \tag{16}$$

Dalgarno and Kingston applied these two inequalities to the dipole velocity calculation of Chandrasekhar (the 11-parameter Henrich function for H^- and the static central field for $H + e$) with the result 16.0 for the left-hand side of (16), thus proving that this cross section is in error. In Table II, several other cross sections which we have discussed are subjected to this test. Again, the plane wave final state and 20-parameter bound state gives too large a value for the $\langle r^2 \rangle$ sum. Geltman's variational free state (and John's Hartree-Fock free state, which gives almost identical results) gives a cross section consistent with the sum rule for dipole length, velocity, and acceleration calculations.

One may now test the experimental cross section of Smith and Burch with the $\langle r^2 \rangle$ sum rule. We may normalize the relative experimental curve to the theoretical calculation of John at 5280 A, extrapolating to zero wavelength by using the theoretical curve, and extending the experimental data from 13,000 A to the threshold by sketching in a smooth curve. The result, shown in Table II, is 14.3, which is approximately equal to the theoretical maximum value. It is interesting to note that the difference between the experimental curve and the theory (which occurs between 10,000 and 15,000 A) contributes only 0.34 to this sum. The major contributions to the integral arise at short wavelengths, 30 % of the sum arising from the wavelength region short of 4000 A. Nonetheless, if one relies on the theory at high energies, one may conclude that the experimental curve may not lie higher than is shown in Fig. 7.

The normalization given here is also consistent with the absolute cross section determination of Branscomb and Smith. On using the experimental absolute value of the integrated cross section (for $\lambda > 4000$ A), we find $\sigma = 3.28 \pm 0.3 \times 10^{-17}$ cm^2, $\lambda = 5280$, while normalization to John's Hartree-Fock value at this wavelength gives 3.08×10^{-17} cm^2 and the variational free state gives 3.00×10^{-17} cm^2. If we attempt to normalize the experimental values any lower than required to fit the variational calculation at 5280, we will exceed the

TABLE II

OSCILLATOR STRENGTH SUM RULES FOR H^- PHOTODETACHMENT

Wave functions: Bound state	Wave functions: Free state	Form of dipole matrix element	$\frac{mc}{\pi e^2} \int_{\nu_0}^{\infty} \sigma_\nu \, d\nu$	$\frac{mc}{\pi h a_0} \int_{\nu_0}^{\infty} \nu^{-1} \sigma_\nu \, d\nu$	Cross section calculated by
—	—	(Theoretical limit)	< 2	< 14.22	
11-parameter Henrich	Static central field	Velocity	1.8	16.0	Chandrasekhar (1945b)
20-parameter Hart-Herzberg	Plane wave	Length	1.5	13.4	Chandrasekhar (1945a)
		Velocity	1.6	15.0	
		Acceleration	2.0	22.1	
20-parameter Hart-Herzberg	Variational	Length	1.6	13.3	Geltman and Krauss (1960)
		Velocity	1.5	13.6	
		Acceleration	1.2	10.3	
20-parameter Hart-Herzberg	Hartree-Fock	Velocity		13.9	John (1960)
Experiment normalized to calculation (John, 1960) at 5280 A	Hartree-Fock			14.3	Smith and Burch (1959)
Experiment normalized to variational calculation (Geltman, 1960) at 5280 A				13.9	
Experiment normalized by experimental absolute integrated cross section (Smith and Burch, 1959)				15.4 ± 1.5	

10 % error estimate on the absolute cross section measurement. Thus, we are almost compelled to set the experimental values as they are shown in Fig. 7. In summary, we conclude that the most recent dipole velocity calculations are correct in energy dependence and absolute value from 4000 to 10,000 A, that these calculations appear to be about 12 % low near 12,000 A, and that the $\langle r^2 \rangle$ oscillator strength sum is almost fully accounted for by photodetachment into the ground state of the atom. The recent calculation by Schwartz (1961) of the polarizability of H^- permits the application of a sum rule which relates the polarizability of H^- to an integral of the cross section divided by the frequency squared. This will provide an improved test of the threshold region (Dalgarno, 1961).

6 Photodetachment Cross Sections for O^-, S^-, and C^-

The photodetachment spectra of only three atomic ions (O^-, S^-, and C^-) other than H^- have been investigated. Of the three, only O^- and C^- have been studied in complete detail. It has been of particular interest to make a precise determination of the wavelength of the absorption threshold for O^-. The shape of the cross section near threshold also provides a test for the possibility that a very weakly bound excited state might produce a resonance peak in the photodetachment continuum (Bates and Massey, 1943), which in turn would strongly affect the radiative attachment rate (Massey, 1950). Just as in the work on H^-, an absolute measurement of the integrated O^- spectrum was followed by a series of relative measurements of the spectrum near threshold, each leading to an increasingly precise measure of the threshold wavelength and hence of the affinity of atomic oxygen (Smith and Branscomb, 1955a; Branscomb and Smith, 1955b; Branscomb *et al.*, 1958).

The transition $O^-\ {}^2P \rightarrow O\ {}^3P + e$ is actually made up of six overlapping continua corresponding to transitions between each of the levels ${}^2P_{1/2}$ and ${}^2P_{3/2}$ of O^- and the three continua of the O ${}^3P_{2,1,0}$ levels. The electron affinity is defined as the minimum energy necessary to remove the electron from the negative ion in its lowest energy level, and hence corresponds to $O^-\ {}^2P_{3/2} \rightarrow O\ {}^3P_2 + e$. This happens also to be the transition with the largest statistical weight. The wavelength of this threshold and the corresponding oxygen electron affinity were found to be

$$\lambda_0 = 8460 \pm 30\ \text{A}; \qquad A(0) = 1.465 \pm 0.005\ \text{ev}. \tag{17}$$

This accuracy could not be achieved with the low resolution available in the broad-band filter combinations. It was achieved by fitting the experimental results using seven filters with radically differing transmissions in the threshold region to a trial cross section introduced into seven integrals like (12), where $I(\lambda)$ is understood to include the transmission of the *i*th filter. The form of this trial cross section is the threshold law of (8), written to include a small contribution due to photodetachment from the $^2P_{1/2}$ component of the initial O^- state:

$$\sigma(\lambda) = \frac{10}{54}\frac{\gamma B}{\lambda}\left[\frac{\lambda_1 - \lambda}{\lambda_1 \lambda}\right]^{1/2} \quad \text{for} \quad \lambda_0 < \lambda < \lambda_1 \tag{18}$$

$$= \frac{\gamma}{\lambda}\left[\frac{\lambda_0 - \lambda}{\lambda_0 \lambda}\right]^{1/2} + \frac{\gamma A}{\lambda}\left[\frac{\lambda_0 - \lambda}{\lambda_0 \lambda}\right]^{3/2} \quad \text{for} \quad \lambda < \lambda_0.$$

Here γ is a constant of proportionality, B is a factor introduced to compensate for the uncertainty in the relative populations of the $^2P_{1/2}$ and $^2P_{3/2}$ states of O^-, and λ_0 and λ_1 are, respectively, the wavelengths of onset of photodetachment from these states. The ground state

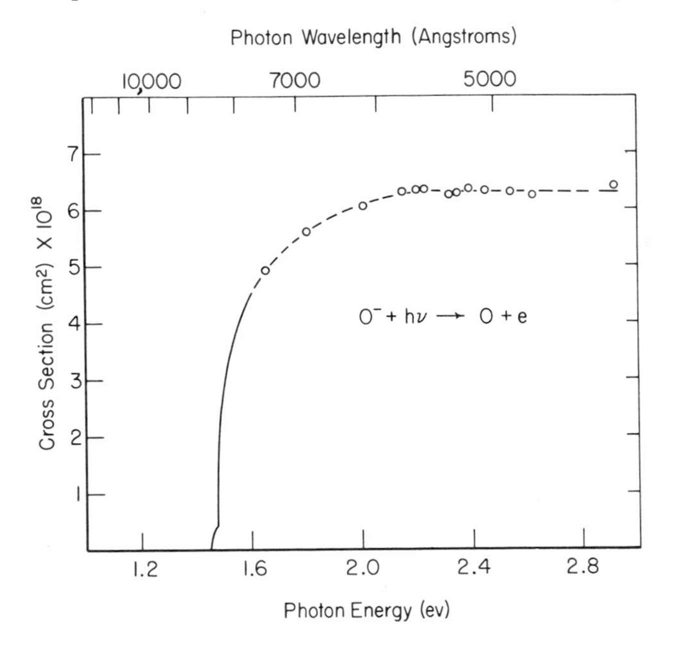

FIG. 10. Experimental cross section (Smith, 1960) for photodetachment of O^- The points were obtained with band-pass filters. The solid line and threshold wavelength were determined by fitting a threshold expansion to data obtained from filters with sharp absorption edges in this region.

splitting of O^- was estimated by isoelectronic extrapolation to be 230 cm^{-1}, so $\lambda_1 - \lambda_0$ is taken as 160 A. Therefore, four unknown parameters must be determined with seven pieces of data, resulting in the value for λ_0 given in (17).

A careful remeasurement of the shape of the O^- cross section above threshold was then made by Smith (1960). These data are combined with the threshold measurements to give in Fig. 10 the best available experimental cross section for O^-. The absolute value, accurate to about 10 %, is such that the cross section at λ5280 A has the value 6.3×10^{-18} cm^2. Relative values are accurate to 2 % or better.

The application of the oscillator strength sum rule (13) to the measured cross section from threshold up to 4000 A, makes N equal to about 0.02. Apparently the cross section remains large to much higher photon energies. To extend the photodetachment cross section for $O^-\ {}^2P \rightarrow O\ {}^3P + e$ to zero wavelengths, Smith calculated the cross section in the Born approximation, using for the O^- bound p state an analytic function approximating the wave function of Klein and Brueckner (1958) and for the continuum wave adopting the asymptotic solutions for $l = 0$ and 2, neglecting phase shifts. Dipole length matrix elements M_S and M_D were used, following Bates and Massey (1943), giving

$$\sigma = \frac{8\pi}{3} \frac{me^2k\omega}{\hbar^2c} (M_S^2 + 2M_D^2) \tag{19}$$

where $\hbar k$ is electron momentum and $\hbar\omega = h\nu$, the photon energy. The resulting cross section (Fig. 11) is, as expected, too large at low energies.

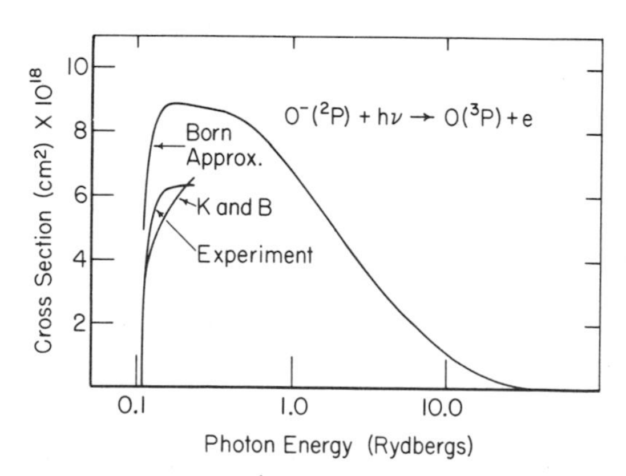

FIG. 11. Estimated cross section for photodetachment of O^- into the ground ${}^3P$ state of atomic oxygen at high energy using a crude Born approximation (Smith, 1960). The curve labeled K and B is due to Klein and Brueckner (1958).

The calculation of Klein and Brueckner, in which distortion of the continuum wave by the atomic field is included, and the experimental curve are shown for comparison. When the experimental curve is extended to shorter wavelengths and joined smoothly on to the Born approximation calculation, the oscillator strength turns out to be equivalent to 4 or 5 electrons.

In order to get an estimate of the total absorption coefficient of O^- in the ultraviolet region, it is necessary to include the effect of absorption processes which leave the oxygen atom in excited ^1D and ^1S terms. Assuming that the matrix elements for transitions to final ^3P, ^1D, and ^1S states are in proportion to their statistical weights (9, 5, and 1), we can adjust (19) for the different threshold energies corresponding to these transitions and sum the three partial cross sections to get the result shown in Fig. 12. The integrated oscillator strength is now found to be

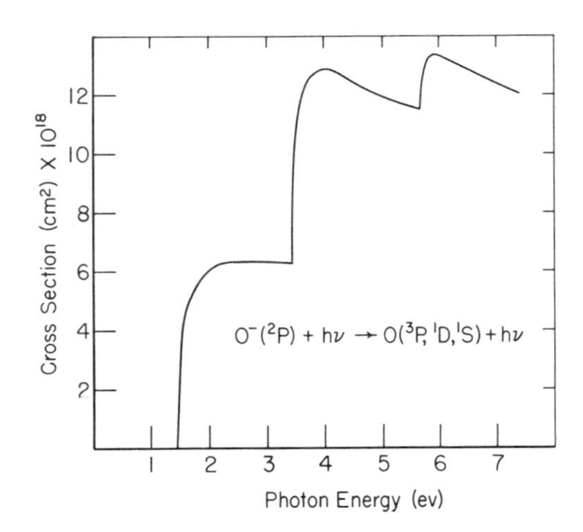

FIG. 12. Predicted form of the total O^- photodetachment cross section using experimental data, the oscillator strength sum rule, and the Born approximation calculation of Fig. 11.

7 or 8, which is consistent with the sum rule for a 9-electron atom, since excited final configurations have still been neglected.

A more accurate calculation of the O^- absorption in the ultraviolet has been completed by Martin and Cooper (1961), repeating with minor improvements and extending the work of Klein and Brueckner (1958). Martin and Cooper also have calculated the cross sections for C^-, F^-, and Cl^-. In all cases for which experimental data is available the cross section appears to peak much nearer threshold than predicted by theory, although the absolute value of the theoretical cross section appears to be quite satisfactory.

This result is of interest not only because it provides the opacity of O^- in the ultraviolet (Meyerott, 1957), but because analogous behavior is expected for the sulfur ion. It is a matter of interest in astrophysics to investigate the role of negative ions other than H^- as sources of opacity in stars of atypical composition (Branscomb and Pagel, 1958). Identification of negative ion absorption by C^-, O^-, Al^-, Si^-, P^-, or S^- is facilitated by the moderately sharp absorption edge associated with photodetachment from these unfilled *p*-shell ions. However, these first absorption edges will generally lie in the near infrared, which handicaps observation of the stellar continuum because of the masking effect of molecular absorption. The existence of second and third absorption thresholds in O^- and S^- of strength comparable with the first threshold suggests the possibility of identifying atomic ions in stellar spectra and measuring the affinities in the laboratory by observation of these higher thresholds.

Experimental data on the photodetachment of atomic ions other than H^-, C^-, and O^- is fragmentary or preliminary. Photodetachment signals have been observed from S^- and P^-. The study of S^- was adequate to determine the wavelength of the threshold, from which an affinity of 2.07 ± 0.07 ev was deduced (Branscomb and Smith, 1956). It was also found that the shape of the cross section was similar to that for O^- and in conformity with the expected threshold law. A study of the C^- spectrum (Seman and Branscomb, 1961) gives an electron affinity of 1.25 ± 0.03 ev, and the cross section has been measured from threshold (10,500 A) to 4000 A. It falls slowly to shorter wavelengths from a maximum around 7000 A of about $1.43 \pm 0.20 \times 10^{-17}$ cm^2. Although the absorption edge associated with the excitation of the final carbon atom in the ^1D state lies at 5080 A, in the middle of the observed spectral range, no additional contribution to the cross section is observed short of 5080 A. This is expected from the spin selection rules which forbid the transition $C^- (^4S) \rightarrow C (^1D) + e$. Analogously, no such discontinuity is expected in the spectrum of Si^-, whose ground state is also ^4S.

In the spectral region between 2.6 μ and the ground state absorption threshold at 0.99 ± 0.02 μ a weak detachment signal is observed from C^-. It appears to arise from a very weakly bound metastable state presumably C^- ^2D. No threshold was observed for this absorption, because of weakness of the signal and long wave length limitation of the apparatus (2.6 μ).

A calculation of the photodetachment cross section for Li^- and for

H^- has been made by Geltman (1956) using a procedure quite different from the calculations of the H^- cross section described in § 5.1. Instead of using a variational bound state wave function, which gives an accurate value of the binding energy but fails to give equality for the dipole length and velocity matrix elements, Geltman constructs simple analytic wave functions which are designed to assure equality of the different matrix element formulations [Eqs. (3) to (5)], the validity of the sum rules [Eqs. (13) and (14)], and a reasonable form for the potential which gives the correct binding energy. The results for hydrogen are in general good agreement with other calculations. There is no theoretical or experimental work with which the Li^- cross section can be compared.

7 *Photodetachment of Negative Molecular Ions*

7.1 General Considerations

As pointed out by Massey (1950), the photodetachment of diatomic molecular ions is governed by the Franck-Condon rule, which requires that the internuclear separation of the molecule is little affected during the electronic transition from the negative ion into the continuum of the neutral molecule. In the event that the molecular ion potential function has a minimum at a larger internuclear separation than the minimum of the neutral molecule potential, the energy required to remove an electron by photodetachment may be quite different (even of different sign) from the electron affinity of the molecule. This is illustrated for H_2^- using the potential function of Dalgarno and McDowell (1956) (Fig. 13). As indicated on the figure, the photodetachment threshold from the lowest vibrational level of H_2^- would be expected at 1.37μ, corresponding to a vertical detachment energy of 0.9 ev; yet the electron affinity derived from this calculation is -3.58 ev.

The photodetachment spectrum of a diatomic molecular ion in a specified initial state will consist of the superposition of a very large number of continua, each corresponding to a particular final rotational, vibrational, and electronic quantum state of the neutral molecule. Each of these continua may be expected to follow the threshold laws described in § 3.2. If the two electronic states involved (one of the negative ion, the other of the neutral molecule) have the same internuclear separation at equilibrium, the intense transitions will be restricted to a few final vibrational states, and the over-all absorption spectrum may be expected to resemble that of a single pair of initial and final states. However, if the equilibrium nuclear separation in the neutral molecule is much different from that of the ion, a relatively large number of final vibrational states take part with comparable intensity, with a resulting smearing out of the spectrum, concealing the true energy dependence of the electronic dipole matrix element.

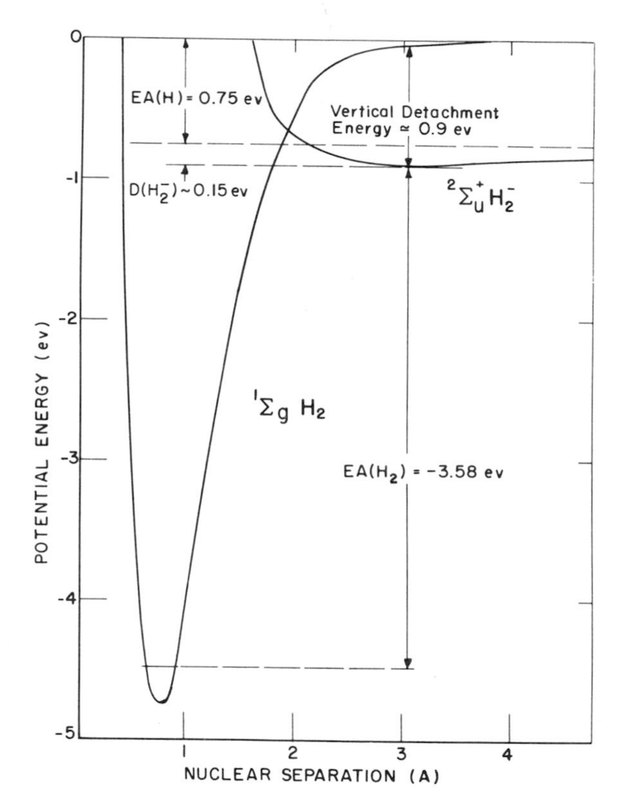

FIG. 13. Potential curves for H_2^- calculated by Dalgarno and McDowell (1956), illustrating the relationship between electron affinity and vertical detachment energy in negative ions of diatomic molecules.

The problem of interpretation of photodetachment spectra of negative ions is further complicated by uncertainty as to the identity of the initial state occupied by the negative ion being observed. This experimental complication is the consequence of the use of discharge ion sources which contain large anode falls of potential (for ion extraction) which might result in the vibrational excitation of the ions during extraction from the source. The transit time of the ions from source to reaction chamber is of the order of 5 μsec, during which some heteronuclear molecule ions might radiate away their vibrational energy, but homonuclear molecules cannot. Again, if the potential functions of both neutral and negatively charged molecules are similar, the influence of initial vibration is reduced by virtue of the relatively great strength of the $\Delta v = 0$ sequence, the members of which fall at roughly the same wavelength and are superposed. When the potential curves are shifted and the ions are vibrationally excited, the effect is greatly to extend the spectral range of the absorption and conceal from the investigator any clear evidence bearing on the shape and location of the negative molecular potential function or the degree of excitation, if any, of the negative ions. Thus, it is apparent that at the present stage of development of apparatus for investigation of spectra of negative molecular ions, a number of conditions should be fulfilled to permit interpretation of the observed spectrum if no *a priori* knowledge of the negative ion potential curve is available:

(a) Initial and final internuclear separation should be similar.

(b) There should be little or no vibrational excitation in the ion beam.

(c) The neutral molecule vibrational spacing should sufficiently exceed the photon energy width of the optical filters to permit resolution of the continua corresponding to different vibrational sequences.

(d) The photodetachment cross section between specified states should exhibit a sufficiently sharp threshold, and

(e) simple over-all shape so that different vibrational sequences can be identified in the total absorption spectrum.

7.2 The Photodetachment Spectrum of OH^- and OD^-

The OH^- ion appeared to meet these five conditions when investigated by Smith and Branscomb (1955b) and Branscomb (1957a) using the low-resolution difference method with glass filters. In confirmation of Geltman's expectations, a sharp absorption edge was observed consistent with the threshold expansion of (9). The threshold energy corresponds to 1.78 ev. Subsequently, this spectrum was reinvestigated with band-pass filters in a search for more than one vibrational sequence and means for identification of the sequences. To prove that the electron affinity of OH is 1.78 ev, one must prove that the one strong threshold which is observed corresponds to $\Delta v = 0$. The higher resolution experiments did not reveal a second threshold one OH vibrational quantum above the first. This might result from the near identity of the internuclear separations of OH and OH^-, which is theoretically anticipated (Ransil and Krauss, 1960). Second, it was shown that the absorption at longer wavelengths than the threshold (7000 A) was less than 10^{-3} of the observed OH^- absorption, which might result from a low degree of vibrational excitation, since absorption in this region presumably results from $\Delta v = -1$. Third, comparisons of OH^- and OD^- were made, and the threshold wavelength was found to be identical to $\pm$ 50 A. This should prove that the observed threshold does indeed result from the (0, 0) vibrational continuum, since any sequence other than $\Delta v = 0$ would shift OD^- relative to OH^- by an amount equal to or greater than the difference in the vibrational quanta of the respective neutral molecules. It appears, however, that condition (e) is not fulfilled. The absorption appears to show a strong maximum near threshold, a deep minimum about 300 A above the threshold, followed by a second maximum. The band-pass filters are too broad to be used in the investigation of this structure, and higher resolution is needed. Until higher resolution is available, the question remains unanswered whether this structure results from (a) a strong, partially resolved absorption (autoionization) line at the threshold, (b) an oscillating photodetachment cross section within 0.2 ev of threshold, or (c) some other cause possibly related to excited electronic or vibrational states of the negative ions.

7.3 The Photodetachment Spectrum of O_2^-

The only other molecule ion whose photodetachment spectrum has been investigated is O_2^-. Burch *et al.* (1958) observed a detachment probability decreasing over the entire range from 4000 to 25,000 A. No threshold could be found, and it appeared that the absorption edge lay below 0.5 ev. The observed points are shown in Fig. 14, together with the O^- cross section for comparison. In this case, conditions (c) and (d) for unequivocal analysis of molecular detachment spectra are not met, and we have no certainty that (a) and (b) are fulfilled. The data were highly reproducible in spite of changes in the ion source conditions which might be expected to produce changes in the population of excited states if such are populated. The threshold law for detachment of a π_g electron from $O_2^-\ {}^2\Pi_g$ is expected to be of the form given in (7) (See Table I). Since the electron affinity of O_2 is in controversy, values from 0.07 to 0.9 having been advanced (Burch *et al.*, 1958), and since the ground state potential function for O_2^- is unknown, it was not possible to predict the energy of onset of the spectrum. Accordingly, the experimental points were fit to the threshold law. A good fit was possible, as is seen from Fig. 14, for the solid line drawn through the points is a plot of (7) with two coefficients in the expansion and a threshold value of 0.15 ± 0.05 ev. The intercept occurred far below the last observed point (0.5 ev).

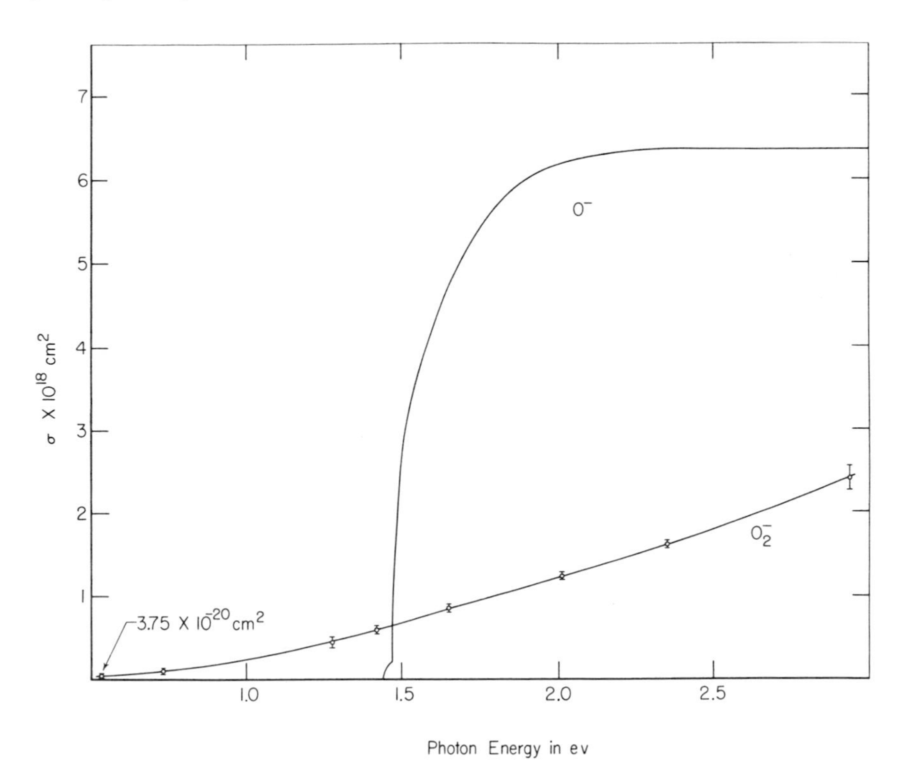

FIG. 14. Observed photodetachment spectrum (Burch *et al.*, 1958) of O_2^-, compared to the cross section for O^-. The solid line is a 3-parameter threshold law expansion fitted to the experimental points. The extrapolated intercept is at a photon energy of 0.15 ev.

In this case it is quite impossible to resolve any vibrational structure in the curve. The resolution of the filters is approximately equal to the vibrational quanta. In addition, the $E^{3/2}$ power law at threshold would cause the continua at each vibrational level of the final O_2 state to merge smoothly into one another.

Recently, additional research bearing on the electron affinity of molecular oxygen has shed doubt on the reliability of the thermochemical value of 0.9 ev (Kazarnovski, 1948; Evans and Uri, 1949). Mulliken has reviewed the qualitative predictions of molecular orbital theory (Mulliken, 1959) with the conclusion that 0.9 ev is unreasonably large. More direct information comes from the swarm experiments of

Phelps and collaborators. Chanin *et al.* (1959) measured the three body attachment coefficient in molecular oxygen under conditions closely approximating thermal equilibrium near room temperature: $e + 2O_2 \rightarrow O_2^- + O_2$. Then Phelps and Pack (1960) measured the collisional detachment rate (the reverse of three body attachment), which they could only observe at temperatures above 425°K. Again the detachment rate is thought to represent thermal equilibrium. By extrapolation of the attachment frequency above 425°K, the electron affinity can be deduced from the law of mass action. The resulting value is

$$A(O_2) = 0.46 \pm 0.02 \text{ ev.} \tag{20}$$

Several of the assumptions upon which the validity of (20) depends would, if unjustified, imply an even higher value of the electron affinity of O_2. Should this value prove correct, one must conclude that the excellent fit of the photodetachment spectrum to the power law given in (7) is fortuitous, unless the observed spectrum arises from an excited state or level, not the lowest level, which is about 0.15 ev below the continuum. That this might be the case was anticipated in the original paper (Burch *et al.*, 1958), where quality of the fit was called "surprising" in view of the very limited region above threshold over which the threshold laws have fit the observed data in other ions.

The only molecular negative ions whose detachment spectra have been carefully studied are OH^- and O_2^-. However, a preliminary attempt to detach electrons from NO_2^- was made using radiation of 5200 A (Branscomb, 1957b). No signal was recorded, and the assumption is made that either the vertical detachment energy of NO_2 exceeds 2.5 ev or that the detachment cross section is unusually small. This matter requires further investigation because of the potential geophysical importance of the NO_2^- ion.

Certain speculations have also been made about the photodetachment spectra of molecular ions such as CN^- and C_2^-, in particular in stellar atmospheres (Branscomb and Pagel, 1958).

8 *Formation of Negative Ions by Radiative Attachment*

8.1 Radiative Attachment to Atoms

The inverse process to photodetachment is radiative attachment:

$$e + X \rightarrow X^- + h\nu. \tag{21}$$

The two cross sections are related by the principle of detailed balancing (Massey, 1950):

$$\sigma_{\text{det}} = (mcv/h\nu)^2 (g_0/g_-)\sigma_{\text{att}} \tag{22}$$

where m and v are electron mass and velocity, $h\nu$ is the photon energy, and g_0 and g_- are statistical weights of the neutral atom and negative ion. Thus, when the photodetachment cross section for an atomic ion has been measured from the threshold to some energy $h\nu$, the radiative attachment cross section can be immediately calculated for electrons whose energy ranges from zero to $h\nu - E_0$, assuming that one knows the ground state term of the negative ion, on which g_- depends.

One can also solve (22) for σ_{att} and substitute the threshold law (6) for σ_{det}, thus deriving the threshold behavior for radiative attachment:

$$\sigma_{\text{att}} \propto \nu^3 k^{2l-1}(b_0 + b_1 k^2 + b_2 k^4 \ldots). \tag{23}$$

For electron attachment into a bound a state, $l = 1$ and

$$\sigma_{kp \to ns} \propto \nu^3 k(b_0 + b_1 k^2 + \ldots) \tag{24}$$

while

$$\sigma_{ks \to np} \propto \nu^3 k^{-1}(b_0 + b_1 k^2 + \ldots). \tag{25}$$

Thus, for electron attachment to atomic hydrogen, σ_{att} vanishes at the threshold according to (24). The cross section calculated from experimental data of Smith and Burch (1959) on the reverse process is shown in

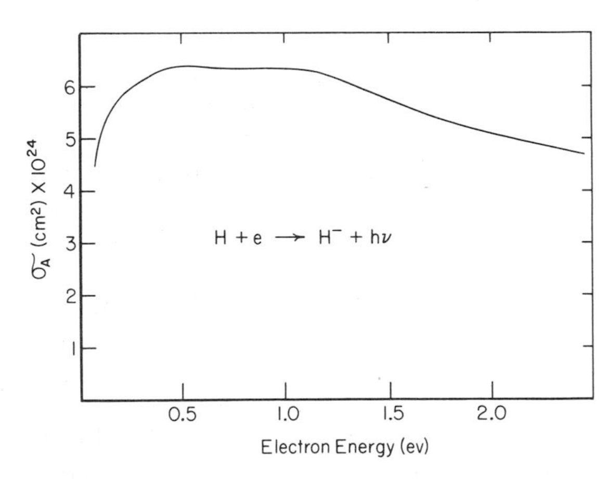

FIG. 15. Radiative attachment cross section for electrons to atomic hydrogen calculated by detailed balancing from the photodetachment spectrum of H^- as determined by Smith and Burch (1959).

Fig. 15. Similarly, Fig. 16 shows the O^- attachment cross section calculated from the experimental photodetachment data (Smith, 1960).

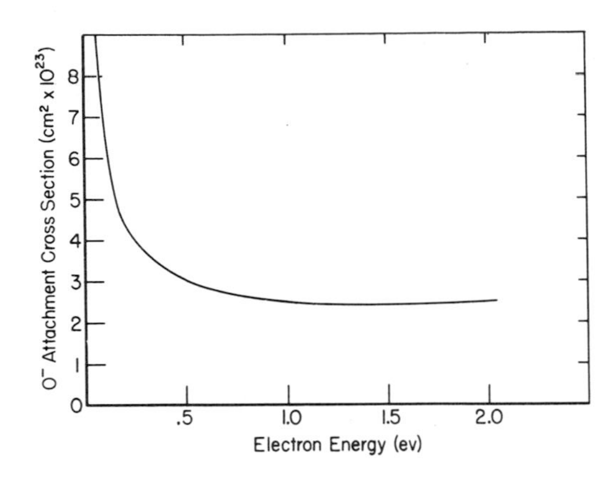

FIG. 16. Radiative attachment cross section of electrons to atomic oxygen calculated by detailed balancing from the photodetachment cross section from Smith (1960).

Here (25) is followed, with the cross section becoming infinite at the origin.

Of more practical interest is the monoenergetic attachment rate

$$\alpha = v\sigma_{\text{att}}. \tag{26}$$

Since k, the electron momentum, is proportional to the velocity v, we see from (24) and (25) that rate for attachment into an s state rises from zero as k^2 (see Fig. 17). For attachment into p states the rate is finite at the threshold, which is seen for O^- in Fig. 18. If one deals with a swarm of electrons with a known distribution of velocities, the average attachment rate is obtained by integration of (26) over this velocity distribution. It is clear by inspection of Figs. 17 and 18 that attachment into s states will have a strong temperature dependence, while the rate for atoms with unfilled p shells will be nearly independent of temperature.

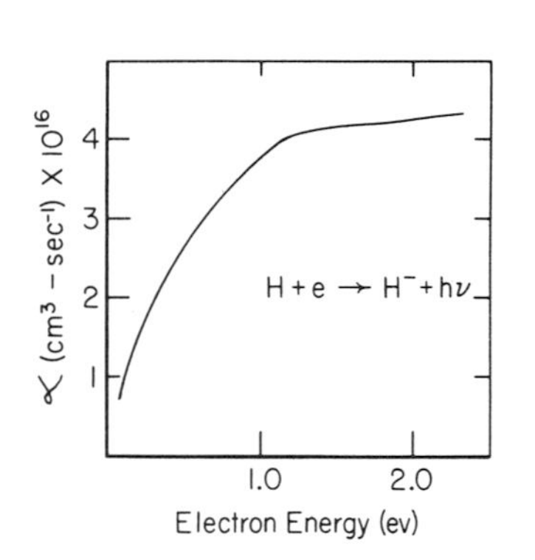

FIG. 17. Rate coefficient for radiative attachment of electrons to atomic hydrogen calculated from Fig. 15.

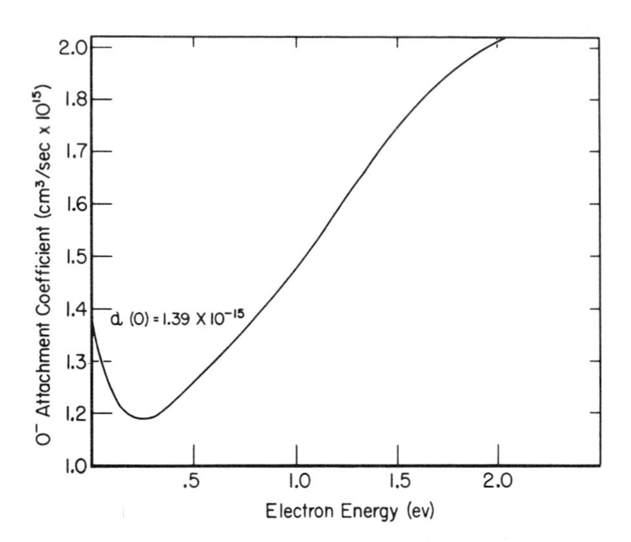

FIG. 18. Rate coefficient for radiative attachment of electrons to atomic oxygen calculated from Fig. 16.

8.2 RADIATIVE ATTACHMENT TO MOLECULES

Equation (22) can also be used to calculate radiative attachment cross sections for the diatomic molecules whose photodetachment spectra have been observed, providing the latter are known to correspond to a single initial and final vibrational state. We have no certainty that this is the case for the spectrum of O_2^- discussed in the previous section. If the absorption spectrum corresponds to a number of transitions to different final vibrational levels, the radiative attachment cross section deduced from the application of (22) will refer to electron attachment to an O_2 gas with this same (highly improbable) vibrational distribution. A further difficulty arises from the fact that we do not observe the threshold in the photodetachment of O_2^-. The effect of the vertical detachment energy on the radiative attachment coefficient can be seen from Fig. 19, which shows the attachment coefficient for O_2 on the above assumption and for $E_0 = 0$, 0.2, and 0.5 ev. These curves are obtained from the power law fit to the photodetachment data, which gives

$$\alpha = 0.78\ 10^{-16}\,(h\nu)^3\,(h\nu - E_0)\,[0.37 - 0.07\,(h\nu - E_0)]\ \text{cm}^3/\text{sec} \qquad (27)$$

where the photon and threshold energies are given in electron volts.

If in fact the photodetachment spectrum does correspond to transitions which leave the O_2 molecule in excited vibrational states, the attachment coefficient will take quite a different form, for one must apply (22) separately to each vibrational transition, and the change in internal energy in each of these transitions will determine the relation between v and ν in (22).

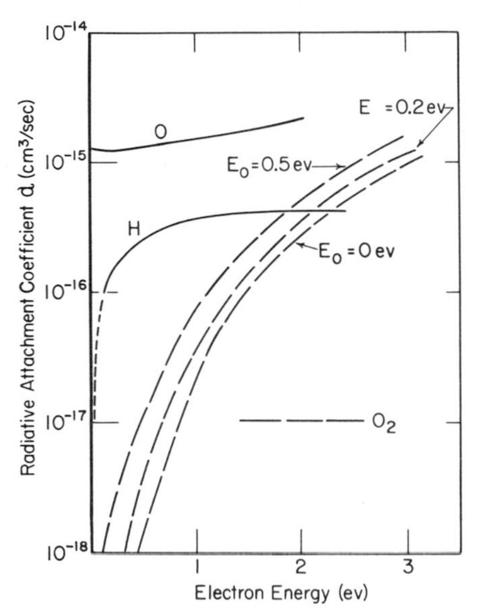

FIG. 19. Radiative attachment coefficients for atomic oxygen and hydrogen and illustrative curves for molecular oxygen. The three curves for O_2 (as a function of vertical detachment energy E_0) are correct only if the experimental photodetachment spectrum from which they were calculated refers to single initial and final vibrational states. The best value for E_0 is thought to be 0.46 ev (Phelps and Pack, 1960).

8.3 Emission Continua Due to Radiative Attachment

Although the radiative attachment continuum (or affinity spectrum) of H^- was identified as the dominant source of solar continuum in the visible by Wildt (1939), a negative ion spectrum was first observed in the laboratory in 1951 by Lochte-Holtgreven (1951) in a water-stabilized hydrogen arc. In these experiments the negative ion spectrum forms only part of the continuum due to a number of different recombination processes. Accordingly, it was not possible to deduce the radiative attachment cross section for atomic hydrogen from these observations. More recent experiments at Kiel have produced much stronger negative ion continua under circumstances more amenable to quantitative evaluation. Weber (1958) has produced a strong H^- attachment spectrum in emission from reflected shock waves of hydrogen expanded into low pressures of krypton. The spectrum was investigated in the blue region in order to minimize the contribution from free-free transitions. The H^- emission is estimated to be 30 times stronger than any other source of continuum in the region immediately to longer wavelengths from the Balmer discontinuity. Relative intensities in the observed spectra are, on the assumption of local thermodynamic equilibrium, in good agreement with the predicted intensity distribution from the theoretical cross section.

The negative atomic oxygen ion continuum has also been studied in emission at Kiel using a wall-stabilized arc at atmospheric pressure. Boldt (1959a) measured the arc temperature as a function of radius using the absolute intensity of an atomic oxygen emission line. He then separated the negative ion continuum from the atomic recombination spectrum by analysis of the temperature dependence of the emission continuum. Again local thermodynamic equilibrium is assumed, and the contribution from free-free transitions is included in the deduced free-bound cross section. The O^- photodetachment cross section was deduced, using Kirchoff's law, and the result differs by about 30 % from the directly measured photodetachment cross section. This is felt to be good agreement, since Boldt quotes a possible 30 % error in its values; the crossed-beam experiments are good to about 10 % in absolute value; and perhaps 10 % of the difference is accounted for by contributions from free-free transitions.

The quite similar investigation by Boldt (1959b) in a nitrogen arc yielded, surprisingly, an even stronger affinity spectrum than the oxygen arc. This result was explained by invoking the metastable ^1D state of N^- suggested by Bates and Moiseiwitsch (1955). For this explanation to hold it is only necessary that the electron collisional destruction rate for $N^-\,{}^1D$ should exceed the spontaneous decomposition rate. At the high pressures and electron densities involved, this is probably the case. Thus, the affinity spectrum in nitrogen may result from radiative attachment of free electrons to metastable atomic nitrogen atoms in the ^2D state.

9 *Atomic Electron Affinities*

Several complete surveys of the available information on atomic electron affinities have recently been published (Buchel'nikova, 1958; Branscomb, 1957a; Pritchard, 1953). The references cited include a summary of molecule affinities. Here we will restrict ourselves to atoms.

TABLE III

BINDING ENERGIES OF SELECTED ATOMIC NEGATIVE IONS

Element	Binding energy (ev)	Method, author, and reference	Binding energies (ev) from empirical extrapolation by	
			Edlen (1960)	Johnson and Rohrlich (1959)
H^- $(1s)^2$	0.75416	Variational calculation (Pekeris, 1958)		
	0.8 ± 0.1	Surface ionization (Khvostenko and Dukel'skii, 1960)		
He^- $1s^2 2s$	< 0			
$1s2s2p$	> 0.075	Variational calculation (Holøien and Midtal, 1955)		
	> 0	Charge exchange (Dukel'skii *et al.*, 1956)		
Li^-	0.616	Configuration interaction (Weiss, 1960)		
Be^- $(2s)^2 (2p)$			—0.19	
B^-			0.33	0.82
C^-	1.33 ± 0.18	Electron impact (Fineman and Petrocelli, 1958)	1.24	1.21
	1.11 ± 0.05	Electron impact (Lagergren, 1955)		
	1.25 ± 0.03	Photodetachment (Seman and Branscomb, 1961)		
N^- ^4S	< 0		0.05	0.54
^4S or ^1D	> 0	Charge exchange (Fogel *et al.*, 1959)		
O^-	1.465 ± 0.005	Photodetachment (Branscomb *et al.*, 1958)	1.47	1.47
F^-	3.62 ± 0.09	Surface ionization (Bailey, 1958)	3.50	3.62
	3.48	Lattice energies (Cubicciotti, 1959)		
	3.47	Surface ionization [assuming A(Br) = 3.50] (Bakulina and Ionov, 1955)		

Na^-	0.84	Statistical theory (Gáspár and Molnar, 1955)	0.47	
Mg^-			—0.32	
Al^-			0.52	1.19
Si^-			1.46	—
P^-	1.12	Statistical theory (Glombas and Ladanyi, 1960)	0.77	1.33
S^-	2.07 ± 0.07	Photodetachment (Branscomb and Smith, 1956)	2.15	2.79
	2.27	Surface ionization [assuming A(Br) = 3.50] (Bakulina and Ionov, 1957)		
Cl^-	3.76 ± 0.09	Surface ionization (Bailey, 1958)	3.70	3.84
	3.69	Lattice energies (Cubicciotti, 1959)		
	3.71	Surface ionization [assuming A(Br) = 3.50] (Bakulina and Ionov, 1955)		
	3.620 ± 0.007	Shock wave absorption (Berry *et al.*, 1961)		
Br^-	3.51 ± 0.06	Surface ionization (Bailey, 1958)		
	3.45	Lattice energies (Cubicciotti, 1959)		
	3.49 ± 0.02	Surface attachment (Doty and Mayer, 1944)		
	3.53 ± 0.12	Photoionization of Br_2 (Morrison *et al.*, 1960)		
	(3.50)	Assumed for other halogen affinities (Bakulina and Ionov, 1955)		
I^-	3.17 ± 0.05	Surface ionization (Bailey, 1958)		
	3.14	Lattice energies (Cubicciotti, 1959)		
	3.23	Surface ionization [assuming A(Br) = 3.50] (Bakulina and Ionov, 1955)		
	3.13 ± 0.12	Photoionization of I_2 (Morrison *et al.*, 1960)		

In Table III are summarized selected experimental and theoretical (or empirical) data. The second column of the table gives the results of experiments or calculations thought to be reliable to within about 0.1 ev (with the exception of data for Na^- and P^- which may be uncertain by a much larger margin). For comparison are given the results of two recent semiempirical extrapolations along isoelectronic sequences. Other forms of extrapolations will be found discussed in (Branscomb, 1957a). The methods of Edlen (1960) and Johnson and Rohrlich (1959) are nearly identical. Edlen included a shorter series and fit to fewer undetermined parameters, using more recent input data on ionization potentials. Neither method fits the results to experimentally determined affinities and both can claim some theoretical justification to the form of the extrapolation. It is very interesting to note that the predictions (particularly of Edlen) are in excellent agreement with the experimental data which exist, and yet quite conflicting predictions are made with regard to Al^-, P^-, and Si^-, which have not yet been studied by the photodetachment method. These differences between the Edlen and Johnson and Rohrlich results demonstrate the extreme sensitivity of the extrapolations to the input data on ionization potentials of isoelectronic positive ions. The apparent excellent agreement of the data for the halogens is upset by the elegant direct observations of Berry *et al.* (1961). All halogen affinities may have to be lowered by about 0.1 ev.

Acknowledgments

I am indebted to my colleagues S. J. Smith and S. Geltman for many helpful suggestions and discussions in the preparation of this chapter. Mr. George Goldenbaum made valuable contributions through his bibliographic research in the preparation of Table III. I am also grateful to a number of scientists who permitted the use of their unpublished data in this manuscript, particularly A. V. Phelps, A. Weiss, T. L. John, R. S. Berry, and my colleagues M. Seman, M. A. Fineman, S. Geltman, J. Cooper, and M. Krauss.

References

Bailey, T. L. (1958) *J. chem. Phys.* **28**, 792.
Bakulina, I. N., and Ionov, N. I. (1955) *Dokl. Akad. Nauk SSSR* **105**, 680.
Bakulina, I. N., and Ionov, N. I. (1957) *Soviet Phys.—JETP* **2**, 423.
Bates, D. R., and Massey, H. S. W. (1940) *Astrophys. J.* **91**, 202.
Bates, D. R., and Massey, H. S. W. (1943) *Trans. Roy. Soc.* **A239**, 269.
Bates, D. R., and Moisciwitsch, B. L. (1955) *Proc. Phys. Soc.* **A68**, 540.
Berry, R. S., Reimann, C. W., and Spokas, G. N. (1961) Private communication, submitted to *J. chem. Phys.*

Boldt, G. (1959a) *Z. Phys.* **154**, 319.
Boldt, G. (1959b) *Z. Phys.* **154**, 330.
Branscomb, L. M. (1957a) *Advances in Electronics and Electron Phys.* **9**, 43.
Branscomb, L. M. (1957b) "Threshold of Space," p. 101. Pergamon, New York.
Branscomb, L. M., and Pagel, B. E. J. (1958) *Monthly Not. Roy. Astron. Soc.* **118**, 258.
Branscomb, L. M., and Smith, S. J. (1955a) *Phys. Rev.* **98**, 1028.
Branscomb, L. M., and Smith, S. J. (1955b) *Phys. Rev.* **98**, 1127.
Branscomb, L. M., and Smith, S. J. (1956) *J. chem. Phys.* **25**, 598.
Branscomb, L. M., Burch, D. S., Smith, S. J., and Geltman, S. (1958) *Phys. Rev.* **111**, 504.
Bucheľnikova, N. S. (1958) *Uspekhi fiz. Nauk* **65**, 351; translated by A. L. Monks, Oak Ridge National Laboratory Library (AEC-tr-3637).
Burch, D. S., Smith, S. J., and Branscomb, L. M. (1958) *Phys. Rev.* **112**, 171; see also erratum (1959) *Phys. Rev.* **114**, 1652.
Chandrasekhar, S. (1945a) *Astrophys. J.* **102**, 223.
Chandrasekhar, S. (1945b) *Astrophys. J.* **102**, 395.
Chandrasekhar, S. (1958) *Astrophys. J.* **128**, 114.
Chandrasekhar, S., and Elbert, D. D. (1958) *Astrophys. J.* **128**, 633.
Chandrasekhar, S., and Krogdahl, M. K. (1943) *Astrophys. J.* **98**, 205.
Chanin, L. M., Phelps, A. V., and Biondi, M. A. (1959) *Phys. Rev. Letters* **2**, 344
Cubicciotti, D. (1959) *J. chem. Phys.* **31**, 1646.
Dalgarno, A. (1961) Private communication.
Dalgarno, A., and Kingston, A. E. (1959) *Proc. Phys. Soc.* **73**, 455.
Dalgarno, A., and McDowell, M. R. C. (1956) *Proc. Phys. Soc.* **A69**, 615; see also the potential function for H_2^- of Inga Fischer-Hjalmers (1959) *Ark. Fys.* **16**, 33.
Doty, P. M., and Mayer, J. E. (1944) *J. chem. Phys.* **12**, 323.
Dukeľskii, V. M., Afrosimov, V. V., and Fedorenko, N. V. (1956) *Zh. eksper. teor. Fiz.* **30**, 792.
Edlén, B. (1960) *J. chem. Phys.* **33**, 98.
Evans, M. G., and Uri, N. (1949) *Trans. Faraday Soc.* **45**, 224.
Fineman, M. A., and Petrocelli, A. (1958) *Bull. Amer. Phys. Soc.* [2] **3**, 258; (1957) AEC. Rept. NYO-AEC-7239.
Fogel, Ya. M., Kozlov, V. F., and Kalmykov, A. A. (1959). *Zh. eksper. teor. Fiz.* **36**(9), 963.
Gáspár, R., and Molnar, B. (1955) *Acta phys. Hungar.* **5**, 75.
Geltman, S. (1956) *Phys. Rev.* **104**, 346.
Geltman, S. (1958) *Phys. Rev.* **112**, 176.
Geltman, S. (1960) Private communication.
Geltman, S., and Krauss, M. (1960) *Bull. Amer. Phys. Soc.* [2] **5**, 339.
Geltman, S., and Martin, J. B. (1960) Private communication.
Glombas, P., and Ladanyi, K. (1960). *Z. Phys.* **158**, 261.
Hart, J. F., and Herzberg, G. (1957) *Phys. Rev.* **106**, 79.
Henrich, L. F. (1944) *Astrophys. J.* **99**, 59, 318.
Hiby, J. W. (1939) *Ann. Phys. (Leipzig)* **34**, 473.
Holøien, E. (1960a) *J. chem. Phys.* **33**, 301.
Holøien, E. (1960b) *J. chem. Phys.* **33**, 310.
Holøien, E., and Midtdal, J. (1955) *Proc. Phys. Soc.* **68**, 815.
John, T. L. (1960) *Astrophys. J.* **131**, 743; (1960) *Monthly Not. Roy. Astron. Soc.* **121**, (1), 41-7.
Johnson, H. R., and Rohrlich, F. (1959) *J. chem. Phys.* **30**, 1608.
Kazarnovski, L. A. (1948) *Dokl. Akad. Nauk SSSR* **59**, 67.
Khvostenko, V. I., and Dukeľskii, V. M. (1960) *Soviet. Phys.—JETP* **10**, 465.

Klein, M. M., and Brueckner, K. A. (1958) *Phys. Rev.* **111**, 1115.
Lagergren, C. R. (1955) Thesis, Univ. of Minnesota, Minneapolis, Minnesota.
Lochte-Holtgreven, W. (1951) *Naturwissenschaften* **38**, 258.
Martin, J. B., and Cooper, J. (1961) Manuscript in preparation.
Massey, H. S. W. (1950) "Negative Ions." Cambridge Univ. Press, London and New York.
Meyerott, R. E. (1957) "The Threshold of Space," p. 259. Pergamon, New York.
Morrison, J. D., Hurzeler, H., Inghram, M. G., and Stanton, H. E. (1960) *J. chem. Phys.* **33**, 821.
Mulliken, R. S. (1959) *Phys. Rev.* **115**, 1225.
Pekeris, C. L. (1958) *Phys. Rev.* **112**, 1649.
Phelps, A. V., and Pack, J. L. (1960) 13th Ann. Gaseous Electronics Conf., Monterey, California, 1960; see abstract to appear in *Bull. Amer. Phys. Soc.* (1961).
Pritchard, H. O. (1953) *Chem. Revs.* **52**, 529.
Ransil, B. J., and Krauss, M. (1960) *J. chem. Phys.* **33**, 840.
Schwartz, C. (1961) *Phys. Rev.* **123**, 1700.
Seman, M., and Branscomb, L. M. (1961) Submitted to *Phys. Rev.*
Seman, M., Fineman, M. A., and Branscomb, L. M. (1961) *Bull. Amer. Phys. Soc.* [2] **6**, 29.
Smith, S. J. (1960) *In* "Proceedings of the Fourth International Conference on Ionization Phenomena in Gases, Uppsala, 1959" (N. R. Nilsson, ed.), p. 219. North-Holland, Amsterdam.
Smith, S. J., and Branscomb, L. M. (1955a) *J. Res. Nat. Bur. Stand.* **55**, 165.
Smith, S. J., and Branscomb, L. M. (1955b) *Phys. Rev.* **99**, 1657 (Abstract).
Smith, S. J., and Branscomb, L. M. (1960) *Rev. sci. Instrum.* **31**, 733.
Smith, S. J., and Burch, D. S. (1959) *Phys. Rev. Letters* **2**, 165; (1959) *Phys. Rev.* **116**, 1125.
Weber, O. (1958) *Z. Phys.* **152**, 281.
Weiss, A. W. (1960) Private communication to S. Geltman; Roothaan, C. C. J., Sachs, L. M., and Weiss, A. W. (1960) *Rev. mod. Phys.* **32**, 186.
Wigner, E. P. (1948) *Phys. Rev.* **73**, 1002.
Wildt, R. (1939) *Astrophys. J.* **89**, 295.
Windham, P. M., Joseph, P. J., and Weinman, J. A. (1958) *Phys. Rev.* **109**, 1193.
See, e.g., Woolley, R.v.d.R., and Stibbs, D. W. N. (1953) "The Outer Layers of a Star." Oxford Univ. Press (Clarendon), London and New York.
Wu, T.-Y. (1953) *Phys. Rev.* **89**, 629.

5.

High-Temperature Shock Waves

Alan C. Kolb and Hans R. Griem

1 Introduction

As a result of a concentrated research effort by numerous workers, it is now generally recognized that the shock tube affords a powerful tool for the investigation of high-temperature, high-velocity gas flow. We will mainly consider the shock tube as a device for the precision determination of atomic properties and will not consider molecular processes in any detail. Included are atomic transition probabilities (f-numbers) and line broadening coefficients (damping constants and shape functions). From the precise measurement of these quantities, one can also obtain information on atomic and ionic wave functions. The various sections which follow are written from the point of view of assessing the applicability of the shock tube for such measurements. We need not dwell on the motivations for carrying out accurate measurements of this kind. However, because of the crucial concern to astrophysicists, we might simply point out, as an illustrative example, that the determination of the structure of stellar atmospheres, of chemical abundances (the hydrogen-helium ratio, for example), and of stellar masses depends sensitively on a knowledge of f-numbers and line profiles.

There is an extensive body of literature on the use of the shock tube to study such things as: the production of plasmas for magnetohydrodynamic studies, boundary layer physics, the measurement of ionization and dissociation energies, various aerodynamic studies, relaxation phenomena, plasma preheating for controlled fusion investigations, structure of collision-free shock waves at low densities, etc. It is not our purpose here to present a comprehensive historical survey of this vast field of physical research; such a review might well fill a book by itself. It is necessary, therefore, to curtail drastically the number of references and to limit the scope of the present chapter quite sharply.

In order to discuss in a critical way the shock tube potentialities, we must also be concerned with the experimental determination of the plasma properties or plasma state. This is necessary so that one can correlate spectroscopic observations with a plasma whose various ionic and atomic concentrations as well as the temperature are known with sufficient accuracy. Since the degree of ionization and also the populations of the atomic and ionic bound excited states depend critically on the temperature, it is imperative that it be measured with a relatively high order of accuracy. Since the concept of temperature is meaningful only if there is local equilibrium, some discussion is necessary of the conditions necessary for equilibrium among the excited states and between the various ionization species, as well as the translational degrees of freedom.

In addition to an emphasis here on the determination of *plasma properties* and *atomic properties*, some discussions of the hydrodynamic aspects are also required in order to appreciate fully the possibilities of the shock tube as a research tool. Since it now appears that for high-temperature shock waves the interaction between radiation and hydrodynamic phenomena is of considerable importance, some attention will be given to this topic, although it is as yet only partly understood. Apparently, radiation effects are important in promoting the observed rapid equilibration behind strong shock waves as well as providing a mechanism for depositing energy ahead of an advancing shock front in magnetic shock tubes. Because of these basic difficulties, only the simplest hydrodynamic considerations will be given here, and it should be emphasized again that in this area much remains to be done from both a theoretical and an experimental point of view. Therefore, most of our remarks concerning deviations from the usual hydrodynamic theory will be of a more or less speculative nature, employing some rather drastic assumptions.

Since one of the major purposes here is to assess the possibilities for developing a high-temperature, thermal light source to obtain information on atomic properties, we will draw heavily on experimental work at the U. S. Naval Research Laboratory, the University of Maryland, the University of Michigan, and Kiel University (with which the authors are most familiar). In any event, it is quite clear that the shock tube is only beginning to emerge as a device useful for precision measurements, and it is the purpose of this brief survey to help delineate those areas which bear immediate promise as well as those areas which are only poorly understood, and whose understanding is necessary for the future development of shock tube techniques.

2 Hydrodynamic Considerations

2.1 Shock-Wave Conditions with Ionization

The state of a plasma behind a strong shock front can be determined, in principle, from the conservation of mass, momentum, and energy. Since detailed accounts (Courant and Friedrichs, 1948; Resler *et al.*, 1952; Turner, 1956; Seay, 1957; Laporte, 1960) of the theory of shock waves can be found elsewhere, we will only summarize here the basic relations which govern the propagation of shock waves which are sufficiently strong to produce ionization. The shock relations take a particularly simple form in case the relaxation times for dissociation, ionization, excitation, and randomization are short enough that equilibrium considerations apply.

As discussed in § 6, these equations accurately correspond to the measured state of the shocked gas (if the densities are high enough) after the relaxation period. This verifies the equilibrium assumption used in the calculation and makes the conventional shock tube useful as a thermal light source (§ 7).

Neglecting for a moment relaxation effects, preheating, and pre-excitation, and assuming that there is an equilibrium state behind the shock front, then the usual conservation equations for a plane shock wave propagating into, e.g., a helium gas initially at rest, are

$$N(V - u) = N_0 V \tag{1}$$

$$NM(V - u)^2 + P = N_0 MV^2 + P_0 \tag{2}$$

$$\tfrac{1}{2} M(V - u)^2 + H = \tfrac{1}{2} MV^2 + H_0, \tag{3}$$

where $N = N^0 + N^+ + N^{++}$ is the (number) density behind the shock front and N_0 the ambient density, V the shock velocity, u the flow velocity, M the helium atomic mass, P the pressure, and H the enthalpy per atom (neutral, singly or doubly ionized) behind the shock. P_0 and H_0 are the corresponding quantities ahead of the shock. The pressure is

$$P = kT(N^0 + 2N^+ + 3N^{++}), \tag{4}$$

and the enthalpy is given by

$$H = \tfrac{5}{2}(1 + \alpha + 2\rho)kT + (1 - \alpha - \rho)\bar{E}_0 + \alpha(\bar{E}_+ + \chi^0) + \rho\chi^+, \tag{5}$$

where $\alpha \equiv N^+/N =$ degree of single ionization, and $\rho \equiv N^{++}/N =$ degree of double ionization. χ^0 and χ^+ are the ionization energies of the neutral and singly ionized atoms.

The excitation energies of the neutral atom and the singly ionized atom are given by the thermal averages, with a suitable cutoff (e.g., Unsöld, 1948) which introduces a slight density dependence.

$$\bar{E}_0 = \frac{\sum_n \chi_n^0 g_n^0 \exp(-\chi_n^0/kT)}{\sum_n g_n^0 \exp(-\chi_n^0/kT)}, \tag{6a}$$

$$\bar{E}_+ = \frac{\sum_n \chi_n^+ g_n^+ \exp(-\chi_n^+/kT)}{\sum_n g_n^+ \exp(-\chi_n^+/kT)}. \tag{6b}$$

These equations have been solved numerically, e.g., for argon (Resler *et al.*, 1952), helium (Faneuff *et al.*, 1958; Seay, 1957), hydrogen (Turner, 1958, 1959), and for the rare gases at low degrees of ionization (Niblett and Kenny, 1957) (see Fig. 1 which is reproduced from that paper).

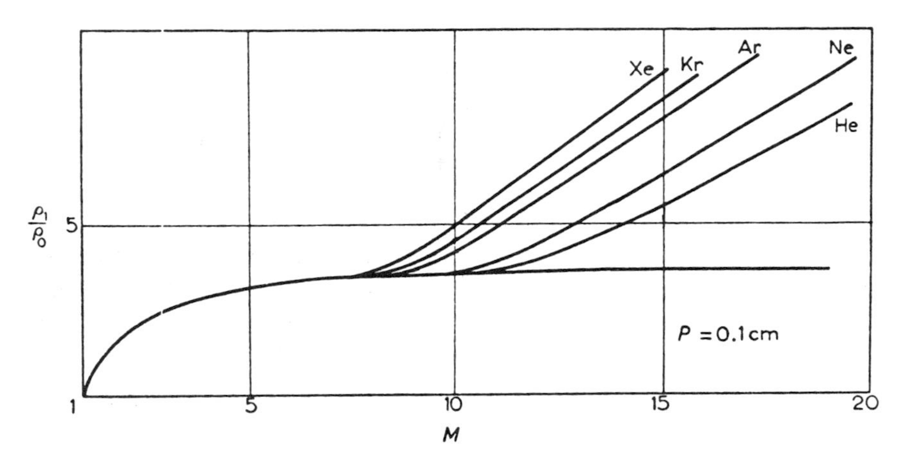

Fig. 1. The density ratio ρ_1/ρ_0 as a function of Mach number for rare gases. $T_0 = 296°K$ and $P_0 = 1$ mm Hg. [Reprinted from Niblett and Kenny (1957).] Ionization results in density ratios greater than 4.

For partial single ionization behind the shock and no ionization ahead, the density ratio is given by $[(4P/P_0) + 1]/[(P/P_0) + 4 - 2\alpha\chi^0/kT]$, and it is seen that for large pressure ratios and low degree of ionization the limiting density ratio is 4. When ionization is important, the density ratio across the front can exceed 4, which is the maximum value for an equilibrated gas with only translational degrees of freedom. Also, the temperature does not rise as steeply as for an ideal gas because energy has to go into excitation, ionization and dissociation in the case of molecular gases.

2.2 Strong Shock Approximation

The situation is further simplified (strong shock limit) if the temperatures and densities behind the shock wave are so large that one may neglect the pressure and enthalpy ahead of the shock front, i.e., in the undisturbed gas. This presupposes that the gas ahead of the shock front has not been preheated (for example, by precursor radiation, as discussed in § 5.3).

If single ionization is nearly complete ($\alpha = 1$) but the number of doubly ionized atoms is small ($\rho \ll 1$), then one obtains with the strong shock approximation, i.e., neglecting H_0 and P_0 and also neglecting the excitation energies

$$V = \left(\frac{2kT}{M}\right)^{1/2} \frac{4 + \frac{\chi^0}{kT}}{\left(3 + \frac{\chi^0}{kT}\right)^{1/2}} \cdot \left[1 + \frac{3 + \frac{1}{2}\frac{\chi^0\chi^+}{(kT)^2} + \frac{\chi^+}{kT} + \frac{5}{4}\frac{\chi^0}{kT}}{\left(4 + \frac{\chi^0}{kT}\right)\left(3 + \frac{\chi^0}{kT}\right)}\rho\right] \tag{7}$$

and

$$\frac{N^+}{N_0} = \left(4 + \frac{\chi^0}{kT}\right)\left(1 + \frac{\frac{\chi^+}{kT} - \frac{3}{2}\frac{\chi^0}{kT} - 4}{4 + \frac{\chi^0}{kT}}\rho\right). \tag{8}$$

The degree of second ionization is determined by Saha's equation (see § 7.2) with $N_e \approx N^+$ and T given by (7) with $\rho = 0$. The correction for second ionization in (7) results in a $\sim 10\ \%$ increase in the shock velocity for temperatures of $\sim$4 ev, whereas the correction

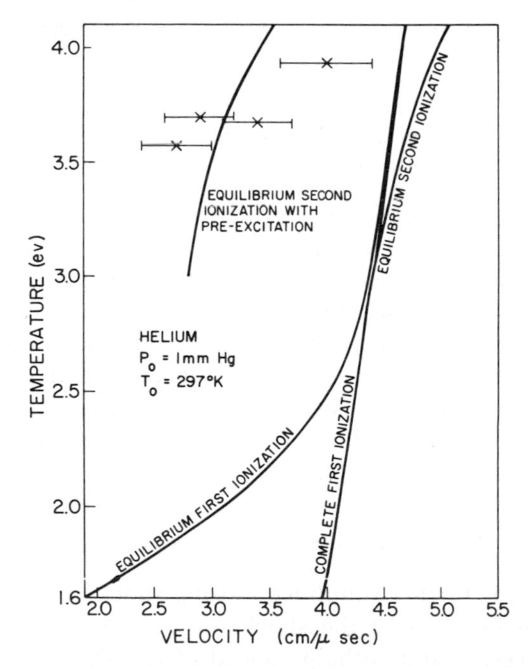

FIG. 2. Temperature vs. velocity obtained from modified Rankine-Hugoniot equations calculated in the strong shock approximation except for the curve labeled "equilibrium first ionization" in which the enthalpy and pressure ahead of the front were taken into account. In the curve labeled "equilibrium second ionization," first ionization is complete. The points refer to experimental determinations. [Reprinted from McLean *et al.* (1960).]

term containing ρ in (8) turns out to be completely negligible. The electron density follows from (8) (without the ρ correction term) and the quasi-neutrality condition.

$$N_e = [4 + (\chi^0/kT)]\,(1 + 2\rho)\,N_0. \tag{9}$$

Figure 2 shows several temperature-velocity curves for helium obtained from the strong shock approximation with and without including the effect of double ionization. Also shown are the results of earlier Rankine-Hugoniot numerical calculations by Faneuff *et al.* (1958) where the strong shock approximation was not used, but the effects due to single ionization were included. Second ionization and the excitation energy were neglected in these calculations. The experimental points refer to experiments (McLean *et al.*, 1960) and the solid curve takes into account the preheating due to precursor radiation (§ 5.3). Temperatures behind the incident and reflected shock waves in hydrogen and deuterium calculated in the strong shock approximation are shown in Fig. 3 (Kolb, 1957a, b).

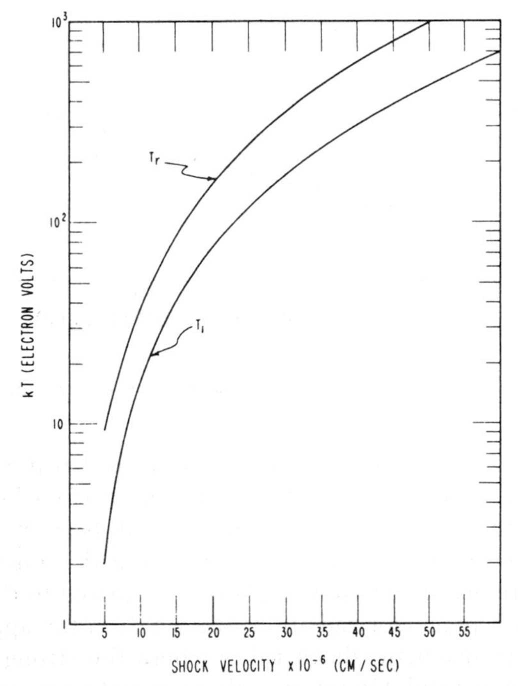

FIG. 3. (a) Temperature behind incident T_i and reflected T_r shocks versus shock velocity for deuterium in the strong shock approximation. [Reprinted from Kolb (1957a).] Pre-excitation of the gas ahead of the front is neglected.

FIG. 3 (b) follows.

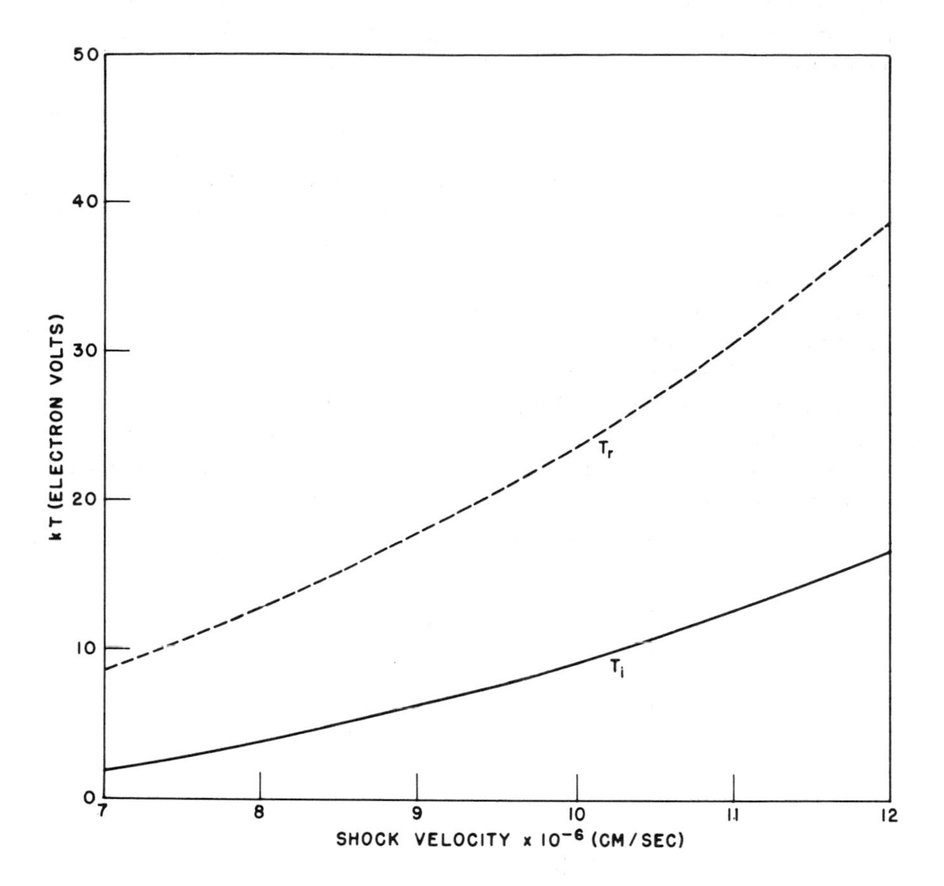

FIG. 3 (b) - Hydrogen. See FIG 3 (a) caption on preceding page.

For the range of temperatures and electron densities normally encountered, the terms in the enthalpy equation (5) which involve the excitation energies $\bar{E}_0$ and $\bar{E}_+$ are usually small (see estimates in Wiese *et al.*, 1960) compared to the terms containing the ionization energies. They are, therefore, completely negligible compared to the uncertainties (connected with precursor radiation, see § 5.3) in the application of the shock theory to magnetic shock tubes where the strong shock approximation should be valid. However, with increased measurement precision these terms should be considered for application to conventional shock tubes, where the usual Rankine-Hugoniot relations seem to apply in many cases (§ 6).

One might also expect for magnetic shock tubes that the usual shock equations, as given above, have validity in the limit of high densities where the mean free path for precursor radiation (§ 5.3) is sufficiently short so that the gas ahead of the front is not pre-excited. This is the situation for strong shock waves propagating in helium or hydrogen at initial pressures of the order 10 mm Hg or greater (Elton, 1961).

In Elton's magnetic shock tube experiments, the electron density is measured to be as high as 10^{19} cm^{-3} behind the reflected shock. The continuum radiation is used for the density determination which agrees within the experimental error with that calculated from the Rankine-Hugoniot relations. At these high densities, the ion-ion relaxation time is of the order 10^{-12} sec, and the electron relaxation time is still shorter. Here, we have a clear case in which one can discuss the shock phenomena from equilibrium considerations.

At lower densities, of the order 10^{17} ions/cm^3 in helium (McLean *et al.*, 1960) and also in hydrogen (Wiese *et al.*, 1960, 1961), it has been shown that an equilibrium state is established in times of the order of 10^{-7} sec. However, here the influence of precursor radiation seems to be of paramount importance since the mean free paths for the resonance lines of He II or atomic hydrogen are comparable to the dimensions of the shock tube. The shock temperatures here were of the order 40,000° K (in helium) or 20,000° K (in hydrogen), and the duration of the high temperature flow was about 1 μsec.

These various experiments seem to show that one can use high-velocity magnetically driven shock waves to establish a plasma whose state can be described from equilibrium considerations and for which the hydrodynamic equations are applicable when the effects of precursor radiation and stray magnetic fields are taken into account properly. The situation concerning the state of a plasma produced by a high-velocity imploding cylindrical shock wave moving into a partially ionized gas containing a magnetic field is much more complicated (see § 4). Here the microscopic processes leading to equilibration must be considered in detail, and much work remains to be done in order to understand the heating and ionization processes for strong shock waves having high current densities, low electron densities, and intermediate relaxation times.

3 *Plane Shock Waves*

3.1 Conventional Shock Tubes

Shock tubes used for the production of high-temperature gases divide themselves naturally into three classes: (1) conventional (energy source is a high-pressure chamber), (2) explosive driven, and (3) electrical. Conventional shock tubes were first used in 1951 to study self-luminous shock waves at Cornell (Resler *et al.*, 1952) and the University of Michigan (Hollyer *et al.*, 1952, 1953; Turner, 1956). In these tubes the high-pressure chamber is separated from a low-pressure expansion tube by a diaphragm (Fig. 4). When the diaphragm is broken, the expansion of the high-pressure gas generates a shock wave in the low-pressure gas. The shock heated gas is, in turn, separated from the driver gas by a contact surface moving with the flow velocity. Relatively large volumes of gas (several liters) can be heated to temperatures up to $\sim 15{,}000°$ K and pressures of ~ 0.1 to 10 atm. The time scale in these experiments generally lies between 10 and 1000 μsec, depending on the shock tube dimensions, temperatures, etc. A detailed review of the theory and operation of conventional shock tubes has been prepared by Glass and Hall (1959). The following comments serve only to point out some of the essential features and our attention in later sections will be mainly concerned with shocks sufficiently strong to produce ionization.

The parameters of the Michigan shock tube are typical (Turner, 1956). It has a cross section of $1\frac{5}{8} \times 2\frac{5}{8}$ in., the expansion chamber is 8 ft long, the high pressure chamber is 4 ft long and is operated with hydrogen at 400 to 700 psi, although pressures up to 10,000 psi have been utilized elsewhere and experiments with 30,000 psi heated-hydrogen drivers are now being prepared at the Cornell Aeronautical Laboratory (Treanor, 1961). With 1 cm Hg of neon in the expansion tube, shock waves with velocities of ~ 2.5 to 3.5 mm/μsec are obtained which traverse the tube in about a millisecond. The shock wave is reflected off the end of the tube with ~ 0.5 the primary velocity. The reflected wave moves into compressed neon, which has a flow speed of ~ 0.75 the primary velocity. The gas behind the reflected wave is brought nearly to rest and is further compressed and heated to temperatures between 8000° and 15,000° K. This gas is now being used for *f*-number and line broadening studies.

These wave motions can be photographed with rotating drum or rotating-mirror streak cameras because of the luminosity of the gases.

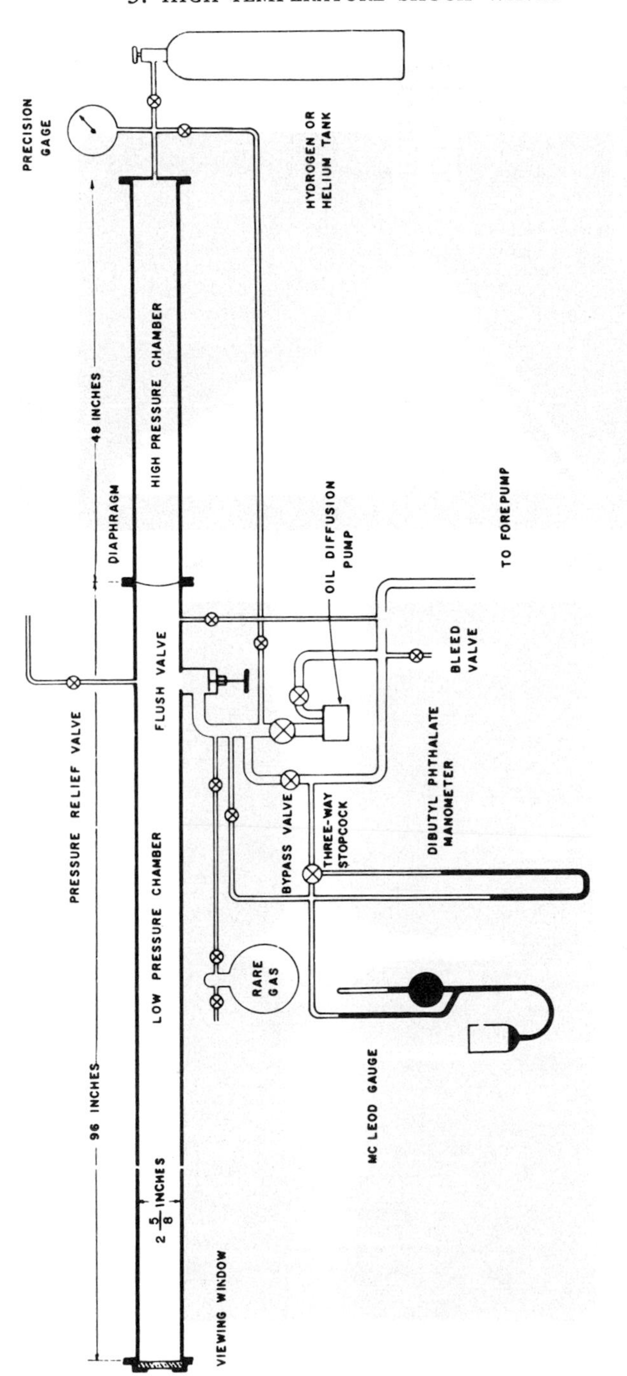

FIG. 4. Schematic of Michigan luminous shock tube. [Reprinted from Turner (1956).]

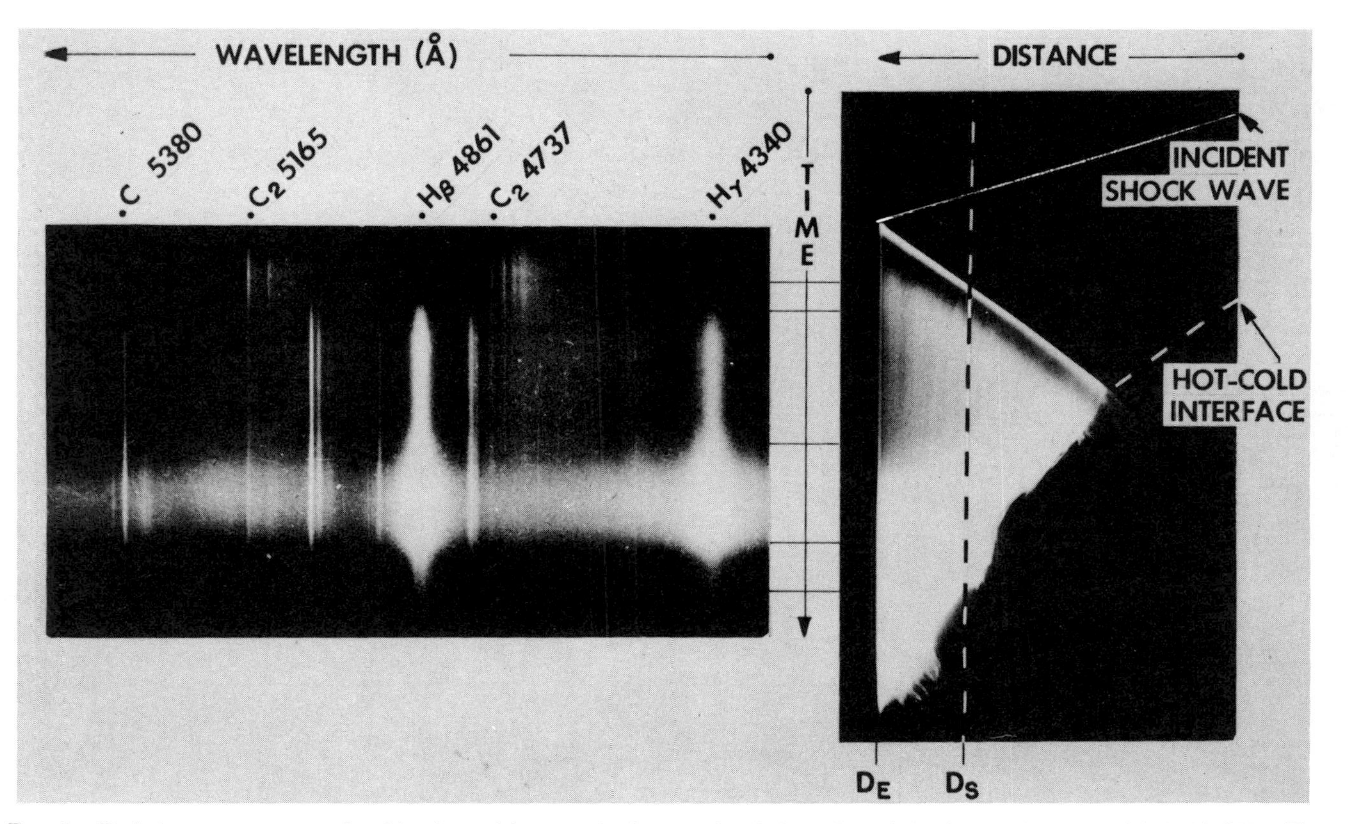

FIG. 5. Emission spectrum correlated in time with a streak photograph of the reflected shock wave in neon with 0.1% CH_4. The Balmer lines have a steady width and intensity for ~50 μsec, and the relaxation zone can be seen behind the reflected wave. The incident shock wave is visible on the original negatives and is drawn in this reproduction. [Reprinted from Laporte and Wilkerson (1960).]

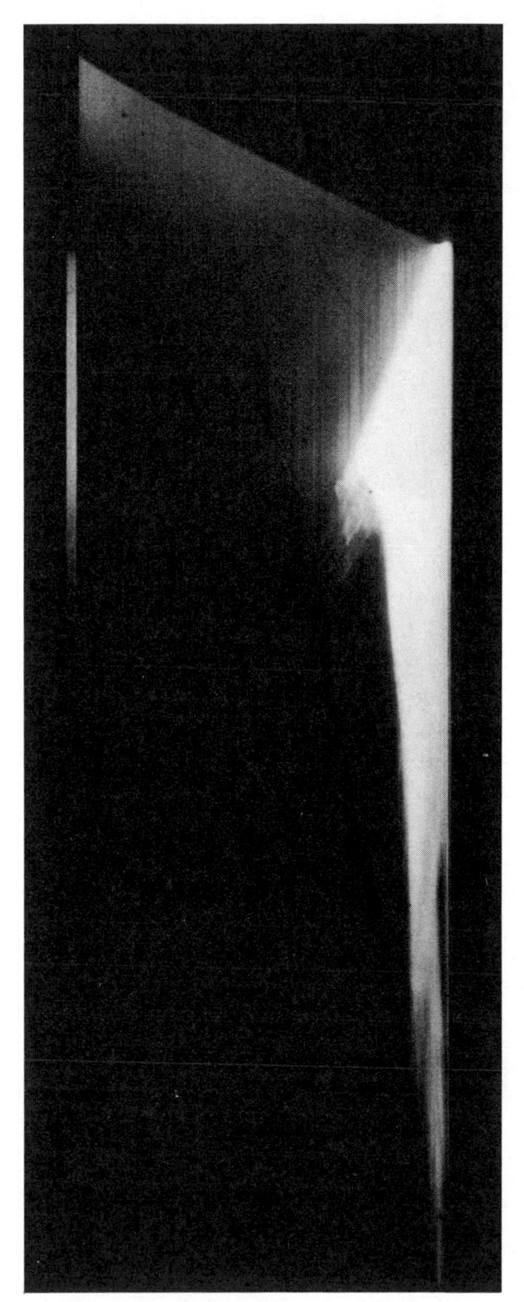

FIG. 6. Incident and reflected shock waves in xenon showing the delayed luminosity behind the incident shock front. Time proceeds from right to left. The reflected shock luminosity appears before the arrival of the incident *luminous* front. The incident shock wave is not visible in the streak photograph. [Courtesy of E. B. Turner.]

A typical streak photograph, correlated with a time-resolved spectrum, shows the essential features, Fig. 5. The radiation from the front of the incident shock wave consists of the bands of C_2 and CN molecules which are formed (Turner, 1955a, b; Rosa, 1955) from organic vapors in the tube. This luminosity sharply defines the front, and precise velocity measurements can be obtained from such photographs. In the Michigan studies with pure neon, these bands are not observed unless CH_4 (normally $\sim 0.1\ \%$) is added. The impurity concentration in these particular experiments has a drastic effect on the observed relaxation times (Greene, 1954; Fairbairn and Gaydon, 1955, 1957; Charatis *et al.*, 1957; Charatis and Wilkerson, 1958).

Behind the reflected shock, there is a dark zone which corresponds to the time for ionization relaxation. This is also particularly evident behind the primary shock wave in xenon, where the continuous radiation is very strong (Fig. 6) (Turner, 1956; Gloersen, 1960). Following multiple

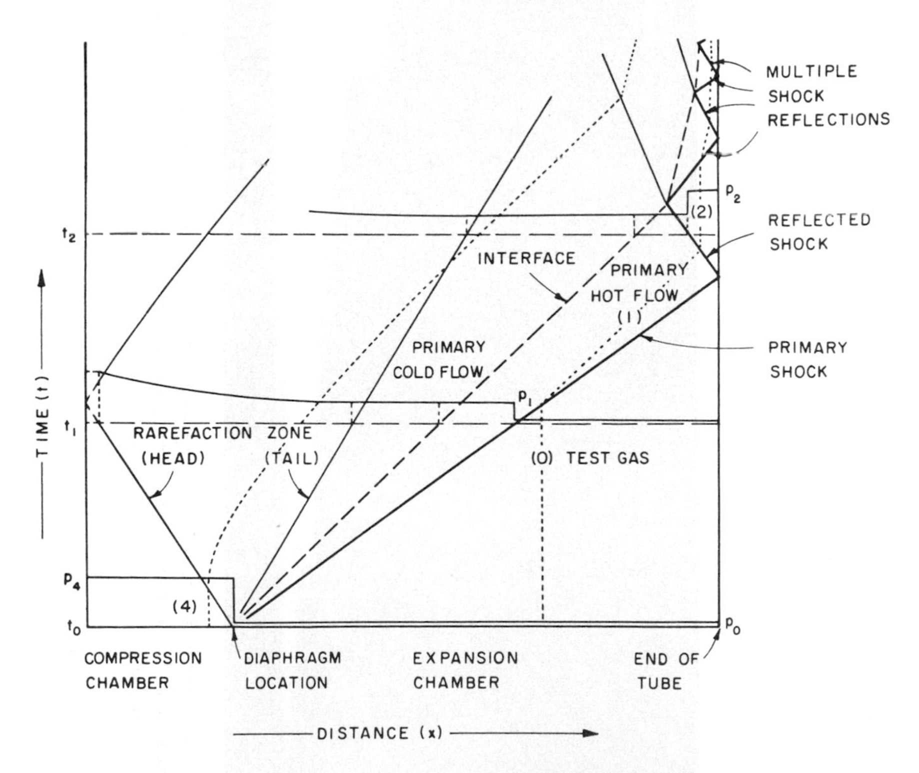

FIG. 7. Motion of shock waves and gas particles according to the ideal theory for a one-dimensional shock tube as described in the text. ρ, p, and u denote density, pressure, and flow velocity, respectively. [Reprinted from Laporte and Wilkerson (1960).]

shock interactions with the contact surface, the high-temperature emission is quenched by the cold hydrogen driver gas.

The conventional shock tube characteristics are summarized in an $x - t$ diagram, Fig. 7. The x-axis represents the length of the shock tube. The motion of a fluid element is shown by the dotted line. The fluid element is overtaken by the primary shock and is brought to rest due to the reflection of the shock at the end. The expanding driver gas is to the left of the interface line. These curves are calculated from ideal theory which neglects boundary layer effects, radiation losses, real gas effects, etc. In practice, the flow lines are not straight and there can be rather large departures from this ideal behavior as discussed in subsequent sections.

3.2 Explosively Driven Shock Tubes

One limitation of conventional shock tubes is that they are only efficient when the ratio of the sound speeds of the driving and driven gas is large. Solid or gaseous (Shreffler and Christian, 1954; Christian *et al.*, 1955; Christian and Yarger, 1955; Seay, 1957) explosives can be used to increase this ratio and produce temperatures of 10,000° to 20,000° at much higher pressures than in conventional tubes. By this method Seay was able to excite the helium spectrum at calculated temperatures of $\sim$ 20,000° K and investigate the broadening of the spectral lines by the Stark effect of surrounding ions and electrons (see Baranger, Chapter 13, and § 7.4). In this experiment, the driver gas and diaphragm were replaced by a block of high explosive which provided the necessary shock strength. The generation of such strong waves in helium with conventional tubes is difficult because of the low atomic weight (high sound speed) of helium and the relatively low sound speed in the driver gas, even with hydrogen. Direct temperature measurements behind the primary and reflected shock waves in explosively driven tubes have not yet been reported, to our knowledge.

3.3 Electrical (Magnetic) Shock Tubes

There are a variety of electric shock tubes now in use for the production of shock-heated plasmas with temperatures above 20,000° K. Mach numbers in the 20 to 200 range can be produced in these tubes. The energy and momentum input into the gas is due to Ohmic dissipation of high pulsed currents and the action of Lorentz forces that are set up by plasma currents and magnetic fields. The pressure associated with a magnetic field of, say, 100,000 gauss is about 1500 psi, and this plays the role of the high-pressure chamber-diaphragm combination or the high explosive discussed previously. We will refer to these tubes in general as magnetic shock tubes because the energy input due to the magnetic acceleration is generally larger than that due to resistive dissipation alone.

3.3.1 *T-Tubes*

Shock waves generated by striking a discharge between two electrodes at one end of a T-shaped tube, with the subsequent expansion of the Ohmic heated gas into the side arm, were first studied by Fowler *et al.* (1951, 1952a, b). Temperatures up to $\sim 30{,}000°$ K could be produced in this way. To generate higher energy plasmas with the T-tube, one uses a current-return backstrap which is perpendicular to the side arm and parallel to the gas-current path (see Fig. 8). This produces a rapidly

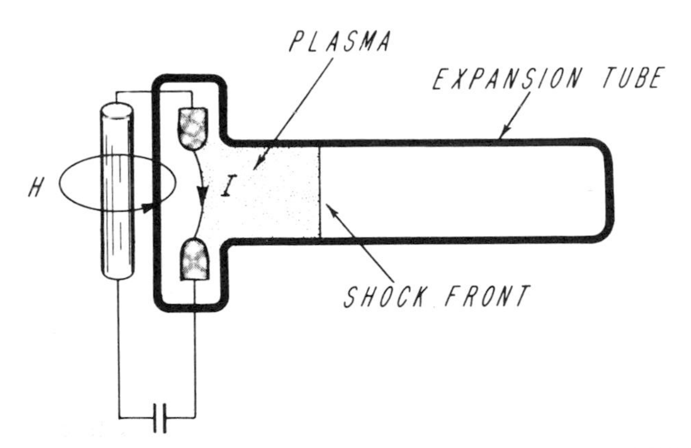

FIG. 8. Magnetic shock tube (T-tube). [Fig. 2 of Kolb (1960b).]

rising magnetic field and $\mathbf{J} \times \mathbf{H}$ force which further accelerates the plasma. With high voltage, and low inductance circuits, a deuterium plasma has been driven with velocities up to ~ 15 cm/μsec, corresponding to ion energies in the 100-ev range (Kolb, 1957a, b).

Following the initial experiments which were designed to demonstrate the feasibility of using magnetic acceleration of plasmas to produce high Mach number flows, these tubes have been used (McLean *et al.*, 1960; Wiese *et al.*, 1960, 1961; Kolb, 1960a, b, 1961) for the production and spectroscopic study of equilibrated plasmas at high densities ($\sim 10^{17}$ electrons/cm^3) with temperatures of $\sim 40{,}000°$ K in helium and $\sim 20{,}000°$ K in hydrogen. The object of these particular experiments was to prove by purely spectroscopic means that the shock-heated plasma is in local thermodynamic equilibrium, LTE (see § 5). Experiments are now in progress which utilize the equilibrated helium and hydrogen plasmas produced by a T-tube as a radiation source for f-number and collision broadening studies.

A typical streak camera photograph of a magnetically driven plasma is shown in Fig. 9. The self-luminous shock front is clearly visible, with

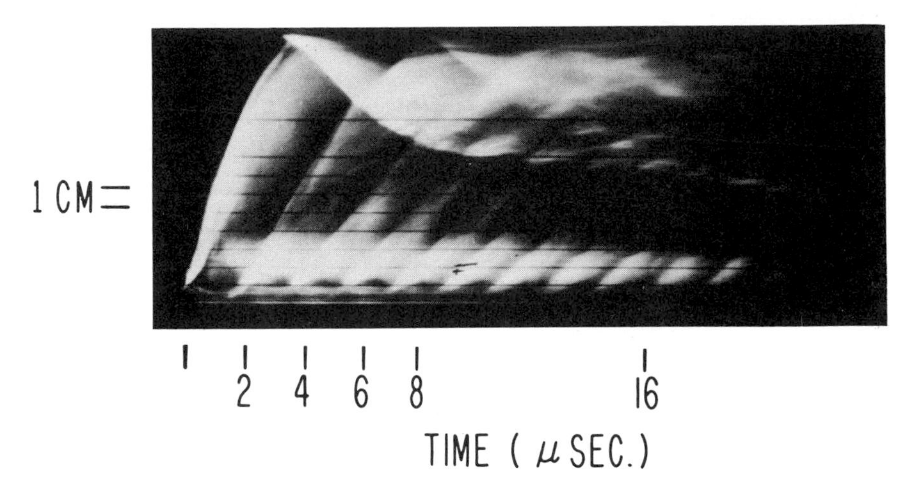

FIG. 9. Typical streak camera photograph of a magnetically driven shock wave in a T-tube (28 kv, 0.8 μf, 500 kc/sec, 3.6 mm Hg hydrogen). [Reprinted from Kolb (1957a).]

the wave reflected off the end of the tube appearing at the top of the photograph. The successive shocks which follow the primary shock are due to the ringing of the electrical LC circuit. The slope of the luminous front increases at first due to the magnetic acceleration and then decreases rapidly due to radiation and wall cooling. The slower rate of attenuation far up the tube is characteristic of a blast wave (Harris, 1956) since the energy is deposited in a narrow (compared to the length of the tube) slab of gas near the electrodes in a time short compared to the transit time of the shock along the tube. A relaxation zone is observed behind the incident shock wave only near the end of the tube.

3.3.2 *Conical Pinch Tubes*

The hydromagnetic implosion of a plasma in a conical pinch tube (Fig. 10) has been used to produce a hydrodynamic flow through a hollow electrode (Scott *et al.*, 1958; Josephson, 1958). Shock velocities up to 12 to 14 cm/μsec in 0.1 mm Hg deuterium have been reported by Scott and Wentzel (1959) and later by Josephson and Hales (1961) who made a parametric study of factors involved in the design of conical

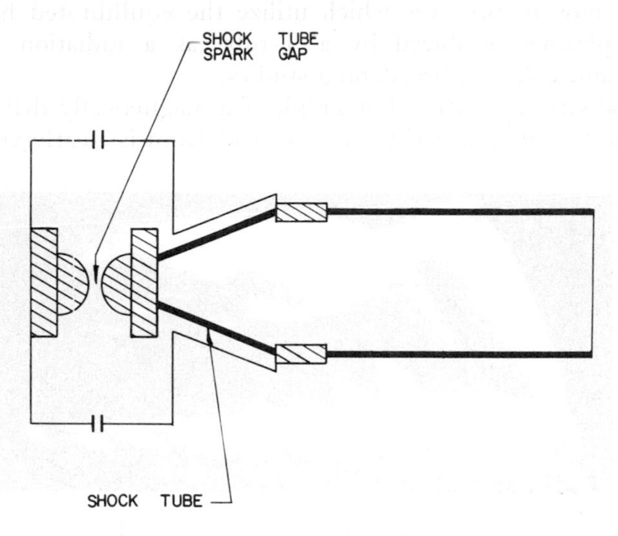

FIG. 10. Schematic of the conical shock tube with electrodes. The B_z coils are used to study the propagation of a shock wave in a magnetic channel. [Photograph provided by F. R. Scott.]

shock tubes. There is a pronounced dependence of the shock velocity on the angle of the cone. This is to be expected (Josephson, 1958) because the pinching occurs at the small end first and produces a plasma which heats the neigboring gas before it is pinched, which then requires a higher current for pinching. This produces an axial flow out the hollow electrode. Scott and Wentzel (1959) have improved the reproducibility and planarity of the shock front by reducing the asymmetry of the current distribution in the tube caused by the erratic breakdown from the solid electrode to a preferred position on the circumference of the hollow electrode. This was accomplished by breaking the return conductor into 8 spiral conductors (45°), which effectively smoothed out the potential distribution and improved the azimuthal symmetry. With this arrangement the shock velocities, for a given capacitor bank energy, were reduced by a factor of 2, but were much more reproducible. It was also found that a magnetic multipole propagated with the shock wave, presumably due to a spiral instability in the conical discharge. However, Scott and Wentzel (private communication from Scott) find that the observed magnetic multipole acts as a coupled piston driving the shock for approximately 20 cm beyond the cone exit, thus extending the region of constant velocity but complicating any analysis of the shocked region (because of the magnetic field). To avoid this they added a coarse (1 cm sq) mesh normal to the axis which "filtered" the multipole. Then the subsequent velocity vs. axial distance corresponded to the blast wave approximation of Harris (1956). Josephson and Hales (1961) also report recently that they achieve the same effect as Scott and Wentzel, i.e., reproducible and uniform breakdown of the gas, without a decrease in shock velocity by use of a 20-turn coil wrapped around the conical tube and connected to the two electrodes. Prior to gas breakdown, the coil current produces a E_θ field which increases the path of the free electrons and the probability of an ionizing collision. Subsequent to breakdown, the current flows mainly through the gas because of its lower impedance. Since breakdown occurs very fast ($\langle 1$ μsec down to pressures of 25 microns D_2), very little energy is wasted in the coil.

One disadvantage of these tubes is that contaminants are introduced into the flow from electrode erosion. Josephson (1958) reports that the contamination can be reduced by coupling the energy inductively by placing coils around the tube (Fig. 11), but that the shock velocity is

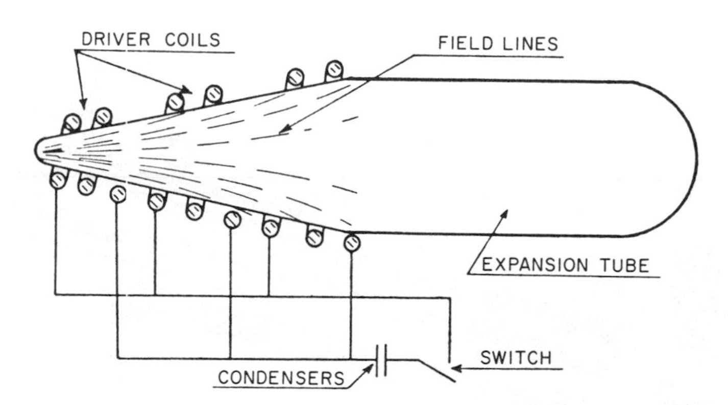

FIG. 11. Schematic of electrodeless conical shock tube after Fig. 7 in Josephson (1958).

$\sim 60\ \%$ of that in the electrode system, using the same energy source. However, later studies by Josephson (private communication) also indicate that when using Pyrex or quartz tubes the major impurities come from the tube walls rather than from the electrodes. It appears, therefore, that from the standpoint of degree of contamination, further quantitative work needs to be done to compare the relative merits of electrode systems and inductively coupled drivers for producing shock waves. Detailed measurements of the temperature, electron density, and impurity concentrations have not been reported, so that a comparison with hydrodynamic theory cannot be made, or the degree of equilibration determined. One advantage of the conical tubes over the T-tubes with fast rising currents is that very little attenuation of the velocity occurs [see Fig. 2 of Scott and Wentzel (1959) and Fig. 4 of Josephson (1958)]. Whether or not this means that there is a steady electron density and temperature behind the front is not yet known, especially when there are stray magnetic fields. This is also the case with T-tubes operated at higher pressures (10-50 mm Hg) with low-voltage ($\sim$ 5 kv) and low-frequency ($\sim$ 100 kc) circuits for magnetic driving (Elton, 1961).

3.3.3 *Magnetic Annular Shock Tubes*

High-speed shock waves with velocities of $\sim 4 \times 10^7$ cm/sec at initial hydrogen pressures of 0.03 mm Hg have been reported (Patrick, 1959) and described theoretically (Kemp and Petschek, 1959). In these shock tubes the gas is confined in an annular region between two coaxial cylinders whose radii are large compared to their annular spacing (Fig. 12). An essentially constant magnetic field is first established

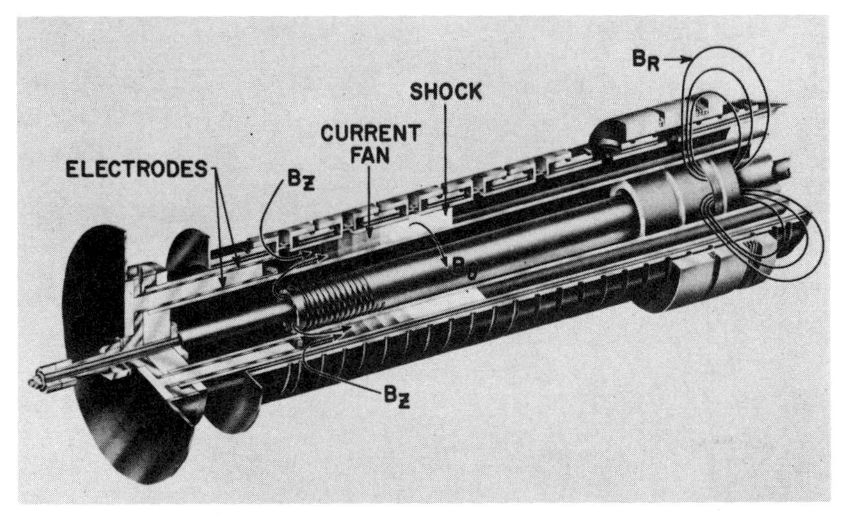

FIG. 12. Annular shock tube for generating high-velocity plasmas for studies of the shock front. Photograph provided by R. M. Patrick, Avco-Everett Research Laboratory.

along the annulus. This aids in the breakdown of the preionized gas when a capacitor bank is discharged across electrodes placed on the cylinder at one end, which causes the plasma to spin with a high velocity.

The preionization is carried out with a low current ($\sim$ 10 amp) high-frequency ($\sim$ 1Mc) discharge across the electrodes.

After the discharge of the main capacitor bank, the plasma is accelerated along the tube by the axial force produced by the interaction of the radial currents with B_θ fields produced by these currents.

The axial bias field also serves to inhibit the radial motion of the shocked gas and reduces the heat conduction to the walls. Furthermore, two concentric cylinders of brass around the annulus prevent the bias field from leaking out during the passage of the shock wave so that it serves to contain the heated plasma. This tube has the advantage that there is a stable acceleration over several microseconds, resulting in substantial flow speeds (here, up to 20 cm/μsec). To increase the velocity to 40 cm/μsec, an azimuthal bias field (in a plane parallel to the shock front) was necessary. These high speeds were observed only if the ion cyclotron radius of heated ions was small compared to the annulus spacing.

The object of these experiments was to study the structure of a shock front when the plasma energy density is comparable to the magnetic energy density, the ion mean free path is large compared to the ion cyclotron radius, and the ion cyclotron radius is small compared to the width of the annulus. The experiments were designed to determine the fundamental question of whether or not in a magnetic field a shock wave could be produced whose thickness was less than the collisional mean free path. The shock thickness was measured from the observation of the hydrogen continuum [only free-free transitions were taken into account in comparing the theory (§ 7.2) with experiment], i.e., the shock velocity and rise time of the radiation defined the shock thickness. This thickness was observed to be less than the mean free path for temperatures calculated from the shock velocity in the initial pressure range $\sim$ 30-60μ. At 30μ, the thickness was about 0.5 the mean free path and 0.1 that calculated according to Mott-Smith (1951). There still seems to be some question about these conclusions since the mean free path is determined from the observed radiation which is a function of the electron density and temperature. Temperatures of 2.5×10^5 to 10^6 °K are calculated from the measured velocities; but whether or not the electrons reach the ion temperature for high velocities and low densities is open to doubt (see the discussion in § 4.6).

This is a critical question since the calculated mean free path is proportional to T^2. A factor of 2 or 3 error in the electron temperature would reduce the electron mean free path to near the observed values. The decision about the relevant dissipative mechanism must await additional measurements which corroborate the calculated temperatures and confirm the continuum analysis which depends on N_e and T_e.

At pressures above 80μ, the calculated shock thickness (Mott-Smith, 1951) is about one-third the measured value and the experimental points have a scatter of about a factor of 1.5 to 2. At the higher pressures and slower shock speeds there could be precursor radiation (see § 5.3) which pre-excites and ionizes the hydrogen. This has already been observed in T-tubes (Wiese *et al.*, 1961) where the resulting temperature in hydrogen is higher than calculated from the usual hydrodynamic considerations. If this were the case, the calculated mean free path could well be too small by a factor up to 4.

Clearly, the whole subject of factors which govern the structure of shock fronts is still in a controversial stage, and further discussions of the difficulties follow in other contexts.

4 *High-Energy Cylindrical Shock Waves*

4.1 Special Problems in Shock Heating of High - Energy Plasmas

In the preceding sections we have been mainly concerned with the physical properties of equilibrium plasmas and with some experimental techniques for their generation and diagnostics. However, for temperatures above 50,000° to 100,000° K the situation becomes much more complicated, especially for rarefied plasmas where the collision frequencies are low. This is so because at high temperatures the energy losses due to radiation and heat conduction can proceed at a rate which may preclude the possibility of establishing an equilibrated plasma by shock waves. One also has the troublesome difficulty connected with the liberation of impurities at the walls, so that the composition of the plasma is, in general, not known precisely. Furthermore, depending on the density and kinetic temperature, the relaxation time for translational equilibrium can be quite long so that the ions and electrons may have different "temperatures."

In order to study high-energy shock waves experimentally, it is desirable to isolate the plasma from surrounding walls. One possible method for accomplishing this with one-dimensional shock waves is to surround the expansion chamber of a T-tube (Fig. 13) with coils to produce an axial magnetic field to compress the plasma (Kolb, 1957c, 1959b, 1960a).

A shock wave is generated in the T-tube and the magnetic field is turned on after the shock enters the coil array. The resultant magnetic pressure drives the plasma radially inward so that the shock-heated plasma propagates in a magnetic channel. With axial fields of 15,000

FIG. 13. T-tube and parallel single-turn coils for generating a pulsed axial magnetic field to isolate the plasma from the tube walls. [Reprinted from Kolb (1960a).]

gauss, kinetic temperatures of $\sim 7 \times 10^5$ °K were calculated from measured shock velocities. However, this technique has the disadvantage that the shock front becomes curved [similar experiments and observations have been carried out by Scott *et al.* (1958) using a conical pinch tube to produce the initial shock wave], and the timing of the axial field relative to the switching of the T-tube discharge is critical because of the high velocities (8-20 cm/μsec in deuterium). The main difficulty is that, even with reproducibility and complete isolation from the walls, the flow field behind the "one-dimensional" shock is really three-dimensional because the radial diffusion of the magnetic field and radial compression behave differently along the axis of the tube because of axial temperature and density gradients.

4.2 Magnetic Acceleration of Cylindrical Shock Waves†

A basically simpler method for studying high-energy (the word temperature needs qualification in this context) shock-heated plasmas is to implode a preionized gas by means of a rapidly rising external magnetic field in a cylindrical geometry. This results in an essentially two-dimensional flow field and the implosion phenomena can be made reproducible from shot to shot (Hintz, 1961), as determined by magnetic probe measurements.

Very little is known experimentally about shock waves with kinetic temperatures in excess of 100,000° K so that the ensuing discussion is more or less speculative and is intended as a prognosis of possible future developments. High-energy plasmas are of interest in many areas of astrophysics, e.g., the solar corona, and controlled fusion research, and afford a distinct possibility for the eventual determination of excitation and ionization cross sections of highly ionized species, e.g., O VIII, Ne X have already been observed (Stratton *et al.*, 1960). This prospect is perhaps the most germane to the subject of this volume, but with careful experiments one can also check on the validity of the ordinary magnetohydrodynamic equations, the corona-formula, relaxation theories, the factors which influence the structure of the shock front, precursor radiation from a plasma where electrode radiation is not present, etc.

Schemes for using fast shock waves as a preheater for the eventual generation of plasma energies greater than 1 kev by subsequent (adiabatic) magnetic compression require a detailed knowledge of the initial implosion (shock) phase so that the initial conditions are known at the beginning of the slower compression. The state of the plasma produced in this fashion is now the subject of much controversial discussion (see for example, Kolb, 1960b; Green, 1960; Griem *et al.*, 1959a; Boyer *et al.*, 1960).

In spite of all the difficulties it is possible to estimate the implosion times and velocities with some accuracy using the simple "snowplow" model of Rosenbluth and Garwin (1954) which was first applied to the pinch effect (I_z current and H_θ field) and later (Kolb, 1959a, 1960b) to the solenoid configuration (I_θ current and H_z field). Because the latter field configuration seems to have the greater stability, we will confine our attention to it. The first experiments of this type were performed

† Attention is brought to several papers on this topic presented recently at the Fifth International Conference on Ionization Phenomena in Gases, Munich (1961) and the International Conference on Plasma Physics and Controlled Nuclear Fusion Research, Salzburg (1961) *Fusion J.* (in press).

by Colgate (1957) in the "collapse" experiment whose object was to produce energies so high that interparticle collisions were not important. At about the same time, experiments with a similar object were begun at AVCO (Janes and Patrick, 1958) (see also § 3.3.3). Other experiments are continuing at Los Alamos (Boyer *et al.*, 1960; Jahoda *et al.*, 1960; Nagle *et al.*, 1960), U. S. Naval Research Laboratory (Kolb *et al.*, 1960; Griem *et al.*, 1961b), A.W.R.E., England (Green, 1960), Jülich, Germany (Fay *et al.*, 1960), and Sukhumi, U.S.S.R. (Kvartskava *et al.*, 1960), to name but a few.

4.3 The Snowplow Model

The snowplow model (Rosenbluth and Garwin, 1954) is instructive because it provides scaling laws for the design of experiments. The basic assumption is that the initial ionization is sufficiently high so that induced currents flow in the plasma so as to exclude the externally applied magnetic field (infinite conductivity model). It is also assumed that all the gas is swept up by the magnetic "piston" in a thin shell whose inertial forces are balanced by the magnetic forces

$$\frac{d}{dt}\left[\mathrm{M_p}\,\frac{dR_\mathrm{p}}{dt}\right] = -\,2\pi\,R_\mathrm{p}\,\frac{H_z^2}{8\pi} \tag{10}$$

where $M_p = \pi\rho[R_0^2 - R_p^2]$ is the mass per unit length swept up at the time t when the plasma, whose initial density is ρ, has a radius R_p. R_0 is the initial radius, and $H_z = (4\pi/c)I_\theta/l$ is the field strength for a coil of length l with a current I_θ. On introducing the dimensionless radius $y = R_p/R_0$, we have

$$\frac{d}{dt}\left[(1-y^2)\,\frac{dy}{dt}\right] = -\,\frac{4\pi\,I_\theta^2}{\rho R_0^2\,c^2\,l^2}\,. \tag{11}$$

The gas (or coil) current I_θ can be found from

$$V = \frac{d}{dt}\,(LI_\theta) \tag{12}$$

where V is the instantaneous voltage around the plasma column, and L is the inductance connected with the volume $\pi(R_0^2 - R_\mathrm{p}^2)l$

$$L = \frac{4\pi}{c^2 l}\,(R_0^2 - R_\mathrm{p}^2), \tag{13}$$

so that

$$I_\theta = \frac{c^2 l}{4\pi}\frac{1}{(R_0^2 - R_p^2)}\int_0^t V(t')\,dt'$$

$$= \frac{c^2 l}{4\pi^2 R_0^2}\frac{1}{(1-y^2)}\int_0^t V(t')\,dt'. \tag{14}$$

The voltage $V(t)$ is found from the circuit equation

$$V(t) = V(0) - L_e\frac{dI_\theta}{dt} - \frac{1}{C}\int_0^t I_\theta\,dt' \tag{15}$$

where L_e is the external inductance and $V(0)$ the charging voltage. It is assumed that the capacitance is sufficiently high so that the charging voltage does not drop appreciably during the implosion; then one can neglect the term $\int_0^t I_\theta\,dt'/C$. This is generally valid for large capacitor

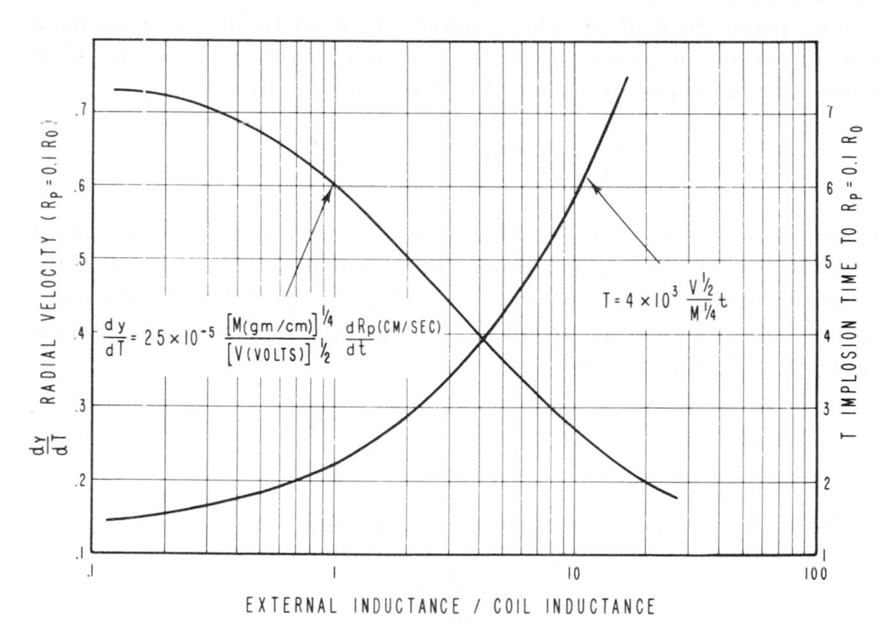

FIG. 14. "Snowplow" calculation of the shock velocity (dR_p/dt) in terms of a dimensionless velocity (dy/dT) when R_p is one-tenth the initial radius R_0. The implosion time t, in terms of a dimensionless time T, is also shown. The dependence of these quantities on the external inductance and the coil inductance is shown for the magnetic compression geometry. [Fig. 15 of Kolb (1960b).] Here T corresponds to τ in the text.

banks where the implosion time is short compared to the quarter period.

Combination of these equations gives finally

$$\left[\frac{d}{d\tau}(1-y^2)\frac{dy}{d\tau}\right] = \frac{-y\tau^2}{[1-y^2+L_{eo}]} \tag{16}$$

where we have introduced $L_{eo} = L_e/L_0$, the ratio of the external induct-

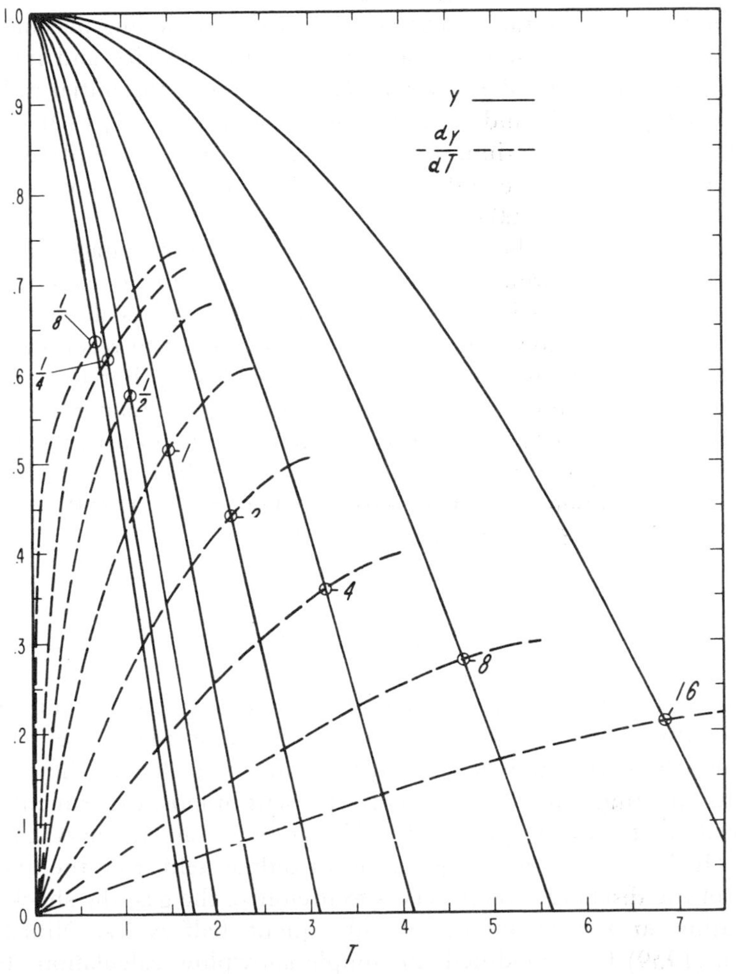

Fig. 15. Radius and velocity of the imploding plasma vs. time (dimensionless variables as discussed in text). [Reprinted from Kolb (1959a), Fig. 19.] The curves are drawn for different values of the parameter $L_{eo} \equiv L_e/L_o$.

ance to the inductance associated with the initial volume $\pi R_0^2 l$ and the characteristic time

$$\tau \equiv \left[\frac{4\pi^2}{Ml}\right]^{1/4} \left[\frac{V(0)}{L(0)}\right]^{1/2} t \tag{17}$$

where $M \equiv \pi \rho^2 l R_0^2$ is the mass of the gas initially present.

Expressed in this way, it is seen that the implosion time and velocity scale with $M^{1/4}$ and $V(0)^{1/2}$, so that the energy imparted to the plasma is proportional to the charging voltage for a given tube diameter. Numerical solutions for the radius and velocity of the imploding plasma versus time with L_{e0} as a parameter and also for the velocity and implosion time T as a function of L_{e0} when R_p is one-tenth the initial radius R_0 are shown in Figs. 14 and 15. For example, with $L_{e0} = 1$, $V(0) = 20$ kv, and for an initial deuterium pressure of 0.1 mm Hg, one finds an implosion time of ~ 0.1 μsec and a velocity of ~ 14 cm/μsec, corresponding to ion energies of ~ 200 ev.

It is evident from these rough considerations that voltages of several hundred kilovolts would be required (for the above conditions) to reach plasma energies in excess of 1 kev by shock waves alone. One can, of course, think of lowering the initial density so as to reach final velocities in deuterium in excess of 30 cm/μsec (1 kev), but the problems connected with forming a thin current sheath in an initially fully ionized, low-density plasma have not received enough attention experimentally to evaluate the possibilities. The highest radial velocity reported to date is 15 cm/μsec in deuterium with $V(0) = 80$ kv, $L_{e0} = 1$, $R(0) = 2.5$ cm and $P_0 = 0.08$ mm Hg (Elmore *et al.*, 1959).

4.4 Further Magnetohydrodynamic Calculations†

The snowplow model gives no information concerning ion and electron temperatures, shock thickness, etc. In the first place, we have not included here the influence of trapped magnetic fields (they tend to increase the implosion time) generally present in most experiments. This trapped field may be provided deliberately by an auxiliary capacitor bank (Kolb *et al.*, 1959) or appear as a residual field from the first half-cycle of the discharge which serves to preionize the gas, the shock waves appearing at the beginning of subsequent half-cycles. Niblett and Green (1959) have modified the simple snowplow calculations to take

† Much of the discussion in § 4.4 to § 4.6 follows from private communication with K. Roberts and K. Hain.

into account the work done in compressing internally trapped fields, but they assume that the mass of the plasma remains constant during the implosion. This assumption has been removed by Kever (1961) recently.

In spite of the rough agreement of the experimental velocities and implosion times with the preceding theory, much more elaborate calculations must be performed to study the influence of the initial temperature and degree of ionization on the shock wave, and to investigate the ohmic heating due to the dissipation of internal reverse fields or the penetration of external fields. The full magnetohydrodynamic equations for a cylindrically symmetric, fully ionized plasma, including electric and thermal conductivity have recently been coded for digital computers (Hain and Roberts, 1960; Hain *et al.*, 1961). The electrons and ions are assigned separate temperatures, with the usual collisional relaxation terms included in the energy equation. Measurements by Köppendörfer of the internal magnetic fields with probes during a stabilized pinch collapse show general agreement to within $\sim 10\ \%$ of the detailed numerical results, taking into account the difference between the electric conductivity perpendicular and parallel to the magnetic fields. In these calculations the electrons are assumed to be heated by Ohmic dissipation and the ions by the shock wave. Conservation laws at the front are satisfied by the von Neuman method of introducing an artificial viscosity in the momentum equation as discussed by Richtmeyer (1957). The initial good success of the comparison between the full nonlinear magnetohydrodynamic theory and the pinch experiments indicates that the theory has some validity under certain circumstances. Furthermore, since the radial distribution of the magnetic field depends on the radial distribution of the conductivity or electron temperature, the experiments tend to confirm the Spitzer conductivity formula (Cohen *et al.*, 1950) as well as the theoretically predicted rate of heating. Also, the close agreement of the time dependence of the fields with the theory indicates that the mass density (or inertial term) is correct, although this does not test the MHD theory in a critical way since even the simple snowplow model is sufficiently accurate to account for the field variation in dynamic pinches. It should now be possible to make more detailed comparisons with the theory by spectroscopic observations of local densities and temperatures.

Recently these calculations have been extended to the B_z pinch with a reverse trapped field. The calculated implosion time, densities, and heating rate are confirmed by experiment and indicate that the field dissipation is determined by the Spitzer resistivity (Hain and Kolb, 1961; Griem *et al.*, 1961b).

4.5 Shock Structure and Precursor Radiation

One of the biggest difficulties which remains in understanding the rapid collapse of a fully ionized plasma is connected with calculating the structure of the shock front and ionization processes in a more realistic manner. This problem is now being investigated by Roberts, Taylor, and Hain by introducing the plasma viscosity instead of the artificial von Neuman viscosity and by considering the ionization of residual neutrals by electron collisions and charge exchanging collisions between ions and neutrals. At this time, there is no clear understanding of the most important dissipative mechanisms which lead to the formation of pressure, density, and temperature jumps at the shock front. But by introducing the various possibilities into the computer code one can hope for the eventual prediction of the results of more and more refined experiments.

As discussed in § 5.2 and § 5.3, the ionization and excitation of neutrals by the propagation of precursor radiation ahead of the advancing shock front in an ordinary magnetic shock tube drastically influences the energy balance, temperatures and densities. It might be expected that for a cylindrical shock propagating in partially ionized, preheated hydrogen or deuterium (no molecules) precursor radiation will not play an important role because of the short mean free path for the atomic resonance lines. However, for very high flow speeds there will be a Doppler shift which shifts the maximum of the emission line away from the absorption maximum, and this tends to increase the mean free path so that volume pre-excitation might be possible.

4.6 Ion and Electron Heating

One might also ask whether or not the shock wave only heats the ions or whether one should take into account the space charge electric fields that are set up by ambipolar diffusion near the front. These fields are in such a direction as to decelerate ions and accelerate electrons as they pass through the advancing front. The problem is complicated still further if the shock front propagates into a magnetic field which leads to complicated trajectories for the charged particles, perhaps exciting plasma waves (Fishman *et al.*, 1960). An experimental approach is first to perform experiments at high plasma densities where the gas pressure is isotropic, the equation of state is known, and collisional dissipation dominates insofar as the shock structure is concerned. If there is any agreement with theoretical predictions for the structure in this regime, the density could be decreased and the velocities increased so as to reduce the importance of collisions, and then one could look for deviations from the magnetohydrodynamic predictions.

In most experimental situations one does not deal with an initial condition where the gas is fully ionized. There are always some impurities present and in general the preheating phases (by rf, low-energy capacitor discharges, etc.) yield a partially ionized gas. In this case, ionization by charge exchanging collisions can be as important as ionization by electron impact (Schlüter and Biermann, 1950, 1958; Cowling, 1956). The charge exchange mechanism provides a kind of high-energy neutral injection through the front and leads to a direct heating of the ion component of the plasma. As the shock wave passes with a high velocity over neutral gas ahead of the front, ions produced by charge exchange are reflected off the advancing sheath and gain momentum from the field, i.e., the injected fast neutral, which becomes a fast ion, is turned by the transverse magnetic field within a Larmor radius and is carried along with the imploding plasma. The energy gained directly is of the order of the flow energy. If it should turn out that the decelerating electric fields due to charge separation at the shock front are important and reduce the ion temperature, then it could be the case that higher ion temperatures will be produced as a result of the charge exchange mechanism for a shock wave moving into a partially ionized gas.

It is also important to have an estimate of the thickness of a shock front moving into a magnetic field because the current density, which governs the electron heating near the front, depends on the thickness. The relative importance of various heating mechanisms in a strong shock moving into a high magnetic field is completely open to question, even at high densities because of the short times involved in most experimental situations (Hain and Kolb, 1961).

4.7 Radiation Cooling

Another area that has been little investigated are the plasma effects caused by radiation cooling. At temperatures of 20 to 100 ev, rough estimates (Knorr, 1958; Post, 1960) of the radiation losses from a hydrogen plasma containing a few per cent impurities show that the electrons could radiate their internal energy in a few tenths of a microsecond. This radiation can be used to estimate local electron temperatures and impurity concentrations, but from the point of view of generating 20 to 100-ev shock waves in low z gases, it is clear that the high z impurity levels should be kept low. Possible exceptions are cases where the density is initially very low and the plasma is accelerated to a high velocity in a time short compared to the relaxation time for multiple ionization and significant radiation loss.

The radiation effects may be calculated by coupling rate equations for the ionization of various species to the equation of state and magneto-hydrodynamic equations, using the appropriate cross sections for ionization by electron collisions, radiative recombination (as in the corona formula) and three-body recombination (leading to Saha equilibrium at high densities). Eventually, one would hope to measure some of these cross sections involving highly ionized species from observations of the time history of the plasma radiation.

In spite of the rather long and incomplete list of unknowns, there seems to be little doubt that shock waves can be used with varying degrees of success to learn more about nonequilibrium plasma physics, the validity of magnetohydrodynamic theory, and radiation from ionized gases at temperatures above 100,000° K.

5 Establishment of Equilibrium Plasmas

If the plasma produced by a shock wave is in local thermal equilibrium, i.e., if its state (the relative population numbers of bound and free particle states) is uniquely described by temperature and chemical composition, it becomes a valuable source for the measurement of oscillator strengths and damping constants. LTE can be achieved even if there is no radiative equilibrium (i.e., the radiation field is not necessarily that of a blackbody which would render the plasma useless for the determination of oscillator strengths and damping constants), provided the various collisional processes are much more frequent than their radiative counterparts would be for the case where there is radiative equilibrium. (For a more detailed discussion see Finkelnburg and Maecker, 1956.) In some cases of purely hydrodynamic interest one may have only local kinetic equilibrium between the translational degrees of freedom. This is sufficient for calculating flow velocities, pressures, etc., in low-temperature experiments where the degree of ionization and dissociation is small; i.e., one may have local kinetic equilibrium without LTE, but not conversely.

5.1 Ionization by Shock Waves

Because of their high velocities, electrons are most effective in establishing collisional equilibrium, especially for ionization and excitation. To attain LTE, it is therefore necessary that the shock rapidly produces a sufficient electron population, which can then relax into a Maxwellian distribution, equilibrate with the ion distribution, and ionize, excite, recombine, or de-excite. If a shock wave propagates into a nonionized gas, it is not at all obvious that the production of electrons will be sufficiently fast (Petschek and Byron, 1957; Alpher and White, 1959b). Atom-atom collisions of the principal gases involved are usually ineffective in providing an initial critical ionization which would suffice to make ionization by electron impact as fast as the measured ionization rates indicate.† Several mechanisms have been invoked to explain the initial ionization rates: ionization of impurity atoms with low ionization potentials by atom-atom impacts (Bond, 1957); production of photoelectrons by precursor radiation (Gloersen, 1959); excitation of atoms in the ambient gas by precursor radiation and subsequent impact ionization from the excited states; two-step collisional processes (Weymann, 1958; Weymann and Troy, 1961); gas ionization by energetic precursor electrons (Weymann, 1960); and production of electrons by the photoeffect on the shock tube walls (Hollyer, 1957; Gloersen, 1960); or from light emitted by the shock-heated gas due to the formation of electronically excited molecules (Roth and Gloersen, 1958; Gloersen, 1960). There does not seem to be a simple way to decide which of these processes dominates (see also § 5.3) or if perhaps some other mechanism is more effective. However, in case of the T-tube it is most likely that rapid ionization is made possible by excitation of the ambient gas by uv radiation from the initial discharge (McLean *et al.*, 1960; Wiese *et al.*, 1960, 1961).

5.2 Role of Precursor Radiation in Equilibration

Significant pre-excitation had been postulated to explain the apparent discrepancy between temperatures and densities calculated from Rankine-Hugoniot relations (assuming an ambient gas unperturbed

† In a recent discussion at the Lebedev Institute (Moscow), N. N. Sobolev and F. S. Faizulov pointed out that there are indications from shock tube studies that excitation by atom-atom collisions, followed by ionizing collisions, may be important in some cases. For example, the excitation of a 2.7-ev level of Ba II by argon (in a 0.2-ev gas) indicates a cross section of 10^{-17} cm^{-3}.

by the discharge and using the measured shock velocities) and those measured spectroscopically. It was found in helium (McLean *et al.*, 1960) that practically all atoms in the ambient gas would have to be excited, and in hydrogen (Wiese *et al.*, 1960) that most molecules would have to be dissociated and the atoms excited to resolve the discrepancy. It should be noted that this effect is not expected to be noticeable at high initial pressures (above ~ 10 mm Hg), because then the mean free path of the uv radiation becomes too small (as in Elton's experiments, § 2).

If this hypothesis is adopted, one can understand how LTE can be established behind shock fronts in times that are short compared with the times characterizing the decay of the shock-heated plasma in a T-tube. First of all, the gas-kinetic cross sections of excited atoms are so large that kinetic equilibrium between them is reached in a negligible time. But because of the small ionization potentials of these excited atoms, also ionization by atom-atom collisions will now be extremely fast. The electrons rapidly assume a Maxwellian distribution (typically in times of 10^{-10} to 10^{-13} sec) which will then relax with the atom-ion distribution in a time of the order of the electron-ion scattering time. All these processes are usually faster than those necessary to establish an equilibrium population of the various bound states of atoms and ions by electron impact or three-body recombination. Even the latter times were found to be of the order of 0.1 μsec as measured from the rise times of spectral lines. (Wiese *et al.*, 1961. In this reference a more quantitative analysis of the relevant relaxation processes was attempted.)

For conventional shock tubes, Petschek and Byron (1957) considered the effect of photoionization to explain the initial build-up of the ionization (to $\sim 10\,\%$ of the final electron density) and its influence on the relaxation times behind argon shocks. At higher densities the ionization rates could be accounted for theoretically by considering electron impacts. Because of the very small absorption coefficients for the continuous bremsstrahlung and recombination radiation emitted from the shock-heated gas, they conclude that this is not an important factor. They then estimated the amount of radiation emitted by the resonance lines and found that it was small compared to the continuum radiation and concluded on this basis that it may also be neglected and that impurities must therefore be responsible for the observed short relaxation times. However, it should be recognized that the absorption coefficient for the resonance lines of the neutral atoms behind the front is much greater than that for the continuous radiation. Because of this, the reabsorption of the neutral resonance lines in a thin zone (a millimeter or less, in general) ahead of the advancing front of ionized gas is capable of raising a fraction of the neutral gas immediately ahead of the front to excited states, for which the probability for subsequent ionization by collisions is several orders of magnitude larger than for atoms in the ground state. In this case, one would not expect that the temperature and density behind the shock front would be very much different from that predicted by the conventional shock equations because the radiative transport by classical diffusion is some 3 or 4 orders of magnitude slower than the shock velocity. Hence, it cannot result in volume preheating, as seems to be the case for low-density ($P_0 < 10$ mm Hg) electromagnetic shock tubes where there is apparently a significant amount of arc radiation.

One might think that radiation on the line wings of the neutral resonance line could cause volume preheating. The width of this line on both sides of the shock front is essentially determined by Doppler broadening, i.e., changes by a factor of the order 10. The mean free path of radiation in the ambient gas is $\sim 10^{-3}$ cm at the line center and reaches 10 cm at a distance of about 3 Doppler widths. Atoms having the corresponding velocities are immediately excited. But at densities below $\sim 10^{17}$ neutrals/cm^3, collisions will, in general, not be sufficiently frequent to replenish the population of ground state atoms in this velocity group, i.e., at large distances only a minute fraction of the atoms in the ambient gas will be raised into the second quantum state by this process.

These conjectures concerning the relaxation processes in conventional shock tubes are not yet supported by detailed numerical calculations and it is only suggested that the whole question of ionization relaxation seems to deserve re-examination, especially in clean baked tubes with high purity gases (see also Gloersen, 1960).

5.3 Influence of Precursor Radiation on Equilibrium Conditions

The existence of precursor signals ahead of a shock wave is already indicated by electrostatic probe measurements made earlier by Hollyer (1957), Weymann (1960) who also used magnetic probes, and Gloersen (1959, 1960). These workers observed measurable electron densities well ahead of shock waves produced in conventional shock tubes. Gloersen has also studied experimentally the correlation between the onset of the electrostatic probe signal and the time at which luminosity appears behind a shock front in xenon. He attributes the precursor signals to the ejection of photoelectrons from the walls of the shock tube as a result of the presence of ultraviolet radiation. Weymann (1960) considers electron diffusion to be the main source of precursor electrons ahead of shock fronts in argon.

Also, Shreffler and Christian (1954) have noticed the presence of optical radiation ahead of high-luminosity shocks in an explosively driven shock tube and Voorhees and Scott (1959) have seen optical radiation ahead of magnetically driven shock waves. Although the existence of precursor radiation is by now well established, it does not seem to have been generally recognized that this radiation can, in certain cases of practical interest, drastically affect the hydrodynamics and the energy balance (magnetic shock tubes) and relaxation times (conventional and magnetic shock tubes). Depending on the particular conditions, either electron diffusion or photoionization can be the main source of electrons ahead of a shock front.

Some rough estimates of the radiative transfer in the resonance lines of ionized helium (McLean *et al.*, 1960) can be given. (Similar estimates for hydrogen can be found in Wiese *et al.*, 1960.) The emitted intensity will be that of a blackbody from the line cores out to points in the wings where the mean free path for radiation is of the order of the depth of the emitting region. The wing broadening is dominated by the Stark effect (Doppler broadening is negligible) caused by the fields produced by ions and electrons in the plasma.

The absorption coefficient is

$$\alpha_{\nu_n} = \pi e^2 \lambda^2 N^+ f_{n1} S_n(\alpha)/mc^2F_0 \tag{18}$$

with $F_0 = 2.61\, eN_e^{2/3}$ and the reduced wavelength $\alpha \equiv \Delta\lambda/F_0$, where $\Delta\lambda$ is measured from the position of the unbroadened line. The equivalent width corresponds to twice that value of α for which $\alpha_{\nu_n} l \sim \frac{1}{2}$, where l (a few centimeters for a T-tube) is the depth of the emitting layer. For the hydrogenlike He II lines one has (using appropriate scaling laws described further in Griem, 1960),

$$S_n(\alpha) \approx 3 \times 10^{-19} \left(\frac{n}{2}\right)^3 \left[1 + 10^4 \left(\frac{n}{2}\right)^2 (\alpha F_0)^{1/2}\right] \alpha^{-5/2} \tag{19}$$

for a line with an upper state of principal quantum number n. This leads to an equivalent width $\Delta\nu_n \sim 10^{14}\,(n/2)$ sec^{-1} for typical T-tube conditions, i.e., $T_e \sim 4$ ev, $N_e \sim 10^{17}$ cm^{-3}, $f_{n1} \sim 0.4\,(2/n)^3$ and a corresponding photon flux

$$F = \sum_n N_{\nu_n} \Delta\nu_n \sim \frac{1}{c^2} \sum_n 2\pi\, \nu_n^2 \exp\left(\frac{-h\nu_n}{kT}\right) (\Delta\nu_n)$$

$$\approx \sum_n 10^{21}\, \frac{n}{2}\ \text{cm}^{-2}\,\text{sec}^{-1} \qquad (\text{with } \nu_n = 1.2 \times 10^{16}\ \text{sec}^{-1}). \tag{20}$$

The sum has to be extended up to n_{max}, where the maximum of the absorption coefficient is smaller than l^{-1}. This occurs when $\alpha \approx 10^{-11}\, n$ corresponding to half the equivalent width $(\Delta\nu_n/2)$ approaches the half-width of the absorption coefficient $\alpha_{1/2} \approx 10^{-15}\, n^5$ for large n, which yields $n_{max} \approx 12$ and a total photon flux $F \approx 3 \times 10^{22}$ cm^{-2} sec^{-1} for an emitting region with an extent $l \sim 3$ cm, for example.

The mean free path l_{ph} of the He II line radiation in the predominantly neutral gas ahead of the front by continuous photoelectric absorption is calculated to be about 10 cm for a neutral helium pressure of 1 mm Hg. This is also of the order of shock tube dimensions for most of the reported work. At higher pressures one does not expect this radiation to have an appreciable influence on the enthalpy of the gas ahead of the shock front, e.g., as in the experiments of Elton (§ 2) for $P_0 \sim$ 10-50 mm Hg where the usual Rankine-Hugoniot relations seem to apply.

From the foregoing estimates one can also estimate the ionization rate in the cold neutral gas ahead of the advancing front,

$$\frac{dN_0}{dt} \approx -F/l_{ph}. \tag{21}$$

However, depending on the distance from the source of the resonance radiation, a large fraction of the photons will hit the walls and not be effective for photoionization. Taking these effects into account, for a shock tube 3 cm in diameter one calculates that (a) if the precursor radiation originates in the plasma immediately behind the shock front ($T_e \sim 4$ ev, $N_e \sim 10^{17}$ cm^{-3}) that only 1-10 % of the gas 6 cm from the arc will be photoionized, and (b) for an arc temperature of ~ 8 ev and again $N_e \sim 10^{17}$ cm^{-3}, that the arc radiation can ionize all the neutral atoms once in about 1 μsec during the initial discharge.

On the basis of these rough considerations, it appears that mainly arc radiation is responsible for any radiation preheating of the gas ahead of the shock front. This has been verified recently in T-tube experiments (McLean *et al.*, 1961) and is also borne out by preliminary experiments at Harwell (Allen and Martin, 1960) and Aldermaston (Niblett *et al.*, 1960) with magnetic shock tubes using an electrodeless, conical discharge (§ 3.3.2) where it is found that the precursor radiation can be photographed in a variety of gases and seems to originate during the initial breakdown and implosion of the gas, when the light output and perhaps the temperature is the highest. In this case, microwave measurements indicate that precursor electron densities substantially greater than 2×10^{13} cm^{-3} are reached ahead of the shock front since the 8 mm cutoff was observed.

Again we emphasize that these effects are only poorly understood and the present remarks are intended only as some kind of guide to the kind of effects one might expect with fast magnetically driven shock waves. Assuming now that such precursor radiation is an important effect, one asks what happens to the ejected photoelectrons. Their mean free path ($P_0 \sim 1$ mm Hg) is a few centimeters and they can, therefore, lose their energy by collisions with the walls in times of $\sim 10^{-7}$ to 10^{-8} sec, depending on the tube diameter. But there does not seem to be a mechanism by which the ionization energy can be lost in times shorter than the observation times. Both radiative recombination and wall recombination following ambipolar diffusion are slower processes.

Most likely, ions and electrons will recombine mainly by three-body interactions (Bates and Kingston, 1961; McWhirter, 1961; D'Angelo, 1961). Only in very rare cases is recombination into the ground state expected to occur because of the required large energy transfer to the third particle. The recombination, therefore, leads usually to an excited state. The excitation energy will then be trapped in the gas, since the corresponding lines are so narrow that it can only leak out by radiative diffusion. Thus, each primary ionization process will provide the cold gas with an energy close to the excitation or ionization energy of the atoms. Even if the ions and electrons recombine slowly, the additional energy would obviously stay in the gas.

The shock equations must accordingly be modified to take into account the absorption of the resonance radiation, i.e., a term $H_0 \sim \chi^0$ be added on the right-hand side of (3). For strong shocks, the density ratio will now be ≈ 4 because little energy has to go into internal degrees of freedom, as long as second ionization is small.

The experiments with magnetically driven shock waves in H_2 and He seem to be in accord with these calculations as can be seen from Fig. 2, where the pre-excitation was taken into account modifying the Rankine-Hugoniot equations. (The observed temperature was about a factor of 2 greater than one would calculate from the ordinary shock equations, and the density is about a factor of 3 smaller.) Extensive observations with magnetic probes seem to rule out the suggestion that stray electric currents are responsible for the effects.

To summarize, for magnetic shock tubes, at the present time it appears that insofar as precursor radiation has any influence on the determination of atomic properties from radiation from an equilibrated, shock-heated plasma, the main influence is to promote rapid equilibration behind the shock front and to result in considerably higher excitation temperatures (and degree of ionization) for a given shock velocity than would be normally expected. This is extremely important because otherwise the relaxation times would be much longer than the flow duration and times for boundary layer growth and wall cooling.

6 *Experimental Verification of the Rankine-Hugoniot Relations in Conventional Shock Tubes*

6.1 Density and Pressure Measurements†

Following the relaxation time for LTE, there is now good evidence that the state of the shock heated gas in a conventional shock tube can be described with good accuracy by the R-H equations. This seems to be the case over a rather wide range of conditions (~ 2000°-20,000° K and 0.5-10 atm) behind the reflected shock wave. In the Michigan work with neon, as well as in studies by Alyamovskii and Kitaeva (1960) using argon, it is shown that with an added trace of hydrogen as an impurity, the Balmer lines are strongly broadened by the Stark effect and have a steady intensity and shape for about 50 to 100 μsec behind the reflected shock [see also Fig. 37 of Turner (1956)]. This implies that the electron density and temperature do not vary appreciably during this time, and that one might expect LTE to be established if collisional processes dominate radiative processes. This is borne out by recent measurements of the electron density from the half-width of H_β (Alyamovskii and Kitaeva, 1960; Doherty, 1961) and the total pressure (Laporte and Wilkerson, 1960) in neon at ~ 10,000° K. In the case of the neon experiments, the temperature was too low for ionization to have any influence on the hydrodynamics (the degree of ionization is $<0.5\%$). Low fractional ionization of the neon prevents the emission of strong recombination radiation and thereby facilitates the measurement of spectral line shapes and intensities, as well as reducing the energy loss due to radiation cooling (Petschek *et al.*, 1955; Gloersen, 1960). However, the ionization can be an important factor in the hydrodynamic equations and must be included (Resler *et al.*, 1952; Turner, 1956).

Since the plasma remains in a steady state, the total pressure can be measured with a quartz piezoelectric transducer having a ringing period of about 7 μsec and mounted so that its active face is flush with the inside wall of the flow channel. For a range of initial pressures between 7 and 30 mm Hg, corresponding to a pressure ratio across the primary shock of 50 to 60, it was found in the Michigan investigations that the pressures behind the primary and reflected shocks predicted by the ideal theory agreed with the measured pressure to within the experimental accuracy

† Much of this section is taken from Laporte and Wilkerson (1960) and from discussions with Laporte, Doherty, and Wilkerson.

of $\pm 5\%$. The pressure remains constant for about 50 μsec behind the primary shock and then begins to fall near the interface. The flow velocity behind the primary and reflected shock waves can also be measured by observing the luminosity due to cesium (cesium nitrate placed on a fine thread in the flow channel) emission. From these flow velocities as well as the reflected shock velocity, one can again compute the pressure ratio across the reflected wave and compare with the transducer measurements. Further, one can compute temperatures and the density ratio across the shock. This procedure assumes only a local validity of the shock conditions and does not depend on the uniformity of the flow fields, which might be influenced by the interaction of the boundary layer (Mark, 1957) and the reflected wave. Mass, momentum, and energy losses were considered by Doherty (1960, 1961) to describe the observed nonideal behavior.

The agreement between the computed and measured pressure (Laporte and Wilkerson, 1960) was again within the experimental accuracy with a residual drift velocity toward the end of the tube that is 2 % of the primary shock velocity, indicating some degree of nonuniformity. The rare gas temperatures computed from the nonzero drift velocity, primary flow velocity, and reflected shock velocity are consistently in agreement with the ideal values (using only the primary shock velocity) within 1.5 %. It appears, therefore, that deviations in the kinetic temperature due to nonideal effects are quite small for temperatures below 10,000° K. However, the use of conventional shock tubes for absolute f-number determinations, employing temperatures calculated from velocity measurements, can still lead to rather large errors because of the sensitive dependence of the Boltzmann factors on the temperature. Accordingly, small temperature corrections are sometimes necessary (see § 7.3).

Errors in f-number determinations due to uncertainties in the temperature can be minimized by choosing a line with known f-number as an intensity standard whose upper level has an excitation potential close to the excitation potentials of the lines to be measured. If the difference of the excitation potentials is ΔE, the relative error in the f-number will be $\Delta E/kT$ times the relative error in the temperature, assuming $\Delta E \ll kT$. It is therefore advantageous to work at high temperatures, which is especially easy with electromagnetic shock tubes.

One might argue that the agreement between the calculated and measured pressure in the Michigan experiment could result from too low a density and too high a temperature as in the case of the magnetic shock tube. However, the measurements of the electron density from Stark broadening of H_β in the same temperature and pressure range seem to rule this out. The electron density behind the primary shock in argon as computed from the Rankine-Hugoniot equations, including the effect of ionization (Saha equation), generally agreed with that inferred from linewidths to within $\sim 20\%$ (Alyamovskii and Kitaeva, 1960). This error is consistent with the accuracy of the half-width determination as well as errors due to neglecting electron broadening and the distortion of the ion fields due to Debye screening and Coulomb interactions (Baranger, Chapter 13). These latter errors are relatively small because the two effects tend to cancel one another (Griem *et al.*, 1959b).

Doherty's (1961) calculations of the electron density behind the reflected shock using ideal theory (neon with 1 % hydrogen at 10,000 to 13,000° K) showed the H_β width to be too large by about 15 %. This implies that the electron pressure is larger than that calculated hydrodynamically by $\sim 25\%$. However, since the degree of ionization is only $\sim 0.5\%$ the total pressure is still in accord with the Rankine-Hugoniot relations within a few per cent. It appears that the slight deviations from the ideal shock tube theory, due to boundary layer effects, etc., which cause the residual gas flow toward the tube end, leads to an additional compression of the gas. An analysis (Doherty, 1961) shows that this leads to a small increase in temperature of about 4 % at 12,000° K (but is negligible at 10,000° K) which accounts for the observed H_β width and electron density (see also the discussion at the end of § 6.2).

Another method for verifying the Rankine-Hugoniot equations and equilibrium assumption is to measure the electron concentration by optical interferometry. Alpher and White (1959a, b) have measured the refractive index of shock-ionized argon (behind the primary shock) with up to 20 % ionization and electron densities of the order 10^{16}-10^{17} cm^{-3}. Since the specific refraction of neutral argon is known and the electron refractivity dominates the ion refractivity (Alpher and White, 1959b), it is possible to deduce the electron contribution by measuring the total index at two wavelengths (the difference in the index of refraction at two wavelengths is almost entirely due to the electrons because of their much greater dispersion). This method does not require the assumption of thermal equilibrium.

For Mach numbers of 10 to 20 it is found that the density ratio across the front is nearly 4, the value expected if the energy appears only in translational energy of neutral argon. As the argon relaxes and the ionization builds up, the total density ratio increases by about 2, as expected for thermal equilibrium. For 10 experiments with electron densities between 10^{16} and 10^{17} cm^{-3}, the ratio of the observed to computed electron density was $1.02 \pm 11\%$. In the calculations the influence of plasma microfields on the ionization was not included and it was stated that the good agreement between the predictions from conventional shock Hugoniots and the measurements suggests (Alpher and White, 1959b) that this effect is not important under the particular conditions of these experiments. However, these corrections are typically 1-10 % in the electron densities (see, for example, Table I of McLean *et al.*, 1960) so that better precision is needed for their observation.

The interferometric measurements show that even though there is only a momentary state of equilibrium, followed by a nonuniform flow, that the degree of ionization increases so rapidly that radiation cooling is negligible and the Rankine-Hugoniot equations apply behind the primary shock, in agreement with the spectroscopic measurements of Alyamovskii and Kitaeva (1960).

Rink *et al.* (1961) have also done some work on the measurement of densities behind shock waves by using an X-ray densitometer. They have compared the results with theoretical density profiles over a wide range of compositions and initial conditions and have made kinetic studies involving O_2, O, and Xe which agree with previous results. It appears that this method will also be useful for observing the kinetics of other reactions.

6.2 Temperature Measurements

The line reversal method (§ 7.2, last paragraph) has been applied to measure temperatures behind shocks in air, oxygen, and nitrogen (Clouston *et al.*, 1958). Around 2000° K this method is accurate within ± 30° C, and, in general, the measured temperatures agree in the equilibrium region with those obtained from hydrodynamic calculations. In air and oxygen the temperature near the interface with the hydrogen driver gas increases by as much as 400° C, probably because of burning. In oxygen, temperatures near the shock front tend to be too high, presumedly because of a delayed dissociation. In nitrogen, there is a steady rise in the temperature, which may be associated with vibrational relaxation. (The excitation of the sodium line used for the temperature measurement seems to follow the vibrational relaxation of the molecules.)

The line reversal method can be extended to higher temperatures using double beams (Clouston *et al.*, 1959). Near 2000° K the accuracy is now ± 20° C and at 3600° K (using a carbon arc) ± 100° C. The agreement with hydrodynamic theory is again satisfactory, only in argon the measured temperature is 140° C low, probably because of the small cross section for excitation by argon atoms.

This slight disagreement seems to disappear at higher temperatures (6000°-9000° K) where rotational temperatures can be obtained from CN impurity radiation (Parkinson and Nicholls, 1960). The accuracy is here $\pm$ 5 %.

Measurements of the temperature behind nitrogen, air, and argon shocks in the temperature range 2000°-5000° K have also been reported recently (Faizullov *et al.*, 1960a) using a generalized method of spectral line reversal (Sobolev *et al.*, 1959; Faizullov *et al.*, 1960b). First attempts to measure the temperature from intensities of nonresonance lines of sodium and lithium (at 4000° K) showed that it was not possible without introducing such large concentrations of salts that the thermodynamic properties were affected by the impurities. There were also difficulties due to distortion of the lines by self-absorption. These complications are avoided by the line reversal technique since strong, self-absorbed resonance lines can be used without introducing enough contaminant to disturb the thermodynamic properties. In fact, the reabsorption only increases the accuracy of the measurements which are based on the observation of three quantities: (1) the radiation flux I_l in the spectral line, (2) the flux I_s from the source at a known temperature T_s, and (3) the flux I_{ls} at the position of the line when irradiated by the comparison source which emits a continuous spectrum [a tungsten lamp ($T_s = 2660°$ K) and dc xenon arc lamp ($T_s = 4750 \pm 50°$ K) were used]. The xenon lamp brightness was calibrated with a standard lamp in the usual way. If τ is the effective absorption coefficient of a uniform source for the line in question and $\delta\lambda$ is the effective line width, then neglecting the continuum at the line (corrections could be made)

$$I_l = B_\nu(T_x)(1 - e^{-\tau})\delta\lambda$$

$$I_s = B_\nu(T_s)Ds$$

$$I_{ls} = B_\nu(T_s)Ds - B_\nu(T_s)(1 - e^{-\tau})\delta\lambda + B_\nu(T_x)(1 - e^{-\tau})\delta\lambda \tag{22}$$

where D is the linear dispersion and s the entrance slit width of the spectrograph, T_x the excitation temperature and $B_\nu(T)$ the Planck function. By knowing D, s, T_s, I_l, and I_{ls} one can solve for τ and T_x, and it is not essential to observe the moment of reversal.

For nitrogen and air, the measured temperature agrees to $\pm$ 70°K (the maximum error of the measurement is $\pm$ 100°K) with the calculated equilibrium values for pressures behind the shock front of the order 0.1 atm and shock speeds from 2 to 4 mm/μsec. The sodium D and Ba II 4554 A lines were used. Special care to remove sodium from the glass windows was necessary to eliminate strong absorption in the boundary layer. It was also found that the measured variations in the temperature behind the shock front agreed within the experimental errors with those calculated from the velocity and attenuation of the shock wave. Each volume element swept up by the shock wave is raised to a slightly different temperature, depending on the instantaneous velocity. These volume elements "remember" the condition of their birth because the times are apparently too short for thermal conduction to smooth out the temperature variations ($\sim$ 10 %).

Temperatures measured behind argon shocks were too low by 1000°K for gas pressures of 0.5 atm and by 400°K at 4 atm. Only for pressures greater than 11 or 12 atm were the calculated kinetic and excitation temperatures the same within the experimental errors. This is a confirmation that for argon the effective cross section for collisions of the second kind with excited barium ions is approximately 2 orders of magnitude smaller than for nitrogen and air (Prigsheim, 1949; Gaydon, 1954). Faizullov *et al.* (1960a) suggest that this seems to be a promising method for the measurement of cross sections of the second kind since the condition of the shock-heated gas can now be measured accurately.† This is also the case for the measurement of vibrational and dissociational relaxation times in the nonequilibrium zone as well as *f*-numbers in the equilibrium zone since the line reversal method is capable of $\sim \pm$ 1.5 % temperature determinations.

For temperatures and electron densities which are high enough to Stark broaden spectral lines (for which the shape functions must be known as a function of density) but not so high that the degree of ionization is appreciable, the line broadening affords a sensitive probe to correct slight deviations from calculated Rankine-Hugoniot temperatures in conventional shock tubes.

This method has been used by Doherty (1961) to determine the temperature in neon with 1 % H (see also § 6.1 and § 7.3), where the degrees of neon and hydrogen ionization are $\sim$ 0.5 and 50 %, respectively. By measuring the shock velocities, the kinetic temperature can be determined to within a few per cent from the Rankine-Hugoniot equations. The excitation and ionization temperature can then be corrected by comparing the calculated and measured width of H_β and asking what increase in temperature is required to bring them into agreement. This method is very sensitive, because a small change in the temperature manifests itself in a large change in the electron pressure (or linewidth) but only a small change in the total pressure, i.e., this iteration procedure is accurate since the degree of ionization is so small that it has a negligible

† See the footnote in § 5.1.

influence on the gross hydrodynamic behavior. In this way, Doherty (1961) has determined ionization temperatures to 100°K at 12,000°K in the Michigan shock tube and concludes that the slight deviations from the Rankine-Hugoniot predictions are due to wall effects which perturb the flow velocities and temperature.

Charatis and Wilkerson (1959) describe an attempt to verify the excitation temperature (in the Michigan shock tube) obtained using the measured set of relative *f*-values for neutral chromium (Hill and King, 1951). They generated the lines of Cr I and Cr II at known abundance by using chromium carbonyl [$Cr(CO)_6$] as an additive vapor in the neon carrier gas. The Cr I excitation temperature was determined essentially by dividing the observed relative line intensities by the relative *f*-values of Hill and King. In the shock tube experiment the Cr I temperature (e.g., 5400°K) was found to be considerably lower than the kinetic temperature (e.g., 9300°K) calculated from the Rankine-Hugoniot equations (including all dissociations, ionizations, etc.). Assuming this to be a real effect, Charatis and Wilkerson have since examined a number of possible causes and found them wanting. Principally, Doherty's (1961) demonstration of LTE for hydrogen is felt to guarantee LTE for the chromium case, due to the greater ease of ionization, higher electron density, etc. Further, the ionization relaxation (mostly Cr I $\rightarrow$ Cr II) behind the reflected shock and the strong effects on the reflected shock speed predicted for the carbonyl additive were observed. Lastly, they fail to observe a strong continuum or diminishing line intensities indicative of radiation cooling.

Proceeding on the assumption that the previously measured *f*-values are in error, Wilkerson (1961) has applied absolute intensity measurements to redetermine the Cr I *f*-values. Charatis (1961) is using a line reversal technique and extending the intensity measurements to more lines, in order to find *f*-values for both Cr I and Cr II. The progress of these chromium experiments demonstrates, in particular, the serious need for reliable *f*-values (§ 7.3) and, in general, the close interdependence between diagnostic methods and the knowledge of atomic constants.

Another check on the validity of the shock theory is to measure the electrical conductivity of shock-heated gases where the degree of ionization is so high that electron-ion collisions dominate electron-neutral collisions. Then the conductivity depends only on the temperature (Cohen *et al.*, 1950). Lin *et al.* (1955) find for temperatures above 8000 to 10,000°K in argon shocks that the measured conductivity compares with the theoretical value within 10 %. This constitutes a temperature measurement with about the same accuracy. It appears that greater precision for temperature determinations is available using spectroscopic techniques, but it should be possible to combine the two methods so as to make an accurate comparison between the experimental and theoretical conductivity. This would be possible if the temperature were determined from an independent measurement. Such experiments are now in progress at the Moscow Power Institute.

Also at lower temperatures, where electron-atom collisions are important in the expression for the conductivity, there is again agreement between the shock theory and experiment. Since the conductivity here depends on both the temperature and the electron-atom elastic cross section, this gives some confirmation of the mobility measurements of Townsend and Bailey (1922a, b), assuming the equilibrium temperature. Similar experiments were carried out later with shock-heated air, which give information on elastic cross sections between electrons and the dissociation products (Lamb and Lin, 1957).

6.3 Conclusions Regarding the Applicability of the Rankine-Hugoniot Relations in Conventional Shock Tubes

From various measurements one can conclude that in conventional shock tubes the mechanisms leading to the large departures from the ideal Rankine-Hugoniot predictions in magnetic shock tubes is not present. This is also consistent with calculations of the influence of precursor radiation on the equilibrium temperature and density (it could still influence the relaxation times), where it is found that such effects should be completely negligible.

Although these various observations show good agreement with local conditions near the fronts of the primary and reflected shock waves, most workers point out that there are several symptoms of nonuniformity in the large. It is generally found that the gas columns behind the primary and reflected shocks are significantly shorter than expected, that the time-dependent primary flow velocities are too high, while the reflected shock velocities are generally too low (by $\sim 8\%$ in the Michigan experiments with neon). In addition, these effects are influenced by the initial pressure and the tube dimensions (Gloersen, 1960) and, their importance must be evaluated in each case. The consequence is that one needs moderately good time resolution so that observations are made only during the limited time of the steady-state plateau behind the primary and reflected waves.

For temperatures near 20,000°K, high explosive drivers can be used, but careful measurements of the temperature, density, and total pressure seem to be lacking (Seay, 1957). However, there seems to be little reason to believe that here also a gas in LTE cannot be generated by shock heating. However, the influence of radiation cooling (Petschek *et al.*, 1955; Gloersen, 1960) might be expected to be of importance, especially for the heavier gases such as argon or xenon where the continuous radiation is very strong. Then, as in the magnetic shock tube, quantitative measurements must be made at a time long compared to the relaxation time but still short compared to significant times for cooling, boundary layer growth, etc.

To summarize the situation with regard to the establishment of equilibrium conditions in conventional shock tubes after the relaxation period (which must be measured): all *reliable* quantitative studies of pressure, density, and temperature reported so far have shown excellent agreement with hydrodynamic theory in the temperature range 2000°-14,000°K if the pressure is sufficiently high. There is certainly a variety of relaxation phenomena, etc., that are not completely understood but, in general, there seems to be little doubt that the shock tube can be used as an equilibrated light source over a wide range of conditions; this will certainly provide essential quantitative information on the character of high temperature radiation and provide an increasingly versatile tool for the measurement of atomic properties.

7 *Shock-Heated Plasmas as Thermal Light Sources*

7.1 General

It is obvious that there is no general method which enables one to predict whether LTE is reached sufficiently fast or not. The following example may serve to illustrate how misleading it would be, e.g., to assume that an increase in density will always further rapid equilibration: Whereas at initial pressures in the mm Hg range one usually does not observe any separation of shock front and luminous front in T-tubes (except far from the discharge), such a separation is quite pronounced at pressures higher than 10 mm Hg (Elton, 1961 and § 2), which is indicative of a relatively long relaxation time at higher initial pressures. Obviously much experimental and theoretical work remains to be done before these phenomena will be understood for any given situation. One has therefore to make sure that LTE exists behind a shock wave, which can be done by checking if various independent measurements and analyses lead to the same temperature. Only then may a shock-heated plasma be safely used as a thermal light source.

Some other classes of phenomena should at least be mentioned in this connection: nonplanarity of the shock front, boundary layer effects, cooling by thermal conduction to the walls or by radiation losses as well as decay of the plasma density and temperature behind the shock front because of the finite extent of the pressure reservoir, mainly in magnetic shock tubes. Few general remarks can be made about these effects, except that one should try to make the times characterizing these effects significantly larger than the relaxation times for LTE, and the various characteristic lengths small compared to the shock tube diameter. All these are necessary requirements for the actual use of a given shock tube for the measurement of oscillator strengths and damping constants, etc. If they are fulfilled, the shock tube offers some decided advantages compared with stabilized arcs: the relative ease of achieving higher temperatures, the production of a fairly homogeneous slab of plasma, and, finally, the absence of appreciable demixing in case of gas mixtures (Mastrup and Wiese, 1958). These advantages might very well off-set the greater difficulties in the instrumentation caused by the required time resolution.

7.2 Spectroscopic Analysis of Shock-Heated Plasmas

Quantitative spectroscopy seems to offer the most powerful tool for the diagnostics of LTE plasmas, and it is therefore appropriate to collect the relevant formulae and discuss their applicability and the accuracy that can be achieved in the laboratory with present techniques. The theory is much simpler than that used by astrophysicists for the analysis of stellar atmospheres, because in applications to plane shock waves one *usually* deals with nearly homogeneous and optically thin layers, i.e., the properties of the emitting layer are independent of the coordinate in the direction of the line of sight and self-absorption is negligible.

The intensity per unit wavelength interval (in erg sec^{-1} cm^{-3} sterad^{-1}) from a homogeneous layer of depth l can be written as

$$\begin{aligned} I_\lambda = \Bigg\{ & \sum_{m,n,s,z} \frac{2\pi e^2 h}{m} (\lambda_{mn}^{s,z})^{-3} f_{mn}^{s,z} \frac{g_n^{s,z}}{Z^{s,z}} N^{s,z} \exp\left(\frac{-\chi_m^{s,z}}{kT}\right) L_{mn}^{s,z}(\lambda) \\ & + \sum_{s,z} \pi \left(\tfrac{2}{3}\right)^{3/2} \frac{e^4 h z^2}{m^2 c^2 \lambda^2} \left(\frac{\chi_\infty^{\mathrm{H}}}{kT}\right)^{1/2} \Bigg[g_{ff} \exp\left(\frac{\Delta\chi^{s,z}}{kT}\right) \\ & + 2z^2 \frac{\chi_\infty^{\mathrm{H}}}{kT} \sum_n \frac{g_{fb}}{n^3} \exp\left(\frac{\Delta\chi_n^{s,z}}{kT}\right)\Bigg] N_e N^{s,z+1} \exp\left(-\frac{hc}{\lambda kT}\right)\Bigg\} \, l \end{aligned} \tag{23}$$

[for details in the derivation of this formula, consult astrophysical texts such as Unsöld (1955) or Aller (1953)]. The first sum represents the line radiation from various atomic species s in ionization stages $z = 1$ (neutral atom), 2 (singly ionized), etc., arising from transitions between upper states m and lower states n with unperturbed wavelengths $\lambda_{mn}^{s,z}$. The $f_{mn}^{s,z}$ are the absorption oscillator strengths, $g_n^{s,z}$ and $Z^{s,z}$ are statistical weights of the lower states and partition functions

$$Z^{s,z} \equiv \sum g_m^{s,z} \exp(-\chi_m^{s,z}/kT),$$

respectively; the $\chi_m^{s,z}$ are the energies of the upper states measured from the ground states, and $L_{mn}^{s,z}(\lambda)$ finally describes the line profiles and is normalized such that the integral over the whole line $\int L_{mn}^{s,z}(\lambda)\, d\lambda = 1$ (the problem of calculating line profiles is treated in Chapter 13 by Baranger).

The second sum gives the continuum intensity. Here g_{ff} and g_{fb} are quantum-mechanical correction (Gaunt) factors which are usually close to unity. Furthermore, $\Delta\chi^{s,z} \equiv \chi_\infty^{s,z} - \chi_{\max}^{s,z}$; $\Delta\chi_n^{s,z} = \chi_\infty^{s,z} - \chi_n^{s,z}$, where $\chi_\infty^{s,z}$ is the ionizational potential and $\chi_{\max}^{s,z}$ the energy of the highest bound state that can be observed. The sum involving g_{fb} must be extended from n defined by $(\chi_{\max}^{s,z} - \chi_n^{s,z})/h \leqslant c/\lambda$ to an n-value corresponding to $\chi_{\max}^{s,z}$. For numerical convenience the ionization potential χ_∞^{H} of hydrogen was introduced, and also the continuum intensity was written in terms of electron density N_e and $N^{s,z+1}$, i.e., the density of the next higher ionization stage of species s.

The formula for the line contribution is exact, but unfortunately oscillator strengths are known only for a small fraction of the lines. Exact calculations can be performed only for hydrogen and hydrogenic ions (Bethe and Salpeter, 1957). Fairly detailed approximate calculations are available for neutral helium ($z = 1$) (Trefftz *et al.*, 1957) and for more complicated systems one can use the relatively simple Coulomb approximation (Bates and Damgaard, 1949). If the coupling scheme is known, relative oscillator strengths can be obtained also in cases where the Coulomb approximation no longer applies (Goldberg *et al.*, 1960). Attempts have been made to treat even more complicated cases, but it is an open question as to whether the large effort is warranted by an increase in accuracy. A compilation of calculated and measured oscillator strengths has been made for astrophysical purposes (Allen, 1955) and presently an effort is being made to provide a complete survey of the literature on the subject (Wiese, private communication).

The formula for the continuum contribution is exact only for hydrogen and hydrogenic ions, where also the Gaunt factors are known

(Karzas and Latter, 1958; Kazachevskaya and Ivanov-Kholodnr, 1959; Berger, 1956). However, the free-free transitions should be described reasonably well by the hydrogenic formula, because they essentially occur in a Coulomb field. But also the bound-free continuum can probably be calculated with a fair accuracy, if one replaces n by effective quantum numbers defined by $n_{\text{eff}}^2 = z^2\chi_\infty^{\text{H}}/(\chi_\infty^{s,z} - \chi_n^{s,z})$, as long as the deviations of n_{eff} from n are small. For neutral helium, e.g., this is the case for most levels, and by using the measured energy levels in the exponentials and for n_{eff}† and employing the hydrogenic Gaunt factors, the calculated continuum intensity is most likely accurate to better than 10 %, unless transitions to the ground state and the $2S$ states are significant [see Goldberg (1939) and Huang (1948) for calculations for helium with 2^3S, 2^1S, 2^3P, and 2^1P as lower states]. Such an accuracy would be quite adequate, since present standards for absolute intensity measurements are only slightly better than that. If transitions into partially occupied shells are important, one must also consider the number of available states in these shells (Elwert, 1954). Biberman and Norman at the Moscow Power Institute inform us that they have obtained corrections to the Unsöld-Kramers formula for radiative recombination using the quantum defect method of Burgess and Seaton (1960).

In addition to the intensity formula, one has in LTE the various Saha equations

$$\frac{N^{s,z+1}N_{\text{e}}}{N^{s,z}} = \frac{2Z^{s,z+1}}{Z^{s,z}}\left(\frac{2\pi mkT}{h^2}\right)^{3/2} \exp\left(-\frac{\chi_{\max}^{s,z}}{kT}\right). \tag{24}$$

Also here the highest observable discrete energy should be used instead of the ionization potential of the isolated atoms or ions, and the sum in the partition function should only be extended to this $\chi_{\max}^{s,z}$. The actual value of $\chi_{\max}^{s,z}$ or the corresponding principal quantum number can be estimated by comparing $\chi_\infty^{s,z} - \chi_{\max}^{s,z}$ with the mean interaction energy of free electrons $e^2N^{1/3}$ (Unsöld, 1948).

Other useful equations are the quasi-neutrality condition

$$N_{\text{e}} = \sum_{s,z} (z-1)\, N^{s,z} \tag{25}$$

which is practically always valid, and the concentrations of the various elements

$$N^s = \sum_z N^{s,z} \tag{26}$$

† The effective quantum numbers depend on l, the orbital quantum numbers, and one has to perform an average of n_{eff}^{-3} over the various l corresponding to one n, using the continuum oscillator strengths for hydrogen as weight factors, which can be found in Bethe and Salpeter (1957).

whose ratios are often known for shock-heated plasmas. With these equations and with intensity measurements at a sufficient number of wavelengths, one can in principle always determine the unknowns T, N_e, and $N^{s,z}$, and by taking more than the minimum number of intensity measurements also check the internal consistency of measurements and results.

This does not necessarily verify the assumption of LTE, however, because quite frequently the intensities of the observed spectral lines might be affected by any deviations from LTE in a very similar way. A more critical test is to measure, e.g., absolute total line intensities, and use them to calculate T and N_e with the LTE assumption, and then also to determine N_e from the continuum intensity or the Stark broadening, which are both practically independent of LTE. Such an experiment

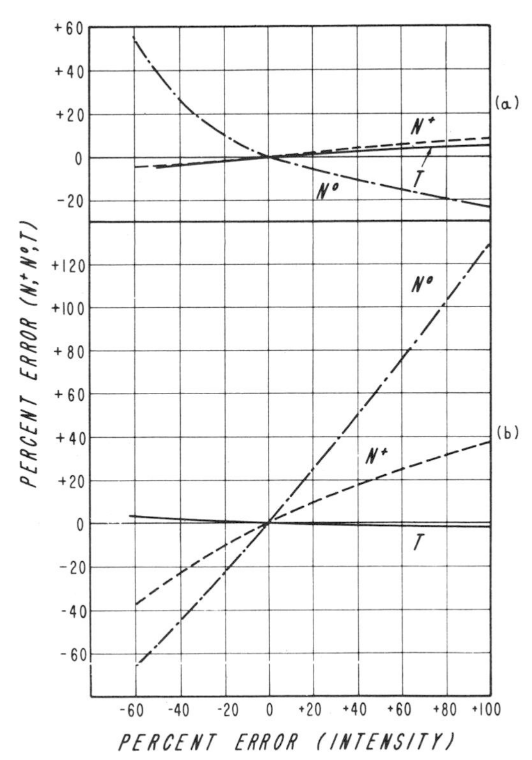

FIG. 16. Effect of errors in the intensity measurement on the calculated temperature, T; ion density, N^+; and neutral atom density, N^0. Set of curves (a) was determined by holding the intensity of He I λ3889 constant and varying the intensity of He II λ4686. Set of curves (b) was determined by holding the intensity of He II λ4686 constant and varying the intensity of He I λ3889. [Reprinted from McLean *et al.* (1960), Fig. 8.]

was performed for a helium plasma in a T-tube (McLean *et al.*, 1960), where one obtains best accuracies by measuring the total line intensities of a He I and a He II line, because of the large difference in the excitation potentials. Figure 16 demonstrates how an intensity error is reflected in temperature and the various densities in this case. Measured absolute intensities can hardly be expected to be more accurate than $\sim 10\,\%$. Temperature errors are therefore of the order of 1 % and density errors of the order of 10 %. Since the theoretical accuracy of the formulae for continuum intensities and linewidths also corresponds to a 5 or 10 % uncertainty in the resulting electron density, one can really check whether or not LTE exists in this case with a precision compatible with the experimental accuracy. (In the experiment discussed here, the error in the linewidth determination was too large to really achieve such a precision because of the poor spectral resolution of the time-resolving spectrograph.)

In situations where the plasma is optically thick, e.g., in the centers of strong lines, one can make another temperature measurement by equating the measured maximum intensities to those of a blackbody in the same wavelength intervals. A similar method is to observe the absorption by the plasma of radiation from a source of known effective temperature. If plasma and source temperatures are equal, the intensity at a given wavelength of the source as seen through the plasma (which absorbs and emits according to Kirchhoff's law) is equal to the source intensity not passing through the plasma (line reversal method, see also § 6.2). To extend method to higher temperatures, i.e., above 5000°K, continuous light sources with higher brightness temperatures than a tungsten lamp or carbon arc are required.

7.3 Measurements of Oscillator Strengths

Oscillator strengths (f-numbers or transition probabilities) are usually determined from lifetimes of excited states or from the intensity of emitted or absorbed radiation. For experiments of the second category, light sources are needed where the relative or absolute occupation numbers for the excited states can be measured or calculated. This is practically impossible, unless LTE exists. There are three classes of thermal light sources: furnaces of the King type (King, 1914, 1922) filled with metallic vapors, high-current arcs in atmospheres consisting of one or several gases, and finally shock tubes. Furnaces are suitable for the measurement of relative oscillator strengths of low-excitation metal lines. Absolute determinations are difficult because of the uncertainties in the vapor pressure curves. Arcs have been used mainly in Kiel to measure oscillator strengths for, e.g., O I (Jürgens, 1954), N I (Motschmann, 1955), O II and N II (Mastrup and Wiese, 1958), C I (Richter, 1958), Si I, Si II, Cl I, and Cl II (Hey, 1959). Also here, only relative oscillator strengths are obtained with good accuracies, and it is again difficult to assess the errors in the absolute oscillator strengths, unless special precautions are taken to prevent demixing of the constituents of the arc gas to occur.

In addition to some work on molecular oscillator strengths (Treanor and Wurster, 1960; Daiber and Williams, 1961; Keck *et al.*, 1959), it appears that only three successful shock tube experiments to measure atomic oscillator strengths have been completed.

Doherty (1961) has measured the absolute f-numbers of 21 neon lines, 12 argon lines, and various CI, NI, and OI lines in the spectral range 5800-8500 A using the Michigan shock tube described in § 3.1. The temperature was determined accurately from velocity measurements and the broadening of H_β (introduced as an impurity) by the method described in § 6.2. The relative plate sensitivity was determined with a carbon arc using a shutter. The characteristic curves at each wavelength were found with a xenon flash lamp and step filter. Both the flash lamp and arc had the same exposure time as the streak spectra of the shock luminosity. The absolute neon and argon f-numbers were found by using the H_β line as a standard since it has a known f-number. Except for a few lines, the relative f-numbers of neon agree to within 10 % with Ladenburg's and Levy's (1934) values, but the absolute values are systematically lower by a factor of about 3, which is within the accuracy of the latter authors' calibration. It is estimated that the absolute values as determined in the shock tube are accurate to $\pm$ 20 % after correcting for the slight deviations ($\sim$ 1-2 %) from the Rankine-Hugoniot kinetic temperature. It should be possible to check the absolute scale by comparing the neon and neutral helium spectra at higher temperatures. Helium and neon both have high excitation energies, and the f-numbers of neutral helium can be measured as described below.

With the same Michigan shock tube, Wilkerson (1961) measured Cr I and some Cr II oscillator strengths using absolute intensities.

Magnetic shock tubes easily produce temperatures significantly higher than those obtainable in arcs or conventional shock tubes. Therefore, it seemed reasonable to measure first the oscillator strengths of He I using a T-tube, especially because helium is most suitable to test the accuracy of the various approximations for calculating oscillator strengths (McLean, 1961). The helium measurements were made using photomultipliers and oscilloscopes with photographic recording, and absolute line and continuum intensities could be obtained by calibrating the system with a tungsten ribbon filament lamp as a radiation standard. It was possible to reduce all errors of the actual experiment to a few per cent, so that one of the major uncertainties in the absolute line intensities is now given by the error in the standard lamp calibration (5 % at 3000 A, 2 % at 6000 A). Intensity ratios could be determined with slightly better accuracies, because here the standard lamp contribution to the error is 3 % or smaller for lines in the visible depending on their wavelength separation.

Some of the *f*-numbers for He I transitions are so close to the corresponding hydrogen values that calculated oscillator strengths for them have uncertainties that are negligible compared with the experimental errors. The oscillator strengths of other transitions were therefore determined relative to them, and their absolute values are expected to have an accuracy of better than 10 %. Similar accuracies should be attainable for lines from other gases, if mixtures with helium or hydrogen are used, and if the reference lines with known oscillator strengths are chosen to have their upper levels close to the upper levels of the transitions whose oscillator strengths are to be measured. This ensures that an error in the measured temperature has only a negligible influence on the resulting oscillator strengths. Furthermore, one should avoid working at too high temperatures, because otherwise most of the line radiation might come from the boundary layer.

This method of mixing several gases is certainly applicable for shock tubes, because there is not enough time for any appreciable de-mixing to occur. Another advantage over arcs is the better homogeneity of the source in the direction of the line of sight, which simplifies the analysis considerably. On the other hand, in a steady-state stabilized arc the whole spectrum can be scanned, whereas with shock tubes one will usually measure only the total intensities in appropriate wavelength intervals containing the spectral lines in question. These must then be corrected for continuum background, and one usually also has to allow for the contribution from the far line wings which are not received by the exit slits. The continuum intensities can be measured near the lines in regions free of impurity radiation, and the wing contribution can either be calculated from theoretical line profiles obtained from measured electron densities and temperatures or from directly measured line profiles (see the next section for details on the measurement of line profiles from shock-heated transient plasmas). In the helium experiment the uncertainties of these corrections gave no significant contribution to the error in the oscillator strengths, because the corrections were usually not much larger than 10 %, and the error in the corrections in turn could be kept below 10 %.

Therefore, in addition to the errors in the standard lamp calibration, only reading errors from the oscilloscope traces seem to give rise to serious errors. Their influence can, however, be reduced by averaging over a large number of measurements. One can accordingly hope to achieve an accuracy of better than 10 % for absolute oscillator strengths measured in shock tubes using the methods described in this section, which should be applicable to a great variety of lines, including ion lines, of interest in astrophysics and laboratory plasmas. A significant improvement of the accuracy of oscillator strengths from intensity measurements will have to wait for better radiation standards.

7.4 Damping Constants and Line Profiles

The principal causes of line broadening in shock-heated plasmas are thermal Doppler effect and Stark effect from the electric microfields produced by ions and electrons in the plasma. Doppler broadening is important at high temperatures and low electron densities and can then be used to measure the kinetic temperature of emitting atoms and ions. But at moderate temperatures and relatively high electron densities (the usual situation) Stark broadening dominates. It is mainly a function of the electron density and can, therefore, be used to measure this quantity provided the theory of Stark broadening has been worked out for the lines whose profiles are observed. The present status of the theory is reviewed in Chapter 13, and here only some of the most important numerical results will be presented.

Table I summarizes the most recent calculations for the hydrogen H_β line (Griem *et al.*, 1961c) which have an estimated error of less then 10 %. The function $S(\alpha)$ describes the normalized line profile ($\int_{-\infty}^{+\infty} S(\alpha)d\alpha = 1$) with α defined by $\alpha \equiv \Delta\lambda/F_0 = \Delta\lambda/(2.61eN_e^{2/3})$. In this approximation the profiles are symmetric $[S(\alpha) = S(-\alpha)]$. Actually some slight asymmetries do occur (Griem, 1954) and one should therefore compare the mean of the measured blue and red wings with the calculated profile. Other hydrogen line profiles (H_α, H_γ, H_δ, Ly_α, and Ly_β) have also been calculated (Griem *et al.*, 1959b) and detailed profiles are available (Griem *et al.*, 1960). In these early calculations the electron impact broadening of the lower states was neglected and also the ion field-strength distribution functions were only approximated, which introduced errors of usually less than 10 % (Griem *et al.*, 1961c). For higher series members of hydrogen and hydrogen-like spectra one can use approximate profiles which depend on two parameters describing the ion broadening and the relative importance of the electron broadening (Griem, 1960).

TABLE I

SHAPE FUNCTIONS S(α) FOR THE H_β PROFILE[a]

N/α	0	0.01	0.02	0.03	0.04	0.05	0.06	0.07	0.08	0.09	0.10	0.12	0.14	0.16	0.18	0.20	0.25	0.30	0.35
									$T = \frac{1}{2} \times 10^4$ °K										
10^{14}	2.27	3.65	5.20	5.52	5.16	4.49	3.75	3.09	2.53	2.05	1.67	1.16	0.820	0.605	0.455	0.350	0.200	0.126	0.086
10^{15}	2.90	3.79	4.79	5.02	4.68	4.11	3.51	2.93	2.41	1.99	1.67	1.19	0.870	0.659	0.502	0.398	0.234	0.157	0.114
10^{16}	3.66	4.02	4.37	4.42	4.13	3.68	3.21	2.73	2.29	1.92	1.64	1.22	0.906	0.703	0.552	0.443	0.272	0.185	0.137
									$T = 1 \times 10^4$ °K										
10^{14}	1.85	3.34	5.13	5.53	5.30	4.64	3.87	3.14	2.57	2.10	1.73	1.18	0.822	0.602	0.457	0.357	0.202	0.119	0.073
10^{15}	2.56	3.52	4.71	5.08	4.83	4.28	3.66	3.05	2.52	2.07	1.74	1.22	0.885	0.662	0.510	0.397	0.232	0.149	0.107
10^{16}	3.09	3.62	4.29	4.53	4.33	3.90	3.42	2.90	3.42	2.04	1.73	1.28	0.950	0.723	0.564	0.445	0.269	0.180	0.136
10^{17}	3.36	3.62	3.91	4.01	3.85	3.53	3.16	2.74	2.34	2.00	1.73	1.31	1.00	0.787	0.622	0.502	0.313	0.218	0.164
									$T = 2 \times 10^4$ °K										
10^{14}	1.55	3.18	5.18	5.57	5.40	4.79	3.95	3.22	2.64	2.18	1.79	1.22	0.860	0.610	0.445	0.338	0.188	0.112	0.070
10^{15}	2.12	3.23	4.69	5.18	5.02	4.45	3.80	3.15	2.61	2.13	1.80	1.23	0.890	0.660	0.500	0.370	0.207	0.130	0.095
10^{16}	2.83	3.44	4.24	4.59	4.45	4.03	3.53	3.00	2.52	2.10	1.78	1.29	0.946	0.724	0.555	0.438	0.262	0.172	0.126
10^{17}	3.08	3.48	3.98	4.20	4.08	3.73	3.32	2.87	2.43	2.07	1.77	1.32	0.995	0.772	0.608	0.487	0.299	0.201	0.149
									$T = 4 \times 10^4$ °K										
10^{14}	1.28	3.05	5.29	5.65	5.49	4.80	3.98	3.24	2.65	2.26	1.88	1.31	0.926	0.665	0.480	0.352	0.178	0.100	0.060
10^{15}	1.87	3.12	4.76	5.29	5.17	4.62	3.90	3.23	2.67	2.21	1.83	1.28	0.898	0.640	0.475	0.352	0.201	0.127	0.087
10^{16}	2.44	3.18	4.25	4.75	4.66	4.10	3.58	3.05	2.55	2.12	1.79	1.27	0.930	0.700	0.540	0.420	0.246	0.158	0.112
10^{17}	2.79	3.15	3.68	3.98	3.94	3.67	3.29	2.86	2.44	2.08	1.79	1.32	1.00	0.776	0.608	0.485	0.297	0.198	0.146

[a] Profiles $S(\alpha)$ of the hydrogen line H_β Stark broadened by ions and electrons as discussed in the text (Griem *et al.*, 1961c).

TABLE II

FULL HALF-WIDTHS OF SOME NEUTRAL HELIUM LINES[a]

Line		Temperature (°K)				
Transition	Wavelength	5000	10,000	20,000	40,000	80,000
$2^3S - 4^3P$	3188	7.6	8.1	8.2	7.7	6.9
$2^3S - 3^3P$	3889	2.4	2.7	2.8	2.8	2.6
$2^1S - 4^1P$	3965	24	22	20	17	14
$2^3P - 5^3S$	4121	19	22	24	25	23
$2^1P - 5^1S$	4438	34	38	39	38	35
$2^3P - 4^3S$	4713	8.0	9.8	11	12	11
$2^1S - 3^1P$	5016	8.9	8.3	7.6	6.8	5.9
$2^1P - 4^1S$	5048	15	17	18	18	17
$2^3P - 3^3D$	5876	3.8	4.0	4.1	4.0	3.8
$2^1P - 3^1D$	6678	9.8	8.7	7.8	6.9	6.0
$2^3P - 3^3S$	7065	4.2	5.1	6.0	6.6	6.8
$2^1P - 3^1S$	7281	7.5	8.9	9.9	10	10

[a] Full half-widths (in angstroms) of neutral helium lines broadened by electrons (mainly) and ions for $N_e = 10^{17}$ cm^{-3} (Griem *et al.*, 1961a). Doppler broadening is not included.

In Table II some calculated half-widths (between half-intensity points) are collected for the stronger isolated neutral helium lines in the visible region of the spectrum. Here the electron contribution dominates, which has been calculated recently with the impact approximation (Griem *et al.*, 1961a). In Table II a 10 % allowance for ion broadening was included. The accuracy should then be better than 20 %, and linewidths will scale linearly with the electron density until they approach the width of a hydrogen line with the same principal quantum number. Recently calculations were also made for He II 4686 A and He II 3203 A (Griem and Shen, 1961). At $N_e = 10^{17}$ cm^{-3} and $T = 40{,}000$°K, e.g., the half-width of He II 4686 A is 4 A. As long as Doppler broadening is negligible or only a small correction, this width is essentially independent of the temperature and scales approximately with $N_e^{2/3}$.

It is planned to calculate damping constants (half-widths) for lines from other than hydrogen or helium atoms using atomic matrix elements derived with the methods used by Bates and Damgaard (1949) for oscillator strengths. This method is expected to be applicable to a large class of spectral lines, because Stark broadening essentially depends only on the wave functions of the upper states, which are more hydrogen-like than those of the lower states.

Measurements of profiles of spectral lines emitted from shock-heated plasmas require rather involved instrumentation because both wavelength and time resolution is needed. If photographic methods are employed, an optically fast spectrograph is essential that still possesses sufficient resolving power. It must be stigmatic so that the dimension in the slit direction can be used as time-axis. One can either use rotating mirrors to sweep the image of the shock tube along the slit or use rotating film drums in conjunction with a pinhole entrance slit. Clearly none of these systems yields spatial resolution. This can be obtained by employing a fast shutter [Kerr-cell or mechanical (Wiese, 1960)] with a stigmatic spectrograph. Such a system then yields a snapshot of the spectrum with spatial resolution in the entrance slit direction.

All photographic methods have the advantage that a number of spectral lines can be observed simultaneously. But normally one has to use the most sensitive films, and one can hardly expect an accuracy of more than 10 % in the half-width determinations. Often measurements on the line wings will be extremely difficult because of the limited range of such films, and it is then next to impossible to determine the continuum background. Finally, the intensity calibration and photometry of the film is extremely laborious, and it is therefore very likely that photoelectric methods will eventually be used almost exclusively for precision measurements of line profiles.

For photoelectric measurements one does not usually require a stigmatic spectrograph. The major problem in this case is to feed the light from different wavelength bands within one line to the various photomultipliers. This has been accomplished by multiple slit and mirror systems in the focal plane. Another possibility is the use of glass fiber optics to split the beam. If shot-to-shot reproducibility is sufficiently good, the line can just be scanned from shot-to-shot with a monochromator. In a stigmatic instrument a larger separation of the various wavelength channels may be achieved with an exit slit that is somewhat inclined with respect to the entrance slit. For very narrow lines one can finally use interferometric techniques, where then a line may be spread so far that the placement of the detectors poses no special difficulty.

The decisive advantage of all photoelectric systems stems from the linearity of the photomultipliers, which practically eliminates the need for intensity calibration if one is interested only in line profile measurements. This linearity is especially important on the line wings, where the line may only contribute a small fraction of the continuum intensity which must be subtracted. It appears therefore that photoelectric systems make possible much more reliable profile determinations.

However, they require more equipment than the photographic method, and also yield only the profile of one line at a time.

As pointed out earlier in this section, there are several motivations for line broadening studies. If Doppler broadening dominates, one can measure the kinetic temperatures. This can be done with a three-channel system (Ramsden *et al.*, 1961) once it is established that the profile is Gaussian. Or, if Stark broadening dominates, one can measure electron densities from lines with calculated Stark profiles. Also, this only requires a small number of channels.

On the other hand, if one wants to check the theory of Stark broadening, one certainly needs more detailed information on the profiles. In this case, the monochromator technique is quite appropriate. Measurements of some hydrogen and helium lines from shock waves in T-tubes have been made with this method (Berg *et al.*, 1961), and the results were compared with calculated profiles using the electron density from absolute continuum intensities. For hydrogen a 5 % agreement and for helium a 10 % agreement in the width was achieved. Presently an attempt is being made by Elton to measure the Stark profile of the Ly_α line emitted by the plasma behind the reflected shock wave in a high pressure T-tube.

But as mentioned before, line profile determinations are also necessary if one wants to measure absolute line intensities precisely, because then wing and continuum corrections must be made. Finally, since the accuracy of, for example, the calculated H_β line profile seems really to be better than 10 %, it is now indeed possible to perform the experimental test of the validity of the LTE assumption described in connection with the measurement of equilibrium temperatures and densities in a helium-filled T-tube.

7.5 Measurement of the H^- Continuum

As an example of a quantity of great interest for atomic physics and astrophysics that can probably be measured directly only in the laboratory with shock-heated gases, one may cite the recombination continuum of the negative hydrogen ion. An experiment designed to measure the intensity of this continuum was performed at Kiel (Weber, 1958) and it may serve here as a demonstration for the flexibility of the shock tube.

The main difficulty in an experiment of this kind is to obtain conditions under which the H^- continuum is much stronger than the usual bound-free and free-free continua of the atom. This is the case for high electron densities, but low temperatures (low degree of ionization), i.e., a set of conditions that is difficult to achieve in arcs.†

Therefore the experiment was done with a diaphragm-type shock tube filled with a mixture of a noble gas (krypton in this case) and hydrogen in the low-pressure section and pure hydrogen in the high-pressure section. The optimum pressure ratio H_2:Kr was found to be 20 %. For this mixture the heavy monatomic gas still determined the shock behavior, but the continuum to be observed behind the reflected shock was already sufficiently strong.

The spectrum of the shock-heated gas was measured photographically and the continuum intensity was then compared with the sum of the various calculated continua, using the temperatures and densities obtained from the Rankine-Hugoniot relations and the measured shock velocities and initial pressures.

The shapes of observed and calculated continua were found to be in good agreement, and also the measured and calculated temperature dependence. Furthermore, the discontinuity at the advanced Balmer limit (the higher series members of the Balmer series merge due to Stark broadening) was measured to be 25 %, which compares favorably with the calculated value using the Chandrasekhar theory for the H^- absorption coefficient. (For these conditions the continuum intensity would change at the Balmer limit by a factor of 30, if H^- would not contribute at all.)

One should compare this method with the direct measurements of the cross sections for photo detachment using negative hydrogen beams (Smith and Burch, 1959) irradiated crosswise with monochromatic light. Probably the latter method is capable of yielding a better accuracy for the absolute cross sections, but it will be difficult to approach the wavelength resolution of the shock tube experiment. As to the relative cross sections, the results of the two methods are in agreement and it is quite likely that a shock tube experiment, in which temperatures and densities are measured spectroscopically, preferably with a photoelectric system, will also give reliable absolute cross sections.

† Because of the small electron affinity of H (0.75 ev), the temperature should not be higher than 7000 °K in the practical pressure range. Continua from negative ions like O^-, however, can be obtained at higher temperatures because of the larger electron affinity. It is for this reason that the O^- continuum could be measured in an arc (Boldt, 1959).

REFERENCES

Allen, C. W. (1955) "Astrophysical Quantities." Univ. London Press (Athlone), London.
Allen, T. K., and Martin, J. (1960) Private communication.
Aller, L. H. (1953) "The Atmospheres of the Sun and the Stars." Ronald Press, New York.
Alpher, R. A., and White, D. R. (1959a) *Phys. of Fluids* **2**, 153.
Alpher, R. A., and White, D. R. (1959b) *Phys. of Fluids* **2**, 162.
Alyamovskii, V. N., and Kitaeva, V. F. (1960) *Optika i Spektrosk.* **8**, 152; see *Opt. Soc. Amer. Translation* **8**, 80.
Bates, D. R., and Damgaard, A. (1949) *Phil. Trans.* **A242**, 101.
Bates, D. R., and Kingston, A. E. (1961) *Nature* **189**, 652.
Berg, H. F., Lincke, R., Ali, A. W., and Griem, H. R. (1961) To be published.
Berger, J. M. (1956) *Astrophys. J.* **124**, 550.
Bethe, H. A., and Salpeter, E. E. (1957) "Quantum Mechanics of One- and Two-Electron Atoms." Academic Press, New York.
Boldt, G. (1959) *Z. Phys.* **154**, 319.
Bond, J. W. (1957) *Phys. Rev.* **105**, 1683; Los Alamos Scientific Laboratory Report LA-1693 (1954).
Boyer, K., Elmore, W. C., Little, E. M., Quinn, W. E., and Tuck, J. L. (1960) *Phys. Rev.* **119**, 831.
Burgess, A., and Seaton M. J. (1960) *Monthly Not. Roy. Astron. Soc.* **120**, 121.
Charatis, G. (1961) Dissertation, University of Michigan, Ann Arbor, Michigan.
Charatis, G., and Wilkerson, T. D. (1958) *Bull. Amer. Phys. Soc.* [2] **3**, 292.
Charatis, G., and Wilkerson, T. D. (1959) *Phys. of Fluids* **2**, 578.
Charatis, G., Doherty, L. R., and Wilkerson, T. D. (1957) *J. chem. Phys.* **27**, 1415.
Christian, R. H., and Yarger, F. L. (1955) *J. chem. Phys.* **23**, 2042.
Christian, R. H., Duff, R. E., and Yarger, F. L. (1955) *J. chem. Phys.* **23**, 2045.
Clouston, J. G., Gaydon, A. G., and Glass, I. I. (1958) *Proc. Roy. Soc.* **A248**, 429.
Clouston, J. G., Gaydon, A. G., and Hurle, I. R. (1959) *Proc. Roy. Soc.* **A252**, 143.
Cohen, R. S., Spitzer, L., and Routly, P. McR. (1950) *Phys. Rev.* **80**, 230.
Colgate, S. (1957) University of California Radiation Laboratory Report UCRL-4829.
Courant, R., and Friedrichs, K. O. (1948) "Supersonic Flow and Shock Waves." Interscience, New York.
Cowling, T. G. (1956) *Monthly Not. Roy. Astron. Soc.* **116**(1).
Daiber, J. W., and Williams, M. J. (1961) *J. Quant. Spect. and Rad. Trans.* (in press).
D'Angelo, N. (1961) *Phys. Rev.* **121**, 505.
Doherty, L. R. (1960) *Bull. Amer. Phys. Soc.* [2] **5**, 131.
Doherty, L. R. (1961) Dissertation, University of Michigan, Ann Arbor, Michigan; private communication in advance of publication.
Elmore, W. C., Little, E. M., and Quinn, W. E. (1959) *Proc. U. N. Internat. Conf. Peaceful Uses Atomic Energy, 2nd Conf. Geneva, 1958* **32**, P/356, p. 337.
Elton, R. C. (1961) Private communication (to be published).
Elwert, G. (1954) *Z. Naturforsch.* **9**a, 632.
Fairbairn, A. R., and Gaydon, A. G. (1955) *Nature* **175**, 253.
Fairbairn, A. R., and Gaydon, A. G. (1957) *Proc. Roy. Soc.* **A239**, 464.
Faizullov, F. S., Sobolev, N. N., and Kudryavtsev, E. M. (1960a) *Optika i Spektrosk.* **8**, 761; see *Opt. Soc. Amer. Translation* **8**, 400.
Faizullov, F. S., Sobolev, N. N., and Kudryavtsev, E. M. (1960b) *Optika i Spektrosk.* **8**, 585; see *Opt. Soc. Amer. Translation* **8**, 311.

Faneuff, C. E., Anderson, A. D., and Kolb, A. C. (1958) U.S. Naval Research Laboratory Report NRL-5200.

Fay, H., Hintz, E., and Jordan, H. (1960) *In* "Proceedings of the Fourth International Conference on Ionization Phenomena in Gases, Uppsala, 1959" (N. R. Nilsson, ed.), Vol. II, p. 1061. North-Holland, Amsterdam.

Finkelnburg, W., and Maecker, H. (1956) *In* "Handbuch der Physik" (S. Flügge, ed.), Vol. 22/II. Springer, Berlin.

Fishman, F. J., Kantrowitz, A., and Petschek, H. E. (1960) *Rev. mod. Phys.* **32**, 959.

Fowler, R. G., Goldstein, J. S., and Clotfelter, B. E. (1951) *Phys. Rev.* **82**, 879.

Fowler, R. G., Atkinson, W. R., and Marks, L. W. (1952a) *Phys. Rev.* **87**, 966.

Fowler, R. G., Atkinson, W. R., Compton, W., and Lee, R. (1952b) *Phys. Rev.* **88**, 137.

Fowler, R. G., Atkinson, W. R., Clotfelter, B. E., and Lee, R. (1952c) University of Oklahoma Research Institute Project Report (unpublished).

Gaydon, A. G. (1954) *Nat. Bur. Stand. (U.S.) Circ. No.* **523**, 1.

Glass, I. I., and Hall, J. G. (1959) Handbook of Supersonic Aerodynamics, Section 18: Shock Tubes, NAVORD Report 1488 (Vol. 6), available from Superintendent of Documents, U. S. Government Printing Office, Washington, D. C.

Gloersen, P. (1959) *Bull. Amer. Phys. Soc.* [2] **4**, 283.

Gloersen, P. (1960) *Phys. of Fluids* **3**, 857.

Goldberg, L. (1939) *Astrophys. J.* **90**, 414.

Goldberg, L., Müller, E. A., and Aller, L. H. (1960) *Astrophys. J. Suppl.* **5**, No. 45, 1.

Green, T. S. (1960) *Phys. Rev. Letters* **5**, 297.

Greene, E. F. (1954) *J. Amer. Chem. Soc.* **76**, 2127.

Griem, H. R. (1954) *Z. Phys.* **137**, 280.

Griem, H. R. (1960) *Astrophys. J.* **132**, 883.

Griem, H. R., and Shen, K. Y. (1961) *Phys. Rev.* **122**, 1490.

Griem, H. R., Kolb, A. C., and Faust, W. R. (1959a) *Phys. Rev. Letters* **2**, 521.

Griem, H. R., Kolb, A. C., and Shen, K. Y. (1959b) *Phys. Rev.* **116**, 4.

Griem, H. R., Kolb, A. C., and Shen, K. Y. (1960) U. S. Naval Research Laboratory Report NRL-5455.

Griem, H. R., Baranger, M., Kolb, A. C., and Shen, K. Y. (1961a) "Stark Broadening of Neutral Helium Lines" (in preparation).

Griem, H. R., Kolb, A. C., Lupton, W. H., and Phillips, D. T. (1961b) International Conference on Plasma Physics and Controlled Nuclear Fusion Research, Salzburg, Austria. *Fusion J.* In press.

Griem, H. R., Kolb, A. C., and Shen, K. Y. (1961c) Stark Broadening of H_β. *Astrophys. J.* In press.

Hain, K., and Roberts, K. V. (1960) Second Annual Meeting of the Division of Plasma Physics of the American Physics Society, Gatlinburg, Tennessee.

Hain, K., Hain, G., Roberts, K. V., Roberts, S. J., and Köppendörfer, W. (1961) *Z. Naturforsch.* **15a**, 1039.

Hain, K., and Kolb, A. C. (1961) International Conference on Plasma Physics and Controlled Nuclear Fusion Research, Salzburg, Austria. *Fusion J.* In press.

Harris, E. G. (1956) U. S. Naval Research Laboratory Report NRL-4858 (unpublished).

Hey, P. (1959) *Z. Phys.* **157**, 79.

Hill, A. J., and King, R. B. (1951) *J. Opt. Soc. Amer.* **41**, 315.

Hintz, E. (1961) Dissertation, Technische Hochschule, Aachen; International Conference on Plasma physics and Controlled Nuclear Fusion Research, Salzburg, Austria. *Fusion J.* In press.

Hollyer, R. N. (1957) Johns Hopkins University Applied Physics Laboratory Report CM-9030.
Hollyer, R. N., Hunting, A. C., Laporte, O., Schwarz, E. H., and Turner, E. B. (1952) *Phys. Rev.* **87**, 911.
Hollyer, R. N., Hunting, A. C., Laporte, O., and Turner, E. B. (1953) *Nature* **171**, 395.
Huang, S.-S. (1948) *Astrophys. J.* **108**, 354.
Jahoda, F. C., Little, E. M., Quinn, W. E., Sawyer, G. A., and Stratton, T. F. (1960) *Phys. Rev.* **119**, 843.
Janes, G. S., and Patrick, R. M. (1958) *Conf. on Extremely High Temp. Boston*, p. 3.
Josephson, V. (1958) *J. appl. Phys.* **29**, 30.
Josephson, V., and Hales, R. W. (1961) *Phys. of Fluids* **4**, 373.
Jürgens, G. (1954) *Z. Phys.* **138**, 613.
Karzas, W. J., and Latter, R. (1958) Rand Corporation Reports AECU-3703 and RM-2091-AEC.
Kazachevskaya, T. V., and Ivanov-Kholodnr, G. S. (1959) *Astron. Zh.* **36**(6).
Keck, J. C., Camm, J. C., Kivel, B., and Wentink, T. (1959) *Ann. Phys.* (*New York*) **7**, 1.
Kemp, N. H., and Petschek, H. E. (1959) *J. appl. Phys.* **2**, 599.
Kever, H. (1961) Dissertation, Technische Hochschule, Aachen; International Conference on Plasma Physics and Controlled Nuclear Fusion Research, Salzburg, Austria. *Fusion J.* In press.
King, A. S. (1914) *Astrophys. J.* **40**, 205.
King, A. S. (1922) *Astrophys. J.* **56**, 318.
Knorr, G. (1958) *Z. Naturforsch.* **13a**, 941.
Kolb, A. C. (1957a) *Phys. Rev.* **107**, 345.
Kolb, A. C. (1957b) *In* "Magnetohydrodynamics" (R. K. M. Landshoff, ed.), p. 76. Stanford Univ. Press, Stanford, California.
Kolb, A. C. (1957c) *Phys. Rev.* **107**, 4 (Letter to the Editor).
Kolb, A. C. (1959a) *Proc. U. N. Internat. Conf. Peaceful Uses Atomic Energy, 2nd Conf. Geneva, 1958* **31**, P/345, p. 332.
Kolb, A. C. (1959b) *Phys. Rev.* **112**, 291.
Kolb, A. C. (1960a) *In* "Plasma Dynamics" (F. H. Clauser, ed.), p. 206. Addison-Wesley, Reading, Massachusetts.
Kolb, A. C. (1960b) *In* "Proceedings of the Fourth International Conference on Ionization Phenomena in Gases, Uppsala, 1959" (N. R. Nilsson, ed.), Vol. II, p. 1021. North-Holland, Amsterdam.
Kolb, A. C. (1961) "Symposium on Optical Spectrometric Measurements of High Temperatures." Univ. Chicago Press, Chicago, Illinois. In press.
Kolb, A. C., Dobbie, C. B., and Griem, H. R. (1959) *Phys. Rev. Letters* **3**, 5.
Kolb, A. C., Griem, H. R., and Faust, W. R. (1960) *In* "Proceedings of the Fourth International Conference on Ionization Phenomena in Gases, Uppsala, 1959" (N. R. Nilsson, ed.), p. 1037. North-Holland, Amsterdam.
Kvartskava, I. F., Kervalidze, K. N., and Gvaladze, J. S. (1960) *In* "Proceedings of the Fourth International Conference on Ionization Phenomena in Gases, Uppsala, 1959" (N. R. Nilsson, ed.), p. 876. North-Holland, Amsterdam.
Ladenburg, R., and Levy, S. (1934) *Z. Phys.* **88**, 461.
Lamb, L., and Lin, S.-C. (1957) *J. appl. Phys.* **28**, 754.
Laporte, O. (1960) *In* "Combustion and Propulsion, Third Agard Colloquium" p. 499. Pergamon, New York.
Laporte, O., and Wilkerson, T. D. (1960) Randall Festschrift. *J. Opt. Soc. Amer.* **50**, 1293.
Lin, S.-C., Resler, E. L., and Kantrowitz, A. (1955) *J. Appl. Phys.* **26**, 95.

McLean, E. A. (1961) The measurement of transition probabilities using a magnetic shock tube. *In* "Third Symposium on Temperature, Its Measurement and Control in Science and Industry, 1961, Columbus, Ohio" (to be published).

McLean, E. A., Faneuff, C. E., Kolb, A. C., and Griem, H. R. (1960) *Phys. of Fluids* **3**, 843.

McLean, E. A., Kolb, A. C., and Griem, H. R. (1961) *Phys. of Fluids* **4**, 1055.

McWhirter, R. W. P. (1961) *Nature* **190**, 902.

Mark, H. (1957) *J. Aeronaut. Sci.* **24**, 304.

Mastrup, F., and Wiese, W. (1958) *Z. Astrophys.* **44**, 259.

Motschmann, H. (1955) *Z. Phys.* **143**, 77.

Mott-Smith, H. M. (1951) *Phys. Rev.* **82**, 885.

Nagle, D. E., Quinn, W. E., Ribe, F. L., and Riesenfeld, W. B. (1960) *Phys. Rev.* **119**, 857.

Niblett, G. B. F., and Green, T. S. (1959) *Proc. Phys. Soc.* **74**, 737.

Niblett, G. B. F., and Kenny, A. (1957) Princeton University Report NR061-020, N6ori-105.

Niblett, G. R. F., and co-workers (1960) Private communication.

Parkinson, W. H., and Nicholls, R. W. (1960) *Canad. J. Phys.* **38**, 715.

Patrick, R. M. (1959) *J. appl. Phys.* **2**, 589.

Petschek, H. E., and Byron, S. (1957) *Ann. Phys.* (*New York*) **1**, 270.

Petschek, H. E., Rose, P. H., Glick, H. S., Kane, A., and Kantrowitz, A. (1955) *J. appl. Phys.* **26**, 83.

Post, R. F. (1960) *In* "Plasma Dynamics" (F. H. Clauser, ed.), p. 30. Addison-Wesley, Reading, Massachusetts; Riso Lectures (to be published).

Prigsheim, P. (1949) "Fluorescence and Phosphorescence." Interscience, New York.

Ramsden, S. A., Hearn, A. G., and Jones, B. B. (1961) *Bull. Amer. Phys. Soc.* **6**, 205.

Resler, E. L., Lin, S.-C., and Kantrowitz, A. (1952) *J. appl. Phys.* **23**, 1390.

Richter, J. (1958) *Z. Phys.* **151**, 114.

Richtmeyer, R. D. (1957) "Difference Methods for Initial Value Problems," p. 208. Interscience, New York.

Rink, J. P., Knight, H. T., and Duff, R. E. (1961) *Bull. Amer. Phys. Soc.* [2] **6**, 213.

Rosa, R. J. (1955) *Phys. Rev.* **99**, 633.

Rosenbluth, M., and Garwin R. (1954) Los Alamos Scientific Laboratory Report LA-1850.

Roth, W., and Gloersen, P. (1958) *J. chem. Phys.* **28**, 820.

Schlüter, A., and Biermann, L. (1950) *Z. Naturforsch.* **5a**, 237.

Schlüter, A., and Biermann, L. (1958) *Rev. mod. Phys.* **30**, 975.

Scott, F. R., and Wentzel, R. F. (1959) *Phys. of Fluids* **2**, 609.

Scott, F. R., Basman, W. P., Little, E. M., and Thompson, D. B. (1958) *In* "The Plasma in a Magnetic Field" (R. K. M. Landshoff, ed.), p. 110. Stanford Univ. Press, Stanford, California.

Seay, G. L. (1957) Dissertation, University of Oklahoma, Norman, Oklahoma; Los Alamos Scientific Laboratory Report LAMS-2125 (unpublished).

Shreffler, G., and Christian, R. H. (1954) *J. appl. Phys.* **25**, 324.

Smith, S. J., and Burch, D. S. (1959) *Phys. Rev.* **116**, 1125.

Sobolev, N. N., Potapov, A. V., Kitaeva, V. F., Faizullov, F. S., Alyamovskii, V. N., Antropov, E. T., and Isaev, I. L. (1959) *Optika i Spektrosk.* **6**, 185, 284.

Stratton, T. F., Sawyer, G. A., and Ribe, F. L. (1960) Private communication; Los Alamos Scientific Laboratory Report 2488.

Townsend, J. S., and Bailey, V. A. (1922a) *Phil. Mag.* **43**, 593.

Townsend, J. S., and Bailey, V. A. (1922b) *Phil. Mag.* **44**, 1033.

Treanor, C. E. (1961) Private communication.

Treanor, C. E., and Wurster, W. H. (1960) *J. chem. Phys.* **32**, 758.

Trefftz, E., Schlüter, A., Dettmar, K. H., and Jörgens, K. (1957) *Z. Astrophys.* **44**, 1.

Turner, E. B. (1955a) *Phys. Rev.* **99**, 633.

Turner, E. B. (1955b) University of Michigan Engineering Research Institute Report 2189-1-T.

Turner, E. B. (1956) Dissertation, University of Michigan, Ann Arbor, Michigan; University of Michigan Engineering Research Institute Report 2189-2-T, AFOSR TN-56-150, ASTIA document No. AD 86309 (unpublished).

Turner, E. B. (1958) Space Technology Laboratory Report GM-TR-0165-00460.

Turner, E. B. (1959) Space Technology Laboratory Report TR-59-0000-00744.

Unsöld, A. (1948) *Z. Pstrophys.* **24**, 355.

Unsöld, A. (1955) "Physik der Sternatmosphären," 2nd ed. Springer, Berlin.

Voorhees, H. G., and Scott, F. R. (1959) *Phys. of Fluids* **2**, 576.

Weber, O. (1958) *Z. Phys.* **152**, 281.

Weymann, H. D. (1958) University of Maryland Institute for Fluid Dynamics, Tech. Note BN-144.

Weymann, H. D. (1960) *Phys. of Fluids* **3**, 545.

Weymann, H. D., and Troy, B. (1961) *Bull. Amer. Phys. Soc.* [2] **6**, 212.

Wiese, W. (1960) *Rev. sci. Instrum.* **31**, 943.

Wiese, W., Berg, H. F., and Griem, H. R. (1960) *Phys. Rev.* **120**, 1079.

Wiese, W., Berg, H. F., and Griem, H. R. (1961) *Phys. of Fluids* **4**, 250.

Wilkerson, T. D. (1961) Dissertation, University of Michigan, Ann Arbor, Michigan; Univ. Michigan Office of Research Administration Rept. O 2822-3-T, AFOSR1115 (June, 1961).

6.

Attachment and Ionization Coefficients

A. N. Prasad and J. D. Craggs

1 *Electron Attachment*

The following chapter gives a summary, necessarily brief and incomplete, of the newer work on electron attachment cross sections and coefficients. To do this, an account of the recent associated work on Townsend's primary and secondary ionization coefficients has been included, because of the importance that these constants have in relation to one of the most important methods of measuring attachment coefficients.

Determination of attachment cross sections. Virtually no absolute measurements of low-pressure, single collision condition, attachment cross sections were made until the middle 1950's. Thus, in "Negative Ions" (Massey, 1950) the only data cited are those of Buchdahl (1941) who used a Lozier ionization chamber (Lozier, 1934) to measure attachment in iodine vapor. The electron path lengths were uncertain and the results were therefore not accurate. Buchdahl's negative ions were predominantly I^- from dissociated I_2^{-*} formed by direct capture.

Despite many studies (e.g., Hagstrum, 1951) of the critical potentials for formation of negative ions by resonance capture (dissociative attachment)

$$AB + e \rightarrow A + B^- \tag{1}$$

and by ion pair formation

$$AB + e \rightarrow A^+ + B^- + e \tag{2}$$

no serious attempt at measuring negative ion currents was made until quite recently.

The early work has been reviewed by Massey (1950) and by Craggs and Massey (1959), but a few specific references will be given below.

Buchel'nikova (1958a) has described an ionization chamber which she has used for attachment studies, i.e., for the determination of critical potentials and cross sections, in single collision conditions. The RPD technique due to Fox *et al.* (1951) was used, together with the pulse technique for ensuring field free ion collection. In a suitably designed chamber the latter should not be necessary since field penetration into the collector region from accelerating electrodes, etc., can usually be made negligible, unless high voltage extraction as with a mass spectrometer is used.

In Buchel'nikova's first paper, some preliminary data for SF_6 were briefly described. The earlier results of Hickam and Fox (1956) for appearance potentials of SF_6^- were confirmed, the critical energy for which capture occurs was found to be 0 ± 0.01 ev, and the resonance capture curve width was taken to be 0-0.5 ev, giving a maximum capture cross section of $1.2 \pm 0.4 \times 10^{-15}$ cm^2. However, SF_6^- ions were observed for electron energies of 3 ev or more. Figure 1 (Hickam and

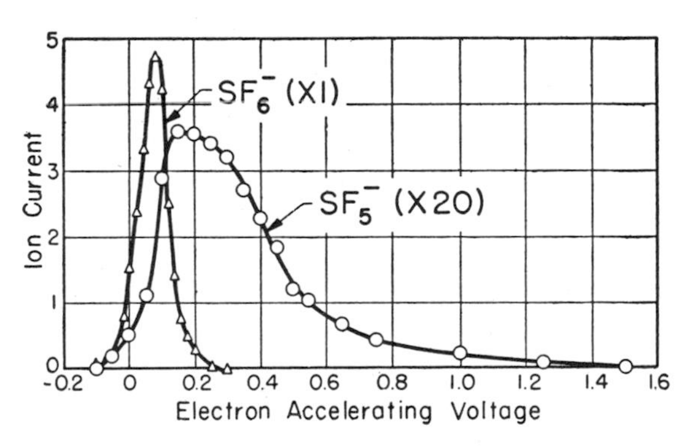

FIG. 1. SF_6^- and SF_5^- ion currents as a function of the electron accelerating voltage. The curves were obtained with a mass spectrometer employing the RPD method. (From Hickam and Fox, 1956.)

Fox, 1956) shows the SF_6^- and much smaller SF_5^- ion currents as a function of electron energy. These data were the first to be taken for low-energy negative ion processes with any degree of precision on the energy scale and demonstrate very clearly the remarkable success of the RPD method.

In her later paper Buchel'nikova (1958b) gives further data for oxygen and water vapor and for certain halogen molecules (SF_6, CCl_4, CF_3I, CCl_2F_2, BCl_3, HCl, and HBr). She did not carry out measurements at higher energies, such as those required for the ion-pair process. Her data are summarized in Table 1.

TABLE I

ATTACHMENT CROSS SECTIONS[a]

Molecule	Cross section at the 1st maximum (cm^2)	Electron energy at the 1st maximum (ev)	Cross section at the 2nd maximum (cm^2)	Electron energy at the 2nd maximum (ev)
SF_6	5.7×10^{-16}	0.00		0.1
CCl_4	1.3×10^{-16}	0.02	1.0×10^{-16}	0.6
CF_3I	7.8×10^{-17}	0.05	3.2×10^{-17}	0.9
CCl_2F_2	5.4×10^{-17}	0.15		
BCl_3	2.8×10^{-17}	0.4		
HBr	5.8×10^{-17}	0.5		
HCl	3.9×10^{-18}	0.6		
H_2O	4.8×10^{-18}	6.4	1.3×10^{-18}	8.6
O_2	1.3×10^{-18}	6.2		

[a] Buchel'nikova (1958b).

Figure 2 shows the oxygen data as obtained by Buchel'nikova and Craggs *et al.* The difference between the two sets of data is greater than the sum of the fairly large experimental errors and further investigations (now being carried out in Liverpool with a new and different form of ionization chamber) are necessary.

The work of Thorburn, Tozer, and Craggs on capture cross section measurements was carried out with a Lozier ionization chamber (Lozier, 1934). The apparatus was later used for cross section measurements in oxygen, carbon monoxide, and carbon dioxide.

A description of the performance of the apparatus has been given by Tozer (1958) who studied, for example, collection efficiencies in some detail. It is necessary to allow for the varying efficiency of collection of ions with varying initial kinetic energy, as Tozer has shown. This is particularly important in the case of oxygen, where the O^- and O_2^+ ions have different kinetic energies, the latter of course have virtually thermal energies while the O^- ions may have initial kinetic energies of more than 1 ev.

The cross sections for attachment in oxygen were measured by Craggs *et al.* (1957) by comparing the O^- current in the Lozier apparatus with

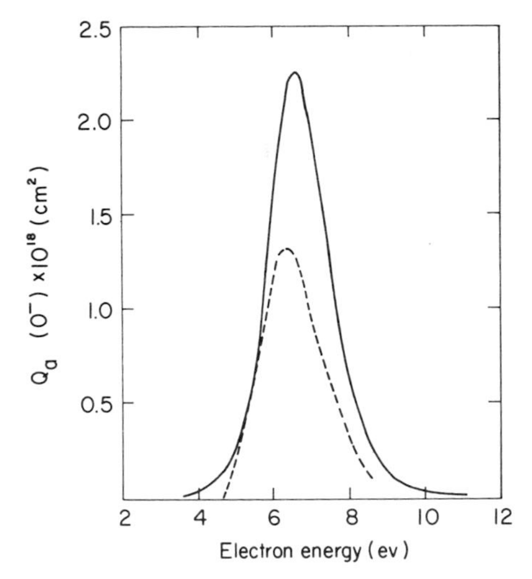

FIG. 2. Resonance capture peak in oxygen: ——, Craggs *et al.* (1957); - - -, Buchel'nikova (1958b).

O_2^+ and with Ar^+ currents. The standard data were those of Tate and Smith (1932) for oxygen and argon, although it was found that if the O_2^+ currents were normalized at, say, 38 ev (in the neighborhood of which the Tate and Smith data follow precisely the same variation with energy as those of Craggs *et al.*), there were discrepancies immediately above the ionization potential. Thus, at 20 ev, the latter data were only 75 % of those due to Tate and Smith, and it was suggested (Craggs *et al.*, 1957) that the difference is due to the neglect by Tate and Smith of any spiraling by the electrons in the magnetic field in their apparatus. This needs further study. After various corrections had been made, the data of Fig. 2 were obtained, with the maximum value of the attachment cross section $Q_a(O^-)$ of 2.25×10^{-18} cm².

The next data were obtained by Craggs and Tozer (1958) on carbon monoxide. Again, no previous values were available and the same procedure was followed. It was found in carbon monoxide that the ion pair process, interpreted as

$$CO \rightarrow C^+ + O^- \text{ (about 21 ev)} \tag{3}$$

and

$$CO \rightarrow C^- + O^+ \text{ (about 23 ev)} \tag{4}$$

is more abundant, relative to the capture peak, than in oxygen and further that the O^- ions in the resonance capture peak which appeared at 9.35 ± 0.1 ev for ions of zero initial kinetic energy had very small initial kinetic energies, cf. O^- in oxygen. A maximum cross section of $2.7 \pm 0.3 \times 10^{-18}$ cm² at 10.1 ev was obtained.

Finally (Craggs and Tozer, 1960), data were obtained for carbon dioxide in which no previous observations of electron attachment in low-pressure single collision or in swarm conditions had been made. The relevant process, checked with a mass spectrometer, is

$$CO_2 + e \rightarrow CO + O^-, \tag{5}$$

i.e., a resonance capture. No ion pair process was observed. The cross sections are shown in Fig. 3 with a maximum value of $5.1 \pm 0.5 \times 10^{-19}$ cm² at 7.8 ev. In the course of these measurements on carbon dioxide, Tozer made observations of the resonance capture peak at various retarding potentials and, as a result, was able to offer an explanation for the high value, about 2.2 ev, for the electron affinity of O, i.e., $EA(O)$, in electron collision experiments, as compared with Branscomb's

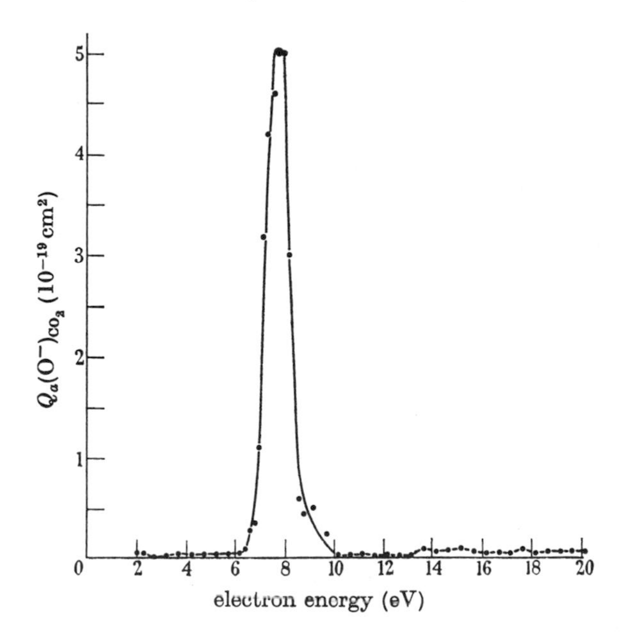

FIG. 3. Attachment cross sections in CO_2. (From Craggs and Tozer, 1960.)

value (Branscomb *et al.*, 1958), by photodetachment, of about 1.45 ev. On making corrections for initial kinetic energy effects the $EA(\mathrm{O})$ values for carbon dioxide, carbon monoxide, and oxygen, respectively, are 1.2 ± 0.3, 1.6 ± 0.2, and 1.5 ± 0.2 ev in agreement with Branscomb's data.

2 *Ionization Coefficients*

It has already been explained (§ 1.1) that it is necessary to describe recent measurements of ionization coefficients in order to give details of attachment coefficients measured by the same methods. Both coefficients are connected, not only in Townsend's method, i.e., the measurement of non-self-maintained field intensified currents in uniform field conditions, but also in Raether's avalanche methods. Further, of course, the ionization coefficients themselves are of interest in connection with mechanisms of electrical breakdown in gases, so for that reason, and for their relation to ionization cross sections (Chapter 12) through energy distributions, they will be discussed here.

2.1 Townsend's Primary Ionization Coefficient

Townsend's primary ionization coefficient, α, is defined as the mean number of ionizing collisions, i.e., ion pairs produced, per electron per centimeter of drift path in the field direction. It was first introduced by Townsend (1915) to describe the growth of prebreakdown ionization currents in uniform fields in gases. The coefficient α is therefore an important parameter in the spark breakdown of gases and the low current discharges. From simple kinetic theory considerations it may be shown that α/p is a function of E/p (p = gas pressure, mm Hg and E = applied field, volts cm^{-1}), if no pressure-dependent ionization is present in the gas (Townsend, 1915). Some investigators (Druyvesteyn and Penning, 1940) have introduced a coefficient η in place of α. The coefficient η is then defined as the mean number of ion pairs produced per electron per volt potential difference. Consequently,

$$\alpha = \eta E. \tag{6}$$

However, since α is more widely employed in the literature, it will be adopted throughout this chapter and η is later utilized to describe electron attachment in gases.

In uniform field conditions, a single electron in traversing a distance d cm in the direction of the field would produce on the average $\exp(\alpha d)$ new electrons in the absence of any secondary effects. This multiplication caused by a single electron is termed an "electron avalanche," and the build-up of ionization in a gas is accordingly produced by a succession of electron avalanches.

2.2 Determination of Ionization Coefficients

2.2.1 *The Method of Townsend*

Following the theory developed by Townsend (1915, 1947) and by subsequent workers (Loeb, 1955), the steady-state growth of prebreakdown currents at any particular value of E/p in uniform fields, due to ionization by electron impact and secondary electron production, can be represented as a function of the electrode separation (d cm) by

$$\frac{I}{I_0} = \frac{\exp(\alpha d)}{[1 - \gamma\{\exp(\alpha d) - 1\}]}.$$

Here I_0 is a small† externally generated initiatory electron current (e.g., photoelectric) and γ is Townsend's secondary coefficient describing electron production due to secondary processes (e.g., secondary emission at the cathode due to incidence of positive ions, or photons formed in the gas). Since in practice several such secondary processes act in concert, γ represents a generalized secondary coefficient describing several secondary processes. In obtaining this equation it has been assumed that the initiatory electrons are in equilibrium with the applied field (i.e., $d \gg$ electron mean free path) and that space charge effects are absent. The latter condition is in general satisfied if the current density corresponding to I_0 is kept $< 10^{-10}$ amp/cm^2 (Crowe *et al.*, 1954). In addition, it is assumed that loss of electrons by attachment and diffusion are negligible.

At any particular value of E/p, a plot of log I against d (or applied potential) is initially linear with an upcurving at large values of d. For $\gamma \sim 10^{-4}$ or less, as would be the case in most molecular gases at low E/p (Loeb, 1955), the mean slope of the initial linear part of the plot yields α. For $\gamma > 10^{-4}$ (e.g., rares gases, hydrogen at high E/p values), the coefficients α and γ have to be evaluated simultaneously from the log I, d plot.

† I_0 — 10^{-12} to 10^{-11} amp.

In electron attaching gases, on the other hand, additional parameters are necessary to account for the removal of electrons by the various attachment processes. Negative ions may result from direct or resonance capture or from ion pair formation. Defining rate coefficients η and δ analogous to α to describe the rates of capture and ion pair formation, respectively, per electron per centimeter of drift in the field direction, it can be shown that Townsend's current growth equation is modified to

$$\frac{I}{I_0} = \frac{\left[\frac{(\alpha + \delta)}{(\alpha - \eta)} \exp(\alpha - \eta)d + \frac{(\eta + \delta)}{(\alpha - \eta)}\right]}{\left[1 - \gamma\left(\frac{\alpha + \delta}{\alpha - \eta}\right)\{\exp(\alpha - \eta)d - 1\}\right]} \tag{8}$$

(Harrison and Geballe, 1953; Howard, 1957). However, since in most molecular gases the δ processes occur at relatively high electron energies (electron energies > ionization energy of the constituent atom or radical) and the cross sections for these processes are in general smaller than those for low-energy processes (§ 1), the contributions to the attachment due to the δ processes can in general be neglected at low values of the mean energy of the electron swarm, i.e., at low values of E/p (Prasad, 1960). Equation (8) is reduced to

$$\frac{I}{I_0} = \frac{\left[\frac{\alpha}{\alpha - \eta} \exp(\alpha - \eta)d - \frac{\eta}{\alpha - \eta}\right]}{\left[1 - \frac{\gamma\alpha}{\alpha - \eta}\{\exp(\alpha - \eta)d - 1\}\right]} \tag{9}$$

(Penning, 1938; Harrison and Geballe, 1953; Prasad, 1959). This approximation is valid only at low values of E/p (§ 3). Under these conditions the log I, d plot would be initially nonlinear (i.e., with a concavity towards the axis representing d) with an upcurving at large values of d. The coefficients α, η (and γ) are then to be evaluated simultaneously by a process of curve fitting. The accuracies attainable have been considered at some length by Prasad (1960) and Bhalla and Craggs (1960), and it would appear that the maximum accuracy is attainable in general only in the case of strongly attaching gases (e.g., oxygen). Further, at high values of E/p where $\alpha/p > \eta/p$, the exponential term in (9) tends to dominate the subtractive term, and accordingly the curvature in these plots would be confined to low values of d. The log I, d plot for large values of d (or for high values of E/p) closely resembles that in a nonattaching gas and only the coefficient $(\alpha - \eta)$ can be obtained with any accuracy (Prasad, 1959). In addition, at very high values of E/p (where mean energies are generally 5-10 ev) contributions due to the δ

processes are likely to be appreciable and a simultaneous evaluation of the coefficients α, η, δ (and γ) is complex and tedious, and approximations regarding the relative significances of these processes have to be employed which in general limit the over-all accuracies in determining α/p.

2.2.2 *Electron Avalanche Method*

While the steady-state experimental method of Townsend described above has hitherto been the accepted method of evaluating α/p in gases, a new technique based on the statistical distribution of electron avalanche sizes has been developed by Raether and his colleagues (Schlumbohm, 1960) for the study of α/p over a limited range of E/p. Although the average multiplication in the gas due to a single electron is given by $\exp(\alpha d)$, Wijsman (1949) and Legler (1955) have pointed out that the multiplication due to individual electrons is distributed statistically. Following Wijsman, the probability for the existence of an "electron avalanche" with a multiplication between n and $(n + dn)$ in uniform fields at any particular E/p can be written as

$$P(n) = \frac{1}{\bar{n}} \exp\left(\frac{-n}{\bar{n}}\right) \tag{10}$$

where $\bar{n} = \exp(\alpha d)$. A semilogarithmic plot of $P(n)$ or the current pulse height due to an avalanche against n would be linear with a slope of $-1/\bar{n}$. From the slope of such a plot α may be obtained.

Frommhold (1956, 1958) and Schlumbohm (1958, 1959) have studied the validity of the above statistical distribution over a wide range of values of n (from $10^{-2}\,\bar{n}$ to $10^{2}\,\bar{n}$) by employing very low values of I_0 ($\sim 10^{-16}$ amp). With an over-all accuracy of better than 10 % in the evaluation of $\bar{n}$, the accuracies in the evaluation of α would be better than 1 %. The log $P(n)$, n plots, however, appear to deviate from the linear form under the following conditions: (a) when gas amplification is $> 10^6$ and an appreciable space charge distortion of the applied field is present, (b) when avalanches are produced by secondary electrons and the apparatus cannot resolve them, and (c) when the parameter $E/V_1\alpha$ (in which V_1 is the first ionization potential of the gas) is very low, i.e., for high values of E/p (> 120).

When electron attachment occurs to an appreciable extent in the gas under investigation (e.g., as in oxygen) the foregoing equation for the statistical distribution of electron avalanches appears to be valid if $\bar{n}$ is written as equal to $[\alpha/(\alpha - \eta)]^2 \exp(\alpha - \eta)d$ assuming that the attachment is due predominantly to the η process discussed above. From measurements of

$\bar{n}$ at two different values of pd for any given value of E/p, the coefficients α and η may be evaluated simultaneously. The accuracy under these conditions is, however, relatively poor and the method requires further detailed studies, particularly at high values of E/p where contributions due to δ processes are apt to be appreciable. Further, unlike the experimental method of Townsend, no clear indication of the existence of electron attachment in the gas under investigation is available in this technique. It therefore requires prior knowledge of the existence and the mechanisms of negative ion formation in the gas under consideration from independent experiments (e.g., mass-spectrometric studies) before investigations can be attempted.

2.3 Ionization Coefficients in Gases

Since the early work of Townsend and his colleagues (Townsend, 1915, 1947) numerous investigations of α/p in various gases have been carried out by several workers. Table II summarizes some of the more important of these investigations.

2.3.1 *Rare Gases*

The early data of Kruithof and Penning in the rare gases argon (1936), neon (1937), neon-argon mixtures (1937), and of Kruithof (1940) in argon, krypton, and xenon constitute the most extensive and probably the

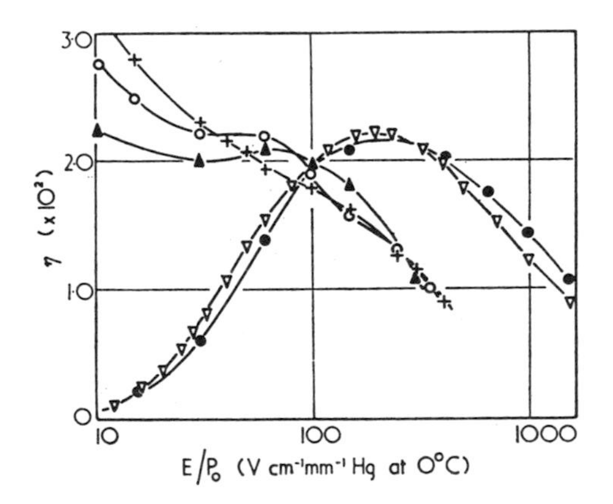

Fig. 4. Ionization coefficient η ($= \alpha/E$) as a function of E/p in Ne and Ar. ● = argon (Davies and Milne, 1959; Davies *et al.*, 1960), ∇ = argon (Kruithof, 1940), ▲ = neon (sample 1), ○ = neon (sample 2) (Davies and Milne, 1959; Davies *et al.*, 1960), + = neon (+ 0.1% argon) (Kruithof and Penning, 1937). (From Davies and Milne, 1959.)

TABLE II

IONIZATION STUDIES

Gas	Observers	Approximate range of E/p	Comments
Helium	(1) Townsend and McCallum (1934)	12-100	(1) These constitute the only available data to date in pure helium ($p < 10$ mm Hg). Agreement with the calculated values is good.
	(2) Peacock (1960)	2-6	(2) Impure helium (0.13% N_2) employed at $p = 50$ to 700 mm Hg.
		2-3	Impurity of 1 in 10^5 to 10^4 of air at $p = 700$ mm Hg. In both samples α/p values are considerably higher than those of Townsend and McCallum, due to ionization of impurities by electron impact and metastable He.
Neon	(1) Townsend and McCallum (1928)	7.5-200	(1) Data deviate slightly from those of Kruithof and Penning at high E/p's indicating minor contamination.
Neon-argon mixtures	(2) Kruithof and Penning (1937)	2-400	(2) Probably the most accurate data available to date.
	(3) Davies and Milne (1959)	6-350	(3) Correspond to 10^{-2}% impurity of argon if compared with data of Kruithof and Penning. Ultrahigh vacuum techniques and spectroscopically pure gas were employed with an all-glass ionization chamber and evaporated copper film electrodes (see text for discussion).
Argon	(1) Kruithof and Penning (1936), Kruithof (1940)	5-1600	(1) Probably the most accurate data available to date ($p = 0.7$ to 160 mm Hg).
	(2) Huxford (1939), Engstrom and Huxford (1940)	5-2000	(2) Data exhibit a pronounced scatter with respect to those of Kruithof and Penning.
	(3) Davies and Milne (1959)	15-1500	(3) Data deviate from those of Kruithof and Penning at high and low values of E/p (see also Davies and Milne for neon).

Krypton and xenon	Kruithof (1940)	5-2000 6-2500	Probably the most accurate data available to date.
Hydrogen	(1) Hale (1938, 1939a, b) (2) Hopwood *et al*, (1956) (3) Crompton *et al.* (1956) (4) Rose (1956) (5) De Bitetto and Fisher (1956) (6) Varnerin and Brown (1950) (7) Jones and Llewellyn-Jones (1958) (8) Davies and Milne (1959), Davies *et al.* (1960) (9) Blevin *et al.* (1957)	(1) 22.7-1500 (2) 15-40 (3) 16-25 (4) 15-1000 (5) 14-22 (6) 15-100 (7) 40-350 (8) 40-500 (9) 50-160	The data of (1) have been superseded by those of (4), (7), and (8). Cylinder H_2 at high pressure was employed by (2), (3), and (5) [99.95% purity by (2) and 99.6% purity by (5)] and the data obtained are in close agreement with those of (4), (7), and (8). The data of (4), (7), and (8) in high purity gas constitute perhaps the best available to date despite the marked discrepancies between the data of (4), on the one hand, and (7) and (8), on the other, at $E/p > 130$. These arise mainly from the neglect of contributions from γ in the evaluation of α. Marked difference exists also between (7) and (8) at high E/p due possibly to small differences in gas purity (see text for a full discussion).
Deuterium	Rose (1956)	18-600	The only available accurate data to date. The gas was prepared from decomposition of uranium deuteride.
Nitrogen	(1) Ayres (1923) (2) Masch (1932) (3) Posin (1936) (4) Bowls (1938) (5) Dutton *et al.* (1952) (6) De Bitetto and Fischer (1956) (7) Harrison (1957) (8) Heylen (1959) (9) Peacock (1960)	(1) 10-3000 (2) 25-500 (3) 20-1000 (4) 59-1000 (5) 41-45 (6) 30-45 (7) 25-80 (8) 30-85 (9) 25-60	The gas samples employed by (1)-(3) were mercury contaminated and hence the data are of little use except perhaps at high E/p. Cylinder N_2 at high pressures was employed by (6) and (9) (99.9 and 99.95% purity, respectively) and the data obtained are in close agreement with those of (8). Three grades of N_2, i.e., (a) sodium azide N_2 (impurity $< 0.01\%$), (b) purified cylinder N_2 (impurity 0.06% O_2 and 0.04% Ar), and (c) cylinder N_2 (99.9% purity) were employed by (7). However, data in all three cases differs widely from those obtained by (8) employing 99.998% pure N_2. Further, the data of (8) are in good agreement with those of the others suggesting insensitivity to impurity. Hence, the differences between (7) and (8) are inexplicable at present.

TABLE II (*continued*)

Gas	Observers	Approximate range of E/p	Comments
Mercury vapor	Badareu and Bratescu (1944)	150-1400	Appear to be the only comprehensive data available to date.
Oxygen	(1) Masch (1932) (2) Harrison and Geballe (1953) (3) Prasad and Craggs (1961) (4) Schlumbohm (1959)	(1) 30-350 (2) 25-75 (3) 30-50 (4) 40-65	The gas employed by (1) was mercury contaminated and further electron attachment was neglected. Hence, the data are of use only at high E/p (> 100). The data obtained by (2) and (3) are in close agreement (to within 5%). The electron avalanche method was employed by (4) and the scatter in the data obtained is pronounced although these are in general agreement with those of (2) and (3). The gas purity in the experiments of (3) was 99.5% while chemically pure gas was manufactured *in situ* by (2).
Dry air	(1) Sanders (1932, 1933) (2) Masch (1932) (3) Townsend (1947) (4) Hochberg and Sandberg (1942) (5) Llewellyn-Jones and Parker (1952) (6) Harrison and Geballe (1953) (7) Prasad (1959, 1960) (8) Dutton *et al.* (1960)	(1) 20-160 (2) 30-500 (3) 100-1000 (4) 38-45 (5) 39-45 (6) 25-65 (7) 25-45 (8) 45-40	The gas employed by (1) to (4) was Hg contaminated and electron attachment was neglected. Although the data of (5) were obtained with care, the results are in error due to neglect of attachment being closely $= (\alpha - \eta)/p$. The η/p observed by (6) appear to be appreciably in error and hence α/p values obtained by these may also be slightly in error. The data of (7) are probably the most comprehensive although the accuracy is relatively poor. The η/p data obtained by (8) differ widely from those of (6) and (7) and no values of α/p are quoted although $(\alpha - \eta)/p$ obtained is shown to agree with the α/p obtained by (5).
Humid air	Prasad and Craggs (1960a)	30-45	α/p and η/p values are given for various concentrations of water vapor. Results indicate that these increase markedly at high concentrations. Values of $(\alpha - \eta)/p$ are given for $E/p = 45$ to 50.

Carbon monoxide	Bhalla and Craggs (1961a)	36-200	Spectroscopically pure gas at $p = 2.5$ to 100 mm Hg was employed and electron attachment was observed. The values of η/p are given for $E/p = 35$ to 60. α/p and η/p in general are considerably smaller than those in dry air.
Hydrogen chloride	(1) Townsend (1915) (2) Bailey and Duncanson (1930)		Gas purities are uncertain and the data require further confirmation.
Water vapor	(1) Townsend (1915) (2) Bailey and Duncanson (1930) (3) Prasad and Craggs (1960a)	(1) 75-100 (3) 25-50	Electron attachment was neglected in (1) and (2). Very strong attachment was observed by (3). Values of η/p are also available.
Carbon dioxide	(1) Bishop (1911) (2) Schlumbohm (1959) (3) Bhalla and Craggs (1960)	(1) 40-1200 (2) 44-70 (3) 26-1200	Gas purity in the experiments of (1) are uncertain, while electron attachment was neglected. The data of (3) are the most accurate and comprehensive available to date. Spectroscopically pure gas at $p = 1$ to 100 mm Hg was used and electron attachment was observed; η/p values are also given. The data of (2) were obtained using the electron avalanche method and are in close agreement with values of $(\alpha - \eta)/p$ obtained by (3).
Methane	Schlumbohm (1959)	40-80	The data were obtained employing the electron avalanche method.
Isopentane	Harrison and Geballe (Loeb, 1955)	70-130	—
Ammonia	Bailey and Duncanson (1930)		—
SF_6	(1) Harrison and Geballe (Loeb, 1955) (2) Hochberg and Sandberg (1946) (3) Bhalla and Craggs (1961b)	80-160	Strongly attaching gases possessing high dielectric strengths. Studies were in general confined to high values of E/p. α/p and η/p were calculated simultaneously. In these gases $\alpha/p < \eta/p$ at low values of E/p. Attachment was neglected in the studies of Hochberg and Sandberg.
CCl_4 CF_3SF_5 CCl_2F_2 $SiCl_4$	Harrison and Geballe (Loeb, 1955)	255-335 175-225 80-210 100-200	

TABLE II (*continued*)

Gas	Observers	Approximate range of E/p	Comments
Carbon tetrafluoride (CF_4)	Howard (1958)	—	Values of both α/p and η/p are given in this highly attaching gas.
C_2H_5Cl C_2H_5Br C_5H_{12} CCl_4 $CHCl_3$	Hochberg and Sandberg (1946)		Strongly attaching gases possessing high dielectric strengths. Attachment was, however, neglected in the evaluation of α/p. Thus, data represent $(\alpha - \eta)/p$.
Ethyl alcohol (C_2H_5OH)	Raether (1937)	40-200	—
Cyclohexane (C_6H_{12})	Badareu and Valeriu (1941)	200-3000	Values of α/p are very large compared with those in molecular gases.
Benzene (C_6H_6)	Valeriu (1943)	—	—
Toluene (C_7H_8)			
Methyl alcohol (CH_3OH)	Schlumbohm (1960)	48-68	Data obtained employing the avalanche method. All these confirm to $\alpha/p = A \exp - B/(E/p)$ over these ranges of E/p. The values of the constants A and B are also given.
Diethylethane ($C_2H_5OC_2H_5$)		72-116	
Acetone [$CO(CH_3)_2$]		74-110	
Methylal [$CH_2(OCH_3)_2$]		60-90	
Cyclohexane (C_6H_{12})		89-108	

most accurate available to date. The data of Kruithof are compared with the more recent data of Davies and Milne (1959; Davies *et al.*, 1960) in commercial spectroscopically pure grade samples of neon and argon in Fig. 4. The latter authors have employed an all-glass ionization chamber with evaporated copper film electrodes and by utilizing stringent ultrahigh vacuum techniques have achieved a background impurity pressure $\sim 10^{-9}$ mm Hg or less. These recent data in argon show deviations of some 15 % compared with those of Kruithof and Penning at high and low E/p's for which no satisfactory explanation is available at present. In neon, on the other hand, the data of Davies and Milne correspond to approximately 10^{-2} % impurity of argon, and this appears to vary between individual flasks of gas samples nominally of the same grade. These latter features thus indicate the necessity of further detailed studies in these gases under ultrahigh purity conditions. In the case of helium, the early data of Townsend and McCallum (1934) appear to be the only available to date over any appreciable range of E/p ($12 < E/p < 100$) and, consequently, further studies in this gas in high purity conditions are urgently required. It is, however, to be noted that these data are in close agreement with the rigorously calculated data (Dunlop, 1949) indicating that contamination of gas if any in the studies of Townsend and McCallum may not be serious.

2.3.2 *Molecular Gases*

Nonattaching Gases. Of the various molecular gases hydrogen and nitrogen appeared to have been investigated by numerous workers in recent studies employing gas samples of varying degrees of purity over different ranges of pressure and E/p's (Table II). In hydrogen, these various studies indicate that α/p is relatively insensitive to traces of impurity at low E/p's (see Fig. 5). At high values of E/p (Fig. 6) the recent measurements of Rose (1956) (accuracy $\pm$ 2 %), Davies and Milne (1959); Davies *et al.*, (1960) (accuracy $\pm$ 2 %), and Jones and Llewellyn-Jones (1958) (accuracy $\pm$ 5 %) indicate that α/p is remarkably sensitive to traces of impurity inherent in the preparation of the gas since all these investigators used ultrahigh vacuum techniques. The gas samples of Rose were prepared from thermal decomposition of uranium hydride (prepared in turn from exposing reactor grade uranium to reagent grade hydrogen) while those of Davies and Milne were obtained from electrolysis of barium hydroxide in distilled water and admitted to the chamber via a palladium thimble, and those employed by Jones and Llewellyn-Jones were doubly purified. Hydrogen from electrolysis of barium hydroxide was transmitted through a palladium thimble on to activated

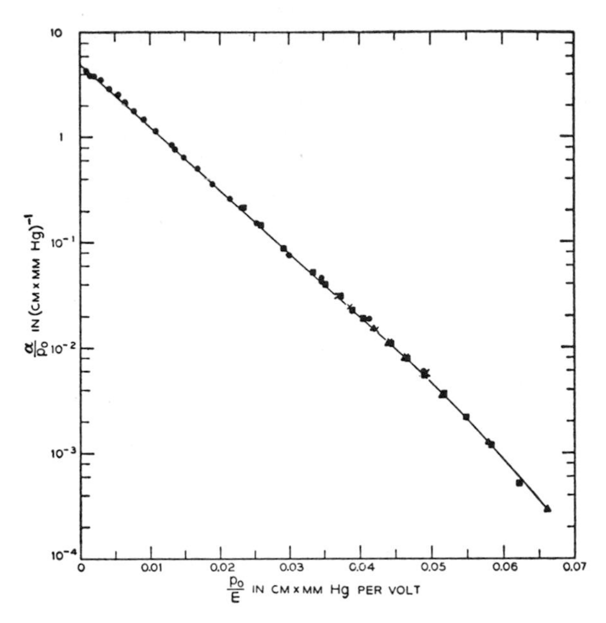

FIG. 5. The ionization coefficient α/p (cm $\times$ mm Hg)$^{-1}$ at 0°C vs. p_0/E cm $\times$ mm Hg/volts at 0°C for H_2, showing the results of several workers. (From Rose, 1956.) ●, Rose (1956); ▲, De Bitetto and Fischer (1956); ■, Hopwood *et al.* (1956); ×, Crompton *et al.* (1956).

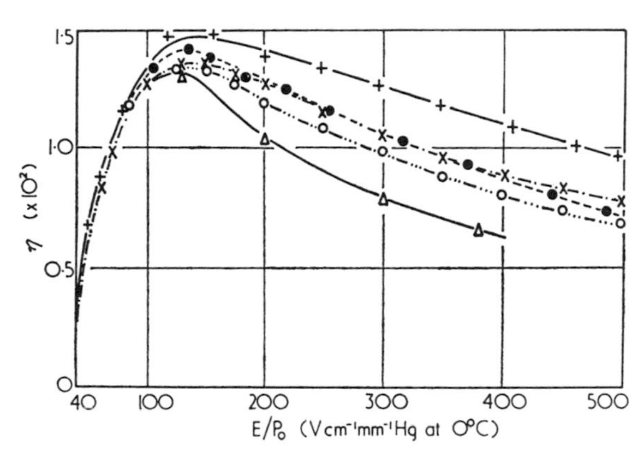

FIG. 6. Ionization coefficient η ($= \alpha/E$) as a function of E/p in hydrogen. +—+—+ = bulk metal tube, uncorrected results (Davies and Milne, 1959; Davies *et al.*, 1960), - - ● - - ● - - = bulk metal tube, corrected results (Davies and Milne, 1959; Davies *et al.*, 1960), × - . - . × = evaporated-film tube, uncorrected results (Davies and Milne, 1959; Davies *et al.*, 1960; also Ayres, 1923; Rose, 1956), ○····-···○ = evaporated-film tube, corrected results (Davies and Milne, 1959; Davies *et al.*, 1960), -△— —△- = Jones and Llewellyn-Jones (1958). (From Davies and Milne, 1959.)

uranium and the uranium hydride so formed was decomposed thermally. In all the cases P_2O_5 and liquid nitrogen traps were employed at various stages to ensure dryness of the gas. Further, Jones and Llewellyn-Jones and Davies and Milne point out that the data of Rose are apt to be appreciably in error at high E/p's ($E/p > 130$) due to the neglect of contributions from γ in the evaluations of α (§ 2.2.1 and Fig. 6). The available data are satisfactory up to $E/p = 100$ to 130, but beyond this further investigations are needed.

In contrast, in nitrogen the position appears to be complex even at low values of E/p. The data of Heylen (1959) in nitrogen of 99.998 % purity over $30 < E/p < 85$ which are in remarkably close agreement with those obtained in nearly all the earlier studies (Table II and Fig. 7) indicating relative insensitivity of α/p in this gas to impurities, differ markedly from the observations of Harrison. The latter has obtained values of α/p in three different grades of nitrogen (of differing purities) which exhibit a marked sensitivity to impurity at variance with the

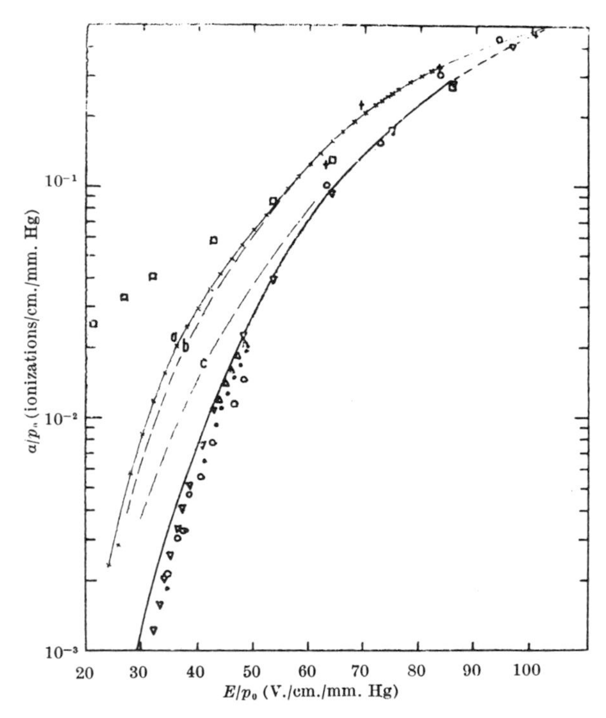

FIG. 7. Ionization coefficient α/p as a function of E/p in nitrogen. □, Ayres (1923); ▽, Masch (1932); ○, Posin (1936); +, Bowls (1938); △, Dutton *et al.* (1952); ●, De Bitetto and Fisher (1956); × a, b, c Harrison, (1957); —, Heylen (1959). (From Heylen, 1959.)

conclusions of Heylen. Further detailed studies in controlled conditions are required.

Attaching Gases. The recent investigations in the common attaching gases, viz., air (Harrison and Geballe, 1953; Prasad, 1959; Prasad and Craggs, 1960b; Dutton *et al.*, 1960), oxygen (Harrison and Geballe, 1953; Prasad and Craggs, 1961; Schlumbohm, 1959), carbon monoxide (Bhalla and Craggs, 1961a), carbon dioxide (Bhalla and Craggs, 1960), and water vapor (Prasad and Craggs, 1960a) are of considerable interest since simultaneous evaluations of both α/p and η/p are available from these studies, and these clearly demonstrate the errors in the evaluations of α/p when attachment is either neglected or is not detected. These various investigators show that the α/p data from the earlier studies in these gases, except in carbon monoxide, approximate closely to $(\alpha - \eta)/p$, particularly at low E/p's, because of the neglect of attachment (Fig. 8). The studies of Prasad and Craggs (1960a) in humid (Fig. 18) air are of some interest in view of the relative importance of this gas in practical open air spark gaps. These measurements clearly demonstrate the marked increase in both α/p and η/p in air with the addition of water vapor. The investigations of Bhalla and Craggs (1960, 1961a) in spectroscopically pure carbon monoxide and carbon dioxide, are of considerable value since no attachment has been observed in these gases in high-pressure electron swarm conditions. Further, in carbon monoxide there appear to have been no earlier investigations of either α/p or η/p.

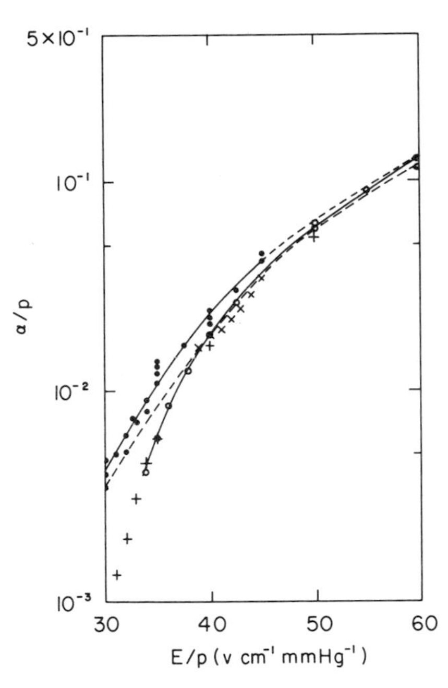

FIG. 8. Values of α/p in dry air as a function of E/p (normalized to 20°C). ●, true values of α/p (Prasad, 1959, 1960); ○, "apparent" values of α/p (Prasad, 1959, 1960); - - -, values of α/p obtained by Harrison and Geballe (1953); ×, values of α/p obtained by Llewellyn-Jones and Parker (1952); +, values of α/p obtained by Sanders (1932, 1933).

The studies of Hochberg and Sandberg (1946) in gases with complex molecules (e.g., CCl_4, SF_6, C_2H_5Cl, etc.) indicate that their (apparent) values of α/p are considerably smaller than in the common molecular gases at low values of E/p (say,

< 100). This feature accounts for the relatively high dielectric strengths. However, since no evaluations of the attachment rates were attempted by Hochberg and Sandberg, these results are of limited value and do not facilitate precise quantitative evaluation.

3 Attachment Coefficients

The present section gives a review of some of the recent work on measurements of attachment in swarm conditions. The variety of experimental techniques in this field is now very wide but, apart from giving appropriate references, little detailed reference to methods is given here. Instead, we have emphasized the experimental results, comparison of which for the different techniques shows now a reasonable agreement for certain common gases, notably oxygen and dry and humid air.

3.1 Basic Relations

If the probability of attachment at a collision is h, then the loss of population in an electron swarm in uniform drift motion and in equilibrium with the applied field E (volts/cm) is given by

$$dn_e = \frac{-h\bar{U}}{\lambda W} n_e dx = \frac{-h\nu_c n_e dx}{W} = -\eta n_e dx \tag{11}$$

where $\bar{U}$ and W are the average random and drift velocities of the electrons, λ the average electron mean free path, ν_c is the electron collision frequency, and $\eta = (h\bar{U}/\lambda W)$ is the attachment coefficient already defined. In the steady state, the number of electrons surviving attachment after traversing a distance d in the direction of the field is given by

$$n_e = n_0 \exp\left(\frac{-h\bar{U}}{\lambda W} d\right) \quad \text{or} \quad n_0 \exp(-\eta d) \tag{12}$$

where n_0 is the initial number of electrons. This equation, or variations based on it, constitutes the basis of the various experimental methods for the study of h or η/p as a function of E/p.

The cross section for attachment for the electron swarm at any particular value of E/p (or mean energy $\bar{\epsilon}$) is given by

$$Q_{\bar{\epsilon}}^{-} = h/\lambda N = hQ \tag{13}$$

where N is the concentration of the captor molecules and Q is the gas kinetic cross section or, substituting for h,

$$Q_{\epsilon}^{-} = (\eta/p)\,(W/\bar{U})\,(1/N_0) \tag{14}$$

where N_0 is the concentration of the captor molecules at $p = 1$ mm Hg and the appropriate temperature. The quantitus W and $\bar{U}$ are related by the expression

$$W = 0.75\, Ee\lambda/m\bar{U}. \tag{15}$$

In the foregoing analysis it has been assumed that ionization of the gas is negligible. If, however, ionization occurs to a limited extent, η in the preceding equations is to be replaced by $(\eta - \alpha)$. In view of this assumption, the experimental methods based on the survival equation above are restricted to conditions where $\eta \gg \alpha$, i.e., to low values of E/p.

3.2 Determination of Attachment Coefficients at Low Energies in Gases

3.2.1 *Microwave Methods*

The microwave deionization method, due originally to Biondi and Brown at M.I.T., has primarily been used to study recombination and diffusion processes. Biondi (1951), however, published a brief note indicating the value of the method for the determination of attachment coefficients for electrons of thermal energies.

If recombination (coefficient α' to avoid confusion with Townsend's α coefficient in this chapter) is the deionizing mechanism, then

$$\frac{dn_e}{dt} = -\,\alpha' n_+ n_e = -\,\alpha' n^2 \tag{16}$$

where n_e and n_+ are the electron and ion densities. Then if $n = n_1$ and n_2 at times t_1 and t_2 after the beginning of the deionization period, then

$$\frac{1}{n_2} - \frac{1}{n_1} = \alpha'(t_2 - t_1) \tag{17}$$

for constant electron temperature. This gives a linear plot of $1/n$ against t. For attachment, we have

$$\frac{dn_e}{dt} = n_e h \nu_c = -\,n_e h N \bar{U} Q. \tag{18}$$

Put $Qh = Q_{\bar{\epsilon}}^{-}$ = attachment cross section, so that

$$Q_{\bar{\epsilon}}^{-} = \frac{-1}{N\bar{U}} \frac{d}{dt} \log n_e$$

and the slope of the linear plot of log n_e with t gives the attachment cross section in the absence of complicating factors, as shown by Sexton and associates (1960), who discussed the situation at the time of the Uppsala conference, and proceed to show, for oxygen as in Biondi's earlier results, an attachment cross section of $\sim 10^{-22}$ cm^2, as compared with the Bloch-Bradbury (1935) value of $\sim 10^{-19}$ cm^2. Sexton *et al.* (1960) discuss this situation in some detail and Biondi (1960) at the same conference also refers to it. A likely explanation of the discrepancy was put forward some years ago by Hopwood in Liverpool (see Craggs, 1957), i.e., that collisions of the negative ions with vibrationally excited neutral molecules would give detachment, and so a production of free electrons tending to balance the loss due to attachment. If this were true, then data taken at very low pressures of oxygen alone, or with oxygen at low pressure mixed with a suitable "neutral" buffer gas at higher pressure to reduce the diffusion loss, would show higher capture

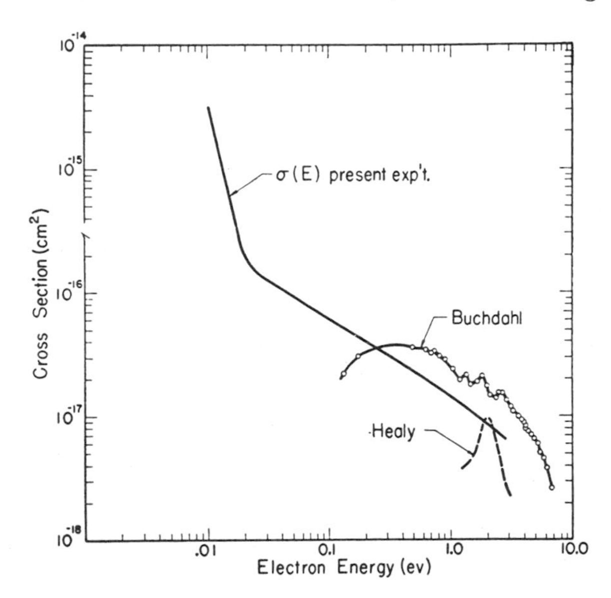

FIG. 9. A comparison of various measurements of the attachment cross section in iodine. The curves of Buchdahl and of Healey represent cross sections averaged over the particular energy distributions used in their measurements; present experiments refer to Biondi and Fox (1958). (From Biondi and Fox, 1958.)

cross sections. Rogers and Biondi (Biondi, 1960) report early results on these lines which show qualitative agreement with the Chanin drift tube data (q.v.). Sexton, Lennon, and their collaborators in Liverpool have been engaged in studies of this kind, for some years and cannot yet report unequivocal data supporting the 10^{-19} cm^2 oxygen capture cross section value, although their recent data approach this value and certainly considerably exceed the 10^{-22} cm^2 value.

Biondi (1958) has reported careful microwave measurements of absolute cross sections for dissociative attachment in iodine at 300°K. The process is

$$I_2 + e \rightarrow I^- + I. \tag{19}$$

The iodine vapor pressure varied over a wide range but was always much less than the pressure of helium (3 mm Hg) present as a buffer gas, i.e., to thermalize the electron energies in the afterglow and to prevent serious diffusion of electrons to the cavity walls. The latter would vitiate the measurements of attachment if the iodine only were present at the pressures used. The attachment (swarm) cross section in iodine is found to be $Q_{\epsilon}^{-} = 3.9 \times 10^{-16}$ cm^2.

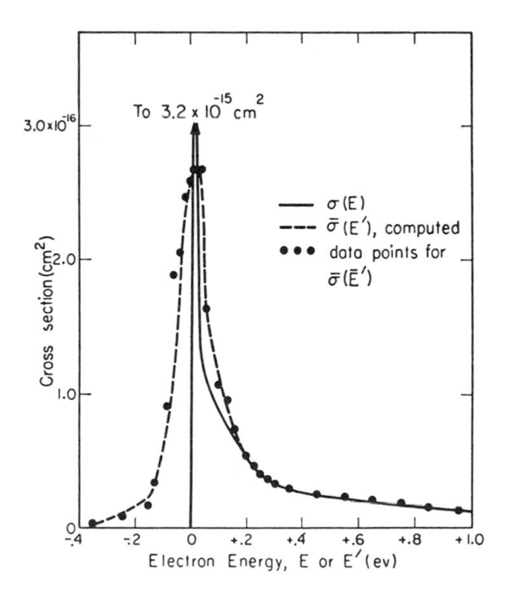

FIG. 10. The attachment cross section as a function of electron energy in iodine. The solid circles represent the measured values of $\bar{\sigma}$ (E') described in Fox (1958), the solid line the derived curve of the actual cross section $\sigma(E)$ obtained by solution of an integral equation. The dashed line represents the fit of the data obtained with this choice of $\sigma(E)$. (From Biondi and Fox, 1958.)

Fox (1958) confirmed the above process in iodine with a mass spectrometer using the RPD technique, etc., and showed that the maximum attachment cross section occurs for electrons with nearly zero energy, with a capture cross section falling to half-value in about 0.03 ev. Some difficulty was experienced in these experiments due to I^- from HI, probably formed by dissociated iodine reacting with water of crystallization in the glass tube. Biondi and Fox (1958) determined, from the above results, the absolute cross sections for electron capture in iodine, allowing for the electron energy distribution in the beam, etc., and the maximum cross section thus derived is about 3×10^{-15} cm^2, a very high value. Various data are shown in Figs. 9 and 10, giving a comparison with Buchdahl's (1941) earlier values, and are discussed at length in the paper cited.

3.2.2 *Drift Tube Methods (Thermal)*

Chanin (1959) has used a drift tube technique for studying attachment in oxygen at low values of E/p and data have been published by Chanin and associates (1959), who give data for (Fig. 11) a three-body, pressure-

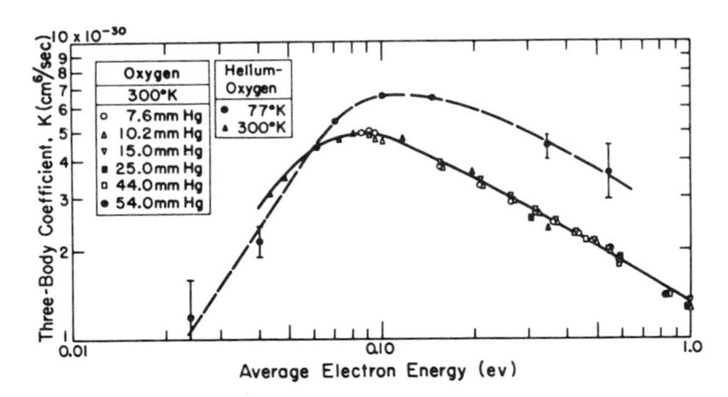

FIG. 11. Energy dependence of the three-body attachment coefficient for oxygen at 300° and 77°K. The solid points were obtained from measurements of attachment coefficients in oxygen-helium mixtures containing 1-5% of oxygen. (From Chanin *et al.*, 1959.)

dependent process of attachment for which the Bloch-Bradbury theory (1935) is presumably applicable, i.e.,

$$O_2 + e \rightarrow O_2^{-*} \rightarrow O_2^- . \tag{20}$$

The second stage, i.e., the stabilization of the excited molecular negative ion is possible if the density of third bodies, in this case oxygen molecules, is sufficiently high. Figure 12 shows the transition to the pressure-independent process, studied by many other workers (cf. § 3.3) at higher E/p's.

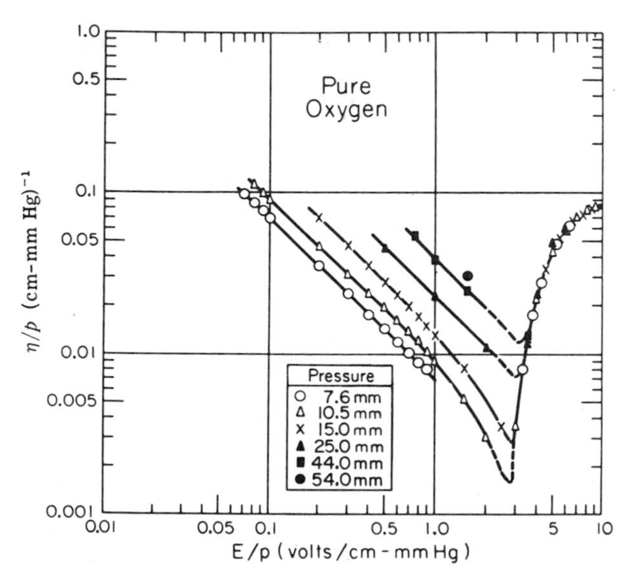

FIG. 12. Measured attachment coefficients (η/p) as a function of E/p for pure O_2 at 300°K. Drift distances of 2.5 and 10 cm were used. (From Chanin *et al.*, 1959.)

The data of Fig. 11 are probably the first on this pressure-dependent process. By defining an attachment frequency $\nu_a = \eta W$ (in which η is a generalized attachment coefficient), Chanin *et al.* describe the two- and three-body processes by

$$\nu_a = \beta N + KN^2. \tag{21}$$

Here β is the two-body attachment coefficient, K is the three-body coefficient, and N is the gas density. Figure 13 shows the β values compared with those due to Craggs *et al.* (1957) (§ 3.3.3 and § 3.3.4), and Fig. 11 gives the K coefficient as a function of mean electron energy. K-values due to Bortner and Hurst (1958) who later confirmed the data of Chanin *et al.* are some 10 % lower. Chatterton in Liverpool (§ 3.3.1) was also able to obtain supporting data with the Bradbury filter (§ 3.3.1) over the energy range 0.6-1.2 ev, 15-30 mm Hg pressure (Fig. 14 and Table III). The agreement here between the two sets of data is satisfactory.

TABLE III

ATTACHMENT COEFFICIENTS IN OXYGEN

K (Chanin *et al.*) $cm^6/sec \times 10^{30}$	K (Chatterton) $\times 10^{30}$	E/p volts/-cm/mm Hg	$\bar{E}$ mean electron energy (ev)	p mm Hg	η/p cm/mm Hg
1.15	1.15	1.97	1.033	15.8	0.0072
1.095	1.02	2.00	1.035	20	0.0075
1.08	1.1	2.06	1.05	30.2	0.0132

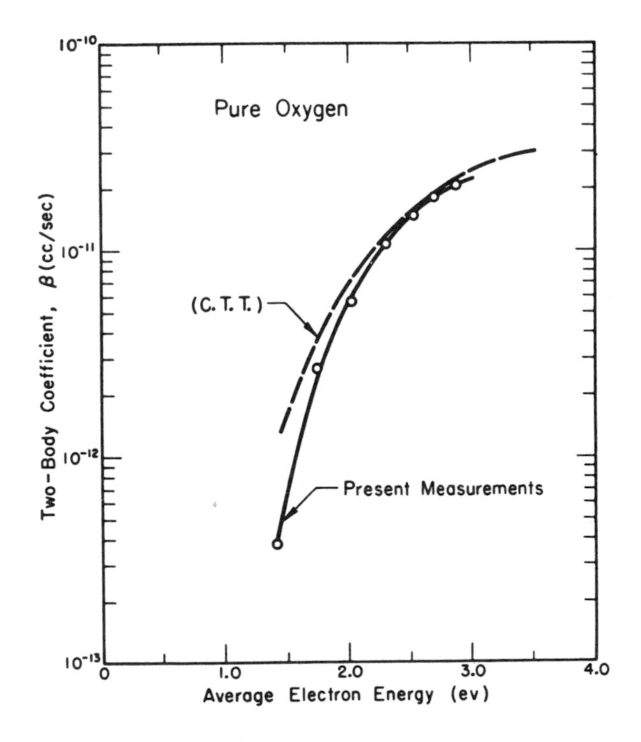

FIG. 13. Energy dependence of the two-body attachment coefficient. The dashed curve was obtained by averaging the cross sections measured by Craggs *et al.* (1957) over a Druyvesteyn electron energy distribution for various average energies. Present measurements refer to Chanin (1959). (From Chanin, 1959.)

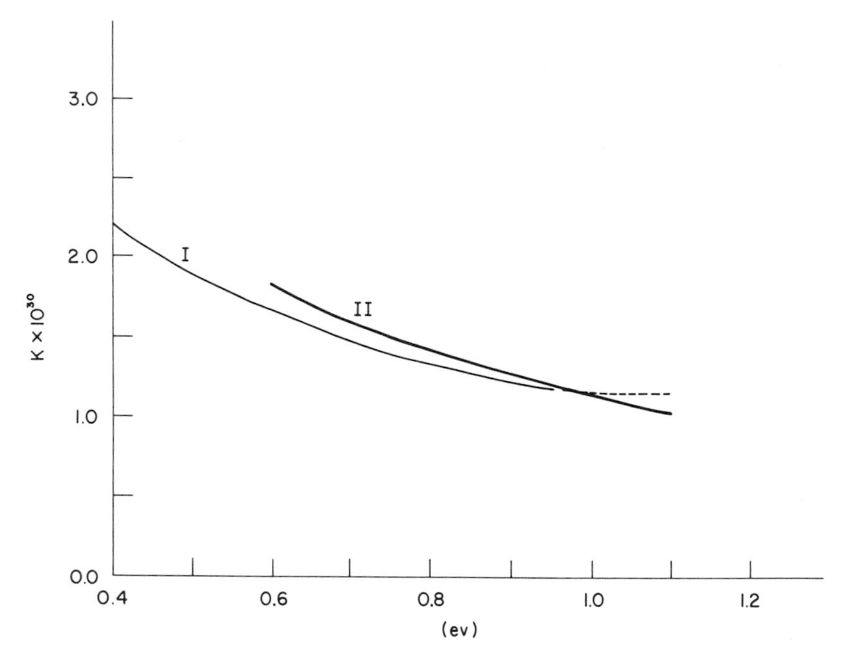

FIG. 14. Three-body attachment in oxygen at various mean electron energies. I, Chanin *et al.* (1959); II, Chatterton (1961).

3.3 Attachment Coefficients at High Energies in Gases

3.3.1 *Method of Bradbury*

The Bradbury electron filter method used to study attachment coefficients at E/p's ~ 4 to 30 is well known (Bradbury, 1933).

Kuffel (1959) working in the A.E.I. (Manchester) Research Laboratories obtained results using a Bradbury filter with oxygen, dry air, humid air, and water vapor. His data (for oxygen) are shown in Fig. 15. They do not agree as closely as may be desired with the data of Chatterton for oxygen who made, apparently, a fuller study of ion loss, etc., and whose data are also in better agreement with those of Prasad and the other later workers.

The most comprehensive study of the Bradbury electron filter for use in measurements of attachment coefficients is, we believe, that due to Chatterton in Liverpool, following earlier work by Twiddy and Hopwood in the same laboratory. Chatterton particularly studied certain possible sources of error, neglected partly or wholly by previous workers, namely,

(1) diffusion losses of electrons in the discharge, (2) collection of negative ions at the rf grids, and (3) current multiplication at the grids, including ionization, secondary emission of electrons, and electron detachment.

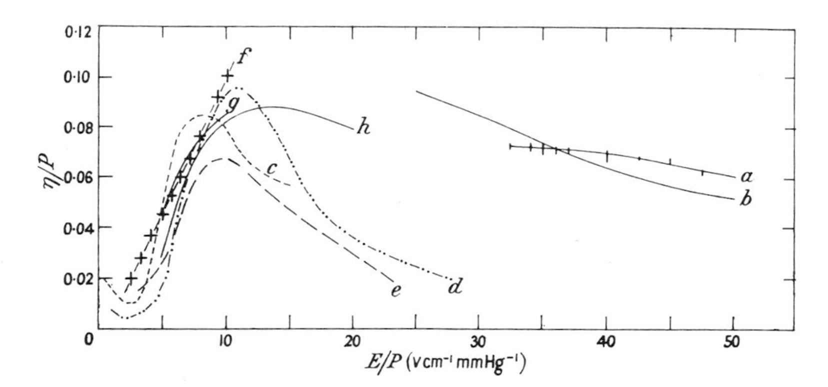

FIG. 15. Values of η/p as a function of E/p in dry oxygen (normalized to 20°C). a, Prasad and Craggs (1961); b, Harrison and Geballe (1953); c, Bradbury (1933); d, Kuffel (1959); e, Herreng (1952); f, Doehring (1952); g, Chanin *et al.* (1959); h, Huxley *et al.* (1959). (From Prasad and Craggs, 1961.)

The falloff in η/p shown in Bradbury's earlier work on oxygen (1933) has now been found (cf. Harrison and Geballe, 1953; Craggs *et al.*, 1957) to be spurious, and to be due to ionization (Townsend's α process) and to diffusion losses, etc.

The results of Bradbury (1933) and Kuffel (1959) are probably not reliable over their whole range, because of neglect of the above sources of error.

3.3.2 *The Diffusion Methods*

The experimental methods of Bailey (1925; Bailey and Duncanson, 1930) and Huxley *et al.* (1959) are based on the measurement of the lateral diffusion of a mixed stream of electrons and negative ions in uniform drift motion and in equilibrium with the applied field, although they differ considerably in detail.

The probabilities of attachment at a collision and attachment coefficients in a wide variety of gases (chlorine, bromine, iodine, hydrogen chloride, water vapor, ammonia, oxygen, nitrous oxide, and dry air) were obtained by Bailey and others and the results of these investigations have been compiled by Healey and Reed (1941). However, in view of the complexity of the equipment and the consequent uncertainty in the gas purity, apart from the errors arising from the loss of electrons by diffusion, etc., particularly at high values of E/p, the results of these investigations are in general somewhat uncertain and require confirmation by independent experiments using modern techniques. In the few instances where such data are available from other experiments (e.g., oxygen, air, water vapor), the results obtained by Bailey *et al.* show marked departures which can be attributed only to errors inherent in this technique (Prasad and Craggs, 1960a, b, 1961).

This method of Bailey, although ingenious, is superseded by the more recent and relatively simpler experimental method of Huxley *et al.* (1959). In these experiments, a mixed stream of photoelectrons and negative ions enters a diffusion chamber through an orifice (sufficiently small to be approximated to a point source). After traversing a known distance (h) under the action of the applied field, the stream is fully intercepted by an anode consisting of a central disk and two outer annuli. Uniformity of the applied field in the drift space is ensured by means of suitable guard rings. Under these conditions, the ratio R of the current received by the inner annulus to the total current received by both annuli is a function ϕ of $(W/D, \eta, h, a, b)$ where D is the diffusion coefficient of electrons, and a and b are the inner and the outer radii of the inner annulus (Huxley, 1959). From diffusion theory (Townsend, 1947; Loeb, 1955), the ratio $W/D = 40.3\, E/Ak_T$, where k_T is Townsend's energy factor defined as the ratio of the average energy of the electrons to the thermal energy of the gas molecules at $T = 15°C$, and A is a numerical constant which depends on the electron energy distribution (Huxley and Zaazou, 1949). Here the lateral diffusion of the ions formed is neglected and the central disk is made large enough to intercept all the ions. A set of theoretical curves are accordingly prepared showing R as a function of η and a parameter involving W/D and for given values of h. At any given E/p, from measurements of R at two values of h, the quantities k_T and η are obtained simultaneously from the theoretical curves by interpolation and successive approximations. The method has the inherent advantage of allowing the simultaneous determination of the mean energy of the electron Swarm and attachment coefficients, as functions of E/p.

However, to date Huxley *et al.* (1959) appeared to have investigated only attachment in oxygen by this method up to $E/p = 20$ (Fig. 15). The results are in excellent agreement with the data obtained from the independent experiments of Chanin (1959). They constitute perhaps the best available data in oxygen at low E/p.

3.3.3 *The Drift Tube or Dynamic Methods*

While the methods mentioned previously are essentially steady-state studies of the electron population, alternative experiments based on the measurement of the time-dependent currents due to the arrival at the anode of ions (formed in the gas) and the (surviving) electrons, when a burst of initiatory electrons are released in the gas (or at the cathode), were developed by Herreng (1952), Doehring (1952), McAfee (1955), and Chanin *et al.* (1959). Of these, the methods of Doehring and Chanin *et al.* differ only in experimental detail and will therefore be discussed together.

In the method of Herreng (1952), a well-collimated thin slab of X-rays was flashed parallel to the anode in a uniform field gap (approximately 12-cm separation) at known distances from the anode. This X-ray ionization was flashed in a pulse of 1 μsec duration at 50 per second. The ions and electrons following any pulse were drawn to the anode by the applied field. On a fast time scale, the resultant current at the anode measured on a cathode-ray oscilloscope represented the arrival of electrons which survived attachment in traversing the distance d, from the X-ray slab to the anode. Herreng (1952) has shown that if the current pulses i_1 and i_2 are measured for two different distances (d_1 and d_2) of the X-ray slab from the anode,

$$\frac{i_1}{i_2} = \exp \frac{-h\bar{U}}{\lambda} \left(\frac{d_1 - d_2}{W}\right), \tag{22}$$

the value of h and thence η/p could be obtained. The method further yielded the drift velocity of the electrons.

In contrast to the above method, Doehring and Chanin *et al.* investigated on a slow time scale the current at the anode, resulting from the arrival of the negative ions created in the gap. If a pulse of n_0 initiatory electrons are released at the cathode (at distance d from the anode) at time $t = 0$, then neglecting the initial component due to the arrival of the surviving electrons, the current at the anode is given by

$$\begin{aligned} I_t &= A \exp(\eta W_i t) \quad &\text{for} \quad t < d/W_i \\ \text{and} \quad I_t &= O \quad &\text{for} \quad t > d/W_i \end{aligned} \tag{23}$$

where A is a constant and W_i the drift velocity of the ions, all of which are assumed to be of the same species. If the latter is not true, the equation is to be modified accordingly. In practice, however, the transition at $t = d/W_i$ is not sharp but has a finite time spread Δt due to diffusion and other secondary effects. From measurements of I_t, η, and W_i may be evaluated as $f(E/p)$.

In these experiments a wire mesh grid incorporated near the anode end of the drift space could be pulsed by means of appropriate square-wave potentials, to act effectively as a "shutter" which could be "opened" for brief periods at known intervals of time after the initiatory electron pulse and thus facilitated the measurements of the ionic currents as functions of time.

In the experiments of Doehring the initiatory electron pulse was itself produced by means of yet another "shutter" in conjunction with a thermionic electron source and synchronized with the "exit shutter" at the anode end.

In the experiments of Chanin *et al.*, on the other hand, the initiatory electron pulse was obtained from a photocathode by pulsing a uv light source in synchronization with the "exit shutter." Further, it is easily seen that by pulsing the shutter on fast time scales the drift velocity of electrons can be obtained simultaneously in both methods.

In the experiments of McAfee (1955) the "shutters" were dispensed with entirely and the total current pulse (ionic and electronic) was measured at the anode. The separation of the electronic and ionic components was then achieved from a study of the transient current flowing immediately after the initiation and the termination of the initiatory electron pulse. The latter was produced as in Chanin's experiments, by the use of a photocathode and a pulsed light source.

As in the case of the diffusion method of Huxley, the dynamic methods of Herreng, Doehring, and Chanin *et al.* outlined previously appear to have been utilized to date for the study of attachment in oxygen only (Fig. 15). From Fig. 15 it is seen that the results obtained by Herreng (1952) deviated markedly from those of Chanin *et al.*, Doehring, or Huxley (1959), and in view of the remarkable consistency in the data obtained by the latter three workers, these deviations can only be attributed at present to experimental errors or errors inherent in the technique. The data of Chanin *et al.* together with those of Huxley (Fig. 12) constitute perhaps the most valuable data available to date in oxygen at low E/p and clearly demonstrate the progressive onset of the pressure-independent two-body attachment for $E/p > 5$, thereby eliminating the inconsistency and the uncertainty of the earlier interpretations of attachment in oxygen (Bradbury, 1933) in these conditions.

In summary, it would appear that the experimental methods of Huxley *et al.* and Chanin are inherently simpler and consequently superior to the other methods described above. However, all the experimental methods described suffer from some common limitations, namely:

(a) The experiments are restricted in general to conditions where $\eta/p > \alpha/p$, i.e.; to low values of E/p and low electron energies. If electron attachment occurs at relatively high energies (cf. the case of carbon monoxide, § 1.2), the simultaneous onset of appreciable ionization would distort the measurements and in some cases prevent the detection of attachment.

(b) The experiments in general are conducted at low pressures ($p \sim 10$ mm Hg) and are affected profoundly by the loss of electrons by diffusion.

(c) In the case of the experiments employing grids as shutters, penetration of the grid fields into the drift space and other secondary effects due to the grids distort the measurements.

3.3.4 *The Method of Townsend and the Avalanche Method*

For the study of attachment at relatively high values of E/p ($E/p > 25$ to 30 in common attaching gases), the experimental method of Townsend and the avalanche method of Raether and others (cf. Schlumbohm, 1959) can be employed with appropriate modifications to the basic equations, as outlined in § 2.2.2. The limitations of these methods have already been discussed.

In addition to the steady-state Townsend method, a pulsed method for the study of ionic drift velocities and attachment has been developed by Burch and Geballe (1957). In principle, the experimental method is somewhat similar to that of McAfee. The results obtained are, however, of an indirect nature and the method is not discussed any further here.

Attachment coefficients have been obtained by Harrison and Geballe (1953; see also Loeb, 1955) at low pressures ($p \sim$ tens of mm Hg) in a wide variety of highly attaching (and consequently high dielectric strength) gases (CCl_2F_2, SF_6, CCl_4, CF_3SF_5) and in oxygen and dry air, and by Howard (1957, 1958) in CF_4 utilizing the modified Townsend method. Similarly, data regarding attachment coefficients have been obtained over a wide range of pressures ($p \sim$ few tens to a few hundreds of mm Hg) in dry air (Prasad, 1959; Dutton *et al.*, 1960), humid air and water vapor (Prasad and Craggs (1960a), oxygen (Prasad and Craggs, 1961), carbon dioxide (Bhalla and Craggs, 1960), carbon monoxide (Bhalla and Craggs, 1961a), and sulfur hexafluoride (Bhalla and Craggs, 1961b) utilizing this method. The avalanche method, on the other hand, appears to have been used so far for the study of attachment coefficients in oxygen only (Schlumbohm, 1959), and the data appear to be in broad agreement with those of Harrison and Geballe and Prasad and Craggs if due allowances are made for the relatively poor accuracies attainable in this technique.

Oxygen. In view of the fact that attachment in oxygen has been studied by numerous investigators using a variety of techniques for $E/p > 5$, it is instructive to compare the results obtained in some of these investigations (Fig. 15).

It is seen that η/p is in general pressure independent over $5 < E/p < 60$ and p = few mm Hg to 600 mm Hg (Prasad and Craggs, 1961). However, apart from the measurements of Huxley *et al.* (1959), Chanin

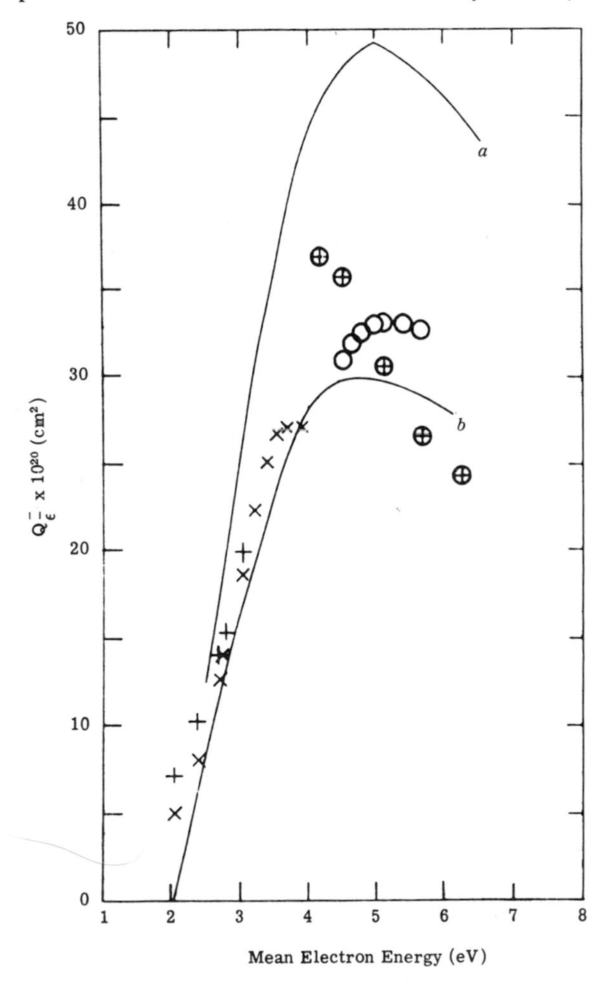

FIG. 16. Mean values of attachment cross sections in oxygen for a Druyvesteyn form of electron energy distribution. a, calculated from the data of Craggs *et al.* (1957); b, calculated from the data of Buchel'nikova (1958b); ○, Prasad and Craggs (1961); ⊕, Harrison and Geballe (1953); ×, Huxley *et al.* (1959); +, Chanin *et al.* (1959). (From Prasad and Craggs, 1961.)

(1959), and Doehring (1952), the agreement in η/p at low values of E/p is relatively poor. Similarly at high E/p, the results of Harrison and Geballe (1953) and Prasad and Craggs (1961) appear to diverge considerably. These differences have been considered at legth by Prasad and Craggs and it has been shown that the results of Huxley *et al.*, Chanin, and Prasad and Craggs were in fact consistent despite their marked deviations from the other measurements. It would therefore appear that the η/p, E/p curve has a broad maximum at $E/p = 10$ to 15 and thereafter declines relatively slowly. Some recent observations of η/p in oxygen obtained by Chatterton (1961) employing the method of Bradbury where the results were corrected for the various intrinsic errors outlined earlier in § 3.3.1 appears further to confirm this conclusion.

Further, the mean cross sections for attachment for electron swarms of various mean energies have been computed by Prasad and Craggs (1961) from the cross sections for the dissociative attachment in oxygen obtained in electron beam experiments by Craggs *et al.* (1957) and Buchel'nikova (1958b). These calculations (Fig. 16) show that the attachment in oxygen in these various high-pressure (from $\sim$ few mm Hg to $\sim$ few hundred mm Hg) swarm experiments is consistent with the two-body dissociative process if the electron energy distribution is assumed to be of a Druyvesteyn form in agreement with Chanin's conclusions (see Fig. 13). Under these conditions, the observed attachment in the swarm experiments has been shown to be consistent with the cross-section data of Buchel'nikova (Fig. 16). In addition, from a comparison of the computed mean cross sections for ionization with those observed in the swarm experiments, Prasad and Craggs postulate that an electron energy distribution of the Druyvesteyn form is adequate to describe ionization in the high-pressure experiments (Fig. 17).

Air. From similar considerations, Prasad and Craggs (1960b) have shown that the observed attachment rates in dry air (Fig. 18) are consistent with the dissociative attachment in the constituent oxygen if the electron energy distribution is assumed to be of a Maxwellian form. The measurements of Kuffel (1959) and Prasad and Craggs (1960a) in humid air (Fig. 18) indicate that attachment in air is profoundly influenced by the presence of even traces of water vapor. Prasad (1960) has put forth quantitative preliminary interpretation for these high rates of attachment in humid air in terms of the attachment in the constituent water vapor and oxygen assuming a Maxwellian electron energy distribution.

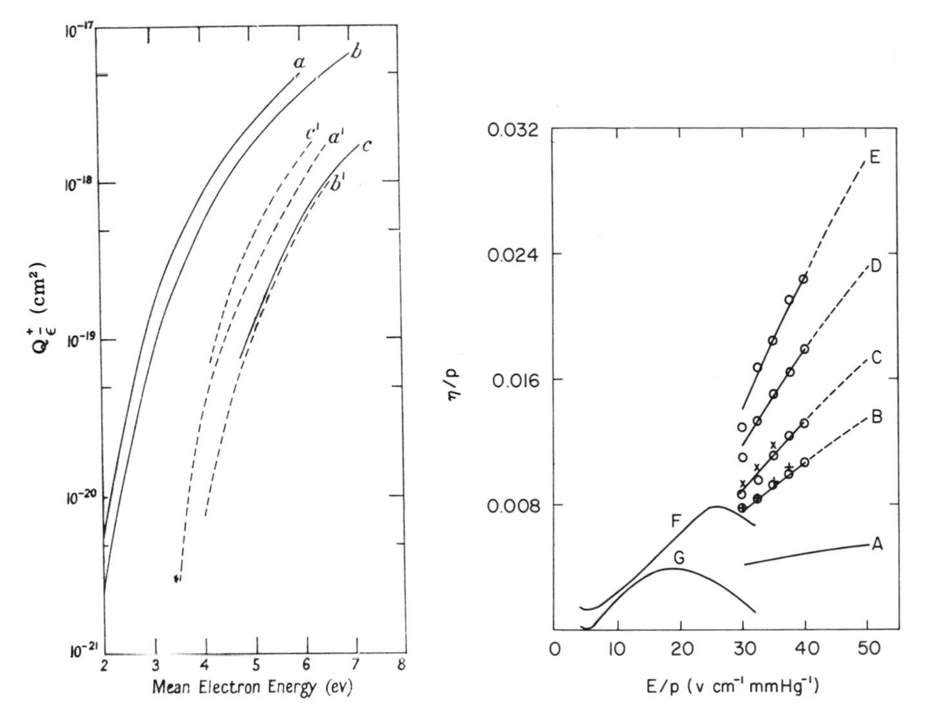

FIG. 17. Mean values of ionization cross sections in oxygen. Maxwellian form of electron energy distribution. a, calculated from the data of Tate and Smith (1932); b, calculated from the data of Craggs *et al.* (1957); c, observed values from Prasad and Craggs (1961) and Harrison and Geballe (1953); a′, b′, and c′ represent the corresponding data for a Druyvesteyn form of electron energy distribution. (From Prasad and Craggs, 1961.)

FIG. 18. Values of η/p as a function of E/p in humid air for various values of partial pressure of water vapor (Prasad and Craggs, 1960a). A, dry air; B, 150/2.5/Pt; +, 300/5/Pt; C, 150/5/D; ×, 300/10/Pt; D, 150/9/D; E, 150/15/D; F, Kuffel (1959); G, Kuffel (1959). (These figures represent the total pressure of humid air and the partial pressure of water vapor; Pt = obtained with a platinum cathode; D = obtained with a dural cathode.)

Carbon dioxide. The data obtained by Bhalla and Craggs in carbon dioxide are of some interest since no attachment had been observed in this gas in the earlier studies, due both to the relatively high energies at which attachment occurs and to the considerably smaller cross sections (Fig. 3). Despite these considerations, Chatterton (1961) has also obtained values of η/p at low E/p utilizing the method of Bradbury with considerable care. The two sets of data are shown in Fig. 19 and appear to be in satisfactory agreement. The correlation of the cross sections for attachment from these data with the corresponding computed cross sections, however, appears to be somewhat difficult (Bhalla and Craggs, 1960) and indicates an unusual electron energy distribution in this gas.

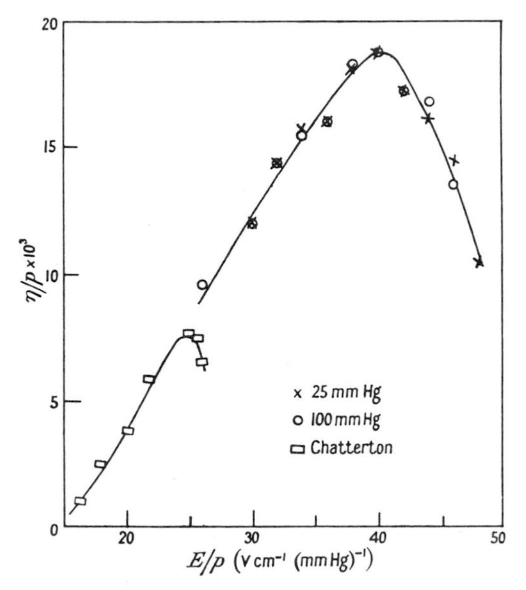

FIG. 19. Values of η/p as a function of E/p in carbon dioxide. The values of Chatterton are from Chatterton (1961). (From Bhalla and Craggs, 1960.)

Carbon Monoxide. Similarly, the data obtained in carbon monoxide are of interest since there appear to be no other quantitative results on attachment in this gas, although Bhalla and Craggs (1961a) show that at high values of E/p the α/p and η/p data obtained are apt to be in error due to appreciable contributions from the ion pair processes which have a cross section comparable with that of the dissociative process (§ 1.2).

Highly Attaching Gases. The observations of Harrison and Geballe (Loeb, 1955) and Howard in the strongly attaching gases already referred to are of considerable interest in view of the high dielectric strength of these gases. These indicate very high attachment rates which dominate the ionization rates up to relatively high values of E/p and thus account for the high dielectric strength of these gases (cf. Meek and Craggs, 1953). The attachment coefficients obtained by Harrison and Geballe (and by Bhalla and Craggs, 1961b) and McAfee (§ 3.3.3) in sulfur hexafluoride show marked deviations for which no satisfactory explanation is available at present. It is possible that the differences in the initial energy of the photoelectrons in the two types of experiments is to some extent responsible for these since the dominant attachment processes in sulfur hexafluoride occur at very low energies.

ACKNOWLEDGMENTS

The authors are grateful to their colleagues and ex-colleagues in Liverpool, notably Professor J. M. Meek, Dr. M. S. Bhalla, and Dr. B. A. Tozer, for discussions on the measurement of ionization and attachment cross sections and coefficients and their implications.

REFERENCES

Ayres, T. L. R. (1923) *Phil. Mag.* **45**, 353.
Badareu, E., and Bratescu, G. G. (1944) *Bull. soc. roumaine phys.* **45**, 9.
Badareu, E., and Valeriu, M. (1941) *Bull. soc. roumaine phys.* **42**, 9.
Bailey, V. A. (1925) *Phil. Mag.* **50**, 825.
Bailey, V. A., and Duncanson, W. E. (1930) *Phil. Mag.* **10**, 145.
Bhalla, M. S., and Craggs, J. D. (1960) *Proc. Phys. Soc.* **76**, 369.
Bhalla, M. S., and Craggs, J. D. (1961a) *Proc. Phys. Soc.* **78**, 438.
Bhalla, M. S., and Craggs, J. D. (1961b) *Proc. Phys. Soc.*, in press.
Biondi, M. A. (1951) *Phys. Rev.* **84**, 1072.
Biondi, M. A. (1958) *Phys. Rev.* **109**, 2005.
Biondi, M. A. (1960) *In* "Proceedings of the Fourth International Conference on Ionization Phenomena in Gases, Uppsala, 1959" (N. R. Nilsson, ed.), Vol. 1, p. 72. North-Holland, Amsterdam.
Biondi, M. A., and Fox, R. E. (1958) *Phys. Rev.* **109**, 2012.
Bishop, E. S. (1911) *Phys. Rev.* **33**, 325.
Blevin, H. A., Haydon, S. C., and Somerville, J. M. (1957) *Nature* **179**, 38.
Bloch, F., and Bradbury, N. E. (1935) *Phys. Rev.* **48**, 689.
Bortner, T. E., and Hurst, G. S. (1958) *Health Phys.* **1**, 39.
Bowls, W. E. (1938) *Phys. Rev.* **53**, 293.
Bradbury, N. E. (1933) *Phys. Rev.* **44**, 883.
Branscomb, L. M., Burch, D. S., Smith, S. J., and Geltman, S. (1958) *Phys. Rev.* **111**, 504.
Buchdahl, R. (1941) *J. chem. Phys.* **9**, 146.
Buchel'nikova, N. S. (1958a) *Instrum. and Measurement Eng.* **6**, 803.
Buchel'nikova, N. S. (1958b) *Zh. eksper. teor. Fiz.* **35**, 1119.
Burch, D. S., and Geballe, R. (1957) *Phys. Rev.* **106**, 183.
Chanin, L. M. (1959) Ph. D. Thesis, University of Pittsburgh, Pittsburgh, Pennsylvania.
Chanin, L. M., Phelps, A. V., and Biondi, M. A. (1959) *Phys. Rev. Letters* **2**, 344.
Chatterton, P. (1961) Ph. D. Thesis, University of Liverpool.
Craggs, J. D. (1957) *In* "Proceedings of the Third International Conference on Ionization Phenomena in Gases, Venice, 1957," p. 207. Italian Phys. Soc., Milan.
Craggs, J. D., and Massey, H. S. W. (1959) *In* "Handbuch der Physik" (S. Flügge, ed.), Vol. 37, p. 314. Springer, Berlin.

Craggs, J. D., and Tozer, B. A. (1958) *Proc. Roy. Soc.* **A247**, 337.
Craggs, J. D., and Tozer, B. A. (1960) *Proc. Roy. Soc.* **A254**, 229.
Craggs, J. D., Thorburn, R., and Tozer, B. A. (1957) *Proc. Roy. Soc.* **A240**, 473.
Crompton, R. W., Dutton, J., and Haydon, S. C. (1956) *Proc. Phys. Soc.* **B69**, 1.
Crowe, R. W., Bragg, J. K., and Thomas, V. G. (1954) *Phys. Rev.* **96**, 10.
Davies, D. E., and Milne, J. G. C. (1959) *Brit. J. appl. Phys.* **10**, 301.
Davies, D. E., Milne, J. G. C., and Myatt, J. (1960) *In* "Proceedings of the Fourth International Conference on Ionization Phenomena in Gases, Uppsala, 1959" (N. R. Nilsson, ed.), p. 161. North-Holland, Amsterdam.
De Bitetto, D. J., and Fisher, L. H. (1956) *Phys. Rev.* **104**, 1213.
Doehring, A. (1952) *Z. Naturforsch.* **79**, 253.
Druyvesteyn, M. J., and Penning, F. M. (1940) *Rev. mod. Phys.* **12**, 87.
Dunlop, S. H. (1949) *Nature* **164**, 452.
Dutton, J., Haydon, S. C., and Llewellyn-Jones, F. (1952) *Proc. Roy. Soc.* **A213**, 203.
Dutton, J., Llewellyn-Jones, F., and Palmer, R. W. (1960) *In* "Proceedings of the Fourth International Conference of Ionization Phenomena in Gases, Uppsala, 1959" (N.R. Nilsson, ed.), p. 137. North-Holland, Amsterdam.
Engstrom, R. W., and Huxford, W. S. (1940) *Phys. Rev.* **58**, 67.
Fox, R. E. (1958) *Phys. Rev.* **109**, 2008.
Fox, R. E., Hickam, W. M., Kjeldaas, T., and Grove, D. J. (1951) *Phys. Rev.* **84**, 985.
Frommhold, L. (1956) *Z. Phys.* **144**, 396.
Frommhold, L. (1958) *Z. Phys.* **150**, 172.
Hagstrum, H. D. (1951) *Rev. mod. Phys.* **23**, 185.
Hale, D. H. (1938) *Phys. Rev.* **54**, 241.
Hale, D. H. (1939a) *Phys. Rev.* **55**, 815.
Hale, D. H. (1939b) *Phys. Rev.* **56**, 1199.
Harrison, M. A. (1957) *Phys. Rev.* **105**, 366.
Harrison, M. A., and Geballe, R. (1953) *Phys. Rev.* **91**, 1.
Healey, R. H., and Reed, J. W. (1941) "The Behaviour of Slow Electrons in Gases." Amalgamated Wireless (Australasia) Ltd., Sydney, Australia.
Herreng, P. (1952) *Cahiers de Phys.* **38**, 7.
Heylen, A. E. D. (1959) *Nature* **183**, 1545.
Hickam, W. M., and Fox, R. E. (1956) *J. chem. Phys.* **25**, 642.
Hochberg, B., and Sandberg, E. (1942) *J. Tech. Phys. U.S.S.R.* **12**, 65.
Hochberg, B., and Sandberg, E. (1946) *C. R. Acad. Sci. U.S.S.R.* **53**, 511.
Hopwood, W., Peacock, N. J., and Wilkes, A. (1956) *Proc. Roy. Soc.* **A235**, 334.
Howard, P. R. (1957) *In* "Proceedings of the Third International Conference on Ionization Phenomena in Gases, Venice, 1957," p. 499. Italian Phys. Soc., Milan.
Howard, P. R. (1958) *Nature* **181**, 645.
Huxford, W. S. (1939) *Phys. Rev.* **55**, 754.
Huxley, L. G. H. (1959) *Austral. J. Phys.* **12**, 171.
Huxley, L. G. H., and Zaazou, A. A. (1949) *Proc. Roy. Soc.* **A196**, 402.
Huxley, L. G. H., Crompton, R. W., and Bagot, C. H. (1959) *Austral. J. Phys.* **12**, 303.
Jones, E., and Llewellyn-Jones, F. (1958) *Proc. Phys. Soc.* **72**, 363.
Kruithof, A. A. (1940) *Physica* **7**, 519.
Kruithof, A. A., and Penning, F. M. (1936) *Physica* **3**, 515.
Kruithof, A. A., and Penning, F. M. (1937) *Physica* **4**, 430.
Kuffel, E., (1959) *Proc. Phys. Soc.* **74**, 297.
Legler, W. (1955) *Z. Phys.* **140**, 221.
Llewellyn-Jones, F., and Parker, A. B. (1952) *Proc. Roy. Soc.* **A213**, 185.

Loeb, L. B. (1955) "Basic Processes of Gaseous Electronics." California Univ. Press, Berkeley, California.
Lozier, W. W. (1934) *Phys. Rev.* **46**, 268.
McAfee, K. B. (1955) *J. chem. Phys.* **23**, 1435.
Masch, U. (1932) *Arch. Elektrotech.* (*Berlin*) **26**, 587, 589.
Massey, H. S. W. (1950) "Negative Ions." Cambridge Univ. Press, London and New York.
Meek, J. M., and Craggs, J. D. (1953) "Electrical Breakdown of Gases." Oxford Univ. Press, London and New York.
Peacock, N. J. (1960) Ph. D. thesis, University of Liverpool.
Penning, F. M. (1938) *Ned. Tijdschr. Natuurkde.* **5**, 33.
Posin, D. Q. (1936) *Phys. Rev.* **50**, 650.
Prasad, A. N. (1959) *Proc. Phys. Soc.* **74**, 33.
Prasad, A. N. (1960) Ph. D. thesis, University of Liverpool.
Prasad, A. N., and Craggs, J. D. (1960a) *Proc. Phys. Soc.* **76**, 223.
Prasad, A. N., and Craggs, J. D. (1960b) *In* "Proceedings of the Fourth International Conference on Ionization Phenomena in Gases, Uppsala, 1959" (N. R. Nilsson, ed.), p. 142. North-Holland, Amsterdam.
Prasad, A. N., and Craggs, J. D. (1961) *Proc. Phys. Soc.* **77**, 385.
Raether, H. (1937) *Z. Phys.* **107**, 91.
Rose, D. J. (1956) *Phys. Rev.* **104**, 273.
Sanders, F. H. (1932) *Phys. Rev.* **41**, 667.
Sanders, F. H. (1933) *Phys. Rev.* **44**, 1020.
Schlumbohm, H. (1958) *Z. Phys.* **151**, 563.
Schlumbohm, H. (1959) *Z. angew. Phys.* **11**, 156.
Schlumbohm, H. (1960) *In* "Proceedings of the Fourth International Conference on Ionization Phenomena in Gases, Uppsala, 1959" (N. R. Nilsson, ed.), p. 121. North-Holland, Amsterdam.
Sexton, M. C., Mulcahy, M. J., and Lennon, J. J. (1960) *In* "Proceedings of the Fourth International Conference on Ionization Phenomena in Gases, Uppsala, 1959" (N. R. Nilsson, ed.), p. 94. North-Holland, Amsterdam.
Tate, J. T., and Smith, P. T. (1932) *Phys. Rev.* **39**, 270.
Townsend, J. S. (1915) "Electricity in Gases." Oxford Univ. Press, London and New York.
Townsend, J. S. (1947) "Electrons in Gases." Hutchinson, London.
Townsend, J. S., and McCallum, S. P. (1928) *Phil. Mag.* **6**, 856.
Townsend, J. S., and McCallum, S. P. (1934) *Phil. Mag.* **17**, 878.
Tozer, B. A. (1958) *J. Electronics and Control* **4**, 149.
Valeriu, M. (1943) *Bull. soc. roumaine phys.* **44**, 3.
Varnerin, L. J., and Brown, S. C. (1950) *Phys. Rev.* **79**, 946.
Wijsman, R. J. (1949) *Phys. Rev.* **75**, 833.

7.

Electronic Recombination

D. R. Bates and A. Dalgarno

Electrons and positive ions may unite by collisional-radiative recombination (a complex process tending to radiative recombination in the low plasma density limit and to collisional recombination in the high plasma density limit) or by a process (such as dielectronic recombination or dissociative recombination) involving a free-bound radiationless transition. These processes are described in § 1 and § 2 and an account is given of the theoretical work relating to them. Accurate calculations on the rate coefficients have been carried out in only a few cases. The experimental results are summarized in § 3. It is often difficult to identify the particular process studied with certainty.

1 Collisional-Radiative Recombination

Recombination between electrons and even the simplest positive ions, those of hydrogen, normally involves a complicated sequence of events.

Energy is liberated in a transition from the free state to a bound state. This energy may be radiated

$$H^+ + e \rightarrow H(p) + h\nu \tag{1}$$

(p denoting the principal quantum number of the level entered); or it may be given to another electron.

$$H^+ + e + e \rightarrow H(p) + e. \tag{2}$$

These processes cannot in general be considered in isolation. The atoms formed (many of which are initially in high quantum levels) may be re-ionized by the inverse of (2)†; that is, by

$$H(p) + e \rightarrow H^+ + e + e; \tag{3}$$

they may suffer collisional de-excitation

$$H(p) + e \rightarrow H(q) + e, \qquad q < p, \tag{4}$$

or excitation

$$H(p) + e \rightarrow H(q) + e, \qquad q > p; \tag{5}$$

or they may take part in line emission

$$H(p) \rightarrow H(q) + h\nu, \qquad q < p, \tag{6}$$

or absorption

$$H(p) + h\nu \rightarrow H(q), \qquad q > p \tag{7}$$

(the radiation concerned originating within the plasma). It is apparent that an electron may flit through many different states before reaching the $1s$ state around the proton to which it ultimately becomes attached.

Bates *et al.* (1962) have suggested that the *net* loss mechanism depending on the interacting processes listed be called *collisional-radiative decay*.

The well-known *radiative recombination* is the limiting form assumed by this general mechanism in plasmas in which the number density of free electrons $n(c)$ is too low for the collisional processes (2) and (3) to have an appreciable effect. In such plasmas the recombination coefficient α, defined by the equation

$$\frac{dn(P^+)}{dt} = -\alpha n(c) n(P^+) \tag{8}$$

where $n(P^+)$ is the number density of positive ions, is naturally independent of $n(c)$ since it describes the rate of (1) which is a two-body process.

Collisional recombination is another limiting form. It controls the initial of stage the decay of plasmas in which $n(c)$ is so high that the

† The inverse of (1) is unimportant in the present connection and will be ignored.

radiative prosesses (1), (6), and (7) may be neglected compared with the competing processes. Since (2) is a three-body process, α is proportional to $n(c)$ in the important early part of the collisional-recombination range.

It should be noted that though processes (6) and (7) do not influence the decay when $n(c)$ is either low or high, they may do so when $n(c)$ is moderate.

1.1 Plasmas of Low Density

References to the main early work on radiative recombination have been given by Massey and Burhop (1952) and Allen (1955).

1.1.1 *Hydrogenic Ions*

The rate coefficient in the case of hydrogenic ions may be calculated exactly, but the labor entailed rises rapidly as the principal quantum number n is increased. Seaton (1959a) has greatly improved the position by making use of the asymptotic expansion for the Kramers-Gaunt g factor:

$$g(n, \epsilon) = 1 + 0.1728n^{-2/3}(u+1)^{-2/3}(u-1) - 0.0496n^{-4/3}(u+1)^{-4/3}\left(u^2 + \frac{4}{3}u + 1\right) + \cdots \tag{9}$$

with

$$u = n^2\epsilon \tag{10}$$

where, if Ze is the nuclear charge, $Z^2\epsilon$ is the energy of the free electron in Rydberg units (Menzel and Pekeris, 1935; Burgess, 1958). The recombination coefficient at temperature T into level n may be written

$$\alpha_n(Z, T) = \mathscr{D}Zx_n^{3/2}\,S_n(\Lambda) \tag{11}$$

where, if α_f is the fine structure constant,

$$\mathscr{D} = \frac{2^6}{3}\left(\frac{\pi}{3}\right)^{1/2}\alpha_f^4 c a_0^4 = 5.197 \times 10^{-14}\ \text{cm}^3/\text{sec}, \tag{12}$$

$$\Lambda = hRcZ^2/kT = 157{,}890\ Z^2/T, \tag{13}$$

$$x_n = \Lambda/n^2, \tag{14}$$

and

$$S_n(\Delta) = \int_0^\infty \frac{g(n, \epsilon) \exp(-x_n u)}{1+u} du. \tag{15}$$

If only the leading term of (9) is retained, which is an approximation frequently employed, (15) reduces to

$$S^0(x_n) = \exp(x_n) \mathscr{E}i(x_n), \tag{16}$$

$\mathscr{E}i(x_n)$ denoting the exponential integral. Seaton has prepared tables from which the contributions from the next two terms may be found. He estimates that the error in the derived recombination coefficient should not exceed 2 % if T is $10^4 Z^2$ °K or less but that it may be more serious if T is of order $10^6 Z^2$ °K.

The computed values of the partial radiative recombination coefficients $\alpha_n(1, T)$ for $n = 1$ to 12, $T = 250$ °K ($\times$ 2) 64,000 °K and the total radiative recombination coefficient

$$\alpha_\Sigma(1, T) = \sum_{n=1}^{\infty} \alpha_n(1, T) \tag{17}$$

are given in Table I (Bates *et al.*, 1960).

When the mean thermal energy is small compared with the ionization potential of the nth level, then

$$\alpha_n(1, T) \propto n^{-1} T^{-1/2} \tag{18}$$

and, when it is large, then

$$\alpha_n(1, T) \propto n^{-3} T^{-3/2} \left\{ \ln \left[\frac{n^2 T}{157,890} \right] - 0.5772 + 8.56 \times 10^{-3} T^{1/3} - 2.3 \times 10^{-5} T^{2/3} \right\}. \tag{19}$$

As would be expected from (18) and (19) the contribution of the excited levels to $\alpha_\Sigma(1, T)$ is greater when T is low than when T is high. The variation of $\alpha_\Sigma(1, T)$ is about as $T^{-0.7}$.

Because of formula (11) Table I may be scaled to apply to the case of nuclei of charge Ze simply by multiplying the entries in the temperature row by Z^2 and those in the recombination coefficient rows by Z. For fixed T, $\alpha_\Sigma(Z, T)$ varies as $Z^{2.4}$ (approximately).

It is sometimes necessary to know the contribution, $\alpha_{nl}(Z, T)$, to the radiative recombination coefficient from each nl level. If T is low it is permissible to write

$$\alpha_{nl}(1, T) = \{a_{nl}/T^{1/2}\} \, 10^{-12} \text{ cm}^3/\text{sec} \tag{20}$$

TABLE I

RADIATIVE RECOMBINATION TO H^+ IONS[a, b]

Principle quantum number n	Rate coefficient (cm³/sec) $\alpha_n(1, T)$								
	250	500	1000	2000	4000	8000	16,000	32,000	64,000 °K
1	1.02^{-12}	7.17^{-13}	5.07^{-13}	3.56^{-13}	2.50^{-13}	1.74^{-13}	1.20^{-13}	8.02^{-14}	5.19^{-14}
2	5.66^{-13}	3.98^{-13}	2.79^{-13}	1.94^{-13}	1.32^{-13}	8.80^{-14}	5.63^{-14}	3.42^{-14}	1.95^{-14}
3	3.90^{-13}	2.72^{-13}	1.88^{-13}	1.28^{-13}	8.44^{-14}	5.33^{-14}	3.19^{-14}	1.80^{-14}	9.46^{-15}
4	2.95^{-13}	2.04^{-13}	1.40^{-13}	9.23^{-14}	5.86^{-14}	3.53^{-14}	2.00^{-14}	1.06^{-14}	5.33^{-15}
5	2.36^{-13}	1.62^{-13}	1.08^{-13}	6.99^{-14}	4.29^{-14}	2.48^{-14}	1.35^{-14}	6.87^{-15}	3.32^{-15}
6	1.96^{-13}	1.33^{-13}	8.70^{-14}	5.48^{-14}	3.26^{-14}	1.82^{-14}	9.53^{-15}	4.71^{-15}	2.22^{-15}
7	1.66^{-13}	1.11^{-13}	7.16^{-14}	4.39^{-14}	2.54^{-14}	1.38^{-14}	7.02^{-15}	3.39^{-15}	1.56^{-15}
8	1.43^{-13}	9.46^{-14}	5.99^{-14}	3.59^{-14}	2.02^{-14}	1.07^{-14}	5.34^{-15}	2.53^{-15}	1.14^{-15}
9	1.25^{-13}	8.17^{-14}	5.08^{-14}	2.98^{-14}	1.64^{-14}	8.51^{-15}	4.16^{-15}	1.93^{-15}	8.66^{-16}
10	1.11^{-13}	7.13^{-14}	4.36^{-14}	2.51^{-14}	1.35^{-14}	6.88^{-15}	3.31^{-15}	1.52^{-15}	6.72^{-16}
11	9.88^{-14}	6.27^{-14}	3.77^{-14}	2.13^{-14}	1.13^{-14}	5.65^{-15}	2.68^{-15}	1.22^{-15}	5.33^{-16}
12	8.87^{-14}	5.56^{-14}	3.29^{-14}	1.83^{-14}	9.53^{-15}	4.71^{-15}	2.21^{-15}	9.89^{-16}	4.30^{-16}
$\alpha_\Sigma(1, T)$	4.84^{-12}	3.12^{-12}	1.99^{-12}	1.26^{-12}	7.85^{-13}	4.83^{-13}	2.93^{-13}	1.73^{-13}	1.00^{-13}

[a] Bates *et al.* (1960).
[b] The indices indicate the powers of 10 by which the entries must be multiplied.

TABLE II

PARAMETERS a_{nl} APPEARING IN EQ. (20) OF TEXT[a]

n	l			
	0	1	2	3
1	16.5			
2	2.4	6.6		
3	0.81	2.5	2.9	
4	0.38	1.2	1.8	1.5

[a] There is a slight discrepancy between the first entry and the first row of Table I (which is less accurate).

where a_{nl} is as given in Table II. The most extensive calculations are those carried out for high T by Burgess (1958). Table III contains the 10,000 °K set of values of a quantity $\mathfrak{f}(n, l)$ which determines $\alpha_{nl}(1, T)$ through the relation

$$\alpha_{nl}(1, T) = \left(\frac{2}{\pi}\right)^{1/2} \left(\frac{8\pi\alpha_{\mathfrak{f}} a_0^2}{3c^2}\right) \left(\frac{kT}{m}\right)^{3/2} \mathfrak{f}(n, l). \tag{21}$$

TABLE III

RELATIVE CONTRIBUTIONS $\mathfrak{f}(n, l)$ TO RADIATIVE RECOMBINATION BETWEEN IONS AND ELECTRONS AT 10,000 °K[a]

n	l											
	0	1	2	3	4	5	6	7	8	9	10	11
1	1757											
2	261	600										
3	87.3	227	193									
4	40.5	108	121	61.9								
5	22.2	58.8	74.5	55.0	19.9							
6	13.5	35.2	47.8	41.8	22.8	6.52						
7	8.85	22.6	32.1	31.4	20.8	9.16	2.23					
8	6.11	15.3	22.3	23.3	17.6	9.57	3.62	0.787				
9	4.39	10.8	16.0	17.4	14.3	8.88	4.18	1.43	0.289			
10	3.27	7.90	11.8	13.4	11.6	7.94	4.27	1.79	0.560	0.111		
11	2.49	5.95	8.85	10.2	9.42	6.91	4.09	1.97	0.767	0.226	0.043	
12	1.94	4.45	6.78	8.05	7.63	5.99	3.80	2.04	0.919	0.339	0.094	0.017

[a] Burgess (1958).

Burgess also gives the 20,000 °K set. The relative importance of levels of low l is an increasing function of T but the effect is not very marked (cf. Bates *et al.*, 1939).

Much theoretical work has been done on hydrogenic recombination spectra in connection with nebular studies (Baker and Menzel, 1938; Searle, 1958; Burgess, 1958; Seaton, 1959b, 1960).

Following Baker and Menzel (1938) it is customary to consider two cases: case A, a plasma which is optically thin towards all radiation; and case B, a plasma which is optically thick towards the Lyman lines but is otherwise thin. The rate of entering an excited level $n'l'$ by radiative recombination and downward cascading together is equated to the rate of leaving the level by radiative transitions to lower levels $n''l''$ with

$$n' > 1, \qquad n' > n'', \qquad \text{case A}$$

and

$$n' > 2, \qquad n' > n'' > 1, \qquad \text{case B.}$$

This gives a set of linear equations from which the populations in the different levels and hence the intensities of the emitted lines may in principle be evaluated.

The intensities I_{n2} of the Balmer lines, $n \rightarrow 2$, are of special interest. Burgess (1958) has carried out calculations on them taking exact† account of levels with $n \leqslant 12$ but taking only approximate account of the higher levels. Table IV compares the computed values of the relative intensities referred to H_β as unity,

$$\mathscr{I}_n = I_{n2}/I_{42} \tag{22}$$

with corresponding observed values for planetary nebulae. The agreement is excellent.

TABLE IV

BALMER DECREMENTS[a]

	Relative intensity $\mathscr{I}_n$—computed				Relative intensity $\mathscr{I}_n$—observed (mean for six planetary nebulae corrected for reddening)
	Case A		Case B		
n	10,000 °K	20,000 °K	10,000 °K	20,000 °K	
3	2.48	2.36	2.62	2.53	2.55
4	1.00	1.00	1.00	1.00	1.00
5	0.501	0.510	0.489	0.493	0.50
6	0.288	0.296	0.276	0.280	0.29
7	0.181	0.188	0.172	0.174	—
8	0.122	0.128	0.114	0.117	—

[a] Burgess (1958).

The integrand of (15) controls the intensity distributions in the recombination continua. These distributions depend on the temperature of the plasma. Measurements on one of them enable T to be determined. Complications may arise from overlap with other recombination continua and from free-free transitions (bremsstrahlung).

1.1.2 *Complex Ions*

If the photoionization cross section $\sigma_{ai}(\nu)$ for the process

$$C_a + h\nu \rightarrow C_i^+ + e, \tag{23}$$

† He avoided the simplifying assumption that all quantum states of the same energy have equal populations (cf. Seaton, 1959b).

where the subscripts denote the levels occupied, is known (Chapter 3) the radiative recombination coefficient $\alpha_{ia}(T)$ for the inverse process

$$C_i^+ + e \rightarrow C_a + h\nu \tag{24}$$

may be computed using the standard relation of Milne (1924) in the form

$$\alpha_{ia}(T) = \frac{\omega_a}{\omega_i}\left\{\frac{2^{1/2}\exp(I_a/kT)}{c^2\pi^{1/2}(mkT)^{3/2}}\right\}\int_{I_a}^{\infty}(h\nu)^2\sigma_{ai}(\nu)\exp\left\{\frac{-h\nu}{kT}\right\}d(h\nu), \tag{25}$$

I_a being the ionization potential, and ω_a and ω_i being the statistical weights of C_a and C_i^+, respectively.

In many cases of interest $kT \ll I_a$ and $\nu^2\sigma_{ai}(\nu)$ does differ appreciably from the value $\nu_0^2\sigma_{ai}(\nu_0)$ at the threshold for energies of ejection up to several times kT so that (25) may be written approximately as

$$\begin{aligned}\alpha_{ia}(T) &= \frac{\omega_a}{\omega_i}\left\{\frac{2^{1/2}I_a^2}{(\pi kT)^{1/2}m^{3/2}c^2}\right\}\sigma_{ai}(\nu_0)\\ &= \frac{1.3\times 10^6 I_a^2}{T^{1/2}}\frac{\omega_a\sigma_{ai}(\nu_0)}{\omega_i}\end{aligned} \tag{26}$$

where I_a is in rydbergs, $\sigma_{ai}(\nu_0)$ is in cm², and T is in °K.

It is best to divide the calculation of the total recombination coefficient $\alpha_{i\Sigma}(T)$ into two parts. Provided the conditions specified in the preceding paragraphs are satisfied, recombination into the lower levels may be treated by formula (26) the threshold photoionization cross sections required being estimated by the quantum defect method (Chapter 3) if they are not already known. The higher levels may be regarded as hydrogenic† so that the coefficients given in Tables I and II may be adopted.

TABLE V

TOTAL RADIATIVE RECOMBINATION COEFFICIENTS AT 250 °K

Ion	H^+	He^+	Li^+	C^+	N^+	O^+	Ne^+	Na^+	K^+
$\alpha_{g\Sigma}$(250 °K)[a]	4.8	4.8	3.7	4.2	3.6	3.7	3.4	3.2	3.0

[a] Units: 10^{-12} cm³/sec.

† Levels for which l is small may remain far from hydrogenic (cf. Burgess and Seaton, 1960) but the contribution from them is not great (cf. Table III).

Table V contains some approximate values of $\alpha_{g\Sigma}(250\ ^\circ\mathrm{K})$, the subscript g indicating that the recombining ion is unexcited. They are based on the assumption that the threshold photoionization cross sections from the ground levels of the atoms concerned are as recommended in Table I of Chapter 3 and on the assumption that all excited levels are hydrogenic. As would be expected the values do not differ much from one singly charged ion to another.

If T is high, formula (25) must be used. Burgess and Seaton (1960) have carried out accurate calculations on recombination to He^+ ions the process being of astrophysical interest. They did not assume any of the S or P levels to be hydrogenic. Their main results are given in Table VI.

TABLE VI

RADIATIVE RECOMBINATION TO He^+ (AND, FOR COMPARISON, H^+) IONS[a]

Atom formed:	Helium				Hydrogen
		Excited			
Level entered:	Ground	Singlets	Triplets	Any	Any
Temperature (°K)	Recombination coefficient[b]				
10,000	1.59	0.63	2.10	4.31	4.17
20,000	1.15	0.35	1.20	2.69	2.51

[a] Burgess and Seaton (1960).
[b] Units: 10^{-13} cm^3/sec.

1.2 PLASMAS OF MODERATE AND HIGH DENSITY

Though it had long been appreciated that the recombination coefficient in plasmas of moderate and high density must be greater than the recombination coefficient in plasmas of low density quantitative estimates of the effect were not made until recently when the problem was attacked independently by D'Angelo (1961) in Princeton, by Bates and Kingston (1961) in Belfast, and by McWhirter (1961) in Harwell-Culham. Since the treatment of Bates and Kingston and of McWhirter, who later combined their efforts (Bates *et al.*, 1962) is more refined than that of D'Angelo, it alone will be outlined here.

The reaction path followed by a particular electron is in general very complicated (p. 246) but the statistical picture is quite simple.

Consider an initially fully ionized plasma composed of bare nuclei of charge Ze and electrons and suppose that it remains optically thin during the period of interest. For a wide range of electron densities $n(c)$ and temperatures T the population $n(p)$ of the pth level of the one-electron system formed by recombination satisfies the conditions

$$n(p) \ll n(c), \qquad p \neq 1 \tag{27}$$

and

$$n(p) \ll n(1). \tag{28}$$

Table VII shows the limitations imposed on $n(c)$ by condition (27).† Condition (28) is met if the mean thermal energy is much less than the excitation energy.

TABLE VII

COLLISIONAL-RADIATIVE RECOMBINATION[a]

Temperature T/Z^2 (°K)	Greatest value of $n(c)/Z^3X$ for which condition (27) is valid in an optically thin plasma (per cm³)[b]
250	10^{12}
1000	10^{14}
4000	10^{16}
16,000	10^{18}
64,000	10^{20}

[b] Bates *et al.* (1962).
[b] X is here the ratio of the number density of electrons to that of positive ions.

Great simplification ensues when conditions (27) and (28) are fulfilled (as they usually are in practice). The relaxation time associated with level p ($\neq 1$) is then very much shorter than the relaxation time of level (1) or of the free electrons. In consequence, the population in level p ($\neq 1$) may be regarded as growing to a quasi-equilibrium value instantaneously; and since $n(p)$, ($p \neq 1$) is relatively minute, $n(c)$ is not appreciably diminished by this growth.

During the subsequent decay of the plasma each $n(p)$, ($p \neq 1$) changes extremely slowly compared with the rates per unit volume at which

† These limitations, though not stringent, are more severe than those imposed by the need to ensure that significance can be attached to the orbitals of the levels of importance in the recombination.

systems in the level concerned are formed and destroyed. Equating these rates yields a set of linear equations from which the $n(p)$'s corresponding to given $n(c)$ and $n(1)$ may be found.

It may be noted that if inelastic or superelastic collisions evacuate or populate a level at a rate comparable with or greater than the rate at which radiative processes evacuate or populate the level, then elastic collisions will maintain a uniform distribution through the states of the level; and if inelastic and superelastic collisions do not do this, the distribution through the states does not affect the rate of collisional-radiative decay.

The set of equations for the $n(p)$'s may be kept finite by taking advantage of the fact that when p is large collisional processes are much more important than radiative processes so that $n(p)$ does not differ significantly from the Saha equilibrium number density

$$n_E(p) = p^2 n(c) n(P^+)\{h^2/2\pi mkT\}^{3/2} \exp\{Z^2\,(\mathrm{Ryd})/p^2kT\}$$

$$= 4.2 \times 10^{-16}\{p^2 n(c) n(P^+)/T^{3/2}\} \exp(157{,}890\, Z^2/p^2T) \qquad (29)$$

[$n(P^+)$ being as before the number density of positive ions].

Having evaluated the $n(p)$'s it is a simple task to calculate the rate at which 1 is populated by downward transitions from all discrete levels. Adding the rate at which it is populated directly from the continuum C and subtracting the rate at which it is evacuated by the totality of upward transitions, $1 \rightarrow p$ and $1 \rightarrow c$ gives

$$\frac{dn(1)}{dt} = -\frac{dn(P^+)}{dt} = \gamma n(c) n(P^+) \qquad (30)$$

where γ is an effective two-body rate coefficient.

Bates, Kingston, and McWhirter took the spontaneous transition probabilities needed for the detailed calculations from the papers of Baker and Menzel (1938) and of Green *et al.* (1957) and derived the radiative recombination coefficients from the tables of Seaton (1959a) already mentioned (§ 1.1). They adopted rate coefficients for collisional excitation and ionization (and the reverse processes) based on the work of Gryzinski (1959) in the case of hydrogen atoms and on the work of Burgess (1961) and Seaton (Chapter 11) in the case of hydrogenic ions.

If the plasma is optically thin, the coefficient defined by (30) may be expressed in the form

$$\gamma = \alpha - S n(1)/n(P^+) \qquad (31)$$

where α and S are functions of $n(c)$ and T only. It is apparent that α is an effective recombination coefficient and that it controls the decay when $n(c)$ is well above the equilibrium value. Substitution of (31) in (8) shows that S may be interpreted as an effective ionization coefficient. Bates, Kingston, and McWhirter suggest that α be called the collisional-radiative *recombination* coefficient, that S be called the collisional-radiative *ionization* coefficient, and that γ be called the collisional-radiative *decay* coefficient.

The computed values of α for H^+ ions are presented in Table VIII. This table is carried to values of $n(c)$ higher than condition (27) allows (cf. Table VII). For the lower of such values the theory in its simple form underestimates α during an initial period when the electron-reservoir formed by the excited levels is being filled but overestimates α during the remainder of the decay when this reservoir is being emptied.

Reference to Table VIII shows that whereas α is a slowly decreasing function of T if $n(c)$ is low, it is a very rapidly decreasing function if $n(c)$ is high; and that whereas it is a rapidly increasing function of $n(c)$ if T is low, it is a very slowly increasing function if T is high.

In order to obtain some indication of what happens in alkali ion plasmas, Bates, Kingston, and McWhirter did calculations similar to those just described but with level 1 made inaccessible (so that the model alkali atom has an excitation potential of 1.9 ev and an ionization potential of 3.4 ev). The results suggested that the rate of collisional-radiative decay is not very sensitive to the species of singly charged ion.

Turning to the case of multiply charged ions, let $\alpha(Z; n(c); T)$ denote the collisional-radiation recombination coefficient for bare nuclei of atomic number Z when the density and temperature of the plasma are as indicated. From the way in which the spontaneous transition probabilities and the radiative recombination, collisional excitation, and collisional ionization coefficients appropriate to hydrogenic ions scale, it may be shown that to a close approximation

$$\frac{1}{Z}\alpha(Z; Z^7 n(c); Z^2 T) = \alpha(1; n(c); T) \tag{32}$$

where the term on the right is what the collisional-radiative recombination coefficient for H^+ ions would be if all the hydrogenic ion formulae applied to H atoms. The relation greatly facilitates the presentation of the results of the computations. A table similar to Table VIII but for the hydrogenic ion case has been given by Bates *et al.* (1962). As would be expected, corresponding entries in the two tables are comparable in magnitude.

TABLE VIII

OPTICALLY THIN H^+ ION PLASMA[a]

$n(c)$	Collisional-radiative recombination coefficient α (cm^3/sec) at temperature T[b]:								
	250	500	1000	2000	4000	8000	16,000	32,000	64,000°K
$Lt\ n(c) \to 0$	4.8^{-12}	3.1^{-12}	2.0^{-12}	1.3^{-12}	7.9^{-13}	4.8^{-13}	2.9^{-13}	1.7^{-13}	1.0^{-13}
10^8	8.8^{-11}	1.4^{-11}	4.1^{-12}	1.8^{-12}	9.2^{-13}	5.1^{-13}	3.0^{-13}	1.8^{-13}	1.0^{-13}
10^9	4.0^{-10}	3.8^{-11}	7.5^{-12}	2.5^{-12}	1.0^{-12}	5.3^{-13}	3.0^{-13}	1.8^{-13}	1.0^{-13}
10^{10}	2.8^{-9}	1.6^{-10}	1.9^{-11}	4.1^{-12}	1.4^{-12}	6.1^{-13}	3.2^{-13}	1.8^{-13}	1.0^{-13}
10^{11}	2.7^{-8}	1.0^{-9}	6.9^{-11}	9.1^{-12}	2.2^{-12}	8.1^{-13}	3.4^{-13}	1.8^{-13}	1.0^{-13}
10^{12}	2.6^{-7}	9.0^{-9}	3.9^{-10}	2.9^{-11}	4.4^{-12}	1.2^{-12}	4.3^{-13}	2.0^{-13}	1.0^{-13}
10^{13}	2.6^{-6}	8.9^{-8}	3.1^{-9}	1.4^{-10}	1.2^{-11}	2.1^{-12}	6.2^{-13}	2.4^{-13}	1.1^{-13}
10^{14}	2.6^{-5}	8.8^{-7}	2.9^{-8}	9.8^{-10}	5.1^{-11}	5.1^{-12}	1.0^{-12}	3.1^{-13}	1.2^{-13}
10^{15}		8.8^{-6}	2.9^{-7}	8.7^{-9}	2.7^{-10}	1.7^{-11}	2.3^{-12}	4.9^{-13}	1.6^{-13}
10^{16}			2.9^{-6}	8.5^{-8}	2.3^{-9}	8.4^{-11}	5.0^{-12}	7.3^{-13}	1.9^{-13}
10^{17}				8.4^{-7}	2.1^{-8}	3.4^{-10}	1.4^{-11}	1.8^{-12}	4.4^{-13}
10^{18}					2.0^{-7}	2.5^{-9}	9.6^{-11}	1.2^{-11}	2.8^{-12}
$Lt\ n(c) \to \infty$	$2.6^{-19}\ n(c)$	$8.8^{-21}\ n(c)$	$2.9^{-22}\ n(c)$	$8.4^{-24}\ n(c)$	$1.9^{-25}\ n(c)$	$2.4^{-27}\ n(c)$	$9.1^{-29}\ n(c)$	$1.1^{-29}\ n(c)$	$2.7^{-30}\ n(c)$

[a] Bates *et al.* (1962).

[b] The indices give the power of 10 by which the entries must be multiplied.

2 *Recombination Involving a Free-Bound Radiationless Transition*

A complex atom or molecule X has series of levels with excitation potentials converging on the ionization potentials corresponds to the different levels of the ion X^+. Suppose that one of the levels d of the series of X associated with the ionization potential I_j lies within the continuum of the system composed of X^+ in level i $(I_i < I_j)$ and a free electron (Fig. 1). Provided certain selection rules (Condon and

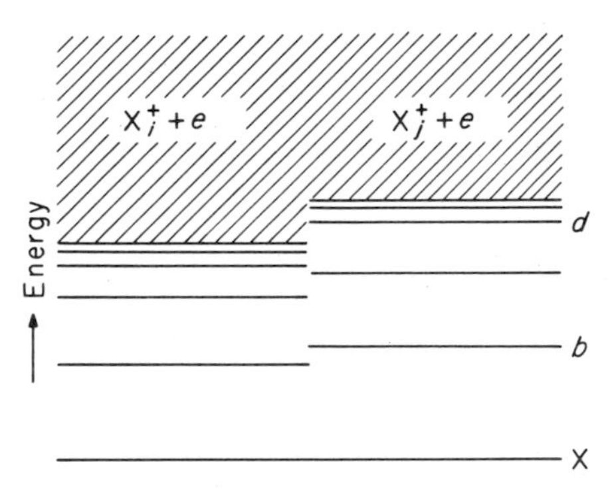

FIG. 1. Illustrative energy level diagram.

Shortley, 1935; Herzberg, 1950) are satisfied, the radiationless transition

$$X_i^+ + e \rightarrow X_d \tag{33}$$

may take place in a plasma. In general this only leads to a transient recombination of little direct physical interest since autoionization by the inverse radiationless transition

$$X_d \rightarrow X_i^+ + e \tag{34}$$

maintains a quasi-equilibrium in which the number density $n(X_d)$ is minute. For true recombination there must be some means by which a neutral system which is formed can become immune to autoionization. If A_S in the probability that such *stabilization* occurs, the recombination rate is $n(X_d)A_S$, and hence the recombination coefficient is

$$\alpha^{rlss} = n(X_d)A_S/n(c)n(X_i^+). \tag{35}$$

It is apparent that

$$n(X_d) = \frac{A_a}{A_s + A_a} n_E(X_d) \tag{36}$$

wher A_a is the probability of autoionization and $n_E(X_d)$ is the number density in Saha equilibrium so that

$$\frac{n_E(X_d)}{n(c)n(X_i^+)} = \frac{\omega_d}{2\omega_i} \frac{h^3}{(2\pi mkT)^{3/2}} \exp\left\{\frac{-\epsilon_{di}}{kT}\right\}, \tag{37}$$

ϵ_{di} being the amount by which the energy of X_d exceeds that of X_i^+ and ω_d and ω_i being the statistical weights of X_d and X_i^+, respectively. Substitution from (36) and (37) in (35) yields

$$\alpha^{rlss} = \frac{A_s A_a}{A_s + A_a} \frac{\omega_d}{2\omega_i} \frac{h^3}{(2\pi mkT)^{3/2}} \exp\left\{\frac{-\epsilon_{di}}{kT}\right\}. \tag{38}$$

Introducing lifetimes, τ_s and τ_a, defined according to

$$\tau_s = A_s^{-1}, \qquad \tau_a = A_a^{-1}, \tag{39}$$

(38) may be written

$$\alpha^{rlss} = \frac{1}{\tau_s + \tau_a} \frac{\omega_d}{2\omega_i} \frac{h^3}{(2\pi mkT)^{3/2}} \exp\left\{\frac{-\epsilon_{di}}{kT}\right\}. \tag{40}$$

2.1 Dielectronic Recombination

In dielectronic recombination,

$$X_i^+ + e \rightleftharpoons X_d \to X_b + h\nu, \tag{41}$$

the stabilization is effected by a radiative transition from the level d to some level b which is not subject to autoionization (Fig. 1). The lifetime $\tau_r(d, b)$ towards emission of the line $d \to b$ is in general much longer than the lifetime $\tau_a(d, i)$ towards autoionization $d \to i$ which may be of the order 10^{-13} second or even less so that (40) reduces to

$$\alpha^{\text{dielect}} = \mathscr{C}_1 T^{-3/2} \frac{\omega_d}{\omega_i} \exp\left\{\frac{-\epsilon_{di}}{kT}\right\} \frac{1}{\tau_r(d, b)}, \tag{42}$$

with

$$\mathscr{C}_1 = h^3/2(2\pi mk)^{3/2} = 2.1 \times 10^{-16} \text{ cm}^3 \text{ deg}^{3/2} \tag{43}$$

(Massey and Bates, 1942). To get the total dielectronic recombination coefficient $\alpha_{\Sigma}^{\text{dielect}}$ for X_i^+ summations over all levels d and all levels b must be performed.

If $\tau_r(d, b)$ is expressed in terms of the absorption oscillator strength f_{bd} for the transition $b \rightarrow d$ formula (42) becomes

$$\alpha^{\text{dielect}} = \mathscr{C}_2 T^{-3/2} \frac{\omega_b}{\omega_i} \tilde{\nu}(b, d)^2 \exp\left\{\frac{-\epsilon_{di}}{kT}\right\} f_{bd} \tag{44}$$

where

$$\mathscr{C}_2 = \frac{(2\pi)^{1/2} e^2 h^3}{ck^{3/2} m^{5/2}} = 1.4 \times 10^{-16} \text{ cm}^5 \text{ deg}^{3/2} \text{ sec}^{-1} \tag{45}$$

and $\tilde{\nu}(b, d)$ is the wave number of the emitted radiation.

It is instructive to note that since the cross section σ_{bi} and oscillator strength f_{bi} for

$$X_b + h\nu \rightarrow X_i^+ + e \tag{46}$$

are related by

$$\sigma_{bi} = \frac{\pi e^2 h}{mc} \frac{df_{bi}}{d\epsilon}, \tag{47}$$

ϵ being the energy of the ejected electron, formula (25) for the radiative recombination coefficient (which will here be denoted by α^{radiat}) may be written

$$\alpha^{\text{radiat}} = \mathscr{C}_2 T^{-3/2} \frac{\omega_b}{\omega_i} \int_{\text{continuum}} \tilde{\nu}^2 \exp\left\{\frac{-\epsilon}{kT}\right\} df_{bi}. \tag{48}$$

The close parallelism of (44) and (48) is revealing.

Dielectronic recombination naturally cannot occur to H^+ ions. Reliable calculations on complex ions have not been carried out but in some cases, for example, in the case of He^+ ions, it may be shown that the process is very slow compared with radiative recombination because any suitable level d lies far enough above the first ionization potential to make the factor $\exp\{-\epsilon_{di}/kT\}$ extremely small.*

The position may be different *if* ϵ_{di} is such that this factor is almost unity. Comparison of (44) and (48) suggests that dielectronic recombination is then likely to be faster than radiative recombination: thus it is

* Dielectronic recombination to normal N^+ and O^+ ions have been investigated by Bates (1962). The computed rate coefficients are less than those for radiative recombination.

faster if the oscillator strength of the line $b \rightarrow d$ exceeds the (usually) small fraction of the oscillator strength of the photoionization continuum $b \rightarrow i$ which arises from within kT of the threshold.

To obtain an indication of how large the total dielectronic recombination coefficient may be in a favorable case suppose that

$$\exp \left\{ \frac{-\epsilon_{di}}{kT} \right\} = 1, \tag{49}$$

$$\tilde{\nu}(d, b) = 10^5 \text{ cm}^{-1}, \tag{50}$$

and

$$\sum_{b,d} \frac{\omega_b}{\omega_i} f_{bd} = 0.3. \tag{51}$$

Substitution in (44) then gives $\alpha_\Sigma^{\text{dielect}}$ to be about $1._1 \times 10^{-10}$ cm^3/sec at 250 °K, $1._3 \times 10^{-11}$ cm^3/sec at 1000 °K, $1._7 \times 10^{-12}$ cm^3/sec at 4000° K, and $2._1 \times 10^{-13}$ cm^3/sec at 16,000 °K. The corresponding values for $\alpha_\Sigma^{\text{radiat}}$ to H^+ ions are 4.8×10^{-12}, 2.0×10^{-12}, 7.8×10^{-13}, and 2.9×10^{-13} cm^3/sec (Table I). In considering the trend it should be borne in mind that the chance that there is a level for which (49) is even approximately true is less if T is low than if T is high. The existence of a case as favorable as that examined has not been established. A suggestion by Garton and associates (1960) that $\alpha_\Sigma^{\text{dielect}}$ for Ar^+ ($^2P_{3/2}$) ions is 5×10^{-11} cm^3/sec or rather greater at 300 °K has been withdrawn by Pery-Thorne and Garton (1960) who adduce evidence that the value of f_{bd} adopted in the original calculations is much too large.

Terms allowing for the effect of dielectronic recombination should of course be introduced into the equations of collisional-radiative recombination. The formal introduction of such terms is a trivial task; but in the present state of knowledge it would be of little service.

2.2 Collisional Stabilization

Stabilization may be brought about by an encounter with a thermal electron

$$X_d + e \rightarrow X_b + e. \tag{52}$$

Formula (42) may be modified to take this possibility into account. All that need be done is to replace $\tau_r(d, b)$ the lifetime towards the parallel radiative process, by $\tau_{rc}(d, b)$ defined by

$$\frac{1}{\tau_{rc}(d, b)} = \frac{1}{\tau_r(d, b)} + \frac{1}{\tau_c(d, b)} \tag{53}$$

$\tau_c(d, b)$ being the lifetime towards (52) which is given by

$$\tau_c(d, b) = 1/\kappa(d, b)\, n(c) \tag{54}$$

where $\kappa(d, b)$ is the de-excitation rate coefficient and $n(c)$ is as usual the number density of electrons.

Examination of the results available on processes involving optically allowed transitions shows that $\kappa(d, b)$ and $\tau_r(d, b)$ are related thus:

$$\kappa(d, b) = \mathrm{p}\, \{10^{-29}\, \lambda^4/\tau_r(d, b)\}\, \mathrm{cm^3/sec} \tag{55}$$

in which p, a parameter characteristic of the process, is of the order of unity and in which the wavelength λ and radiative lifetime $\tau_r(d, b)$ are measured in angstroms and seconds, respectively. Substitution in (54) gives

$$\tau_c(d, b)/\tau_r(d, b) = 10^{29}/\mathrm{p}\lambda^4 n(c). \tag{56}$$

Hence collisional stabilization only becomes as effective as radiative stabilization when $n(c)$ is about as high as the following values (with the assumed λ in brackets): $1 \times 10^{17}/\mathrm{cm}^3$ (1000 A), $6 \times 10^{15}/\mathrm{cm}^3$ (2000 A), $4 \times 10^{14}/\mathrm{cm}^3$ (4000 A), $2 \times 10^{13}/\mathrm{cm}^3$ (8000 A). Neither form of stabilization is likely to be important in such cases.

2.3 Dissociative Recombination

Dissociative recombination

$$AB^+ + e \rightleftharpoons AB' \rightarrow A' + B' \tag{57}$$

(the primes indicating possible excitation) occurs as a result of a radiationless transition to some state of the molecule in which the constituent atoms move apart and gain kinetic energy under the action of their mutual repulsion so that the neutralization is rendered permanent by virtue of the Franck-Condon principle (Bates, 1950). The lifetime τ_p towards this stabilization process is extremely brief: for example, if the repulsive force between the atoms is 5 ev/A and if their reduced mass is 8 on the ^{16}O scale, their separation is increased by 0.15 A and their kinetic energy of relative motion is increased by 0.75 ev in only about 7×10^{-15} sec. It appears likely that in most cases the rate-limiting step is the radiationless transition.

Denote the lifetime towards autoionization when the nuclei are held fixed by τ_a and assume this to be almost independent of R, the internuclear distance. For simplicity, suppose in the first instance that only the zeroth vibrational level of the molecular ion need be taken into account, let $\psi_0(R)$ be its normalized vibrational wave function and put

$$|\psi_0(R)|^2 \frac{dR}{d\epsilon} = f_0(\epsilon) \tag{58}$$

where ϵ is the energy supplied by the electron in a vertical transition (Fig. 2).

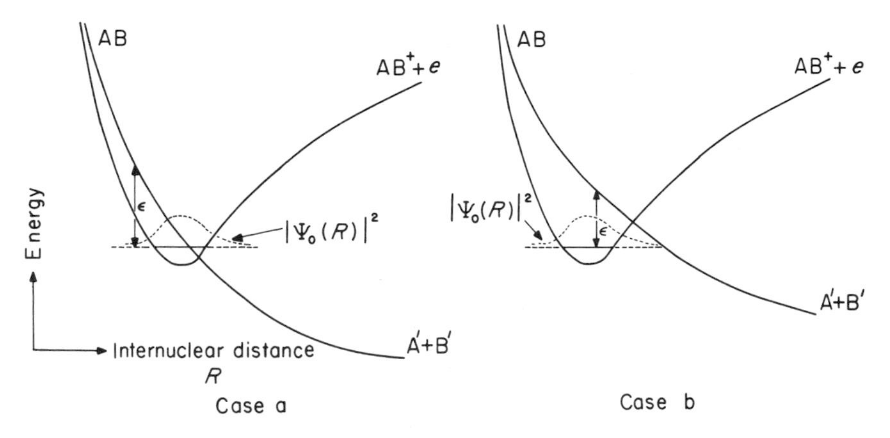

FIG. 2. Potential energy curves of states concerned in dissociative recombination.

Using the Winans-Stueckelberg (1928) approximation for the vibrational wave functions of the neutral molecule it may be seen from (40) that the rate coefficient for dissociative recombination is

$$\alpha^{\text{diss}} = \mathscr{C}_1 T^{-3/2} \frac{\omega_{\text{AB}}}{\omega_{\text{AB}^+}} \mathscr{A}_0(T) \tag{59}$$

in which $\mathscr{C}_1$ is as defined in (43), ω_{AB^+} and ω_{AB} are the statistical weights of the relevant electronic states of the molecular ion and of the neutral molecule, and

$$\mathscr{A}_0(T) = \frac{g_0(T)}{\tau_a + g_0(T)\tau_p} \tag{60}$$

with

$$g_0(T) = \int_0^\infty f_0(\epsilon) \exp\left\{\frac{-\epsilon}{kT}\right\} d\epsilon. \tag{61}$$

If the stabilization is the rate-limiting step, (60) reduces to

$$\mathscr{A}_0(T) = 1/\tau_{\mathrm{p}}; \tag{62}$$

but if, as is probably more common, the radiationless transition is the rate-limiting step, it reduces to

$$\mathscr{A}_0(T) = g_0(T)/\tau_{\mathrm{a}}. \tag{63}$$

Several vibrational levels v may be important if the temperature is high. Assuming that the distribution amongst them is thermal, it may be shown that $g_0(T)$ in (60) should be replaced by

$$\bar{g}(T) = \sum_v g_v(T) \exp\left\{\frac{-\mathrm{E}_v}{kT}\right\} \Big/ \sum_v \exp\left\{\frac{-\mathrm{E}_v}{kT}\right\} \tag{64}$$

where $g_v(T)$ is an analogously defined function and E_v is the energy of the vth level (with E_0 chosen to be zero).

Account should strictly be taken of the possible dependence of τ_{p} on $\epsilon + \mathrm{E}_v$ and of several other minor effects but it would be unrealistic to seek to introduce such refinements at present.

Though proper calculations have not been performed Bates (1950) has made estimates which show that in certain, by no means unusual circumstances, α^{diss} may be some 10^{-7} cm^3/sec.

The predicted dependence of α^{diss} on T is occasionally stated incorrectly in the literature. Some writers have wrongly inferred a $T^{-3/2}$ law from (59) overlooking the fact that $\mathscr{A}_0(T)$ itself depends on T unless the rate-limiting step is the stabilization. There is no simple general law when the rate-limiting step is the radiationless transition. Thus it is apparent from (61) and (63) that if $f_0(\epsilon)$ varies little over the effective range of integration (e.g., case (a) of Fig. 2, low T), then $\mathscr{A}_0(T)$ is proportional to T giving a $T^{-1/2}$ law; but if $f_0(\epsilon)$ is initially a rapidly increasing function of ϵ [e.g., case (b) of Fig. 2], α^{diss} may rise to a maximum. The position is likely to be complicated by more than one unbound state being involved.

Its great rapidity may make dissociative recombination important even in plasmas in which atomic ions predominate since these may be converted into molecular ions by ion-atom interchange or by other collision processes. For example the charges carried by the O^+ ions in the upper atmosphere are neutralized through sequences like

$$\left.\begin{aligned} O^+ + O_2 &\rightarrow O_2^+ + O \\ O_2^+ + e &\rightarrow O' + O'' \end{aligned}\right\} \tag{64}$$

or

$$\left.\begin{aligned} O^+ + N_2 &\rightarrow NO^+ + N \\ NO^+ + e &\rightarrow N' + O' \end{aligned}\right\} \tag{65}$$

(cf. Bates and Massey, 1947; Bates, 1955; Bates and McDowell, 1957; Krassovsky, 1957; Bates and Nicolet, 1960).

3 *Experimental Studies*

In experimental studies of electron recombination, measurements are made of the decay of electron density in gases which have been partially ionized. The interpretation of the measurements is complicated because electrons are removed not only by recombination but also by diffusion to the walls, because electron production may continue after the termination of the ionizing discharge and because the identification of the ionic species may be ambiguous. Arising from these complications is the possibility that the behavior of the plasma may be very sensitive to the presence of small amounts of impurities.

3.1 Plasmas of Low Densities

The decay of electron concentration in a low density plasma is conveniently determined using the microwave method of Biondi and Brown (Biondi and Brown, 1949; Biondi, 1951a; Brown and Rose, 1952; Goldstein, 1955; Sexton *et al.*, 1959) in which the shift in frequency Δf of a resonant cavity containing the plasma is measured relative to the frequency f of the cavity without the plasma. The frequency shift is related to the electron density n by the formula

$$\frac{\Delta f}{f} = \frac{1}{2} \frac{\bar{n}}{1 + (\nu/\omega)^2 n_p} \tag{65}$$

where ν is the electron collision frequency, ω is the radian frequency of the probing microwave signal, n_p denotes $m\omega^2/e^2$, and

$$\bar{n} = \int_V nE^2 dv \Big/ \int_V E^2 dv, \tag{66}$$

E being the electric field of the probing signal and V being the volume of the microwave cavity. Because of macroscopic polarization of the plasma, (65) is not valid at high electron densities (Persson, 1957) and because of secondary production and heating effects, (65) is not valid at high power levels of the probing pulse (Oskam, 1958).

Electron densities may also be determined from measurements of the transmission of microwave signals through a discharge (Whitmer, 1956; Takeda and Holt, 1959).

The electron density $n(\mathbf{r}, t)$ satisfies the equation

$$\frac{\partial n}{\partial t} = -\alpha n^2 + D_a \nabla^2 n \tag{67}$$

where α is the recombination coefficient, D_a is the ambipolar diffusion coefficient (inversely proportional to the gas pressure), and it is assumed that electron production has ceased. If diffusion is negligible and n_1 and n_2 are the electron concentrations at times t_1 and t_2, respectively, then

$$\frac{1}{n_2} - \frac{1}{n_1} = (t_2 - t_1)\alpha \tag{68}$$

and the slope of a plot of $1/n$ against t is the recombination coefficient. However, the plasma is always diffusion-controlled near the cavity walls and the interpretation of a plot of $1/\bar{n}$ against t must be made with caution (Gray and Kerr, 1960). Persson and Brown (1955) have argued indeed that the experimental results for molecular hydrogen (Biondi and Brown, 1949; Richardson and Holt, 1951; Varnerin, 1951) are due to higher diffusion modes and to attachment to impurities and not to a large recombination coefficient ($\sim 10^{-6}$ cm^3 sec^{-1}) as was originally supposed.

3.1.1 *Helium*

Recombination in a helium plasma was investigated by Biondi and Brown (1949) and by Johnson *et al.* (1950) who observed that at pressures greater than 20mm-/Hg electron removal is apparently controlled by recombination and at pressures below 5 mm Hg electron removal is controlled by diffusion. They derive a recombination coefficient of about 10^{-8} cm^3 sec^{-1}. Although this numerical value is called into question by the further measurements of Oskam (1958) and the analysis of Gray and Kerr (1960), which suggests that the importance of diffusion was underestimated, the identification of the main recombination process as that of dissociative recombination

$$\mathrm{He}_2^+ + e \rightarrow \mathrm{He}' + \mathrm{He}'' \tag{69}$$

(Bates, 1950) has been widely accepted.

The phenomenon of afterglow quenching has been used by Chen *et al.* (1961) to determine the dependence on electron temperature T of the recombination coefficient α in a helium plasma. They observe the intensity variations of the total visible light and of the two helium lines at 5876 A and 3888 A consequent upon selective heating of the electrons by a pulsed microwave and deduce that α varies as $T^{-3/2}$ for $300\ °K < T < 1500\ °K$, in harmony with an earlier investigation by Anderson (1958). In addition, measurements at 300 °K of n by microwave and by optical methods yield a value of $(8.9 \pm 0.5) \times 10^{-9}$ cm^3 sec^{-1} for α. There is little doubt that the main recombination mechanism is (69) but measurements of the widths of the helium lines (cf. Biondi, 1961) would be of interest.

3.1.2 *Other Inert Gases*

Recombination in a low density neon plasma has been studied by Biondi and Brown (1949), by Holt *et al.* (1950), and by Oskam (1958). A recombination coefficient of about 2×10^{-7} cm^3 sec^{-1} has been derived which is apparently independent of electron temperature for $195\ °K < T < 410\ °K$.

Of the remaining inert gases, argon has been studied by Biondi and Brown (1949), by Biondi (1951b), by Redfield and Holt (1951), and by Sexton and Craggs (1958), krypton by Richardson (1952), by Lennon and Sexton (1959), and by Popov and Afanaśeva (1960), and xenon by Lennon and Sexton (1959). There are some uncertainties as to the role of impurities in these experiments but from them recombination coefficients of 7×10^{-7} cm^3 sec^{-1} for argon, 3×10^{-7} cm^3 sec^{-1} for krypton, and 2×10^{-6} cm^3 sec^{-1} for xenon have been derived.

There seems little doubt that all these rates refer to dissociative recombination (57), the singly charged diatomic molecular ions occurring readily in the laboratory plasmas (cf. Pahl, 1959).

3.1.3 *Hydrogen*

Persson and Brown (1955) have concluded that the recombination coefficient in the hydrogen afterglow is less than 3×10^{-8} cm^3 sec^{-1} and this conclusion is in harmony with the measurements of Popov and Afanaśeva (1960). The predominant ion in the experimental hydrogen afterglows is not identified with certainty but it may be H_3^+.

3.1.4 *Nitrogen*

Because of its importance in the earth's atmosphere recombination in nitrogen has been the subject of many investigations (cf. Dalgarno,

1961), the most complete being that by Faire and Champion (1959). They derive a recombination coefficient of $(4.0 \pm 0.3) \times 10^{-7}$ cm^3 sec^{-1} at a temperature of about 400 °K but there is some doubt over the identity of the recombining ion. Indeed, several ionic species may be present in their plasma and a more general analysis (Kunkel, 1951; Loeb and Kunkel, 1952) of the measurements may be necessary. Kasner, *et al.* (1961) have employed a mass spectrometer to identify the ions and they find that for N_2^+ the recombination coefficient is $(5.9 \pm 1) \times 10^{-7}$ cm^3 sec^{-1} and that for N_3^+ and N_4^+ it is about 2×10^{-6} cm^3 sec^{-1}.

3.1.5 *Other Gases*

Experiments on low-density plasmas in mercury and in oxygen have been carried out (cf. Massey, 1952; Loeb, 1956) but the interpretation is complicated by the removal of electrons through attachment processes and no meaningful values have been obtained for recombination coefficients, except possibly for the recombination of O_2^+ ions and electrons (Kasner *et al.*, 1961). The value obtained—$(3.8 \pm 1) \times 10^{-7}$ cm^3 sec^{-1}—is unexpectedly large in view of the evidence provided by upper atmosphere data (cf. Dalgarno, 1961).

3.2 Plasmas of Moderate and High Densities

The earliest investigations of electron-ion recombination were carried out by Kenty, Mohler, Boeckner, and Sayers using interrupted arcs in argon, caesium, and mercury, the electron density being measured by probes and by photometric methods. This work has been reviewed by Massey and Burhop (1952), by Massey (1952), and by Loeb (1956): for electron densities of about 10^{12} cm^{-3} and electron temperatures in the range 1000 °K-4000 °K, apparent recombination coefficients of two or three times 10^{-10} cm^3 sec^{-1} were derived for all three gases. Since ambipolar diffusion has been ignored in the analysis and since the identity of the ions is uncertain (some of the ions will be diatomic molecular ions), the numerical value of the recombination coefficient may not be significant. It is of interest to note, however, that the values obtained are of the order of magnitude expected for collisional-radiative recombination (cf. Table VIII). Larger values of α are obtained in the more recent microwave studies of caesium and mercury at similar electron densities and temperatures (Dandurand and Holt, 1951) but again their significance is obscure. Much greater attention must be given to establishing the identity of the ions before the complex processes which occur in plasmas of moderate densities can be understood.

Electron-ion recombination in high current spark channels in hydrogen, argon, and helium has been investigated by Craggs and his associates (Craggs and Meek, 1946; Craggs and Hopwood, 1947; Tsui-Fang and Craggs, 1958; McChesney and Craggs, 1958; Mitchell, 1960) using photomultipliers and fast-sweep oscilloscopes. Detailed measurements have been made of line profiles and ion densities of about 10^{17} cm^{-3} have been derived from analysis of the Stark broadening (Griem, *et al.*, 1959; Griem, 1960; Griem and Shen, 1961). Considerations based on the Saha equilibrium formula then lead to electron temperatures of the order of 10,000 °K. The decay curves observed are complex and it is clear that recombination is not a simple two-body process. The apparent recombination coefficients vary between 10^{-11} and 10^{-12} cm^3 sec^{-1} depending upon the pressure.

Similar investigations have been carried out by Olsen and Huxford (1952) of highly condensed discharges in argon and neon, the ion densities being determined from the Stark broadening of small amounts of hydrogen introduced into the gases. The apparent recombination coefficients are near 10^{-13} cm^3 sec^{-1} for $n \sim 10^{17}$ cm^{-3} and $T \sim 10{,}000$ °K, but the interpretation of the measurements has been criticized by Fowler and Atkinson (1959).

Fowler and Atkinson (1959) measured the absolute intensity of the continuum associated with the Balmer lines in a shock tube containing an expanding hydrogen plasma in a field-free region. An ion density of about 6×10^{16} cm^{-3} was derived from the Stark broadening of H_β and a recombination coefficient of about 10^{-12} cm^3 sec^{-1} at a temperature of 4500 °K was obtained. The derived coefficient is that appropriate to radiative recombination and it has the predicted order of magnitude.

Recently Kuckes *et al.* (1961) have studied the plasma losses in the afterglow of discharges in the B-1 stellarator. For a helium plasma, they obtain a recombination coefficient proportional to $n^{1/2}$ for $10^{11} < n < 5 \times 10^{13}$, thereby providing clear evidence in support of the collisional-radiative recombination process.

REFERENCES

Allen, C. W. (1955) "Astrophysical Quantities," Univ. London Press (Athlone), London.
Anderson, J.M. (1958) *Phys. Rev.* **108**, 898.
Baker, J. G., and Menzel, D.H. (1938) *Astrophys. J.* **88**, 52.
Bates, D. R. (1950) *Phys. Rev.* **77**, 718; **78**, 492.
Bates, D. R. (1962) *Planet. Space Sci.* **9** (in press).
Bates, D. R. (1955) *Proc. Phys. Soc.* A**68**, 344.
Bates, D. R., and Kingston, A.E. (1961) *Nature* **189**, 652.

Bates, D. R., and McDowell, M. R. C. (1957) *J. atmos. terrest. Phys.* **10**, 96.
Bates, D. R., and Massey, H. S. W. (1947) *Proc. Roy. Soc.* **A192**, 1.
Bates, D. R., and Nicolet, M. (1960) *J. atmos. terrest. Phys.* **18**, 65.
Bates, D. R., Buckingham, R. A., Massey, H. S. W., and Unwin, J. J. (1939) *Proc. Roy. Soc.* **A170**, 322.
Bates, D. R., Kingston, A. E., and McWhirter, R. W. P. (1960) Unpublished data.
Bates, D. R., Kingston, A. E., and McWhirter, R.W. P. (1962) *Proc. Roy. Soc.* **A** (in press).
Biondi, M. A. (1951a) *Rev. sci. Instrum.* **22**, 500.
Biondi, M. A. (1951b) *Phys. Rev.* **83**, 1078.
Biondi, M. A. (1961) *Planet. Space Sci.* **3**, 104.
Biondi, M. A., and Brown, S. C. (1949) *Phys. Rev.* **75**, 1700.
Brown, S. C., and Rose, D. J. (1952) *J. appl. Phys.* **23**, 711, 719, 1028.
Burgess, A. (1958) *Monthly Not. Roy. Astron. Soc.* **118**, 477.
Burgess, A. (1961) *Mem. Soc. Roy. Sci. Liège* [5] **4**, 299.
Burgess, A., and Seaton, M. J. (1960) *Monthly Not. Roy. Astron. Soc.* **121**, 471.
Chen, C. L., Leiby, C. C., and Goldstein, L. (1961) *Phys. Rev.* **121**, 1391.
Condon, E. U., and Shortley, G. H. (1935) "The Theory of Atomic Spectra," Cambridge Univ. Press, London and New York.
Craggs, J. D., and Hopwood, W. (1947) *Proc. Phys. Soc.* **59**, 771.
Craggs, J. D., and Meek, J. M. (1946) *Proc. Roy. Soc.* **A186**, 241.
Dalgarno, A. (1961) *Ann. Geophys.* **17**, 16.
Dandurand, P., and Holt, R. B. (1951) *Phys. Rev.* **82**, 278, 868.
D'Angelo, N. (1961) *Phys. Rev.* **121**, 501.
Faire, A. C., and Champion, K. S. W. (1959) *Phys. Rev.* **113**, 1.
Fowler, R. G., and Atkinson, W. R. (1959) *Phys. Rev.* **113**, 1268.
Garton, W. R. S., Pery, A., and Codling, K. (1960) *In* "Proceedings of the Fourth International Conference on Ionization Phenomena in Gases, Uppsala, 1959" (N. R. Nilsson, ed.), Vol. I, p. 206. North-Holland, Amsterdam.
Goldstein, L. (1955) *Advances in Electronics and Electron Phys.* **7**, 399.
Gray, E. P., and Kerr, D. E. (1960) *In* "Proceedings of the Fourth International Conference on Ionization Phenomena in Gases, Uppsala, 1959" (N. R. Nilsson, ed.), Vol. I, p. 184. North-Holland, Amsterdam.
Green, L. C., Rush, P. P., and Chandler, C. D. (1957) *Astrophys. J. Suppl.* **3**, 37.
Griem, H. R. (1960) *Astrophys. J.* **132**, 883.
Griem, H. R., and Shen, K. Y. (1961) *Phys. Rev.* **122**, 1490.
Griem, H. R., Kolb, A. C., and Shen, K. Y. (1959) *Phys. Rev.* **116**, 4.
Gryzinski, M. (1959) *Phys. Rev.* **115**, 374.
Herzberg, G. (1950) "Spectra of Diatomic Molecules," 2nd ed. Van Nostrand, Princeton, New Jersey.
Holt, R. B., Richardson, J. M., Haviland, B., and McClure, B. T. (1950) *Phys. Rev.* **77**, 239.
Johnson, R. A., McClure, B. T., and Holt, R. B. (1950) *Phys. Rev.* **80**, 376.
Kasner, W. H., Rogers, W. A., and Biondi, M. A. (1961) *Phys. Rev. Letters.* **7**, 321.
Krassovsky, V. I. (1957) *Izv. Akad. Nauk SSSR., Ser. geofiz.* **4**, 504.
Kuckes, A. F., Motley, R. W., Hinnov, E., and Hirschberg, J. G. (1961) *Phys. Rev. Letters* **6**, 337.
Kunkel, W. B. (1951) *Phys. Rev.* **84**, 218.
Lennon, J. J., and Sexton, M. C. (1959) *J. Electronics and Control* **7**, 123.
Loeb, L. B. (1956) *In* "Handbuch der Physik" (S. Flügge, ed.) Vol. 21, p. 471. Springer, Berlin.

Loeb, L. B., and Kunkel, W. B. (1952) *Phys. Rev.* **85**, 493.
McChesney, M., and Craggs, J. D. (1958) *J. Electronics and Control* **4**, 481.
McWhirter, R. W. P. (1961) *Nature* **190**, 902.
Massey, H. S. W. (1952) *Advances in Phys. (Phil. Mag. Suppl.)* **1**, 395.
Massey, H. S. W., and Bates, D. R. (1942) *Rep. Progr. Phys.* **9**, 62.
Massey, H. S. W., Burhop, E. H. S. (1952) "Electronic and Ionic Impact Phenomena," Oxford Univ. Press, London and New York.
Menzel, D. H., and Pekeris, C. L. (1935) *Monthly Not. Roy. Astron. Soc.* **96**, 77.
Milne, E. A. (1924) *Phil. Mag.* **47**, 209.
Mitchell, E. E. L. (1960) Thesis, University of Liverpool.
Olsen, H. N., and Huxford, W. S. (1952) *Phys. Rev.* **87**, 922.
Oskam, H. J. (1958) *Philips Res. Rep.* **13**, 419.
Pahl, M. (1959) *Z. Naturforsch.* **14a**, 239.
Persson, K. B. (1957) *Phys. Rev.* **106**, 191.
Persson, K. B., and Brown, S. C. (1955) *Phys. Rev.* **100**, 729.
Pery-Thorne, A. and Garton, W. R. S. (1960) *Proc. Phys. Soc.* **76**, 833.
Popov, N. A., and Afaneśeva, E. A. (1960) *Soviet Phys.—Tech. Phys.* **4**, 764.
Redfield, A., and Holt, R. B. (1951) *Phys. Rev.* **82**, 874.
Richardson, J. M. (1952) *Phys. Rev.* **88**, 895.
Richardson, J. M., and Holt, R. B. (1951) *Phys. Rev.* **81**, 153.
Searle, L. (1958) *Astrophys. J.* **128**, 489.
Seaton, M. J. (1959a) *Monthly Not. Roy. Astron. Soc.* **119**, 81.
Seaton, M. J. (1959b) *Monthly Not. Roy. Astron. Soc.* **119**, 90.
Seaton, M. J. (1960) *Rep. Progr. Phys.* **23**, 313.
Sexton, M. C., and Craggs, J. D. (1958) *J. Electronics and Control* **4**, 493.
Sexton, M. C., Lennon, J. J., and Mulcahy, M. J. (1959) *Brit. J. appl. Phys.* **10**, 356.
Takeda, S., and Holt, E. H. (1959) *Rev. sci. Instrum.* **30**, 722.
Tsui-Fang, and Craggs, J. D. (1958) *J. Electronics and Control* **4**, 385.
Varnerin, L. J. (1951) *Phys. Rev.* **84**, 563.
Whitmer, R. F. (1956) *Phys. Rev.* **104**, 572.
Winans, J. G., and Stueckelberg, E. C. G. (1928) *Proc. Nat. Acad. Sci. U.S.A.* **14**, 867.

8.

Ionic Recombination

J. Sayers

1 Three Body

In 1896 Thomson and Rutherford conducted an illuminating series of experiments on the nature of the conductivity in gases irradiated by Roentgen rays. By examining the electrical characteristics of the gas enclosed between two metal plates, they were the first to discover that for small electric fields the current flowing was proportional to the field strength, but as the field was increased the current approached a limiting value. They showed, moreover, that this behavior could be explained on the assumption that the ions produced by the rays were capable of recombining at a rate proportional to the product of the population densities of the positive and negative ions. Very soon after this discovery Rutherford (1897) devised a means of measuring the rate at which this recombination proceeds, and he further verified that the recombination law is obeyed, i.e.,

$$\frac{dN}{dt} = -\alpha N^2$$

where N is the population density of ions of either sign and α is the recombination coefficient. In 1903 Langevin not only measured the recombination coefficient by an important new method, but also developed a theory to explain the physical processes involved.

Since these pioneer experiments many other investigators have taken up the problems of elucidating all the various aspects of ionic recombination. McClung (1902), Hendren (1905), and Thirkill (1913) were among the early investigators of the variation of the recombination coefficient with gas pressure. In the variation of the recombination coefficient with temperature, the most comprehensive early study has been made by Erikson (1909) whose experiments include air, carbon dioxide, and hydrogen at constant density and ionized by α- and β-rays. Further experiments were conducted by Phillips (1910) on air at constant pressure, ionized by X-rays.

All the early experiments had one unfortunate—though at the time unavoidable—feature in common; the gases used were always far from pure and the ions were in many cases complex clusters of molecules, as has been demonstrated by Loeb (1928, 1932) and by Tyndall and Powell (1930). It is, therefore, not surprising that a poor measure of agreement between the results of different investigators was recorded.

With the invention of the diffusion pump more significant and consistent results were obtained for the rates of ion-ion recombination. Outstanding among this more thorough experimental work was that of Loeb and his associates, notably Gardner (1938) who carried out an extensive study of recombination in chemically pure oxygen. A study of recombination, under similar experimental standards, was made in dust free dry air by Sayers (1938). In most of this work the pressure range explored did not extend upwards above about 2 atm. An important exception was the work of Machler (1936) whose investigations of ionic recombination in air extended over the pressure range 5-15 atm.

On the theoretical side the two outstanding contributions of the period were those of Thomson (1924) in the case of pressures below 1 atm and Langevin (1903) in the case of high pressures. These theories are so frequently dealt with in the subsequent literature that it will be necessary here only to remind the reader of the underlying principles and assumptions.

In the Thomson theory it is assumed that the random diffusive motion of the positive and negative ions will result in pair encounters in which, in the absence of the interaction of a third body, the energy will be conserved between the two ions and they will execute open orbits in their mutual Coulomb field. The presence of a third body with which one of the ions collides will result in stabilization of the encounter and the formation of closed orbits and eventual recombination. The Thomson process is often referred to as three-body ionic recombination.

In the Langevin theory it is assumed that the approach of the positive and negative ions in pairs is governed by mobility laws and the Coulomb attractive forces. This leads to a recombination rate which is a function of the mobility of the ions and therefore inversely proportional to gas pressure.

For many years from the late 1930's onwards very little progress was made in the study of ion-ion recombination at intermediate and high pressures. The next important advance was made by Natason (1959) as a result of theoretical study. In Natason's paper a unified theory has been developed which gives excellent agreement with previous experiments including the low and very high pressure ranges. Natason's theory yields an expression for the recombination coefficient which is very similar to the Thomson expression at low pressure and coincides with Langevin's theory at high pressure.

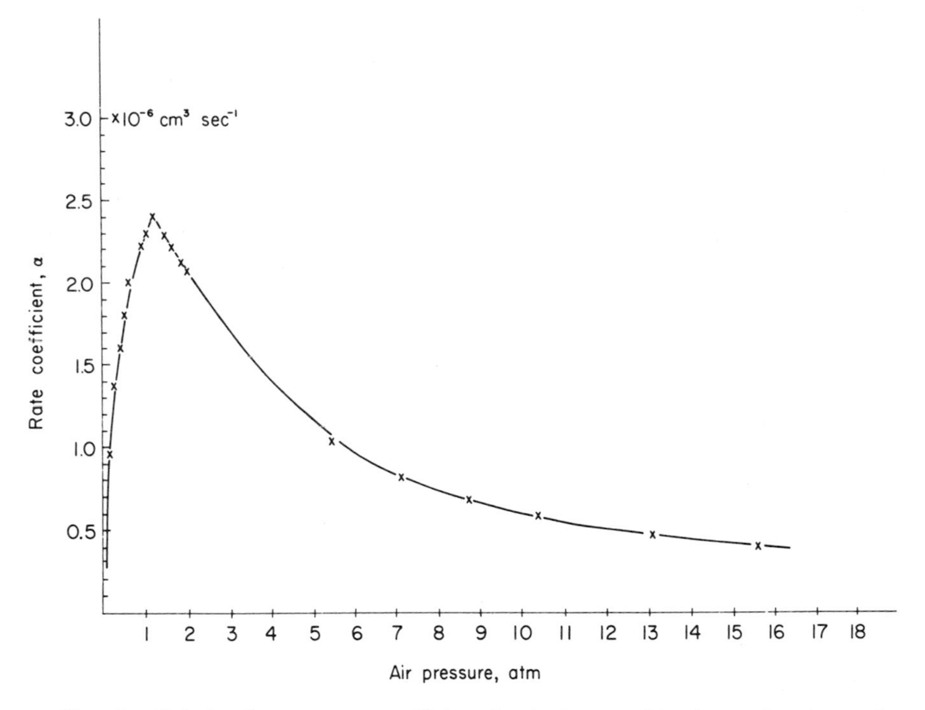

FIG. 1. Relation between rate coefficient for ionic recombination and pressure for dry air at normal temperature.

Figure 1 shows the variation of the recombination coefficient in dry air. The experimental points represent the results obtained by Sayers (1938) for the pressure range up to just over 1 atm and by Machler (1936) for pressures above 5 atm. The accuracy of the fit of Natason's theory to these experimental values is as good as the line in Fig. 1 drawn through the points.

All the early work on ion-ion recombination was concerned with pressures of 100 mm Hg and upwards. There are, however, a number of applications, in the theory of the ionosphere for example, in which it is necessary to have some knowledge of ionic recombination at very low pressures. It is not possible to investigate experimentally the three-body recombination process in the millimeter range of pressures because, the coefficient for this process being proportional to gas pressure, its rate becomes small by comparison with ionization loss by diffusion or other recombination processes of the two-body type. Nevertheless, the three-body process is always present and may have to be taken into account as a correction term to calculated or measured rates of recombination by two-body processes.

The three-body ion-ion recombination process may be a significant though minor process of ion loss in the lower regions of the ionosphere where diffusion loss of ionization may be ignored, although it is very probable that two-body recombination processes will predominate. If it is desired to calculate the contribution to ion decay by the three-body process at very low pressures, the Thomson theory may be used without the introduction of any adjustable constants. There are, however, some approximations inherent in the Thomson theory and it is more accurate to use the Natason expression. Alternatively, the following expression, derived from the Thomson theory but incorporating an adjusted constant, may be used for either air or oxygen:

$$\alpha_i = CT^{-5/2}p \text{ cm}^3 \text{ sec}^{-1}$$

where T is the temperature in degrees Kelvin, p is the gas pressure in mm Hg, and C is an adjustable constant having a value of 1.5×10^{-2} in terms of these units.

2 *Two Body*

2.1 General Considerations

When a positive ion and a negative ion in the gas phase collide, electron transfer between the two colliding particles will result in electrical recombination. Such electron transfer will be energetically possible since the electron affinity of the negative ion will be, in general, less than the ionization potential of the positive ion. The two ions of opposite sign may, individually, be of either atomic or molecular form. In the simplest case of two atomic ions the energy balance in the reaction, being the excess of ionization potential of the positive ions over the electron affinity of the negative ion, is taken up in energy of the two neutral atoms after the collision. This energy may be kinetic energy or electronic excitation energy or a combination of the two. The probability of excess energy being radiated in the collision process is very small. In the case where one or both of the colliding ions is molecular there is the additional probability of chemical dissociation of either molecule.

Even in the simplest case above, namely, recombination of a pair of atomic ions, it is not possible to calculate the recombination cross section on purely theoretical considerations without the aid of drastic approximations which may be acceptable only for a positive and negative ion pair of atomic hydrogen. For example, as shown by Bates (1960), the approximations involved in the application of the Landau-Zener formula cannot be justified.

Though the details of the early calculations of Bates and Lewis (1955) and of Bates and Boyd (1956) are unreliable, it is still believed that the crossing of the curves which relate potential energy to nuclear separation in the two states is an important theoretical concept and that the rate coefficients are likely to be high, perhaps of the order of 10^{-7} cm^3 sec^{-1}

2.2 Experimental Investigations

Experimental study of this reaction necessitates the use of low gas pressures in order to reduce, as far as possible, recombination by the three-body process. This requires a gas discharge as the source of ionization and a study of ionization decay in the afterglow. In addition, the methods of measuring ion population densities at higher pressures are invalid in the millimeter pressure range and new methods have therefore had to be developed. The foregoing experimental problems have been solved by Yeung (1958a, b) who has carried out a preliminary study of ion-ion recombination in the afterglow of a discharge in iodine. Iodine was selected because of the rapid electron attachment in this gas which, as he has shown, rapidly removes the free electrons in the early stages of the afterglow. This produces a plasma consisting of positive and negative ions in equal concentrations. For the measurement of the positive and negative ion densities in the decaying plasma Yeung employed a radio-frequency probing technique. This depends on the fact that the dielectric properties of the plasma are related to the ion density. The probing electrodes consisted of an internal cylindrical gauze at the center of the discharge tube and a sleeve electrode on the outside surface. By using these two electrodes the dielectric constant of the decaying plasma was measured as a function of time, using a probing frequency in the region of 15 Mc/sec. The changes in dielectric constant were interpreted in terms of changes in ionization density and from the results the recombination coefficient was determined. A satisfactory linear relation was found between the reciprocal of ionization density and time in the afterglow thus establishing that the decay of ionization followed a recombination law.

2.2.1 *Two-Body Recombination at Various Gas Pressures*

The two-body process of ionic recombination should proceed at a rate which is independent of gas pressure, and if this fact is established it renders the identification of the loss process more certain. Yeung succeeded in measuring the rate of ionic recombination in iodine over a substantial pressure range. He repeated the whole series of experiments in bromine and in each case he found a recombination rate independent of pressure in the range studied. The results for these two gases are tabulated in Table I. It was concluded that, within experimental error, the recombination rate was pressure independent.

TABLE I

IONIC RECOMBINATION RATES IN IODINE AND BROMINE

Iodine		Bromine	
p (mm Hg)	α_i (relative values)[a]	p (mm Hg)	α_i (relative values)[a]
0.07	1.4	0.05	1.9
0.095	1.4	0.07	1.8
0.15	1.6	0.1	1.8
0.185	1.4	0.5	1.8
0.32	1.5	0.7	1.8
0.45	1.5	1.0	2.0
0.62	1.5		
1.00	1.55		

[a] For absolute values see later discussion.

2.2.2 *Variation of Ionic Recombination with Temperature*

Yeung has explored the variation of two-body ionic recombination with temperature. Owing to the experimental difficulties associated with this work and the many new technical problems encountered, it was not possible to measure recombination over a sufficient temperature range, this being limited to the range 0°-100°C. While the results did not justify the fitting of any empirical law to the temperature dependence, the conclusion was that the recombination rate decreased as the temperature increased.

While the work of Yeung represents substantially the pioneer experimental work in the field of two-body ionic recombination, the numerical measurements must be taken as preliminary and the work leaves the question of the identity of the ions—as between atomic and molecular forms—still unsettled.

Very recently Greaves (1959), working in the same laboratory, has taken up the study of ionic recombination by similar methods to those employed by Yeung. Greaves has, however, included a miniature mass spectrometer in his experimental tube in order to identify the ions whose recombination rates are being studied. In the course of the work it became necessary to recalibrate some of the electronic equipment used previously by Yeung and it was discovered that a fault had developed in one of the amplifiers since the time of an earlier calibration. There are strong reasons for believing that the fault had existed throughout the use of the equipment in Yeung's work; if so, the revised calibration can be readily applied to Yeung's results. When this is done the multiplying factor to be applied to all the relative values of the recombination coefficient given in Table I is 10^{-7} and the figures are then absolute values in $cm^3 \ sec^{-1}$ units.

Yeung's value of the recombination coefficient, modified in this way, would then be about 50 % larger than a new value found by Greaves, being approximately

$$\alpha_i = 10^{-7} \ cm^3 \ sec^{-1}$$

at room temperature in an iodine afterglow.

On the assumption that the negative ions are I^- and the positive ions I_2^+, the corresponding recombination cross section is

$$\sigma_i = 3 \times 10^{-13} \ cm^2.$$

REFERENCES

Bates, D. R. (1960) *Proc. Roy. Soc.* **A257**, 22.
Bates, D. R., and Boyd, T. J. M. (1956) *Proc. Phys. Soc.* **A69**, 910.
Bates, D. R., and Lewis, J. T. (1955) *Proc. Phys. Soc.* **A68**, 173.
Erikson, H. A. (1909) *Phil. Mag.* **18**, 327.
Erikson, H. A. (1912) *Phil. Mag.* **23**, 747.
Gardner, M. E. (1938) *Phys. Rev.* **53**, 75.
Greaves, C. (1959) Thesis, University of Birmingham, England.
Hendren, L. L. (1905) *Phys. Rev.* **21**, 314.
Langevin, P. (1903) *Ann. chim. phys.* **28**, 433.

Loeb, L. B. (1928) *Phys. Rev.* **32**, 81.
Loeb, L. B. (1932) *Phys. Rev.* **38**, 549.
McClung, R. (1902) *Phil. Mag.* **3**, 283.
Machler, W. (1936) *Z. Phys.* **164**, 1.
Natason, G. L. (1959) *Zh. tekh. Fiz.* **29**, 1373.
Phillips, P. (1910) *Proc. Roy. Soc.* **83**, 246.
Rutherford, E. (1897) *Phil. Mag.* **44**, 422.
Sayers, J. (1938) *Proc. Roy. Soc.* **A169**, 83.
Thirkill, H. (1913) *Proc. Roy. Soc.* **88**, 488.
Thomson, J. J. (1924) *Phil. Mag.* **47**, 337.
Thomson, J. J., and Rutherford, E. (1896) *Phil. Mag.* **42**, 392.
Tyndall, A. M., and Powell, C. F. (1930) *Proc. Roy. Soc.* **129**, 162.
Yeung, H. Y. (1958a) *Proc. Phys. Soc.* **71**, 341.
Yeung, H. Y. (1958b) *J. Electronics and Control* **5**, 307.

9.

Elastic Scattering of Electrons

B. L. Moiseiwitsch

In the case of elastic collisions, no energy is transferred from the motion of the incident electron to the internal motion of the target atomic or molecular system. However, the electron does lose some of its energy in providing kinetic energy of translational motion of the target system. Since this loss of energy is only of the order of the ratio of the mass of the electron to the mass of the atom or molecule times the kinetic energy of the incident electron, it can be generally neglected.

Further, in the investigation of elastic collisions between electrons and atoms, it is frequently permissible to regard the target system as a fixed scattering center without any structure. Consequently, to begin with, we shall develop an elementary quantum theory of elastic scattering in which this simplifying assumption is made. At a later stage we shall discuss the effects which arise from allowance for the structure of the atom or molecule.

1 Scattering by a Potential Field

1.1 Differential and Total Cross Sections

We consider a parallel beam of particles moving in the direction of the z-axis of a frame of reference with N particles crossing unit area perpendicular to the beam per second. If we take the scattering center to be located at the origin, the number of particles deflected per second through polar angles θ and ϕ into the element of solid angle $d\omega = \sin\theta\, d\theta\, d\phi$ can be expressed in the form $NI(\theta, \phi)d\omega$, where $I(\theta, \phi)$ has the dimensions of area and is termed the *differential cross section.*

The total number of particles scattered per second from an incident beam of unit flux density is defined to be the *total cross section* and is given by

$$Q = \int_0^\pi \int_0^{2\pi} I(\theta, \phi) \sin\theta\, d\theta\, d\phi. \tag{1}$$

1.2 Asymptotic Behavior

Let us suppose that the incident particles all have the same mass m and the same speed v. Then for large values of the distance r from the origin, the asymptotic form of the wave function $\psi(\mathbf{r})$ describing the scattering is given by

$$\psi(\mathbf{r}) \sim A[e^{ikz} + r^{-1}e^{ikr}f(\theta, \phi)] \tag{2}$$

where $f(\theta, \phi)$ is called the *scattering amplitude.*

The first term e^{ikz} is a plane wave and represents a particle belonging to the incident stream moving in the direction of the z-axis with linear momentum $\hbar k = mv$. The second term $r^{-1}e^{ikr}f(\theta, \phi)$ is an outgoing spherical wave and represents a scattered particle moving away from the scattering center in the radial direction.

Since the current density associated with a wave function ψ is given by

$$\mathbf{j} = \frac{\hbar}{2mi}(\psi^* \operatorname{grad} \psi - \psi \operatorname{grad} \psi^*), \tag{3}$$

it follows that the flux density of incident particles is $v \mid A \mid^2$ and so $A = \sqrt{N/v}$. Now the leading term of the radial flux density of scattered particles is given by

$$v \mid A \mid^2 \mid f(\theta, \phi) \mid^2 r^{-2} \tag{4}$$

and therefore

$$I(\theta, \phi) = \mid f(\theta, \phi) \mid^2 \tag{5}$$

which provides an expression for the differential cross section in terms of the scattering amplitude.

1.3 Partial Waves Method

In the present section we consider the scattering of particles by a spherically symmetrical potential $V(r)$. This is determined by the wave equation

$$-\frac{\hbar^2}{2m}\nabla^2\psi + V(r)\psi = E\psi \tag{6}$$

where $E = \hbar^2k^2/2m$ is the energy of the particles and $\psi(\mathbf{r})$ is a continuous, bounded, single-valued function of position having the asymptotic form for large r,

$$\psi \sim e^{ikz} + r^{-1}e^{ikr}f(\theta, \phi), \tag{7}$$

corresponding to an incident flux density of magnitude v.

Owing to the symmetry about the z-axis, we may expand the wave function ψ in terms of Legendre polynomials $P_l(\cos\theta)$. On putting

$$\psi(\mathbf{r}) = \frac{1}{r}\sum_{l=0}^{\infty} A_l\phi_l(r)\, P_l(\cos\theta) \tag{8}$$

and substituting into (6), we obtain the radial equation

$$\frac{d^2\phi_l}{dr^2} + \left\{k^2 - U(r) - \frac{l(l+1)}{r^2}\right\} \phi_l = 0 \tag{9}$$

where

$$U(r) = \frac{2m}{\hbar^2} V(r). \tag{10}$$

Since ψ must be bounded at the origin, it follows that $\phi_l(0) = 0$.

Provided $U(r)$ falls off more rapidly than $1/r^2$ as $r \to \infty$, we have

$$\phi_l \sim r[a_l j_l(kr) + b_l n_l(kr)] \tag{11}$$

for large r, where j_l and n_l are spherical Bessel and spherical Neumann functions, respectively, with asymptotic forms

$$j_l(kr) \sim \frac{1}{kr} \sin (kr - \tfrac{1}{2} l\pi) \tag{12}$$

$$n_l(kr) \sim -\frac{1}{kr} \cos (kr - \tfrac{1}{2} l\pi) \tag{13}$$

as $r \to \infty$.

Putting $a_l = \cos \eta_l$ and $b_l = -\sin \eta_l$, we see that for large r

$$\phi_l \sim \frac{1}{k} \sin (kr - \tfrac{1}{2} l\pi + \eta_l). \tag{14}$$

η_l is termed the *phase shift* or *eigenphase*.

Expanding the scattering amplitude in the form

$$f(\theta) = \sum_{l=0}^{\infty} c_l P_l(\cos \theta) \tag{15}$$

and using the formula (Watson, 1944)

$$e^{ikz} = \sum_{l=0}^{\infty} i^l(2l+1) j_l(kr) P_l(\cos \theta), \tag{16}$$

it can be readily shown, by equating the coefficients of e^{ikr} and of e^{-ikr} on both sides of (7), that

$$A_l = i^l e^{i\eta_l}(2l+1) \tag{17}$$

and

$$c_l = \frac{1}{2ik} (2l+1) (e^{2i\eta_l} - 1). \tag{18}$$

Hence, the scattering amplitude is given by

$$f(\theta) = \frac{1}{k} \sum_{l=0}^{\infty} (2l + 1) e^{i\eta_l} \sin \eta_l P_l(\cos \theta). \tag{19}$$

Integrating over θ and ϕ and using the orthonomal property of Legendre polynomials, we see that the total cross section can be written in the form

$$Q = \sum_{l=0}^{\infty} Q_l \tag{20}$$

where

$$Q_l = \frac{4\pi}{k^2} (2l + 1) \sin^2 \eta_l. \tag{21}$$

Q_l is called the lth order *partial cross section.*

From (15) and (18) it also follows that

$$Q = \frac{4\pi}{k} \operatorname{Im} f(0). \tag{22}$$

This result is known as the *cross section theorem* and gives the total cross section in terms of the imaginary part of the forward scattering amplitude $f(0)$.

The foregoing method is due to Lord Rayleigh and was first applied to the elastic scattering of electrons by atoms by Faxén and Holtsmark (1927).

1.4 Integral Equations

The solution of the equation

$$(\nabla^2 + k^2) \psi(\mathbf{r}) = F(\mathbf{r}) \tag{23}$$

where ψ is an outgoing spherical wave for large r, can be expressed in the form

$$\psi(\mathbf{r}) = \int G(\mathbf{r}, \mathbf{r}') F(\mathbf{r}') \, d\mathbf{r}' \tag{24}$$

where the *Green's function* G satisfies the equation

$$(\nabla^2 + k^2) G(\mathbf{r}, \mathbf{r}') = \delta(\mathbf{r} - \mathbf{r}') \tag{25}$$

and is given by

$$G(\mathbf{r}, \mathbf{r}') = -\frac{\exp ik \mid \mathbf{r} - \mathbf{r}' \mid}{4\pi \mid \mathbf{r} - \mathbf{r}' \mid}. \tag{26}$$

On writing the wave equation (6) in the form

$$(\nabla^2 + k^2)\,\psi(\mathbf{r}) = U(r)\,\psi(\mathbf{r}) \tag{27}$$

we see that

$$\psi(\mathbf{r}) = e^{ikz} - \frac{1}{4\pi}\int \frac{\exp ik\,|\,\mathbf{r} - \mathbf{r}'\,|}{|\,\mathbf{r} - \mathbf{r}'\,|}\,U(r')\,\psi(\mathbf{r}')\,d\mathbf{r}'. \tag{28}$$

Denoting the unit vector in the direction of $\mathbf{r}$ by $\mathbf{n}$, we have for large r

$$|\,\mathbf{r} - \mathbf{r}'\,| \sim r - \mathbf{n}\cdot\mathbf{r}' \tag{29}$$

and thus we may write the scattering amplitude in the integral equation form

$$f(\theta) = -\frac{1}{4\pi}\int \exp(-ik\mathbf{n}\cdot\mathbf{r}')\,U(r')\,\psi(\mathbf{r}')\,d\mathbf{r}'. \tag{30}$$

Expanding both sides of this equation in terms of Legendre polynomials we obtain an integral equation for the phase shift η_l. It is given by

$$\sin \eta_l = -k\int_0^\infty r\,j_l(kr)\,\phi_l(r)\,U(r)\,dr. \tag{31}$$

1.5 Born's Approximation

At sufficiently high impact energies, the incident particles will be only slightly affected by the scattering center. In this case, we may replace the exact wave function $\psi(\mathbf{r})$ on the right-hand side of (30) by the plane wave $\exp ik\mathbf{n}_0\cdot\mathbf{r}$ where $\mathbf{n}_0$ is a unit vector in the direction of the incident beam of particles (Born, 1926). Then we have

$$f(\theta) = -\frac{1}{4\pi}\int \exp(i\mathbf{K}\cdot\mathbf{r})\,U(r)\,d\mathbf{r} \tag{32}$$

where $\mathbf{K} = k(\mathbf{n}_0 - \mathbf{n})$. Since the angle of scattering θ is the angle between $\mathbf{n}$ and $\mathbf{n}_0$, it follows that

$$K = 2k\sin\frac{\theta}{2}\,. \tag{33}$$

By choosing the polar axis of the coordinate system in the direction of the momentum change vector $\hbar\mathbf{K}$, we obtain

$$f(\theta) = -\frac{1}{K}\int_0^\infty r\sin Kr\,U(r)\,dr. \tag{34}$$

This is the *first Born approximation* to the scattering amplitude. To the same approximation we may substitute $rj_l(kr)$ for $\phi_l(r)$ in the right-hand side of (31). We then get

$$\sin \eta_l = -k \int_0^\infty r^2 \{j_l(kr)\}^2 U(r)\, dr. \tag{35}$$

Now for $(kr)^2 \ll 4l + 6$, we have

$$j_l(kr) \simeq \frac{l!}{(2l+1)!} (2kr)^l \tag{36}$$

and so it follows from the first Born approximation expression (35) that

$$\sin \eta_l \cong -k^{2l+1} \left\{ \frac{2^l l!}{(2l+1)!} \right\}^2 \int_0^\infty r^{2l+2} U(r)\, dr. \tag{37}$$

Hence, if $U(r)$ is negligible for $r > R$ and if $| r^2 U(r) | \leqq C$ for $0 < r < R$, we get

$$| \sin \eta_l | < \frac{C(kR)^{2l+1}}{2l+1} \left\{ \frac{2^l l!}{(2l+1)!} \right\}^2 \tag{38}$$

and therefore η_l is a rapidly decreasing function of l for $(kR)^2 \ll 4l + 6$.

Higher order Born approximation expressions can be obtained by iteration. To the first order in the interaction energy U, we have

$$\psi(\mathbf{r}) = \exp(ik\mathbf{n}_0 \cdot \mathbf{r}) - \frac{1}{4\pi} \int \frac{\exp ik\, | \mathbf{r} - \mathbf{r}' |}{| \mathbf{r} - \mathbf{r}' |} U(r') \exp(ik\mathbf{n}_0 \cdot \mathbf{r}')\, d\mathbf{r}'. \tag{39}$$

If this is substituted into the right-hand side of (30), we obtain the *second Born approximation* to the scattering amplitude given by

$$f(\theta) = f_{\mathrm{B1}}(\theta) + f_{\mathrm{B2}}(\theta),$$

where $f_{\mathrm{B1}}(\theta)$ is just the first Born approximation (34) and

$$f_{\mathrm{B2}}(\theta) = \frac{1}{(4\pi)^2} \iint \exp(-ik\mathbf{n} \cdot \mathbf{r})\, U(\mathbf{r}) \frac{\exp ik\, | \mathbf{r} - \mathbf{r}' |}{| \mathbf{r} - \mathbf{r}' |} \times U(\mathbf{r}') \exp(ik\mathbf{n}_0 \cdot \mathbf{r}')\, d\mathbf{r}\, d\mathbf{r}'. \tag{41}$$

The convergence of Born expansions has been discussed in some detail by Kohn (1954).

1.6 Scattering Length

Let $u_k(r)$ be the solution of the radial equation for the zero order partial wave with wave number k. Then

$$\frac{d^2u_k}{dr^2} + \{k^2 - U(r)\}\, u_k = 0, \tag{42}$$

where $u_k(0) = 0$, $u_k(r) \sim v_k(r)$ for large r and

$$v_k(r) = \cos kr + \cot \eta \sin kr, \tag{43}$$

η being the zero order phase shift.

Defining the *scattering length* a to be the quantity

$$a = -\lim_{k\to 0} \frac{1}{k} \tan \eta \tag{44}$$

it can be readily shown that (Bethe, 1949)

$$k \cot \eta = -\frac{1}{a} + k^2 \int_0^\infty (v_k v_0 - u_k u_0)\, dr, \tag{45}$$

where u_0 is the solution of (42) for zero wave number k and

$$v_0 = 1 - \frac{r}{a}. \tag{46}$$

Equation (45) is exact. However, if we replace v_k and u_k by v_0 and u_0, respectively, it reduces to the result†

$$k \cot \eta = -\frac{1}{a} + \tfrac{1}{2} r_0 k^2 \tag{47}$$

where

$$r_0 = 2 \int_0^\infty \left[\left(1 - \frac{r}{a}\right)^2 - u_0^2\right] dr \tag{48}$$

is called the *effective range*. Equation (47) is correct to the second order in k.

It follows from the definition of the scattering length that in the limit of vanishing energy the zero order partial cross section Q_0 is given by

$$\lim_{k\to 0} Q_0 = 4\pi a^2. \tag{49}$$

† This formula is valid only for short range potentials. The effective range theory has to be modified for long range potentials such as the r^{-4} potential (cf. O'Malley *et al.*, 1961).

In general, the scattering length is a finite quantity so that tan η must tend to zero in the limit of vanishing energy. Hence $\eta(0)$, the phase shift for zero wave number k, must be an integer multiple of π. In certain cases the scattering length may be infinite and then $\eta(0)$ must be a half odd integer multiple of π. In fact, it can be shown (Weiss, 1952; Swan, 1954) that $\eta(0) = n\pi$ for an attractive potential having n bound states of negative energy and zero angular momentum, while $\eta(0) = (n + \frac{1}{2})\pi$ if there exists an additional bound state of zero energy. The presence of the bound state of zero energy results in a resonance effect which gives rise to an infinite value for the zero order partial cross section as the energy of the incident particles becomes very small.

For the lth order partial wave it can be shown without difficulty that (47) generalizes to the form†

$$k^{2l+1} \cot \eta_l = -\frac{1}{a_l} + \tfrac{1}{2} r_{0,l} k^2, \tag{50}$$

where a_l and $r_{0,l}$ are independent of k. It follows that for small k the lth order partial cross section is given by

$$Q_l = 4\pi a_l^2 k^{4l}. \tag{51}$$

Hence, Q_l vanishes in the limit of zero energy for $l > 0$.

1.7 Variational Methods

An analogous procedure to that used for deriving the Rayleigh-Ritz variational method for stationary states can be employed to obtain variational methods for the evaluation of phase shifts and scattering amplitudes. In this section we shall be concerned with the variational methods derived by Hulthén (1944) and Kohn (1948) for determining phase shifts. We write the radial equation (9) in the form

$$\mathscr{L}\phi = 0, \tag{52}$$

where

$$\mathscr{L} \equiv \frac{d^2}{dr^2} + k^2 - \frac{2m}{\hbar^2} V(r) - \frac{l(l+1)}{r^2}. \tag{53}$$

$\phi(r)$ satisfies the boundary conditions $\phi(0) = 0$ and

$$\phi(r) \sim \sin (kr - \tfrac{1}{2} l\pi + \eta) \tag{54}$$

† This formula is valid only for short range potentials. (cf. O'Malley *et al.*, 1961).

for large r, if $V(r)$ falls off more rapidly than $1/r$ as $r \to \infty$. Defining the functional

$$L[\chi] = \int_0^\infty \chi \mathscr{L} \chi \, dr \tag{55}$$

where χ is an arbitrary function of r, we see that $L[\phi] = 0$ where ϕ is the solution of (52) satisfying the correct boundary conditions.

Suppose that $\phi + \delta\phi$ is a function, differing infinitesimally from ϕ, which vanishes at the origin and has the asymptotic form

$$\sin\,(kr - \tfrac{1}{2} l\pi + \eta + \delta\eta).$$

Then $\delta\phi = 0$ at $r = 0$ and for large r

$$\delta\phi \sim \cos\,(kr - \tfrac{1}{2}\, l\pi + \eta)\, \delta\eta. \tag{56}$$

Owing to the infinitesimal change $\delta\phi$, $L[\phi]$ changes by the amount

$$\delta L[\phi] = \int_0^\infty \delta\phi \, \mathscr{L} \, \phi \, dr + \int_0^\infty \phi \, \mathscr{L} \, \delta\phi \, dr + \int_0^\infty \delta\phi \, \mathscr{L} \, \delta\phi \, dr. \tag{57}$$

On integrating by parts and using the boundary conditions imposed on ϕ and $\delta\phi$, we see that

$$\int_0^\infty \phi \, \frac{d^2}{dr^2}\,(\delta\phi)\, dr = \int_0^\infty \delta\phi \, \frac{d^2\phi}{dr^2}\, dr - k\,\delta\eta \tag{58}$$

and so

$$\delta L[\phi] = 2\int_0^\infty \delta\phi \, \mathscr{L} \, \phi \, dr - k\,\delta\eta \tag{59}$$

to the first order of small quantities. Now $\mathscr{L}\phi = 0$ and so

$$\delta L[\phi] = -\,k\,\delta\eta. \tag{60}$$

This variational principle is due to Hulthén.

Let us first consider only those variations of ϕ for which L remains equal to zero. Then δL and $\delta\eta$ both vanish. This stationary behavior of η may be employed to obtain an approximate value for the phase shift η and an approximation to the radial wave function $\phi(r)$ by choosing a trial function $\phi_t(r)$ which is continuous, satisfies the boundary conditions that $\phi_t(0) = 0$ and

$$\phi_t(r) \sim \sin\,(kr - \tfrac{1}{2}\, l\pi + \zeta) \tag{61}$$

for large r, and depends upon n arbitrary parameters c_i ($i = 1, ..., n$) as as well as the phase shift parameter ζ. In Hulthén's variational method, these parameters are determined by solving the set of equations

$$L_t = 0 \tag{62}$$

$$\frac{\partial L_t}{\partial c_i} = 0 \qquad (i = 1, ..., n) \tag{63}$$

where

$$L_t = L[\phi_t]. \tag{64}$$

An approximate value for the phase shift is then given by ζ.

We now consider arbitrary variations of ϕ so that $\delta\eta$ is no longer necessarily zero. Then we may use the variational method due to Kohn which determines the parameters ζ and c_i ($i = 1, ..., n$) by solving the set of equations

$$\frac{\partial L_t}{\partial \zeta} = -k \tag{65}$$

$$\frac{\partial L_t}{\partial c_i} = 0 \qquad (i = 1, ..., n). \tag{66}$$

Since L_t is not necessarily equal to zero in this case, it follows from (60) that the value of the phase shift is given by $\zeta + L_t/k$.

In general, the variational principle (60) does not provide a bound to the phase shift. However, it has been shown by Spruch and Rosenberg (1960) that in the limit of zero energy a bound to the scattering length can be obtained. This is to be expected as it is the limiting case of the Rayleigh-Ritz variational method for stationary states which gives a bound to the eigenenergy.

Consider the zero order partial wave. For particles of zero energy we have

$$\mathscr{L} \equiv \frac{d^2}{dr^2} - \frac{2m}{\hbar^2} V(r). \tag{67}$$

The radial wave function $\phi(r)$ satisfies the boundary conditions $\phi(0) = 0$ and

$$\phi(r) \sim a - r \tag{68}$$

for large r, where a is the scattering length.

We define the function

$$\psi(r) = \phi_t(r) - \phi(r) \tag{69}$$

where $\phi_t(r)$ is a trial function satisfying the boundary conditions $\phi_t(0) = 0$ and

$$\phi_t(r) \sim a_t - r \tag{70}$$

for large r. Then

$$\int_0^\infty \phi_t \mathscr{L} \phi_t \, dr = \left[\phi \frac{d\psi}{dr} - \psi \frac{d\phi}{dr}\right]_0^\infty + \int_0^\infty \psi \mathscr{L} \psi \, dr \tag{71}$$

and so

$$a = a_t - \int_0^\infty \phi_t \mathscr{L} \phi_t \, dr + \int_0^\infty \psi \mathscr{L} \psi \, dr. \tag{72}$$

For simplicity, we suppose that the potential $V(r)$ is too weak to have states of negative energy. It then follows from the variational principle for stationary states that

$$\int_0^\infty \psi \mathscr{L} \psi \, dr \leqq 0 \tag{73}$$

and therefore

$$a \leqq a_t - \int_0^\infty \phi_t \mathscr{L} \phi_t \, dr \tag{74}$$

which provides an upper bound to the scattering length a.

In Kohn's variational method the parameters upon which the trial function ϕ_t depends are varied so as to give the least value for $a_t - \int_0^\infty \phi_t \mathscr{L} \phi_t dr$ and thus provides a least upper bound to the scattering length a. In Hulthén's method, the value of $a_t - \int_0^\infty \phi_t \mathscr{L} \phi_t dr$ which is obtained is not in general the least upper bound to a, and so for the zero energy case, Kohn's method is superior to Hulthén's method. However, this is not necessarily true at nonzero energies.

An analogous result to (74) can be derived when the potential $V(r)$ is sufficiently strong to have states of negative energy and zero angular momentum. In this case, in order to obtain an upper bound to the scattering length, variational wave functions for all the states of negative energy and zero angular momentum are required (Rosenberg, *et al.*, 1960).

2 Measurement of Collision Cross Sections

2.1 Absorption Coefficient

Consider a parallel beam of electrons having the same velocity v passing through a gas consisting of n identical atoms or molecules per unit volume. Let us suppose that the intensity I of the beam is reduced by an amount δI in traversing a distance δx. Then, assuming that any collision between an electron and a gas particle results in the removal of the electron from the beam, we have

$$\delta I = nQ\delta x \tag{75}$$

where Q is the total collision cross section for encounters between electrons of velocity v and the gas particles. Hence, it follows that the intensity $I(x)$ of the beam at a distance x from a fixed point O is given by

$$I(x) = I(O)e^{-\alpha x} \tag{76}$$

where $\alpha = nQ$ is called the *absorption coefficient*.

If the gas is at a pressure p mm Hg and at a temperature 0°C, then

$$Q = 2.81 \times 10^{-17} \frac{\alpha}{p} \text{ cm}^2. \tag{77}$$

The absorption coefficient at 1 mm Hg pressure and 0°C temperature is sometimes referred to as the *probability coefficient of collision* P_c. The numerical value of P_c in units of cm^2/cm^3 (0°C, 1 mm Hg) is by chance very nearly the same as the total collision cross section Q in units of $a_0^2 = 2.80 \times 10^{-17}$ cm^2.

Provided the energy of the incident electrons is less than the minimum energy required to excite a gas particle (e.g., 10.2 ev for atomic hydrogen and 19.7 ev for helium), the total collision cross section is the same as the total elastic cross section. At electron energies above the threshold energy for excitation, the total collision cross section will include a contribution from inelastic collisions.

The earliest measurements of probability coefficients of collision were made using the method of Ramsauer (1921).

2.2 Diffusion Through Gases

We consider a swarm of electrons diffusing through a gas, the mean kinetic energy of the electrons being assumed so small that only elastic collisions between the electrons and gas atoms can occur.

If m is the mass of an electron and M is the mass of a gas atom, the fractional energy lost by an electron deflected through an angle θ due to a collision with an atom is given by

$$\frac{2m}{M}(1 - \cos\theta) \tag{78}$$

neglecting terms of order $(m/M)^2$. The mean fractional energy loss per collision is therefore

$$\frac{2m}{M}\frac{Q_D}{Q} \tag{79}$$

where

$$Q_D = \int_0^\pi \int_0^{2\pi} (1 - \cos\theta)\, I(\theta) \sin\theta \, d\theta \, d\phi \tag{80}$$

is called the *momentum transfer cross section* or the *diffusion cross section.* At very low impact energies the differential cross section $I(\theta)$ is independent of the scattering angle θ and then Q_D is identical to the total elastic cross section Q.

The *collision frequency* for momentum transfer for electrons having velocity v is given by

$$\nu_c = vP_c p \tag{81}$$

where P_c is the probability coefficient of collision for momentum transfer and p is the gas pressure normalized to 0°C temperature.

We now consider a swarm of electrons of density n_e diffusing through a gas in the presence of an electric field $E \exp i\omega t$. Then the equation of motion of a typical electron is

$$m\frac{dv}{dt} = eE \exp i\omega t - mv\,\nu_c \tag{82}$$

and so, making the simplifying assumption that ν_c is independent of v, we obtain

$$v = \frac{eE \exp i\omega t}{m(\nu_c + i\omega)}. \tag{83}$$

Hence, the current density $j = n_e ev$ is given by

$$j = \frac{n_e e^2 E \exp i\omega t}{m(\nu_c + i\omega)}. \tag{84}$$

Now $j = \sigma E \exp i\omega t$, where $\sigma = \sigma_r + i\sigma_i$ is the complex conductivity, and so it follows that

$$\frac{\sigma_r}{\sigma_i} = -\frac{\nu_c}{\omega}. \tag{85}$$

In the case of a constant electric field E, we see that the mean velocity with which the swarm of electrons diffuses through the gas in the direction of the field is given by

$$u = eE/m\nu_c. \tag{86}$$

This is called the *drift velocity*.

2.3 Microwave Measurements

Phelps *et al.* (1951) and Gould and Brown (1954) have developed a microwave method for determining the probability coefficient of collision for momentum transfer by measuring the conductivity of a decaying plasma after the electrons have attained thermal equilibrium with the gas atoms.

Consider an ionized gas in thermal equilibrium under the action of an applied high-frequency electric field $E \cos \omega t$. For low gas pressures, we have $\nu_c^2 \ll \omega^2$, and then the electron velocity distribution function has the Maxwellian form

$$f(v) = A \exp(-\tfrac{1}{2} mv^2/kT_e) \tag{87}$$

with

$$kT_e = kT_g + e^2ME^2/6m^2\omega^2 \tag{88}$$

where T_g is the temperature of the gas atoms and T_e is the electron temperature.

It follows from (85) that at low gas pressures

$$\frac{\sigma_r}{\sigma_i} = -\int_0^\infty \frac{\nu_c}{\omega} \epsilon^{3/2} \exp(-\epsilon/kT_e)\, d\epsilon \bigg/ \int_0^\infty \epsilon^{3/2} \exp(-\epsilon/kT_e)\, d\epsilon. \tag{89}$$

A general proof of this result has been given by Margenau (1946) who made no assumption regarding the dependence of ν_c on v. In the method used by Gould and Brown (1954) to measure the electron conductivity ratio $\rho = p^{-1}(\sigma_r/\sigma_i)$, the gas being investigated was contained in a copper cavity constructed in the form of a rectangular parallelepiped resonating in its three fundamental modes at wavelengths 9.5, 10.0, and 10.5 cm.

A 9.5 cm pulsed magnetron was employed to break down the gas for a time duration varying between 0.1 and 5 msec and at a repetition rate varying from 20 to 120 cps. The time constant for the electrons to cool down from the high mean energy during the discharge to the gas temperature is of the order of $M/2m\nu_c$ which for helium is about $20/p$ μsec.

A 10.0 cm continuous-wave magnetron was used to increase the average energy of the electrons in the afterglow above the gas temperature. Now, in order that the condition $\nu_c^2 \leqq 0.04\,\omega^2$ should be satisfied at a frequency of 3000 Mc, we require that $p \leqq 350\,T_e^{-1/2}$. Hence, for pressures of 10 mm Hg, the maximum allowed electron temperature within the plasma is about 2000°K. At pressures below 10 mm Hg, the effect of energy gradients within the cavity becomes important. However, by placing the plasma within a quartz bottle contained inside the cavity, pressures as low as 1 mm Hg could be used without producing significant energy gradients so that electron temperatures up to 25,000°K were permissible.

The presence of a plasma results in a change in the resonant frequency of the cavity and also in the conductance of the cavity at resonance. The ratio of the microwave power transmitted through the cavity to the incident power as a function of frequency near to the cavity resonance enables these changes to be determined and these yield the conductivity ratio ρ. The incident power was provided by a continuous-wave tunable magnetron operating in the neighborhood of the 10.5-cm mode.

The temperature of the gas within the cavity could be varied by cooling the cavity to the temperature 195° and 77°K using dry ice and liquid air, respectively, and by heating the cavity from room temperature to 400°C.

If the collision probability coefficient P_c is assumed constant, (89) gives

$$\rho = 1.505\,\frac{P_c}{\omega}\left(\frac{2kT_e}{m}\right)^{1/2} \tag{90}$$

so that P_c can then be readily calculated from the measurements of ρ. More accurate values for the collision probability coefficient can be obtained by expanding in the form $P_c = a + bv + cv^2$.

2.4 Drift Velocity Measurements

The earliest measurements of drift velocity were carried out by the crossed electric and magnetic fields method introduced by Townsend (1925) and by the electrical shutter method devised by Bradbury and Nielsen (1936).

A more recent method for measuring drift velocities has been used by Hornbeck (1951). In this method a 0.1 μsec light pulse from a spark source strikes a cathode plate releasing a short burst of photoelectrons which drift through the gas to the anode under the influence of an electric field. The drift velocity can then be determined from the time of flight of the electron burst.

2.5 Atomic Beam Experiments

In this type of experiment a chopped atomic beam is crossed by a dc electron beam. The signal derived from the electrons scattered by the particles of the neutral beam can be separated from the dc signal due to electrons scattered by particles of the residual gas in the vacuum chamber by utilizing the frequency and phase of the former electrons. The most successful attempt to measure a total elastic cross section using this approach has been carried out by Brackmann *et al.* (1958) who were concerned with scattering by atomic hydrogen. The apparatus used by them is illustrated in Fig. 1. An atomic hydrogen beam pro-

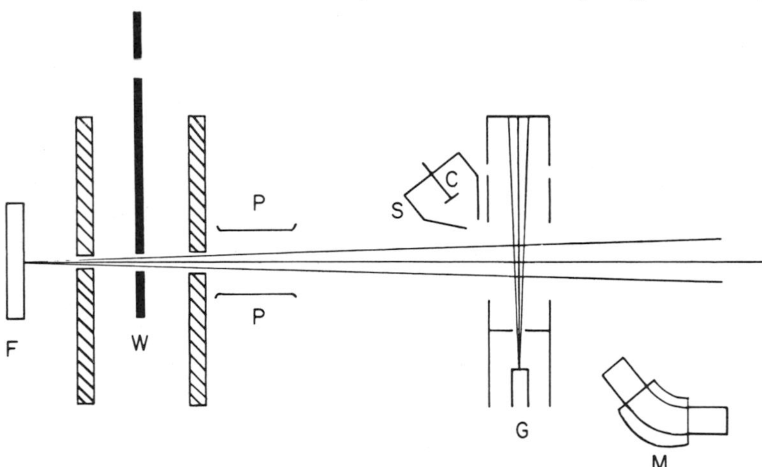

Fig. 1. Atomic beam apparatus used by Brackmann *et al.* (1958) to investigate the elastic scattering of electrons by hydrogen atoms.

duced by the tungsten furnace F was modulated at 100 cps by a mechanical chopper wheel W. Electrostatic deflection plates P were used to prevent electrons and ions from the furnace from reaching the scattered electron collector C. A mass spectrometer M was employed to determine the proportions of hydrogen atoms and molecules in the beam, the dissociation fraction being usually maintained at about 0.90 to 0.96. The dc electron beam was produced by the electron gun G. The observed electrons were scattered into a cone with a semiangle of 45° and with its axis perpendicular to the electron beam. The scattered electron collector C was placed within a metal shield S.

The absolute cross section for scattering of electrons by hydrogen molecules into the cone observation was obtained by integrating the absolute differential cross section data of Ramsauer and Kollath (1932). When the measurements of Brackmann, Fite, and Neynaber for scattering by a purely molecular beam were normalized to this absolute cross section agreement was obtained to within 5 %. Hence, knowing the degree of dissociation in the atomic beam the absolute cross section Q_A for scattering by hydrogen atoms into the cone of observation could be derived.

2.6 Measurements of Angular Distribution

The angular distribution of electrons scattered by elastic collisions with hydrogen atoms has been investigated by Gilbody *et al.* (1961). A modulated atomic hydrogen beam was passed through a circular hole situated at the center of a large disk. An electron gun was placed on the disk so that the electron beam intersected the neutral beam at right angles. The scattered electrons passed through a hole in a plate and were focused on to an electron multiplier. The variation of the scattered electron signal with scattering angle was investigated by rotating the electron gun. The energy of the incident electrons was determined by a retarding potential analysis of the electron current to a small collector placed in front of the gun. The half-width of the energy spread was about 0.4 ev for electron energies below 10 ev.

A second electron gun was employed to ionize the beam after leaving the collision volume. From the ratio of the peak heights of H^+ to H_2^+ as measured by a mass spectrometer, the dissociation fraction in the neutral beam could be determined.

3 *Scattering by Hydrogen Atoms*

3.1 Direct Scattering

The simplest case of collisions between electrons and atoms is that involving atomic hydrogen. The wave function $\Psi(\mathbf{r}_1, \mathbf{r}_2)$ of the system of two electrons moving in the field of a proton satisfies the wave equation

$$\left[-\frac{\hbar^2}{2m}(\nabla_1^2 + \nabla_2^2) - \frac{e^2}{r_1} - \frac{e^2}{r_2} + \frac{e^2}{r_{12}} - E\right]\Psi(\mathbf{r}_1, \mathbf{r}_2) = 0 \tag{91}$$

where E is the total energy of the system, $\mathbf{r}_1$ and $\mathbf{r}_2$ are the position vectors of electrons 1 and 2 relative to the proton and $r_{12} = |\mathbf{r}_1 - \mathbf{r}_2|$.

To solve this equation we expand Ψ in terms of the orthogonal and normalized set of atomic hydrogen wave functions $\psi_n(\mathbf{r})$ which satisfy

$$\left[-\frac{\hbar^2}{2m}\nabla^2 - \frac{e^2}{r} - \epsilon_n\right]\psi_n(\mathbf{r}) = 0, \tag{92}$$

ϵ_n being the eigenenergy corresponding to the nth state of a hydrogen atom.

Regarding the two electrons as being distinguishable, we may take electron 1 as the incident electron and electron 2 as the atomic electron. Then in the case of direct scattering for which no rearrangement of electrons occurs, we may put

$$\Psi(\mathbf{r}_1, \mathbf{r}_2) = \underset{m}{S}\, F_m(\mathbf{r}_1)\, \psi_m(\mathbf{r}_2). \tag{93}$$

On substituting into (91), multiplying by $\psi_n^*(\mathbf{r}_2)$ and integrating with respect to $\mathbf{r}_2$ we obtain

$$[\nabla_1^2 + k_n^2]\, F_n(\mathbf{r}_1) = \frac{2m}{\hbar^2}\underset{m}{S}\, V_{nm}(\mathbf{r}_1)\, F_m(\mathbf{r}_1) \tag{94}$$

where

$$k_n^2 = \frac{2m}{\hbar^2}(E - \epsilon_n) \tag{95}$$

and

$$V_{nm}(\mathbf{r}_1) = \int \psi_n^*(\mathbf{r}_2)\left(\frac{e^2}{r_{12}} - \frac{e^2}{r_1}\right)\psi_m(\mathbf{r}_2)\, d\mathbf{r}_2. \tag{96}$$

In order to proceed further we require to know the boundary conditions satisfied by $F_n(\mathbf{r})$. If the atom is initially in its ground state ($n = 0$) we have for large r

$$F_0(\mathbf{r}) \sim \exp ik_0\mathbf{n}_0 \cdot \mathbf{r} + r^{-1} \exp ik_0 r\, f_0(\theta, \phi) \tag{97}$$

and

$$F_n(\mathbf{r}) \sim r^{-1} \exp ik_n r\, f_n(\theta, \phi) \qquad (n \neq 0) \tag{98}$$

where $\mathbf{n}_0$ is a unit vector in the direction of the incident electron.

3.2 Born Approximations

Clearly it is impossible to solve the infinite set of coupled differential equations given by (94). Thus we must resort to approximation. The simplest is the first Born approximation and is obtained by putting

$$F_0(\mathbf{r}) = \exp ik_0\mathbf{n}_0 \cdot \mathbf{r} \tag{99}$$

$$F_m(\mathbf{r}) = 0 \qquad (m \neq 0) \tag{100}$$

on the right-hand side of (94). Then we have

$$[\nabla_1^2 + k_n^2]\, F_n(\mathbf{r}_1) = \frac{2m}{\hbar^2} V_{n0}(\mathbf{r}_1) \exp ik_0\mathbf{n}_0 \cdot \mathbf{r}_1 \tag{101}$$

and so using the Green's function (26) we get

$$F_0(\mathbf{r}_1) = \exp ik_0\mathbf{n}_0 \cdot \mathbf{r}_1 - \frac{1}{4\pi}\left(\frac{2m}{\hbar^2}\right) \times \int \frac{\exp ik_0 \,|\, \mathbf{r}_1 - \mathbf{r}_2 \,|}{|\, \mathbf{r}_1 - \mathbf{r}_2 \,|} V_{00}(r_2) \exp ik_0\mathbf{n}_0 \cdot \mathbf{r}_2 \, d\mathbf{r}_2 \tag{102}$$

and

$$F_n(\mathbf{r}_1) = -\frac{1}{4\pi}\left(\frac{2m}{\hbar^2}\right) \int \frac{\exp ik_n \,|\, \mathbf{r}_1 - \mathbf{r}_2 \,|}{|\, \mathbf{r}_1 - \mathbf{r}_2 \,|} V_{n0}(\mathbf{r}_2) \exp ik_0\mathbf{n}_0 \cdot \mathbf{r}_2 \, d\mathbf{r}_2. \tag{103}$$

Hence, to the first Born approximation the elastic scattering amplitude is given by

$$f_{B1}(\theta) = -\frac{1}{4\pi}\left(\frac{2m}{\hbar^2}\right) \int \exp ik_0(\mathbf{n}_0 - \mathbf{n}) \cdot \mathbf{r}\, V_{00}(r)\, d\mathbf{r} \tag{104}$$

where $\mathbf{n} \cdot \mathbf{n}_0 = \cos\theta$.

$V_{00}(r)$ is the potential of an electron in the static field of a hydrogen atom in the ground $1s$ state. It can be readily evaluated and is given by

$$V_{00}(r) = -e^2\left(\frac{1}{r} + \frac{1}{a_0}\right) e^{-2r/a_0}. \tag{105}$$

On carrying out the integration in (104) we obtain

$$f_{B1}(\theta) = \frac{\dfrac{1}{2a_0}\left(\dfrac{2}{a_0^2} + k_0^2 \sin^2\dfrac{\theta}{2}\right)}{\left(\dfrac{1}{a_0^2} + k_0^2 \sin^2\dfrac{\theta}{2}\right)^2}. \tag{106}$$

For forward scattering we have $f_{B1} = a_0$ and so as θ becomes small the differential cross section tends to a_0^2 and is therefore independent of the energy of the incident electron in this limit.

If $a_0 k_0 \sin(\theta/2) \gg 1$, we have $f_{B1} \sim (1/2\, a_0 k_0^2) \csc^2(\theta/2)$ and therefore the differential cross section is given by

$$I(\theta) \sim (1/4\, a_0^2 k_0^4) \csc^4 \frac{\theta}{2}. \tag{107}$$

This is identical to the differential cross section for the elastic scattering of electrons by the Coulomb field of a proton and so at high energies the effect of the screening due to the atomic electron is negligible except at small angles of scattering.

The second Born approximation to the elastic scattering amplitude can be obtained by substituting (102) and (103) into the right-hand side of (94) with $n = 0$ and solving for F_0. This gives

$$f_0(\theta) = f_{B1}(\theta) + f_{B2}(\theta) \tag{108}$$

where

$$f_{B2}(\theta) = \underset{m}{S} f_{B2}^m(\theta) \tag{109}$$

and

$$f_{B2}^m(\theta) = \left(\frac{2\pi m}{h^2}\right)^2 \iint \exp ik_0(\mathbf{n}_0 \cdot \mathbf{r}_1' - \mathbf{n} \cdot \mathbf{r}_1) \times \frac{\exp ik_m |\mathbf{r}_1 - \mathbf{r}_1'|}{|\mathbf{r}_1 - \mathbf{r}_1'|} V_{0m}(\mathbf{r}_1) V_{m0}(\mathbf{r}_1')\, d\mathbf{r}_1\, d\mathbf{r}_1'. \tag{110}$$

As this expression for f_{B2} involves a summation over an infinite number of intermediate states, it cannot be evaluated without great computational effort unless the series of terms is cut off. Since about two-thirds of the polarizability of a hydrogen atom arises from the $2p$ states a satisfactory approximation to f_{B2} can be obtained by assuming that the terms corresponding to the $1s$, $2s$, and $2p$ intermediate states provide the major contribution to f_{B2} and that the terms corresponding to all other intermediate states can be neglected. The effect of the $1s$ intermediate state on elastic scattering is termed *distortion* while the effect of all the other intermediate states is termed *polarization.*

On introducing the quantities $\alpha = \mathrm{Re}\, f_{B2}/f_{B1}$ and $\beta = \mathrm{Im}\, f_{B2}/f_{B1}$ we see that the differential cross section given by the second Born approximation is

$$I(\theta) = |f_{B1}|^2 \{(1 + \alpha)^2 + \beta^2\}. \tag{111}$$

However, in the derivation of this expression, terms of the fourth order in the interaction energy between the incident electron and the hydrogen atom have been neglected. Thus, defining $\alpha' = \mathrm{Re}\, f_{B3}/f_{B1}$ where f_{B3} is the third Born approximation correction to the scattering amplitude, we find that to the fourth order

$$I(\theta) = |f_{B1}|^2 \{(1 + \alpha)^2 + 2\alpha' + \beta^2\}. \tag{112}$$

Since α' is neglected in (111) it is necessary to ignore all terms of the fourth order and to employ the expression

$$I(\theta) = |f_{B1}|^2 (1 + 2\alpha) \tag{113}$$

which is exact to the third order in the interaction energy.

TABLE I

ELASTIC SCATTERING OF ELECTRONS BY HYDROGEN ATOMS[a]

Wave number k_0 (in a_0^{-1})	First Born approximation	Third order approximation		
		Distortion	Polarization	Distortion-polarization
0.5	2.90	6.55	4.43	8.09
1.0	1.54	2.74	2.12	3.32
2.0	0.523	0.674	0.555	0.705
3.0	0.247	0.281	0.253	0.287
4.0	0.142	0.153	0.144	0.156

[a] Total cross sections (in units of πa_0^2).

In Table I the cross sections for the elastic scattering of electrons by hydrogen atoms obtained by Kingston and Skinner (1961) with the first Born approximation and the third order approximation (113) are tabulated against the wave number k_0 of the incident electron. We refer to the results obtained with $f_{B2} = f_{B2}^{1s}$, $f_{B2} = f_{B2}^{2s} + f_{B2}^{2p}$ and $f_{B2} = f_{B2}^{1s} + f_{B2}^{2s} + f_{B2}^{2p}$ as the *distortion*, *polarization*, and *distortion-polarization* approximations, respectively.

All the values obtained with the third order expression (113) are larger than the first Born cross section, the largest being that given by the distortion-polarization approximation. For $k_0 \lesssim 1.5a_0^{-1}$ the values of α are so large that it is not permissible to neglect fourth order terms. However, for $k_0 \gtrsim 2a_0^{-1}$ (i.e., for electron impact energies $\gtrsim$ 50 ev) the fourth order correction is small and consequently the cross section given by the third order distortion-polarization approximation should be reliable.

Unfortunately, there are no available laboratory data on the elastic scattering of electrons by atomic hydrogen at sufficiently high electron impact energies for making a comparison between the third order approximation calculations and experiment.

3.3 Static Field Approximation

Consider the elastic scattering of electrons by hydrogen atoms in the state n. Neglecting the coupling between this state and all other states (94) reduces to

$$\left[\nabla^2 + k_n^2 - \frac{2m}{\hbar^2} V_{nn}(\mathbf{r})\right] F_n(\mathbf{r}) = 0 \tag{114}$$

with

$$F_n(\mathbf{r}) \sim \exp ik_n\mathbf{n}_0 \cdot \mathbf{r} + r^{-1} \exp ik_n r\, f_n(\theta) \tag{115}$$

for large r.

In the case of a spherically symmetrical potential V_{nn}, such as for an s state, the solution of (114) may be obtained by employing the method of partial waves described in § 1.3. On expanding F_n in the form

$$F_n(\mathbf{r}) = \frac{1}{r} \sum_{l=0}^{\infty} i^l(2l+1)\, e^{i\eta_{n,l}}\, \phi_{n,l}(r)\, P_l(\cos\theta) \tag{116}$$

we obtain the radial equation

$$\left\{\frac{d^2}{dr^2} + k_n^2 - \frac{2m}{\hbar^2} V_{nn}(r) - \frac{l(l+1)}{r^2}\right\} \phi_{n,l}(r) = 0 \tag{117}$$

with

$$\phi_{n,l}(r) \sim \frac{1}{k_n} \sin\left(k_n r - \tfrac{1}{2} l\pi + \eta_{n,l}\right) \tag{118}$$

for large r.

Consider the elastic scattering of electrons by the static field $V_{00}(r)$ of a hydrogen atom in the $1s$ state given by (105). At sufficiently low electron energies the dominant contribution to the total elastic cross section comes from the zero order partial wave. The radial equation (117) with $l = 0$ and $n = 0$ has been solved by numerical integration by

McDougall (1932) and by Chandrasekhar and Breen (1946). Their values of the phase shift η_0 are given in Table II. Seaton (1957a) has calculated the scattering length by numerical integration. He finds that it has the value $-9.44a_0$ which gives $356\pi a_0^2$ for the total elastic cross section in the limit of zero energy incident electrons. Also given in Table II are the values of the phase shift η_0 obtained by Massey and Moiseiwitsch (1951) using Hulthén's and Kohn's variational methods with the trial function

$$\phi_t(r) = k_0^{-1}(1 + a^2)^{-1/2} \times \{\sin k_0 r + (a + b\, e^{-r/a_0})(1 - e^{-r/a_0}) \cos k_0 r\}. \tag{119}$$

Both variational methods result in essentially the same values for the phase shift η_0 and are in good agreement with the exact calculations.

TABLE II

ELASTIC SCATTERING OF ELECTRONS BY THE STATIC FIELD OF A HYDROGEN ATOM IN THE 1s STATE[a]

Wave number k_0 (in a_0^{-1})	Phase shift η_0 (in radians)	
	Numerical integration	Hulthén's and Kohn's variational methods
0	(−9.44)[b]	(−9.44)[b]
0.1	0.730	0.721
0.2	0.973	0.972
0.3	1.046	1.045
0.4	—	1.057
0.5	1.045	1.044
0.6	—	1.020
0.8	—	0.962
1.0	0.906	0.904
1.2	—	0.851
1.5	—	0.783
2.0	—	0.694

[a] Zero order partial wave.
[b] Scattering length in a_0.

Chandrasekhar and Breen (1946) have also carried out a numerical integration of the radial equation (117) with $l = 1$. Their phase shift η_1 behaves like $0.269\, k_0^3$ for small k_0.

3.4 Exchange Scattering

In the previous sections we have taken into account only the possibility of direct scattering in which the incident electron 1 is scattered by the target hydrogen atom and electron 2 remains bound to the proton. There remains the possibility of exchange scattering in which electron 2 is ejected from the hydrogen atom while electron 1 is captured by the proton.

If electron 1 is initially moving with insufficient energy to excite the target hydrogen atom, the boundary conditions on the wave function $\Psi(\mathbf{r}_1, \mathbf{r}_2)$ are

$$\Psi(\mathbf{r}_1, \mathbf{r}_2) \sim \{\exp ik_0\mathbf{n}_0 \cdot \mathbf{r}_1 + r_1^{-1} \exp ik_0 r_1 f(\theta_1, \phi_1)\} \psi_0(\mathbf{r}_2) \quad \text{as} \quad r_1 \to \infty \tag{120}$$

$$\sim r_2^{-1} \exp ik_0 r_2 g(\theta_2, \phi_2) \psi_0(\mathbf{r}_1) \quad \text{as} \quad r_2 \to \infty \tag{121}$$

These boundary conditions are independent of each other and so (121) cannot be deduced from (120) as has been shown by Foldy and Tobocman (1957) and by Epstein (1957).

Consider the expansion of Ψ in the form

$$\Psi(\mathbf{r}_1, \mathbf{r}_2) = \sum_m \psi_m(\mathbf{r}_2) F_m(\mathbf{r}_1) + \int \psi_\kappa(\mathbf{r}_2) F_\kappa(\mathbf{r}_1) \, d\kappa \tag{122}$$

where ψ_m is a hydrogen atom wave function belonging to the discrete spectrum and ψ_κ is a hydrogen atom wave function belonging to the continuous spectrum with energy $\hbar^2\kappa^2/2m$. Lippman (1956) has proved that F_κ has a singularity at the point $\kappa = k_0$. The path of integration around the singularity must be chosen so that the asymptotic behavior (121) is fulfilled. By carrying out the integration over κ with an appropriate contour, Castillejo *et al.* (1960) have shown that

$$g(\theta_2, \phi_2) = -\frac{1}{2\pi a_0} \int \psi_0(\mathbf{r}_1') \psi_{-\mathbf{k}_s}(\mathbf{r}_2') (1/|\mathbf{r}_1' - \mathbf{r}_2'|) \Psi(\mathbf{r}_1', \mathbf{r}_2') \, d\mathbf{r}_1' \, d\mathbf{r}_2' \tag{123}$$

where $\psi_{\mathbf{k}_s}(\mathbf{r})$ is the wave function of a free electron moving in the Coulomb field $-(e^2/r)$ and having the asymptotic form

$$\psi_{\mathbf{k}_s}(\mathbf{r}) \sim \exp ik_0\mathbf{n} \cdot \mathbf{r} + r^{-1} \exp ik_0 r f_s(\theta, \phi), \tag{124}$$

θ_2, ϕ_2 being the polar angles of $\mathbf{n}$ relative to $\mathbf{n}_0$.

In order to take account of the inherent indistinguishability of the two electrons, the total electronic wave function of the system must be antisymmetrical with respect to interchange of both the space and spin coordinates of the two electrons. Hence, the space part of the wave function is taken to be

$$\Psi^{\pm}(\mathbf{r}_1, \mathbf{r}_2) = \Psi(\mathbf{r}_1, \mathbf{r}_2) \pm \Psi(\mathbf{r}_2, \mathbf{r}_1) \tag{125}$$

where Ψ^+ and Ψ^- are associated with the singlet and triplet spin functions, respectively. Then for large r_1

$$\Psi^{\pm}(\mathbf{r}_1, \mathbf{r}_2) \sim \{\exp ik_0\mathbf{n}_0 \cdot \mathbf{r}_1 + r_1^{-1} \exp ik_0 r_1(f \pm g)\} \psi_0(r_2). \tag{126}$$

It follows that the differential cross section is given by

$$I(\theta, \phi) = \tfrac{1}{4}\{3\,|f - g\,|^2 + |f + g\,|^2\}. \tag{127}$$

3.5 Exchange Approximation

On writing Ψ in the form

$$\Psi(\mathbf{r}_1, \mathbf{r}_2) = \underset{m}{S}\, F_m^{\pm}(\mathbf{r}_1)\, \psi_m(\mathbf{r}_2) \tag{128}$$

we get

$$\Psi^{\pm}(\mathbf{r}_1, \mathbf{r}_2) = \underset{m}{S}\, F_m^{\pm}(\mathbf{r}_1)\, \psi_m(\mathbf{r}_2) \pm \underset{m}{S}\, F_m^{\pm}(\mathbf{r}_2)\, \psi_m(\mathbf{r}_1). \tag{129}$$

Castillejo *et al.* (1960) have shown that in such a symmetrized expansion the integration over the continuous spectrum does not involve a singularity.

Substituting into (91), multiplying by $\psi^*(\mathbf{r}_2)$, and integrating with respect to $\mathbf{r}_2$ we obtain

$$[\nabla_1^2 + k_n^2]\, F_n^{\pm}(\mathbf{r}_1)$$

$$= \frac{2m}{\hbar^2} \underset{m}{S} \left\{ V_{nm}(\mathbf{r}_1)\, F_m^{\pm}(\mathbf{r}_1) \mp \int K_{nm}(\mathbf{r}_1, \mathbf{r}_2)\, F_m^{\pm}(\mathbf{r}_2)\, d\mathbf{r}_2 \right\} \tag{130}$$

where

$$K_{nm}(\mathbf{r}_1, \mathbf{r}_2) = \psi_n^*(\mathbf{r}_2)\, \psi_m(\mathbf{r}_1) \left\{ E - \epsilon_n - \epsilon_m - \frac{e^2}{r_{12}} \right\}. \tag{131}$$

Consider the elastic scattering of electrons by hydrogen atoms in the state n. If we neglect the effect of all states other than the state n the infinite set of coupled integro-differential equations (130) reduce to the pair of uncoupled equations corresponding to positive and negative symmetry

$$\left[\nabla_1^2 + k_n^2 - \frac{2m}{\hbar^2} V_{nn}(\mathbf{r}_1)\right] F_n^{\pm}(\mathbf{r}_1) \pm \frac{2m}{\hbar^2} \int K_{nn}(\mathbf{r}_1, \mathbf{r}_2)\, F_n^{\pm}(\mathbf{r}_2)\, d\mathbf{r}_2 = 0 \tag{132}$$

where for large r

$$F_n^{\pm}(\mathbf{r}) \sim \exp ik_n\mathbf{n}_0 \cdot \mathbf{r} + r^{-1} \exp ik_n r \, f_n^{\pm}(\theta, \phi). \tag{133}$$

The differential cross section for elastic scattering by a hydrogen atom in the state n is then given by

$$I(\theta, \phi) = \tfrac{1}{4}\{3 \, | f_n^- |^2 + | f_n^+ |^2\}. \tag{134}$$

Expanding $F_n^{\pm}(\mathbf{r})$ in the form

$$F_n^{\pm}(\mathbf{r}) = \frac{1}{r} \sum_{l=0}^{\infty} i^l(2l+1) \exp(i\eta_{n,l}^{\pm}) \, \phi_{n,l}^{\pm}(r) \, P_l(\cos\theta) \tag{135}$$

we obtain in the case of scattering by an s state atom

$$\left\{\frac{d^2}{dr^2} + k_n^2 - \frac{2m}{\hbar^2} V_{nn}(r) - \frac{l(l+1)}{r^2}\right\} \phi_{n,l}^{\pm}(r) \pm \frac{2m}{\hbar^2} \int_0^{\infty} \kappa_{nn}^l(r, r') \, \phi_{n,l}^{\pm}(r') \, dr' = 0 \tag{136}$$

where

$$\kappa_{nn}^l(r, r') = \frac{4\pi}{2l+1} rr' \, \psi_n(r) \, \psi_n(r') \, \{(E - 2\epsilon_n)\delta_{0n} - e^2\gamma_l(r, r')\} \tag{137}$$

and

$$\gamma_l(r, r') = \begin{cases} r^l/r'^{l+1} & (r < r') \\ r'^l/r^{l+1} & (r' < r), \end{cases} \tag{138}$$

the asymptotic form of $\phi_{n,l}^{\pm}(r)$ for large r being

$$\phi_{n,l}^{\pm}(r) \sim \frac{1}{k_n} \sin(k_n r - \tfrac{1}{2} l\pi + \eta_{n,l}^{\pm}). \tag{139}$$

The solutions of the pair of equations (136) for the $l = 0$ partial wave corresponding to the case of elastic scattering by hydrogen atoms in the ground $1s$ state was first attempted by Morse and Allis (1933) by numerical integration. Unfortunately, it seems that they made an error in their derivation of the kernal κ_{nn}^l. This was pointed out by Seaton (1957a) who carried out the numerical integration of the correct radial equations in the limit of zero energy obtaining the value $a_+ = 8.06\, a_0$, $a_- = 2.35\, a_0$ for the singlet and triplet scattering lengths, respectively.

The total elastic cross section in the limit of vanishing energy is given by

$$Q(k_0 = 0) = \pi(3a_-^2 + a_+^2) \tag{140}$$

the value of $Q(k_0 = 0)$ obtained by Seaton being $81.5\pi a_0^2$ which is much smaller than the corresponding value of $356\pi a_0^2$ calculated by Seaton neglecting exchange (cf. § 3.3).

For nonvanishing values of the electron energy, calculations have been carried our for the zero order partial wave by Massey and Moiseiwitsch (1951) using Hulthén's and Kohn's variational methods with the trial function (119) for $\phi_0^{\pm}(r)$. The values of their phase shifts $\eta_0^{\pm}$ are given in Table III. Whereas the zero order phase shift η_0 corresponding to the static field approximation tends to zero in the limit of vanishing electron energy, the zero order phase shifts $\eta_0^{\pm}$ corresponding to the exchange approximation tend to π.

TABLE III

ELASTIC SCATTERING OF ELECTRONS BY HYDROGEN ATOMS IN THE 1s STATE[a]

Wave number k_0 (in a_0^{-1})	Phase shift (in radians): Exchange approximation η_0^+	Exchange approximation η_0^-	Correlation approximation η_0	Exchange-correlation approximation η_0^+	Exchange-correlation approximation η_0^-
0	(+9.03)[b]	(+2.35)[b]	—	(+7.4)[b]	(2.34)[b]
0.1	2.403	2.908	2.386	2.484	2.909
0.2	1.819	2.679	1.963	2.003	2.680
0.3	1.486	2.463	1.716	1.649	2.447
0.4	1.250	2.257	1.545	1.425	2.248
0.5	1.074	2.070	1.413	1.250	2.039
0.6	0.940	1.901	1.307	1.095	1.909
0.8	0.756	1.614	1.143	0.857	1.621
1.0	0.645	1.390	1.027	0.708	1.398
1.2	0.583	1.217	0.938	0.630	1.225
1.5	0.546	1.027	0.840	0.599	1.036
2.0	0.542	0.826	0.730	0.600	0.838

[a] Zero order phase shifts calculated by Hulthén's variational method.
[b] Scattering length in a_0.

Bransden *et al.* (1958) have also calculated $\eta_0^{\pm}$ to the exchange approximation using variational methods with a different form of trial function to (119). Their values of η_0^- are only slightly different from those obtained by Massey and Moiseiwitsch while their values of η_0^+, though more accurate at low energies, are less accurate at the higher energies than those calculated by Massey and Moiseiwitsch. Bransden *et al.* (1958) also calculated the phase shifts $\eta_1^\pm$ to the exchange approximation using variational methods.

The elastic scattering of electrons by hydrogen atoms in states other than the ground state have also been investigated. Thus, Erskine and Massey (1952) calculated the zero order phase shifts $\eta_0^\pm$ for scattering by 2*s* hydrogen atoms and Massey and Khashaba (1958) calculated the zero order and first order phase shifts $\eta_0^\pm$, $\eta_1^\pm$, respectively, for scattering by 2*p* hydrogen atoms.

3.6 Correlation

In the static field approximation to the elastic scattering of electrons by hydrogen atoms in the ground state, it was assumed that the zero order partial wave function can be written in the separable form

$$\Psi_0(\mathbf{r}_1, \mathbf{r}_2) = \psi_0(\mathbf{r}_2)\, F_0(\mathbf{r}_1). \tag{141}$$

However, this makes no explicit allowance for the mutual interaction of the two electrons involved in the collision. For the zero order partial wave it has been shown (Hylleraas, 1929) that the wave function depends on the three distances r_1, r_2, and r_{12} but not on any of the angular coordinates. Hence, we may write the wave function of the $l = 0$ partial wave in the form $\Psi_0(r_1, r_2, r_{12})$.

Massey and Moiseiwitsch (1951) have used Hulthén's and Kohn's variational methods to investigate the effect of the correlation between the two electrons by introducing a term depending on r_{12} in the trial function. They chose a trial function first suggested by Huang (1949) having the form

$$\Psi_t(r_1, r_2, r_{12}) = e^{i\eta_0}\, r_1^{-1}\, \phi_t(r_1, r_{12})\, \psi_0(r_2) \tag{142}$$

where

$$\phi_t(r_1, r_{12}) = k_0^{-1}(1 + a^2)^{-1/2} \times [\sin k_0 r_1 + \{a + (b + c r_{12})\, e^{-r_1/a_0}\}\, (1 - e^{-r_1/a_0}) \cos k_0 r_1]. \tag{143}$$

We shall refer to this as the *correlation approximation* and the values of the phase shift η_0 obtained with its use are given in Table III. Clearly the effect of correlation is very important. Thus, in the static field approximation the phase shift η_0 tends to zero as k_0 becomes small whereas in the correlation approximation η_0 tends to π.

3.7 Positron Scattering

Massey and Moussa (1958) have used Hulthén's and Kohn's variational methods with the trial function given by (142) and (143) to calculate the zero order phase shift for the elastic scattering of positrons by hydrogen atoms. They found that the effect of the correlation term cr_{12} in (143) was small. Thus, they obtained the values 0.512 a_0 and 0.582 a_0 for the scattering length with and without allowance for correlation, respectively.

However, Spruch and Rosenberg (1960) have re-examined the problem in the zero energy limit using the trial function

$$\Psi_t(r_1, r_2, r_{12}) = r_1^{-1}[e^{-r_2/a_0} \{A(1 - e^{-r_1/a_0}) - r_1\} + Br_1 \, e^{-(sr_1+qr_2)/a_0}$$
$$+ Cr_1 \, e^{-(vr_2+tr_{12})/a_0} + Dr_1 \, e^{-(2r_1+r_{12}/4)/a_0}]. \qquad (144)$$

They obtained the value $-$ 1.397 a_0 for the scattering length which is opposite in sign and larger in magnitude than the value calculated by Massey and Moussa. Hence, the effect of correlation is actually very important in the case of the elastic scattering of positrons by hydrogen atoms. Spruch and Rosenberg also point out that the value of the scattering length obtained with the variational method is an upper bound to the exact scattering length (cf. § 1.7) and that since it is negative it provides a lower bound to the zero energy limit of the elastic cross section of magnitude 7.80 πa_0^2.

3.8 Exchange-Correlation Approximation

For the elastic scattering of electrons by hydrogen atoms a more accurate approximation than either the exchange approximation or the correlation approximation is the exchange-correlation approximation which was first used by Massey and Moiseiwitsch (1951). They took a trial function having the form

$$\Psi_t^{\pm}(r_1, r_2, r_{12}) = \Psi_t(r_1, r_2, r_{12}) \pm \Psi_t(r_2, r_1, r_{12}) \qquad (145)$$

where $\Psi_t(r_1, r_2, r_{12})$ is given by (142) and (143). This function allows for both exchange and correlation. The values of the phase shifts $\eta_0^{\pm}$ obtained with it are given in Table III. In the antisymmetrical case the introduction of correlation has a negligible effect upon η_0^-. This is to be expected as the antisymmetry of the space wave function gives a small probability of finding the two electrons close together. For the symmetrical case the effect of introducing the correlation term cr_{12} is to increase the phase shift η_0^+ somewhat. For comparison purposes the phase shifts obtained with the use of the various approximation are illustrated in Fig. 2.

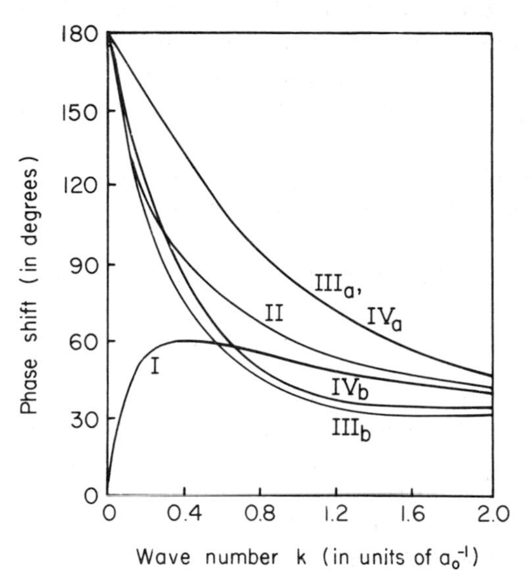

FIG. 2. Zero order phase shifts for the elastic scattering of electrons by hydrogen atoms calculated by Massey and Moiseiwitsch (1951) using the variational method. Curve I, static field approximation; curve II, polarization approximation; curve IIIa, exchange approximation, antisymmetric case; curve IIIb, exchange approximation, symmetric case; curve IVa, exchange-correlation approximation, antisymmetric case; curve IVb, exchange-correlation approximation, symmetric case.

Rosenberg *et al.* (1960) have investigated the case of zero energy scattering using the trial functions given by (145) with

$$\Psi_t(r_1, r_2, r_{12}) = r_1^{-1}[e^{-r_2/a_0}\{A(1 - e^{-br_1/a_0}) - r_1\} + Br_1\, e^{-(cr_1+dr_2)/a_0} + Cr_1\, e^{-(fr_1+gr_2+hr_{12})/a_0}]. \tag{146}$$

They obtained the values $a_+ = 6.23\, a_0$ and $a_- = 1.93\, a_0$ for the singlet and triplet scattering lengths. These values for $a_\pm$ are upper bounds to the scattering lengths and so the corresponding value $49.8\, \pi a_0^2$ of the zero energy limit to the elastic cross section is also an upper bound.

It has been shown by Ohmura *et al.* (1958) and also by O'Malley *et al.* (1961) that

$$k_0 \cot \eta_0^+ = -\gamma + \frac{\rho}{2}(\gamma^2 + k_0^2) + O\{(\gamma^2 + k_0^2)^2\} \tag{147}$$

where $\hbar^2\gamma^2/2m$ is the electron affinity of atomic hydrogen and

$$\rho = \frac{1}{\gamma} - \frac{1}{4\pi^2 c^2} \tag{148}$$

is the effective range for the singlet case, c being the asymptotic amplitude of the normalized wave function of H^-.

Ohmura and Ohmura (1960) have used the wave function for H^- calculated by Pekeris (1958) with a variational trial function having 202 adjustable parameters to calculate γ and c. They find that $\gamma = 0.235588\ a_0^{-1}$ and $\rho = (2.646 \pm 0.004)a_0$ which gives good agreement with the exchange-correlation approximation calculation of Massey and Moiseiwitsch (1951) in the singlet case for values of $k_0 \lesssim 1.0\ a_0^{-1}$.

If the third term on the right-hand side of (147) is neglected we obtain $a_+ = 6.167\ a_0$ for the scattering length in the singlet case. This is somewhat smaller than the value given in Table III but is in good agreement with the value obtained by Rosenberg *et al.* (1960).

3.9 Polarized Orbitals

An electron situated at a distance r_1 from the nucleus of an atom induces a dipole moment in the atom which gives rise to an induced dipole potential $V_P(r_1)$ behaving like $-\alpha/2r_1^4$ for large r_1, where α is the polarizability of the atom and is given by $\alpha = (9/2)\ e^2 a_0^3$ in the case of hydrogen. The corresponding first order perturbed hydrogen atom wave function is (Dalgarno and Lewis, 1955)

$$\psi_0(r_2)\left[1 - \sum_{m=1}^{\infty} \frac{1}{r_1^{m+1}} \left\{\frac{r_2^{m+1}}{(m+1)} + a_0 \frac{r_2^m}{m}\right\} P_m(\cos\Theta)\right] \tag{149}$$

where $\hat{\mathbf{r}}_1 \cdot \hat{\mathbf{r}}_2 = \cos\Theta$.

The method of polarized orbitals employs a trial wave function having the form

$$\Psi^{\pm}(\mathbf{r}_1, \mathbf{r}_2) = F_0(\mathbf{r}_1)\{\psi_0(\mathbf{r}_2) + \psi_{\mathrm{POL}}(\mathbf{r}_1, \mathbf{r}_2)\}$$
$$\pm F_0(\mathbf{r}_2)\{\psi_0(\mathbf{r}_1) + \psi_{\mathrm{POL}}(\mathbf{r}_2, \mathbf{r}_1)\} \tag{150}$$

where

$$\psi_{\mathrm{POL}}(\mathbf{r}_1, \mathbf{r}_2) \sim -\psi_0(r_2) \sum_{m=1}^{\infty} \frac{1}{r_1^{m+1}} \left\{\frac{r_2^{m+1}}{(m+1)} + a_0 \frac{r_2^m}{m}\right\} P_m(\cos\Theta). \tag{151}$$

TABLE IV

ELASTIC SCATTERING OF ELECTRONS BY HYDROGEN ATOMS IN THE 1s STATE[a]

Wave number k_0 (in a_0^{-1})	Phase shift (in radians)											
	η_0^+		η_0^-		η_1^+		η_1^-		η_2^+		η_2^-	
	P.O.	E.	P.O.	E.	P.O.	E.	P.O.	E.	P.O.	E.	P.O.	E.
0	(5.7)[b]	(8.10)[b]	(1.7)[b]	(2.35)[b]	—	—	—	—	—	—	—	—
0.01	3.085	3.061	3.125	3.118	—	—	—	—	—	—	—	—
0.05	2.855	2.746	3.049	3.024	—	—	—	—	—	—	—	—
0.1	2.583	2.396	2.946	2.907	0.00481	−0.0012	0.0090	0.00214	—	—	—	—
0.2	2.114	1.870	2.732	2.679	—	—	—	—	—	—	—	—
0.3	1.750	1.508	2.519	2.461	0.0322	−0.0241	0.098	0.0511	0.0113	−0.00057	0.0118	0.000763
0.4	1.469	1.239	2.320	2.257	—	—	—	—	—	—	—	—
0.5	1.251	1.031	2.133	2.070	0.0392	−0.0703	0.245	0.169	0.266	−0.00397	0.0350	0.00698
0.75	0.908	0.694	1.745	1.679	0.0347	−0.1126	0.390	0.304	0.0456	−0.0123	0.0746	0.0274
1.0	0.758	0.543	1.460	1.391	0.0428	−0.1059	0.453	0.358	0.0627	−0.0176	0.112	0.0555

[a] Phase shifts obtained with the method of polarized orbitals (P.O.) and with the exchange approximation (E.) by numerical integration.
[b] Scattering length in a_0.

This method has been used by Temkin and Lamkin (1961) who made allowance for the dipole term of ψ_{POL} only by putting

$$\psi_{POL}(\mathbf{r}_1, \mathbf{r}_2) = - \epsilon(r_1, r_2)\, \psi_0(r_2) \frac{1}{r_1^2} \left(\frac{r_2^2}{2} + a_0 r_2 \right) P_1(\cos \Theta) \tag{152}$$

where

$$\epsilon(r_1, r_2) = \begin{cases} 1 & (r_1 > r_2) \\ 0 & (r_1 < r_2). \end{cases} \tag{153}$$

Substitution of (150) with (152) into the wave equation (91), multiplication by $\psi_0(r_2)\, P_l(\cos \theta_1)$ and integration with respect to all coordinates except r_1 results in an infinite set of radial equations, one for each angular momentum value of the incident electron.

Temkin and Lamkin (1961) have solved the radial equations for the $l = 0, 1, 2$ partial waves by numerical integration. The values of the phase shifts obtained by them are given in Table IV where they are compared with the phase shifts obtained by the exact numerical integration of the radial equations (136) corresponding to the exchange approximation in which ψ_{POL} is neglected.

The partial cross sections for $l = 0, 1, 2$ given by

$$Q_l = \tfrac{1}{4}(3Q_l^- + Q_l^+) \tag{154}$$

where

$$Q_l^{\pm} = \frac{4\pi}{k_0^2}(2l + 1) \sin^2 \eta_l^{\pm} \tag{155}$$

are displayed in Table V as is the total elastic cross section

$$Q = \sum_{l=0}^{\infty} Q_l \, .$$

TABLE V

CROSS SECTIONS FOR THE ELASTIC SCATTERING OF ELECTRONS BY HYDROGEN ATOMS IN THE 1s STATE

Wave number k_0 (in a_0^{-1})	Partial cross sections (in πa_0^2) Q_0	Q_1	Q_2	Total cross section (in πa_0^2) $Q = \sum_{l=0}^{\infty} Q_l$
0	41	0	0	41
0.3	22.1	0.99	0.030	23.1
0.5	12.2	2.14	0.088	14.4
0.75	6.28	2.31	0.166	8.8
1.0	3.44	1.73	0.207	5.4

3.10 Comparison between Theory and Experiment

Until the atomic beam experiments of Fite and his collaborators, there was scanty and uncertain experimental information about the elastic scattering of electrons by hydrogen atoms.

The earliest investigations were carried out by Kruithof and Ornstein (1935) and by Lindemann (1935) who were concerned with the measurement of the total cross section including both elastic and inelastic collisions. Kruithof and Ornstein directed an electron beam through hydrogen gas at pressures of 0.01 to 0.05 mm Hg. From the broadening of the beam due to elastic collisions they estimated the total cross section for scattering of electrons by a mixture of atomic and molecular hydrogen. Since the degree of dissociation of the hydrogen gas was known and the total cross section for the scattering of electrons by molecular hydrogen had been measured by previous investigators (cf. § 5.1), they were able to determine the total cross section for scattering of electrons by atomic hydrogen. For electron energies in the range 17-28 ev, they found a total cross section of about 7 to 8 πa_0^2. In their paper they also give some results obtained by Lindemann. They make no reference to his experimental method but quote values of about 5 πa_0^2 for 13 ev electrons rising to about 9 πa_0^2 for 50 ev electrons. At 13 ev the total cross section obtained by Lindemann is slightly lower than the calculated total elastic cross section for $k_0 = 1$ given in Table V.

Maecker *et al.* (1955) have obtained a mean value of about 150 πa_0^2 for the total cross section at an electron temperature of about 12,000°K from measurements of the conductivity of a water-stabilized high-pressure arc. This value is very much greater than the calculated total elastic cross section at the corresponding mean electron energy of 1 ev (i.e., $k_0 = 0.3\ a_0^{-1}$). Bederson *et al.* (1957) have measured the total cross section by a method employing a chopped atomic hydrogen beam crossed by a dc electron beam (cf. § 2. 5). They found a resonance in the total cross section with a maximum value of about 100 πa_0^2 occurring at an electron impact energy of 3 ev decreasing to about 6 πa_0^2 at 11 ev. The latter value is in satisfactory accord with theory, but at 3 ev their total elastic cross section is very much larger than that calculated by Temkin and Lamkin (1961) with allowance made for both exchange and polarization for the *s*, *p*, and *d* waves.

The most recent experimental data has been obtained by Brackmann *et al.* (1958) employing a similar experimental approach to that used by Bederson *et al.* and already described in § 2.5. Brackmann *et al.* show that if partial cross sections of higher order than unity are neglected, the total elastic cross section Q and the first order partial cross section Q_1 are related to their absolute cross section Q_A for scattering of electrons by hydrogen atoms into the cone of observation by the formula

$$Q = 6.85\, Q_A + 0.60\, Q_1. \tag{156}$$

Choosing Q_1 to be that calculated by Temkin and Lamkin (1961) and given in Table V, the experimental measurements of Q_A result in the values of the total elastic cross section depicted by crosses in Fig. 3.†

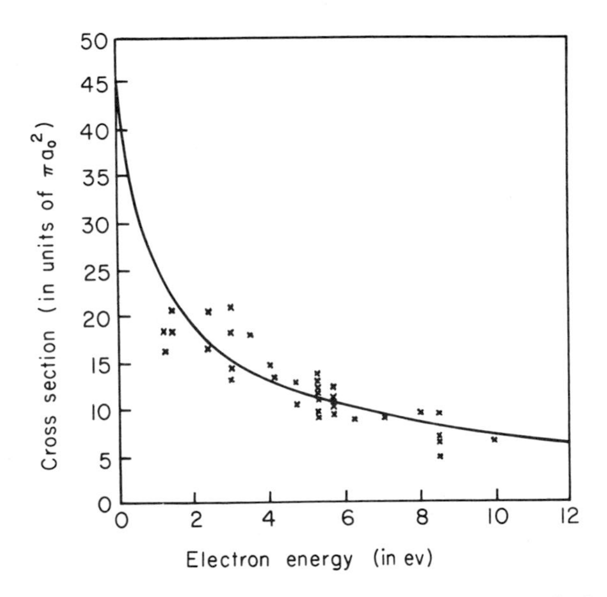

FIG. 3. Total cross sections for the elastic scattering of electrons by hydrogen atoms. Curve represents total cross section calculated by Temkin and Lamkin (1961) using the method of polarized orbitals. Crosses denote total cross sections derived from the experimental data of Brackmann *et al.* (1958).

Very good agreement between the experimental observations by Brackmann *et al.* (1958) and the polarized orbital calculations of Temkin and Lamkin (1961) is apparent.

Gilbody *et al.* (1961) have used the method given in § 2.6 to obtain the differential cross section for the elastic scattering of electrons by hydrogen atoms. Their results are displayed in Fig. 4. At 7.1 ev they obtain the value 0.25 ($\pm$ 20 %) for the ratio Q_1/Q_0 which is somewhat

† Recent measurements by Neynaber *et al.* (1961b) suggest a smaller total elastic cross section.

smaller than the value 0.37 calculated by Temkin and Lamkin (1961) using the method of polarized orbitals at 7.6 ev.

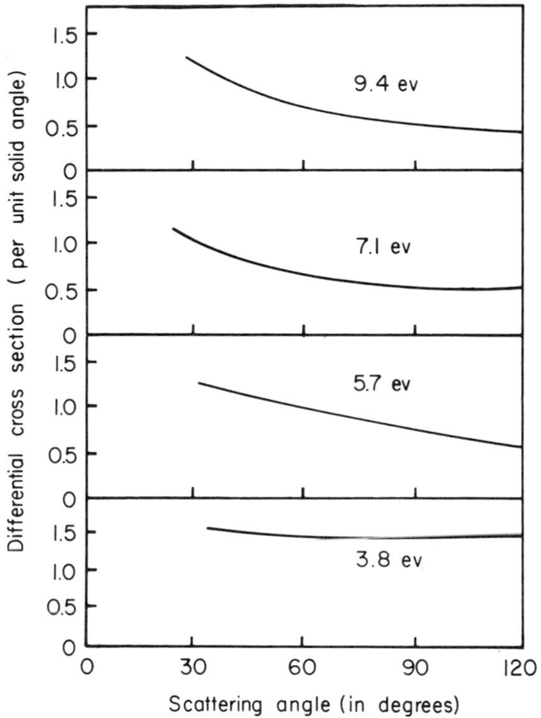

FIG. 4. Differential cross sections for the elastic scattering of electrons by hydrogen atoms obtained experimentally by Gilbody *et al.* (1961).

4 *Scattering by Complex Atoms and Ions*

4.1 HELIUM: CALCULATIONS

The elastic scattering of electrons by the Hartree field of a helium atom in the ground state has been investigated by McDougall (1932) who solved the appropriate radial equations for the $l = 0, 1, 2$ partial waves by numerical integration. The values of the phase shifts obtained by him are given in Table VI. Subsequently, Morse and Allis (1933) solved the integro-differential equations corresponding to the exchange approximation for the $l = 0,1$ partial waves by numerical integration. The wave function of the system was taken to have the symmetrized form

$$\Psi(\mathbf{r}_1, \mathbf{r}_2, \mathbf{r}_3) = \sum_{1,2,3} F_0(\mathbf{r}_1)\, \psi_0(\mathbf{r}_2, \mathbf{r}_3)\, \chi(1; 2, 3) \tag{157}$$

where $\chi(1; 2, 3) = (1/\sqrt{2})\, \alpha_1(\alpha_2\beta_3 - \beta_2\alpha_3)$ is a doublet spin function, ψ_0 is the Hartree wave function for the ground state of the helium atom, and the summation is cyclical over 1, 2, 3. The scattering lengths

TABLE VI

ELASTIC SCATTERING OF ELECTRONS BY THE HARTREE FIELD OF HELIUM[a]

Wave number k (in a_0^{-1})	Phase shift (in radians)		
	η_0	η_1	η_2
0	(5.266)[b]	0	0
0.136	2.507	—	—
0.272	2.108	—	—
0.608	1.659	0.020	—
0.859	1.481	—	—
1.053	1.360	0.070	0.006
1.359	1.256	—	—
1.922	1.093	0.186	0.041
3	0.898	0.272	0.095
4	0.784	0.301	0.130
5	0.696	0.308	0.152

[a] Phase shifts calculated by numerical integration.
[b] Scattering length in a_0.

for the static field approximation and the exchange approximation have been calculated by Moiseiwitsch (1961) who obtained the values 5.27 a_0 and 1.442 a_0 respectively. These give for the total elastic cross section in the limit of zero energy the values 111 πa_0^2 and 8.32 πa_0^2, respectively. Evidently the effect of exchange is very important. Moiseiwitsch (1953) has also carried out a variational method calculation for the zero order partial wave employing the trial function:

$$F_0(r) = (kr)^{-1}(1 + a^2)^{-1/2} \times \{\sin kr + [a + b \exp(-Zr/a_0)][1 - \exp(-Zr/a_0)] \cos kr\} \tag{158}$$

where k is the wave number of the incident electron, and taking for the ground state wave function of the helium atom

$$\psi_0(r_2, r_3) = \frac{Z^3}{\pi a_0^3} \exp[-Z(r_2 + r_3)/a_0] \tag{159}$$

with $Z = 27/16$. Even with such a simple form of trial function the resulting phase shifts which are displayed in Table VII are quite accurate particularly in the case of the exchange approximation for which a scattering length of 1.46 a_0 is obtained.

TABLE VII

ELASTIC SCATTERING OF ELECTRONS BY HELIUM ATOMS IN THE GROUND STATE[a]

Wave number k (in a_0^{-1})	Phase shift η_0 (in radians)	
	Static field approximation	Exchange approximation
0	—	(1.46)[b]
0.136	2.34	2.95
0.272	1.96	2.75
0.608	1.58	2.31
0.859	1.43	2.03
1.053	1.34	1.84
1.359	1.22	1.60
1.922	1.07	1.29

[a] Zero order phase shift calculated by Hulthén's variational method.
[b] Scattering length in a_0.

In the limit of zero energy incident electrons the phase shift η_0 is equal to π radians. However, this does not mean that a stable helium negative ion exists (cf. § 1.6) due to the fact that the Pauli principle excludes the possibility of the third electron being bound in a $1s$ state and the potential field is insufficiently strong to bind the additional electron in a $2s$ state.

A third order distortion approximation calculation for the elastic scattering of electrons by helium atoms (cf. § 3.2) has been performed by Moiseiwitsch and Shields (1962). Their partial cross sections are in satisfactory agreement with those derived from the phase shifts calculated by McDougall (1932) except for the zero order partial wave.

4.2 HELIUM: COMPARISON BETWEEN THEORY AND EXPERIMENT

In Fig. 5 a comparison is made between the total elastic cross sections calculated by Morse and Allis (1933) and the laboratory data obtained by Ramsauer and Kollath (1932) and by Normand (1930) for slow electrons. The differential cross sections predicted by the calculations of Morse and Allis (1933) are compared with the experimental data of Ramsauer and Kollath (1932) and of Bullard and Massey (1931) in Fig. 6. Since Bullard and Massey do not give absolute values, their data have been multiplied by an arbitrary constant in order to obtain the best fit between experiment and theory.

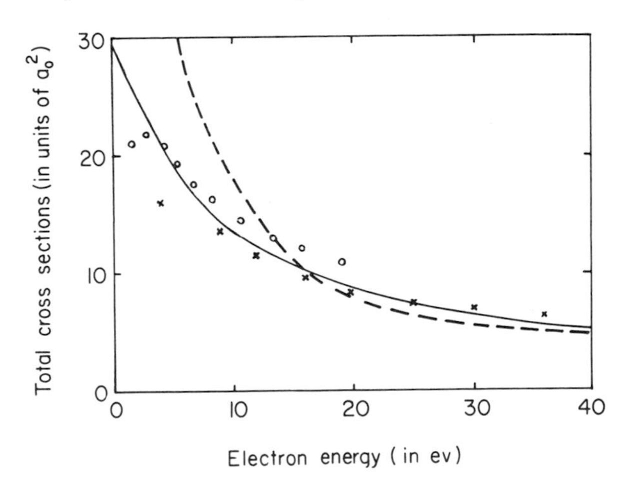

FIG. 5. Total cross sections for the elastic scattering of electrons by helium atoms. The full line and the broken line represent cross sections calculated by Morse and Allis (1933) with and without allowance for exchange, respectively. The circles and crosses denote the experimental data of Ramsauer and Kollath (1932) and of Normand (1930), respectively.

Evidently, good agreement between the calculations using the exchange approximation and the experimental data is obtained for both the total and differential cross sections. However, when exchange is neglected in the theory the agreement is rather poor at the lower energies.

Gabriel and Heddle (1960) have investigated the excitation of helium atoms by electrons. If their excitation cross sections are added to the ionization cross section of Smith (1930) and the sum subtracted from the total cross section data of Normand (1930), the total elastic cross section can be determined. It is found that if Normand's data are increased by 20 %, good agreement between the calculations of McDougall (1932) and experiment is obtained for electron energies greater than 20 ev.

For electron energies below about 2 ev, a considerable amount of experimental information has been obtained by the investigation of the behavior of electron swarms in helium gas. From microwave measurements of electrical conductivity (cf. § 2.4) Phelps *et al.* (1951) obtained the value 19 cm²/cm³ (0°C 1 mm Hg) for the probability coefficient of collision P_c for momentum transfer at the mean electron energy 0.04 ev, Gould and Brown (1954) obtained 18 ± 2 cm²/cm³ for electron energies

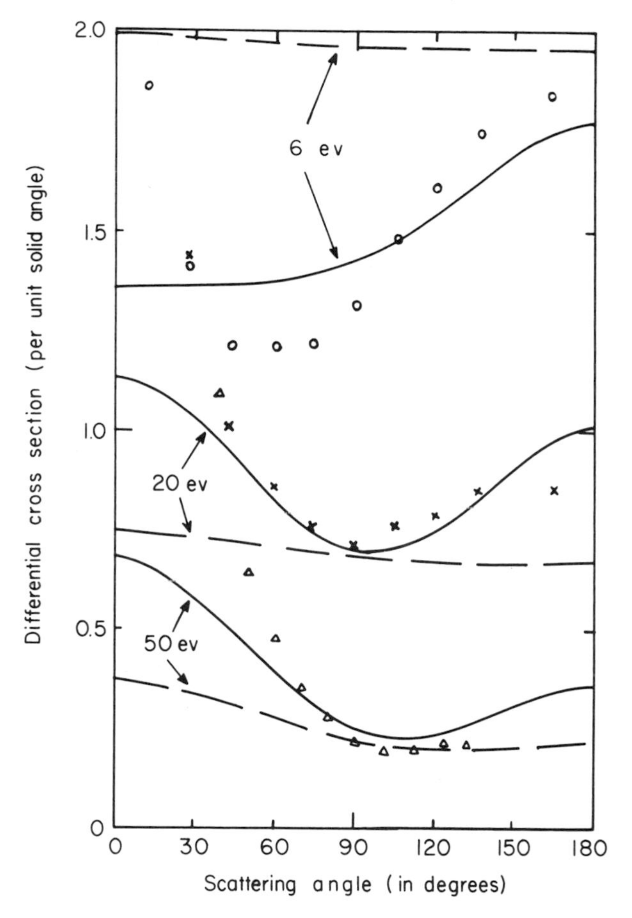

FIG. 6. Differential cross sections for the elastic scattering of electrons by helium atoms. The full line and the broken line represent cross sections calculated by Morse and Allis (1933) with and without allowance for exchange, respectively. The circles and crosses denote the experimental data of Ramsauer and Kollath (1932) and the triangles denote the data of Bullard and Massey (1931).

between 0 and 0.75 ev, increasing to a maximum value of 19 ± 2 cm²/cm³ at 2.2 ev, and Anderson and Goldstein (1956) obtained 23-26 cm²/cm³ in the range of electron energies between 0.04 and 0.4 ev. By measuring drift velocities Bowe (1960) obtained 28 ± 1 cm²/cm³ for P_c for electron energies between 0.13 and 4 ev, and Phelps *et al.* (1960) obtained

22 cm^2/cm^3 for electron energies below about 2 ev.† The latter authors found good agreement with the drift velocity measurements of Townsend and Bailey (1923), Nielsen (1936), and Hornbeck (1951) for E/p values greater than 0.05 volt cm^{-1}/mm Hg, but not with the measurements of Townsend and Bailey for E/p in the range 0.01-0.05 volt cm^{-1}/mm Hg. Apart from this early work of Townsend and Bailey for low values of E/p, the experimental data quoted above is consistent with a constant value of P_c for electrons with energies less than about 2 ev.

From Fig. 5 we see that the total elastic cross section for electrons of zero energy calculated by Morse and Allis (1933) using the exchange approximation is equivalent to a probability coefficient of collision P_c of magnitude 30 cm^2/cm^3 which is somewhat larger than most of the experimental data for P_c. The total elastic cross section for zero energy electrons given by the numerical calculation of the scattering length by Moiseiwitsch (1961) is equivalent to a probability coefficient of collision equal to 26 cm^2/cm^3. This is in fairly satisfactory agreement with the experimental data at low electron energies, the remaining discrepancy being probably due to the neglect of polarization in the calculations.

4.3 Inert Gases

A detailed discussion of the early experimental and theoretical work on the elastic scattering of electrons by the inert gases has been given by Massey and Burhop (1952) and need not be repeated here.

A recent source of information concerning the scattering of electrons by the inert gases comes from the observations of drift velocities of electrons carried out by Bowe (1960) using the technique introduced by Hornbeck (1951) (cf. § 2.4). The momentum transfer cross sections Q_D derived by him are in satisfactory accord with the calculated values of Q_D obtained from the measurements of the total cross section carried out by Ramsauer and Kollath (1932).

Microwave measurements of the collision probability coefficient for electrons with a mean energy of 0.04 ev have been made by Phelps *et al.* (1951) who obtained the values 3.3, 2.1, 54, and 180 cm^2/cm^3 (0°C 1 mm Hg) for neon, argon, krypton, and xenon, respectively.

Kivel (1959) has calculated the total elastic cross section for the case of argon employing essentially the same method as Holtsmark (1929) involving the use of a Hartree atomic field modified by the inclu-

† Pack and Phelps (1961) have reported the value 19 cm^2/cm^3 for P_c at mean electron energies in the range 0.003-0.05 ev.

sion of a polarization term behaving as $-\alpha/2r^4$ for large distances r from the atomic nucleus, α being the polarizability of the atom. He obtained good agreement with the total cross section measurements of Ramsauer and Kollath (1929). However, at very low energies he found that the effect of the tail of the polarization potential results in a large increase in the cross section. Thus, he obtained 27, 14, 2, and $5a_0^2$ for the zero order partial cross section Q_0 at the electron energies 0, 0.02, 0.55, and 1.2 ev, respectively. His cross section curve lies considerably above the value for the collision probability obtained by Phelps *et al.* (1951) for argon at 0.04 ev.

4.4 Positron Scattering by Inert Gases

Teutsch and Hughes (1956) have derived cross sections for momentum transfer for positrons in helium, neon, and argon by analyzing the data of Marder *et al.* 1956) on the slowing down of positrons in the inert gases. At a mean positron energy of approximately 18 ev they found 0.023, 0.12, and $1.5\pi a_0^2 \pm 25\%$ for helium, neon, and argon, respectively.

These values are much smaller than were obtained by Massey and Moussa (1958) who calculated the elastic cross sections for the scattering of positrons by helium, neon, and argon by numerical integration of the appropriate differential equations employing the static field approximation with self-consistent atomic fields. Thus, for positrons having 13.6 ev energy, Massey and Moussa obtained 0.53, 2.25, and $5.21\pi a_0^2$, respectively. It appears therefore that polarization effects are very important. In the case of the scattering of positrons by hydrogen atoms, it was found by Spruch and Rosenburg (1960) that the effect of polarization changed the sign of the scattering length and at the same time increased its absolute magnitude. However, Allison *et al.* (1961), using the variational method, have shown that for the scattering by helium atoms the introduction of a term corresponding to dipole polarizability decreases the absolute magnitude of the scattering length though changing its sign. They find a total elastic cross section for zero energy positrons of $0.1\pi a_0^2$ which is still rather larger than that found experimentally at 18 ev.

4.5 Oxygen and Nitrogen

The first detailed calculations on the elastic scattering of electrons by atomic oxygen were carried out by Bates and Massey (1943, 1947). They considered the case of a free electron with zero angular momentum moving in the field of an oxygen atom in the lowest ^3P term of the ground $2p^4$ configuration. Thus, there arise two states of the whole system, a ^2P state and a ^4P state. Bates and Massey solved the relevant Hartree-Fock equations by numerical integration allowing for polarization by introducing an empirical polarization potential term having the form

$$V_P(r) = -\tfrac{1}{2}\sum_{n,l} \alpha_{n,l}/(r_{n,l}^2 + r^2)^2 \tag{160}$$

where $\alpha_{n,l}$ and $r_{n,l}$ are the polarizability and radius of the shell of the oxygen atom with principal quantum number n and azimuthal quantum number l. For large r this term behaves like $-\alpha/2r^4$ where $\alpha = \Sigma_{n,l}\,\alpha_{n,l}$ is the polarizability of an oxygen atom. Bates and Massey chose $\alpha = 5.69e^2a_0^3$ which they obtained by using the method described by Buckingham (1937).

A more recent calculation has been carried out by Temkin (1957) who employed the method of polarized orbitals (cf. § 3.9). He made allowance for the d orbitals of the oxygen atom which contribute $3.6e^2a_0^3$ to the polarizability. This is about 70 % of the semiempirical value $5.2e^2a_0^3$ for the polarizability of atomic oxygen (Alpher and White, 1959).

The zero order partial cross section for the elastic scattering of electrons by atomic oxygen in the ^3P term of the ground configuration is given by

$$Q_0 = \tfrac{1}{3}\{Q(^2P) + 2Q(^4P)\} \tag{161}$$

where $Q(^2P)$ and $Q(^4P)$ are the cross sections corresponding to doublet and quartet scattering. The zero order partial cross sections obtained by Bates and Massey (1947) and by Temkin (1957) are illustrated in Fig. 7. The exchange-polarization terms which are included in the method of polarized orbitals increase the zero order partial cross section by only 15 %. However, the cross section is very sensitive to the value of the polarizability α which is assumed. Thus, a large increase in the zero order partial cross section results from using $\alpha = 3.6e^2a_0^3$ as chosen by Temkin (1957) rather than $\alpha = 5.7e^2a_0^3$ as chosen by Bates and Massey.

Klein and Brueckner (1958) have used a similar approach to that employed by Bates and Massey to investigate the elastic scattering of electrons by atomic oxygen. They chose $\alpha = 5.589\, e^2a_0^3$ so as to give the value 1.45 ev for the electron affinity of an oxygen atom determined experimentally by Branscomb and Smith (1955). With this value of the polarizability they obtained a scattering length $a = 1.613a_0$ and an effective range $r_0 = 0.860a_0$ for the doublet case of the elastic scattering of electrons by oxygen atoms.

From simultaneous measurements of the dc conductivity and the microwave absorption coefficient of atomic oxygen in a shock tube, Lin and Kivel (1959) have been able to determine the collision frequency

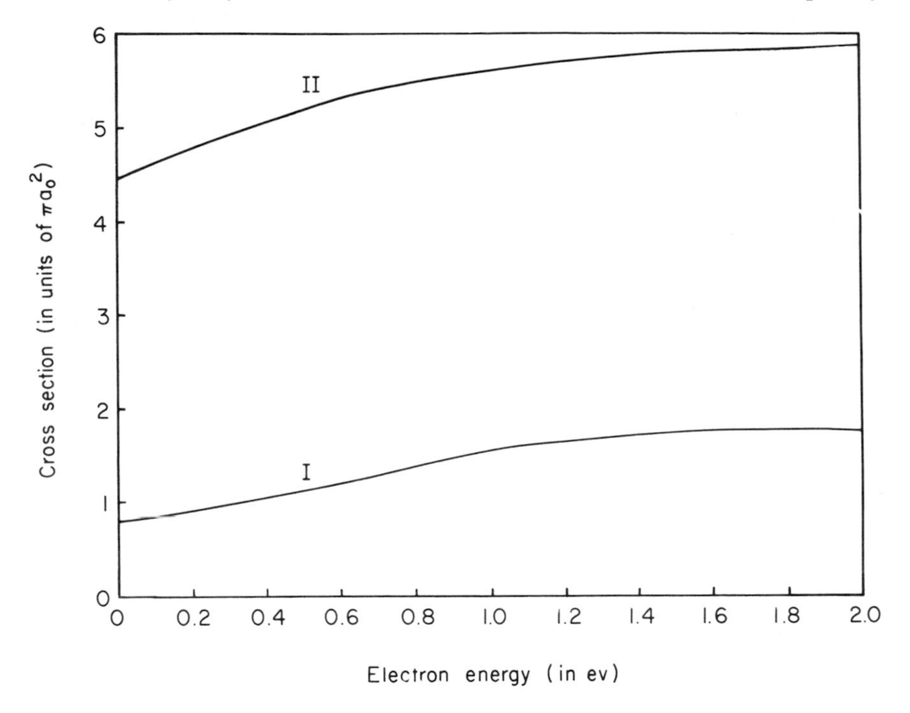

FIG. 7. Zero order partial cross sections for the elastic scattering of electrons by oxygen atoms. Curve I, cross section calculated by Bates and Massey (1947); curve II, cross section calculated by Temkin (1957).

of electrons in atomic oxygen giving a cross section of about $2\pi a_0^2$ for a mean electron energy of 0.5 ev. This is in fairly satisfactory agreement with the calculations of Bates and Massey but not with those of Klein and Brueckner.

Neynaber *et al.* (1961a) have measured the total cross section for the elastic scattering of electrons by oxygen atoms using a crossed beam technique similar to that described in § 2.5. They find a cross section which is nearly constant at $6.2 \pm 0.5\pi a_0^2$ over the range of energies from 2.3 to 11.6 ev. This value is in satisfactory agreement with an extrapolation of the cross section calculated by Bates and Massey (1947) if allowance is made for the p wave contribution.

Klein and Brueckner have also investigated the elastic scattering of electrons by atomic nitrogen. With a polarizability $\alpha = 7.084e^2a_0^3$

(which is slightly less than the value $7.6e^2a_0^3$ measured by Alpher and White, 1959), they obtained a scattering length $a = 1.857a_0$ and an effective range $r_0 = 0.845$ for the triplet scattering case.

4.6 Atomic Ions

In the case of the elastic scattering of electrons by positive ions with net charge Z, it can be shown (Massey, 1956) that the asymptotic form of the radial wave function $\phi_l(r)$ is given by

$$\phi_l(r) \sim \sin\left(kr + \frac{Z}{k}\log 2kr - \tfrac{1}{2}l\pi + \sigma_l + \eta_l\right) \tag{162}$$

where

$$\sigma_l = \arg \Gamma(1 + l - iZ/k), \tag{163}$$

η_l being the phase shift and k the wave number.

Consider two incident electrons with wave numbers k and k' and zero order phase shifts η_0 and η_0', respectively. Then it has been shown by Bethe (1949), using a similar procedure to that employed in deriving (47), that to the second order in k and k'

$$\frac{\cot \eta_0}{1 - \exp(-2\pi Z/k)} - \frac{\cot \eta_0'}{1 - \exp(-2\pi Z/k')} + \frac{1}{\pi}\left(\log\frac{k'}{k} + Q\right) = \tfrac{1}{2}(k'^2 - k^2)\rho \tag{164}$$

where

$$Q = \frac{Z^2}{k'^2}\sum_{\nu=1}^{\infty}\frac{1}{\nu}\frac{1}{\nu^2 + Z^2/k'^2} - \frac{Z^2}{k^2}\sum_{\nu=1}^{\infty}\frac{1}{\nu}\frac{1}{\nu^2 + Z^2/k^2} \tag{165}$$

and ρ is the effective range.

The continuum state with wave number k' can be converted to a bound state with eigenenergy $-Z^2/n'^2$ by putting $\cot \eta_0' = i$ and $k' = iZ/n'$ where n' is the effective principle quantum number of the state.

On defining the quantum defect $\delta(n)$ by the equation

$$n' = n - \delta(n) \tag{166}$$

where n is the principle quantum number, we obtain

$$\frac{\cot \eta_0}{1 - \exp(-2\pi Z/k)} = \cot \pi\delta(n) \tag{167}$$

for small k and large n.

It follows (Seaton, 1955) that

$$\eta_0(k = 0) = \pi\delta(n = \infty). \tag{168}$$

By considering the $1sns\,^1S$ and $1sns\,^3S$ series of He, Seaton (1957b) has shown that the phase shifts for the elastic scattering of electrons by He^+ ions are given by

$$\eta_0^+(k) = 0.437 - 0.109k^2$$

$$\eta_0^-(k) = 0.932 - 0.126k^2 \tag{169}$$

where the "+" and "−" refer to the symmetric and antisymmetric space wave function cases, respectively. These values are in satisfactory agreement with the phase shifts calculated by Bransden and Dalgarno (1953) using Hulthén's variational method, who obtained the values $\eta_0^+ = 0.419$ and $\eta_0^- = 0.959$ for $k^2 = 0.241$ using the exchange approximation.

5 *Scattering by Molecules*

5.1 Hydrogen

The scattering of slow electrons by hydrogen molecules is similar to the scattering of slow electrons by helium atoms since in both cases the target system consists of two electrons bound in a singlet spin state. It follows that in the molecular case we may expand the wave function of the total system in the symmetrized form (157) where now ψ_0 is the ground $^1\Sigma$ state wave function of the hydrogen molecule.

For the investigation of the scattering of electrons by homonuclear diatomic molecules, it is convenient to employ spheroidal coordinates λ, μ, ϕ where

$$\lambda = \frac{r_A + r_B}{d}, \qquad \mu = \frac{r_A - r_B}{d}, \tag{170}$$

r_A and r_B being the distances of the incident electron from the nuclei A and B, respectively, and d being the distance between the nulcei.

A plane wave $\exp ik\mathbf{n}_0 \cdot \mathbf{r}$ representing an electron moving with linear momentum $\hbar k$ in the direction of the unit vector $\mathbf{n}_0$ with polar angles

θ_0, ϕ_0 relative to the internuclear axis AB can be expanded in the form (Morse, 1935, Stratton *et al.*, 1956)

$$\exp ik\mathbf{n}_0 \cdot \mathbf{r} = \sum_m \sum_l a_{ml}\, S_{ml}(c, \cos\theta_0)\, S_{ml}(c, \mu)\, R_{ml}(c, \lambda) \cos m(\phi - \phi_0) \tag{171}$$

where $c = \frac{1}{2}kd$, m and l are zero or positive integers, and the functions S_{ml} are spheroidal harmonics satisfying the equation

$$\frac{d}{d\mu}\left\{(\mu^2 - 1)\frac{d}{d\mu} S_{ml}\right\} + \left\{A_{ml} + c^2\mu^2 - \frac{m^2}{\mu^2 - 1}\right\} S_{ml} = 0. \tag{172}$$

If θ is the angle between the internuclear axis AB and the line joining the midpoint of AB to the incident electron, then

$$S_{ml}(c, \mu) \to P^m_{m+l}(\cos\theta) \qquad \text{as} \qquad d \to 0.$$

The asymptotic behavior of R_{ml} for large λ is given by

$$R_{ml}(c, \lambda) \sim \frac{1}{c\lambda} \sin\{c\lambda - \tfrac{1}{2}(m + l)\pi\}. \tag{173}$$

If exchange is neglected and if the static potential due to the molecule is chosen to have the form

$$V(\mathbf{r}) = \frac{\hbar^2}{2m}\frac{f(\lambda)}{\lambda^2 - \mu^2}, \tag{174}$$

the wave function $F_0(\mathbf{r})$ of the incident electron may be expanded in the form

$$F_0(\mathbf{r}) = \sum_m \sum_l a_{ml} \exp(i\eta_l^m)\, S_{ml}(c, \cos\theta_0)\, S_{ml}(c, \mu)\, T_{ml}(c, \lambda) \cos m(\phi - \phi_0) \tag{175}$$

where T_{ml} satisfies the equation

$$\frac{d}{d\lambda}\left\{(\lambda^2 - 1)\frac{d}{d\lambda} T_{ml}\right\} + \left\{A_{ml} + c^2\lambda^2 - \frac{m^2}{\lambda^2 - 1} - \frac{d^2}{4} f(\lambda)\right\} T_{ml} = 0 \tag{176}$$

and has the asymptotic form for large λ

$$T_{ml}(c, \lambda) \sim \frac{1}{c\lambda} \sin\{c\lambda - \tfrac{1}{2}(m + l)\pi + \eta_l^m\}. \tag{177}$$

Then

$$F_0(\mathbf{r}) \sim \exp ik\mathbf{n}_0 \cdot \mathbf{r} + \frac{2}{d\lambda} e^{ic\lambda} f_0(\mu, \phi; \theta_0, \phi_0) \tag{178}$$

for large λ, where

$$f_0(\mu, \phi; \theta_0, \phi_0) = \frac{1}{2ik} \sum_m \sum_l a_{ml}(-i)^{m+l} [\exp(2i\eta_l^m) - 1] \times S_{ml}(c, \cos\theta_0)\, S_{ml}(c, \mu) \cos m(\phi - \phi_0). \tag{179}$$

On averaging over all orientations of the molecular axis and integrating the resulting mean differential cross section over all elements of solid angle into which the incident electron is deflected, it is found that the total elastic cross section is given by (Stier, 1932; Fisk, 1936)

$$Q_0 = \sum_m \sum_l q_{ml} \tag{180}$$

where

$$q_{ml} = \frac{4\pi}{k^2} (2 - \delta_{0m}) \sin^2 \eta_l^m. \tag{181}$$

As $k \to 0$ all the partial cross sections q_{ml} vanish except for q_{00}.

Massey and Ridley (1956) have calculated the phase shift η_0^0 for the scattering of slow electrons by hydrogen molecules both with and without exchange included. They employed Hulthén's and Kohn's variational methods with the trial function

$$F_t(\lambda, \mu, \phi) = (1 + a^2)^{-1/2} \frac{1}{c(\lambda - 1)} [\sin c(\lambda - 1) + \{a + b\, e^{-\gamma(\lambda-1)}\} \times \{1 - e^{-\gamma(\lambda-1)}\} \cos c(\lambda - 1)]. \tag{182}$$

This trial function is independent of μ which is a satisfactory approximation at low incident electron energies since $S_{00}(c, \mu) \cong 1$ for $c < 1$. As $\lambda \to \infty$ we see that

$$F_t \sim \frac{1}{c\lambda} \sin(c\lambda + \eta_0^0) \tag{183}$$

where

$$\eta_0^0 = \tan^{-1} a - c. \tag{184}$$

For the ground state wave function of the hydrogen molecule, Massey and Ridley (1956) used the form derived by Coulson (1938)

$$\psi_0(\mathbf{r}_2, \mathbf{r}_3) = \phi(\mathbf{r}_2)\, \phi(\mathbf{r}_3) \tag{185}$$

where

$$\phi(\mathbf{r}) = N e^{-\gamma\lambda} (1 + \alpha\mu^2 + \beta\lambda) \tag{186}$$

with $\alpha = 0.2195$, $\beta = 0.07957$, $\gamma = 0.75$, and $N = 0.693$.

The values of the phase shift η_0^0 and the cross section q_{00} calculated by Massey and Ridley (1956) are given in Table VIII.

TABLE VIII

ELASTIC SCATTERING OF ELECTRONS BY HYDROGEN MOLECULES IN THE GROUND STATE[a]

	Static field approximation		Exchange approximation	
Wave number k (in a_0^{-1})	Phase shift η_0^0 (in radians)	Cross section q_{00} (in πa_0^2)	Phase shift η_0^0 (in radians)	Cross section q_{00} (in πa_0^2)
0.143	2.103	145	2.941	7.77
0.286	1.778	46.8	—	—
0.429	1.514	21.7	2.352	10.96
0.714	1.228	6.96	1.830	7.33
1.429	0.748	0.906	1.056	1.48

[a] Zero order phase shift and cross section calculated by Hulthén's variational method.

By expanding the nuclear potential of the hydrogen molecule in spherical harmonics retaining only the spherically symmetric term, Carter *et al.* (1958) reduced the determination of the ground state hydrogen molecular wave function to a self-consistent field problem for two electrons moving under the action of a uniform surface distribution of charge on a sphere of radius $d/2$. The phase shifts and partial cross sections for the elastic scattering of electrons by the hydrogen molecule calculated by Carter *et al.* (1958) employing their molecular wave function together with (a) the static field approximation and (b) approximate allowance for exchange, are given in Table IX. Their zero order phase shifts are in good agreement with those obtained by Massey and Ridley (1956).† Both calculations show that the effect of exchange substantially decreases the total elastic cross section at low energies, a feature which also occurred in the case of the elastic scattering of electrons by helium atoms (cf. § 4.1). In fact, the phase shifts displayed in Table VII for helium are in fair accord with the zero order phase shifts obtained by Massey and Ridley (1956) and by Carter *et al.* (1958) for the hydrogen molecule. This is to be expected, owing to the similarity between the helium atom and the hydrogen molecule in their respective ground states.

† This is because orientation-dependent terms have little effect on the total elastic cross section for low energy electrons (cf. Arthurs and Dalgarno, 1960).

TABLE IX

ELASTIC SCATTERING OF ELECTRONS BY HYDROGEN MOLECULES IN THE GROUND STATE[a]

Static field approximation

Wave number k (in a_0^{-1})	Phase shifts (in radians)			Partial cross sections (in πa_0^2)			Total cross section (in πa_0^2)
	η_0	η_1	η_2	Q_0	Q_1	Q_2	
0.3	1.763			42.82			42.8
0.4	1.561			25.00			25.0
0.5	1.435	0.048		15.71	0.11		15.8
0.7	1.241	0.108		7.31	0.28		7.6
1.0	1.026	0.219	0.021	2.93	0.56	0.01	3.5
1.3	0.854	0.311	0.055	1.35	0.66	0.04	2.1

Exchange approximation

Wave number k (in a_0^{-1})	Phase shift η_0 (in radians)	Total cross section (in πa_0^2)
0.3	2.366	13.2
0.4	2.376	12.0
0.5	2.190	10.7

[a] Phase shifts and cross sections calculated by Carter *et al.* (1958).

A comparison between the calculated total elastic cross sections and those obtained in the laboratory is made in Fig. 8. It can be seen that when allowance is made for exchange, the agreement between theory and experiment is fairly satisfactory.

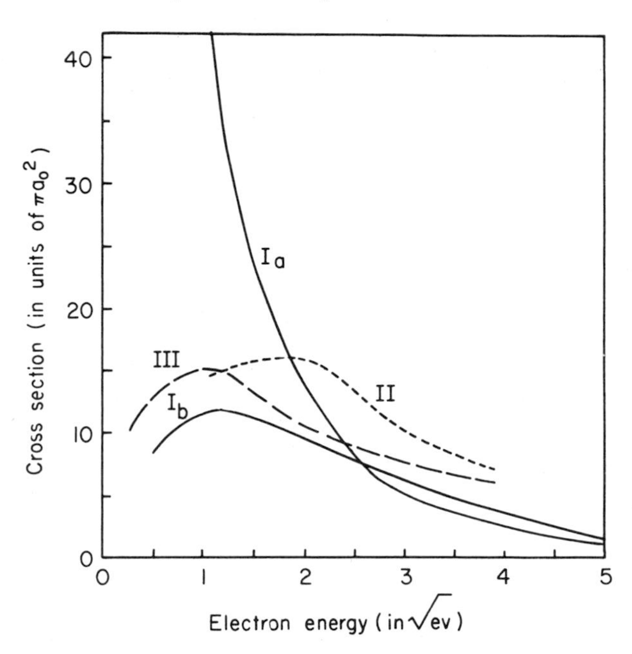

FIG. 8. Zero order partial cross sections for the elastic scattering of electrons by hydrogen molecules. Curves Ia and Ib represent cross sections calculated by Massey and Ridley (1956) using the static field approximation and the exchange approximation, respectively. Curves II and III represent the experimental data obtained using the Ramsauer method and the electron swarm diffusion method, respectively.

5.2 Complex Molecules

The early work on the elastic scattering of electrons by molecules has been described by Massey and Burhop (1952) and need not be given here

There has been no recent theoretical investigation of the scattering of electrons by molecules other than hydrogen. The only recent experimental work has been confined to the case of the N_2 molecule. By using microwave techniques (cf. § 2.3) Phelps *et al.* (1951) obtained 15 cm^2/cm^3 (0°C 1 mm Hg) for the probability coefficient of collision P_c for electrons with a mean energy of 0.04 ev while Anderson and Goldstein (1956) obtained 60 cm^2/cm^3 for electrons with a mean energy of 0.04 ev falling to 28 cm^2/cm^3 at 0.46 ev. Drift velocity measurements have been carried out by Chrompton and Sutton (1952) using the Townsend method (cf. § 2.4). At 0.04 ev their value of P_c is in accord with that obtained by Phelps *et al.* (1951), while at higher electron energies their results are in fair agreement with the early investigations of Brüche (1927), Ramsauer and Kollath (1930), and Normand (1930).

References

Allison, D. C. S., McIntyre, H. A.J., and Moiseiwitsch, B. L. (1961) *Proc. Phys. Soc.* **78**, 1169.

Alpher, R. A., and White, D. R. (1959) *Phys. of Fluids* **2**, 159.

Anderson, J. M., and Goldstein, L. (1956) *Phys. Rev.* **102**, 388, 933.

Arthurs, A. M., and Dalgarno, A. (1960) *Proc. Roy. Soc.* **A256**, 540.

Bates, D. R., and Massey, H. S. W. (1943) *Phil. Trans.* **A239**, 269.

Bates, D. R., and Massey, H. S. W. (1947) *Proc. Roy. Soc.* **A192**, 1.

Bederson, B., Malamud, H., and Hammer, J. (1957) *Bull. Amer. Phys. Soc.* [2] **2**, 172.

Bethe, H. A. (1949) *Phys. Rev.* **76**, 38.

Born, M. (1926) *Z. Phys.* **37**, 863; **38**, 803.

Bowe, J. C. (1960) *Phys. Rev.* **117**, 1416.

Brackmann, R. T., Fite, W. L., and Neynaber, R. H. (1958) *Phys. Rev.* **112**, 1157.

Bradbury, N. E., and Nielsen, R. A. (1936) *Phys. Rev.* **49**, 388.

Branscomb, L. M., and Smith, S. J. (1955) *Phys. Rev.* **98**, 1127.

Bransden, B. H., and Dalgarno, A. (1953) *Proc. Phys. Soc.* **A66**, 268.

Bransden, B. H., Dalgarno, A., John, T. L., and Seaton, M. J. (1958) *Proc. Phys. Soc.* **71**, 877.

Brüche, E. (1927) *Ann. Phys. (Leipzig)* **84**, 279.

Buckingham, R. A. (1937) *Proc. Roy. Soc.* **A160**, 94.

Bullard, E. C., and Massey, H. S. W. (1931) *Proc. Roy. Soc.* **A130**, 579.

Carter, C., March, N. H., and Vincent, D. (1958) *Proc. Phys. Soc.* **71**, 2.

Castillejo, L., Percival, I. C., and Seaton, M. J. (1960) *Proc. Roy. Soc.* **A254**, 259.

Chandrasekhar, S., and Breen, F. H. (1946) *Astrophys. J.* **103**, 41.

Chrompton, R. W., and Sutton, D. J. (1952) *Proc. Roy. Soc.* **A215**, 467.

Coulson, C. A. (1938) *Proc. Cambridge Phil. Soc.* **34**, 204.

Dalgarno, A., and Lewis, J. T. (1955) *Proc. Roy. Soc.* **A233**, 70.

Epstein, S. T. (1957) *Phys. Rev.* **106**, 598.

Erskine, G. A., and Massey, H. S. W. (1952) *Proc. Roy. Soc.* **A212**, 521.

Faxén, H., and Holtsmark, J. (1927) *Z. Phys.* **45**, 307.

Fisk, J. B. (1936) *Phys. Rev.* **49**, 167.

Foldy, L. L., and Tobocman, W. (1957) *Phys. Rev.* **105**, 1099.

Gabriel A. H., and Heddle, D. W. O. (1960) *Proc. Roy. Soc.* **A258**, 124.

Gilbody, H. B., Stebbings, R. F., and Fite, W. L. (1961) *Phys. Rev.* **121**, 794.

Gould, L., and Brown, S. C. (1954) *Phys. Rev.* **95**, 897.

Holtsmark, J. (1929) *Z. Phys.* **55**, 437.

Hornbeck, J. A. (1951) *Phys. Rev.* **83**, 374.

Huang, S.-S. (1949) *Phys. Rev.* **76**, 477.

Hulthén, L. (1944) *K. Fysiogr. Sällsk. Lund Förhandl.* **14**, 21.
Hylleraas, E. A. (1929) *Z. Phys.* **54**, 347.
Kingston, A. E., and Skinner, B. G. (1961) *Proc. Phys. Soc.* **77**, 724.
Kivel, B. (1959) *Phys. Rev.* **116**, 926.
Klein, M. M., and Brueckner, K. A. (1958) *Phys. Rev.* **111**, 1115.
Kohn, W. (1948) *Phys. Rev.* **74**, 1763.
Kohn, W. (1954) *Rev. mod. Phys.* **26**, 292.
Kruithof, A. A., and Ornstein, L. S. (1935) *Physica* **2**, 611.
Lin, S.-C., and Kivel, B. (1959) *Phys. Rev.* **114**, 1026.
Lindemann, H. (1935) Quoted by Kruithof and Ornstein (1935).
Lippman, B. A. (1956) *Phys. Rev.* **102**, 264.
McDougall, J. (1932) *Proc. Roy. Soc.* **A136**, 549.
Maecker, H., Peters, T., and Schenk, H. (1955) *Z. Phys.* **140**, 119.
Marder, S., Hughes, V. W., Wu, C. S., and Bennett, W. (1956) *Phys. Rev.* **103**, 1258.
Margenau, H. (1946) *Phys. Rev.* **69**, 508.
Massey, H. S. W. (1956) *In* "Handbuch der Physik" (S. Flügge, ed.), Vol. 36, II. Springer, Berlin.
Massey, H. S. W., and Burhop, E. H. S. (1952) "Electronic and Ionic Impact Phenomena." Oxford Univ. Press, London and New York.
Massey, H. S. W., and Khashaba, S. (1958) *Proc. Phys. Soc.* **71**, 574.
Massey, H. S. W., and Moiseiwitsch, B. L. (1951) *Proc. Roy. Soc.* **A205**, 483.
Massey, H. S. W., and Moussa, A. H. A. (1958) *Proc. Phys. Soc.* **71**, 38.
Massey, H. S. W., and Ridley, R. O. (1956) *Proc. Phys. Soc.* **A69**, 659.
Moiseiwitsch, B. L. (1953) *Proc. Roy. Soc.* **A219**, 102.
Moiseiwitsch, B. L. (1961) *Proc. Phys. Soc.* **77**, 721.
Moiseiwitsch, B. L., and Shields, D. B. (1962). To be published.
Morse, P. M. (1935) *Proc. Nat. Acad. Sci. U.S.A.* **21**, 56.
Morse, P. M., and Allis, W. P. (1933) *Phys. Rev.* **44**, 269.
Neynaber, R. H., Marino, L. L., Rothe, E.W., and Trujillo, S. M. (1961a) *Phys. Rev.* **123**, 148.
Neynaber, R. H., Marino, L. L., Rothe, E. W., and Trujillo, S. M. (1961b) *Phys. Rev.* **124**, 135.
Nielsen, R. A. (1936) *Phys. Rev.* **50**, 950.
Normand, C. E. (1930) *Phys. Rev.* **35**, 1217.
Ohmura, T., and Ohmura, H. (1960) *Phys. Rev.* **118**, 154.
Ohmura, T., Hara, Y., and Yamanouchi, T. (1958) *Progr. theor. Phys.* **20**, 82.
O'Malley, T. F., Spruch, L., and Rosenberg, L. (1961) *J. Math. Phys.* **2**, 491.
Pack, J. L., and Phelps, A. V. (1961) *Phys. Rev.* **121**, 798.
Pekeris, C. L. (1958) *Phys. Rev.* **112**, 1649.
Phelps, A. V., Fundingsland, O. T., and Brown, S. C. (1951) *Phys. Rev.* **84**, 559.
Phelps, A. V., Pack, J. L., and Frost, L. S. (1960) *Phys. Rev.* **117**, 470.
Ramsauer, C. (1921) *Ann. Phys. (Leipzig)* **64**, 513; **66**, 546.
Ramsauer, C., and Kollath, R. (1929) *Ann. Phys. (Leipzig)* **3**, 536.
Ramsauer, C., and Kollath, R. (1930) *Ann. Phys. (Leipzig)* **4**, 91.
Ramsauer, C., and Kollath, R. (1932) *Ann. Phys. (Leipzig)* **12**, 529.
Rosenberg, L., Spruch, L., and O'Malley, T. F. (1960) *Phys. Rev.* **119**, 164.
Seaton, M. J. (1955) *C. R. Acad. Sci. (Paris)* **240**, 1317.
Seaton, M. J. (1957a) *Proc. Roy. Soc.* **A241**, 522.
Seaton, M. J. (1957b) *Proc. Phys. Soc.* **A70**, 620.
Smith, P. T. (1930) *Phys. Rev.* **36**, 1293.

Spruch, L., and Rosenberg, L. (1960) *Phys. Rev.* **117**, 143.
Stier, H. (1932) *Z. Phys.* **76**, 439.
Stratton, J. A., Morse, P. M., Chu, L. J., Little, J. D. C., and Corbató, F. J. (1956) "Spheroidal Wave Functions." Wiley, New York.
Swan, P. (1954) *Proc. Roy. Soc.* **A228**, 10.
Temkin, A. (1957) *Phys. Rev.* **107**, 1004.
Temkin, A., and Lamkin, J. C. (1961) *Phys. Rev.* **121**, 788.
Teutsch, W. B., and Hughes, V. W. (1956) *Phys. Rev.* **103**, 1266.
Townsend, J. S. (1925) "Motion of Electrons in Gases." Oxford Univ. Press (Clarendon), London and New York.
Townsend, J. S., and Bailey, V. A. (1923) *Phil. Mag.* **46**, 657.
Watson, G. N. (1944) "Theory of Bessel Functions." Cambridge Univ. Press, London and New York.
Weiss, P. (1952) *Phys. Rev.* **87**, 226.

Note added in proof: Recent work on the elastic scattering of electrons has been reported in Abstracts of Papers, 2nd International Conference on the Physics of Electronic and Atomic Collisions, 1961. Benjamin, Inc., New York.

10.

The Motions of Slow Electrons in Gases

L. G. H. Huxley and R. W. Crompton

1 *Introduction*

In what follows we shall be concerned with an effective method, introduced by Townsend, for investigating the motions of slow electrons in gases, and some physical quantities associated with collisions between electrons and gas molecules.

In this method a stream of electrons, having entered a diffusion chamber through an aperture (usually small) in its cathode, proceeds through the gas (usually at pressure p of a few millimeters of mercury) to the anode in a uniform electric field $\mathbf{E}$. The mean speed of drift W of the electrons in the electric field $\mathbf{E}$ is much less than the average value of their speeds of random motion c.

As the stream proceeds to the anode, it also spreads by diffusion, with a coefficient of diffusion D, so that the current received by the anode is distributed broadly, in general, over its surface. By dividing the anode symmetrically (a central disk and one or more concentric annuli, or a central strip, flanked by parallel strips) into insulated sections, the ratio of the currents to any two sections can be measured and from this ratio the ratio W/D for the stream can be deduced.

It was proved by Townsend (1899) that for ions whose masses m are of the same order of magnitude as the masses M of molecules of a gas,

$$W/D = Ee/\kappa T = EN_0e/RT \tag{1}$$

in which N_0 is Loschmidt's number, e is the ionic charge, κ is Boltzmann's constant, R is the gas constant, and T is the absolute temperature of the ions.

Townsend (1899, 1900) used this formula to show, from the experimental values of W/D, that N_0e was the same for gaseous ions as for monovalent ions in electrolytes; at that time this was an important step in establishing the atomic nature of electricity.

In 1908 Townsend investigated the quantity W/D in streams of diffusing electrons (for which $m \ll M$) and found that abnormally small values of N_0e were obtained in dry gases. Since N_0e is constant, it follows that when electrons drift through a gas in a steady state of motion in a uniform electric field $\mathbf{E}$ their effective temperature T_e exceeds the temperature T of the gas molecules that appears in (1). The mean energy of agitation of an electron may therefore be written

$$\tfrac{1}{2}m\overline{c^2} = k \cdot \tfrac{3}{2}\kappa T = k \cdot \tfrac{1}{2}M\overline{C^2} \qquad (k \geqslant 1)$$

so that for electrons (1) becomes

$$\frac{W}{D} = \frac{EN_0e}{kRT} = \frac{3}{2}\frac{Ee}{(\frac{1}{2}m\overline{c^2})}. \tag{2}$$

If E is expressed in volts cm^{-1} and other quantities in cgs units, then on substituting the known value of N_0e, and adopting a standard temperature $T = 288°\text{K} = 15°C$, (2) becomes

$$W/D = 40 \cdot 3\,(E/k) \tag{3}$$

so that from measurements of W/D in a given field E, in a gas at a fixed pressure p, k may be deduced.

If in addition to the measurement of W/D, at some combination of values of E and p, the drift speed W is separately determined, then D can be found. As will appear in the sequel, (2) and (3) require correction.

The theory of the motions of ions and electrons in gases, confirmed by innumerable measurements, shows that (at a standard temperature T) the fundamental experimental parameter is not E or p individually, but their ratio E/p. For instance, the following quantities are functions of (E/p) only: W, pD, k, and W/pD, that is to say, the same values of these quantities are found with an electric field E and a pressure p as are found at fields rE with pressures rp. Experimental data are therefore conveniently represented, after correction to the standard temperature of 15°C, in the form of tables or curves in terms of the universal parameter E/p. Any one quantity may be expressed in terms of any other through their common dependence on E/p. Moreover, the dependence of a quantity, such as W, upon E/p provides a convenient means for checking the inner consistency of measurements made at different combinations of rE with rp.

An important quantity is $(\overline{c^2})^{1/2}$ which is related to k as follows. From

$$\tfrac{1}{2}m\overline{c^2} = k \cdot \tfrac{3}{2}\kappa T \qquad (T = 288°\text{K})$$

$$(\overline{c^2})^{1/2} = \left(\frac{3\kappa T}{m}k\right)^{1/2} = 1.15 \times 10^7\, k^{1/2}\ \text{cm sec}^{-1}. \tag{4}$$

The coefficient of diffusion D and speed of drift W of electrons in gases are given by the following theoretical formulae (e.g., Huxley 1957, 1960a, b):

$$D = \tfrac{1}{3}(\overline{lc}) = \frac{1}{3N}\overline{\left(\frac{c}{A}\right)} \tag{5}$$

$$W = \frac{Ee}{3m}\overline{c^{-2}\frac{d}{dc}(lc^2)} = \frac{e}{3}\left(\frac{E}{Nm}\right)\overline{c^{-2}\frac{d}{dc}\left(\frac{c^2}{A}\right)} \tag{6}$$

in which $l \equiv l(c)$ is an equivalent mean free path of an electron with speed c, and $A \equiv A(c)$ is the equivalent collision cross section of a molecule in a collision with an electron with speed c. The averages are taken with respect to the distribution of the speeds c. From these formulae the dependence of $A(c)$ upon c may be inferred from the experimental results.

We proceed, next, to a summary of the theory of the motions of ions and electrons in gases. In what follows the term ion is taken to include electrons, but in the case of electrons it may be assumed that m is always much less than M.

2 *Theory of Motion of Ions and Electrons in Gases*

It is assumed that the number n of ions in unit volume is much smaller than the number N of gas molecules in unit volume so that the mutual interactions of the ions are unimportant.

2.1 Random Motion and Mean Motion of a Group of Ions

In the absence of an electric field, the centroid of a group of ions in a gas is at rest and the directions of the agitational velocities of the ions are distributed isotropically, that is to say, in velocity space their representative points are distributed with spherical symmetry about the origin. Let the group comprise n_0 ions and let the number of velocity points residing within an element $du\,dv\,dw$ at the vector point $c(u, v, w)$ in velocity space be $n_0 f(c)\,du\,dv\,dw$ so that $n_0 f(c)$ is the density of points in velocity space. In this case, where $E = 0$, the distribution function $f(c)$ is given by Maxwell's formula

$$f(c) = \frac{1}{\alpha^3 \sqrt{\pi^3}} \exp\left(-\frac{c^2}{\alpha^2}\right) \tag{7}$$

and the number of points in a spherical shell with radius c and thickness dc with its center at the origin is

$$n_0 f(c) \cdot 4\pi c^2 dc = n_0 \frac{4}{\alpha^3 \sqrt{\pi}} \exp\left(-\frac{c^2}{\alpha^2}\right) \cdot c^2 dc. \tag{8}$$

Moreover, $\frac{1}{2} m\overline{c^2} = \frac{1}{2} M\overline{C^2}$ (equipartition of energy).

In the presence of an electric field **E** the motion is modified in important respects:

(a) $\frac{1}{2}m\overline{c^2} = k \cdot \frac{1}{2}M\overline{C^2}$; $k > 1$.

(b) The centroid of the group drifts through the gas with a velocity $\mathbf{W}(W \ll \bar{c})$ parallel to **E** on which is superimposed a random agitational motion whose distribution function $f(c)$ no longer conforms to (7).

It will be supposed that the distribution of points in velocity space is as follows. When $E = 0$, the density of points $n_0 f(c)$ is the same throughout a thin spherical shell with radius c with its center at the origin. In the modified distribution of points when $E \neq 0$, it will be supposed that the density of points is uniform within a thin displaced shell with radius c, whose center is not the origin but the end of a vector **V** parallel to **E** such that $V \equiv V(c)$. The resultant velocity $\mathbf{c} = \mathbf{c}_1 + \mathbf{V}$. Without loss of generality it may be assumed that **E** is directed along Ox so that **V** is along the u-axis in velocity space. Let the density of points in this shell be $n_0 f(c_1)$ so that the number of velocity points within an element $d\tau$ of velocity space at the end of the vector $\mathbf{c}_1$ is $n_0 f(c_1)d\tau$. But the element $d\tau$ is also at the end of the vector **c** from the origin with respect to which the distribution function is of the form $F(c, u)$. But,

$$c_1^2 = c^2 + V^2 - 2cV\cos\theta,$$

where $\cos\theta = u/c$. Let $V/c \ll 1$, then

$$c_1 \approx c\left(1 - \frac{V}{c}\cos\theta\right) = c - \frac{Vu}{c}.$$

It follows that

$$F(c, u) = f(c_1) = f\left(c - \frac{uV}{c}\right) \approx f(c) - \frac{uV}{c}\frac{df}{dc}. \tag{9}$$

Thus, the density of points in a thin spherical shell with radius c and center at the origin is not constant but of the form of a constant term modulated by a term proportional to $\cos\theta = u/c$. The total number of points within such a shell of thickness dc is $4\pi n_0 c^2 f(c)dc$, but the mean value of their u-component is obtained from

$$\bar{u}f(c) = c\frac{f(c)}{2}\int_0^{\pi}\cos\theta\sin\theta\, d\theta - \frac{cV}{2}\frac{df}{dc}\int_0^{\pi}\cos^2\theta\sin\theta\, d\theta = -\frac{cV}{3}\frac{df}{dc}. \tag{10}$$

The mean speed of the whole group, which is the drift speed W, is

$$W = -\frac{4\pi}{3}\int_0^{\infty} V(c)\frac{df}{dc}c^3 dc = \frac{4\pi}{3}\left[-fVc^3\Big|_0^{\infty} + \int_0^{\infty} c^{-2}\frac{d}{dc}(c^3V)fc^2 dc\right]$$

or

$$W = \tfrac{1}{3} c^{-2} \frac{d}{dc} \overline{(c^3 V)} \tag{11}$$

since $-fVc^3$ vanishes at both limits. Thus, in order to calculate W by this method it is necessary first to find the velocity $\mathbf{V}$ in terms of $\mathbf{E}$, c, and the equivalent collision cross section.

2.2 Specification of an Encounter

Consider two interacting particles with a relative speed g. If their masses are m and M, their speeds relative to their centroid X are, respectively, $r = [M/(M + m)]\, g$ and $R = [m/(M + m)]\, g$. In consequence of the interaction at close approach, the velocities $\mathbf{r}$ and $\mathbf{R}$ become $\mathbf{r}'$ and $\mathbf{R}'$ after the encounter, the angle between $\mathbf{r}$ and $\mathbf{r}'$ being θ (Fig. 1). The angle of deflection θ is a function $\theta(b)$ of the impact

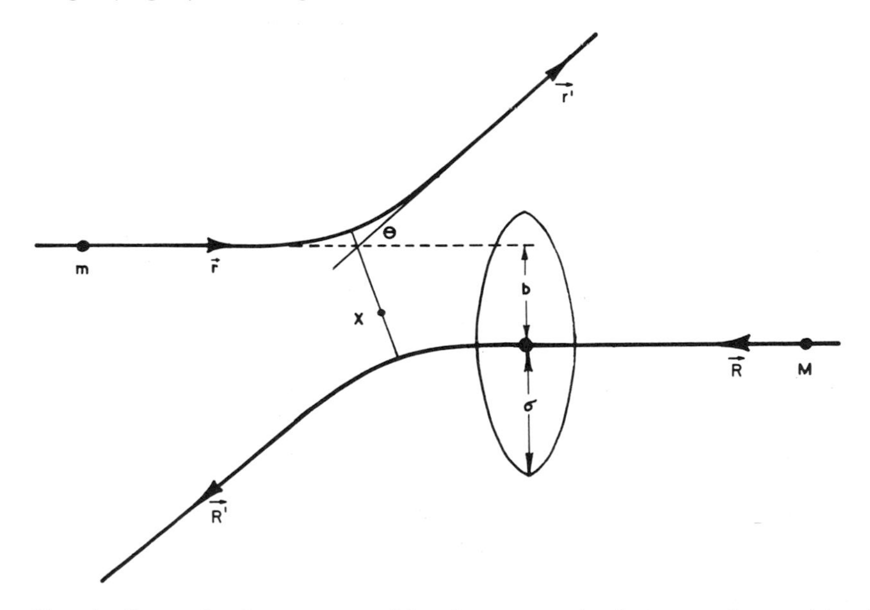

Fig. 1. Interaction between a particle of mass m and velocity $\mathbf{r}$ and a particle of mass M and velocity $\mathbf{R}$.

parameter b which is the perpendicular distance between the extension of their trajectories in the center of mass system before the encounter.

Suppose that $\theta(b) = 0$ when $b \geqslant \sigma$, then the mean value of $\cos\theta(b)$ is

$$\overline{\cos\theta} = \frac{2}{\sigma^2}\int_0^\sigma \cos\theta(b) b\, db. \tag{12}$$

Alternatively, if the proportion of particles m scattered within the elementary solid angle $2\pi\sin\theta\, d\theta$ is $\frac{1}{2}\phi(\theta)\sin\theta\, d\theta$ such that

$$\tfrac{1}{2}\int_0^\pi \phi(\theta)\sin\theta\, d\theta = 1,$$

then

$$\overline{\cos\theta} = \tfrac{1}{2}\int_0^\pi \phi(\theta)\cos\theta\sin\theta\, d\theta.$$

If $\phi(\theta) = 1$, then $\overline{\cos\theta} = 0$ and the scattering is termed isotropic. If g is the mean speed, relative to the molecules, of an ion with speed c, the mean time between encounters is l_0/g and the mean free path is $l \equiv l(c) = (c/g)\, l_0$, where $l_0 = 1/N\pi\sigma^2$.

3 *Diffusion and Drift*

3.1 Drift: Derivation of Formulae for V and W

Figure 2 is the vector diagram of an encounter between an ion with velocity $\mathbf{c}(\overrightarrow{OP})$ and a molecule with velocity $\mathbf{C}(\overrightarrow{OQ})$, their masses being m and M as before.

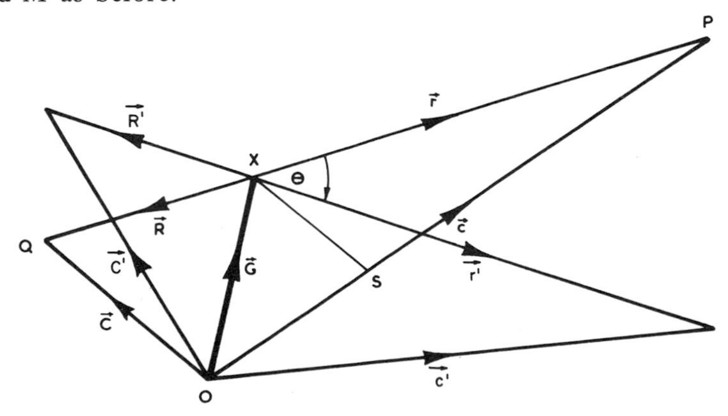

FIG. 2. Vector diagram of an encounter between an ion of mass m and velocity $\mathbf{c}$ and a molecule of mass M and velocity $\mathbf{C}$.

The velocity of the centroid is $\mathbf{G}(\overrightarrow{\mathrm{OX}})$ and $\mathbf{r}$ and $\mathbf{R}$ are the velocities of m and M relative to the centroid before the encounter. Thus, $\mathbf{r} + \mathbf{R} = \mathbf{g} = \overrightarrow{\mathrm{QP}}$; $r = [M/(M + m)]\, g$. In the encounter $\mathbf{r}$ and $\mathbf{R}$ are deflected through the angle θ to become $\mathbf{r}'$ and $\mathbf{R}'$ with $r' = r$ and $R' = R$, and the velocities of m and M become $\mathbf{c}'$and $\mathbf{C}'$. The vectors $\mathbf{r}'$ are distributed with axial symmetry about the direction of $\mathbf{r}$ so that the mean value of $\mathbf{r}'$ in a large number of encounters in which $\mathbf{c}$ and $\mathbf{C}$ are specified is $\mathbf{r}\,\overline{\cos\theta} = \alpha\mathbf{r}$, where $\alpha = \overline{\cos\theta}$ and $\overline{\cos\theta}$ is defined in (12).

Thus, the mean velocity of ions after encounters of this type is $\mathbf{G} + \alpha\mathbf{r}$. Draw XS parallel to QO to meet OP in S. Then OS $= [m/(M + m)] \cdot$ OP and S is a fixed point. Consequently, as $\mathbf{C}$ is allowed to assume all directions, X moves over the surface of a sphere with S as center. The mean value of

$$\mathbf{G} + \alpha\mathbf{r} = \mathbf{G} + \alpha[\overrightarrow{\mathrm{SP}} - \overrightarrow{\mathrm{SX}}] \text{ is } \overrightarrow{\mathrm{OS}} + \alpha\overrightarrow{\mathrm{SP}} = [(m + \alpha M)/(m + M)]\mathbf{c}.$$

The mean momentum residing in the ions after encounters when they possess momentum $m\mathbf{c}$ before an encounter is $[(m + \alpha M)/(M + m)]\, m\mathbf{c}$. and the u-component of this is $[(m + \alpha M)/(M + m)]\, mu$, where $c \equiv c(u, v, w)$.

The mean value of the residual momentum for all directions of c is, from (9),

$$-\left(\frac{m + \alpha M}{M + m}\right)\frac{mcV}{3}\frac{df}{dc}.$$

The mean momentum before the encounters is

$$-\frac{mcV}{3}\frac{df}{dc}.$$

Consequently, the mean momentum imparted to the molecules of the gas in encounters is

$$-\left[1 - \frac{m + \alpha M}{M + m}\right]\frac{mcV}{3}\frac{df}{dc} = -\frac{Mm}{M + m}(1 - \alpha)\frac{cV}{3}\frac{df}{dc}.$$

The total momentum imparted to the gas in unit time is

$$-\frac{4\pi n_0}{3}\frac{Mm}{(M + m)}\int_0^\infty \frac{g(1 - \alpha)}{l_0} c^3 V \frac{df}{dc}\, dc$$

$$= \frac{4\pi}{3} n_0 \left(\frac{Mm}{M + m}\right)\int_0^\infty c^{-2}\frac{d}{dc}\left(\frac{gc^3 V}{l'}\right) f\, c^2 dc,$$

where $l' = l_0/(1 - \alpha)$. In a steady state of motion this momentum is equal to the total momentum given to the group in unit time by the field, namely, $n_0 Ee$, whence,

$$\frac{4\pi}{3}\frac{Mm}{(M+m)}\int_0^\infty c^{-2}\frac{d}{dc}\left(\frac{gc^3V}{l'}\right)fc^2dc = Ee. \tag{13}$$

Since $4\pi\int_0^\infty fc^2dc = 1$, (13) is satisfied by

$$V = \frac{Eel'}{g}\left(\frac{1}{m}+\frac{1}{M}\right) = \frac{Eel}{mc}$$

where

$$l = \frac{l'(M+m)c}{Mg} = \frac{l_0(M+m)}{M(1-\alpha)}\frac{c}{g}\,. \tag{14}$$

When $m \ll M$ and $\alpha = 0$, $l = l_0$ and the velocities $\mathbf{c}'$ are scattered isotropically. Thus l is an equivalent mean free path in a model gas in which the ions are scattered isotropically in encounters and V is the same as in the actual gas. It follows from (11) and (14) that the drift speed is

$$W = \frac{Ee}{3m}\overline{c^{-2}\frac{d}{dc}(lc^2)}\,. \tag{15}$$

In the case of electrons ($m \ll M$) for which $g \to c$, according to (14) l in (15) becomes

$$l = l_0/(1-\alpha).$$

The equivalent collision cross section that corresponds to l as defined in (14) is

$$A \equiv A(c) = \frac{1}{Nl} = \frac{gM(1-\alpha)\pi\sigma^2}{c(M+m)} = \frac{gM}{c(M+m)}\cdot 2\pi\int_0^\sigma [1-\cos\theta(b)]b\,db \tag{16}$$

which, for electrons, becomes

$$A(c) = 2\pi\int_0^\sigma [1-\cos\theta(b)]b\,db.$$

When electrons or ions interact with molecules as point centers of repulsive force according to a law $P = k/r^\nu$, then, if $\nu > 2$, the integral $2\pi\int_0^\infty [1-\cos\theta(b)]b\,db$ is convergent and serves to define a cross section $A(c)$ through (16) such that $A(c) \propto c^{-4/(\nu-1)}$, (Chapman and Cowling, 1952, p. 171); for instance, when $\nu = 5$, $A(c) \propto c^{-1}$.

3.2 Diffusion

The random motion of the ions operates to diminish inequalities in the concentration n (number of particles per unit volume) and to disperse a group of ions throughout the gas. At a point on an elementary geometrical surface $d\mathbf{S}$ where grad n is not zero, there is a net flux of ions across $d\mathbf{S}$ which is a function of the components of grad n and is proportional to dS. When (grad n)/n is small, the flux in unit time (when no magnetic field is present) is $-D\,(\text{grad}\,n)\cdot d\mathbf{S}$. The coefficient of proportionality D is the coefficient of diffusion of the ions. The net number of ions that pass across $d\mathbf{S}$ in time dt, being the difference between those that cross in opposite senses, is therefore $-D(\text{grad}\,n)\cdot d\mathbf{S}dt$. It is convenient to introduce an equivalent convective velocity $\mathbf{w}$ that would cause the same number of ions to cross $d\mathbf{S}$ if grad n were zero, or diffusion were inhibited; thus,

$$n\mathbf{w}\cdot d\mathbf{S}dt = -D\,(\text{grad}\,n)\cdot d\mathbf{S}dt.$$

3.3 Formula for Coefficient of Diffusion D

Let the direction of grad n be $+Ox$, then

$$w = -\frac{D}{n}\frac{dn}{dx}.$$

Let $n_c = 4\pi n f(c)c^2dc$ be the number of ions in unit volume whose speeds lie in value between c and $c + dc$ and write

$$w_c = -\frac{D(c)}{n_c}\frac{dn_c}{dx}.$$

Consider a volume bounded by planes of unit area, normal to Ox, and at a separation dx. The mean u-component of the ions $n_c dx$ is w_c.

It was shown in § 2.2 that the mean residual momentum in the direction Ox carried by an ion after a collision is $[(m + \alpha M)/(M + m)]\,mu$, so that the mean momentum lost by an ion in an encounter is

$$mu\left[1 - \frac{m + \alpha M}{m + M}\right] = \frac{mM(1-\alpha)}{(m+M)}u.$$

Consequently, the mean momentum lost in time dt by the ions $n_c dx$ within the specified volume is

$$\frac{mM(1-\alpha)g}{(m+M)l_0}n_c w_c dt dx.$$

This loss of momentum to the volume is restored by the total net momentum in the direction $+ Ox$ transported to the volume by the ions that cross its faces in time dt. That is,

$$-\frac{m}{3} c^2 \frac{dn_c}{dx} dtdx,$$

so that on equating these expressions it follows that

$$D(c) = -\frac{nw_c}{dn_c/dx} = \tfrac{1}{3} lc \tag{17}$$

where, as in (14)

$$l = \frac{l_0(M+m)}{M(1-\alpha)} \cdot \frac{c}{g}.$$

For electrons, $l = l_0/(1-\alpha)$. The diffusion coefficient for the whole group of ions is the mean value of $D(c)$ taken with respect to the distribution of speeds c, namely,

$$D = \tfrac{1}{3}(\overline{lc}) = \frac{1}{3N}\overline{\left(\frac{c}{A}\right)}. \tag{18}$$

This formula can also be derived by use of the method of free paths (§ 1).

3.4 Diffusion in a Magnetic Field

Let the ions diffuse in a uniform and constant magnetic field parallel to $+ Oy$, and let $|\,\text{grad}\, n\,| = dn/dx$. The Lorentz force on the ion is $\mathbf{F} = e\mathbf{w}_c \times \mathbf{B}$; consequently, $\mathbf{w}_c$ is not parallel to $-$ grad n. The equations describing the balance of momentum are:

$$\begin{aligned} n_c\left[-\frac{mw_{cz}c}{l} + w_{cx}eB\right] &= 0 \\ -\frac{n_c m w_{cx} c}{l} - n_c w_{cz} eB - \frac{mc^2}{3}\frac{dn_c}{dx} &= 0 \end{aligned} \tag{19}$$

whence, with $\omega = -eB/m$,

$$n_c w_{cx} = -\frac{lc}{3(1+\omega^2 l^2/c^2)}\frac{dn_c}{dx} = -D_B(c)\frac{dn_c}{dx}.$$

The coefficient of diffusion in the direction $+ Ox$, therefore, is

$$D_B = \tfrac{1}{3}\overline{\left(\frac{lc}{1+\omega^2 T^2}\right)}; \qquad T = \frac{l}{c}. \tag{20}$$

Also,

$$n w_z = -\overline{\omega T} \cdot n w_x = \overline{\omega T D_B(c)} \frac{dn}{dx} . \tag{21}$$

The more general case in which grad n is arbitrarily directed, and B is directed along $+ Oy$, is described in matrix notation as follows:

$$-n \begin{bmatrix} w_x \\ \\ w_y \\ \\ w_z \end{bmatrix} = \begin{bmatrix} D_B & 0 & \overline{\omega T D_B(c)} \\ \\ 0 & D & 0 \\ \\ \overline{-\omega T D_B(c)} & 0 & D_B \end{bmatrix} \begin{bmatrix} \frac{dn}{dx} \\ \\ \frac{dn}{dy} \\ \\ \frac{dn}{dz} \end{bmatrix} . \tag{22}$$

3.5 Drift in an Alternating Electric Field with a Constant Magnetic Field

It will suffice to state a general formula without proof. Let the electric field vector rotate in the XOY plane with a phase such that at time $t = 0$ it is directed along $+ Ox$ whereas $\mathbf{B}$ is directed along $+ Oz$. The electric field is therefore designated by $E = E_0 \exp ipt$. It may be established, by an extension of the methods used previously, or by free path methods that the formulae for the speed V and the drift speed W (§ 3.1) are (Huxley, 1960a, b)

$$V = \frac{E_0 e \cdot \exp ipt}{m[\nu - i(\omega - p)]}$$

$$W = \frac{E_0 e \exp ipt}{3m} \cdot \overline{c^{-2} \frac{d}{dc} \frac{c^3}{\nu - i(\omega - p)}} \tag{23}$$

where

$$\nu = \frac{c}{l} = \frac{M(1 - \alpha)}{l_0(M + m)} g ; \qquad \omega = - \frac{Be}{m} .$$

This general formula contains some special cases of interest.

Case 1 : $p = 0$; $\omega = 0$. Formulae (23) reduce to (14) and (15), respectively.

Case 2 : *Magnetic deflection of a stream of ions.*
Put $p = 0$ and write $W = W_x + iW_y = W_0 \exp i\theta$.

Then,

$$W_x = \frac{E_0 e}{3m} \overline{c^{-2} \frac{d}{dc} \frac{\nu c^3}{\nu^2 + \omega^2}}$$

$$W_y = \frac{E_0 e}{3m} \overline{\omega c^{-2} \frac{d}{dc} \frac{c^3}{\nu^2 + \omega^2}}$$

$$\tan \theta = \frac{W_y}{W_x} = \omega \cdot \overline{c^{-2} \frac{d}{dc} \left(\frac{c^3}{\nu^2 + \omega^2}\right)} \Big/ \overline{c^{-2} \frac{d}{dc} \left(\frac{\nu c^3}{\nu^2 + \omega^2}\right)} \tag{24}$$

where θ is the angle of deflection of the stream.

In laboratory experiments where the pressure of the gas is of the order of 1 mm Hg and the field B a few tens of gauss, $\nu^2 \ll \omega^2$. In that event,

$$W_x \approx \frac{Ee}{3m} \overline{c^{-2} \frac{d}{dc} (lc^2)} = W$$

$$W_y \approx \frac{Ee}{3m} \overline{\omega c^{-2} \frac{d}{dc} (l^2 c)}$$

$$W = C \frac{E}{B} \tan \theta \tag{25}$$

where

$$3C = \left\{ \overline{\left[c^{-2} \frac{d}{dc} (lc^2) \right]} \right\}^2 \Big/ \overline{\left[c^{-2} \frac{d}{dc} (l^2 c) \right]},$$

a dimensionless quantity. It is possible to measure $\tan \theta$ and thus determine W if C can be calculated (Hall, 1955a).

Case 3 : High frequency conductivity of a weakly ionized gas.

This case is discussed by Huxley (1959) where it is pointed out that the usual Appleton-Hartree magneto-ionic formulae are inexact. In effect these formulae assume that the formula for W is that which gives V in (23). When $\nu = 0$ it can be seen that $W \rightarrow V$ and there is no error.

4 Equation of Continuity

In the present notation the equation of continuity for the concentration n is

$$\frac{dn}{dt} = \frac{\partial n}{\partial t} - \operatorname{div} n(\mathbf{w} + \mathbf{W}).$$

In the special case in which **B** is directed along $+ Oy$ and **E** is constant uniform, and directed along $+ Oz$, it can be deduced from § 3 that

$$\frac{dn}{dt} = \frac{\partial n}{\partial t} - W_x \frac{\partial n}{\partial x} - W_z \frac{\partial n}{\partial z} + D_B \left(\frac{\partial^2 n}{\partial x^2} + \frac{\partial^2 n}{\partial z^2} \right) + D \frac{\partial^2 n}{\partial y^2} \cdot \quad (26)$$

Thus, cross diffusion represented by the coefficient $\omega \overline{TD_B(c)}$ in (22) plays no part in the dispersion of ions by diffusion or drift.

When $\partial n / \partial t = 0$, as in a steady stream of ions free from the processes of electron attachment or ionization by collision and with $B = 0$, $W = W_z$, $D_B = D$,

$$\nabla^2 n = 2\lambda \frac{\partial n}{\partial z} \quad (27)$$

where $2\lambda = W/D$. If the substitution $n = e^{\lambda z} V$ is made, (27) reduces to

$$\nabla^2 V = \lambda^2 V$$

of which a useful solution corresponding to a pole source at the origin is

$$V = \text{const} \cdot e^{-\lambda r}/r.$$

The distribution due to a pole source at the origin emitting ions at the rate S in unit time is

$$n = \frac{S}{4\pi D} \exp - \lambda(r - z) \quad (28)$$

where $r^2 = x^2 + y^2 + z^2$. If an image source is placed at the point $(O, O, 2h)$ and such that the distribution due to it is

$$n' = - \frac{S}{4\pi D} \exp - \lambda(r' - z)$$

where

$$r'^2 = x^2 + y^2 + (z - 2h)^2,$$

the combined distribution

$$n = \frac{S}{4\pi D} (\exp \lambda z) \left[\frac{\exp - \lambda r}{r} - \frac{\exp - \lambda r'}{r'} \right] \quad (29)$$

is such that $n = 0$ over the plane $z = h$ where $r' = r$.

This distribution was shown (Huxley and Crompton, 1955) to correspond accurately to the distribution in a stream of electrons which enters the cathode of a diffusion chamber through a small hole and proceeds through the gas under the action of a uniform field **E** to a plane-parallel anode over which $n = 0$.

5 Mean Loss of Energy in Elastic Encounters

The velocities in Fig. 2 are related as follows:

$$\mathbf{c} = \mathbf{G} + \mathbf{r}; \qquad \mathbf{c}' = \mathbf{G} + \mathbf{r}'$$

whence,

$$c^2 = G^2 + r^2 + 2\mathbf{G} \cdot \mathbf{r}; \qquad c'^2 = G^2 + r'^2 + 2\mathbf{G} \cdot \mathbf{r}'$$

and,

$$c^2 - c'^2 = 2\mathbf{G} \cdot (\mathbf{r} - \mathbf{r}') + r^2 - r'^2.$$

We consider elastic collisions to be those in which $r' = r$, in which event

$$c^2 - c'^2 = 2\mathbf{G} \cdot (\mathbf{r} - \mathbf{r}').$$

Let $\mathbf{r}$ be allowed to assume all permissible directions, then its mean value (§ 3.1) is $\alpha\mathbf{r}$, and the mean value of $c^2 - c'^2 = 2(1 - \alpha)\, \mathbf{G} \cdot \mathbf{r}$. But

$$\mathbf{G} \cdot \mathbf{r} = \left(\frac{m\mathbf{c} + M\mathbf{C}}{m + M}\right) \frac{M}{m + M} (\mathbf{c} - \mathbf{C})$$

and it will be found, after reduction, that

$$c^2 - c'^2 = \frac{2M}{(m + M)^2}(1 - \alpha)\, [mc^2 - MC^2 + (M - m)\mathbf{c} \cdot \mathbf{C}].$$

The term $\mathbf{c} \cdot \mathbf{C}$ vanishes in the mean since the velocities $\mathbf{C}$ are distributed isotropically. Consequently, the mean loss of energy in elastic collisions, when c is specified, is

$$\tfrac{1}{2}\, mc^2 - \tfrac{1}{2}\, mc'^2 = \frac{2Mm(1 - \alpha)}{(m + M)^2}\, [\tfrac{1}{2}\, mc^2 - \tfrac{1}{2}\, M\overline{C^2}]. \tag{30}$$

6 The Distribution Function f(c)

When the electric field **E** is zero, the distribution function $f(c)$ is that of Maxwell as given in (7), in which

$$\alpha^2 = \frac{2\overline{c^2}}{3} = \frac{2M}{3m}\, \overline{C^2}$$

so that

$$M\overline{C^2}\frac{df}{dc} = -3mcf. \tag{31}$$

Consider another extreme case where E is not zero, but the molecules are at rest. The rate at which the group of ions $n_c = n4\pi f(c)c^2dc$ transfers energy to the molecules through elastic encounters is, from (30),

$$n_c \frac{g}{l_0} \cdot \frac{Mm}{(m+M)^2}(1-\alpha)mc^2 = 4\frac{c^3}{l}\frac{m^2}{(M+m)}\cdot fc^2dc \qquad (\overline{C^2}=0)$$

and the rate at which this is restored by the field is, from (9),

$$-4\pi Ee\frac{\overline{u^2}}{c}V\frac{df}{dc}c^2dc = -\frac{4\pi}{3}EeVc\frac{df}{dc}c^2dc$$

whence

$$EeV\frac{df}{dc} = -3\frac{m^2c^2}{l(M+m)}f$$

or, using (14),

$$(M+m)V^2\frac{df}{dc} = -3mcf. \tag{32}$$

Thus, the full equation, of which (31) and (32) are special cases, may be inferred to be

$$[(M+m)V^2 + M\overline{C^2}]\frac{df}{dc} = -3mcf \tag{33}$$

whence

$$f(c) = \text{const}\cdot\exp\left\{-3m\int^c \frac{c\,dc}{[(M+m)V^2 + M\overline{C^2}]}\right\}. \tag{34}$$

More complete and rigorous discussions will be found in Chapman and Cowling (1952), Loeb (1955), Ginzburg and Gurevič (1959), and Huxley (1960b).

In the case of electrons ($m \ll M$), $f(c)$ assumes the limiting form

$$f(c) = \text{const}\cdot\exp\left\{-\frac{3m}{M}\int^c \frac{c\,dc}{(V^2+\overline{C^2})}\right\} \tag{35}$$

and when $V^2 \gg \overline{C^2}$, as is usually the case,

$$f(c) = \text{const}\cdot\exp\left\{-\frac{3m}{M}\int^c \frac{cdc}{(Eel/mc)^2}\right\}. \tag{36}$$

If $l(c)$ is of the form $l(c) \propto c^r$, (36) with $4\pi \int_0^\infty f(c)c^2dc = 1$ is of the general type

$$4\pi f(c)c^2dc = \frac{4c^2}{\alpha^3 \Gamma(3/n)} \exp\left(-\frac{c^n}{\alpha^n}\right) dc \tag{37}$$

where $n = 4 - 2r$. If use is made of the standard integral

$$\int_0^\infty \exp(-y^n)y^m dy = \frac{1}{n}\Gamma\left(\frac{m+1}{n}\right),$$

it can be seen that the mean value of c^s is

$$\overline{c^s} = \alpha^s \Gamma\left(\frac{s+3}{n}\right) \Big/ \Gamma(3/n). \tag{38}$$

The form of $f(c)$ in a high-frequency field is obtained by using the following expression for V^2:

$$V^2 = \left(\frac{eE_0}{2m}\right)^2 \left[\frac{1}{\nu^2 + (\omega - p)^2} + \frac{1}{\nu^2 + (\omega + p)^2}\right].$$

When some encounters are inelastic and with $c > c_1$, V is replaced by aV, where $a = 1$ when $c < c_1$ and $a < 1$ when $c > c_1$.

7 *The Ratio W/D*

Attention was drawn to this important experimental quantity in § 1. From (15) and (18) it follows that

$$\frac{W}{D} = \frac{Ee}{m} \overline{\left[c^{-2}\frac{d}{dc}(lc^2)\right]} \Big/ [\overline{lc}] \tag{39}$$

and when the distribution function is that of Maxwell it can be seen, using integration by parts, that the factor

$$\overline{c^{-2}\frac{d}{dc}(lc^2)} = 3\overline{(lc)}/\overline{c^2}.$$

Consequently, in this case,

$$\frac{W}{D} = \tfrac{3}{2} Ee/(\tfrac{1}{2} m\overline{c^2}) = Ee/k\kappa T$$

$$= EN_0 e/kRT = 40 \cdot 3E/k \qquad (E \text{ in volt cm}^{-1};\ T = 288°\text{K}) \tag{40}$$

as was stated in § 1 [Eqs. (2) and (3)]. Formula (40) is also correct when the equivalent cross section $A(c) = 1/Nl$ is inversely proportional to c (inverse fifth-power repulsion) as may be easily demonstrated from (39).

In general, however, (40) is inexact since (39) is equivalent to

$$\frac{W}{D} = \frac{Ee}{\frac{1}{2}\,m\overline{c^2}} \cdot F = (\tfrac{2}{3}F)\frac{EN_0 e}{kRT} = \frac{40 \cdot 3E}{k_1}$$

where F is the dimensionless factor,

$$F = \frac{\overline{c^2} \cdot \overline{\left[c^{-2}\frac{d}{dc}(lc^2)\right]}}{2\overline{(lc)}} \qquad \text{and} \qquad k_1 = \frac{3k}{2F}. \tag{41}$$

The value of F depends upon the distribution function $f(c)$. If, for instance, in (37) n has the values $n = 2$, 4, and 6, then the corresponding values of F are, respectively, $\frac{3}{2}$ [Maxwell's distribution, cf. (40)], $1 \cdot 312$ (Druyvesteyn's distribution), and 1. The experimental quantity is therefore k_1, from which the true energy factor k can be deduced if F is known.

8 Measurement of Mean Energy Lost in an Encounter

In order to estimate the mean loss of energy in an encounter from measurements of the speed of drift W, it is necessary to know the distribution function $f(c)$ and the manner in which the equivalent mean free path $l(c)$ depends upon c.

With electrons the equivalent free path l is, according to (14), $l_0/(1 - \alpha)$ and if it is assumed that the scattering is isotropic ($\alpha = 0$), then $l = l_0$.

With this assumption the collision frequency is

$$4\pi \int_0^\infty \frac{c}{l} fc^2 dc = \overline{(c/l)}.$$

Let $Q = \frac{1}{2}\,m\overline{c^2}$ be the mean kinetic energy of the electrons and ΔQ the mean energy lost in an encounter. The mean energy transferred to the gas by an electron in unit time is, therefore, $\overline{(c/l)}\,\Delta Q$.

This energy, in a steady state of motion, is equal to the energy received in unit time in drifting in the electric field $\mathbf{E}$. If follows that

$$\Delta Q \overline{\left(\frac{c}{l}\right)} = EeW = \frac{3mW^2}{\overline{c^{-2}\frac{d}{dc}(lc^2)}}$$

whence

$$\Delta Q = \tfrac{1}{2} mW^2K \qquad \text{and} \qquad 1/K = \frac{1}{6}\left[\overline{\left(\frac{c}{l}\right)} \cdot \overline{c^{-2}\frac{d}{dc}(lc^2)}\right], \tag{42}$$

a dimensionless quantity. The fraction of the mean energy lost on the average in an encounter is

$$\frac{\Delta Q}{Q} = K\frac{W^2}{\overline{c^2}} = 7.65 \times 10^{-15}\frac{KW^2}{k}. \tag{43}$$

If, for example, $l \propto c$, then $K = 2$; if $l =$ constant, $K = 3/(\bar{c} \cdot \overline{c^{-1}})$ so that if $n = 4$ and 2 in (38), $K = 2 \cdot 54$ and $3\pi/4 = 2.37$; if $l \propto c^{-1}$, $K = 6/(\overline{c^{-2}} \cdot \overline{c^2})$ and with $n = 6$, $K = 3$.

The mean rate w at which energy is lost by an electron in collisions may be written

$$w/p = e(E/p)W = \psi(E/p) \tag{44}$$

where p is the pressure, since W is a function of E/p.

9 Experimental Procedure

The two macroscopic quantities that are measured (§ 1) are W/D and W from which D can also be derived. From measurements of W and pD as functions of the parameter E/p (§ 1), it is possible to derive k_1 (§ 7), $\bar{c}$, $\Delta Q/Q$ (§ 8), and $A(c)$ [(5) and (6)] all as functions of E/p. By eliminating E/p between two quantities, the one can be represented as a function of the other as, for instance, $\bar{A}$ as a function of $\bar{c}$.

9.1 Measurement of W/D

This quantity is derived from measurements of the divergence of a stream of electrons moving in a steady state of motion in a uniform electric field **E**. The form of apparatus that has been used in modern investigations (Fig. 3a) comprises a diffusion chamber bounded by a plane gold-plated cathode with a small central hole about 1 mm in diameter whose center is taken as the origin of coordinates, the cathode being the XOY plane. The other end of the diffusion chamber, the anode (also gold plated), is the plane $z = h$ and the anode is subdivided into insulated sections (Fig. 3b). In the more usual form of subdivision, there is a central disk and one or more annular sections all with the common center, the point $(0, 0, h)$. In an alternative method of division there is a central parallel-sided strip flanked symmetrically by similar

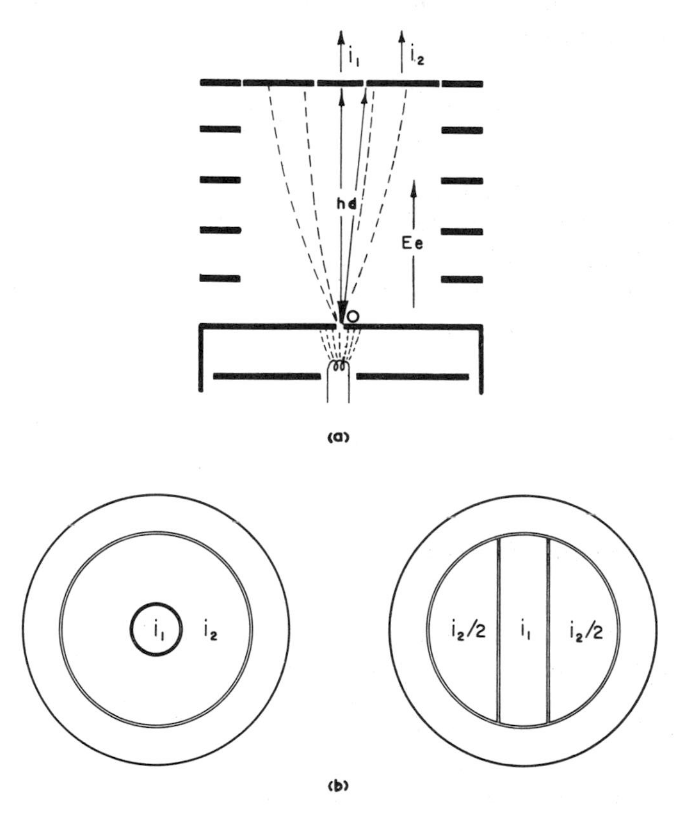

FIG. 3. Schematic diagram of a typical diffusion apparatus showing alternative modes of division of the anode.

strips, with the point $(0, 0, h)$ as the center of symmetry of the system. The stream of electrons is derived from a small filament above the hole in the cathode towards which the electrons are impelled by the same uniform field $\mathbf{E}$ as prevails in the diffusion chamber. Thus, the hole in the cathode acts as a point source of electrons and it is found that the distribution of the concentration n in the stream in the diffusion chamber is accurately represented by (29). The divergence of the stream is measured by the ratio $R = i_1/(i_1 + i_2)$ where i_1 and i_2 are the steady currents to two insulated sections of the anode. In the annular mode of division the currents most commonly measured are i_1, the current to the central disk, and i_2, the current to the remainder of the anode. The formula for the ratio is, in that instance,

$$R = 1 - \frac{h}{d} \exp - \lambda h \left(\frac{d}{h} - 1 \right) \tag{45}$$

in which h is the length of the diffusion chamber, d the distance from the origin (hole) to midway across the narrow gap separating the central disk and the surrounding annulus, and $2\lambda = W/D$.

Since (41), $W/D = (40 \cdot 3/k_1)E$ (at 15°C with E in volt cm^{-1}), it follows that (45) can be represented by a family of curves, each member of which displays R as a function of λh or Eh/k_1 for a given value of h/d. Thus, in an apparatus in which h can be altered, a family of curves can be prepared for each value of h. k_1 can thus be obtained directly from the measured value of R.

Similar curves can be prepared for other modes of division of the anode. The current $(i_1 + i_2)$ is kept small, less than 10^{-11} amp, in order to avoid a spurious spreading of the stream from the field of space charge.

There is good agreement between the measurements of k_1 as a function of E/p at different pressures, with different values of h and different modes of division of the anode.

9.2 Measurement of W

9.2.1 *Direct Method by Use of Electrical Shutters*

This method was first used to measure the drift velocities of ions but was modified by Bradbury and Nielsen (1936) to measure the drift velocities of electrons in a number of gases. Electrons released either from a hot filamant or by ultraviolet light from a metal surface move in a uniform electric field **E** to an electrode G_1 (Fig. 4a) with an aperture across which is stretched a grid of closely spaced and parallel wires. Alternate wires of the grid are connected, the whole forming an interleaved system of two insulated conductors (Fig. 4b). The application of a potential difference of a few volts between the two sets of wires creates a transverse electrical field across which the electrons cannot pass. If the field is alternating, the electrons are admitted to the drift chamber, in which the same field **E** is maintained, only during a short interval about the times when the alternating fields across the grid are zero. Groups of electrons thus enter the drift chamber every half-cycle of the alternating field. Each group traverses the length l of the drift chamber in time $t = l/W$ in a uniform field $\mathbf{E}$. The other electrode G_2 also has a central aperture and a grid system to which is applied the same alternating voltage. If the time $t = l/W$ is equal to an integral

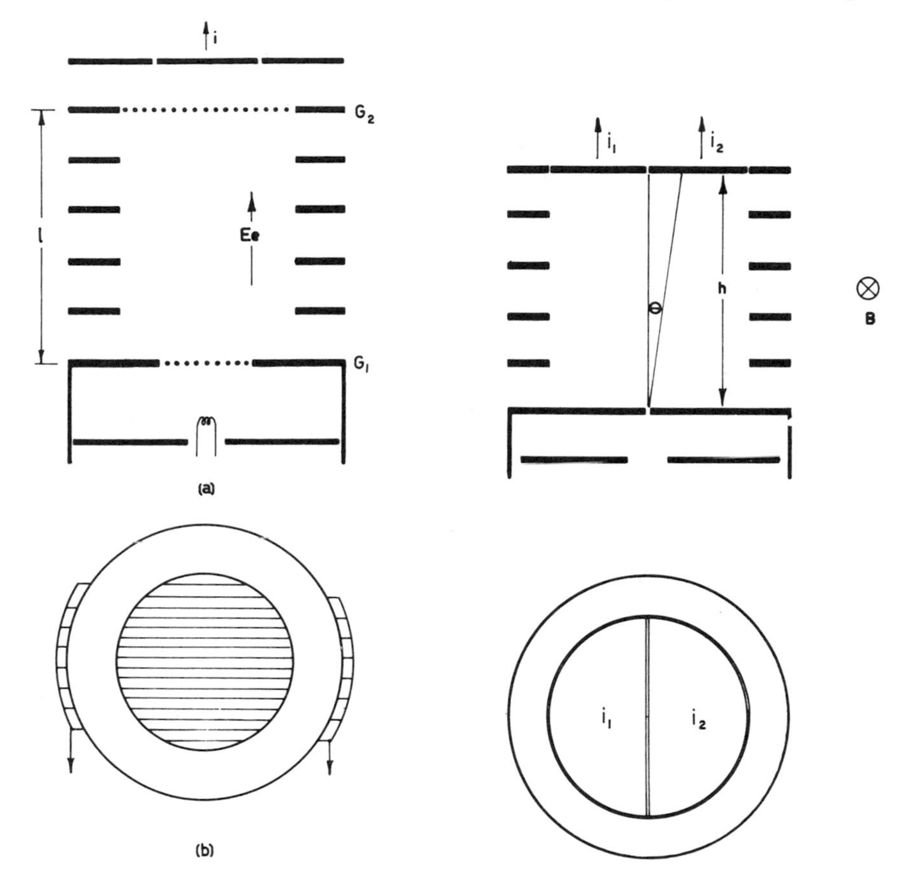

FIG. 4. Schematic diagram of a typical electrical shutter apparatus.

FIG. 5. Schematic diagram of an apparatus for measuring W by magnetic deflection.

multiple n of the half-period T of alternation, the group is able to pass through the second grid to a collecting electrode. Thus, $l/W = nT/2$; $W = 2l/nT = 2lf/n$. If the field $\mathbf{E}$ is held at a fixed value while f is continuously increased from some convenient small value, the current to the receiving electrode remains small until f reaches a value f_1 corresponding to $n = 1$. With further increase in f it exhibits a sequence of minima and maxima, the nth maximum corresponding to a frequency f_n.

Thus, in general, $W = 2lf_n/n$. In this way the internal consistency of the measurements can be checked from the degree of concordance of the values of W corresponding to different values of n.

W is found as a function of E/p (reduced to 15°C). It has recently become apparent that the precision of the measurements can be impaired by the diffusion of the groups as they travel within the diffusion space (Duncan, 1957; Crompton *et al.*, 1957). The greatest inaccuracies were shown to occur in short drift chambers and at small values of E/p. When, however, the appropriate precautions are taken, this method yields measurements of greater precision than any other. Experimental results are given in Brown (1959).

9.2.2 *Method of Magnetic Deflection*

Equation (25), $W = (CE/B) \tan \theta$, relates the drift speed W to the angle θ of deflection of an electron stream by a magnetic field $\mathbf{B}$ at right angles to $\mathbf{E}$ where C is a numerical factor.

This formula, with $C = 1$, was used by Townsend and Tizard (1912) to make the first measurements of W for electrons.

A modification of their procedure which is more flexible was used by Huxley and Zaazou (1949) and by Hall (1955a). The principle of the modified method is shown in Fig. 5. The receiving electrode of a diffusion chamber is bisected by a slit. When the magnetic field B is zero the currents i_1 and i_2 to the halves are equal but become unequal when $\mathbf{B}$ is applied parallel to the slit. The ratio $R_B = i_1/i_2$ can be expressed as a function of $\tan \theta$ and B so that $\tan \theta$ may be deduced from R_B and W from $\tan \theta$ by means of (25) if a value is assumed for C. Since W can be found reliably by the method of the electrical shutter, it would be a more profitable procedure to use W and $\tan \theta$ to determine C and thus obtain information about the distribution function $f(c)$. For instance, with Maxwell's distribution $C = 0.85$, but with that of Druyvesteyn $C = 0.943$.

10 Experimental Results

10.1 Law of Equipartition of Energy

According to the law of equipartition of energy, a particle moving at random in a gas in equilibrium and subject to no forces other than those experienced in encounters with molecules acquires an energy of random motion whose mean value is the same as that of a molecule of a gas.

The random speeds in a group of identical particles are distributed in accordance with Maxwell's formula [(7) and (8)] and, in the notation of (40) and (41), $k = k_1 = 1$.

It was found from measurements of W/D (§ 9.1) for a stream of monovalent ions of potassium moving in hydrogen, that when the ratio E/p was made very small, then $k_1 \to 1$.

For instance,

E/p volt cm^{-1} mm Hg^{-1}	0.3	0.6
k_1	1.002	1.004

The increase in k_1 with E/p actually represents a rise in the temperature of the ions above that of the gas, so that as $E/p \to 0$, $k_1 \to 1$.†

In order to obtain measurements of this precision, especial care was taken to obtain fields E of great uniformity and also to extrapolate the measured values of k_1 to zero current.

Conversely, if the validity of the law of equipartition is assumed, then the measurements show that the experimental methods and the associated mathematical analysis are trustworthy.

10.2 Townsend's Energy Factor k§

Monatomic gases

As no recent measurements have been made, the reader is referred to those made by Townsend and his school about 40 years ago and given in the treatises of Healey and Reed (1941) and Brown (1959). Reference to these works may be made for other parameters of electronic motion in monatomic gases.

Diatomic Gases

As explained in § 7 the measured quantity is k_1, which is the same as k when the speeds of agitation are distributed according to Maxwell's formula. In other cases (41) the true energy factor k is $k = \frac{2}{3} F k_1$. For example, in a Druyvesteyn distribution $k = \frac{2}{3} \times 1.312 k_1 = 0.875 k_1$. The results of recent measurements in air, hydrogen, deuterium, nitrogen, and oxygen are given in the form of curves in Figs. 6 and 7.

† The experiment to which reference is made was performed to check the validity of the experimental methods where great uniformity of the electric field could be assured. Since the construction of the apparatus was such that it is unlikely that the hydrogen used was pure, the results are quoted only as an illustration of the validity of the methods. In pure hydrogen the rise of k_1 with increasing E/p may differ from that indicated by the results in the table.

§ Equations (3), (4), and (40), (41).

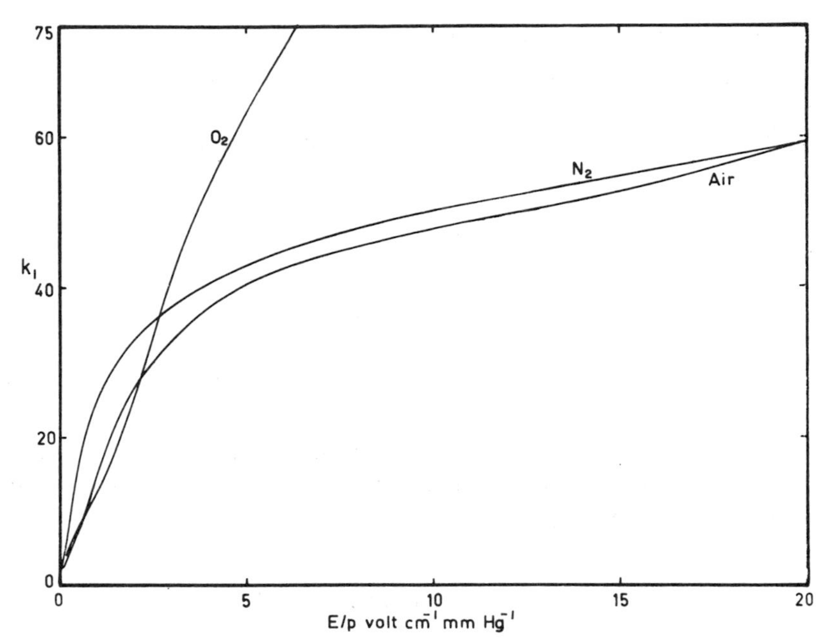

FIG. 6. Curves showing the variation of k_1 with E/p in air, nitrogen, and oxygen (O_2—Huxley *et al.*, 1959; N_2—Crompton and Sutton, 1952; air—Crompton *et al.*, 1953).

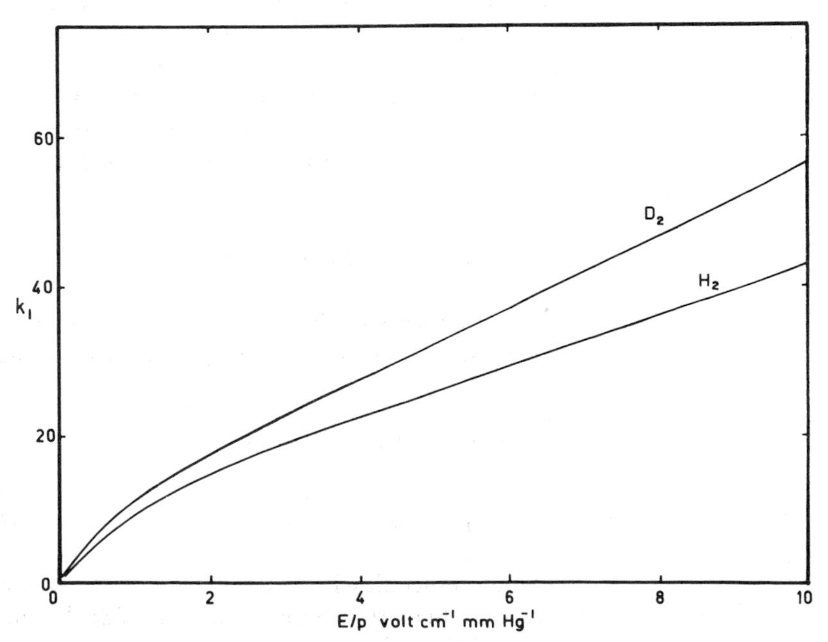

FIG. 7. Curves showing the variation of k_1 with E/p in deuterium and hydrogen (D_2—Hall, 1955; H_2—Crompton and Sutton, 1952).

The root-mean-square speed is (4)

$$(\overline{c^2})^{1/2} = 1.15 \times 10^7\, k^{1/2} = 1.15 \times 10^7 \times (\tfrac{2}{3} Fk_1)^{1/2} \text{ cm sec}^{-1}.$$

With the law of distribution of (37) and from formula (38) with $s = 2$,

$$(\overline{c^2})^{1/2} = \alpha \left\{ \Gamma\left(\frac{5}{n}\right) \Big/ \Gamma\left(\frac{3}{n}\right) \right\}^{1/2}$$

whereas

$$\bar{c} = \alpha \Gamma\left(\frac{4}{n}\right) \Big/ \Gamma\left(\frac{3}{n}\right) = \Gamma\left(\frac{4}{n}\right) \left\{ \Gamma\left(\frac{3}{n}\right) \Gamma\left(\frac{5}{n}\right) \right\}^{-1/2} (\overline{c^2})^{1/2}$$

$$= \Gamma\left(\frac{4}{n}\right) \left\{ \Gamma\left(\frac{3}{n}\right) \Gamma\left(\frac{5}{n}\right) \right\}^{-1/2} (\tfrac{2}{3} Fk_1)^{1/2} \times 1.15 \times 10^7 \text{ cm sec}^{-1}. \quad (46)$$

Thus, $(\overline{c^2})^{1/2}$ and $\bar{c}$ may be derived from k_1 and represented as functions of E/p when the law of distribution is known or specified. For instance, with Maxwell's distribution ($n = 2$) $\bar{c} = (8/3\pi)^{1/2} (\overline{c^2})^{1/2}$.

10.3 Drift Velocity

Diatomic Gases

The experimental results for air, hydrogen, deuterium, nitrogen, and oxygen are shown in Figs. 8 and 9. No results are available for drift velocities in deuterium measured by the electrical shutter method; values reported by Hall (1955b) were obtained by the method of magnetic deflection, the two curves shown in Fig. 9 for this gas resulting from the use of (25) with values of C appropriate to the distributions of Maxwell and Druyvesteyn.

10.4 Collision Cross Section

The collision cross section A appears in the formulae for the speed of drift W and the coefficient of diffusion D [(15), (16), and (18)]. Thus,

$$W = \frac{Ee}{3mN} \overline{c^{-2} \frac{d}{dc}\left(\frac{c^2}{A}\right)} = \frac{e}{3mN_1} \left[\overline{c^{-2} \frac{d}{dc}\left(\frac{c^2}{A}\right)}\right] \frac{E}{p} \quad (47)$$

$$D = \frac{1}{3N}\left(\overline{\frac{c}{A}}\right); \qquad pD = \frac{1}{3N_1}\left(\overline{\frac{c}{A}}\right) \quad (48)$$

in which N_1 is the number of molecules in unit volume of the gas at unit pressure. For instance, when the temperature is 288°K and $p = 1$ mm Hg, $N_1 = 3.35 \times 10^{16}$ cm^{-3}.

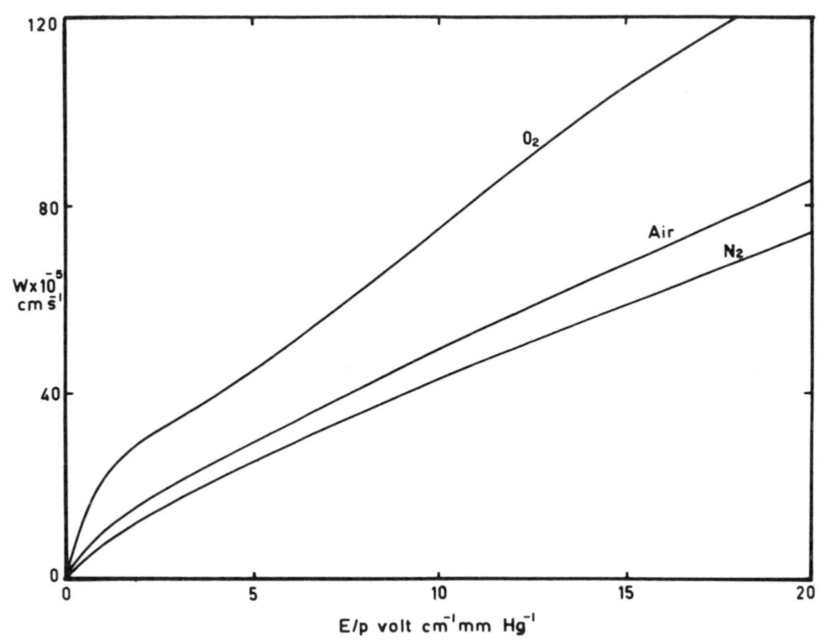

FIG. 8. Curves showing the variation of W with E/p in air, nitrogen, and oxygen (air and O_2—Bradbury and Nielsen, 1937; N_2—Nielsen, 1936).

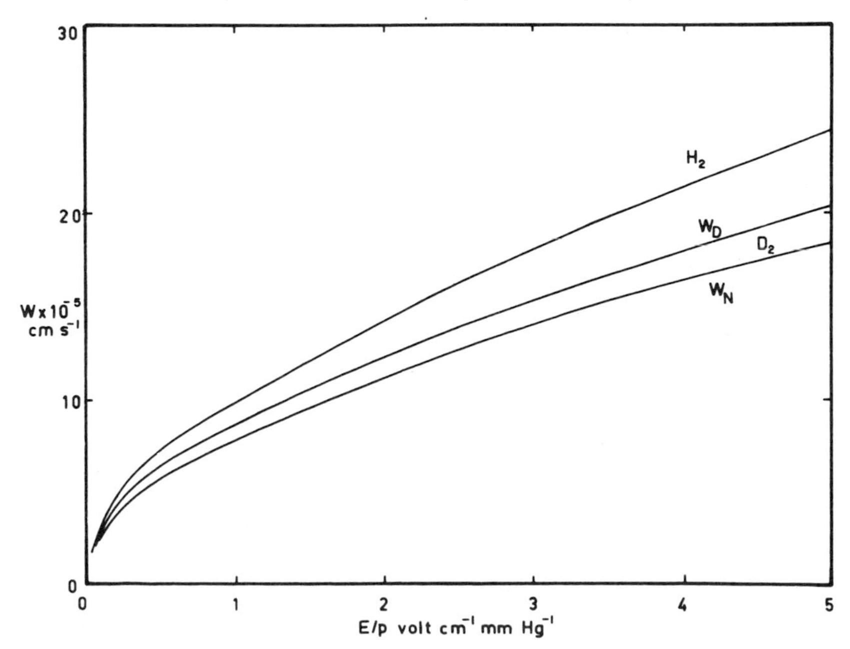

FIG. 9. Curves showing the variation of W with E/p in deuterium and hydrogen (H_2—Bradbury and Nielsen, 1936; D_2—Hall, 1955b).

In practice, the dependence of A upon c has usually been deduced from W by using simplifying assumptions as follows:

In the first instance it is supposed that A may be replaced by its mean value $\bar{A}$ for the given value of E/p ($T = 288°K$) so that (47) becomes

$$W = \frac{2e}{3mN_1\bar{A}} (\overline{c^{-1}}) (E/p), \tag{49}$$

From (38),

$$(\overline{c^{-1}}) = \Gamma\left(\frac{2}{n}\right) \Gamma\left(\frac{4}{n}\right) \Big/ \left\{\Gamma\left(\frac{3}{n}\right)\right\}^2 \bar{c},$$

so that $(\overline{c^{-1}})$ may be replaced by $(\bar{c})^{-1}$ in the formula for W and

$$\bar{A} = \frac{2e}{3mN_1} \times \frac{\Gamma(2/n)\,\Gamma(4/n)}{\{\Gamma(3/n)\}^2} \times \frac{(E/p)}{W} \times \frac{1}{\bar{c}}.$$

But since at each value of (E/p), $(E/p)/W$ is a measured quantity and $\bar{c}$ is known as a function of E/p from k_1, when n is specified, it follows that $\bar{A}$ can be represented as a function of $\bar{c}$ when the form of the distribution law is specified through n.

In practice, the values $n = 2$ (Maxwell) and $n = 4$ (Druyvesteyn) are usually adopted. It is found that $\bar{A}$ as a function of $\bar{c}$ is not sensitive to the choice of n. The dependence of A upon c is then assumed to be the same as that of $\bar{A}$ upon $\bar{c}$.

Figure 10 shows the dependence of $\bar{A}$ upon $\bar{c}$ for the diatomic gases air, oxygen, and nitrogen. In all cases, A diminishes as $\bar{c} \to 0$. In Fig. 11 the results for molecular hydrogen and deuterium (Hall, 1955b) are shown. It can be seen that the collision cross sections are almost the same at each value of c as would be expected from the identity of the electron configurations of the molecules, and it may be concluded that the approximate treatment of formula (47) does not lead to gross inaccuracies.

The simpler form of (48) suggests that the dependence of A upon c could be more conveniently derived frop pD than from W. pD itself is obtained from the measured values of W and W/pD which are both functions of E/p.

If in (48) $(\overline{c/A})$ is replaced by $\bar{c}/\bar{A}$, then $\bar{A} = \bar{c}/3N_1pD$; since $\bar{c}$ and pD are known from the measurements (when the distribution function is specified) for each value of E/p, then $\bar{A}$ can be found as a function of $\bar{c}$. In cgs units with $N_1 = 3.35 \times 10^{16}$, it follows that (48) becomes $pD = 10^{-17} \times (\overline{c/A})$.

Equation (48) can be used without modification when A is proportional to c. Put $A = ac$, then $pD = 10^{-17}/a = $ constant.

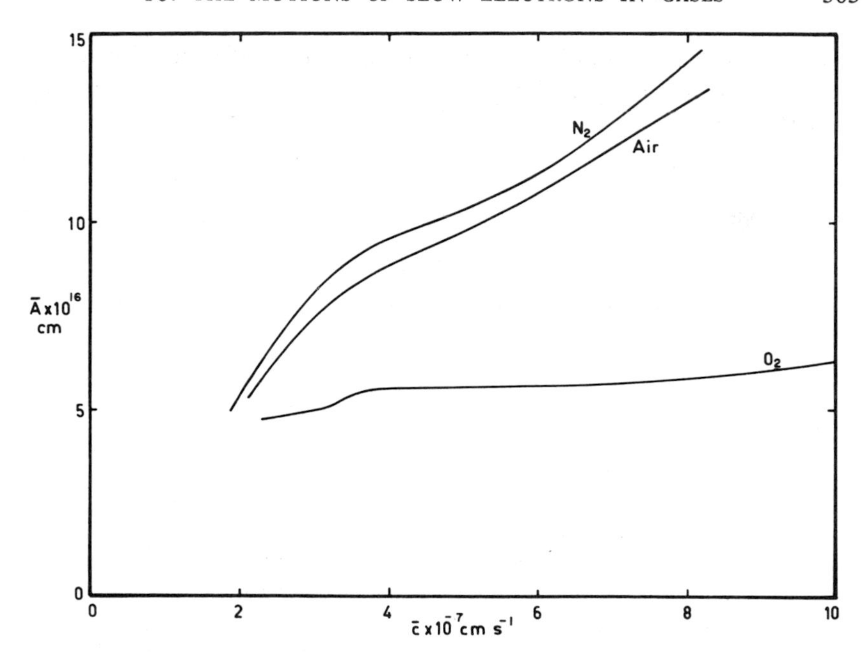

FIG. 10. Curves showing the variation of $\bar{A}$ with $\bar{c}$ in air, nitrogen, and oxygen (N_2—Crompton and Sutton, 1952; air—Crompton *et al.*, 1953; O_2—Crompton, 1953).

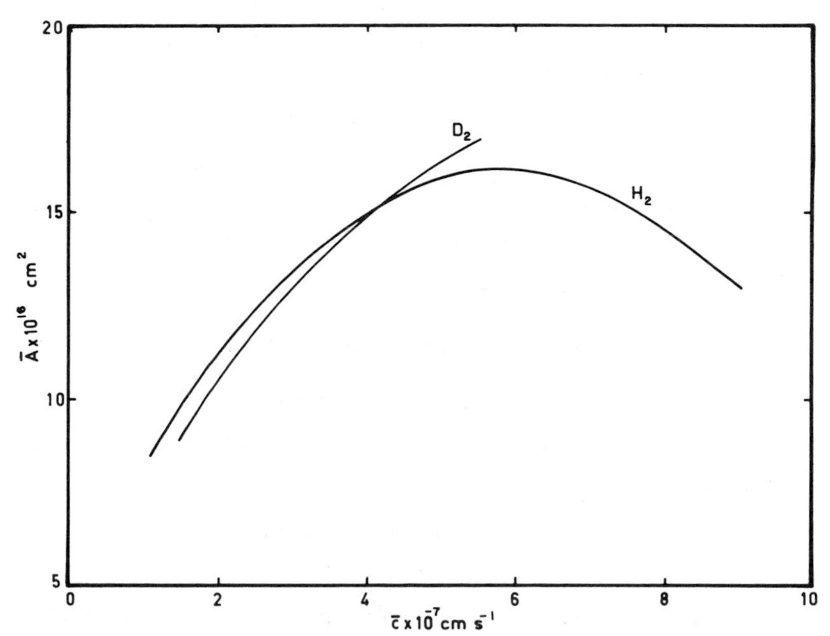

FIG. 11. Curves showing the variation of $\bar{A}$ with $\bar{c}$ in deuterium and hydrogen (D_2—Hall, 1955b; H_2—Crompton and Sutton, 1952).

An important example of this special case is afforded by nitrogen when k_1 is small as is demonstrated by the following experimental measurements:

(E/p) volt cm^{-1} mm Hg^{-1}	0.02	0.03	0.04	0.05	0.06	0.07	0.08	0.09
k_1	1.29	1.56	1.87	2.20	2.56	2.90	3.26	3.65
$pD \times 10^{-5}$ cm^2 sec^{-1}	3.04	2.99	3.04	2.98	3.11	3.02	3.05	3.10

Thus, when $k_1 < 4$, pD remains constant as k_1 is diminished, with $pD = 3.02 \times 10^5$ from which it follows that $A = 3.29 \times 10^{-23} c$ cm^2.

10.5 Collision Frequency

The collision frequency of electrons with speed c is $\nu = NAc$ and the mean collision frequency for all speeds c is $\bar{\nu} = N(\overline{Ac})$. At constant temperature T, N is proportional to the pressure p of the gas and $\overline{Ac}$ is a function of E/p; consequently $\bar{\nu}/p$ is a function E/p. The special case in which A is proportional to c is relevant to studies of the ionosphere. Consider the motion of electrons in nitrogen when $k_1 \leqslant 4$ where it was found (§ 10.4) that $A = 3.29 \times 10^{-23}\, c$ cm^2. It follows that

$$\bar{\nu} = 3.29 \times 10^{-23} N \overline{c^2};$$

but the mean energy of agitation of an electron is $Q = \frac{1}{2} m\overline{c^2}$ from which it may be deduced that in nitrogen ($k_1 \leqslant 4$),

$$\bar{\nu} = 7.22 \times 10^4 NQ \text{ sec}^{-1}.$$

Preliminary measurements indicate (Huxley, 1959) that in this range of energies the collision cross section A of molecules of oxygen are much smaller than those of nitrogen so that in air the collision frequency becomes

$$\bar{\nu} = 0.8 \times 7.22 \times 10^4 NQ = 5.8 \times 10^4 NQ.$$

When electrons are in thermal equilibrium with air, then

$$Q = \tfrac{1}{2} m\overline{c^2} = Q_0 = \tfrac{1}{2} M\overline{C^2}.$$

But the pressure p of the air is $p = \frac{2}{3} NQ$; consequently, in thermal equilibrium (Huxley, 1959)

$$\begin{aligned} \bar{\nu} &= 8.7 \times 10^4 p \; (p \text{ dyne cm}^{-2}) \\ &= 1.2 \times 10^8 p \; (p \text{ mm Hg}). \end{aligned} \tag{50}$$

For instance, in a region of the ionosphere where the pressure is 10^{-3} mm Hg, $\bar{\nu} = 1.2 \times 10^5$ sec^{-1}.

This formula gives a value of the collision frequency in the lower D-region of the ionosphere, at a height of 66 km, that is of the order of magnitude of the values found by Gardner and Pawsey (1953) and Fejer and Vice (1959) from the absorption of radio waves. It has been mentioned in § 3.5 that the magneto-ionic equations of Appleton and Hartree are not precise when the collision frequency $\bar{\nu}$ is important, with the result that the value of $\bar{\nu}$ derived by radio methods could involve "errors, by factors of several units" (Huxley, 1959, 1951). This matter has been investigated in more detail (Phelps and Pack, 1959).

Formula (50) has recently been shown (Seddon, 1960) to accord well with rocket measurements of $\bar{\nu}$ by radio means and with the improved theory.

Measurements of the dependence of A upon c have also proved to be of value in studies of the electromagnetic properties of gases at high temperatures (Shkarofsky *et al.*, 1961).

10.6 A Formula for W in Terms of k_1 and E/p

In the following identity, $W = (W/D)\,[(pD)/p]$, replace W/D by its value given by (41) to find

$$W = \frac{40 \cdot 3}{k_1}\left(\frac{E}{p}\right)(pD).$$

When pD is constant over a range of values of k_1 then within this range W is proportional to $1/k_1 \times (E/p)$. In nitrogen when $k_1 \leqslant 4$

$$pD = 3.02 \times 10^5 \qquad (\S\ 10.4),$$

consequently,

$$W = 1.22 \times 10^7\,(E/p)k_1$$

so that if k_1 can be measured at smaller values of E/p than can W, the values of W can be derived from the formula (Huxley, 1959).

11 *The Losses of Energy by Electrons in Colliding with Gas Molecules*

11.1 Elastic Loss

According to (30), the mean energy lost by an electron with speed c in elastic encounters with molecules of a gas is

$$\tfrac{1}{2}\,mc^2 - \tfrac{1}{2}\,mc'^2 = \frac{2Mm}{(m+M)^2}(1-\alpha)\left(\tfrac{1}{2}\,mc^2 - \tfrac{1}{2}\,M\overline{C^2}\right)$$

so that when $\frac{1}{2}mc^2 \gg \frac{1}{2}M\overline{C^2}$, the loss becomes $[2Mm/(M+m)^2] \times (1-\alpha)\frac{1}{2}mc^2$. The mean rate at which electrons in a group lose energy in encounters when $Q = \frac{1}{2}m\overline{c^2} \gg \frac{1}{2}M\overline{C^2}$ and the scattering is isotropic and l_0 is independent of c (rigid spherical molecules) is,

$$w = \frac{Mm^2}{(M+m)^2 l_0} 4\pi \int_0^\infty fc^5 dc = \frac{Mm^2}{(M+m)^2 l_0} \overline{c^3}$$

so that if the distribution function is of the type defined in (37), the mean rate of loss, according to (38), becomes

$$w = \frac{M}{l_0}\left(\frac{m}{M+m}\right)^2 \frac{\alpha^3 \Gamma(6/n)}{\Gamma(3/n)}$$

But

$$\bar{c} = \frac{\alpha\Gamma(4/n)}{\Gamma(3/n)} \quad \text{and} \quad \overline{c^2} = \frac{\alpha^2\Gamma(5/n)}{\Gamma(3/n)};$$

consequently, the mean energy lost in a single encounter is

$$\Delta Q = \frac{l_0 w}{\bar{c}} = \frac{2\Gamma(6/n)\,\Gamma(3/n)}{\Gamma(5/n)\,\Gamma(4/n)} \times \frac{Mm}{(M+m)^2} \times \tfrac{1}{2}m\overline{c^2}. \tag{51}$$

The mean proportion of the mean energy lost in a collision is

$$\eta = \frac{\Delta Q}{Q} = S\frac{Mm}{(M+m)^2}; \qquad S = \frac{2\Gamma(6/n)\,\Gamma(3/n)}{\Gamma(5/n)\,\Gamma(4/n)}. \tag{52}$$

In order to take into account the influence of the thermal motions of the molecules we note that when no electric field is present $\Delta Q = 0$ and $\frac{1}{2}m\overline{c^2} = \frac{1}{2}M\overline{C^2}$ so that the appropriate generalizations of (51) and (52) are

$$\Delta Q = S\frac{Mm}{(M+m)^2}(\tfrac{1}{2}m\overline{c^2} - \tfrac{1}{2}M\overline{C^2})$$

$$\eta = S\frac{Mm}{(M+m)^2}\left(1 - \frac{1}{k}\right). \tag{53}$$

With Maxwell's distribution, $n = 2$, $S = \frac{8}{3}$ and with Druyvesteyn's distribution ($n = 4$), $S = 2.397$. Also $Mm/(M+m)^2 \approx m/M$. In monatomic gases the values for η predicted by (53) and the measured values derived by means of (43) are in reasonable agreement as is illustrated by the following measurements for helium (Townsend, 1928):

E/p volt cm^{-1} mm Hg^{-1}	0.013	0.02	0.05	0.2	1.0	1.5	2.5	5.0
k_1	1.77	2.12	3.68	11.3	53.0	79	124	172
$\eta \times 10^4$	1.3	1.56	2.3	2.55	2.4	2.6	3.4	9.8
$\frac{2.4m}{M}\left(1 - \frac{1}{k}\right) \times 10^4$†	1.16	1.51	2.25	2.94	3.20	3.22	3.24	3.26

Thus, the majority of collisions are elastic when $k_1 < 124$ ($k_1 = 27 \approx 1$ ev).

11.2 Inelastic Losses

If in (43) Maxwell's distribution is postulated and it is assumed that l is independent of c, then $k = k_1$; $K = 2.37$; consequently,

$$\eta = \frac{\Delta Q}{Q} \approx 1.8 \times 10^{-14} \frac{w^2}{k_1}.$$

When η is derived from this formula by using pairs of measured values of W and k_1 for the same value of E/p, it is found that the values of η so derived are much greater than those predicted for elastic collisions from the formula $\eta = (2.66m/M)(1 - 1/k_1)$. For instance, in hydrogen, when $E/p = 0.2$ volt cm^{-1} mm Hg^{-1}, $W = 4.8 \times 10^5$ cm sec^{-1}, and $k_1 = 2.5$. The measured value of η is therefore $\eta = 1.66 \times 10^{-3}$, whereas

$$\frac{2.66m}{M}\left(1 - \frac{1}{k_1}\right) = \frac{1.6m}{M} = 4.4 \times 10^{-4}.$$

Similarly, when $E/p = 1$; $W = 9.9 \times 10^5$; $k_1 = 9$ and the measured value of η is $\eta = 1.96 \times 10^{-3}$ whereas the value for elastic collisions is 6.6×10^{-4}.

In other diatomic gases, with greater molecular weights, the disparity is even greater; for instance, in nitrogen; $E/p = 0.08$; $k_1 = 3.26$; $\eta = 4.22 \times 10^{-4}$; $(2.66m/M)(1 - 1/k_1) = 3.62 \times 10^{-5}$.

It is apparent, therefore, that in a small proportion of a large number of encounters an electron loses a large proportion of its energy although the majority of encounters are elastic in which the proportion of energy lost is small. Evidence about the nature of these inelastic encounters in diatomic gases at the smaller values of k is provided by measurements in

† The value of S appropriate to Druyvesteyn's distribution has been chosen since this distribution will be a closer approximation to the true distribution for the larger values of k_1.

deuterium (Hall, 1955b) which when compared with corresponding measurements in hydrogen (Huxley, 1956) show that within a wide range of values of $k_1(1 < k_1 < 25)$ the value of η in deuterium is half that in hydrogen when k_1 has the same value in each gas.

Thus, as for elastic collisions, η is inversely proportional to the mass of a molecule of the gas and it may be supposed that in these inelastic collisions the rotational energy state of the molecule is changed by the collision (Huxley, 1956; Gerjuoy and Stein, 1955).

The angular momentum associated with a rotational quantum number J is

$$P_J = \left(\frac{h^2 I}{8\pi^2}\right) \{J(J+1)\}^{1/2}$$

and the corresponding energy is

$$\epsilon_J = \left(\frac{h^2}{8\pi^2 I}\right) J(J+1)$$

where h in Planck's constant and I is the moment of inertia of the molecule.

The angular momentum, relative to the centroid of the molecule imparted by the colliding electron when the molecule is changed from rotational state J to J', is

$$\Delta P_{J'J} = [\{J'(J'+1)\}^{1/2} - \{J(J+1)\}^{1/2}] \frac{h^2 I}{8\pi^2}$$

and the corresponding energy lost by the electron is

$$\Delta\epsilon_{J'J} = \left(\frac{h^2}{8\pi^2 I}\right)(J'-J)(J'+J+1).$$

Since I is proportional to the mass of the molecule, $\Delta\epsilon_{J'J}$ is inversely proportional to the molecular mass. In many encounters the impact parameter of the colliding electron is such that its angular momentum is less than $\Delta P_{J'J}$ although its energy may exceed $\Delta\epsilon_{J'J}$; thus, the majority of collisions are elastic, and the measured collisions cross section A refers to elastic collisions.

On the basis of these principles it is possible to derive a semiempirical formula for η (Huxley, 1956). It is more useful, however, to consider the mean rate at which an electron loses energy in collisions rather than η.

11.3 Rate of Energy Loss in Encounters

Elastic Encounters

The rate at which energy is lost in such encounters is

$$w_e = \bar{\nu}Q \times \frac{2.66m}{M}\left(1 - \frac{1}{k}\right) = \frac{2.66m}{M}\bar{\nu}(Q - Q_0) \tag{54}$$

where Q_0 is the mean energy of agitation of a molecule. In nitrogen when $k_1 \leqslant 4$, $\bar{\nu} = 7.22 \times 10^4 NQ$ (§ 10.5) and it follows that

$$w_e = 3.78\,Q(Q - Q_0)N \text{ erg sec}^{-1}. \tag{55}$$

Inelastic encounters

The theoretical approach outlined in § 11.2 (Huxley, 1956) leads to the following semiempirical formula for the rate of loss of energy in exciting changes in the rotational states of molecules when the law of distribution is that of Maxwell:

$$\begin{aligned} w_r &= \alpha N Q^{1/2}[\exp(-\beta/Q) - \exp(-\beta/Q_0)] \\ &= aNk^{1/2}[\exp(-b/k) - \exp(-b)]. \end{aligned} \tag{56}$$

in which α, β, a, and b are constants independent of k, whose values are determined by experiment.

The measured total rate of loss is $w = EeW$, whence $w/p = e(E/p)W$ which is a function of E/p. In nitrogen at $T = 288°$K and when $k_1 \leqslant 4$ (§ 10.6)

$$W = \frac{1.22 \times 10^7}{k_1} \times \frac{E}{p};$$

consequently, in this gas,

$$w = 5.82 \times 10^{-22}\frac{N}{k_1}\left(\frac{E}{p}\right)^2 \text{ erg sec}^{-1} \qquad (E \text{ in volt cm}^{-1}). \tag{57}$$

From $w_r = w - w_e$ it is possible to deduce the values of the constants α, β, a, and b in (56). It is found that $\alpha = 7.29 \times 10^{-18}$ and $\beta = 4.41 \times 10^{-14}$ and, at $T = 288°$K, $a = 1.78 \times 10^{-24}$ and $b = 0.74$ (Huxley, 1956).

Table I indicates the concordance of (56) and the measurements.

TABLE I[a]

COMPARISON OF THEORETICAL AND EXPERIMENTAL ENERGY LOSSES

$\frac{E}{p}$	0.02	0.03	0.04	0.05	0.06	0.07	0.08
k	1.29	1.56	1.87	2.20	2.56	2.90	3.26
$\frac{w}{N} \times 10^{26}$ [from (57)]	18.0	33.6	49.8	66.1	81.8	98.4	114
$\frac{w_e}{N} \times 10^{26}$ [from (54)]	0.50	1.16	2.16	3.51	5.31	7.33	9.80
$\frac{w_r}{N} \times 10^{26} = \frac{w - w_e}{N} \times 10^{26}$	17.5	32.4	47.6	62.6	75.5	91.0	104
$\frac{w_r}{N} \times 10^{26}$ [from (56)]	17.5	32.3	47.8	62.6	77.4	90.8	103

[a] The experimental results from which the values recorded in this table are derived have not been published as further experimental verification is in progress. However, it is unlikely that the results of the more rigorous investigation will differ greatly from those given here.

Thus, in nitrogen ($k \leqslant 4$),

$$\frac{w}{N} = 7.29 \times 10^{-18} Q^{1/2} \left[\exp\left(-\frac{4.41 \times 10^{-14}}{Q}\right) - \exp\left(-\frac{4.41 \times 10^{-14}}{Q_0}\right)\right] + 3.78\, Q(Q - Q_0) \qquad (58)$$

and at $T = 288°\text{K}$ ($Q_0 = 5.96 \times 10^{-14}$ erg)

$$\frac{w}{N} = 1.78 \times 10^{-24} k^{1/2} \left[\exp\left(\frac{-0.74}{k}\right) - \exp(-0.74)\right] + 1.33 \times 10^{-26} k(k-1). \qquad (59)$$

When $k \to 1$, (58) tends to the form

$$\frac{w}{N} = \left[\alpha\beta Q_0^{-3/2} \exp\left(-\frac{\beta}{Q_0}\right)\right] \left(\frac{Q_0}{Q}\right)^{1/2} (Q - Q_0) + 3.78\, Q(Q - Q_0) \qquad (60)$$

with $\alpha = 7.29 \times 10^{18}$ and $\beta = 4.41 \times 10^{-14}$ as before and (59) becomes

$$\frac{w}{N} = (6.28 \times 10^{-25} k^{-1/2} + 1.33 \times 10^{-26} k)(k - 1). \qquad (61)$$

When w/N is eliminated from formulae (57) and (59) the following relationship between k and E/p at $T = 288°K$ is found:

$$\left(\frac{E}{p}\right)^2 = 3.06 \times 10^{-3} k^{3/2} \left[\exp\left(-\frac{0.74}{k}\right) - \exp(-0.74)\right] + 2.29 \times 10^{-5} k^2 (k-1) \quad (62)$$

and when $k \to 1$,

$$\left(\frac{E}{p}\right)^2 = 1.065 \times 10^{-3} k^{1/2} (k-1) + 2.29 \times 10^{-5} k^2 (k-1). \quad (63)$$

11.4 Rate of Loss of Energy in Air

There are reasons in support of the assumption that formulae (58)-(61) are applicable not only to nitrogen but also to air (Huxley, 1956) so that both in air and nitrogen when $k \to 1$

$$\frac{w}{N} = B(Q - Q_0)$$

with

$$B = \alpha\beta Q_0^{-3/2} \exp(-\beta/Q_0) + 3.78\, Q_0$$

and

$$\alpha = 7.29 \times 10^{-18}; \qquad \beta = 4.41 \times 10^{-14}. \quad (64)$$

It follows that B, through Q_0, depends upon the temperature of the gas; for instance,

$T°K$	288	196
$B \times 10^{11}$	1.07	1.34

11.5 The Interaction of Radio Waves in the Ionosphere

The theory of the interaction of radio waves is based upon an expression for the fluctuation of the mean energy of agitation Q of the electrons about a mean value $\bar{Q}$ under the influence of a modulated disturbing radio wave. This basic equation is

$$\frac{dQ}{dt} + R = w(t)$$

where $w(t)$ is the power supplied in the mean to an electron by the disturbing wave and R is the mean rate at which an electron with mean energy Q loses energy in collisions with the molecules of the atmosphere.

According to (64), $R = BN(Q - Q_0)$ if it is assumed that $k \sim 1$. Measurements of radio-wave interaction lead to values of BN such that in most cases $10^3 < BN < 3 \times 10^3$ with $BN = 2 \times 10^3$ as typical (Huxley, 1952, where BN is written $G\bar{\nu}$).

Other measurements indicate that the cross modulation takes place at heights above the ground of 82 to 90 km.

Rocket measurements indicate that at these heights the atmospheric composition is essentially the same as at the ground and that the temperature is about $T = 196°$K.

The value of B is therefore 1.34×10^{-11} and the values of the molecular concentration associated with the values 10^3, 2×10^3, and 3×10^3 of BN are, respectively, $N = 7.45 \times 10^{13}$, 1.49×10^{14}, and 2.23×10^{14}. The heights corresponding to these values of N in the A.R.D.C. model atmosphere are 90.7, 86.6, and 84.2 km. It may be concluded that the laboratory measurements and the ionospheric measurements are mutually consistent.

A more exact theory of radio-wave interaction and of nonlinear effects in plasmas in general is given by Ginzburg and Gurevič (1960).

References

Bradbury, N. E., and Nielsen, R. A. (1936) *Phys. Rev.* **49**, 388.
Bradbury, N. E., and Nielsen, R. A. (1937) *Phys. Rev.* **51**, 69.
Brown, S. C. (1959) "Basic Data of Plasma Physics." Technology Press, Cambridge, Massachusetts; Wiley, New York.
Chapman, S., and Cowling, T. G. (1952) "The Mathematical Theory of Non-Uniform Gases," 2nd ed., p. 171. Cambridge Univ. Press, London and New York.
Crompton, R. W. (1953) Unpublished Ph. D. thesis, University of Adelaide, Adelaide, Australia.
Crompton, R. W., and Sutton, D. J. (1952) *Proc. Roy. Soc.* **A215**, 467.
Crompton, R. W., Huxley, L. G. H., and Sutton, D. J. (1953) *Proc. Roy. Soc.* **A218**, 507.
Crompton, R. W., Hall, B. I. H., and Macklin, W. C. (1957) *Austral. J. Phys.* **10**, 366.
Duncan, R. A. (1957) *Austral. J. Phys.* **10**, 54.
Fejer, J. A., and Vice, R. W. (1959) *J. atmos. terrest. Phys.* **16**, 291.
Gardner, F. F., and Pawsey, J. L. (1953) *J. atmos. terrest. Phys.* **3**, 321.
Gerjuoy, S., and Stein, S. (1955) *Phys. Rev.* **97**, 1671.
Ginzburg, V. L., and Gurevič, A. V. (1960) *Uspekhi Fiz. Nauk* **70**, 393; translation (1960) *Soviet Phys.—Uspekhi* **3** (2), 175.
Hall, B. I. H. (1955a) *Proc. Phys. Soc.* **B68**, 334.
Hall, B. I. H. (1955b) *Austral. J. Phys.* **8**, 468.
Healey, R. H., and Reed, J. W. (1941) "The Behaviour of Slow Electrons in Gases." Amalgamated Wireless (Australasia) Ltd., Sydney, Australia.
Huxley, L. G. H. (1951) *Proc. Phys. Soc.* **B64**, 844.
Huxley, L. G. H. (1952) *Nuovo cimento* (*Suppl.*) **9**.
Huxley, L. G. H. (1956) *Austral. J. Phys.* **9**, 44.

Huxley, L. G. H. (1957) *Austral. J. Phys.* **10**, 118.
Huxley, L. G. H. (1959) *J. atmos. terrest. Phys.* **16**, 46.
Huxley, L. G. H. (1960a) *Austral. J. Phys.* **13**, 578.
Huxley, L. G. H. (1960b) *Austral. J. Phys.* **13**, 718.
Huxley, L. G. H., and Crompton, R. W. (1955) *Proc. Phys. Soc.* **B68**, 381.
Huxley, L. G. H., and Zaazou, A. A. (1949) *Proc. Roy. Soc.* **A196**, 402.
Huxley, L. G. H., Crompton, R. W., and Bagot, C. H. (1959) *Austral. J. Phys.* **12**, 303.
Loeb, L. B. (1955) "Basic Processes of Gaseous Electronics." Univ. Calif. Press, Los Angeles and Berkeley, California.
Nielsen, R. A. (1936) *Phys. Rev.* **50**, 950.
Phelps, A. V., and Pack, J. L. (1959) *Phys. Rev. Letters* **3**, 340.
Seddon, J. C. (1960) Summary of Rocket and Satellite Observations Related to the Ionosphere. Paper read at London General Assembly of U.R.S.I.
Shkarofsky, I. P., Bachynski, M. P., and Johnston, T. W. (1961) *Planetary and Space Sci.* **6**, 24.
Townsend, J. S. (1899) *Phil. Trans.* **193**, 129.
Townsend, J. S. (1900) *Phil. Trans.* **195**, 259.
Townsend, J. S. (1908) *Proc. Roy. Soc.* **A80**, 207; **81**, 464.
Townsend, J. S. (1928) *Proc. Roy. Soc.* **A120**, 511.
Townsend, J. S., and Tizard, H. T. (1912) *Proc. Roy. Soc.* **A87**, 357.

11.

The Theory of Excitation and Ionization by Electron Impact

M. J. Seaton

In this chapter we survey the theory of excitation and ionization by electron impact and consider the results obtained by calculation and by experiment. Since the main emphasis is on recent developments, we do not dwell on topics discussed adequately in previous reviews (Mott and Massey, 1949; Bates *et al.*, 1950; Massey and Burhop, 1952; Massey, 1956a, b; Seaton, 1958).

1 Classical Theory

Let an electron in collision with an atom have a classical orbit with impact parameter R (Fig. 1). Let the atom be initially in state i and sup-

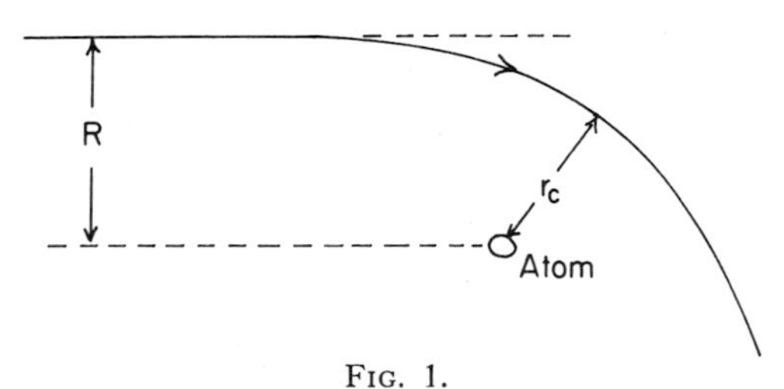

FIG. 1.

pose that there is a probability $P_{ij}(R)$ that a transition occurs to state j; evidently $P_{ij} \leqslant 1$. The cross section for the transition is

$$Q(i \to j) = \int P_{ij}(R)\, 2\pi R dR. \tag{1}$$

Let the colliding electron have initial velocity v_i and angular momentum $L = mv_i R$. Then

$$Q(i \to j) = \frac{2\pi}{(mv_i)^2} \int P_{ij}(L)\, L dL. \tag{2}$$

For a reversible classical process one would expect to have† $P_{ij}(L) = P_{ji}(L)$ and

$$Q(j \to i) = \frac{2\pi}{(mv_j)^2} \int P_{ij}(L)\, L dL \tag{3}$$

where

$$\tfrac{1}{2} mv_i^2 + E_i = \tfrac{1}{2} mv_j^2 + E_j, \tag{4}$$

E_i and E_j being the atom energies.

† More exactly, this would be expected when P_{ij} is a function of total angular momentum, which is conserved during the collision.

The magnitude of P_{ij} may be expected to depend on r_c, the distance of closest approach. According to classical orbit theory, r_c is obtained on solving

$$\frac{L^2}{2mr_c^2} + V(r_c) = \tfrac{1}{2} m v_i^2, \tag{5}$$

$V(r)$ being the potential for the colliding electron. When $V(r_c)$ is small, one has

$$r_c \simeq R = L/mv_i. \tag{6}$$

Consider the $i \to j$ collision to be of de-excitation type ($v_i < v_j$). According to (6), when v_i is small, r_c becomes large for all finite L. One would expect $P_{ij}(L)$ to go to zero as v_i goes to zero. In this case the excitation cross section $Q(j \to i)$ is zero at the excitation threshold, $\frac{1}{2} m v_j^2 = (E_i - E_j)$. Furthermore, when v_i is small one would expect $P_{ij}(L)$ to decrease rapidly as L increases.

The prediction of zero cross sections at excitation thresholds is in agreement with experiment for neutral atoms. For positive ions a different law is to be expected. With $V(r) = -ze^2/r$, from (5) one obtains $r_c = L^2/2zme^2$ in the limit of $v_i \to 0$. It follows that cross sections for positive ion excitation will remain finite at threshold.

The simplest classical estimate of the transition probability P_{ij} is due to Thomson (1912). He considered a collision between two free electrons, one with an initial kinetic energy $W = \frac{1}{2} m v_i^2$ and the other initially at rest. With impact parameter R the energy transferred is

$$\epsilon = \frac{W}{1 + (RW/e^2)^2}. \tag{7}$$

For ionizing collisions one takes $P_{ij} = 1$ for all values of R giving $\epsilon \geqslant I$ where I is the ionization potential. The ionization cross section is $Q_{cl} = n\pi R_0^2$ where $R_0^2 = (e^2/W)^2 [(W/I) - 1]$ and where n is the number of atomic electrons (or the number in the outer shells). In terms of the Bohr radius, $a_0 = \hbar^2/me^2$, and the hydrogen ionization potential, $I_H = me^4/2\hbar^2$, this gives

$$Q_{cl} = 4n \left(\frac{I_H}{I}\right)^2 \left(\frac{I}{W}\right) \left(1 - \frac{I}{W}\right) \pi a_0^2. \tag{8}$$

Figure 2 shows Q/Q_{cl} against $\log_{10}(W/I)$ for H, He, Ne, and Ar, Q being taken from experimental results [Fite and Brackmann (1958), and Boksenberg (1961) for H, Smith (1930) for He, Ne, and Ar]. The classical theory is seen to give the correct order of magnitude, but at low energies it overestimates by a factor of about 5 while at high energies the ratio Q/Q_{cl} increases as log W. In a quantum treatment one finds that P rarely approaches unity, even for small impact parameters; this explains why the classical theory overestimates the low-energy cross

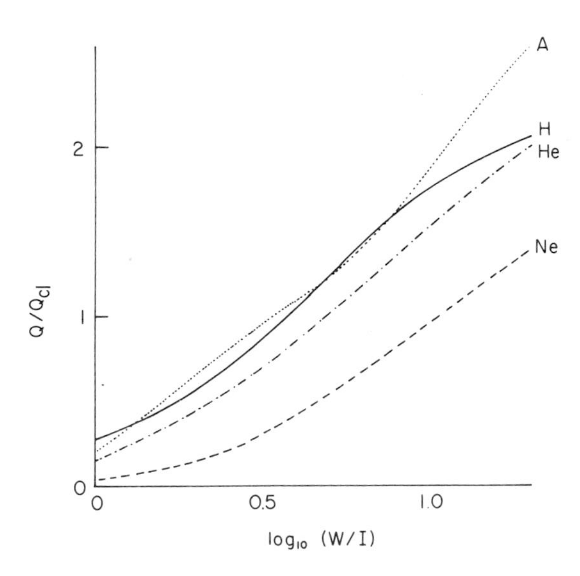

FIG. 2. The classical theory for neutral atom ionization. Q/Q_{cl} is plotted against $\log_{10}(W/I)$ where Q is the experimental cross section, Q_{cl} that given by (8), W is the incident electron energy, and I the ionization potential.

section. Another feature of a quantum treatment is that one obtains finite contributions from impact parameters much larger than those which contribute in the classical theory. This is important at high energies and explains the fact that Q varies as $(1/W) \log W$. This high-energy behavior is obtained if one treats collisional excitation, or ionization, as an induced radiative process (§ 5).

Elwert (1952) has used classical theory as a guide for the extrapolation of experimental data. His method is based on the fact that, according to (8), $(I/I_H)^2(1/n)Q$ is a universal function of (W/I). The attempt is made to determine this function, not from classical theory, but from experimental data. Figure 2 shows to what extent the method is valid for ionization of neutral atoms. If there were some such universal function one would obtain, in Fig. 2, a single curve for all elements. It is seen that the differences between the curves for H, He, and Ar are quite small—smaller than the differences between the cross sections themselves—but that the Ne curve is well below the others. This behavior of Ne depends on detailed quantum-mechanical properties of the atom (§ 5) and cannot be explained in a classical theory.

The Elwert theory has been used to estimate cross sections for ionization of posiitve ions, particularly those of importance in the solar corona. Because the theory does not allow for the distortion of the incident electron orbit by the attractive field of the ion, it tends to underestimate these cross sections by a factor of about 2 or 3 (Burgess, 1960).

Attempts have been made to use classical theory for atomic excitation. The theory gives cross sections for excitation of definite energy intervals in place of discrete energy levels. The usual procedure is to take intervals equal to the separations of levels. It is known that the magnitude of excitation cross sections depends a great deal on the type of transition, whether or not it is optically allowed and whether it involves a change of atom spin. Classical theory takes no account of these differences.

2 *General Quantum Theory*

2.1 Electron-Hydrogen Collisions Neglecting Spin

We use atomic units ($e = m = \hbar = 1$). For the hydrogen atom quantum numbers we use the notation

$$a = nlm \tag{9}$$

and for the wave functions we use

$$\psi(a|\mathbf{r}) = Y_{lm}(\hat{\mathbf{r}}) \frac{1}{r} P_{nl}(r) \tag{10}$$

where Y_{lm} is a normalized spherical harmonic and $\hat{\mathbf{r}}$ is a unit vector specifying the polar angles θ, φ. The atom Hamiltonian is

$$H(\mathbf{r}) = -\tfrac{1}{2}\nabla^2 - \frac{1}{r} \tag{11}$$

and the atom Schrödinger equation is

$$H(\mathbf{r})\,\psi(a|\mathbf{r}) = E_a\,\psi(a|\mathbf{r}) \tag{12}$$

where

$$E_a = -\frac{1}{2n^2}\,.$$

The Hamiltonian for electron-hydrogen collisions is

$$H(\mathbf{r}_1, \mathbf{r}_2) = H(\mathbf{r}_1) + H(\mathbf{r}_2) + \frac{1}{r_{12}} \tag{13}$$

where $r_{12} = | \mathbf{r}_1 - \mathbf{r}_2 |$. The complete wave function, $\Psi(\mathbf{r}_1, \mathbf{r}_2)$, satisfies

$$[H(\mathbf{r}_1, \mathbf{r}_2) - E]\, \Psi(\mathbf{r}_1, \mathbf{r}_2) = 0, \tag{14}$$

E being the total energy. Let the colliding electron have a velocity $\mathbf{k}_a$ when far from the atom and let the atom be in state a; then

$$E = E_a + \tfrac{1}{2} k_a^2. \tag{15}$$

When this gives $k_a^2 < 0$, we define k_a to be $i[2(E_a - E)]^{1/2}$.

Before the collision occurs, let $\mathbf{r}_1$ be the coordinate of the atomic electron, $\mathbf{r}_2$ that of the incident electron, and let $a\mathbf{k}_a$ be the initial state of the system. The wave function $\Psi(a\mathbf{k}_a \mid \mathbf{r}_1, \mathbf{r}_2)$ will have asymptotic form

$$\Psi(a\mathbf{k}_a|\mathbf{r}_1, \mathbf{r}_2) \underset{r_2\to\infty}{\sim} \psi(a|\mathbf{r}_1) \exp (i\mathbf{k}_a \cdot \mathbf{r}_2) + \sum_{a'} \psi(a'|r_1)\, f(a'\hat{\mathbf{r}}_2, a\hat{\mathbf{k}}_a) \frac{\exp ik_{a'}r_2}{r_2}, \tag{16}$$

$f(a'\hat{\mathbf{k}}_{a'}, a\hat{\mathbf{k}}_a)$ is the amplitude for direct scattering from $a\mathbf{k}_a$ to $a'\mathbf{k}_{a'}$. There is also a possibility of exchange scattering, the incident electron being captured and the atomic electron ejected. We have

$$\Psi(a\mathbf{k}_a|\mathbf{r}_1, \mathbf{r}_2) \underset{r_1\to\infty}{\sim} \sum_{a'} \psi(a'|\mathbf{r}_2)\, g(a'\hat{\mathbf{r}}_1, a\hat{\mathbf{k}}_a) \frac{\exp ik_{a'}r_1}{r_1} \tag{17}$$

where $g(a'\hat{\mathbf{k}}_{a'}, a\hat{\mathbf{k}}_a)$ is the amplitude for exchange scattering.

The cross section for direct scattering is

$$I(a\mathbf{k}_a \to a'\mathbf{k}_{a'}) = \frac{k_{a'}}{k_a} | f(a'\hat{\mathbf{k}}_{a'}, a\hat{\mathbf{k}}_a) |^2 \tag{18}$$

per unit solid angle. For the transition $nl \to n'l'$, the total cross section is

$$Q(nl \to n'l') = \frac{k_{n'}}{k_n} \frac{1}{(2l+1)} \sum_{mm'} \int | f(n'l'm'\hat{\mathbf{k}}_{n'}, nlm\hat{\mathbf{k}}_n) |^2 \, d\hat{\mathbf{k}}_{n'} \tag{19}$$

where $d\hat{\mathbf{k}}_{n'}$ is the element of solid angle about the direction $\hat{\mathbf{k}}_{n'}$. In (19) we have summed over final states (m') and averaged over initial states (m). Exchange cross sections are obtained on replacing f by g.

2.2 Spin Variables

The foregoing theory is unsatisfactory in that the two electrons are treated as distinguishable. To correct the theory we must first include spin variables.

For one electron we use a spin quantum number $\mu = \pm \frac{1}{2}$, a spin coordinate $\sigma = \pm \frac{1}{2}$, and a spin function $\delta(\mu|\sigma)$ which is unity for $\sigma = \mu$ and zero for $\sigma \neq \mu$. For the space and spin coordinates we use $\mathbf{x} = (\mathbf{r}, \sigma)$ and for the atom functions we use

$$\psi(a\mu|\mathbf{x}) = \delta(\mu|\sigma)\, Y_{lm}(\hat{\mathbf{r}})\, \frac{1}{r}\, P_{nl}(r). \tag{20}$$

To satisfy the Pauli principle the complete wave function must be antisymmetric in the coordinates:

$$\Psi(\mathbf{x}_1, \mathbf{x}_2) = -\,\Psi(\mathbf{x}_2, \mathbf{x}_1). \tag{21}$$

With this condition we may no longer distinguish between the coordinates of the atomic electron and of the colliding electron, but we may specify their quantum numbers. Let μ_1 be the initial spin of the atomic-atomic electron, and μ_2 that of the colliding electron. For r_2 large the complete wave function will have asymptotic form

$$\Psi(a\mu_1\mu_2\,\mathbf{k}_a|\mathbf{x}_1, \mathbf{x}_2) \underset{r_2\to\infty}{\sim} \frac{1}{\sqrt{2}} \Big\{ \psi(a\mu_1|\mathbf{x}_1)\, \delta(\mu_2|\sigma_2) \exp\,(i\mathbf{k}_a \cdot \mathbf{r}_2)$$

$$+ \sum_{a'\mu_1'\mu_2'} \psi(a'\mu_1'|\mathbf{x}_1)\, \delta(\mu_2'|\sigma_2)\, f(a'\mu_1'\mu_2'\hat{\mathbf{r}}_2, a\mu_1\mu_2\hat{\mathbf{k}}_a)\, \frac{\exp ik_{a'}r_2}{r_2} \Big\} \tag{22}$$

and the corresponding form for r_1 large follows using (21). In (22), the factor $(1/\sqrt{2})$ is inserted for convenience in later work. From the amplitudes $f(a'\mu_1'\mu_2'\hat{\mathbf{k}}_{a'}, a\mu_1\mu_2\hat{\mathbf{k}}_a)$, we obtain the cross sections for transitions in which the spins are specified. If we average over initial spins and sum over final spins, we obtain

$$I(a\mathbf{k}_a \to a'\mathbf{k}_{a'}) = \frac{k_{a'}}{k_a} \cdot \frac{1}{4} \cdot \sum_{\mu_1\mu_1'\mu_2\mu_2'} |f(a'\mu_1'\mu_2'\hat{\mathbf{k}}_{a'}, a\mu_1\mu_2\hat{\mathbf{k}}_a)|^2. \tag{23}$$

On neglecting spin terms in H, we have the spin conservation condition $\mu_1 + \mu_2 = \mu_1' + \mu_2'$.

2.3 Spin Coupling

For two electrons the resultant spin is $S = 0$ or 1 and the functions are

$$\chi(SM_s|\sigma_1, \sigma_2) = \sum_{\mu_1\mu_2} C^{\frac{1}{2}\frac{1}{2}S}_{\mu_1\mu_2M_s} \delta(\mu_1|\sigma_1)\delta(\mu_2|\sigma_2) \tag{24}$$

where $C^{abc}_{\alpha\beta\gamma}$ is a vector-coupling coefficient.

Explicit expressions are:

$$\chi(SM_s|\sigma_1, \sigma_2) =$$

$$\begin{cases} \delta(\frac{1}{2}|\sigma_1)\,\delta(\frac{1}{2}|\sigma_2) & S = 1 \quad M_s = 1 \\ \frac{1}{\sqrt{2}}\{\delta(\frac{1}{2}|\sigma_1)\,\delta(-\frac{1}{2}|\sigma_2) + \delta(-\frac{1}{2}|\sigma_1)\,\delta(\frac{1}{2}|\sigma_2)\} & S = 1 \quad M_s = 0 \\ \delta(-\frac{1}{2}|\sigma_1)\,\delta(-\frac{1}{2}|\sigma_2) & S = 1 \quad M_s = -1 \\ \frac{1}{\sqrt{2}}\{\delta(\frac{1}{2}|\sigma_1)\,\delta(-\frac{1}{2}|\sigma_2) - \delta(-\frac{1}{2}|\sigma_1)\,\delta(\frac{1}{2}|\sigma_2)\} & S = 0 \quad M_s = 0. \end{cases} \tag{25}$$

The complete wave function is

$$\Psi(SM_s|\mathbf{x}_1, \mathbf{x}_2) = \chi(SM_s|\sigma_1, \sigma_2)\, \Psi(S|\mathbf{r}_1, \mathbf{r}_2). \tag{26}$$

Since χ is symmetric for $S = 1$ and antisymmetric for $S = 0$, for the space function we may put $\Psi(S = 1) = \Psi^-$ and $\Psi(S = 0) = \Psi^+$ where

$$\Psi^\pm(\mathbf{r}_1, \mathbf{r}_2) = \pm\, \Psi^\pm(\mathbf{r}_2, \mathbf{r}_1). \tag{27}$$

Let the asymptotic forms be

$$\Psi^\pm(a\mathbf{k}_a|\mathbf{r}_1, \mathbf{r}_2) \underset{r_2 \to \infty}{\sim} \frac{1}{\sqrt{2}} \Big\{ \psi(a|\mathbf{r}_1) \exp(i\mathbf{k}_a \cdot \mathbf{r}_2) + \sum_{a'} \psi(a'|\mathbf{r}_1) f^{\pm}(a'\hat{\mathbf{r}}_2, a\hat{\mathbf{k}}_a) \frac{\exp ik_{a'}r_2}{r_2} \Big\}. \tag{28}$$

The amplitudes for uncoupled spins are given by

$$f(\mu_1'\mu_2', \mu_1\mu_2) = \sum_{SM_s} C^{\frac{1}{2}\frac{1}{2}S}_{\mu_1'\mu_2'M_s} f(S)\, C^{\frac{1}{2}\frac{1}{2}S}_{\mu_1\mu_2M_s} \tag{29}$$

where $f(S = 0) = f^+$, $f(S = 1) = f^-$. For the cross section (23),

$$I = \frac{k_{a'}}{k_a} \cdot \frac{1}{4}\{|f^+|^2 + 3\,|f^-|^2\}. \tag{30}$$

The unsymmetrized function is

$$\Psi = \frac{1}{\sqrt{2}}[\Psi^{+} + \Psi^{-}] \tag{31}$$

and

$$f = \tfrac{1}{2}(f^{+} + f^{-}), \qquad g = \tfrac{1}{2}(f^{+} - f^{-}). \tag{32}$$

2.4 Expansions of the Complete Wave Function

We consider the nature of the expansion

$$\Psi(\mathbf{r}_1, \mathbf{r}_2) = \sum_a \psi(a|\mathbf{r}_1)\, F(a|\mathbf{r}_2). \tag{33}$$

If a is a bound state, $\psi(a|\mathbf{r}_1)$ goes to zero exponentially for $r_1 \to \infty$. To obtain an asymptotic form such as (17), it is therefore necessary to include positive energy states, for which we put

$$a = \kappa lm$$

where the energy is

$$E_\kappa = \tfrac{1}{2}\kappa^2.$$

We take the functions $\psi(\kappa lm|\mathbf{r})$ to be normalized to

$$\int \psi^*(\kappa lm|\mathbf{r})\, \psi(\kappa' lm|\mathbf{r})\, d^3\mathbf{r} = \delta(E_\kappa - E_{\kappa'}). \tag{34}$$

The radial functions $P_{\kappa l}(r)$ will then have asymptotic form

$$P_{\kappa l}(r) \underset{r\to\infty}{\sim} \left(\frac{2}{\pi\kappa}\right)^{1/2} \sin\left[\kappa r - \tfrac{1}{2} l\pi + \frac{1}{\kappa}\log(2\kappa r) + \arg\Gamma\left(l + 1 - \frac{i}{\kappa}\right)\right].$$

The expansion (33) is

$$\Psi = \sum_{lm}\left\{\sum_{n=1}^{\infty} \psi(nlm|\mathbf{r}_1)\, F(nlm|\mathbf{r}_2) + \int_0^{\infty} \psi(\kappa lm|\mathbf{r}_1)\, F(\kappa lm|\mathbf{r}_2)\, dE_\kappa\right\}. \tag{35}$$

It is shown by Castillejo and associates (1960) that $F(\kappa lm)$ contains singularities at all points $\kappa = k_a$ for which $k_a^2 \geqslant 0$. The asymptotic form (17) for r_1 large is obtained on choosing a path of integration which passes around these singularities below the real κ axis.

For $\Psi^{\pm}$ we may use explicitly symmetrized expansions:

$$\Psi^{\pm}(\mathbf{r}_1, \mathbf{r}_2) = \frac{1}{\sqrt{2}} \sum_a \{\psi(a|\mathbf{r}_1)\, F^{\pm}(a|\mathbf{r}_2) \pm \psi(a|\mathbf{r}_2)\, F^{\pm}(a|\mathbf{r}_1)\}. \tag{36}$$

For a given function $\Psi^{\pm}$, (36) does not define the functions $F^{\pm}$ uniquely; thus, (36) is unaltered if one adds $\alpha\psi(a')$ to $F^{\pm}(a)$ and $(\mp)\alpha\psi(a)$ to $F^{\pm}(a')$. One can choose the functions $F^{\pm}$ to be such that there are no singularities in the integral over E_{κ}. The asymptotic form of $\Psi^{\pm}$ is then determined by the asymptotic form of $F^{\pm}$:

$$\Psi^{\pm}(\mathbf{r}_1, \mathbf{r}_2) \underset{r_2\to\infty}{\sim} \sum_a \psi(a|r_1)\, F^{\pm}(a|r_2). \tag{37}$$

These considerations are of importance for practical calculations. In calculations for excitation of atoms it is usual to employ expansions containing only a few terms and to neglect the integrals over E_{κ}. Such expansions can have correct asymptotic forms, in both variables, only when the explicitly symmetrized expansions are used. But for this advantage there is a price to pay: with the symmetrized functions we have to solve coupled integro-differential equations in place of coupled differential equations.

2.5 Coupled Equations

The equations satisfied by the functions $F(a|\mathbf{r})$ in (33) are obtained from

$$\int \psi^*(a|\mathbf{r}_1)\,[H - E]\,\Psi(\mathbf{r}_1, \mathbf{r}_2)\, d^3\mathbf{r}_1 = 0. \tag{38}$$

This gives

$$[\nabla^2 + k_a^2]\, F(a|\mathbf{r}) = 2 \sum_{a'} V_{aa'}(\mathbf{r})\, F(a'|\mathbf{r}) \tag{39}$$

where

$$V_{aa'}(\mathbf{r}_2) = \int \psi(a|\mathbf{r}_1)\left[-\frac{1}{r_2} + \frac{1}{r_{12}}\right] \psi(a'|\mathbf{r}_1)\, d^3\mathbf{r}_1. \tag{40}$$

With symmetrized functions we obtain in a similar way

$$[\nabla^2 + k_a^2]\, F^{\pm}(a|\mathbf{r}) = 2 \sum_{a'} [V_{aa'}(\mathbf{r}) \pm W_{aa'}(\mathbf{r})]\, F^{\pm}(a'|\mathbf{r}) \tag{41}$$

where $W_{aa'}$ is an exchange operator defined by

$$W_{aa'}(\mathbf{r}_2)\, F(a'|\mathbf{r}_2) = \int \psi^*(a|\mathbf{r}_1)\,[H(\mathbf{r}_1, \mathbf{r}_2) - E]\, F(a'|\mathbf{r}_1)\, d^3\mathbf{r}_1\, \psi(a'|\mathbf{r}_2). \tag{42}$$

2.6 Integrals for Scattering Amplitudes: Threshold Laws

The following theorems are proved by Mott and Massey (1949):

I. Let $u(\mathbf{r})$ be such that $ru \to 0$ for $r \to \infty$. Then, functions $\mathscr{F}(\mathbf{k}, \mathbf{r})$ exist which satisfy

$$[\nabla^2 - u + k^2]\mathscr{F}(\mathbf{k}, \mathbf{r}) = 0 \tag{43}$$

$$\mathscr{F}(\mathbf{k}, \mathbf{r}) \underset{r\to\infty}{\sim} \exp(i\mathbf{k}\cdot\mathbf{r}) + p(\hat{\mathbf{r}})\frac{\exp ikr}{r}. \tag{44}$$

II. If $ru \to 0$ and $rA \to 0$ for $r \to \infty$, functions $\mathscr{G}(\mathbf{r})$ exist which satisfy

$$[\nabla^2 - u + k^2]\,\mathscr{G}(\mathbf{r}) = A(\mathbf{r}), \tag{45}$$

$$\mathscr{G}(\mathbf{r}) \underset{r\to\infty}{\sim} h(\hat{\mathbf{r}})\frac{\exp ikr}{r}. \tag{46}$$

Furthermore,

$$h(\hat{\mathbf{k}}) = -\frac{1}{4\pi}\int \mathscr{F}(-\mathbf{k}, \mathbf{r})\,A(\mathbf{r})\,d^3\mathbf{r}. \tag{47}$$

If a and a' are bound states, or if either one of them is a bound state, it may be shown that $rV_{aa'} \to 0$ for $r \to \infty$. Theorems I and II may therefore be used to justify our asymptotic forms for excitation of neutral atoms.

On writing (39) in the form

$$[\nabla^2 + k_{a'}^2]\,F(a'|\mathbf{r}_2) = 2\int \psi^*(a'|\mathbf{r}_1)\left[-\frac{1}{r_2} + \frac{1}{r_{12}}\right]\Psi(\mathbf{r}_1, \mathbf{r}_2)\,d^3\mathbf{r}_1 \tag{48}$$

and using (47) we obtain

$$f(a'\hat{\mathbf{k}}_{a'}, a\mathbf{k}_a)$$

$$= -\frac{1}{2\pi}\int \psi^*(a'|\mathbf{r}_1)\exp(-i\mathbf{k}_{a'}\cdot\mathbf{r}_2)\left[-\frac{1}{r_2} + \frac{1}{r_{12}}\right]\Psi(a\mathbf{k}_a|\mathbf{r}_1, \mathbf{r}_2)\,d^3\mathbf{r}_1\,d^3\mathbf{r}_2. \tag{49}$$

Let $I(a \to a')$ be an excitation cross section. At threshold $k_{a'} \to 0$, but the complete wave function Ψ remains finite and therefore the scattering amplitude remains finite. Since $I(a \to a') = (k_{a'}/k_a) \mid f(a', a) \mid^2$, I is proportional to $k_{a'}$ just above threshold and is independent of the direction of $\mathbf{k}_{a'}$. Figure 3 shows experimental results, for $1s \to 2p$ transitions in hydrogen, which appear to be in satisfactory agreement with the theory. Threshold laws are discussed further in § 3.7.

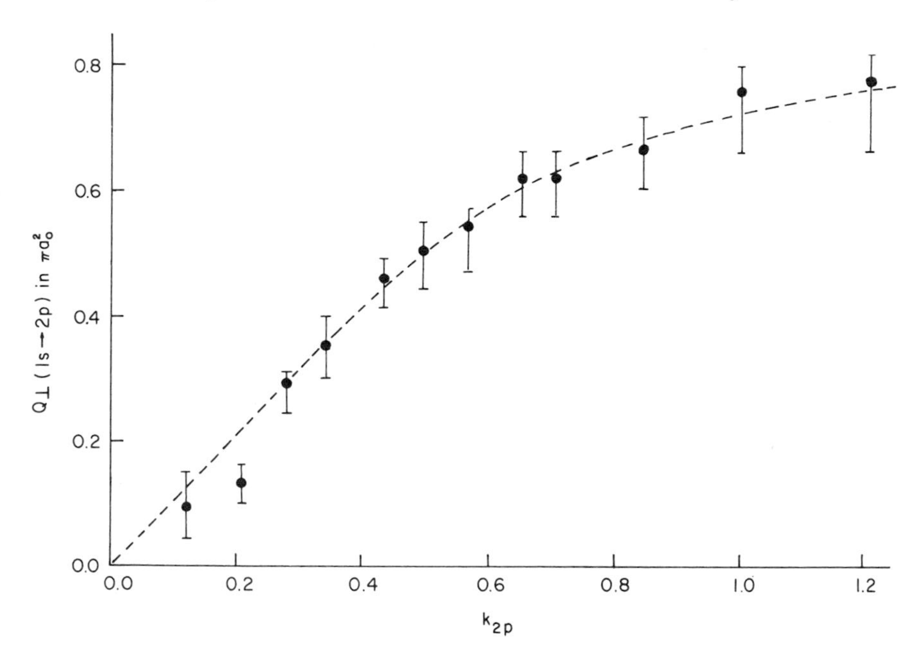

FIG. 3. The threshold law for H $1s \to 2p$. The cross section $Q_\perp(1s \to 2p)$ is obtained (Fite *et al.*, 1959) on counting photons emitted perpendicular to the electron beam and assuming an isotropic photon distribution. This is plotted against k_{2p}, the final electron velocity in atomic units; the final electron energy is k_{2p}^2 in 13.60 ev units.

2.7 Weak Coupling Approximations

We consider the equation

$$[\nabla^2 + k_{a'}^2] F(a', a|\mathbf{r}) = 2 \sum_{a''} V_{a', a''} F(a'', a|\mathbf{r}) \tag{50}$$

where a is the initial state,

$$F(a', a) \underset{r \to \infty}{\sim} \exp(i\mathbf{k}_a \cdot \mathbf{r})\, \delta(a', a) + \frac{\exp ik_{a'}r}{r} f(a', a). \tag{51}$$

Let the kinetic energies, $\frac{1}{2}k_{a'}^2$, be so large that the right-hand side of (50) can be treated as a small perturbation. We then have

$$F(a', a) \simeq \exp(i\mathbf{k}_{a'} \cdot \mathbf{r})\, \delta(a', a). \tag{52}$$

On substituting this approximation on the right-hand side, we have the equations of the first Born approximation,

$$[\nabla^2 + k_{a'}^2]\, F_{B1}(a',a) = 2V_{a'a} \exp i\mathbf{k}_a \cdot \mathbf{r} \tag{53}$$

from which one obtains

$$f_{B1}(a', a) = -\frac{1}{2\pi} \int \exp(-i\mathbf{k}_{a'} \cdot \mathbf{r})\, V_{a'a} \exp(i\mathbf{k}_a \cdot \mathbf{r})\, d^3\mathbf{r}. \tag{54}$$

To obtain the distorted wave approximation we suppose that $V_{aa''}$ is small for all $a'' \neq a$ and that $V_{a'a''}$ is small for all $a'' \neq a'$. We then have, in place of (50),

$$[\nabla^2 + k_{a'}^2]\, F(a, a) = 2\, V_{aa}\, F(a, a) \tag{55}$$

$$[\nabla^2 + k_{a'}^2]\, F(a', a) = 2\, V_{a'a'}\, F(a', a) + 2\, V_{a'a}\, F(a, a), \tag{56}$$

and from this we obtain

$$f_{DW}(a', a) = -\frac{1}{2\pi} \int \mathscr{F}(a', -\mathbf{k}_{a'})\, V_{a'a}\, \mathscr{F}(a, \mathbf{k}_a)\, d^3\mathbf{r} \tag{57}$$

where

$$[\nabla^2 - 2\, V_{a''a''} + k_{a''}^2]\, \mathscr{F}(a'', \mathbf{k}_{a''}) = 0, \tag{58}$$

$$\mathscr{F}(a'', \mathbf{k}_{a''}) \sim \exp(i\mathbf{k}_{a''} \cdot \mathbf{r}) + f_{a''} \frac{\exp ik_{a''}r}{r}. \tag{59}$$

The exchange distorted wave approximation is obtained from the foregoing equations on replacing V by $(V \pm W)$. The Born-Oppenheimer approximation is

$$f_{BO}^{\pm} = -\frac{1}{2\pi} \int \exp(-i\mathbf{k}_{a'} \cdot \mathbf{r})\, [V_{a'a} \pm W_{a'a}] \exp(i\mathbf{k}_a \cdot \mathbf{r})\, d^3\mathbf{r}. \tag{60}$$

In § 3 we shall give more precise criteria for determining the validity of these approximations.

2.8 The Second Born Approximation

The Born approximation may be treated as an iterative procedure. In the first Born approximation the wave functions, solutions of (53), are

$$F_{B1}(a', a|\mathbf{r}) = \exp(i\mathbf{k}_a \cdot \mathbf{r})\, \delta(a', a) - \frac{1}{2\pi} \int \frac{\exp ik_{a'} |\mathbf{r} - \mathbf{r}'|}{|\mathbf{r} - \mathbf{r}'|} \times V_{a'a}(\mathbf{r}') \exp(i\mathbf{k}_a \cdot \mathbf{r}')\, d^3\mathbf{r}'. \quad (61)$$

On substituting on the right-hand side of (50), in the second Born approximation

$$f_{B2}(a', a) = -\frac{1}{2\pi} \sum_{a''} \int \exp(-i\mathbf{k}_{a'} \cdot \mathbf{r})\, V_{a'a''} F_{B1}(a'', a|\mathbf{r})\, d^3\mathbf{r}. \quad (62)$$

On writing the potential, $[-(1/r_2) + (1/r_{12})]$, as λv_0, the Born approximations may be considered to be an expansion in powers of λ. In the first Born approximation, f is proportional to λ and I to λ^2, while in the second approximation f contains a term in λ^2 and I a term in λ^4. In the third approximation, one has a λ^3 term in f which also contributes to the λ^4 term in I and it is shown by Kingston *et al.* (1960) that this may partially cancel the λ^4 term in the second approximation. In using the second Born approximation it is therefore best to omit the λ^4 term in I.

2.9 Collisions with Positive Ions

We consider electron collisions with hydrogenic positive ions, containing one electron and having a nucleus of charge Z. In place of (39) we now obtain

$$\left[\nabla^2 + \frac{2(Z-1)}{r} + k_a^2\right] F(a|\mathbf{r}) = 2 \sum_{a'} V_{aa'}(\mathbf{r})\, F(a'|\mathbf{r}) \quad (63)$$

where $V_{aa'}$ is defined by (40) and the $\psi(a)$ are wave functions for the ion. The term $2(Z-1)/r$ on the left-hand side introduces some modifications into the asymptotic forms. For scattering by a pure Coulomb potential we have

$$\left[\nabla^2 + \frac{2z}{r} + k^2\right] \chi(z, \mathbf{k}, \mathbf{r}) = 0 \quad (64)$$

with solution (Mott and Massey, 1949, p. 46)

$$\chi = \exp(\pi\gamma/2)\, \Gamma(1 - i\gamma) \exp(i\mathbf{k} \cdot \mathbf{r})\, \Phi[i\gamma, 1; i(kr - \mathbf{k} \cdot \mathbf{r})] \tag{65}$$

where $\gamma = z/k$ and where

$$\Phi(a, b; x) = 1 + \frac{ax}{b1!} + \frac{a(a+1)x^2}{b(b+1)2!} + \dots. \tag{66}$$

The function χ has asymptotic form

$$\chi \underset{r\to\infty}{\sim} P(\mathbf{k}, \mathbf{r}) + S(k, r) f_c(\theta) \tag{67}$$

where $\mathbf{k} \cdot \mathbf{r} = kr \cos\theta$ and where

$$P(\mathbf{k}, \mathbf{r}) = (\exp i\mathbf{k} \cdot \mathbf{r}) \times \exp - i\gamma \log [kr - \mathbf{k} \cdot \mathbf{r}]$$

$$S(k, r) = \left(\frac{\exp ikr}{r}\right) \times \exp i\gamma \log(2kr)$$

$$f_c(\theta) = \left(\frac{\gamma}{k(1 - \cos\theta)}\right) \times \exp i\{\gamma \log [\tfrac{1}{2}(1 - \cos\theta)] + 2 \arg \Gamma(1 - i\gamma)\}. \tag{86}$$

The Coulomb differential cross section is $I_c = |f_c|^2$. The function $P(\mathbf{k}, \mathbf{r})$ is a modified plane wave, replacing $\exp i\mathbf{k} \cdot \mathbf{r}$ and $S(k, r)$ is a modified scattered wave, replacing $(\exp ikr)/r$. For the electron-ion collision problem the asymptotic forms must be expressed in terms of P and S. Similar modifications occur in the integrals for the scattering amplitudes. Thus, in place of (49), one has

$$f(a'\hat{\mathbf{k}}_{a'}, a\hat{\mathbf{k}}_a)$$
$$= -\frac{1}{2\pi} \int \psi^*(a'|\mathbf{r}_1)\, \chi(Z - 1, -\mathbf{k}_{a'}, \mathbf{r}_2) \left[-\frac{1}{r_2} + \frac{1}{r_{12}}\right] \Psi(a\mathbf{k}_a|\mathbf{r}_1, \mathbf{r}_2)\, d^3\mathbf{r}_1\, d^3\mathbf{r}_2. \tag{69}$$

To obtain the threshold law one requires the behavior of $\chi(z, \mathbf{k}, \mathbf{r})$ for $k \to 0$. From (65) it may be shown that

$$\chi(z, \mathbf{k}, \mathbf{r}) \underset{k\to 0}{\sim} (2\pi\gamma)^{1/2} \exp(i[\gamma - \gamma \log\gamma - \tfrac{1}{4}\pi])\, J_0[(8zr)^{1/2} \sin\theta/2] \tag{70}$$

where $\gamma = z/k$ and J_0 is a Bessel function. It follows that $|f|$ behaves as $k_{a'}^{-1/2}$ for $k_{a'} \to 0$ and that $I(a \to a')$ remains finite at threshold.

In the Coulomb-Born approximation, the right-hand side of (63) is treated as a perturbation. If we include exchange terms on the right-

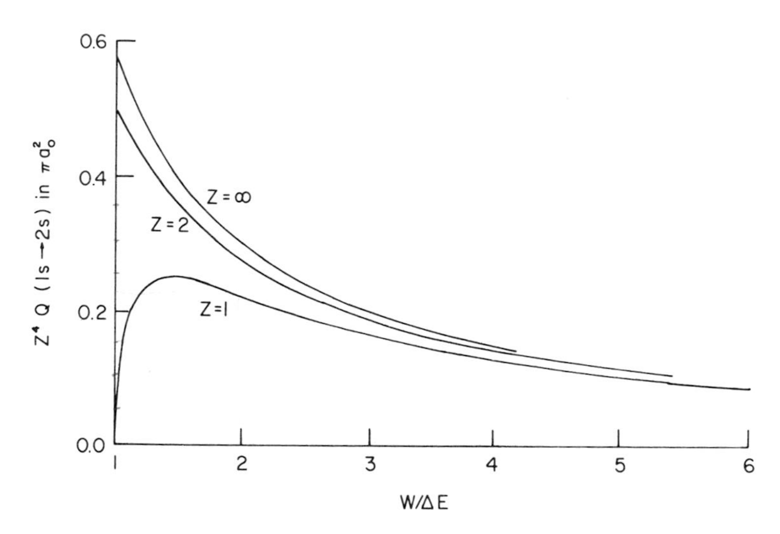

FIG. 4. Cross sections for transitions in hydrogenic ions calculated in the Coulomb-Born approximation (Born for $Z = 1$). Z^4Q is plotted against $(W/\Delta E)$, W being the incident electron energy and ΔE the transition energy difference. (a) $1s \rightarrow 2s$ calculations by Tully (1960); (b) $1s \rightarrow 2p$ calculations by Burgess (1961).

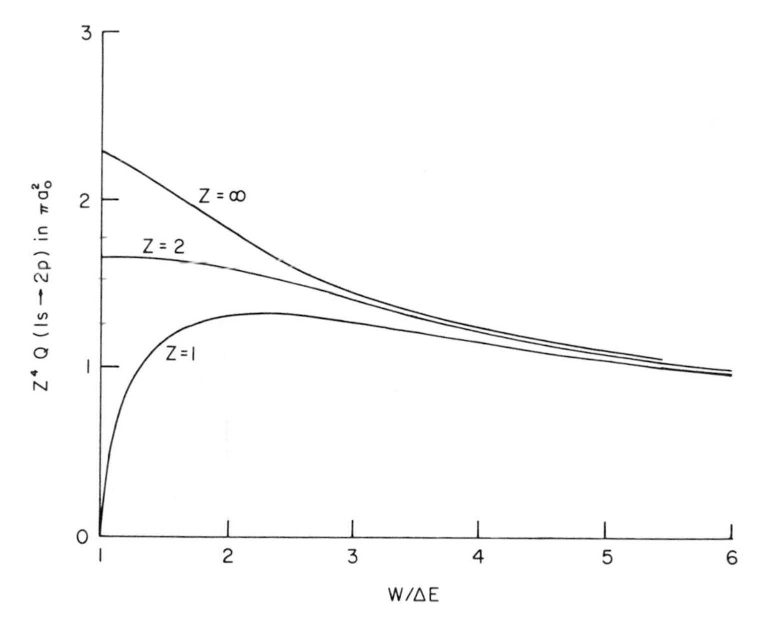

hand side we have the Coulomb-Born-Oppenheimer (CBO) approximation,

$$f^{\pm}_{\mathrm{CBO}}(a'\hat{\mathbf{k}}_{a'}, a\hat{\mathbf{k}}_a) = -\frac{1}{2\pi}\int \chi(Z-1, -\mathbf{k}_{a'}, \mathbf{r})\,[V_{a'a} \pm W_{a'a}]\,\chi(Z-1, \mathbf{k}_a, \mathbf{r})\,d^3\mathbf{r}. \tag{71}$$

In the limit of $Z \to \infty$, the terms treated as perturbations are vanishingly small compared with the terms treated exactly and the CBO approximation is exact even at threshold.

It is of interest to consider the cross sections $Q(a \to a')$ for a definite transition in the isoelectronic sequence obtained when Z varies. In the limit of large k the Coulomb wave χ approximates to a plane wave and the Born approximation may be used. It may then be shown that $Z^4 Q(a \to a')$, as a function of $k_a^2/(k_a^2 - k_{a'}^2)$, does not depend on Z; thus, the high energy cross sections behave as Z^{-4}, in agreement with classical theory [in (8), $(I/I_{\mathrm{H}}) = Z^2$ for the hydrogen isoelectronic sequence].

Figure 4 shows the results obtained in the CB approximation for $1s \to 2s$ (Tully, 1960) and $1s \to 2p$ (Burgess, 1961) transitions in the hydrogen isoelectronic sequence.

2.10 Ionization

In § 2.4 we introduced the functions $\psi(\kappa lm)$, normalized to $\delta(E_\kappa - E_{\kappa'})$ for the positive energy atom states. When ionization can occur, to the asymptotic form of the wave functions (16) one must add

$$\sum_{lm}\int \psi(\kappa lm|\mathbf{r}_1)\,f(\kappa lm\hat{\mathbf{r}}_2, a\hat{\mathbf{k}}_a)\,\frac{\exp ikr_2}{r_2}\,dE \tag{72}$$

where the total energy is

$$E = \tfrac{1}{2}(k^2 + \kappa^2). \tag{73}$$

The total cross section for ionization from state a is

$$Q_{\mathrm{ion}}(a) = \int_0^E Q(a \to E_\kappa)\,dE_\kappa \tag{74}$$

where

$$Q(a \to E_\kappa) = \frac{k}{k_a}\sum_{lm}\int |f(\kappa lm\hat{\mathbf{k}}, a\hat{\mathbf{k}}_a)|^2\,d\hat{\mathbf{k}}. \tag{75}$$

When one uses symmetrized functions one has, in place of $|f(\kappa)|^2$,

$$\tfrac{1}{2}\{\tfrac{1}{4}|f^{+}(\kappa)|^2 + \tfrac{3}{4}|f^{-}(\kappa)|^2\}. \tag{76}$$

The factor of $\frac{1}{2}$ in (76) occurs because the ionization cross section is defined in terms of the number of ionizations, not in terms of the total number of electrons either scattered or ejected (Peterkop, 1961).

It is of importance to consider the effect of screening in the final state for ionization problems. For ionization of hydrogen one has

$$[\nabla^2 + k^2]\, F(\kappa|\mathbf{r}) = 2\left\{\sum_a V_{\kappa a}\, F(a|\mathbf{r}) + \int V_{\kappa\kappa'}\, F(\kappa'|\mathbf{r})\, dE_{\kappa'}\right\} \tag{77}$$

where $E = \frac{1}{2}(k^2 + \kappa^2)$. For $r \to \infty$ the $V_{\kappa a}$ go to zero at least as fast as r^{-2}, but it is shown in the appendix that

$$\int V_{\kappa\kappa'}\, F(\kappa')\, dE_{\kappa'} \underset{r\to\infty}{\sim} \frac{\beta(\kappa, k)}{r}\, F(\kappa) \tag{78}$$

where

$$\beta(\kappa, k) = \begin{cases} 0 & \text{for} \quad \kappa < k \\ [1 - (k/\kappa)] & \text{for} \quad \kappa > k. \end{cases} \tag{79}$$

Thus, in considering the potential for the scattered electron, one can say that the ejected electron screens the nucleus completely only when its velocity is less than that of the scattered electron.

We have defined E as the total energy, so that $E \to 0$ at the ionization threshold. It may be shown that the functions $\psi(\kappa lm)$ remain finite for $E_\kappa \to 0$ and therefore, for finite k, the cross sections $Q(a \to E_\kappa)$ are always finite for $E_\kappa \to 0$. When one uses Coulomb functions for the scattered electron (velocity k) the $Q(a \to E_\kappa)$ remain finite for $k \to 0$ and, for small E, Q_{ion} behaves as E. When allowance is made for incomplete screening it is seen that this is the correct threshold law for ionization of neutral atoms† as well as positive ions. The Born approximation, which assumes complete screening, gives an incorrect threshold law for ionization, Q_{ion} behaving as $E^{3/2}$.

† Geltman (1956) used Coulomb waves (with $\beta = 1$) for the scattered electron and hence obtained the correct threshold law, but his results give only a lower limit for the threshold slope since he considered only s-waves (see § 3.7).

Figure 5 shows experimental results for near-threshold ionization of hydrogen together with the calculated Born cross section and Fig. 6 shows ionization cross sections calculated in the CB approximation for the hydrogenic ions.

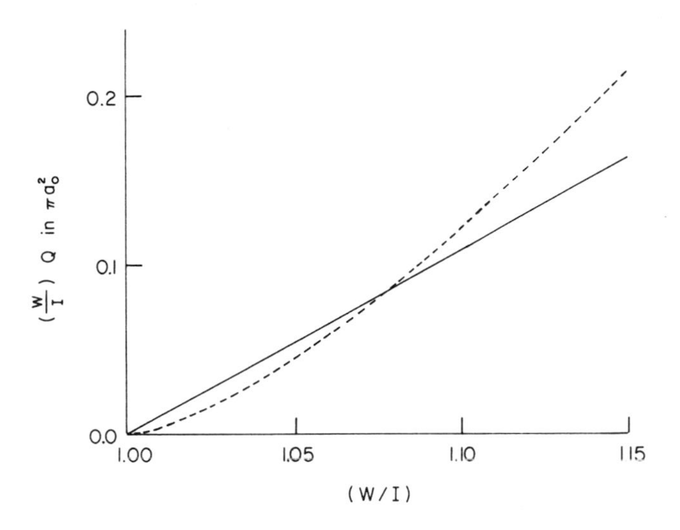

FIG. 5. Cross sections for near-threshold ionization of hydrogen; $(W/I)Q$ against (W/I). Full line, curve experimental; dashed curve, Born approximation. The experimental results give a variation as $[(W/I) - 1]$ compared with $[(W/I) - 1]^{3/2}$ in the Born approximation.

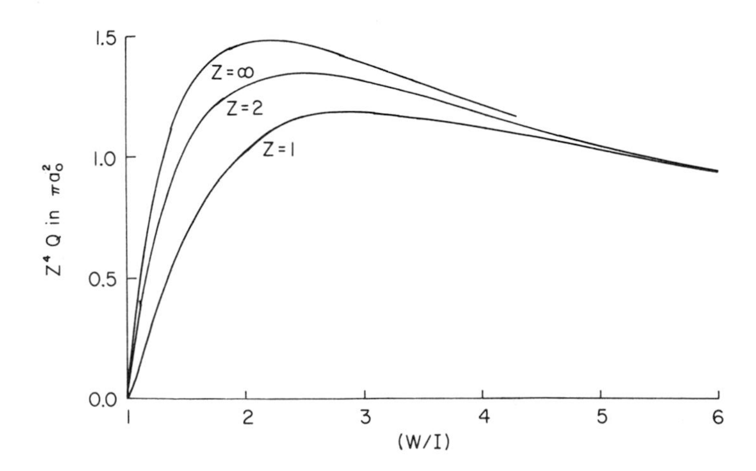

FIG. 6. Cross sections for ionization of hydrogenic ions calculated by Burgess (1960) in the CB approximation (Born for $Z = 1$); Z^4Q against (W/I).

2.11 Transitions in Complex Atoms

When one neglects spin terms in the Hamiltonian, a transition involving a change in atom spin can be brought about only through electron exchange. For hydrogen a change in spin can involve only μ_1, which describes the orientation of the atomic electron, but for complex atoms one may also have changes in S_1, the quantum number for the total spin of the atomic electrons. A change in S_1 can be brought about only through electron exchange. For transitions not involving a change in S_1, exchange may still be important, but one can obtain finite cross sections in approximations which neglect exchange.

Calculations for complex atoms differ from those for hydrogen in another important respect. For complex atoms it is necessary to use approximate atomic wave functions and this may introduce additional errors.

2.12 Polarization of Impact Radiation

Let the transition $a \rightarrow a'$ be excited by an electron beam directed along the Oz axis and assume, to take the simplest case, that a is an S state. Radiation subsequently emitted in the $a' \rightarrow a''$ transition will be partially polarized and will have an intensity $\mathscr{I}(\theta)$ which varies with the angle θ between Oz and the direction of observation. When viewed perpendicular to Oz, let the intensities be $\mathscr{I}^{||}$ and $\mathscr{I}^{\perp}$ for radiation with electric vectors parallel and perpendicular to Oz. On defining the polarization fraction as

$$P = (\mathscr{I}^{||} - \mathscr{I}^{\perp})/(\mathscr{I}^{||} + \mathscr{I}^{\perp}) \tag{80}$$

one has

$$\mathscr{I}(\theta) = \mathscr{I} \left[\frac{1 - P\cos^2\theta}{1 - P/3}\right] \tag{81}$$

where $\mathscr{I}$ is the mean intensity.

The theoretical expressions for P (Percival and Seaton, 1958) involve the cross sections $Q_{M'}$ for exciting the M' states of the upper level. Thus, for a $P \rightarrow S$ radiative transition one has, neglecting spin,

$$P = (Q_0 - Q_1)/(Q_0 + Q_1). \tag{82}$$

To obtain P as a function of energy one must calculate the ratio (Q_1/Q_0). At threshold this ratio is determined by angular momentum conservation. With an initial S state the entire system has zero angular momentum about Oz. The scattered electron cannot have any angular momentum when its velocity tends to zero, and in this limit only $M' = 0$ can be excited. Hence, at threshold, $(Q_1/Q_0) = 0$ and $P = 1$.

The polarization is reduced when fine structure is included and is further reduced by hyperfine structure when the natural linewidth is small compared with the hfs separations. Percival and Seaton consider the case of the Ly-α line of hydrogen for which the width is comparable with the hfs separations.

Skinner and Appleyard (1927) have measured P for a number of Hg lines. At energies not close to thresholds, the general pattern of their results is in agreement with what theory would lead one to expect, but as the excitation thresholds are approached the measurements appear to give P tending to zero. This is illustrated in Fig. 7. There is no satisfactory explanation of this near-threshold behavior of the experimental results.

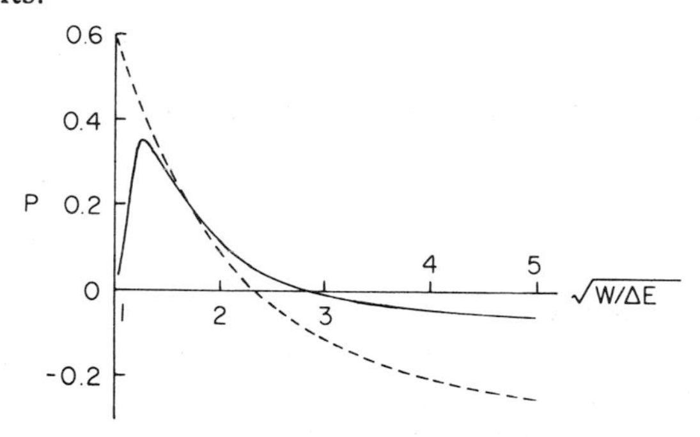

FIG. 7. The polarization fraction for radiative ${}^1D \rightarrow {}^1P$ transitions without hyperfine structure; P against $\sqrt{W/\Delta E}$. Dashed curve calculated for He $3^1D \rightarrow {}^1P$ (Percival and Seaton, 1958), full line curve, experimental for Hg $7^1D \rightarrow 6^1P$ (Skinner and Appleyard, 1927).

3 *Partial Wave Theory*

The partial wave theory, in which one considers states of definite angular momentum for the colliding electron, is needed for all practical calculations other than those using the Born approximation. In place of the partial differential equations of § 2.5 one obtains differential equations in one variable which may be solved by numerical methods.

The partial wave theory also enables us to obtain much more precise estimates of the reliability of different approximations.

3.1 The Scattering Matrix

The functions

$$\varphi_{\pm}(klm|\mathbf{r}) = k^{-1/2}\, Y_{lm}(\hat{\mathbf{r}})\, \frac{\exp \pm i(kr - \frac{1}{2}l\pi)}{r} \tag{83}$$

represent waves with angular momentum lm and a total current of one particle per unit time. The function φ_- represents an incoming wave and φ_+ represents an outgoing wave.

We use quantum numbers $a = nl_1m_1$ for the atomic electron and $k_al_2m_2$ for the colliding electron, and we consider complete wave functions $\Psi_{\mathrm{S}}(al_2m_2|\mathbf{r}_1, \mathbf{r}_2)$ having asymptotic form

$$\Psi_{\mathrm{S}}(al_2m_2|\mathbf{r}_1, \mathbf{r}_2) \underset{r_2\to\infty}{\sim} \psi(a|\mathbf{r}_1)\, \varphi_-(k_al_2m_2|\mathbf{r}_2) - \sum_{a'\, l_2'm_2'} \psi(a'|\mathbf{r}_1)\, \varphi_+(k_{a'}l_2'm_2'|\mathbf{r}_2)\, S(a'l_2'm_2', al_2m_2) \tag{84}$$

where $\mathbf{S}$ is the scattering matrix. For elastic scattering $S(l) = e^{2i\eta_l}$. The condition for conservation of current is†

$$\sum_{\alpha'} |\, S(\alpha, \alpha')\, |^2 = 1. \tag{85}$$

When the coupling is weak for $\alpha \to \alpha'$ one must have

$$|\, S(\alpha, \alpha')\, | \ll 1.$$

To obtain the scattering amplitude in terms of $\mathbf{S}$ we require the relations:

$$\exp i\mathbf{k}\cdot\mathbf{r} = \sum_l (2l+1)\, P_l(\hat{\mathbf{k}}\cdot\hat{\mathbf{r}})\, i^l \frac{1}{kr}\, j_l(kr), \tag{86}$$

$$j_l(x) = \left(\frac{\pi x}{2}\right)^{1/2} J_{l+1/2}(x) \underset{x\to\infty}{\sim} \sin(x - \tfrac{1}{2}l\pi), \tag{87}$$

and

$$P_l(\hat{\mathbf{k}}\cdot\hat{\mathbf{r}}) = \left(\frac{4\pi}{2l+1}\right) \sum_m Y_{lm}^*(\hat{\mathbf{k}})\, Y_{lm}(\hat{\mathbf{r}}). \tag{88}$$

† When unsymmetrized functions are used, in (85) we must include contributions from exchange amplitudes.

One obtains

$$f(a'\hat{\mathbf{k}}_{a'}, a\hat{\mathbf{k}}_a)$$

$$= \frac{2\pi i}{(k_a k_{a'})^{1/2}} \sum_{l_2 m_2 l_2' m_2'} Y^*_{l_2 m_2}(\hat{\mathbf{k}}_a)\, Y_{l_2' m_2'}(\hat{\mathbf{k}}_{a'})\, i^{l_2 - l_2'}\, T(a' l_2' m_2', a l_2 m_2) \tag{89}$$

where

$$T(\alpha', \alpha) = \delta(\alpha', \alpha) - S(\alpha', \alpha) \tag{90}$$

defines an element of the transmission matrix, $\mathbf{T}$. On substituting in (19) we obtain†

$$Q(nl_1 \to n'l_1') = \frac{\pi}{k_n^2} \cdot \frac{1}{(2l_1 + 1)} \sum_{m_1 m_1'} \sum_{l_2 l_2'} \sum_{m_2 m_2'} |\, T(n'l_1'm_1'l_2'm_2', nl_1 m_1 l_2 m_2)\,|^2. \tag{91}$$

A simpler expression is obtained on introducing the total angular momentum, LM. Define

$$\Phi_{\pm}(nl_1 l_2 LM|\mathbf{r}_1, \mathbf{r}_2) = \sum_{m_1 m_2} C^{l_1 l_2 L}_{m_1 m_2 M}\, \psi(nl_1 m_1|\mathbf{r}_1)\, \varphi_{\pm}(k_n l_2 m_2|\mathbf{r}_2) \tag{92}$$

and let

$$\Psi_{\mathbf{S}}(\alpha|\mathbf{r}_1, \mathbf{r}_2) \underset{r_2 \to \infty}{\sim} \Phi_{-}(\alpha|\mathbf{r}_1, \mathbf{r}_2) - \sum_{\alpha'} \Phi_{+}(\alpha'|\mathbf{r}_1, \mathbf{r}_2)\, S(\alpha', \alpha). \tag{93}$$

The matrix $S(n'l_1'l_2'L'M', nl_1 l_2 LM)$ is diagonal with respect to LM, since the total angular momentum does not change, and is independent of M, this quantum number serving only to define the orientation of the entire system with respect to some arbitrary direction. The $\mathbf{S}$ matrix satisfies transformation relations of the type

$$S(l_1'm_1'l_2'm_2', l_1 m_1 l_2 m_2) = \sum_{LM} C^{l_1' l_2' L}_{m_1' m_2' M}\, S(l_1' l_2' L, l_1 l_2 L)\, C^{l_1 l_2 L}_{m_1 m_2 M}. \tag{94}$$

Equation (91) becomes

$$Q(nl_1 \to n'l_1') = \frac{\pi}{k_n^2} \frac{1}{(2l_1 + 1)} \sum_{l_2 l_2' L} (2L + 1)\, |\, T(n'l_1'l_2'L, nl_1 l_2 L)\,|^2. \tag{95}$$

† Expression (19) is independent of the direction of $\mathbf{k}_n$. The simplest derivation of (91) is obtained on integrating (19) over $d\hat{\mathbf{k}}_n$ and dividing by 4π.

When we include spin variables and use antisymmetric functions $\Psi(\mathbf{x}_1, \mathbf{x}_2)$, we obtain

$$Q(nl_1 \rightarrow n'l_1') = \frac{\pi}{k_n^2} \cdot \frac{1}{4(2l_1+1)} \sum_{l_2 l_2' SL} (2L+1)(2S+1) \mid T(n'l_1'l_2'SL, nl_1l_2SL) \mid^2. \tag{96}$$

These expressions may be compared with the classical theory (§ 1); $\mid T(\alpha', \alpha) \mid^2$ is the quantity corresponding to $P_{\alpha\alpha'}$.

3.2 The Reactance Matrix

It is usually convenient to work with real functions. We define

$$\varphi_{\binom{s}{c}}(klm|\mathbf{r}) = k^{-1/2}\, Y_{lm}(\hat{\mathbf{r}}) \frac{1}{r} \binom{\sin}{\cos} (kr - \tfrac{1}{2} l\pi) \tag{97}$$

and Φ_s, Φ_c by relations similar to (92). The $\mathbf{R}$ matrix is defined by a function $\Psi_{\mathbf{R}}$ having asymptotic form

$$\Psi_{\mathbf{R}}(\alpha) \sim \Phi_s(\alpha) + \sum_{\alpha'} \Phi_c(\alpha')\, R(\alpha', \alpha). \tag{98}$$

On using

$$2i\Phi_s = \Phi_+ - \Phi_-, \qquad 2\Phi_c = \Phi_+ + \Phi_- \tag{99}$$

one obtains

$$\mathbf{S} = \frac{1 + i\mathbf{R}}{1 - i\mathbf{R}}. \tag{100}$$

For elastic scattering, $R(l) = \tan \eta_l$.

3.3 Properties of the R and S Matrices

In the representation

$$\alpha = nl_1l_2SL \tag{101}$$

and with the usual phase conventions (Edmonds, 1957) it may be shown that $\mathbf{R}$ is real and symmetric,

$$\mathbf{R} = \mathbf{R}^* = \tilde{\mathbf{R}}, \tag{102}$$

and that $\mathbf{S}$ is unitary and symmetric,

$$\mathbf{S}^{\dagger}\mathbf{S} = 1, \qquad \mathbf{S} = \tilde{\mathbf{S}} \tag{103}$$

The conservation relation (85) is a special case of the unitary property. From $S = \tilde{S}$ we obtain the reciprocity relations, such as

$$(2l_1 + 1)\, k_n^2\, Q(nl_1 \to n'l_1') = (2l_1' + 1)\, k_{n'}^2\, Q(n'l_1' \to nl_1). \tag{104}$$

3.4 Radial Equations

Let us define

$$\Phi(nl_1l_2L|\mathbf{r}_1, \mathbf{r}_2) = \sum_{m_1m_2} C^{l_1l_2L}_{m_1m_2M}\, \psi(nl_1m_1|\mathbf{r}_1)\, Y_{l_2m_2}(\hat{\mathbf{r}}_2)\, \frac{1}{r_2}\, F(nl_1l_2L|r_2) \tag{105}$$

and let the complete wave function be

$$\Psi^{\pm}(\alpha) = \frac{1}{\sqrt{2}} \sum_{\alpha'} \{\Phi(\alpha'|\mathbf{r}_1, \mathbf{r}_2) \pm \Phi(\alpha'|\mathbf{r}_2, \mathbf{r}_1)\}. \tag{106}$$

For the radial functions $F(\alpha)$ one obtains equations of the form

$$\left\{\frac{d^2}{dr^2} - \frac{l_2(l_2+1)}{r^2} + k^2\right\} F(\alpha) = 2 \sum_{\alpha'} \{V_{\alpha\alpha'} \pm W_{\alpha\alpha'}\}\, F(\alpha'). \tag{107}$$

These are discussed in detail by Percival and Seaton (1957).

3.5 Variational Principles

For any symmetrized function $\Psi_t\ (\equiv \Psi_t^{\pm})$ with asymptotic form

$$\Psi_t(\alpha) \underset{r_2 \to \infty}{\sim} \frac{1}{\sqrt{2}} \left\{\Phi_s(\alpha|\mathbf{r}_1, \mathbf{r}_2) + \sum_{\alpha'} \Phi_c(\alpha'|\mathbf{r}_1, \mathbf{r}_2)\, R_t(\alpha', \alpha)\right\}, \tag{108}$$

define

$$L_t(\alpha, \alpha') = \int \Psi_t^*(\alpha)\, [H - E]\, \Psi_t(\alpha')\, d^3\mathbf{r}_1\, d^3\mathbf{r}_2. \tag{109}$$

By a generalization of the argument giving the Hulthén and Kohn variational principles for elastic scattering (Chapter 9, § 1.7) it may be shown that

$$R(\alpha, \alpha') = R_t(\alpha, \alpha') - 2L_t(\alpha, \alpha') + 2\int \delta\Psi^*(\alpha)\,[H - E]\,\delta\Psi(\alpha')\,d^3\mathbf{r}_1\,d^3\mathbf{r}_2 \quad (110)$$

where $\delta\Psi = \Psi_t - \Psi$, Ψ being the exact wave function and $\mathbf{R}$ the exact reactance matrix. Given any approximate wave function Ψ_t, with corresponding R_t, an improved estimate for R is

$$\mathbf{R}_{\mathrm{K}} = \mathbf{R}_t - 2\mathbf{L}_t. \quad (111)$$

3.6 Weak Coupling Approximations

In weak coupling approximations one first obtains functions

$$\Psi_t(\alpha) = \frac{1}{\sqrt{2}}\{\Phi_t(\alpha|\mathbf{r}_1, \mathbf{r}_2) \pm \Phi_t(\alpha|\mathbf{r}_2, \mathbf{r}_1)\} \quad (112)$$

neglecting the coupling, so that $R_t(\alpha, \alpha') = 0$ for $\alpha \neq \alpha'$, and then calculates

$$R_{\mathrm{K}}(\alpha, \alpha') = -\,2L_t(\alpha, \alpha'). \quad (113)$$

In terms of the radial functions this gives

$$R_{\mathrm{K}}(\alpha, \alpha') = -\,2\int F(\alpha|r)\,[V_{\alpha\alpha'} \pm W_{\alpha\alpha'}]\,F(\alpha'|r)\,dr. \quad (114)$$

In the exchange distorted wave method one would have

$$\left\{\frac{d^2}{dr^2} - \frac{l_2(l_2+1)}{r^2} - 2\,(V_{\alpha\alpha} \pm W_{\alpha\alpha}) + k_\alpha^2\right\} F\,(\alpha|r) = 0, \quad (115)$$

$$F(\alpha|r) \sim k_\alpha^{-1/2}\{\sin\,(k_\alpha r - \tfrac{1}{2}\,l_2\pi) + \cos\,(k_\alpha r - \tfrac{1}{2}\,l_2\pi)\,R_t(\alpha, \alpha)\}. \quad (116)$$

In the Born approximation one uses†

$$F_{\mathrm{B}}(\alpha|r) = k_\alpha^{-1/2} j_{l_2}(k_\alpha r) \quad (117)$$

and neglects $W_{\alpha\alpha'}$ in (114).

† j_l is defined by (87).

Given a real symmetric $\mathbf{R}$, $\mathbf{S}$ calculated from $\mathbf{S} = (\mathbf{1} + i\mathbf{R})/(\mathbf{1} - i\mathbf{R})$ will be unitary and symmetric. For a 2×2 matrix,

$$S_{\alpha\alpha'} = \frac{2iR_{\alpha\alpha'}}{[(1 - iR_{\alpha\alpha})(1 - iR_{\alpha'\alpha'}) + R^2_{\alpha\alpha'}]}. \tag{118}$$

For the weak coupling approximation to be fully justified it is necessary that $R_{\alpha\alpha'}$ should be much less than unity. With $R_{\alpha\alpha'} \ll 1$, (118) can be replaced by

$$S_{\alpha\alpha'} = \frac{2iR_{\alpha\alpha'}}{(1 - iR_{\alpha\alpha})(1 - iR_{\alpha'\alpha'})}. \tag{119}$$

It is the last expression which corresponds to the formulation of § 2.7 and it is this expression which has been used in most calculations. In using (119) one does not automatically satisfy the condition that $|S(\alpha, \alpha')|$ should be less than unity and, in practice, values greater than unity have often been obtained. When $R(\alpha, \alpha')$ is of order unity, a better approximation is to use (118) (Seaton, 1961).

In the Born approximation one has $R_t = 0$ and

$$R_B(\alpha, \alpha') = -2 \int F_B(\alpha)\, V_{\alpha\alpha'}\, F_B(\alpha')\, dr. \tag{120}$$

Consider the use of the Born approximation for the calculation of $S(\alpha, \alpha')$. The approximation should give good results when $R_B(\alpha, \alpha'')$ and $R_B(\alpha', \alpha'')$ are small for all α''. For the calculation of $S(\alpha, \alpha')$ one may then replace $\mathbf{S} = (\mathbf{1} + i\mathbf{R}_B)/(\mathbf{1} - i\mathbf{R}_B)$ by

$$\mathbf{S} \simeq \mathbf{1} + 2i\mathbf{R}_B \qquad \text{(approximation B I)}. \tag{121}$$

This is the usual form of the first Born approximation. An alternative form of the first Born approximation is obtained using

$$\mathbf{S} = (\mathbf{1} + i\mathbf{R}_B)/(\mathbf{1} - i\mathbf{R}_B) \qquad \text{(approximation B II)} \tag{122}$$

which ensures that the conservation conditions are satisfied. When the coupling becomes moderately strong $[R_B(\alpha, \alpha') \simeq 1]$, approximation B II will generally be better than B I (Seaton, 1961).

If the distortion potentials $V_{\alpha\alpha}$ and $V_{\alpha'\alpha'}$ are not small, $S(\alpha, \alpha')$ calculated by method B II will be smaller than that calculated by method B I. However, this decrease, which results from imposing conservation conditions, may be offset by another effect; attractive potentials tend to pull the wave functions in towards the nucleus and this tends to increase $S(\alpha, \alpha')$. This effect of distortion can be calculated only by going beyond the first Born approximation.

3.7 Threshold Laws

These may be obtained by arguments similar to those of § 2.6. For excitation of neutral atoms

$$|\, T(a'l_2', al_2)\,|^2$$

behaves as $k_{\alpha'}^{2l_2'+1}$ for $k_{a'} \to 0$. At low energies the partial wave summations therefore converge rapidly. Close to threshold the sum is dominated by the contribution from $l_2' = 0$ and one recovers the law that $Q(a \to a')$ behaves as $k_{a'}$.

In some cases this law has little practical significance. In the transition

$$2p^4\,{}^3P \to 2p^4\,{}^1D$$

in atomic oxygen the partial cross section for $l_2' = 0$ is always very small but the contribution from $l_2' = 1$ is large. For all practical purposes the threshold law is that Q behaves as $k_{a'}^3$ (Seaton, 1953).

For transitions in positive ions, $T(\alpha', \alpha)$ remains finite at threshold for all values of the angular momenta.

3.8 Types of Transitions

We classify transitions according to the range of the interaction, $(V_{\alpha\alpha'} \pm W_{\alpha\alpha'})$.

Let the radial function† $F(al|r)$ be small for $r < r_l$; r_l corresponds to the classical distance of closest approach. For spherical Bessel functions, $j_l(kr)$, one has

$$r_l = [l(l+1)]^{1/2}/k. \tag{123}$$

For collisions with neutral atoms this expression gives an estimate of r_l for the distorted wave functions, but for collisions with positive ions of charge z one has, for small k^2,

$$r_l \simeq l(l+1)/2z.$$

† We here use l in place of l_2 for the angular momentum of the colliding electron, and a for the states of the atom. In using al for the entire system we omit specification of the total angular momentum.

The magnitude of the element of the transmission matrix, $T(al, a'l')$, depends on the size of the interaction, $(V_{aa'} \pm W_{aa'})$, in the region for which the wave functions are large; $T(al, a'l')$ will be small if $(V_{aa'} \pm W_{aa'})$ is small for $r > r_l$ or r'_l, whichever is the larger.

Transitions involving a change in atom spin depend entirely on $W_{aa'}$. According to (42), this is small for $r > r_a$ where r_a is of order of the atomic dimensions; T will be small for $r_l > r_a$. The colliding electron must penetrate the atom if exchange is to occur. At low energies only a few values of l contribute significantly, often only one or two, but for these quite elaborate calculations are required. The important partial waves are usually quite different from spherical Bessel functions (plane wave components) and in consequence the BO approximation rarely gives useful low-energy results. This is discussed by Bates *et al.* (1950). At high energies the spin-change transitions have cross sections which become very small due to interference effects.

For transitions not involving a spin change we consider the potentials $V_{\alpha\alpha'}$. We use the expansion

$$\frac{1}{r_{12}} = \sum_{\lambda} P_{\lambda}(\hat{\mathbf{r}}_1 \cdot \hat{\mathbf{r}}_2) \frac{r_<^{\lambda}}{r_>^{\lambda+1}} \tag{124}$$

where $r_<$ is the smaller of r_1, r_2 and $r_>$ is the greater. For transitions between atomic s states only the $\lambda = 0$ part of $(1/r_{12})$ contributes, and it may then be shown that $V_{aa'}$ goes to zero exponentially for large r. For these transitions one has the selection rule (for the colliding electron) $l' = l$. At low energies the partial wave expansions converge rapidly, as for the spin-change case, but interference effects are usually less serious. At higher energies the Born approximation may be used. In the high-energy limit the cross sections behave as $(1/W)$.

The $\lambda = 1$ part of $(1/r_{12})$, which may be written $(\mathbf{r}_1 \cdot \mathbf{r}_2/r_>^3)$, contributes only to optically allowed transitions. For the colliding electron, $l' = l \pm 1$. The asymptotic form of $V_{aa'}$ is

$$V_{\alpha\alpha'}^{(\lambda=1)}(\mathbf{r}_2) \underset{r_2 \to \infty}{\sim} \frac{\mathbf{r}_2 \cdot (a|\mathbf{r}_1|a')}{r_2^3}. \tag{125}$$

This interaction, behaving as $(1/r^2)$, has a range which is longer than that of any other interaction for $a \neq a'$. The long-range part is particularly important when the dipole matrix element, $(a \mid \mathbf{r}_1 \mid a')$, is large. In general, many values of l, l' contribute to optically allowed transitions and this makes calculations easier. Simple approximations may be used for the larger values of l, l'. Another consequence of the long-range interaction is that the high-energy cross sections behave as $(1/W) \log W$.

Figure 8 illustrates the convergence of partial wave expansions for H $1s \rightarrow 2s$ and $1s \rightarrow 2p$, the latter being optically allowed.

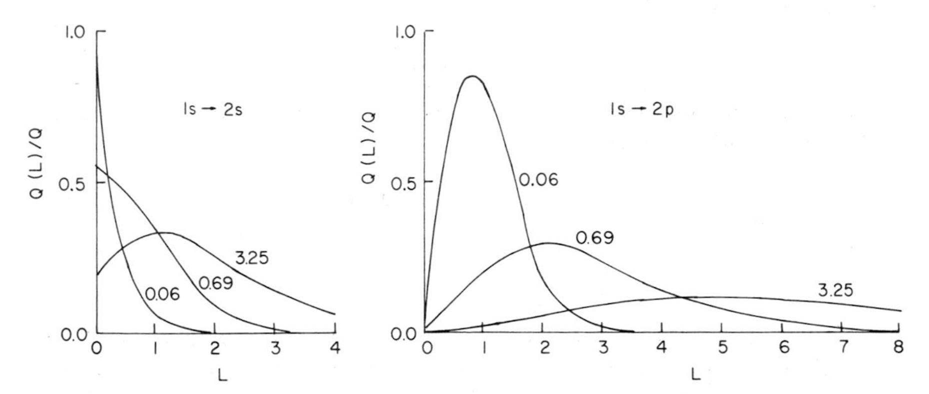

FIG. 8. Partial cross sections for H $1s \rightarrow 2s$ and $1s \rightarrow 2p$; $Q(L)/Q$, in the Born approximation, against total angular momentum L for $k_2^2 = 0.06$, 0.69, and 3.25. For $1s \rightarrow 2p$, zero angular momentum for the scattered electron ($l_2' = 0$) occurs only with $L = 1$; this contribution therefore dominates at low energies. $L = 0$ gives small but finite contributions.

For transitions with quadrupole moments (such as $s - d$ or $p - p$ transitions with no spin change) the $\lambda = 2$ terms in (124) contribute and $V_{aa'}$ then behaves as $(1/r^3)$ for large r. The high-energy cross sections behave as $(1/W)$.

To obtain accurate results for neutral atom cross sections very close to thresholds, it is necessary to make rather elaborate calculations for the one or two angular momenta giving the dominant contributions. For positive ions accurate threshold results are easier to obtain. Even at threshold all angular momenta give finite contributions, and it often happens—at least when there is no spin change—that the dominant contributions come from values of l sufficiently large for simple approximations to be used.

4 *Calculated and Measured Cross Sections*

4.1 TRANSITIONS IN HYDROGEN

4.1.1 *The $1s \rightarrow 2p$ Transition*

Cross sections have been measured by Fite and Brackmann (1958) and, at low energies, by Fite and associates (1959). The low-energy work gave a cross section $Q_\perp$ obtained on counting photons emitted perpendicular to the electron beam and assuming an isotropic photon distribution.† According to (81),

$$Q = (1 - \tfrac{1}{3}P)Q_\perp \tag{126}$$

where P is the polarization fraction. On using the theory of Percival and Seaton (1958) for P,

$$Q_\perp = 0.918Q + 0.246Q_0 \tag{127}$$

where

$$Q_0 = Q(1s \rightarrow 2p\ m_l = 0).$$

The results of Fig. 9 for energies up to 54.4 ev show that the Born approximation gives a cross section $Q_\perp$ which is a good deal too large. The exchange distorted wave method (Khashaba and Massey, 1958) gives some reduction but still gives a result considerably larger than that obtained experimentally. The coupling is not strong and the EDW method should give a good approximation to the results which would be

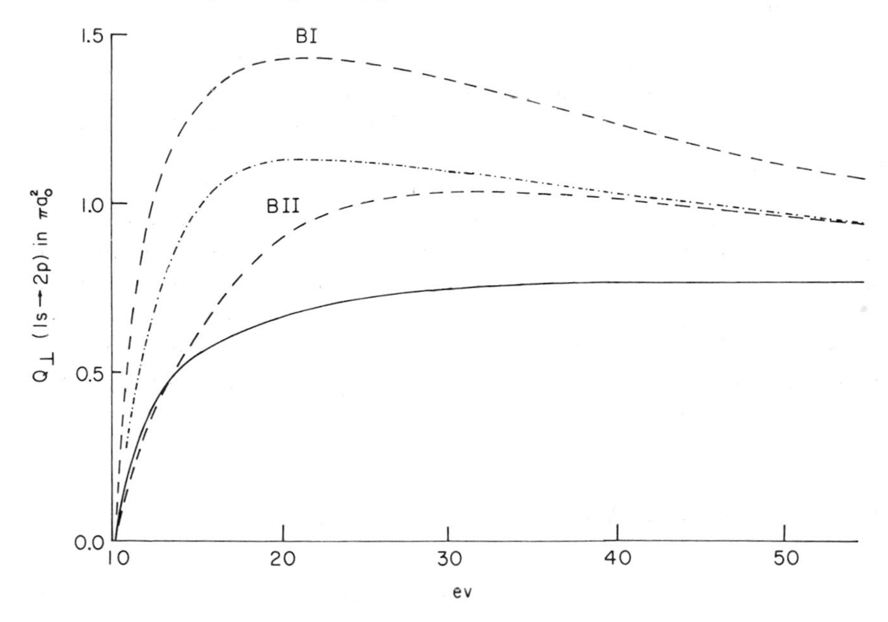

FIG. 9. Cross sections $Q_\perp(1s \rightarrow 2p)$. Full line curve, experimental (Fite *et al.*, 1959); - - -, first Born approximation using methods B I and B II (Burke and Seaton, 1961); -.-.-. EDW approximation (Khashaba and Massey, 1958).

† The experimental results have been shown in Fig. 3.

obtained from exact solutions of the $1s - 2p$ coupled exchange equations. For all but the lowest energies the dominant contributions come from large values of L, for which the EDW method gives results not differing significantly from those obtained in the Born approximation.

Lawson *et al.* (1961) have calculated the Born **R**-matrix, $\mathbf{R}_B$, for $1s$, $2s$, and $2p$. They find that the elements connecting $2s$ and $2p$ are sufficiently large to make this coupling of importance for the calculation of the $1s - 2p$ cross section and since the $2s - 2p$ interaction is of long range, it can have an important effect even for large values of L. Burke and Seaton (1961) calculate the **S**-matrix in two ways: $\mathbf{S}_{\mathrm{B\,I}} = \mathbf{1} + 2i\mathbf{R}_\mathrm{B}$ and $\mathbf{S}_{\mathrm{B\,II}} = (\mathbf{1} + i\mathbf{R}_\mathrm{B})/(\mathbf{1} - i\mathbf{R}_\mathrm{B})$ (see § 3.6). Method B I is the usual first Born approximation. The main difference between $Q(1s \rightarrow 2p)$ calculated by these two methods comes from allowing for $2s - 2p$ coupling. The effect is to reduce Q and to give improved agreement with experiment. Further improvement would be expected in an approximation which allowed both for distortion and exchange and for $2s - 2p$ coupling.

At the lower energies we obtain a cross section Q on multiplying the measured $Q_\perp$ by $(Q/Q_\perp)$ as calculated in method B II; this factor never differs from unity by more than 16 %. At higher energies Fite and Brackmann obtained Q using measurements of the angular distribution of the Ly-α radiation. Figure 10 shows the exeprimental results and the Born curve over an extended energy range.

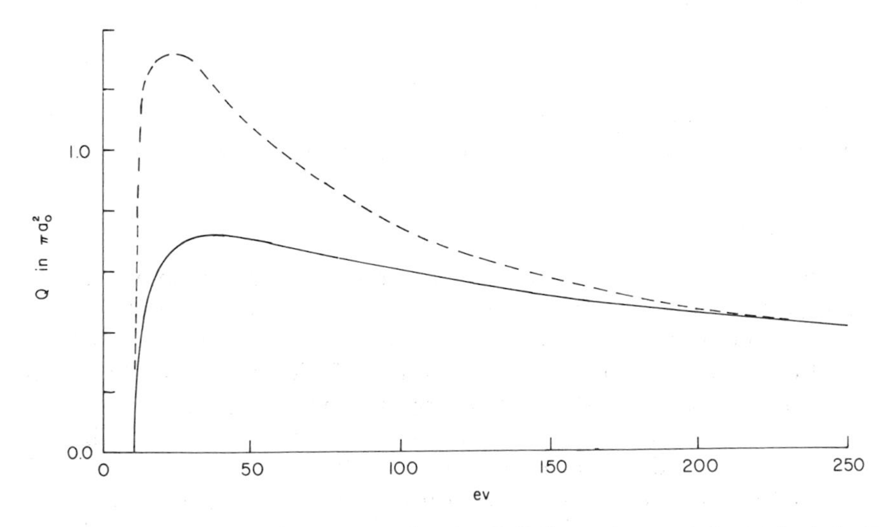

FIG. 10. The cross section for H $1s \rightarrow 2p$. Full line curve, experimental; dashed curve, Born approximation.

4.1.2 *Ionization*

The measured cross section has been compared with classical theory in § 1 and, for low energies, has been compared with the Born approximation in § 2.10. Figure 11 gives experimental and Born cross sections over

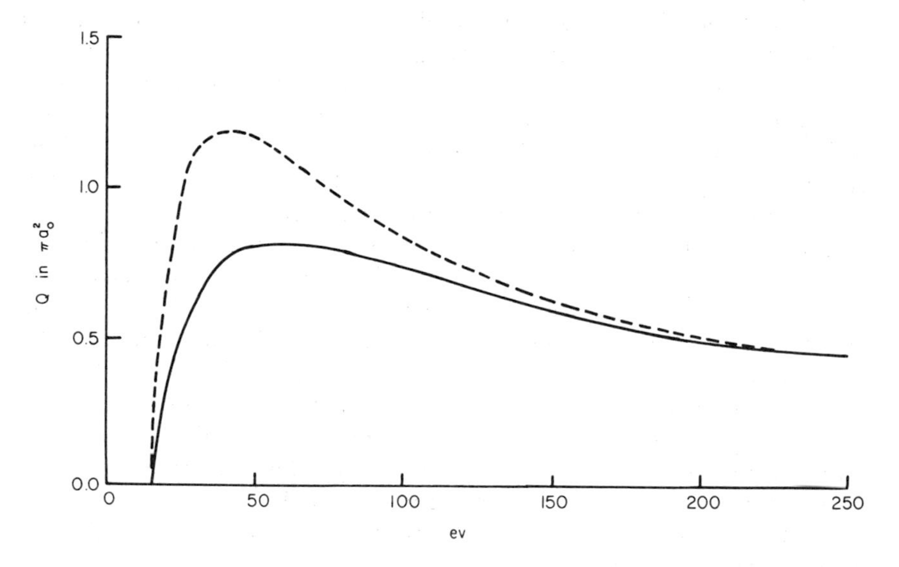

FIG. 11. The cross section for ionization of hydrogen. Full line curve, experimental; dashed curve, Born approximation.

an extended energy range. The comparison is seen to be similar to that for $1s \rightarrow 2p$. This would be expected since the main contribution to ionization comes from the optically allowed $1s \rightarrow \kappa p$ transitions.

4.1.3 *The* $1s \rightarrow 2s$ *transition*

The total cross section, $Q_T(2s)$, for production of $2s$ atoms has been measured by Lichten and Schultz (1959) and by Stebbings and associates (1960). The results obtained have been discussed by Lichten (1961) and by Hummer and Seaton (1961). The results of Lichten and Schultz should give an accurate curve shape for energies up to 45 ev, but their absolute calibration is much more uncertain. Their procedure of normalizing to the Born approximation at energies of 30 to 40 ev is not justified. Stebbings *et al.* made absolute measurements for energies up to 700 ev. Their published results should be multiplied by a factor of 1.5 (Lichten, 1961).

To obtain the cross section $Q(1s \rightarrow 2s)$ for direct excitation one must subtract from $Q_T(2s)$ a cascade cross section $Q_c(2s)$. This is given by

$$Q_c(2s) = \sum_{n=3}^{\infty} \sum_{l} Q(nl)\, C_{nl,2s} \tag{128}$$

where $C_{nl,2s}$ is the probability of cascade from nl to $2s$. The sum, which is dominated by the np states, may be evaluated using the Born approximation. This is a good approximation at high energies; at lower energies a correction factor is obtained using the measured $2p$ cross section. Figure 12 shows $Q(1s \rightarrow 2s)$ obtained from the measurements of Stebbings *et al.* and also the curve of Lichten and Schultz normalized to agree with Stebbings *et al.*

Figure 12 shows that the Born approximation gives Q to be too large for energies less than 100 ev. The $1s - 2s$ coupled equations, with

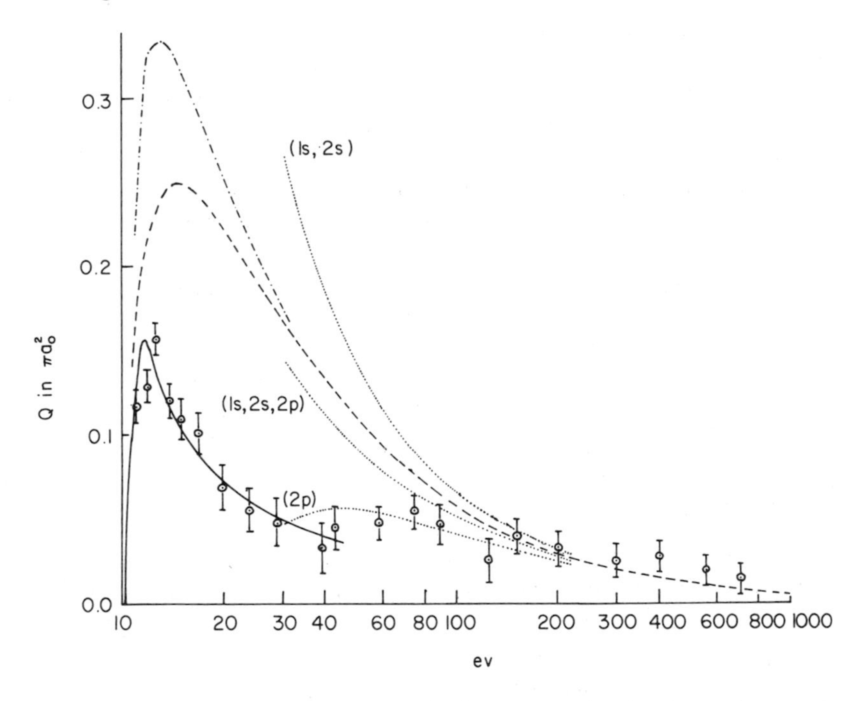

FIG. 12. The cross section for H $1s \rightarrow 2s$. The experimental points with error bars are from the measurements of Stebbings *et al.* (1960) and the full line curve from the relative measurements of Lichten and Schultz (1959) fitted to the data of Stebbings *et al.* - - -, the first Born approximation; ..., the second Born approximation for various terms included in the second Born summation (Kingston *et al.*, 1960); -·-·-·, the solution of the $1s - 2s$ exchange equations (Smith, 1960).

exchange, have been solved by Marriott (1958) for $L = 0$ and by Smith (1960) for $L = 0$, 1, and 2. The cross section obtained† (Fig. 12) is even larger than that obtained in the Born approximation. To explain the experimental results intermediate states must therefore be taken into account.

The B II calculations of Burke and Seaton (1961) show that $2s - 2p$ coupling reduces the cross section but a better way of calculating the effect is provided by the second Born calculations of Kingston *et al.* (1960), who use the method described in § 2.8. In the sum

$$f_{B2} = f_{B1} + \sum_{a''} f_{B2}^{(a'')} \tag{129}$$

[see (62)], they include the states $a'' = (1s,\ 2s)$, $(2p)$, and $(1s,\ 2s,\ 2p)$. The effect of including $(1s, 2s)$ is to increase Q and the effect of $2p$ is to reduce it (fig. 12). In the final curve, $(1s, 2s, 2p)$, these effects tend to cancel. The $(1s, 2s)$ case represents an approximation to the solution of the $1s - 2s$ nonexchange equations. Exact solutions of these equations (Smith *et al.*, 1960) show that the second Born approximation overestimates the effect of distortion in increasing Q. In the energy range of the second Born calculations (30.6-218 ev) inclusion of exchange in these equations would further reduce Q to values not far different from those obtained in the first Born approximation. It is therefore not surprising that the $(1s, 2s, 2p)$ curve is too high and that the $(2p)$ curve gives the best agreement with experiment.

4.2 Transitions in Helium

4.2.1 *High energies*

Experimental high-energy work at Ohio (Lassetre *et al.*, 1953; Silverman and Lassetre, 1957) is of interest in testing the Born approximation. On using

$$\int \frac{\exp i\mathbf{K} \cdot \mathbf{r}_1}{|\mathbf{r}_1 - \mathbf{r}_2|} d^3\mathbf{r}_1 = \frac{4\pi}{K^2} \exp i\mathbf{K} \cdot \mathbf{r}_2 \tag{130}$$

† At the lowest energies solutions were obtained for $L = 0$ and 1 only; this should not introduce significant error. At the higher energies we have introduced a small correction, calculated in the Born approximation, for $L > 2$.

we obtain from (54)

$$f_B(a', a) = -\frac{2}{K^2}(a' \mid \exp i\mathbf{K}\cdot\mathbf{r} \mid a) \qquad (\mathbf{K} = \mathbf{k}_a - \mathbf{k}_{a'}) \tag{131}$$

for $a' \neq a$ and hence

$$\frac{k_a}{k_{a'}} I(a \to a') = 4\left|\frac{(a' \mid \exp i\mathbf{K}\cdot\mathbf{r} \mid a)}{K^2}\right|^2. \tag{132}$$

When summed over degenerate states this depends only on $K = |\mathbf{K}|$ or

$$K^2 = k_a^2 + k_{a'}^2 - 2k_a k_{a'} \cos\omega. \tag{133}$$

This was tested experimentally. For different electron energies in the range 400-600 ev the angular distributions were appreciably different when plotted against ω but all points lay on a single smooth curve when $(k_a/k_{a'})I$ was plotted against K^2. The apparatus was calibrated in absolute units using a Born cross section for He $1\,{}^1S \to 2\,{}^1P$ calculated using accurate wave functions. With one adjustable constant it was found that, between the observed and calculated values of $(k_1/k_2)I\,(1\,{}^1S \to 2\,{}^1P)$, there was a maximum deviation of 6.3 % and a mean deviation of 2.2 %, K^2 being in the range 0.2-2.0 a_0^{-2}.

The use of approximate He wave functions may lead to cross sections being in error by factors of 2 or more (Altshuler, 1952, 1953; Miller and Platzman, 1957). From the high-energy work Born total cross sections may be deduced. With $k_a k_{a'} d\cos\omega = -KdK$, one has

$$Q(a \to a') = \frac{8\pi}{k_a^2}\int_{|k_a - k_{a'}|}^{k_a + k_{a'}} \left|\frac{(a' \mid \exp i\mathbf{K}\cdot\mathbf{r} \mid a)}{K^2}\right|^2 KdK. \tag{134}$$

The Born cross section at all energies may therefore be deduced from experimental angular distributions at high energies.

4.2.2 *The $1\,{}^1S \to 3\,{}^1P$ Transition*

Figure 13 shows the Born cross section calculated from (134) using Ohio high-energy data. We include the experimental cross section, measured by Thieme (1932) in arbitrary units, fitted to the Born curve at high energies and the result of an absolute measurement by Gabriel and Heddle (1960) at 108 ev. The Born approximation for He $3\,{}^1P$

appears to be better than that for H $2p$. This may be due to weaker coupling between excited s and p states in He, a result of the removal of l-degeneracy.

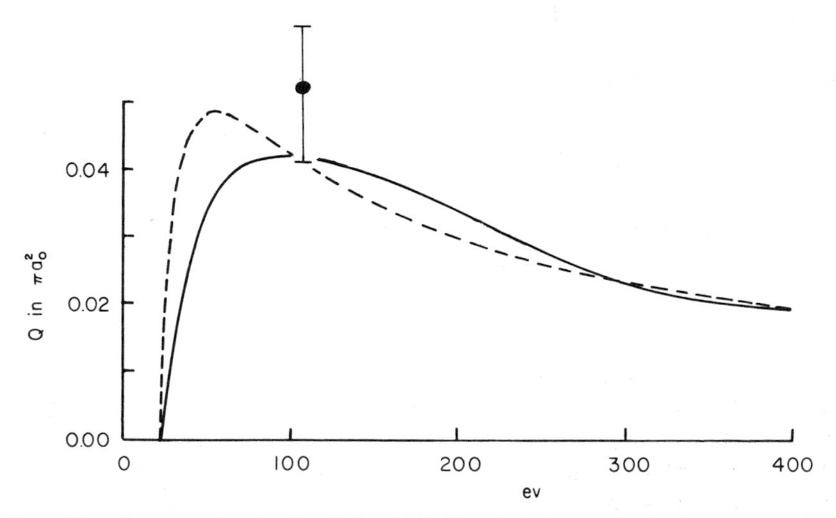

FIG. 13. Cross sections for He $1^1S \rightarrow 3^1P$. The dashed curve is the Born approximation calculated using experimental high-energy data. The full line curve is obtained from the measurements of Thieme (1932) fitted to the Born approximation at high energies. The point at 108 ev is from the absolute measurements of Gabriel and Heddle (1960).

4.2.3 $1\,^1S \rightarrow n\,^1S$ Transitions

The Born approximation for He $n\,^1S$ is also better than that for H $2s$. In Table I Born cross sections calculated by M. Fox (unpublished), using reasonably accurate wave functions, are compared with those measured by Gabriel and Heddle (1960) at† 108 ev.

TABLE I

CROSS SECTIONS FOR HE $1^1S \rightarrow n^1S$ AT 108 EV[a]

	n	3	4	5	6
Q	Experiment	33	18.4	8.0	4.5
	Born	49	18.6	9.2	5.3

[a] Units $10^{-3}\,\pi a_0^2$.

† Taking the significant energy parameter to be $(W/\Delta E)$, the He results at 108 ev should be compared with H results at about 50 ev.

4.2.4 *The* $1\,^1S \rightarrow 3\,^1D$ *Transition*

For this transition the Born approximation seems to be much poorer. At 108 ev Fox obtains $Q = 0.8 \times 10^{-3}\,\pi a_0^2$ in the Born approximation compared with $2.7 \times 10^{-3}\,\pi a_0^2$ measured by Gabriel and Heddle. The discrepancy is unlikely to be due to errors in the He wave functions and B II calculations show that the coupling between $3\,^1D$ and $3\,^1P$ is not important.

4.2.5 *The* $1\,^1S \rightarrow 2\,^1S$ *and* $2\,^3S$ *Transitions*

The cross sections have sharp peaks within 1 ev of their thresholds. For $2\,^3S$, Fig. 14 shows the experimental results of Schulz and Fox (1957) and the results of EDW calculations by Massey and Moiseiwitsch (1954). The effect of exchange distortion is to bring about a large reduction in the magnitude of the cross section and to introduce the low energy peak. The agreement with experiment, although not perfect, is a remarkable success for the EDW method.

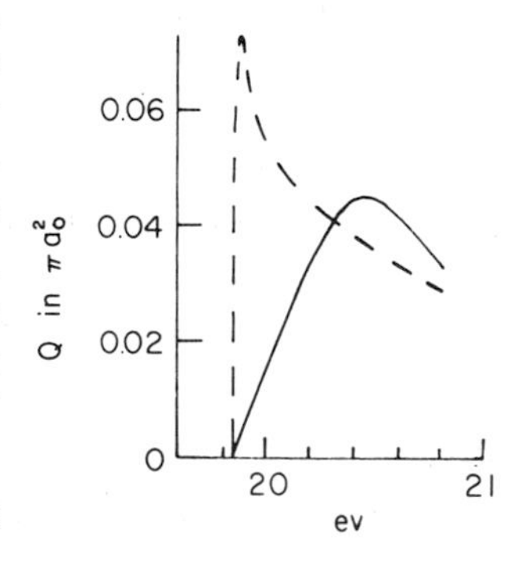

FIG. 14. Cross sections for He $1^1S \rightarrow 2^3S$. Full line curve, experimental (Schulz and Fox, 1957); dashed curve, calculated by the EDW method (Massey and Moiseiwitsch, 1954).

A similar calculation for $2\,^1S$ does not give a peak although one is obtained experimentally. Massey and Moiseiwitsch consider that a peak might be obtained if allowance were made for coupling between $2\,^1S$ and $2\,^3S$.

4.2.6 *The* $2\,^1S \rightarrow 2\,^3S$ *Transition*

The cross section has been calculated by Marriott (1957) on solving the coupled exchange equations. A mean value of $Q(2\,^1S \rightarrow 2\,^3S)$ has been measured by Phelps (1955) for electrons at 300°K. The experimental value is $\bar{Q} = 300\,\pi a_0^2$ compared with the theoretical value of $\bar{Q} = 60\,\pi a_0^2$. The discrepancy may be due to the use of a poor $2\,^1S$ wave function (Marriott and Seaton, 1957) or may be due to coupling with $2\,^1P$ and $2\,^3P$.

4.2.7 *The* $1\,^1S \rightarrow 2\,^3P$ *Transition*

The cross section has been calculated by Massey and Moiseiwitsch (1960) using the EDW method. Exchange distortion produces a considerable reduction but the calculated cross section is still a good deal larger than the best estimates from experimental data. Coupling with the other $n = 2$ states may be particularly important at low energies where only one or two partial waves contribute.

4.3 Strong Coupling in Optically Allowed Transitions

Equation (125) shows that, for an optically allowed transition, the asymptotic form of the interaction potential is proportional to the dipole matrix element. The oscillator strength, $f(a', a)$, therefore provides a measure of the coupling strength. We consider transitions with f of order unity. On putting

$$Q = \sum_L Q(L), \tag{135}$$

the conservation condition gives

$$Q(L) \leqslant \frac{\pi}{k_a^2}(2L+1). \tag{136}$$

When the coupling is strong, $Q_{\mathrm{B\,I}}(L)$ violates (136) for small L. One may obtain a value L_0 of L such that

$$Q_{\mathrm{B\,I}}(L_0) \simeq \tfrac{1}{2}\cdot\frac{\pi}{k_a^2}(2L_0+1). \tag{137}$$

An improved approximation will be

$$Q(L) = \begin{cases} \tfrac{1}{2}\cdot\frac{\pi}{k_a^2}(2L+1) & \text{for} \quad L < L_0 \\ Q_{\mathrm{B\,I}}(L) & \text{for} \quad L \geqslant L_0. \end{cases} \tag{138}$$

In the limit of large L the calculation of $Q_{\mathrm{B\,I}}(L)$ may be simplified on replacing the potential by its asymptotic form. This gives the Bethe, or B′, approximation. It is discussed further in § 5. When L_0 is sufficiently large the difference between $Q_{\mathrm{B\,I}}(L)$ and $Q_{\mathrm{B'\,I}}(L)$ will be small for $L \geqslant L_0$ and one may then use $Q_{\mathrm{B'\,I}}(L)$ in (137) and (138) (Seaton, 1955). Similar results are obtained if one uses approximation B′ II (cf. § 3.6) for all values of L. This is the method to be preferred when there are competing strong-coupling processes (Seaton, 1961). For collisions with positive ions one may use the approximations CB, CB′ in which Coulomb waves replace plane waves.

Figure 15 shows some results for Na $3s \rightarrow 3p$. The experimental cross section is obtained from the relative measurements of Haft (1933) fitted to the absolute measurements of Christoph (1935) as corrected by Bates *et al.* (1950). The failure of the B I approximation at low energies

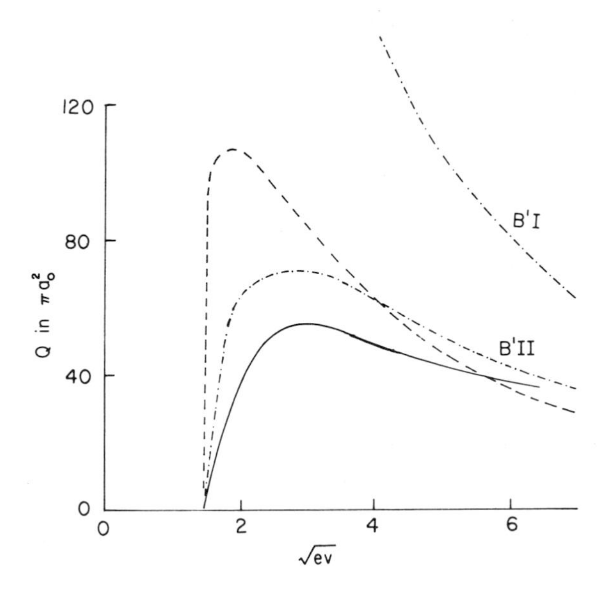

FIG. 15. Cross sections for Na $3s \rightarrow 3p$. Full line, curve, experimental (see text); - - -, Born approximation (B I); -·-·-, Bethe approximation using methods B′ I and B′ II (Salmona and Seaton, 1961).

is largely due to violations of the conservation condition. The cross section $Q_{B'\,I}$ is much too large due to a gross overestimation of the contributions from smaller values of L, but $Q_{B'\,II}$ is in closer agreement with experiment.

Van Regemorter (1960) has calculated cross sections for Ca^+ $4s \rightarrow 4p$ and $3d \rightarrow 4p$ allowing for close coupling in both transitions. He uses methods CB I, CB II, CB′ I, and CB′ II. For the threshold $4s \rightarrow 4p$ cross section he obtains:

Approximation	CB I	CB II	CB′ I	CB′ II
$Q(4s \rightarrow 4p)$ in πa_0^2	126.5	58.2	354.5	63.8

It is seen that $Q_{CB\,II}$ is only a little larger than $Q_{CB'\,II}$.

5 *Collisional Excitation Treated as a Radiative Process*

Approximate cross-section estimates are required in many physical and astrophysical problems. We seek to obtain such estimates for optically allowed transitions.

In the distorted wave expression,

$$Q(i \to j) = \frac{1}{4\pi^2 \omega_i} \cdot \frac{k_j}{k_i} \int |(\mathbf{k}_j \mid V_{ji} \mid \mathbf{k}_i)|^2 \, d\hat{\mathbf{k}}_j \tag{139}$$

where ω_i is the statistical weight of level i, we replace V_{ji} by its asymptotic form (125) to obtain

$$Q(i \to j) = \frac{1}{4\pi^2} \cdot \frac{k_j}{k_i} \int \left| \left(\mathbf{k}_j \,\middle|\, \frac{\mathbf{r}}{r^3} \,\middle|\, \mathbf{k}_i\right) \right|^2 d\hat{\mathbf{k}}_j \cdot \frac{1}{3\omega_i} |(j \mid \mathbf{r} \mid i)|^2. \tag{140}$$

This has the interpretation that, due to the field of the atomic electron, the colliding electron emits a photon which is subsequently absorbed by the atom in the $i \to j$ transition; $(\mathbf{k}_j \mid \mathbf{r}/r^3 \mid \mathbf{k}_i)$ is the dipole acceleration matrix element for emission of a photon in a free-free transition and $(j \mid \mathbf{r} \mid i)$ is the dipole length atomic matrix element. On introducing the Kramers-Gaunt g-factor defined by

$$\int \left| \left(\mathbf{k}_j \,\middle|\, \frac{\mathbf{r}}{r^3} \,\middle|\, \mathbf{k}_i\right) \right|^2 d\hat{\mathbf{k}}_j = \frac{32\pi^4}{k_i k_j \sqrt{3}} g(k_j, k_i) \tag{141}$$

and the oscillator strength defined by

$$\frac{1}{3\omega_i} |(j \mid \mathbf{r} \mid i)|^2 = \frac{f(j, i)}{2(E_j - E_i)} \tag{142}$$

where E_j, E_i are in atomic units (27.20 ev), we have

$$Q(i \to j) = \frac{4\pi^2}{k_i^2 \sqrt{3}} \cdot \frac{f(j, i)}{(E_j - E_i)} g(k_j, k_i). \tag{143}$$

Similarly, for ionization,

$$Q_{\text{ion}}(i) = \frac{2}{k_i^2 \alpha \sqrt{3}} \int_0^{\frac{1}{2}k_i^2 - I} \frac{a(E_\kappa)\, g(k, k_i)}{(I + E_\kappa)} \, dE_\kappa \tag{144}$$

where $\frac{1}{2} k_i^2 = I + \frac{1}{2} k^2 + E_\kappa$ and where I is the ionization potential, E_κ is the ejected electron energy, $a(E_\kappa)$ the photoionization cross section and $\alpha(\simeq 1/137)$ is the fine structure constant.

On evaluating (141) using plane waves,

$$g = \frac{\sqrt{3}}{\pi} \log \left(\frac{k_i + k_j}{| k_i - k_j |} \right). \tag{145}$$

This gives the Bethe approximation; an alternative derivation is to replace $\exp(i\mathbf{K} \cdot \mathbf{r})$ by $(1 + i\mathbf{K} \cdot \mathbf{r})$ in the Born approximation. On

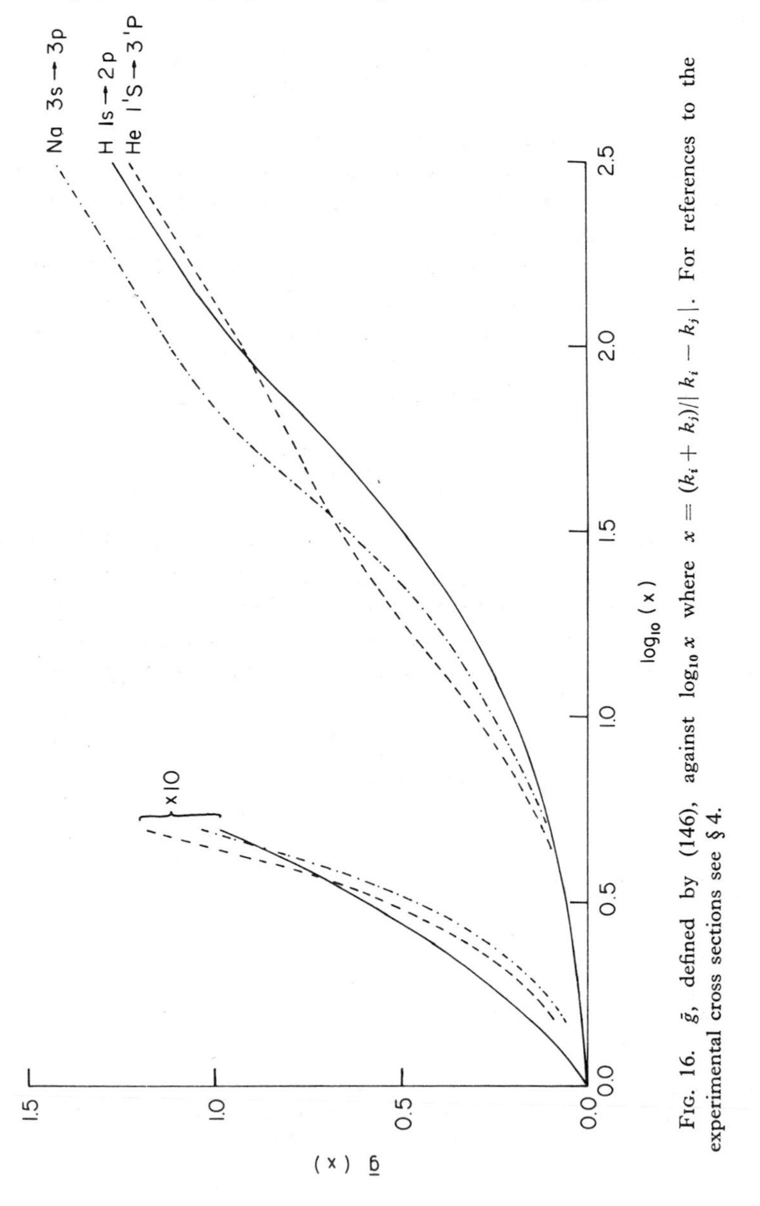

FIG. 16. $\bar{g}$, defined by (146), against $\log_{10} x$ where $x = (k_i + k_j)/| k_i - k_j |$. For references to the experimental cross sections see § 4.

evaluating (141) using Coulomb waves one obtains $g \simeq 1$ at low energies and the expression (145) at high energies (Grant, 1958).

As they stand, expressions (143) and (144) give only very crude estimates. Attempts to improve them have usually been concerned with introducing cut-off factors in the integrations. Our approach is more empirical. Assuming Q to be known we define an effective Gaunt factor, $\bar{g}$, by

$$Q(i \to j) = \frac{4\pi^2}{k_i^2 \sqrt{3}} \cdot \frac{f(j, i)}{(E_j - E_i)} \cdot \bar{g}(k_j, k_i). \tag{146}$$

Figure 16 shows values of $\bar{g}$ against $x = (k_i + k_j)/| k_i - k_j |$ as obtained from the experimental cross sections: H, $1s \to 2p$; He, $1\,^1S \to 3\,^1P$; Na, $3s \to 3p$ (see § 4). Similar curves are obtained for these three cases despite the fact that the maximum cross sections range over three orders of magnitude (maximum values of 0.72 for H, 0.042 for He, and 53 for Na, all in πa_0^2). The curves of Fig. 16 could be used to estimate cross sections for other optically allowed transitions in neutral atoms and would probably give results correct to a factor of 2.

For near-threshold collisional ionization,

$$Q_{\text{ion}}(i) = \frac{2}{k_i^2 \alpha \sqrt{3}} \cdot a(0) \cdot \frac{(\frac{1}{2} k_i^2 - I)}{I} \bar{g}, \tag{147}$$

$a(0)$ being the threshold photoionization cross section. Table II gives values of $\bar{g}$ obtained using experimental cross sections† for neutral atoms and CB calculations for hydrogenic ions. The anomaly for Ne noted in § 1 is now cleared up; the cross section for near-threshold collisional ionization is small because the threshold photoionization cross section is small.

TABLE II

VALUES OF $\bar{g}$ FOR NEAR-THRESHOLD IONIZATION

Neutral atoms		Hydrogenic ions	
Atom	$\bar{g}$	Z	$\bar{g}$
H	0.099	2	0.23
He	0.046	3	0.26
Ne	0.070	∞	0.31
Ar	0.064		

† References for the neutral atom cross sections are given in § 1. For the photoionization cross sections see Burgess and Seaton (1960) and Lee and Weissler (1953, 1955) and for the CB calculations see Burgess (1960).

For positive ion excitation $\bar{g}$ is finite at threshold. Table III gives some threshold values calculated for Ca^+ and Mg^+ by Van Regemorter (1960 and further unpublished material) and for hydrogenic ions by Burgess (1961). Close to threshold, $\bar{g}$ for positive ions varies slowly with energy. The approximation of taking $\bar{g} \simeq 0.2$ should give nearthreshold positive ion excitation cross section estimates generally correct to within a factor of 2. At higher energies, the behavior of $\bar{g}$ for positive ions is similar to that for neutral atoms.

TABLE III

Ion		Transition	$(E_j - E_i)$	f	$\bar{g}$	
					CB I	CB II
Ca^+		$4s - 4p$	0.116	1.19	0.39	0.18
		$3d - 4p$	0.053	0.10	0.75	0.46
		$3d - 4f$	0.248	0.29	0.08	—
Mg^+		$3s - 3p$	0.163	0.91	0.45	0.31
Hydrogenic	$Z = 2$	$1s - 2p$	1.500	0.42	0.16	—
	$Z \to \infty$	$1s - 2p$	$0.375Z^2$	0.42	0.21	—

Distorted wave calculations have been carried out by Vainshtein (1960) for a large number of transitions in neutral atoms and positive ions. For many transitions in neutral atoms the effect of distortion is to increase the cross sections (see § 3.6) and hence to give a discrepancy with experiment greater than that obtained using the first Born approximation.

The results of Vainshtein confirm our statement that taking $\bar{g} = 0.2$ gives near-threshold positive ion cross sections correct to within a factor of 2.

ACKNOWLEDGMENTS

I should like to thank Dr. H. Van Regemorter, Dr. A. Burgess, Mr. J. Tully, Mr. M. Fox, and Mr. A. Boksenberg for permission to quote results of their unpublished work.

APPENDIX

We consider

$$\int V_{\kappa\kappa'}(\mathbf{r})\, F(\kappa'|\mathbf{r})\, dE_{\kappa'} \underset{r\to\infty}{\sim} \frac{\beta(\kappa, k)}{r} F(\kappa|\mathbf{r}) \tag{148}$$

where

$$E_{\kappa'} = \tfrac{1}{2}\kappa'^2, \qquad \kappa'^2 + k'^2 = \kappa^2 + k^2,$$

$$F(\kappa'|\mathbf{r}) \sim f(\kappa') \frac{\exp ik'r}{r}, \tag{149}$$

and where $V_{\kappa\kappa'}$ is defined by (40). Considering only the $\lambda = 0$ term in (124),

$$V_{\kappa\kappa'}(\mathbf{r}) = \int_r^\infty \left(\frac{1}{r_1} - \frac{1}{r}\right) P_\kappa(r_1)\, P_{\kappa'}(r_1)\, dr_1 \tag{150}$$

where (§ 2.4)

$$P_\kappa(r_1) \underset{r_1\to\infty}{\sim} \left(\frac{2}{\pi\kappa}\right)^{1/2} \sin(\kappa r_1 + \eta). \tag{151}$$

Owing to interference, $V_{\kappa\kappa'}$ is small for r large and $\kappa \neq \kappa'$. For $\kappa \simeq \kappa'$, $k' \simeq k + (\kappa/k)(\kappa - \kappa')$ and, for r large,

$$F(\kappa'|\mathbf{r}) \simeq F(\kappa|\mathbf{r}) \exp i(\kappa/k)(\kappa - \kappa')r. \tag{152}$$

Therefore

$$\beta(\kappa, k) = r \int V_{\kappa\kappa'}(r) \exp[i(\kappa/k)(\kappa - \kappa')r]\, dE_{\kappa'} \tag{153}$$

evaluated in the limit of r large. On using

$$P_\kappa(r_1)\, P_{\kappa'}(r_1) \sim \frac{1}{\pi(\kappa\kappa')^{1/2}} \times \{\cos[(\kappa - \kappa')r_1 + \eta - \eta'] - \cos[(\kappa + \kappa')r_1 + \eta + \eta']\} \tag{154}$$

we have, for $\kappa \simeq \kappa'$,

$$V_{\kappa\kappa'} \sim \frac{1}{\pi\kappa} \int_r^\infty \left(\frac{1}{r_1} - \frac{1}{r}\right) \cos(\kappa - \kappa')r_1\, dr_1 \tag{155}$$

and, integrating by parts,

$$V_{\kappa\kappa'} \sim \frac{1}{\pi\kappa} \int_r^\infty \frac{1}{r_1^2} \frac{\sin(\kappa - \kappa')r_1}{(\kappa - \kappa')}\, dr_1 - \frac{1}{r}\, \delta(E_\kappa - E_{\kappa'}) \tag{156}$$

where

$$\delta(E_\kappa - E_{\kappa'}) = \lim_{r_1\to\infty} \left\{\frac{\sin(\kappa - \kappa')r_1}{\pi\kappa(\kappa - \kappa')}\right\}. \tag{157}$$

Expression (153) may now be evaluated using

$$\int_{-\infty}^{+\infty} \frac{\sin t}{t} \exp(ipt)\, dt = \begin{cases} 1 & \text{for} \quad |p| < 1 \\ 0 & \text{for} \quad |p| > 1. \end{cases} \tag{158}$$

One obtains (79).

REFERENCES

Altshuler, S. (1952) *Phys. Rev.* **87**, 992.
Altshuler, S. (1953) *Phys. Rev.* **89**, 1093.
Bates, D. R., Fundaminsky, A., Leech, J. W., and Massey, H. S. W. (1950) *Phil. Trans.* **A243**, 93.
Boksenberg, A. (1961) Thesis, London.
Burgess, A. (1960) *Astrophys. J.* **132**, 503.
Burgess, A. (1961) *Mém. Soc. Roy. Sci. Liège* **4**, 299.
Burgess, A., and Seaton, M. J. (1960) *Monthly Not. Roy. Astron. Soc.* **120**, 121.
Burke, V. M., and Seaton, M. J. (1961) *Proc. Phys. Soc.* **77**, 199.
Castillejo, L., Percival, I. C., and Seaton, M. J. (1960) *Proc. Roy. Soc.* **A254**, 259.
Christoph, W. (1935) *Ann. Phys. (Leipzig)* **23**, 51.
Edmonds, A. R. (1957) "Angular Momentum in Quantum Mechanics." Princeton Univ. Press, Princeton, New Jersey.
Elwert, G. (1952) *Z. Naturforsch.* **7a**, 432.
Fite, W. L., and Brackmann, R. T. (1958) *Phys. Rev.* **112**, 1141; (1958) *Phys. Rev.* **112**, 1151.
Fite, W. L., Stebbings, R. F., and Brackmann, R. T. (1959) *Phys. Rev.* **116**, 356.
Gabriel, A. H., and Heddle, D. W. O. (1960) *Proc. Roy. Soc.* **A258**, 124.
Geltman, S. (1956) *Phys. Rev.* **102**, 171.
Grant, I. P. (1958) *Monthly Not. Roy. Astron. Soc.* **118**, 241.
Haft, G. (1933) *Z. Phys.* **82**, 73.
Hummer, D. G., and Seaton, M. J. (1961) *Phys. Rev. Letters* **6**, 471.
Khashaba, S., and Massey, H. S. W. (1958) *Proc. Phys. Soc.* **71**, 574.
Kingston, A. E., Moiseiwitsch, B. L., and Skinner, B. G. (1960) *Proc. Roy. Soc.* **A258**, 245.
Lassetre, E. N., Krasnow, M. E., and Silverman, S. (1953) Scientific Report No. 3, Contract No. AF 19(122)-642, Department of Chemistry, Ohio State University, Columbus, Ohio.
Lawson, J., Lawson, W., and Seaton, M. J. (1961) *Proc. Phys. Soc.* **77**, 192.
Lee, P., and Weissler, G. L. (1953) *Proc. Roy. Soc.* **A219**, 71.
Lee, P., and Weissler, G. L. (1955) *Phys. Rev.* **99**, 540.
Lichten, W. (1961) *Phys. Rev. Letters* **6**, 12.
Lichten, W., and Schultz, S. (1959) *Phys. Rev.* **116**, 1132.
Marriott, R. (1957) *Proc. Phys. Soc.* **A70**, 288.
Marriott, R. (1958) *Proc. Phys. Soc.* **72**, 121.
Marriott, R., and Seaton, M. J. (1957) *Proc. Phys. Soc.* **A70**, 296.
Massey, H. S. W. (1956a) *In* "Handbuch der Physik" (S. Flügge, ed.), Vol. 36, p. 307. Springer, Berlin.
Massey, H. S. W. (1956b) *Rev. Mod. Phys.* **28**, 199.
Massey, H. S. W., and Burhop, E. H. S. (1952) "Electronic and Ionic Impact Phenomena." Oxford Univ. Press, London and New York.
Massey, H. S. W., and Moiseiwitsch, B. L. (1954) *Proc. Roy. Soc.* **A227**, 38.
Massey, H. S. W., and Moiseiwitsch, B. L. (1960) *Proc. Roy. Soc.* **A258**, 147.
Miller, W. F., and Platzman, R. L. (1957) *Proc. Phys. Soc.* **A70**, 299.
Mott, N. F., and Massey, H. S. W. (1949) "The Theory of Atomic Collisions." Oxford Univ. Press, London and New York.
Percival, I. C., and Seaton, M. J. (1957) *Proc. Cambridge Phil. Soc.* **53**, 654.
Percival, I. C., and Seaton, M. J. (1958) *Phil. Trans.* **A251**, 113.
Peterkop, R. (1961) *Proc. Phys. Soc.* **A77**, 1220.

Phelps, A. V. (1955) *Phys. Rev.* **99**, 1307.
Salmona, A., and Seaton, M. J. (1961) *Proc. Phys. Soc.* **77**, 617.
Schulz, G. J., and Fox, R. E. (1957) *Phys. Rev.* **106**, 1179.
Seaton, M. J. (1953) *Phil. Trans.* **A245**, 469.
Seaton, M. J. (1955) *Proc. Phys. Soc.* **A68**, 457.
Seaton, M. J. (1958) *Rev. Mod. Phys.* **30**, 979.
Seaton, M. J. (1961) *Proc. Phys. Soc.* **77**, 174.
Silverman, S., and Lassetre, E. N. (1957) Scientific Report No. 9, Contract No. AF 19(122)-642, Department of Chemistry Ohio State University, Columbia, Ohio.
Skinner, H. W. B., and Appleyard, E. T. S. (1927) *Proc. Roy. Soc.* **A117**, 224.
Smith, K. (1960) *Phys. Rev.* **120**, 845.
Smith, K., Miller, W. F., and Mumford, A. J. P. (1960) *Proc. Phys. Soc.* **76**, 559.
Smith, P. T. (1930) *Phys. Rev.* **36**, 1293.
Stebbings, R. F., Fite, W. L., Hummer, D. G., and Brackmann, R. T. (1960) *Phys. Rev.* **119**, 1939.
Thieme, O. (1932) *Z. Phys.* **78**, 412.
Thomson, J. J. (1912) *Phil. Mag.* **23**, 419.
Tully, J. (1960) M. Sc. Dissertation, London.
Vainshtein, L. (1960), Effective cross sections for the excitation of atoms by electron impact. I. Neglecting exchange. Report A-33, Lebedev Physics Institute, Moscow.
Van Regemorter, H. (1960) *Monthly Not. Roy. Astron. Soc.* **121**, 213.

12.

The Measurement of Collisional Excitation and Ionization Cross Sections

Wade L. Fite

1 *Introduction*

Concurrent with the appearance of new needs for information on inelastic collision processes, advances in laboratory technique have provided the capability to obtain much of the required data. Improvements in vacuum technique, electronic circuitry, and electron multiplication devices have enabled making superior measurements using previously established experimental designs; and the introduction of experimental methods new to atomic collision physics, such as modulation techniques and optical pumping, has made possible the study of a number of collision phenomena previously inaccessible to measurement.

Of course, notwithstanding the application of improved and new experimental techniques, the basic operational definitions of cross sections for excitation and ionization have remained unaltered, and the large majority of experiments has continued to be designed about the equation

$$dS = nQi\,dx \tag{1}$$

where n is the number density of "target particles," Q is the cross section for the particular process under study, i is the current of the bombarding particles, dx is the distance traversed by the bombarding particles through the target particles, and dS is a signal which is characteristic of the ionization or excitation process. While the foregoing equation defines the absolute cross section, the practical difficulties of making absolute measurements remain quite severe. The problem of determination of mass-spectrometer collection efficiency in ionization studies and the problem of accurate calibration of photon and metastable atom detectors in excitation experiments has caused most of the recent electron-impact studies to concentrate on relative cross-section measurements and excitation functions. Where absolute values have been required, the recent general pattern has been to compare an unknown cross section with one whose absolute value has been determined previously (and usually in the 1930's) and so derive an absolute value for the cross section to be determined. This practice emphasizes the need for more precise remeasurement of certain cross sections for use as standards.

In the case of ion-impact experimentation on inelastic collisions, the situation is somewhat different, for here most of the recent experiments have been designed to determine absolute cross-section values directly. In part this has been done to provide a body of absolute cross sections, such as already exists in part for electron-impact collision phenomena, which can be used as standards for further comparison experiments. Perhaps a stronger reason, however, is that for controlled thermonuclear research it is essential to know absolute cross-section values for processes operative in a number of experimental devices.

This chapter will be concerned with some recent developments in the *measurement* of collisional excitation and ionization cross sections, and interpretation of results will be greatly subordinated to the discussion of experimental techniques.

2 *Electron Impact Studies*

2.1 Ionization

In studies of ionization, the most characteristic signal for detection is the current of positive ions produced in an electron-atom or electron molecule collision. Two methods of ion current detection have been employed and continue to form the basis of ionization experimentation. The first measures total ion current, irrespective of the mass or charge of the ion. The apparent cross section Q' for this type of detection is given by $Q' = Q_1 + 2Q_2 + 3Q_3 + ...$, where Q_n is the sum of the cross sections for all processes leading to ions which are n-times ionized, i.e., irrespective of the final ion state. While lack of knowledge of the ions formed is a disadvantage, the fact that in nondiscriminating detectors ions may be collected with 100% efficiency lends to such detectors the excellent property that absolute measurements of Q' may be made easily.

The more commonly used detector for ionization cross-section measurements is the mass spectrometer with which determination of the cross section for production of ions with a given charge-to-mass ratio may be made. A severe disadvantage with the mass spectrometer is that measurement of absolute currents of ions formed in the ionizer is quite difficult to make, for the collection efficiency of the particular instrument for the particular ion under study must be known accurately. As a result, most recent mass-spectrometer studies of ionization have been content with obtaining only relative cross-section curves.

However, this type of relative data is important for investigating two aspects of ionization phenomena which have been of considerable recent interest. The first is the electronic state in which an ion is left after its formation and the second is the energy dependence of ionization cross sections just above threshold, a subject on which several recent theoretical developments have occurred.

Both of these aspects entail the measurement of relative cross sections for ionization just above threshold, and the principal advances over the earlier threshold ionization work of Lawrence (1926) and Nottingham (1939) have come about as a result of improved energy resolution of ionizing electron beams.

2.1.1 *Techniques*

Energy Resolution Techniques. The first method recently developed to obtain improved information on ionization in the near-threshold energy region is the retarding potential difference method (RPD) described by Fox and associates (1951). Figure 1 illustrates the mass-spectrometer ion source which they devised and shows the essential features of the RPD method. Electrons emitted from the filament are accelerated into the grounded ionization enclosure, 5, by the potential V_1. These electrons would normally have a distribution of energies as indicated by the dashed curve near the filament. If electrode 4 is interposed and is placed at a negative potential, V_R, with respect to the filament, the less energetic electrons in the distribution will be prevented from reaching the ionization region while the electrons with kinetic energy in the direction toward the enclosure 5 greater than eV_R will not be stopped by electrode 4. The energy distribution of electrons actually entering the ionization enclosure will be as illustrated by the distribution curve inside 5 on the figure; i.e., it will be cut off sharply. If now the potential

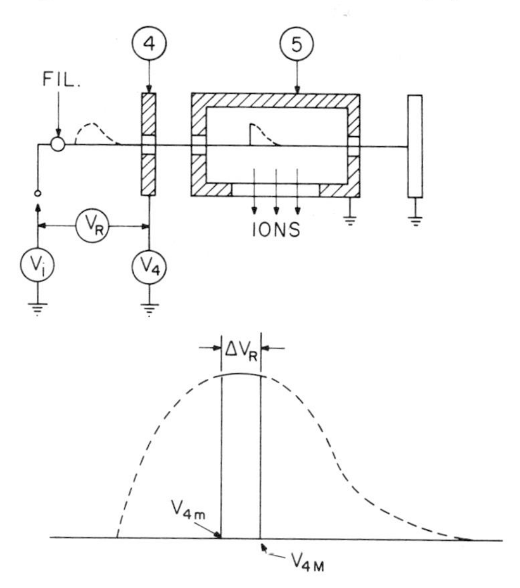

FIG. 1. The elements of the retarding potential difference method of Fox *et al.* (1951). See text for description of the use of the RPD method for production of electron beams of very narrow effective energy spread.

V_4 is changed by a small amount from V_{4M} to V_{4m}, the retarding potential will change by an amount $\Delta V_R = V_{4M} - V_{4m}$, and the distribution cutoff will change so as to permit electrons in the energy range $e\Delta V_R$ to reach the ionization enclosure. The increase in observed ion current can then be attributed to the increase in electron current which arises from the inclusion of electrons with the very narrow energy spread $e\Delta V_R$. A longitudinal magnetic field is used to collimate the electrons. If the aperture in electrode 4 is made small compared to the Larmor radius of the electrons, then those electrons with appreciable velocity components transverse to the magnetic field will be excluded by a mechanical as well as an electrical barrier. As a result, the only electrons eligible to reach enclosure 5 are those whose initial directions of motion are very nearly parallel to the axis of the ionizer.

Since Fox *et al.* used a mass spectrometer as the ion detector, it was necessary to apply an electric field to extract the ions from the ionizing enclosure. To avoid the generation of an electron energy spread by the extraction field, Fox *et al.* adopted a pulsing technique in which a pulse ΔV_R of short time duration was first applied to the gun at a time when no extraction field existed in the ionizing enclosure. Only after the ionization had been completed was a pulse of ion extraction field applied. Under these conditions the maximum energy spread of the effective electron beam was 0.06 ev and the kinetic energy of the electrons was known to a precision of 0.1 ev, after proper consideration of contact potentials (Fox *et al.*, 1953).

A second and more straightforward approach to the problem of electron energy resolution is through the production of monoenergetic electron beams by electrostatic energy analysis. Both Yarnold and Bolton (1949) and Harrower (1955) have examined the properties of a parallel-plate electrostatic energy analyzer, which provides direction focusing but not energy focusing when charged particles are admitted between the plates at an angle of 45°. This type of analyzer has recently been employed by Foner and Nall (1961) for the study of threshold ionization.

Clarke (1954) also used an electrostatic electron energy selector in his mass spectrometric studies of threshold ionization. His device was a 127° sector of a cylindrical condenser of the type described by Hughes and Rojansky (1929) and used in the precision mass spectrograph of Bainbridge and Jordan (1936). The electron analyzer of Clarke was capable of producing beams with an energy spread of down to 0.2 ev.

Further development of the 127° sector electrostatic electron selector to reduce space charge effects has been carried on by Marmet and Kerwin (1960). By substituting curved tungsten mesh grids for the curved plates of Clarke's selector, reflection of electrons at these plates was reduced and therefore electron space charge within the analyzer proper also. In order to reduce space charge in the ionization chamber arising from the reflection of electrons at the chamber walls, Marmet and Kerwin replaced the normally solid walls with a surface of very small and finely packed gold-plated copper tubes viewed end-on by an electron. Their method of manufacture of this wall material, to which the appropriate name "electron velvet" has been applied, is described in their article. With the improved electron selector and ionization chamber Marmet and Kerwin have produced electron beams with an energy spread measured at the half-height of the distribution of 0.04 ev.

Measurements of Derivatives of Cross-Section Curves. Both the RPD method and the electrostatic analyzer method have made it possible to examine fine details in ionization cross section curves. Yet another method has been used to study structure in ionization cross section curves which relies not upon measuring the cross section with very fine energy resolution but rather by directly measuring derivatives of the ionization cross section curve.

This work has been carried out by Morrison (1954) who used a type of modulation technique. Electrons from a conventional ionizer are accelerated by a voltage $V_0 + v \sin \omega t$, where V_0 is the mean electron energy and v is the amplitude of a very small superposed voltage at the modulation angular frequency ω. It may be shown that if such an energy-modulated beam is used in a mass-spectrometer ion source, then the ion current will also be modulated, at frequencies which are harmonics of the electron energy modulation frequency. The amplitude, A_m, of the mth harmonic of the ion current is, in the first approximation, proportional to the mth derivative of the ionization cross section with respect to electron energy, averaged over the distribution of energies in the electron beam. By examination of the second harmonic, Morrison studied the changes in slope of ionization cross section curves without reference to the curves themselves. This technique is quite attractive since it considerably eases the problem of stability of experimental parameters which must always be a prime concern for reproducibility of results in experiments studying the ionization cross section curve itself.

Modulated-Crossed-Beam Techniques. Complementing the development of the foregoing techniques for examination of the fine structure in collision cross section curves, the past few years has also witnessed the advent of modulated-crossed-beam experiments for cross-section measurement. The principal advantage of such experimental techniques in ionization studies is the fact that they can be applied to chemically unstable atomic and molecular systems as well as to systems which are chemically stable. Among the earliest applications of these techniques were the measurement of the cross section for electron impact ionization of H_2 (Boyd and Green, 1958) and for ionization of the free hydrogen atom (Fite and Brackmann, 1958a). Indeed, a major incentive to the development of modulated-crossed-beam experimentation was the need to measure the hydrogen atom ionization cross section.

The logical development of these techniques, framed in terms of the H atom ionization cross-section measurement, begins by recognizing that conventional methods of studying this process would be difficult because of the requirement that the gas must be kept dissociated by means which would not interfere with the electrical measurement of the ionization currents. An obvious way to circumvent these difficulties is to dissociate the gas in a suitable atomic beam source (either a gas discharge or a hot (3000°K) low-pressure furnace, where thermal dissociation occurs), permit a beam of atoms to flow into high vacuum, and then cross the atom beam by an electron beam some distance downstream from the atom source. Ions produced in the volume of intersection of the two beams can then be detected either by a mass spectrometer or by a current detector which does not discriminate according to ion mass. While the number density of atoms in a highly dissociated beam is small ($\sim 10^{8\text{-}9}$ atoms/cm^3) and ion currents are consequently also quite small, they can be measured by sensitive current measuring devices now commonly available.

The difficulty of a crossed-beam experiment arises from the fact that electrons also produce ions in collisions with the residual gases in the vacuum chamber into which the atom beam is admitted. A pressure of 2×10^{-7} mm Hg corresponds to a number density of about 10^{10} molecules per cm^3, and since much of this residual gas will be H and H_2, it is clear that in a normal dc beam experiment considerable difficulty will be encountered in distinguishing ions formed at the intersection of the two beams from those formed in the background gas in the vacuum. While this would not be a severe difficulty if the background gas pressure were steady (say to within 1% or better), pressure drifts and fluctuations arising from unsteady operation of diffusion pumps would seem to make dc experiments and subtractive procedures unattractive.

It is a property of vacuum systems which restores the feasibility of crossed-beam experiments. Every vacuum system possesses a characteristic time constant, which is equal to the volume of the system divided by the speed of the pumps of the system. As a result, the vacuum system itself will integrate pressure fluctuations in a manner completely analogous to integration by an *RC* electrical network, with the integration improving with increasing frequency. It is thus clear that a crossed-beam experiment which is done dc, i.e., where the dc signals with the atom beam crossing and not crossing the electron beam are subtracted, is inferior to an experiment in which the atom beam is interrupted periodically at a sufficiently high frequency that the background gas fluctuations at that frequency are smoothed out by the vacuum system. Under circumstances where the ionizing electron beam is run dc, those ions formed by the intersecting beams occur at the atom beam modulation frequency and in a specified phase, while background gas effects are dc ion currents subject to very little more than the unexludable shot effect noise. Rudimentary electronics permit ready separation of the two types of ion signals.

A block diagram of the experiment of Fite and Brackmann is shown in Fig. 2. The hydrogen atoms were produced by thermal dissociation in a tungsten oven located in the first of three differentially pumped vacuum chambers. The beam was first collimated by a slit in the wall separating the first and second vacuum chambers. A rotating toothed

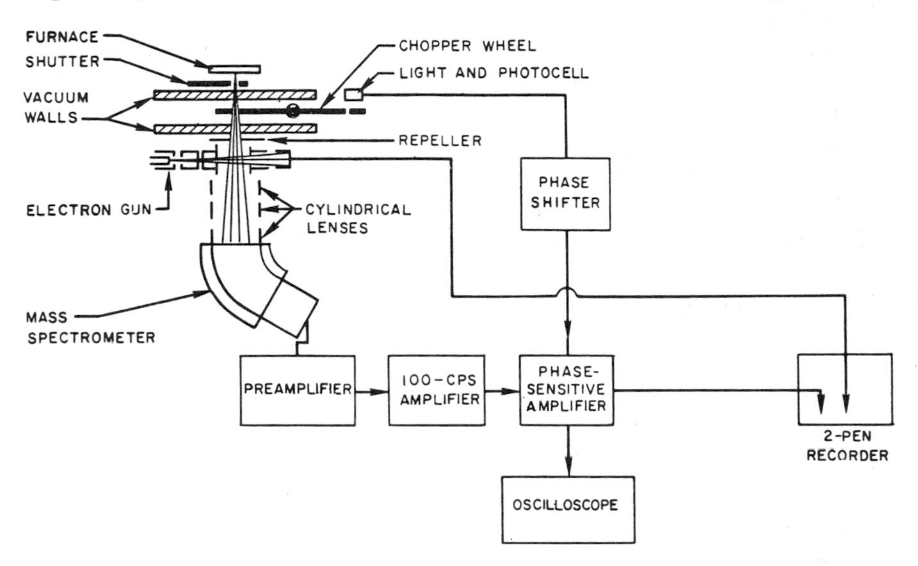

FIG. 2. Modulated atomic beam method used by Fite and Brackmann (1958a) to measure the cross section for $e + H \rightarrow 2e + H^+$ in a crossed-beam experiment.

chopper wheel interrupted the beam at a frequency of 100 cps, and it was this modulated beam which proceeded through a second collimating slit in the wall between the second and third vacuum chambers, after which it was crossed by the dc electron beam. The ions formed by the intersection of the two beams were magnetically analyzed.

The analyzed ion currents were converted to voltage signals by IR drop at the input of a preamplifier, the principal purposes of which were to separate the ac and dc ion signals and to change the ac impedance down to a sufficiently low value that the signal could be taken from the vacuum chamber without excessive signal losses due to stray capacity effects. The ac signal then was passed through a tuned amplifier and into a phase-sensitive detector whose reference signal was taken directly from the chopper wheel by means of a light-and-photocell monitor. An oscilloscope monitored the detected and rectified ion signal for proper phasing before integration and display on a dc recorder.

While it is clear that relative cross sections may be readily taken in this experimental configuration from measuring the signal per unit ionizing electron current as a function of electron energy, it is also possible to obtain absolute cross section values. By maintaining a constant mass flow in the neutral beam and by varying the degree of dissociation of the beam, observation of both the atomic and molecular ion signals can be made to yield a ratio of cross sections for ionization of the atoms and of the molecules. Absolute cross sections for the atom can then be obtained from this ratio and from the known molecular ionization cross section.

2.1.2 *Ionization of Atoms*

Threshold Ionization. The problem which has received most of the attention in recent ionization studies has been the behavior of the ionization cross section near threshold. Not only has the structure in ionization efficiency curves been of concern, but also added incentive to threshold ionization studies has been given by two recent theoretical developments. First, Wannier (1953) developed a theory for single ionization which predicted that near threshold the cross section should increase as the 1.127th power of the electron energy in excess of the threshold energy. Geltman (1956), using a modified Born approximation argument, developed an alternate theory which gave the general prediction that near threshold the cross section should increase as the nth power of the excess energy, where ne is the charge on the resulting ion. The energy range over which these threshold laws should apply was not specified.

Turning first to the question of the dependence of ionization cross sections upon excess energy just above threshold, a number of experiments have now been completed which appear to confirm the threshold laws of Geltman. In atomic hydrogen, Fite and Brackmann (1958a) found a linear threshold law for 5 or 6 ev above threshold, although the experimental value of the slope (0.078 πa_0^2/ev) was somewhat higher than the value predicted by Geltman when he considered only the s-wave component of the incident electrons (0.044 πa_0^2/ev). In studying the ionization of helium, Fox (1959) found that in single ionization, a linear threshold law was obeyed for at least 5 ev above threshold, and that for double ionization the quadratic threshold law was operative for about 20 ev.† In further experimental tests, Dibeler and Reese (1959) studied formation of Na^+, Na^{++}, and Na^{3+}, for which the respective excess energy dependences were linear, quadratic, and cubic.

† It is interesting to note that much of the success of Fox's double ionization experiment may be credited to his use of the mass-3 isotope of helium. By using this isotope, confusion between He^{++} and H_2^+ was completely eliminated.

It is pertinent to note that the above-named ions share a common feature—the absence of excited states lying near the ion ground state. H^+ and H^{++} have no electrons and therefore no excited states, and the first excited state of He^+ is some 40 ev above the ground state. For both Na^+ and Na^{++}, the first excited state is some 33 ev above the ground state, although for the Na^{3+} ion the energy gap is considerably less. It is, however, the absence of near-lying states which has made these ions particularly attractive for the study of the energy dependence of ionization cross sections and has given the most clearly defined confirmation of Geltman's predictions.

In more complicated ions, the situation has not been as clear. In studying single ionization of Kr and Xe, Fox and his collaborators (Fox *et al.*, 1953) used the retarding potential difference method and found that the ionization cross section curve near threshold could be interpreted as consisting of a number of linear segments punctuated by rather well-defined breaks. This is illustrated for the case of Xe in Fig. 3, in which a break occurs at an excess energy of 1.27 ± 0.03 ev. It is interesting to note that the first excited state of Xe^+ is found from spectroscopic studies to be 1.31 ev, which led Fox *et al.* to interpret

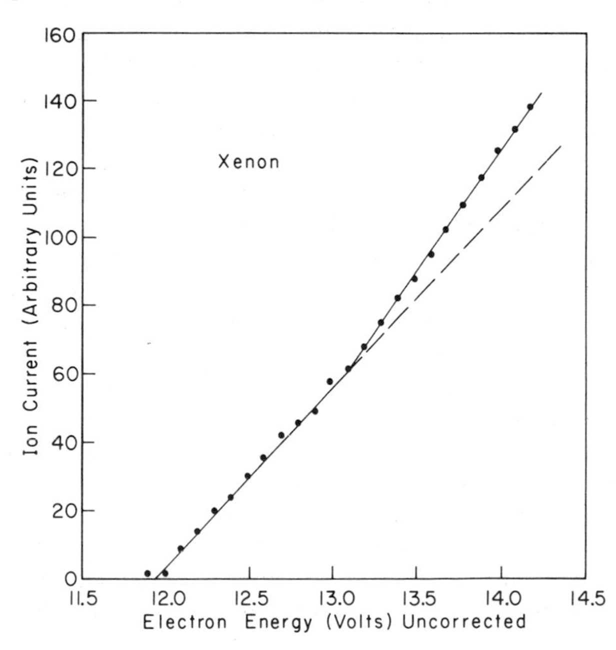

FIG. 3. Relative cross section for single-ionization of xenon near threshold (Fox *et al.*, 1953). The "break" in the curve at 1.27 ev above threshold is identified as the onset of production of ions in the $^2P_{1/2}$ state as well as in the $^2P_{3/2}$ ground state.

the first segment of the curve in Fig. 3 as corresponding to ionization to the ground state of the ion and the second segment as reflecting ionization to both the ground state and the first excited state. The linear segments are, of course, what would be expected from Geltman's threshold laws.

In going on to study multiple ionization, Fox (1959) has found that data using the RPD method in multiply ionized Xenon could also be fitted to straight line segments. This would appear to indicate that multiple ionization is more complex than considered in deriving Geltman's threshold laws or else that the RPD method is subject to error. That this latter possibility may be the case is indicated from recent experiments of Morrison and his collaborators (Morrison and Nicholson, 1959; Dorman *et al.*, 1959) and Krauss *et al* (1959). In studying double ionization, Morrison made measurements of the first derivative of the ionization curve with respect to excess electron energy for formation of Ne^{++}, Ar^{++}, and Xe^{++} and found that in these three cases the first derivative was linear with excess energy. This result implies a quadratic energy dependence, in conflict with Fox's interpretation and in agreement with Geltman's prediction. Further by measuring the zeroth and first derivatives for up to sixfold ionization in xenon, Morrison and his collaborators concluded that Geltman's threshold laws held true for up to at least fourfold ionization and probably up to sixfold ionization as well. It must be noted, however, that the relatively large electron energy spread in these experiments may have obscured some detail in the cross-section curve and its derivatives. Krauss *et al.*, using more conventional mass spectrometric techniques without a monoenergetic electron beam, also obtained data indicating that n-fold ionization occurs with an n-power threshold ionization law in the inert gases.

A second source of complication has arisen in connection with threshold ionization and the interpretation which Fox *et al.* (1953) used in connection with their experiments on Xe^+, holding that structure in the ionization efficiency curves corresponded to the formation of ions in excited states. In studying ionization to form Kr^+, the same authors found that the break between the linear segments corresponding to ionization to the ground and first excited states of Kr^+ was not as distinct as in the case of Xe^+. Figure 4 shows their experimental results. While for a very short range above threshold a linear segment appears to be formed, a rather large transition region occurs before the well-established second linear segment is formed. Extrapolation of the second segment back to the base line gives an apparent threshold for the second process at 0.66 ± 0.01 ev above the ground state ion threshold which is in good agreement with the spectroscopic value for the first excited ion state of 0.666 ev. Subtraction of the observed ion signals from those expected from only the two linear segments gives ion contribution which must be ascribed to processes in addition to those which their early data in Xe^+ formation required.

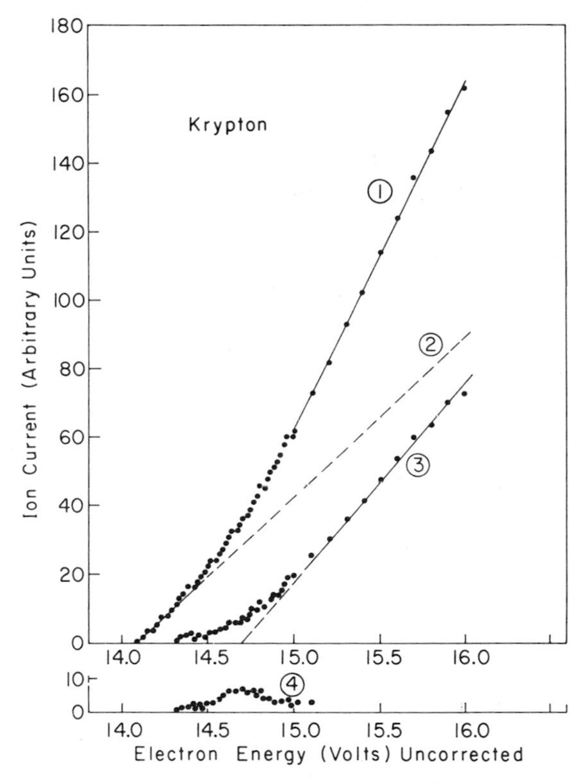

FIG. 4. Relative cross section curve for single-ionization of Krypton near threshold (Fox *et al.*, 1953) and its analysis. Curve ① is a typical experimental curve (where probable error is indicated by the height of the data points), curve ② is the straight line extrapolation of curve ①, curve ③ is obtained by subtracting curve ② from curve ①, and curve ④ is the curve obtained by subtracting the linear portion of curve ③ and its extrapolation, from curve ③.

Fox *et al.* suggest that a possible additional process is autoionization such as was proposed by White (1931), in which the atom is first excited to a discrete state which is energetically higher than required for producing a ground state ion, after which a radiationless transition occurs in which the energy is carried off by an atomic electron, leaving the ground state ion.

Whether this is a correct interpretation remains to be seen. That some such process may be occurring even in the relatively clear case of Xe^+ is suggested by the experiments of Clarke (1954) and a very recent experiment of Foner and Nall (1961). Clarke, using his 127° electron energy selector, observed a cross-section curve which consisted of linear segments but with a break at about 0.8 ev, considerably below the first break observed by Fox *et al.*; and the appearance of a small break at 0.70 ev in the data of Foner and Nall seems to confirm the structure detected by Clarke at an energy intermediate between the threshold for formation of the two lowest-state ions. Foner and Nall also find that identification of breaks with spectroscopic energy levels is not as simple as the early work of Fox *et al.* indicated.

Considerably more confusion accrues to the area of threshold ionization of atoms from the fact that very poor agreement is found between investigators on the relative probabilities of ionization into different ion states, as indicated by the sharpness of breaks in the threshold curves.

At the present time there are many features of threshold ionization which remain unclarified. While it would appear that the threshold laws of Geltman seem unequivocally confirmed for those few ions investigated which do not possess low-lying excited states, the situation for more complicated ions seems in doubt. Not only is there experimental conflict of results, but the full consideration of processes such as autoionization and Auger transitions caused by ionization of inner shell electrons (Burhop, 1952) has yet to be made. In addition, it seems possible that an evaporation model of ionization, such as has been proposed by Russek and Thomas (1958, 1959) for ion-impact ionization phenomena, may be of value for electron-impact ionization as well.

Ionization of Chemically Unstable Atoms. The free hydrogen atom has, for many years, been a major subject for theoretical studies of atomic collision phenomena. The fact that its wave functions are not only completely and exactly known, but are also of simple analytical form, makes it particularly attractive for theoretical studies. However, cross-section predictions even for this simple atom are subject to the uncertainty which arises from inherent deficiencies of the scattering approximations used in any given theoretical treatment. Indeed, a major aim of recent experiments on atomic hydrogen collisions has been to ascertain the ranges of validity of a number of scattering approximations by comparing experimental cross sections with predicted values derived using the approximations.

In studying ionization of ground state atomic hydrogen, the principal aim of Fite and Brackmann (1958a) was to obtain experimental information on the first Born approximation. They used a modulated-crossed-beam experiment, the schematic for which is shown in Fig. 2 and measured the ratio of ionization cross sections of H and H_2. By using their measured ratios and the absolute values for the cross section for ionization of H_2 obtained by Tate and Smith (1932), absolute values

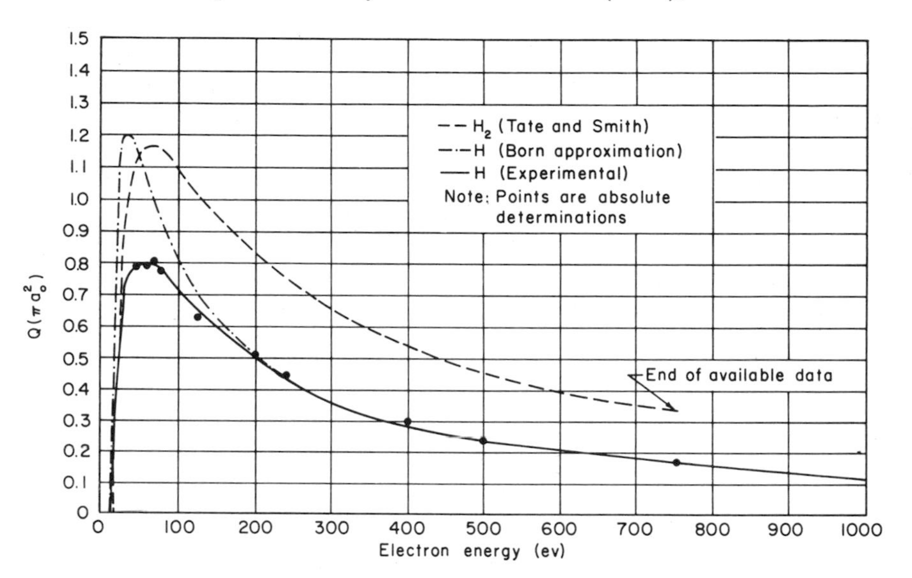

FIG. 5. Cross section for ionization of atomic hydrogen compared with the Born approximation predictions and the cross section for ionization of H_2 (Fite and Brackmann, 1958a).

for the cross section of H were obtained at a number of electron energies. The cross-section behavior at energies between which absolute values were obtained were mapped out by using relative cross section data. Their results are shown in Fig. 5, which shows the Tate and Smith data for H_2 ionization, the experimental atomic hydrogen ionization cross section, and the Born approximation predictions. The agreement of the experimental curve with the Born approximation at electron energies in excess of about 250 ev appears to justify their procedure in obtaining absolute values. Below 250 ev, the departure of the experimental results from theory is not unexpected in view of the basic assumptions used in the Born treatment.

In a more recent experimental study of this process, Boyd and Boksenberg (1960) used a similar experimental approach, but the ion detector was a mass spectrometer using crossed electric and magnetic fields rather than a magnetic sector instrument such as was used by

Fite and Brackmann. The two sets of results are in fairly satisfactory agreement.

In studying the ionization of atomic oxygen in a crossed-beam experiment, Fite and Brackmann (1959) used a gas discharge source for the oxygen atom beam source. Their experimental procedure was again to measure a ratio of cross sections, Q_1/Q_2, where Q_1 is the cross section for $e + O \rightarrow O^+ + 2e$ and Q_2 is that for $e + O_2 \rightarrow O_2^+ + 2e$. A complication arose in obtaining absolute values for Q_1 because of the absence of knowledge of Q_2. The absolute cross section Q_T which was determined by Tate and Smith was for the sum of all processes leading to positive ion production (i.e., including multiple and dissociative ionization) with no analysis of the ions produced. It was thus necessary to combine mass-spectrometric ion detection with a detector of the type used by Boyd and Green (1958) which indiscriminately collected all ions produced by the intersecting beams with 100% efficiency. By using argon beams with both detectors alternately, Fite and Brackmann first determined the collection efficiency of their mass spectrometer for ions produced with no initial kinetic energy. They then ran a beam of O_2 and divided their mass spectrometer O_2^+ signal by this efficiency to ascertain the total number of O_2^+ ions being produced. The ratio of this calculated molecular ion current and the observed signal in their nondiscriminating detector was used as a value for Q_2/Q_T. By combining the two ratios, the ratio Q_1/Q_T was obtained which was used in conjunction with the Tate and Smith data to find absolute values for Q_1. The results of this experiment were in satisfactory agreement at moderate and high electron energies with values predicted by Seaton (1959), although at lower energies substantial deviations between theory and experiment were apparent. It may be added that in this experiment the concern that some of the oxygen atoms from the gas discharge source might not be in the ground state was set aside on the basis of threshold ionization studies. It was found that no-mass-16 signals could be detected below the ground state appearance potential of about 13.6 ev. From the observed noise level below this energy, Fite and Brackmann estimated that at least 97% of their atoms were in the ground state.†

† Professor E. Lindholm has pointed out (private communication) that the absence of ionization of atomic oxygen at electron energies below the appearance potential for ionization of ground state oxygen atoms is not a good test for excited metastable atoms. For oxygen the three lowest states of the neutral atom are ^3P, ^1D, and ^1S and those of the singly charged ion are ^4S, ^2D, and ^2P. Very remarkably, the energies of these levels are such that any ionization process in which optical selection rules are valid yields appearance potentials for excited atom ionization which *exceed* that for ionization of groundstate atoms. Thus to test for the presence of excited atoms one should seek the

They also observed that the ionization cross section for single ionization of atomic oxygen appears linear with excess electron energy.

2.1.3 *Ionization of Molecules*

Ionizing collisions between molecules and electrons has continued to be an active area of experimentation in recent years. Because the experimental techniques are similar to those which have been discussed in § 2.1.1, and in "Electronic and Ionic Impact Phenomena" (Massey and Burhop, 1952), and in view of the existence of the recent review article on electron-molecule collisions by Craggs and Massey (1959), a general discussion of this subject will not be made part of this chapter.

It is pertinent to mention one feature of recent work on ionization of molecules on electron impact: that much of the work now taking place is being done on commercial instrumentation. As representative of work done with commercially available equipment, it is perhaps adequate to cite here the experiments of Lampe *et al.* (1957) who used a Consolidated Electrodynamics Corporation Model 21-620 mass spectrometer, employing crossed electric and magnetic fields and cycloidal ion orbits, to measure ionization cross sections at a fixed electron energy (75 ev) of the inert gases, a number of the lighter diatomic and triatomic gases, and some of the lighter hydrocarbons. In their experiments, ion signals were compared and absolute values for the cross sections were obtained by taking the values for ionization of argon to be those reported in "Electronic and Ionic Impact Phenomenon" (Massey and Burhop, 1952, p. 38).

2.2 Excitation

In comparing the measurement of ionization and excitation cross sections from an experimental point of view, and in terms of the basic operational definition of collision cross section (1), it is evident that the only fundamental difference resides in the method of detection of signals associated with the processes. With ionization, the most characteristic

appearance of "breaks" in the ionization efficiency curve just above threshold rather than the appearance of ions at electron energies below the groundstate threshold.

The possibility of observation of ionization below the ground state threshold energy is probably not excluded, however, for at these low energies electron-exchange effects might be expected to occur and weaken the selection rules. However, it is not presently possible, because of the lack of cross section values for exchange ionization processes, to relate the absence of ionization below the ground state threshold to the maximum fraction of neutral atoms in other than the ground state; therefore the 97% figure quoted in the text should be taken with considerable reserve.

signal derives from the ion formed in the collision, but, as is seen in the foregoing sections, questions of the state of the ion formed remain unanswered using ion detection alone, except in the immediate vicinity of threshold. Offsetting this disadvantage is the experimental advantage that ion currents are simply detectable and readily measurable.

In experiments on the electron-impact excitation of atoms and molecules to discrete states the situation is reversed. While in almost every case there are one or more signals which are unique to the process under study and which clearly identify the state of the excited system, these signals are of a very diverse nature and are much less easily measured than a simple charged particle current is. Indeed, the history of experimentation on excitation processes is, in large part, a history of ingenuity in devising detectors for the products of the excitation process; but, unfortunately, very few of these detectors have been developed to a point of reliability and accuracy such that they can be used to determine absolute excitation cross sections to the precision achievable in processes yielding a charged particle as a reaction product.

2.2.1 *Techniques*

Radiation Detectors. For the study of excitation to states from which radiation to lower states occurs, the radiation itself is the most obvious signal on which to base a cross section measurement. Because of the simultaneous occurrence of many excitation processes, it has long been common practice to use prism or grating monochromators to select a particular spectral line for detection of a specific process. Massey and Burhop cite a number of experiments in which this technique has been used. Perhaps the most significant technical advance for studying excitation processes yielding the emission of visible and near-ultraviolet radiation has been the growing availability of commercial photocells and photomultipliers whereby the sensitivity of monochromator detectors can be greatly increased.

In the far- and vacuum-ultraviolet regions of the spectrum, the experimenter is faced with a loss of sensitivity due to the opacity of the materials of which optical components and photomultiplier envelopes are made, and also with a diminishing reflectivity of grating surfaces. On the other hand, in these spectral regions, the large energy available in each photon restores experimental feasibility for a wide variety of experiments. In the past few years, a number of novel and significant advances have been made which enable improved experimentation in the far- and vacuum-ultraviolet.

In order to extend the range of photomultipliers to wavelengths shorter than the cutoff of their envelope materials, fluorescent detectors are sometimes used. One such detector is constructed by applying sodium salicylate and other materials to the outer surface of a glass-enclosed photomultiplier. Ultraviolet photons, to which the glass is opaque, incident upon the sodium salicylate will produce visible photons which then actuate the multiplier. Johnson *et al.* (1951) and Watanabe *et al.* (1953) have found that this photon-conversion method gives a nearly flat wavelength response down to 540 A. This feature was of particular value in the studies of Corrigan and von Engel (1958) who calibrated such a detector absolutely and then studied, using electron swarm techniques, the excitation and dissociation of molecular hydrogen. Among their results was a comparison between excitation and ionization coefficients in this gas. A disadvantage of this method is that the photomultiplier, having a photocathode of low work function, has a substantial dark current at room temperature, and the noise using this method of detection may be excessive in experiments of low signal level.

Another detection method for ultraviolet photons is the direct photoelectric effect. The high photon energy permits greatly relaxing the care which must be given to the photocathode surface compared to photoelectric detection of visible radiation. Indeed, virtually any metal treated in almost any manner is suitable as a photocathode. A very important recent contribution to the measurement of cross sections for excitation of ultraviolet radiation has been the measurement of the photoelectric efficiency of various materials throughout the entire ultraviolet spectrum (Walker *et al.*, 1955; Hinteregger and Watanabe, 1953) and it may be expected that these measurements will ultimately permit a wider range of measurement of absolute excitation cross sections.

In experiments of low signal level, the direct photoelectron currents are often inadequate. Two methods are used to increase the electrical signal level. The first is electron multiplication using multiplier stages similar to those of the commercial photomultiplier. A disadvantage of this is that in detecting vacuum ultraviolet radiation, the photocathode and therefore the multiplying surfaces must be exposed to a vacuum environment which is somewhat less desirable than that in a sealed photomultiplier tube. Unfortunately, most electron multiplying surfaces are subject to some deterioration under ordinary vacuum conditions and the sensitivity of such devices is not as stable as is desired. The dark currents and noise in such multipliers are much less than in photomultipliers, however, because of the absence of any low-work-function materials.

The second method of enhancing photoelectric signals involves accelerating the photoelectrons to high energies and then detecting the fast electrons using scintillators. The noise associated with dark currents of the photomultiplier used to detect the scintillations can be reduced by pulse-height discrimination techniques.

Another class of detectors which has been used in excitation studies operates by photoionization of gases. Since these detectors employ fairly high pressures of gases, they must be enclosed in solid materials The short-wavelength cutoff is then the transmission limit of whatever window material is used in the envelope; e.g., the cutoff is about 1050 A when using lithium fluoride. The long-wavelength cutoff is the photoionization threshold of the working gas. By judicious selection of windows and working gases, sensitive detectors of radiation of limited wavelength response can be constructed.

Both ionization chambers and photon counters have been constructed using NO (Chubb and Friedman, 1955) and I_2 (Brackmann *et al.*, 1958) for which the ionization thresholds are about 9.3 and 9.7 ev, respectively. When LiF windows are used, these devices have wavelength response bands of 200 to 300 A, depending on the quality of the LiF. While this range is restrictive, a number of interesting spectral lines occur in this wavelength band, not the least of which is the Ly-α line of atomic hydrogen. In photon counters, the practice has been to place small amounts of the working gas into an inert gas filling and use the devices as Geiger-Müller counters. The facts that both NO and I_2 have large uv absorption coefficients and are electronegative accord such counters the property of being self-quenching.

Another major step in detection of photon signals has been the measurement of absorption coefficients of several gases throughout the ultraviolet spectrum by Watanabe *et al.* (1953). By using their data, it is now possible to select combinations of gases for use in gas filters which offer high transmission and excellent selectivity at several places in the uv spectrum. An example of gas filtering for increased selectivity is the use of an O_2 absorption cell (with LiF entrance and exit windows) before an iodine-vapor-filled photon counter for Ly-α detection (Fite and Brackmann, 1958b). In the wavelength band of the I_2 counter, O_2 is a good absorber of radiation except at seven narrow "windows," one of which coincides with the Ly-α lines. By using O_2 filtering, the wavelength response range of the photon counter is reduced by about an order of magnitude. With additional CO_2 filtering, the response range could be further reduced, for this gas is much more transparent at Ly-α than at several of the oxygen windows which admit radiation other than Ly-α (Chubb and Friedman, 1955).

Gas filtering before photoelectric detectors is a possibility which has not yet been fully exploited.

Metastable Atom Detectors. For the detection of metastable atoms, detectors using three principles of operation have been used.

The first and most frequently used method is applicable to those metastable atoms whose excitation energy exceeds the work function of some metal. In this case it is energetically possible for a metastable atom impinging on the metal to give up its excitation energy and eject an electron from the metal, so that an electrical current measurement can then be made. The efficiency of detection of the metastable states of helium at gold surfaces has recently been measured by Stebbings (1957) who found the efficiency to be 29%. A less precise measurement of the efficiency of electron ejection from a platinum surface by the metastable $2S_{1/2}$ state of atomic hydrogen (of excitation energy 10.15 ev) yielded a value of about 6% (Lichten and Schultz, 1959).

A second method of detection relies upon there being near the metastable state another state from which radiation to a third state is permitted, and upon perturbing the metastable so that an admixture of the first two states is produced. This method has been applied to detection of the $2S_{1/2}$ state of atomic hydrogen and He^+ (Novick *et al.*, 1955). Since the $2P_{1/2}$ states has an energy separation from the metastable state of only the Lamb shift ($\sim 4 \times 10^{-6}$ ev in H), the application of a dc electric field causes the metastable atom to radiate Ly-α radiation which can then be detected with a radiation detector.

A third method, which has been successfully applied to detection of metastable helium atom employs absorption spectroscopy (Woudenberg and Milatz, 1941). In particular, the 2^3S state will absorb radiation (10,830 A) in photoexcitation to the 2^3P state, and both attenuation and fluorescence of this radiation can indicate the presence of metastable helium atoms. Phelps and Molnar (1953) detected both the 2^1S and 2^3S states of helium by absorption of radiation at 5016 and 3889 A corresponding to transitions to the 3^1P and 3^3P states, respectively. A number of metastable atom densities in other gases have also been monitored by absorption spectroscopy, e.g., Phelps' (1959) work on de-excitation of excited neon atoms. Clearly, variations of this type of radiation absorption technique can be applied to the detection of metastable states of less excitation energy than necessary for using either of the two methods given previously, although the efficiency of detection using photon absorption is considerably less than necessary for many experiments.

Scattered Electron Detection. Since an electron in exciting an atomic system to a discrete state loses a definite amount of energy, it is possible to monitor the excitation process by detecting those scattered electrons which have lost an amount of energy equal to the excitation energy. Using this approach to excitation studies has the advantage that the basic measurement is a current measurement, and therefore is quite straightforward to carry out. The disadvantage of this method is that it requires good energy resolution in the scattered electron detector, for many of the most interesting cases for study involve atoms and molecules in which the energy separation between states may be quite small. If excitation to a particular state is to be examined, the electron detector must be capable of discriminating between two near-lying states.

Early applications of scattered electron detection were made by Hughes and McMillen (1932) and Mohr and Nicoll (1932, 1933) in which electrostatic analysis of the electrons scattered at a fixed angle was made using a 127° electrostatic analyzer. Massey and Burhop (1952) and Massey (1956) discuss these early measurements and show schematic representations of these experiments. More recent experiments of this type taking advantage of newer experimental techniques have

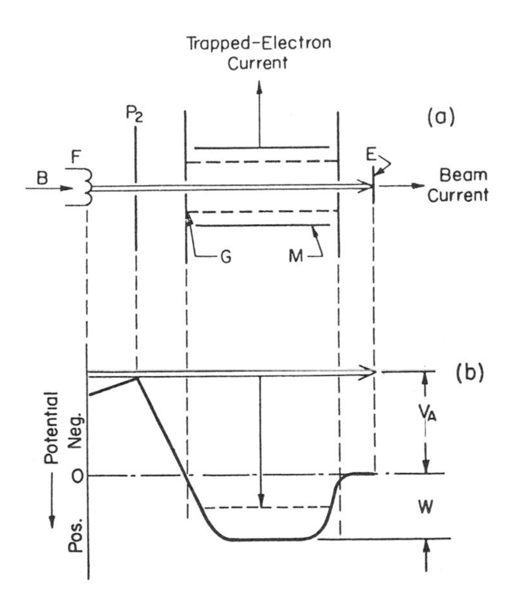

FIG. 6. Schematic diagram of Schulz' trapped electron method used in studying inelastic electron collision processes, showing the tube (a) and the potential distribution at the axis of the tube (b). F is the filament, P_2 is the retarding electrode, G is the cylindrical grid forming the collision chamber, M is the cylinder for collection of trapped electrons, and E is the electron beam collector. V_A is the accelerating potential and W is the well depth. The electrons have energy $V_A + W$ in the collision chamber. Those which lose more energy than V_A in collisions become trapped.

been done by Lassettre and his associates at the Ohio State University, but unfortunately their work has not been published in the open literature.

A second type of experimental method utilizing scattered electron detection is the "trapped electron method" recently developed by Schulz (1959). Just as electrostatic analysis using the 127° analyzer is analogous to the monoenergetic electron studies of Clarke, the trapped electron method might be considered the detection analog of the retarding potential difference method of Fox.

Figure 6 shows a schematic of a trapped electron experiment, in which a magnetic field and narrow apertures confine a beam of electrons to motion in a straight line. The beam passes through potential regions as shown. Those electrons which suffer inelastic collisions and lose energy between V_A and $V_A + W$ are trapped in an electrostatic well. In the absence of subsequent collisions, the electrons have no choice but to arrive ultimately at the collector, M. Since small variation of the potential P_2 permits the use of monoenergetic incident electrons in the sense of RPD method, and since W may also be varied, remarkable versatility accrues to this experimental method. A disadvantage of this method at its present stage of development is that the trapped electron current is the sum of the currents produced by excitation to all states whose energies lie in the interval $\mathrm{V_A}$ to $\mathrm{V_A} + W$. As a result, a clear definition of the state to which the target atom or molecule is excited is obtained only near the lowest excitation threshold. Nonetheless, the facts that this method is possessed of excellent energy resolution and measures total scattering cross sections directly makes it a very important complementary technique to the electrostatic analyzer method described previously, which measures differential scattering and is best suited to high incident electron energies.

2.2.2 *Excitation of Atoms*

Atomic Hydrogen. Study of collisions in which electrons excite atomic hydrogen to discrete states has an unusual appeal because of the large variety of mathematical approximations which have been applied to these problems. Until very recently, however, measurements of hydrogen atom collision cross sections had not been attempted because the chemical instability of this atom precluded the use of conventional experimental techniques. The beginnings of experimentation in this area were the experiments of Lamb and Retherford (1950, 1951) on hydrogen atom fine structure in which they used electron-impact excitation to produce atoms in the $2S$ state. While they made only relative measurements near threshold of the cross section of the excitation process which was used in those experiments, their work dramatically called attention to the fact that thermal dissociation sources could be used to produce groundstate hydrogen atom beams which were both highly pure in atom content and sufficiently intense that electron-impact excitation experiments using crossed-beam configurations were possible.

a. Excitation of Lyman-α radiation. In studying the excitation of Lyman-α radiation produced in collisions between electrons and free hydrogen atoms, Fite *et al.* (Fite and Brackmann, 1958b; Fite *et al.*, 1959) used crossed-beam techniques. Their detectors were oxygen-filtered iodine-vapor-filled uv photon counters which viewed the region of intersection of the two beams.

The basic experimental problem in this experiment was to obtain an adequate signal-to-noise ratio, where the "noise" arose from two principal sources. The first was soft ($\sim$ 10 ev) X-rays which were produced by the electron beam upon its arrival at the anode of the electron gun. The second and more important source of noise was the radiation emitted following collision of the electrons with molecules in the residual gas in their vacuum system. The principal gas responsible for this radiation was molecular hydrogen which was formed when the hydrogen atom beam struck the end of the beam apparatus and reassociated. Because of irregular action of diffusion pumps, drifts and fluctuations in the background pressure prohibited making dc measurements, and modulated-beam techniques used in the study of electron-impact ionization of hydrogen atoms were again necessary.

The most unusual feature of these experiments was the use of counters in ac circuits, a situation made possible since the dead time of the counters (which are basically Geiger-Müller counters) was small compared to the period of the modulation (0.01 sec). In the earlier experiments, the output of the counter tube was fed to a pulse-monitoring oscilloscope and the output of the amplifier of the oscilloscope was fed into their ac circuitry (see Fig. 2). In the later experiments, the pulses were shaped to give a larger Fourier amplitude at the frequency of the modulation in order to give improved signal-to-noise ratio and eliminate the high peaks at the beginning of the counter pulses.

In this experiment, the directly measured quantity was the relative signal per unit electron current as a function of electron energy, i.e., a relative cross-section measurement. In order to assign absolute values, the relative cross section curve was to be normalized to Born approximation values at energies above 250 ev. The justifying arguments for this procedure were (1) the Born approximation appeared to be correct above this energy in the case of ionization of H on electron impact, and (2) the relative cross section curve had the same shape as the Born approximation curve above 250 ev.

The assignment of absolute values, however, encounters two complications. The first is that Ly-α radiation arises from not only the direct $1S$-$2P$ excitation process, but from (a) cascade processes in which the atom is excited to states for which $n > 2$ and radiates back to the $2P$ states and (b) production of $2S$ atoms which could be quenched by stray electric fields. While care to exclude stray fields could minimize the contributions from this latter cause, the former effects cannot be eliminated. At the higher energies, estimation of these contributions to the detected Ly-α radiation on the basis of the Born approximation cross sections (Massey, 1956) and the pertinent transition probabilities (Bethe and Salpeter, 1957) indicate that the error made in normalizing relative cross section data at high energy to only the direct $1S$-$2P$ excitation cross section is considerably less than the experimental uncertainty.

The second complication is that radiation excited on electron impact has an angular distribution which is not generally isotropic and is energy dependent. In order to compare theoretical predictions, which for the most part have considered only total excitation cross sections, with experimental results obtained by observing photons emitted over only a limited solid angle, it becomes necessary to be concerned with the angular distribution of the radiation.

This requirement led to a two-step experiment. The first part was a measurement of the relative signal per unit electron current with the photon counter observing radiation emitted at 90° with respect to the electron beam, a situation in which the best signal-to-noise ratio could be achieved. The second part was the examination of the angular distribution of the radiation, made by comparing signals with the photon counter observing at 45°, 90°, and 135° with respect to the electron beam, at the same electron energy. Since Ly-α is an electric dipole radiation, its angular distribution is given in the form

$$q(\theta) = \frac{3Q}{4\pi} \frac{(1 - P\cos^2\theta)}{(3 - P)} \tag{2}$$

where Q is the total cross section, and P, the polarization fraction, is an energy-dependent parameter in the range $-1 \leqslant P \leqslant 1$. Observations at only two angles are required to determine the constant P and the entire angular distribution at any electron energy. By using a third angle, a check on symmetry of the radiation was obtained.

These angular distribution measurements were subject to much less experimental accuracy than the relative measurements taken when detecting photons emitted at 90° with respect to the electron beam. While the angular distribution measurements did indicate that at low energies the values of P are positive, at higher energies the most probable values of P were less than the experimental uncertainty. As a result, in normalizing their high-energy relative data to the Born approximation, Fite and Brackmann elected to use the value $P = 0$. By correcting their lower energy data taken at the 90° observation angle by the measured most probable values of P, they obtained a total cross section curve.

A check on the absolute values obtained in this way was made by examining the radiation emitted at an angle of 54.5° with respect to the direction of the electron beam. It may be noted from (2) that at the angle $\theta = \cos^{-1}(1/\sqrt{3})$, relative measurements are proportional to the total cross section for excitation, irrespective of the angular distribution. These measurements were subject to considerably more noise than the 90° measurements because of soft X-rays from the electrodes of the electron gun which were now in view of the photon counter; however, they appeared to confirm the results obtained by the procedure of correction of the 90° data to within the experimental accuracy.

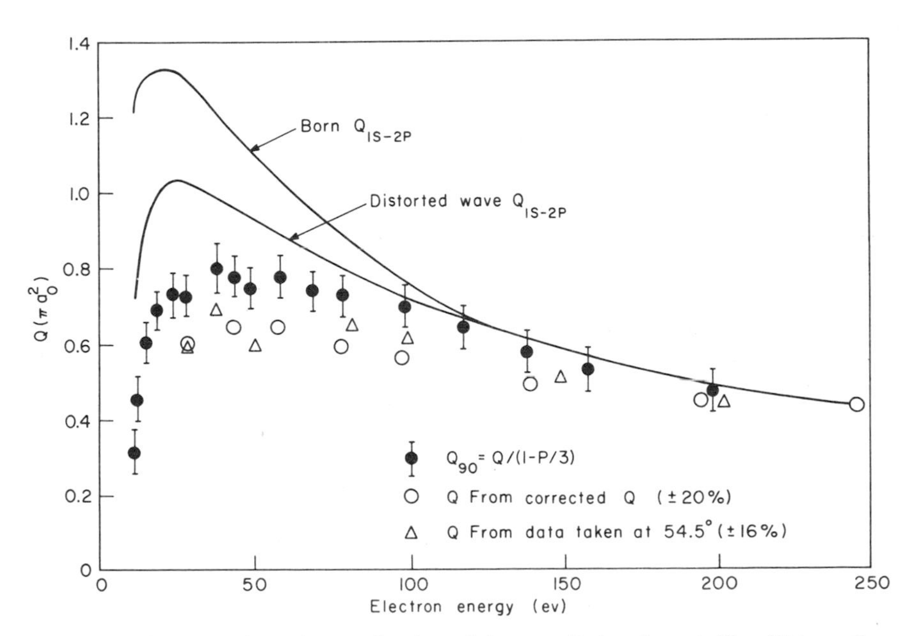

FIG. 7. Cross sections for production of Ly-α radiation in $e + H$ collisions. See text for explanation.

Figure 7 shows a plot of a fictitious cross section, $Q_{90} = Q/(1 - P/3)$, which is the apparent total cross section assuming isotropic angular distribution of the radiation at an intensity equal to that observed at 90°; the total cross section, Q, obtained by correcting the 90° data for measured angular distribution; and the total cross section obtained from data taken at an angle of 54.5°. This figure also shows for comparison the predictions of the Born approximation (Massey, 1956) and the distorted wave approximation (Khashaba and Massey, 1958) for direct excitation to the $2P$ state only.

It is interesting to note that near threshold the cross section Q_{90} appears proportional to the square root of the excess energy. This energy dependence is predicted for the total cross section, Q, from the work of Wigner (1948).

b. Excitation to the metastable $2S_{1/2}$ state. The excitation processes leading to the production of a metastable hydrogen atom in the $2S$ state, which Lamb and Retherford used in their fine-structure experiments, have been studied in two subsequent experiments. The first was one part of an experiment by Lichten and Schultz (1959). This experiment was a dc crossed-beam experiment in which a beam of ground state atoms was produced in a thermal dissociation source of the type developed by Hendrie (1954), which was an improved version of the oven originally used in the experiments of Lamb and Retherford (1950, 1951). A dc electron beam crossed the atom beam and excited metastable atoms, which were slightly deflected in the excitation process and were detected downstream by a platinum surface electron ejection detector. This detector was made of sufficiently large area to intercept all metastables within a recoil angle of 22° off the direction of the ground state atom beam and $\pm$ 7° in the direction perpendicular to the plane of the atom and electron beams.†

Intermediate between the excitation volume where the two beams crossed and the detector, an electrostatic quench field could be applied. When the quench field was removed, all metastable atoms struck the detector and ejected electrons, the current of which was measured. When the quench field was turned on to a field of about 30 volts/cm, the metastables were quenched, radiating Ly-α. Only a small fraction of the quench radiation could reach the detector, and give rise to an electron current through photoelectric effect, because the angular

† Subsequent experiments (Stebbings *et al.*, 1960) on angular distribution of H(2S) atoms showed that this solid angle was adequate to intercept all the metastables produced in the experiment of Lichten and Schultz.

distribution of the radiation is isotropic and the detector surface subtended only a small solid angle. By using a removeable wire stop to block the beam from entering the excitation region of their apparatus, Lichten and Schultz could assess the contributions to their detected currents from electron-impact excitation of the background gas in their vacuum. By proper combination of galvanometer readings with the quench field on and off and with the wire beam stop both in and out of place, the signal associated only with production of $2S$ atoms from electron collisions with the groundstate atoms in their beam could be obtained. Figure 8 shows their relative cross section data for production

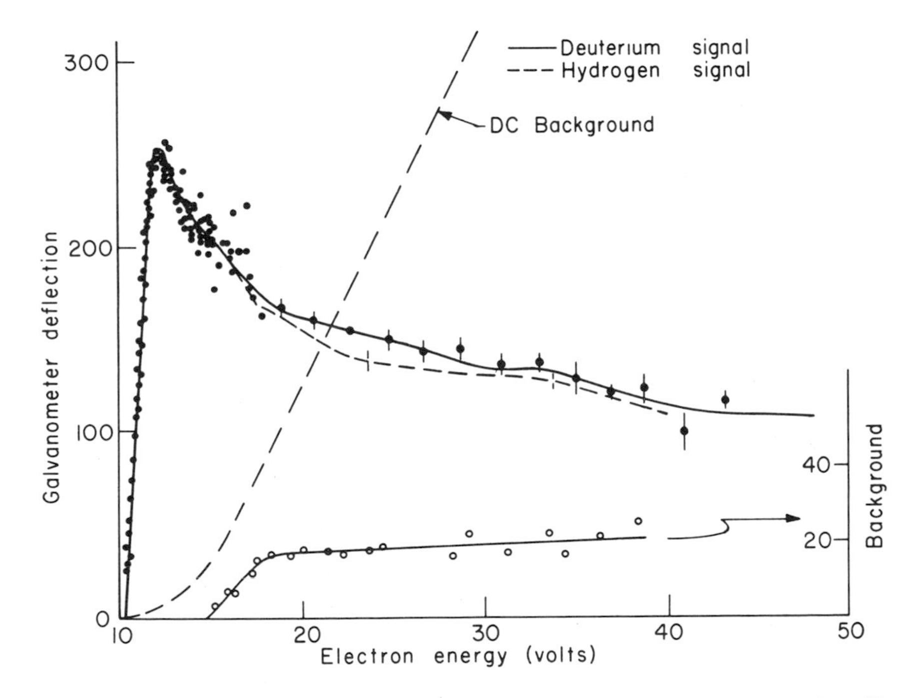

FIG. 8. Relative cross section for production of metastable hydrogen atoms in collisions between electrons and ground state hydrogen atoms (Lichten and Schultz, 1959). The "background" is indicative of metastable atoms produced in collisions of electrons with H_2 in the residual gas in the vacuum system and the "dc background" was the galvanometer deflection when metastable atoms, irrespective of origin, were electrostatically quenched.

of $2S$ atoms. The dc background in this figure is the galvanometer deflection with the atom beam running and with the quench field turned on. The "background" is the difference in galvanometer deflections with the quench field off and on, with the wire stop blocking the atom beam. The background gives an indication of metastable atoms produced by electron collisions with molecular hydrogen in the background gas.

Although not essential to that part of the work of Lichten and Schultz dealing with total excitation cross sections, their apparatus also contained an inhomogeneous magnetic field to permit their using a spin-polarized beam of atoms and a magnetic field rotator. A field of 575 gauss was used to confine their electron beam while also quencing metastable atoms with a given direction of their electron spin. These additional features were used in their measurements of exchange excitation which will be discussed in § 2.3.

The first type of data obtained in the experiment of Lichten and Schultz was a relative cross section, or excitation function curve for the production of H(2*S*). While below about 20 ev, background effects were less than effects associated with the crossed beams, at higher energies the background increased strongly. Because of deteriorating signal-to-noise ratio at higher energies, data were presented only up to about 43 ev. (See Fig. 8.)

In addition to obtaining a relative cross-section curve, Lichten and Schulz took advantage of the relative positions of their quench plates and their detector to make an absolute measurement of the cross section. Since they could estimate the number density in the atom beam from their furnace pressure and the geometry of their apparatus, and measure their electron current, electron path length through the neutral beam, and current at their detector, the use of (1) to determine the absolute cross section required only a knowledge of the efficiency of their detector for ejecting electrons when struck by H(2*S*). In order to determine this efficiency, they compared the detector signal when struck by metastable atoms with the signal produced when the atoms were electrically quenched by the quench field. Since quenching produces Ly-α radiation for which the angular distribution is isotropic, by knowledge of the solid angle subtended to the electric quench region and observing the photoelectric signal, they measured the ratio of the efficiency of electron ejection under H(2*S*) bombardment to the Ly-α photoelectric efficiency. Since the photoelectric efficiency was known from earlier experiments (Walker *et al.*, 1955 ; Hinteregger and Watanabe, 1953), they could determine their detector's efficiency for metastable hydrogen detection. This figure was 0.065 ± 0.025. By using this procedure Lichten and Schultz determined that the cross section for production of H(2*S*) at its peak (at about 11.7 ev) is $0.28 \pm 0.14\ \pi a_0^2$.

An alternate procedure used by Lichten and Schultz to assign absolute values followed the observation that their relative cross section data had the same shape as that predicted by the Born approximation in the higher part of their energy range. By normalizing their relative data to the Born approximation in the 30 to 40-ev range, the peak of the curve occurs at a value of $0.36 \pm 0.05\ \pi a_0^2$ which is within the range specified in their absolute measurement. The peak cross section value which they adopt is a weighted average of $0.35 \pm 0.05\ \pi a_0^2$.

The second experiment bearing on excitation of ground state atomic hydrogen to the $2S_{1/2}$ state is that of Stebbings and associates (1960) in which a comparison of the cross section for this process to the cross section for excitation of Ly-α radiation was made. Their experimental arrangement is shown in Fig. 9. A modulated atom beam was crossed

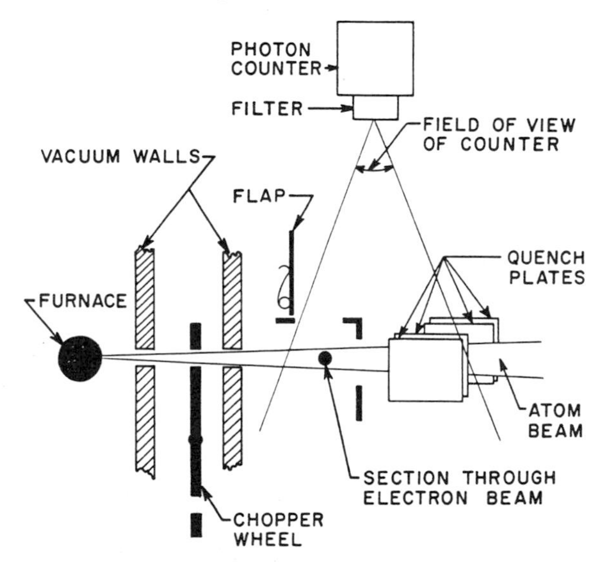

FIG. 9. Schematic representation for measuring the ratio of cross sections for production of metastable atoms and Ly-α radiation (Stebbings *et al.*, 1960).

by an electron beam and the interaction region was viewed by a photon counter, so that the ac counter signal recorded the direct excitation of the Ly-α radiation. A flap could then be lowered to block the counter's view of the interaction region and prevent detection of direct excitation of Ly-α.† The counter could then be moved slightly to view a region in which an electric field could be placed. The metastable $2S$ atoms produced in the interaction region, whose lifetime in the vacuum was determined to be of the order of a millisecond, continued on and upon

† Since the lifetime of the $2P$ state is of the order of 10^{-9} sec, the decay length of $H(2P)$ is of the order of 10^{-3} cm.

entering the region between the electric field plates could be quenched to radiate Ly-α. By comparing the photon counter signals produced by quenching H(2S) and from direct excitation of Ly-α, under conditions of constant atom and electron beams, and with the same solid angle subtended in both parts of the experiment and using the same photon counter, a ratio of cross sections for the two processes could be obtained.

The use of modulation techniques and the discriminating oxygen-filtered photon counters permitted measurements to 700 ev.

Since the counter was intercepting photons emitted at 90° with respect to the electron beam, the appropriate Ly-α excitation cross section to use is Q_{90} discussed previously. Since Q_{90} has been assigned absolute values through normalization to the Born approximation at very high energies (250-700 ev), absolute values of $Q(2S)$ the total cross section for production of 2S atoms could also be obtained.

TABLE I

MOST PROBABLE VALUES FOR THE RATIO OF THE TOTAL CROSS SECTION FOR PRODUCTION OF 2S ATOMS TO Q_{90}[a]

Electron energy (ev)	$Q(2S)/Q_{90}$
11.3	0.45
12	0.36
13	0.36
14	0.28
15	0.22
17.5	0.20
19	0.14
25	0.11
30	0.096
40	0.08
57	0.095
75	0.11
90	0.10
125	0.08
150	0.11
200	0.11
300	0.11
400	0.13
550	0.13
700	0.12

[a] The apparent total cross section for excitation of Lyman-α radiation if it were to be assumed that the radiation angular distribution is isotropic at an intensity equal to that observed at 90° with respect to the direction of the exciting electron beam.

In assigning absolute values to Q_{2S} on the basis of the data of this experiment in the original report of it, a correction was made appropriate to having the radiation produced by the quench field angularly distributed as dipole radiation, with the dipole oriented parallel to the quench field. Lichten (1961) pointed out that this was an erroneous correction, and that the induced Ly-α radiation is in fact isotropic. The direct data of this experiment gives the values of $Q(2S)/Q_{90}(\text{Ly-}\alpha)$ shown in Table I.

Combining the results of this table with values of Q_{90} (see Fig. 7) yields a cross section curve for Q_{2S} given in Fig. 10. For comparison, the Born approximation for this cross section is shown.

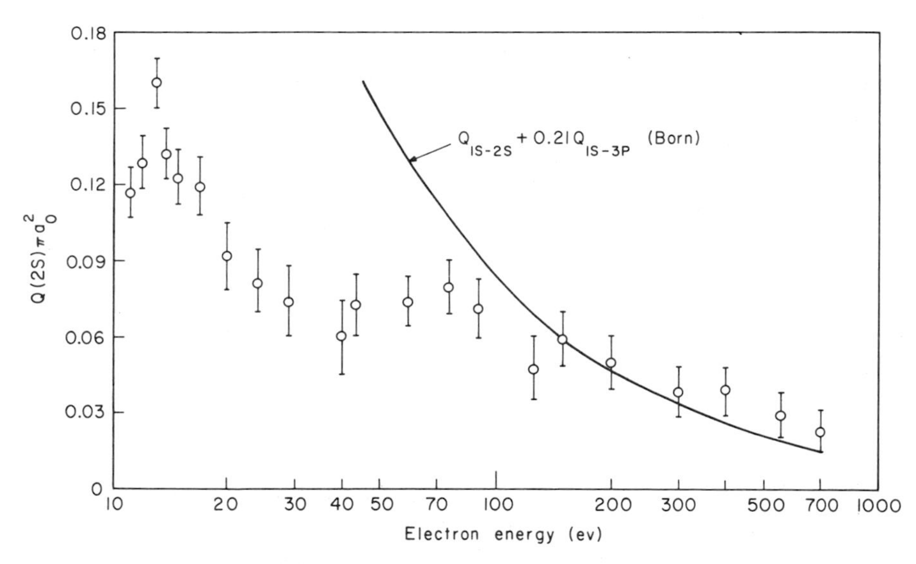

FIG. 10. Cross section for production of metastable $2S$ hydrogen atoms as obtained from measurements of $Q(2S)/Q(\text{Ly-}\alpha)$, and results given in Fig. 7. The Born approximation predictions against which this result is compared includes a term to account for the substantial contributions to the $2S$ production by cascade processes of the type $1S - nP - 2S$.

In considering both the data of this experiment and that of Lichten and Schultz, it may first be noted that the shape below 40 ev of the two curves is in quite good agreement. A point of disagreement exists on the absolute value assignment however. In the experiment of Stebbings *et al.*, Born approximation values fall within the experimental uncertainty only above 150 ev, and substantial deviations occur in the 30 to 40-ev range where Lichten and Schultz normalized their relative data to the Born approximation. While this leads to a discrepancy of about a factor of 2 between the peak values adopted by the two sets of authors, it is to be noted that the most probable peak value taken by Stebbings *et al.* ($0.16\,\pi a_0^2$) falls within the range of uncertainty in the direct experimental absolute cross section determination of Lichten and Schultz ($0.28 \pm 0.14\ \pi a_0^2$).

With reference again to Fig. 10, it is to be noted that in comparing either of the above-described experiments with theory it is necessary to consider cascade contributions as well as contributions from the direct $1S$-$2S$ excitation process. Particularly important are excitation processes $1S$-nP, where $n \geqslant 3$, followed by a radiative transition to the $2S$ state. To assess these contributions, Lichten and Schultz assumed that the shape of relative cross section curves for all $1S$-nP excitation processes is similar to that observed for the excitation of Ly-α radiation (Fite and Brackmann, 1958b) and that the absolute magnitudes of the cross sections are proportional to the squares of the respective dipole matrix elements $\langle 1S \mid z \mid nP \rangle$. On the basis of this assumption and the fact that the branching ratios between the $1S$ and $2S$ states are approximately the same for all nP states (Bethe and Salpeter, 1957, Table 15), an approximate formula for the production of H($2S$) is given by $Q_{2S} = Q_{1S-2S} + 0.21 Q_{1S-3P}$. The first term is the direct excitation process while the second estimates the cascade contributions. It is this formula that is used by both experimental groups in comparison of their data with theory.

Helium

a. Excitation of metastable levels. The fact that helium is both chemically stable and relatively simple from a theoretical point of view has led to its being a favorite atomic system for study for many years. As is, unfortunately, not uncommon in atomic collision physics, there has existed some disagreement between different workers. Recently some of the newer techniques have been applied to unresolved problems, including excitation to the 2^3S and 2^1S levels.

In past studies of excitation to these metastable states a number of experimental approaches have been used. Maier-Leibnitz (1936) used retarding curves on electrons in an electron swarm experiment to determine the inelastic collision cross section in helium. Dorrestein (1942) used an electron beam method with detection being made by electron ejection when the metastable atoms strike metal surfaces. Woudenberg and Milatz (1941) detected He(2^3S) by observing absorption of light at 10,830 A, corresponding to the transition from the 2^3S state to the 2^3P state.

Recently, additional experiments on electron-impact excitation to the lowest metastable states have been carried out near threshold, with particular care being taken in the experiments to increase the electron energy resolution. Schulz and Fox (1957) conducted an experiment of similar concept to that of Dorrestein (1942) in which the retarding potential difference method was employed under conditions that their effective electron energy spread was about 0.1 ev. Their tube is illustrated in Fig. 11. The electrodes P_1, P_2, and P_3 were used to select a

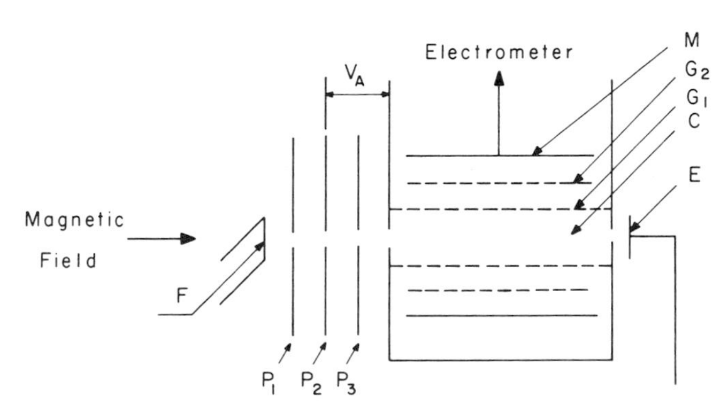

FIG. 11. Electrode assembly for the study of excitation of helium to its metastable states in which the RPD method was used to improve energy resolution (Schulz and Fox, 1957). Metastable atoms were detected by electron ejection at the detector surface, M.

narrow band of energies in a manner described earlier (§ 2.1). Some of the metastable atoms produced in the collision chamber C, pass through the two grids G_1 and G_2, arrive at the gold-plated metastable detection surface M, and eject electrons. These electrons are accelerated to grid G_2, and the electron current leaving M is measured. The data in this experiment were the differences in current at the metastable atom detector when the potential of the filament with respect to P_2, the retarding electrode, was varied. A magnetic field of about 100 gauss was used to weakly confine the electron beam. By using the 0.29 value of Stebbings (1957) for the efficiency of electron ejection of gold surfaces upon metastable helium atom impact, they obtained absolute cross sections as well as highly resolved relative cross section curves. They place the value of the cross section at the peak of the 2^3S excitation process at 4×10^{-18} cm$^2 \pm 30\%$, which is in good agreement with Maier-Leibnitz' value of 5×10^{-18} cm^2.

In a second study of excitation of helium to the lowest metastable states, Schulz (1959) applied the trapped electron method discussed in § 2.2. The data on excitation of helium using both metastable atom detection and the trapped electron method are shown in Fig. 12. The dip following the peak of the 2^3S excitation cross section is not as well pronounced in the trapped electron method as in the metastable detection experiment. Such a diminishing of the dip would arise if the

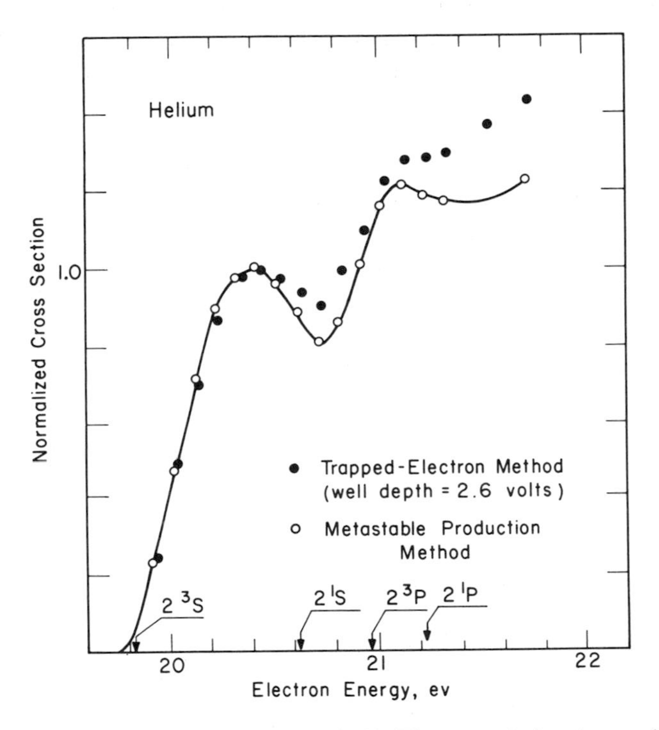

FIG. 12. Excitation of helium near threshold. The open circles give results obtained using metastable atom detection (Fig. 11) and the closed circles represent data obtained using the trapped electron method (Fig. 6).

apparent effective energy spread of the exciting electrons is larger than the true effective energy spread associated with using the RPD method in the electron gun. That such an increase in the effective energy spread should occur was pointed out by Schulz; its cause is the nonrectangular shape of the potential well which traps the scattered electrons.

The second difference between the data obtained using the two methods of studying helium excitation occurs just over 21 ev where excitation to the P states can occur. While the solid surface used to detect metastables in the earlier experiment would not respond to photons radiated by the 2^3P-2^3S transition, photoelectric effect by the energetic photons from the 2^1P-1^1S transition will add to the apparent metastable atom signal. In the trapped electron method, excitation to both the *P* states is detected with the same efficiency as in the case of excitation to the metastable levels.

2.2.3 *Excitation of Molecules*

The recent excellent review article of Craggs and Massey (1959) on electron-molecule collisions makes it unnecessary here to discuss in detail the excitation of molecules on electron impact. It is appropriate, however, to present some unusually interesting results published since Craggs and Massey's review.

In studying excitation of CO and N_2 on electron impact, Schulz (1959) demonstrated the remarkable utility of the trapped electron method and also convincingly substantiated the existence of a scattering process in N_2 at 2.3 ev which had been found by Haas (1957) in a swarm experiment and interpreted as representing the formation of an unstable negative ion of N_2.

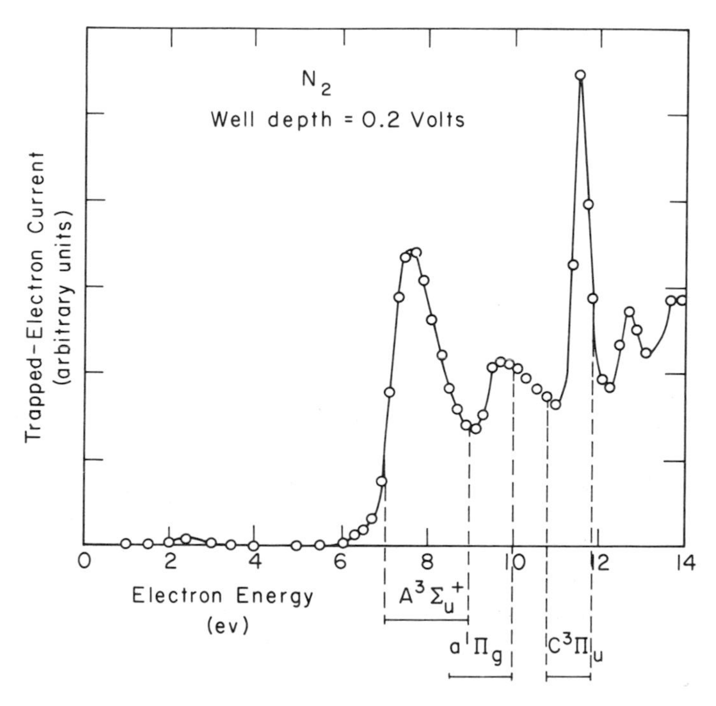

FIG. 13. Excitation of N_2, using the trapped electron method, with a well depth of 0.2 ev (Schulz, 1959). The small peak at 2.3 ev and the large sharp peak at 11.5 ev are to be particularly noted.

Figures 13 and 14 show curves of the trapped electron current as a function of electron energy using two well depths, 0.2 and 0.8 ev, respectively. In interpreting the differences between these two curves or the curves for any molecular excitation, it is necessary to be aware

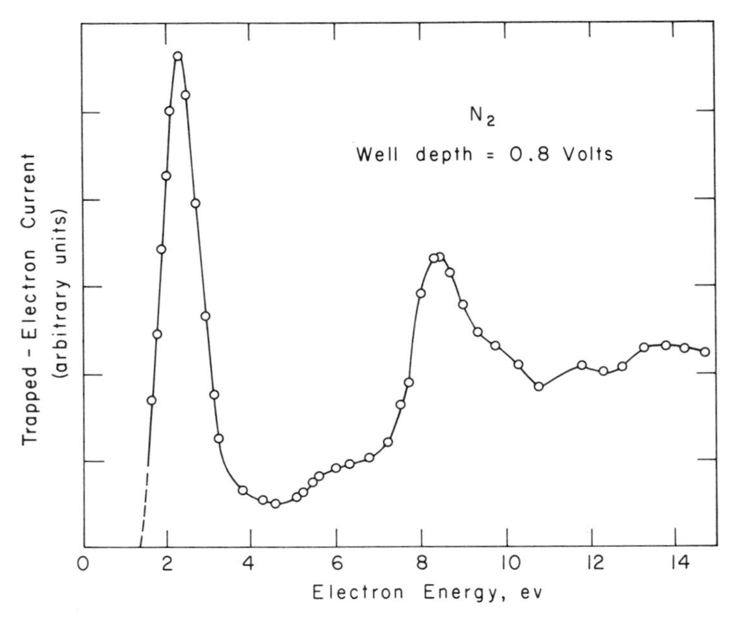

FIG. 14. Excitation of N_2, using the trapped electron method, with a well depth of 0.8 ev (Schulz, 1959). See text for comparison of the peaks at 2.3 and 11.5 ev using the different well depths.

that there are two limiting cases when using the trapped electron method. The first involves transitions to repulsive-type states in which the potential curve for the state in the Franck-Condon range of nuclear separations has an energy range much larger than the well depth. In this case, the trapped electron current traces out the integral of the cross section for production of electrons with kinetic energy between 0 and W (the well depth), resulting from transitions from the ground state to the repulsive state within the Franck-Condon range. The second limiting case is for transitions to bound states, where a given vibrational state is preferentially excited, which resembles excitation to discrete atomic levels.

On recalling that the energy of the electrons before collision is $V_A + W$, where V_A is the accelerating potential and W is the well depth, and electrons which lose energy between V_A and $V_A + W$

constitute the trapped electron current, it is evident that in order to observe a situation represented by the second limiting case W must not be large compared to the range of energies given up in exciting the state. If W is larger than the energy range between the threshold and the peak of the process, the peak will not be clearly discerned.

The peak occurring at 11.5 ev is an example of the second limiting case. With the 0.2-ev well depth, this peak, which Schulz identifies with excitation to the $C^3\Pi_u$ state, is quite distinct, but when using $W = 0.8$, the peak is obscured.

The peak at 2.3 ev represents an opposite situation, where the integrated cross section for producing electron energy loss is measured. Because there are no known electronically excited states below 6.0 ev and because excitation of vibrational states in N_2 on electron impact is improbable, Schulz interprets this peak as arising from formation of a negative ion N_2^- which is unstable, and which then decays to a free electron and a vibrationally excited N_2 molecule. With a shallow well, only decay of N_2^- giving the highest vibrational states of N_2 is detected, and as the well depth is increased, decays yielding more vibrational states, working downward in energy, are included in the trapped electron current.

By (1) holding V_A constant at several values and varying the electron energy by varying the well depth (so as to include only those final N_2 vibrational states with energy above V_A), and (2) holding W fixed at several values and varying electron energy by varying V_A, Schulz obtained data which he interprets as indicating that the N_2 formed by N_2^- decay is formed preferentially in the lower vibrational states. This is consistent with the fact that in Fig. 13 the peak at 2.3 is quite small, for with the 0.2 ev well depth, only those processes leaving the N_2 highly vibrationally excited would be detected.

2.3 Electron Exchange Collisions

It is well known that in collisions at low electron energies it becomes necessary to consider the fact that an incident electron and an atomic electron are identical particles of spin $\frac{1}{2}$. This requirement has been embodied in many of the low-energy calculations for collision cross sections through appropriate symmetrization of the wave functions. In the case of atomic hydrogen, for example, two scattering amplitudes are introduced, $f(\theta)$, for collisions in which the incident electron is scattered, and $g(\theta)$, for collisions in which the incident electron and the atomic electron exchange.

The differential scattering cross section is then given by

$$\sigma(\theta) = \frac{v_n}{v}\{\tfrac{1}{4}|f+g|^2 + \tfrac{3}{4}|f-g|^2\}, \tag{3}$$

where v is the velocity of the incident electron and v_n is the velocity of the electron leaving the atom which has been excited to the nth state in the collision. An alternative expression for the differential cross section is obtained algebraically to be

$$\sigma(\theta) = \frac{v_n}{v}\{\tfrac{1}{2}|f|^2 + \tfrac{1}{2}|g|^2 + \tfrac{1}{4}|f-g|^2\}. \tag{4}$$

It is common to speak of the cross section as being composed of three constituent cross sections corresponding to the three terms in the brackets. They are, respectively, the "direct," "exchange," and "mixed" differential cross sections.

While these concepts have been part of the theory for many years, experimentation has until recently been confined to study those exchange processes in which no competing direct processes occur; e.g., excitation of helium to the 2^3S state can be accomplished only by an exchange collision if spin-orbit coupling is negligible. In the past few years, the scope of experimentation on exchange collisions has been immensely broadened by the development of techniques which can distinguish between direct and exchange collisions in processes where both are allowed. Since sources of polarized electrons are not available, these techniques depend upon the detection of the change of spin of the atomic electron following the collision.

2.3.1 *Optical Absorption*

Phelps and his associates (Phelps and Molnar, 1953; Phelps, 1955; Phelps and Pack, 1955) used optical absorption to detect a change of state from the 2^1S to the 2^3S states of helium. Their experiment involved the study of the decay of the afterglow in a pulsed gas discharge in a helium absorption cell. A monochromator introduced light of 5016 A to detect the 2^1S metastables (2^1S-3^1P transition) and light at 3889 A was used to detect the similar transition in the triplet system and measure the concentration of 2^3S metastables. An interference filter was placed between the cell and the photomultiplier detecting the monochromatic radiation in order to reduce unwanted light from the afterglow itself. The pulse rate of the discharge was sufficiently slow that the afterglow completely decayed between pulses. By observing the transmitted radiation as a function of time, the rate of decay of the two types of metastable atoms could be followed.

In their first experiments Phelps and Molnar displayed the transmitted radiation signals on an oscilloscope which traced out the entire time decay of the metastables. The difficulty of this method of data taking is that small amounts of absorption could not be measured accurately because of poor signal-to-noise ratio. That this method should be limited by signal-to-noise ratio is clear because multiplier shot noise will be accepted by the oscilloscope over its very broad frequency response range.

The second method of taking data traded imprecise knowledge of the absorption at *all* times for more accurate knowledge at a given time following cessation of the pulses. In this second method, gating circuitry was used to accept signals produced only during two very small time intervals within each interval between successive discharge pulses. The gating of the detector occurred at twice the frequency of the discharge pulsing, so that signals were taken at times t following the discharge pulse and $t + T/2$, where T is the discharge pulse period. The gated photomultiplier signals were fed to a narrow-band tuned amplifier operating at the discharge pulse frequency and the ac component was

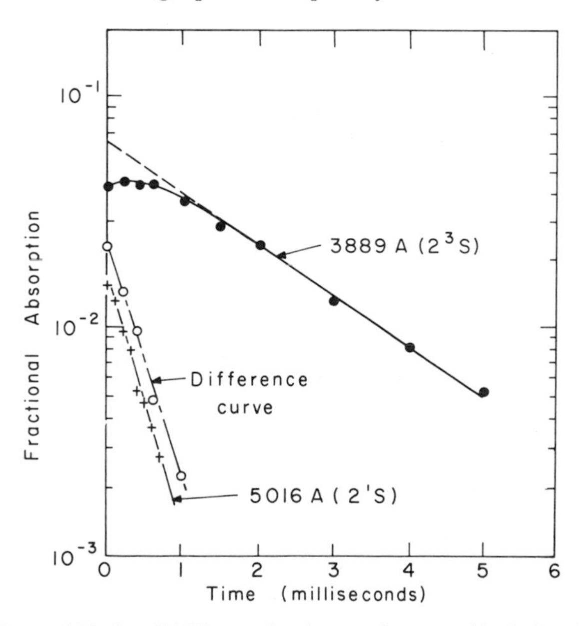

FIG. 15. Data of Phelps (1955) on the decay of metastable helium in a discharge afterglow, from which the value of the cross section for the exchange collision $e + \text{He}(2^1S) \rightarrow e + \text{He}(2^3S)$ was obtained.

amplified. This ac component is indicative of the difference of absorption at the two times at which the detector gate was open. By using narrow-band detection and integration, signal-to-noise ratio was greatly improved and permitted more accurate study of the rates of decay of the metastables.

The results of Phelps are shown in Fig. 15. It is seen that the 2^1S component diminishes much more rapidly than the 2^3S. In considering the loss of the 2^1S component, three effects were taken into account: diffusion, de-excitation in collision with neutral helium atoms, and de-excitation in collisions with free electrons. By varying pressure and electron density (which is very nearly constant during the time interval when 2^1S atoms are present), the authors found that the rate of decay of the atoms was linear with the electron density and determined the coefficient deactivation of the 2^1S atoms.

To interpret the origin of the deactivation, it is appropriate to note that highly pure helium was used so that impurity effects are presumably minor, and that in the afterglow the electron energy is effectively thermal. This latter fact prohibits depopulation of the 2^1S state by excitation processes, and strongly suggests that the loss mechanism is a superelastic collision where a free electron exchanges with the atomic electron, leaving the atom in the 2^3S state. If this is the process, it would be expected that during the decay of the 2^1S population an increase would be seen in the 2^3S concentration. This is, indeed, to be seen in Fig. 15.

If it is assumed that the deactivation occurs only as a result of electron exchange collisions and that the electron energy is the mean thermal energy at room temperature, then the cross section for the process at this energy is 3×10^{-14} cm^2.

2.3.2 *Optical Pumping*

A striking new method for studying exchange collisions is that of "optical pumping." The first measurement using this method, that of the exchange elastic collision cross section in sodium, was carried out by Dehmelt (1958) in the course of his work on the gyromagnetic ratio of the free electron.

Dehmelt's experiment is shown schematically in Fig. 16. A beam of circularly polarized sodium light (D lines) is directed through a bulb containing sodium vapor and an inert buffer gas (argon), and thence onto a photocell. The bulb is contained in a uniform magnetic field, $H_0 = 21.4$ gauss, parallel to the direction of the light beam, so that weak Zeeman splitting of the levels of the sodium occurs. Since the light is circularly polarized, absorption of the radiation can occur only if $\Delta m = +1$ in the transition. These transitions produce excited atoms in only part of the excited substates, but collisions with the buffer gas will alter these state populations and tend to make the excited substates equally populated prior to reradiation. When reradiation occurs, transitions for which $\Delta m = 0, \pm 1$ take place. These effects tend to repopulate

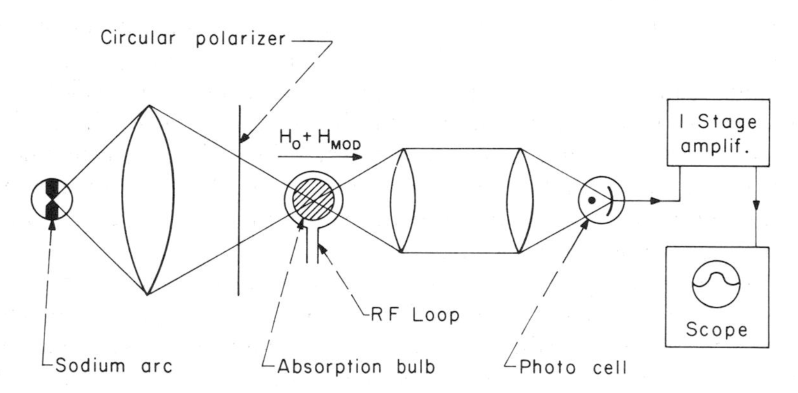

FIG. 16. Schematic representation of the optical pumping experiment of Dehmelt (1958) to measure the *g*-factor of the free electron and the cross section for elastic exchange collisions between electrons and sodium atoms.

the ground substates equally. But since the circularly polarized light is depopulating them unequally (for the $m = -\frac{1}{2}$ state is a better absorber than the $m = +\frac{1}{2}$ state), the result is that ground state atoms are "pumped" from one substate to the other, and a steady state polarization of the sodium atoms is produced. In practice, relaxation effects as well as radiation patterns govern the degree of polarization producible by optical pumping (Dehmelt, 1957).

If anything should occur in the bulb to upset the steady state polarization of the sodium atoms, the number of atoms with $m = -\frac{1}{2}$ will change. Since these are the most effective in absorbing the circularly polarized light, the transmitted light intensity will also change. Such an upset of the steady-state polarization of the sodium can be produced by applying an rf magnetic field, H_{mod}, at the proper frequency to induce transitions between the magnetic sublevels of the $F = 2$ sodium hyperfine structure level. Indeed, by using these transitions Dehmelt calibrated the magnetic field H_0.

In addition, Dehmelt provided for the addition of free electrons to the gas in the bulb by using a gas discharge. Radio-frequency (25 Mcps) pulses of about 1-msec duration were applied to a pair of condenser plates located outside the bulb (not shown in Fig. 16), at a repetition rate of about 10 per second. Observations were made after the electrons had become thermalized (about 50 μsec).

When unpolarized electrons are added to the bulb in which polarized sodium atoms are present, they undergo exchange collisions with the sodium. Not only is a new steady-state polarization of the atoms achieved, but the electron cloud also becomes polarized.

The principal purpose of Dehmelt's experiment was to take advantage of the polarized electrons to measure the free electron spin g-factor. With the field H_{mod} at the gyromagnetic frequency, the polarization of the electrons was changed, thus producing a change in the sodium atom polarization, which in turn was detected by a change in the photocell current. His actual measurement compared the g-factor for free electrons and g_J for ground state sodium atoms, in the same sample and with the same magnetic field H_0.

Of more interest for the present purposes is the fact that by an analysis of the signal strength and line shape of the electron resonance, Dehmelt was able to deduce that the cross section for elastic exchange collisions between free thermal electrons and sodium atoms is $> 2.3 \times 10^{-14}$ cm^2.

In an experiment similar to Dehmelt's optical pumping experiment, Franken *et al.* (1958) examined the electron exchange cross section in potassium and set an upper limit on the cross section of 3×10^{-14} cm^2.

2.3.3 *Preferential Quenching*

To study exchange collisions of the type which leave an atom in an excited state, Lichten and Schultz (1959) experimented with excitation of the hydrogen atom to the metastable $2S_{1/2}$ state. The attribute of this state which permits their experiment to be performed is seen from Fig. 17. When a magnetic field of about 575 gauss is applied to this excited atom its $m_s = -\frac{1}{2}$ substate coincides energetically with the $m_s = \frac{1}{2}$ substate of the $2P_{1/2}$ state, and very small electric fields perpendicular to the magnetic field cause admixing of the P state, so that the atom decays with Ly-α emission. This magnetic quenching removes $2S$ atoms with $m_s = -\frac{1}{2}$ but does not quench metastable atoms with $m_s = +\frac{1}{2}$. Only those atoms of a given polarization survive magnetic quenching.

Lichten and Schultz' experimental plan is given in Fig. 18. Hydrogen atoms produced by thermal dissociation are passed through an inhomogeneous magnetic field in order to produce a beam of polarized ground state atoms. By altering the position of their source, they could select a beam of atoms of either polarization, i.e., $m_s = +\frac{1}{2}$ or $m_s = -\frac{1}{2}$.

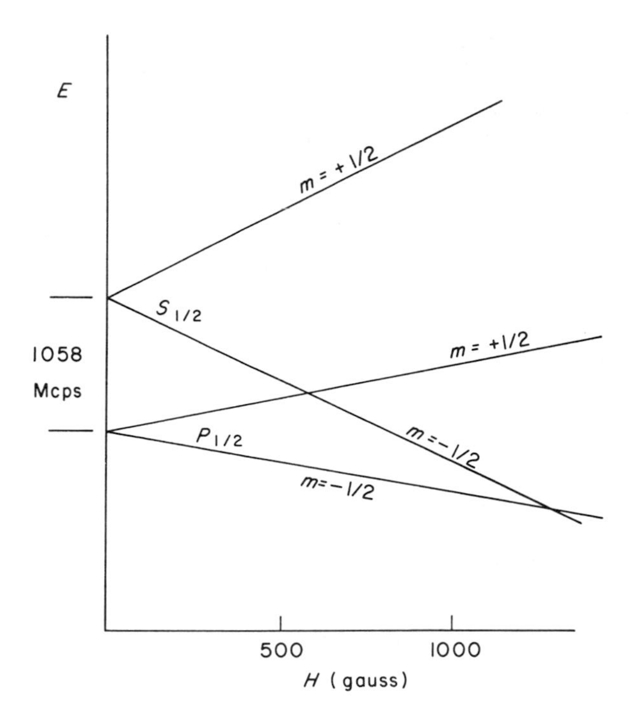

FIG. 17. Zeeman effect on the levels with $n = 2, j = \frac{1}{2}$ in atomic hydrogen in weak magnetic fields. The energy coincidence of the $m = -\frac{1}{2}$ substate of the $S_{1/2}$ state and the $m = +\frac{1}{2}$ substate of the $P_{1/2}$ state at a field of 575 gauss permits magnetic quenching of metastable atoms with $m = -\frac{1}{2}$.

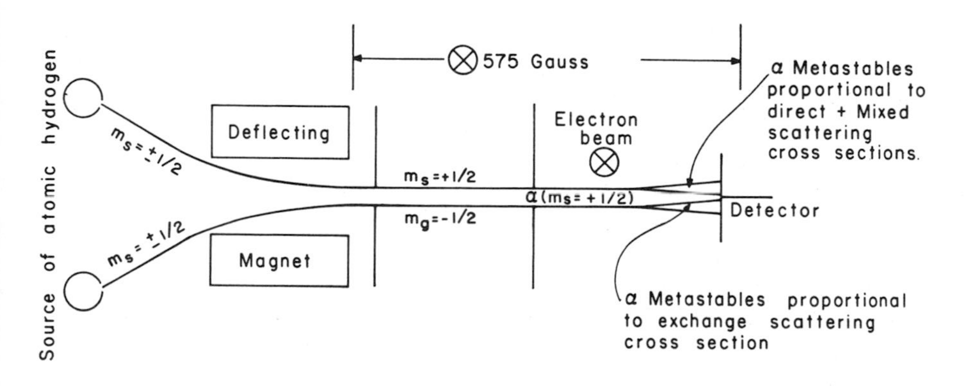

FIG. 18. Experimental plan of Lichten and Schultz (1959) for measuring the ratio of the exchange to total cross sections in excitation of atomic hydrogen to the metastable $2^2S_{1/2}$ state.

These polarized ground state atoms entered a region of uniform magnetic field at 575 gauss, in which their electron gun crossed the ground state atom beam and produced $2S$ atoms. Those metastable atoms with $m_s = -\frac{1}{2}$ were magnetically quenched while those with $m_s = \frac{1}{2}$ continued on to the surface electron ejection detector. In this experiment, since the electrons were unpolarized and metastables of only one polarization were detected, the variable was the polarization of the original ground state atom beam.

In order to deduce information on the exchange cross section for this excitation process, it is helpful to consider Table II from Lichten

TABLE II

Electron Scattering by a Hydrogenic Atom

	Component of electron spin along magnetic field				
	Before collision		After collision		
Case	Atomic electron	Incident electron	Atomic electron	Scattered electron	Partial cross section
A	$\frac{1}{2}$	$\frac{1}{2}$	$\frac{1}{2}$	$\frac{1}{2}$	$\lvert f - g \rvert^2/2$ (mixed)
	$\frac{1}{2}$	$-\frac{1}{2}$	$\frac{1}{2}$	$-\frac{1}{2}$	$\lvert f \rvert^2/2$ (direct)
	$\frac{1}{2}$	$-\frac{1}{2}$	$-\frac{1}{2}$	$\frac{1}{2}$	$\lvert g \rvert^2/2$ (exchange)
B	$-\frac{1}{2}$	$\frac{1}{2}$	$\frac{1}{2}$	$-\frac{1}{2}$	$\lvert g \rvert^2/2$ (exchange)
	$-\frac{1}{2}$	$\frac{1}{2}$	$-\frac{1}{2}$	$\frac{1}{2}$	$\lvert f \rvert^2/2$ (direct)
	$-\frac{1}{2}$	$-\frac{1}{2}$	$-\frac{1}{2}$	$-\frac{1}{2}$	$\lvert f - g \rvert^2/2$ (mixed)

and Schultz' paper. This table lists the pertinent cross sections for all polarizations of both the incident and atomic electrons before the collision and the scattered and atomic electrons after the collision. Since only metastables with $m_s = \frac{1}{2}$ state are detected, it is evident from this table that when the ground state atom beam is in the $m_s = \frac{1}{2}$ state the signal, S_+, is proportional to the sum of the direct and the mixed cross sections. With the ground state atom beam in the $m_s = -\frac{1}{2}$ state, the signal S_- is proportional to the exchange cross section. Therefore, the quantity, $S_-/(S_+ + S_-)$ is equal to the ratio of the exchange cross section to the total cross section. This ratio as determined by Lichten and Schultz is given in Fig. 19.

Again it is important to point out that in this measurement as in those experiments to measure the total cross section for excitation to the

$2S$ state, it is not only the direct $1S$-$2S$ process which is under study. Cascade processes of the type $1S$-nP-$2S$ contribute a quite significant fraction of the observed metastable atoms and care must be exercised in comparing experiment with theory.

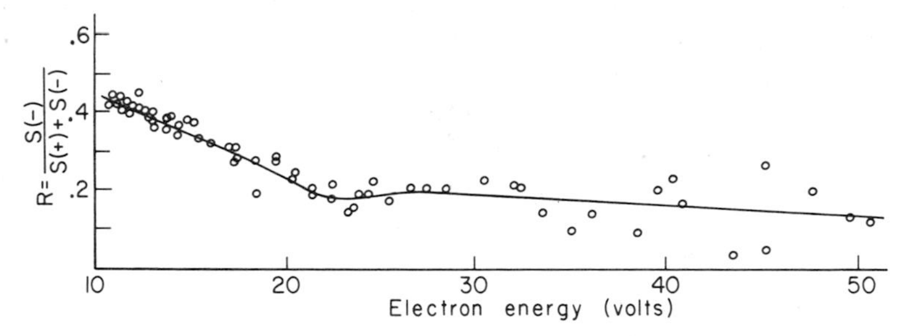

FIG. 19. The ratio of the exchange cross section to the total cross section in excitation of atomic hydrogen to the $2S$ metastable state (Lichten and Schultz, 1959).

2.3.4 *Inhomogeneous Magnetic Field Analysis*

In Lichten and Schultz' work, an inhomogeneous magnetic field was used to select a single spin state in the ground state atoms with which

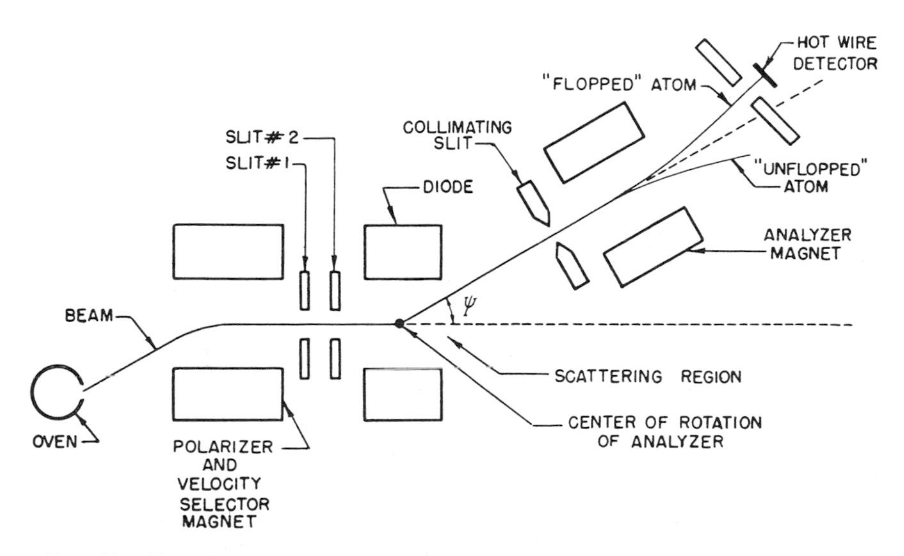

FIG. 20. Experimental plan of Rubin *et al.* (1959) for studying electron exchange collisions. Inhomogeneous magnetic fields are used to produce a polarized and velocity-selected incident atom beam and to analyze the atom beam after collision for its state of polarization.

electrons were to collide and their analyzer for the polarization of the excited atoms was a system taking advantage of the particular excitation process under study. Clearly a more general method to obtain measurements of the ratio of exchange to total cross sections, if the atom after collision is in a relatively long-lived and moderately indestructible state, is to determine the degree of polarization by using a second inhomogeneous magnetic field.

This type of experiment has been done by Rubin *et al.* (1959) who, in their first experiments, studied exchange in elastic collisions of slow electrons with potassium. Their experiment, using a crossed-beam configuration, is illustrated in Fig. 20. Atoms were produced in a furnace which was offset from the axis of their beam apparatus. These atoms, in passing the polarizing inhomogeneous magnetic field, were selected according to their spin state, and this magnet also acted as a velocity selector for atoms of the desired polarization. The atoms were then crossed with a beam of electrons. Conservation of momentum requires that the atoms which have suffered collisions with electrons are deflected slightly away from the main beam of atoms, and to this deflected beam the analyzing inhomogeneous magnetic field is applied. In this experiment, the analyzed beam of atoms was detected by a surface ionization detector (Ramsey, 1956, p. 379).

Since this experiment is very similar in concept to that of Lichten and Schultz, except for the detector of polarization following the collision, the information output of the two experiments is similar; i.e., it measures a ratio of exchange to total cross sections. In the case of exchange collisions in potassium, Rubin, Perel, and Bederson have found that the ratio is about 0.3 at electron energies in the range 0.5-4 ev. Since in the energy range 1.5-4 ev the total cross section as measured by Brode (1929) varies from about 6 to 3×10^{-14} cm^2, exchange cross sections of between 2 and 1×10^{-14} cm^2 are implied by these measurements. The similarity of these values with the values obtained for thermal electrons from the optical pumping experiment of Franken *et al.* (1958) is to be noted.

Both the experiments of Lichten and Schultz and of Rubin, Perel, and Bederson are capable in principle of measuring differential scattering cross sections as well as the cross sections integrated over all angles. In both experiments the ground state atom beam can be made to have a very narrow spread of energies and when a monoenergetic atom beam is struck by a monoenergetic electron beam, the angular distribution of the atoms following the collision may be related to the angular distribution of the scattered electrons. Measurements of this kind are now in progress (Bederson, private communication).

3 *Ion Impact Studies*

In comparing electron-impact and ion-impact inelastic collision phenomena, the simplicity of interpretation of the former in comparison to the latter is striking.

Even in the simplest of ion-atom collisions, collisions of protons with hydrogen atoms, careful distinction must be made between ionization and charge transfer (or electron transfer or capture), and if excitation is under study the desirability of discerning between direct excitation and charge transfer into excited states is obvious. With more complicated atomic ions colliding with atoms, questions of the state of the incident ion and both final products arise and an experimenter must be aware that a large variety of ionization and charge transfer processes may occur in the collisions.

If the target system is a molecule, then dissociation, dissociative ionization and excitation, and ion exchange must be added to the collision processes. If the incident ion is a molecular ion, breakup of that ion as well as of the target molecule must be considered and complicated rearrangement collisions can take place.

Lastly, taking into consideration that the number of producible ion species vastly exceeds the number of neutral chemical species and that these ions may be multiply charged (and some of them negatively charged), the scope of heavy particle collision research begins to become evident.

In addition to the difference in complexity of the fields of electron-impact and ion-impact excitation and ionization, two other basic differences exist which are consequences of the mass difference between electrons and ions. The first difference is that the quantum-mechanical wavelength of ions is very small compared to both electron wavelengths and atomic dimensions. Diffraction effects are therefore absent except at the smallest angles and the ion trajectories are very nearly those predicted by classical mechanics.

The second consequence of large ion mass is that ion excitation experiments are high-energy experiments. At low kinetic energies, ions have velocities small compared to the velocity of atomic electrons. The low energy collisions are therefore very gradual and the atomic electrons have ample time to adiabatically accommodate themselves to the slowly changing environment presented by a passing ion. The near-adiabatic theory of Massey (1949) holds that the cross section will be small when the ion velocity $v \ll a \mid \Delta E \mid / h$, where a is a length of atomic dimensions

($\sim 10^{-8}$ cm), ΔE is the energy defect of the process, and h is Planck's constant. For ionization by the lighter ions, peaks of cross section curves occur at energies of the order of tens of kilovolts, and at correspondingly higher energies for heavier ions. Complete study of inelastic ion collision processes requires energy sources of at least several hundred kev. The unavailability of inexpensive voltage supplies for these high energies has been a major factor causing ion collision experimentation to lag behind electron collision work.

3.1 Definitions and Methods of Measurement

It is convenient to enumerate the processes in ion-neutral collisions in which charge rearrangement occurs and define the appropriate cross sections. For simplicity the ion will be assumed to be an atomic ion I^{n+} of charge ne, and the initial neutral particle will be taken as an atom, A. The cross section for the process

$$I^{n+} + A \rightarrow I^{i+} + A^{j+} + (i + j - n)\, e$$

will be denoted by $Q_{n,0}^{i,j}$.

Special designations for four cases are commonly used. If $i > n$, the term "ion stripping" is often applied, and if $i < n$, reference is made to "electron capture." "Simple ionization" refers to the special case where $i = n$ and $j > 0$; and if $i + j = n$, so that no electrons are produced, the term "simple charge transfer" is appropriate.

In carrying out experiments to measure ion-neutral collisions, a "fast" ion normally is made to collide with a "slow" neutral particle. The ion has a nearly classical trajectory and usually is deflected but slightly in a collision. Thus, an ion beam traversing a gas remains a rather well-defined beam even though inelastic collisions occur to alter the nature of both the incident ion and the target gas particle. The following particles are available for straightforward detection: (1) fast ions, (2) fast neutrals, (3) slow ions, and (4) electrons. Slow neutrals may be easily detected if they are in excited states, but this is usually not done in studies of charge rearranging collisions.

The first type of experimental method, originally formulated by Wien (1908), is called the "equilibrium beam method" or the "multiple collision method." In these experiments a beam of ions of known species is passed through a gas at sufficiently high pressure that each injected ion undergoes many collisions with the target gas before emerging at the end of the collision chamber. Charge-changing collisions cause the original ion beam to become converted into a mixed beam containing fast neutrals and ions of various charges. Analysis of the emergent mixed beam is made to determine the numbers of the variously charged components.

The quantities deduced from this type of measurement, as the pressure in the collision chamber is varied (so that both partially and completely equilibrated fast beams can be examined), are electron capture and loss cross sections, $\sigma_{n,i}$, where n and i denote the charge of the fast particle before and after the collision with the slow neutral. In terms of the generalized cross sections defined previously,

$$\sigma_{n,i} = \sum_j Q_{n,0}^{i,j}. \tag{5}$$

Clearly, this method alone reveals nothing as to the state of the original slow neutral after the collision has occurred or whether the stripped electrons were left free or were captured by the target gas to form slow negative ions. The use of this experimental method and the information derived therefrom are treated fully in the review article by Allison (1958).

The second experimental method might be called the "single collision method." This approach utilizes target gas pressures sufficiently low that any incident ion undergoes not more than one collision with a neutral particle in the collision chamber. Slow ions formed in the collision chamber are collected by electric and/or magnetic fields sufficiently weak that their influence on the motion of the fast incident ions is small, and the quantity determined is the "cross section for the production of slow ions." If mass analysis of the slow ions is used, then the cross section for production of the slow ion A^{j+} is given by

$$Q(\mathrm{A}^j) = \sum_i Q_{n,0}^{i,j}. \tag{6}$$

This cross section obviously includes charge transfer as well as ionization processes.

The single collision method also permits the collection of ionization electrons which, because of their relatively low energies and their light mass, can be readily collected by weak electric and magnetic fields and separated from any slow negative ions formed in the collisions. The often-measured "cross section for free electron production," $Q(e)$, is given in terms of the generalized cross sections by

$$Q(e) = \sum_i \sum_j (i + j - n) Q_{n,0}^{i,j} \tag{7}$$

and does not delineate generally whether the electrons were stripped from the fast incident ions or removed from the slow neutrals.

The low pressure used in the single collision method does, however, exclude the possibility of secondary processes occurring in the collision chamber between ions, A^{j+}, or electrons and the target gas, A.

Almost all experiments on charge-rearranging collisions have been designed to measure the total cross sections for charge capture or loss, slow ion production, or slow electron production. In only rare cases have provisions been made to measure all these quantities simultaneously, notably in the equipment of Fedorenko (1959) which also permits direction, mass, and energy analysis of both the fast and slow particles.

3.1.1 *Collisions of Protons with Hydrogen Molecules*

In this chapter where emphasis is placed on the *measurement* of ionization and excitation cross sections, it is appropriate to consider first the collisions $p + H_2$. Not only is this the simplest ion-molecule collision, but it is one which possesses all of the major experimental problems encountered in more complicated collision processes. In addition, these collisions have been studied by practically everyone engaged in ion-neutral collision work; a review of work on this collision is in large part a review of methods of measurement.

When a proton collides with a neutral hydrogen molecule, the following charge-rearranging processes are allowed by conservation of mass and charge:

$$\begin{aligned} p + H_2 \rightarrow\ & p + H_2^+ + e && \text{(i)} \\ & p + H + H^+ + e && \text{(ii)} \\ & p + H^+ + H^+ + 2e && \text{(iii)} \\ & p + H^+ + H^- && \text{(iv)} \\ & H + H_2^+ && \text{(v)} \\ & H + H + H^+ && \text{(vi)} \\ & H + H^+ + H^+ + e && \text{(vii)} \\ & H^- + H^+ + H^+ && \text{(viii).} \end{aligned}$$

With experimental methods commonly employed to date, this problem is not fully determinate since the measureables indicating charge rearrangement are six in number (fast H and H^- currents and slow H_2^+, H^+, H^-, and electron currents) while there are eight processes of interest.

The cross section for process (viii) is the double capture cross section $\sigma_{1,-1}$ and the sum of the cross sections for (v)-(vii) is the single capture cross section $\sigma_{1,0}$. Both of these cross sections have been measured using the equilibrium beam method (see Allison, 1958).

Of more concern for the study of ionization phenomena analogous to electron-impact ionization processes are the cross sections $Q(e)$ and the cross sections for slow ion production.

Measurements of $Q(e)$. The earliest studies of the slow charged particles produced in $p + H_2$ collisions have been discussed by Massey and Burhop (1952). In 1949 Keene turned to the study of these collisions at energies up to 35 kev and made more careful and complete measurements. In Keene's work a beam of ions was extracted from a low-voltage arc source, and the proton component of this beam was selected by magnetic analysis. With reference to Fig. 21, the proton beam I was

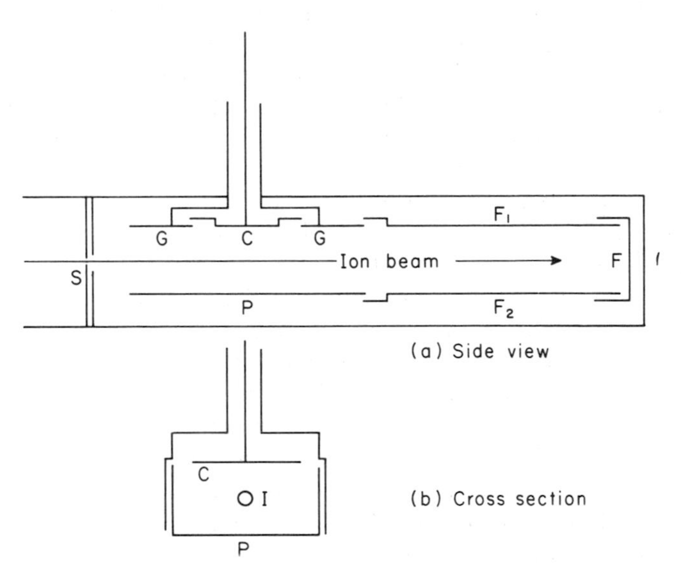

FIG. 21. Collision chamber used by Keene (1949) to measure cross sections for production of slow ions and slow electrons in ion-neutral collisions.

then shot into a collision chamber through an aperture, S, passed through the low-pressure H_2 in the collision chamber, and terminated at a Faraday cage (F, F_1, F_2) where the fast ion current was measured. Slow ions and electrons formed along the length of the fast ion beam were swept by an electric field either to plates C and G, or to electrode P, depending on the charge of the slow particle. The electric sweep field was maintained sufficiently strong that these slow particle currents were saturated.

The slow particle currents were measured at the collector plate C. The guard plates, G, were maintained at the same potential as C to minimize effects of fringing fields which, in the absence of the guard plates, would give uncertainty in the length of the effective source of the slow particles. By measuring the saturated slow particle currents, the fast particle current, the temperature and pressure of the gas in the collision chamber (so that its number density is determinable), and the length of the collector plate C, all data are available for use in (1) to determine the absolute total cross sections for production of electrons and slow ions.

Keene also performed experiments in which a small hole was put in the collector plate so that slow ions could be drawn through the plate. Stopping potential experiments on this slow ion beam indicated that the ions did indeed originate at the position of the fast ion beam and that the energy spread of the slow ions emerging through the hole in the plate was quite small. He also performed mass analysis of these ions and found that only a few per cent of them were protons; the bulk of the ions were H_2^+, suggesting that the major collision processes occurring were (i) and (v).

Several experimental questions may be raised in regard to Keene's experiments. The first and principal question is the role played by secondary electron effects. Such effects can arise from (1) secondary electron emission on slow ion impact, (2) photoelectric effect (caused principally by radiation produced in the $p + H_2$ collisions and by soft X-rays produced at the electron collector), (3) electron ejection by slow metastable atoms produced in the collisions, and (4) secondary electrons produced by fast ions which are scattered from the primary ion beam. All these effects should make the apparent cross sections larger than they should be. When collecting electrons at C, secondary electrons produced at P add to the gas-phase-collection-produced electrons, and when collecting ions at C the loss of secondary electrons would make the measured current larger than the true slow ion current.

The second question is the reliability of the slow ion analysis. If protons were produced as a result of collisions in which H_2 molecules or H_2^+ ions were left in excited repulsive states which subsequently dissociate, the protons would be expected to be formed with kinetic energies of several ev. The initial kinetic energies of ions produced by processes (i) and (v) would be small by comparison. When electrostatic fields are used to extract the slow ions and form them into a beam for subsequent mass or energy analysis, the slower molecular ions are collected more efficiently than the faster protons formed by dissociative processes. Consequently, the measured ratio of atomic to molecular ions should be but a lower limit of the true ratio of these ions. And the measured small energy spread of the slow ions might indicate only that the protons produced in the collisions were simply not being efficiently collected.

In making an improved series of measurements of the $p + H_2$ collisions, Fogel' *et al.* (1955) used a more refined apparatus in which secondary electron emission and photoelectric effect at the collection electrodes were assessed. Their apparatus, which was built for measuring electron capture and loss cross sections as well as slow particle production cross sections, consisted of a series of parallel-plate condensers. Any of these could be used for the measurement of the slow particles with all the others acting as guard plates. In addition, they provided a magnetic field of about 300 oersteds parallel to the fast ion beam. This magnetic field not only prevented the fast ion beam from spreading as it traversed the collision chamber but also permitted the assessment of spurious electron currents at the condenser plates.

The procedure used in the experiment of Fogel' *et al.* was to measure the positive and negative saturation currents at a collector plate first without and then with the longitudinal magnetic field. Without the magnetic field, the current measured when positive ions were collected was

$$i_+ = i_1 + i$$

where i_1 is the true ion current and i_3 is the secondary electron current between the condenser plates. When collecting negatively charged particles, the current was

$$i_- = i_2 + i_3$$

where i_2 is the current of electrons produced in ionizing collisions in the gas phase. Upon applying the magnetic field in which the electron Larmor radius was small compared to the physical dimensions of the collector plates, secondary electrons were returned to the electrode of their origin, and electrons produced in the gas-phase ion-molecule collisions would be trapped along the fast ion beam and constrained from reaching the condenser plates. Thus, with the magnetic field,

$$i_+^H = i_1$$

and

$$i_-^H = 0.$$

By combination of the current readings, i_3 could be evaluated, so permitting both the true slow ion and ionization electron currents to be determined.

Another method of combating the difficulties of secondary electron effects is to use electrostatic suppression of electrons produced at the ion collecting plate. In the measurement of $Q(e)$ with Keene's apparatus, this modification would replace the plate P with a highly transparent wire grid below which the new ion collector plate would be placed. By biasing the plate about 20-volts positive to the grid, secondary electrons produced at the new ion collector plate would be prevented from leaving it. Secondary electrons produced at C would, of course, be returned to that electrode by the electric sweep field when collecting the electrons produced in the $p + H_2$ collisions. If only this single grid is incorporated into Keene's apparatus, slow electron currents must be measured at C and that slow ion currents must be measured at the new plate below the grid.

A single grid was incorporated into the design of the equipment used by Kistemaker and his collaborators (de Heer *et al.*, 1957; Sluyters *et al.*, 1959; Sluyters and Kistemaker, 1959) to measure electron capture and loss cross sections and electron production cross sections in a number of gases. Fedorenko and associates (1956) used grids over both plates of their condenser in order to permit measuring slow ion and slow electron currents at only one plate (upon reversal of the direction of the collection field).

In the very recent high-energy experiments (up to 1.1 Mev) of Hooper and associates (1961) a series of segmented condensers similar to those of Fogel' *et al.* were used, with grids covering both plates.

Gilbody and Hasted (1957) used two configurations in which grids were placed before one and both plates, respectively, of their condenser, and introduced another innovation, namely, a magnetic field *parallel* to the electric field used to collect the slow particles. The use of this type of magnetic field offers several advantages not the least of which is that by magnetically confining electrons which can reach the collector to the volume below the electron collector, the effective length of the line source of slow particles is the linear dimension of the fast ion beam below the electron collector. The magnetic field accomplished the same thing as the guard plates G of Fig. 21. In addition, the magnetic field can be made to contain all electrons produced in the collisions and permit somewhat weaker electric collection fields to be used. The possibility that energetic electrons might escape collection by passage through the open sides of a simple parallel plate condenser such as was used by Fogel' *et al.* is also clearly obviated.

The results of various recent measurements of the cross section for free electron production in $p + H_2$ collisions are shown in Fig. 22. While the general pattern of the cross section is well defined from the work of various experimenters, the not insubstantial discrepancies attest to the difficulty of making this type of cross section measurement.

Other Ionization Measurements. The value for the total cross section for electron production $Q(e)$ is the sum of the cross sections for processes (i), (ii), and (vii) given previously, plus twice the cross section for process (iii). It is interesting to attempt to ascertain the relative importance of these processes in producing free electrons. Several methods have been used.

The first study to this end has already been mentioned. Keene (1949) upon mass analyzing his slow ions found that for ion energies up to 35 kev, only a few per cent of the slow ions he could detect were protons. Although the major portion of the slow H_2^+ ions were produced through process (v), as implied by the smallness of electron signals compared to total slow ion signals, it could be concluded that in his energy range the major electron-producing process was (i). It is probably not safe to make more quantitative deductions from Keene's mass spectrometer measurements, however, because of the fact that the H^+ ions may have been formed with substantial initial kinetic energy by dissociation of highly excited H_2^+. It is probable that the collection efficiency of Keene's mass spectrometer for the protons was not as high as for the H_2^+ ions.

More recent studies have been guided by broader considerations than Keene's and have attempted to exploit several avenues of approach. If one assumes that processes yielding protons [e.g., (ii) and (iii)] are in fact processes in which the incident proton ionizes the molecule and leaves it in an excited state, following which the excited molecular ion breaks up, then several predictions can be made.

The first begins by noting that associated with each excited molecular ion state there is an energy defect ΔE which is usually taken as the energy to make a Franck-Condon transition from the ion ground state to the excited state (less the ionization potential of the hydrogen atom in the case of charge transfer collisions). The near-adiabatic theory of heavy particle collisions predicts that the maximum cross section value should occur at an ion energy which is related to the energy defect. Thus, by examination of the shape of the cross section curve for production of a given fragment ion, particularly noting the energy at which the peaks of the curve occur, the energy defect in the process can be estimated and the most likely states of the molecular ion prior to breakup may be determined. Lindholm and his associates (von Koch and

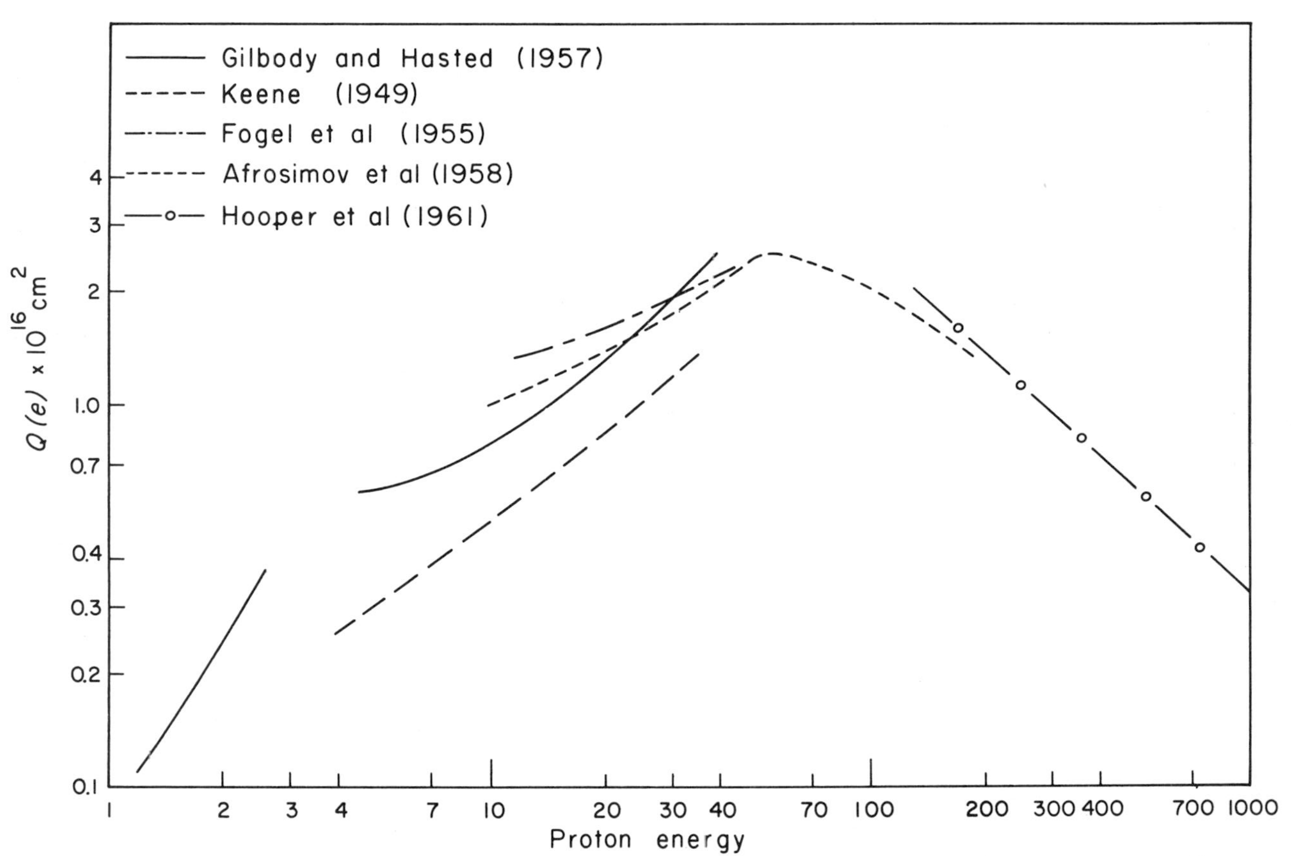

FIG. 22. Compilation of results for the cross section for slow electron production in collisions between protons and hydrogen molecules.

Lindholm, 1961; Gustafsson and Lindholm, 1960) have contributed particularly to the development of this type of study.

While this approach is quite appealing in principle, it is possessed of several practical difficulties. First, the cross-section curve for any given excitation process varies slowly with ion energy and accurate location of the energy at which the maximum occurs is often difficult. Second, several simultaneous processes generally contribute to the production of a given fragment ion and detail is further smeared out. Third, the near-adiabatic theory can at present be regarded as only semiquantitative. Nonetheless, considerable insight into processes yielding slow ions has been obtained from such studies.

A second method of obtaining information on the separate ionization processes is to study the energy and angular distribution of the slow

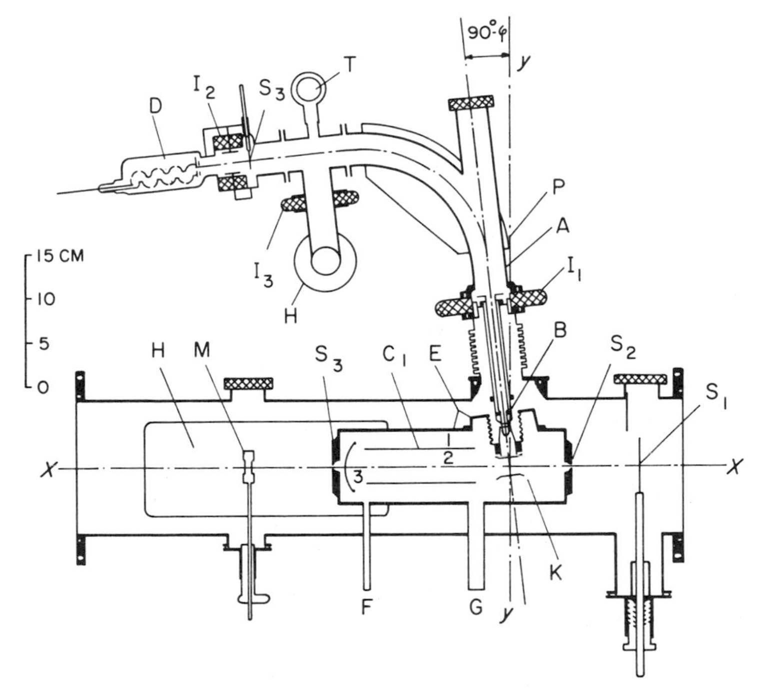

FIG. 23. Apparatus of Afrosimov and Fedorenko (1957) for investigating mass, kinetic energy, and angular distribution of secondary ions formed in ion-neutral collisions. A, mass analyzer chamber; P, pole pieces; D, electron multiplier; I_1, I_2, I_3, insulators; T, liquid air trap; E, magnetic shielding; M and K, auxiliary electrodes; H, diffusion pumps; F and G, tubes for admission of gas to the collision chamber and vacuum gage connection; C_1, collector for primary ions; B, collimator and electrode system for retarding potential studies on the ions; S_1, S_2, S_3, slits. The angle θ ranges from $-2°$ to $+14°$.

secondary ions. The most complete study of this kind is that of Afrosimov *et al.* (1958) who used the apparatus shown in Fig. 23 (Afrosimov and Fedorenko, 1957). The fast ion beam entered from the right and after passage through the slits S_1 and S_2 underwent collisions with the target gas. A probe introduced at the side of the collision chamber sampled the slow ions produced in the collisions. This probe contained a collimator, so that only a very small length along the fast ion beam was viewed and the slow ions originated from an apparent "point" source. A set of three grids was included in the probe with which retarding potential data could be taken to determine the energy of the ions. The entire probe was affixed to the apparatus with flexible bellows to permit observation of the slow ions emitted over angles from 76° to 92° with respect to the fast ion beam. In studying angular distribution of the slow ions, it was necessary that no ion extraction fields be used in the collision chamber, for such fields would alter the direction of ion motion. Omission of extraction fields also removes any question of the relative efficiency of collection of the various slow ion species, a problem which has been discussed earlier in regard to the experiments of Keene (1949).

Of interest to consider is the slow proton energy distribution, since their energy indicates the state of excited H_2^+ which broke up to produce them. As a case in point, protons with zero energy can be produced by H_2^+ excited to the $^2\Sigma_g$ state, while $H_2^+(^2\Sigma_u)$ will dissociate to produce protons in the 5 to 7-ev range. If the molecule is sufficiently excited in a Franck-Condon transition that it will break up to yield two protons, their energies will be in the range 7.5-10 ev. The finding of Afrosimov *et al.* (1958) that at a primary proton energy of 75 kev only an insignificant fraction of the protons carried energies from 7 to 12 ev indicates that at this energy processes (iii) and (vii) contribute negligibly to the electron production.

It is worth noting that extreme care must be applied to the interpretation of measurements of the energy and direction of motion of secondary ions. From the requirement of simultaneous conservation of energy and momentum in an ion-neutral collision, it can be shown, for example, that in a simple ionization process such as process (i), the secondary ion *must* receive some kinetic energy for incident ion energies less than $(1 + M/m) I$, where m is the electron mass, M is the incident ion mass, and I is the ionization threshold. The actual energy, direction of motion, and state of the secondary ion are uniquely defined only if the energies and directions of motion of both the primary ion and the electron are known. Thus, while in the experiments of Afrosimov *et al* the absence of protons above 7 ev implies negligible contributions from processes (iii) and (vii), the presence of lower energy protons does not necessarily imply excitation of H_2^+ to say, the $^2\Sigma_u$ repulsive state. Even though the breakup of a molecule at rest in this state would produce ions in the 5- to 7-ev range, dissociation of a moving molecule in the $^2\Sigma_g$ state can also produce energetic protons even though the proton energy in the center-of-mass system of coordinates is very small.

Information on the elementary electron producing processes can also be obtained by studying the energies of the electrons. Experiments of Moe and Petsch (1958) determined by magnetic analysis the electron energy spectra in collisions between potassium ions and inert gases at ion energies up to 900 ev and revealed quite complex spectra. Particularly interesting is the fact that their results were in qualitative agreement with the Weizel-Beeck theory of ionization by positive ions, which holds that an electron may be ejected by an Auger process when the ion-atom internuclear separation reaches a value at which the energy of the ion-atom system equals that of the ion-ion system. Several peaks appearing in the electron energy spectra suggest that if this interpretation is correct then the mechanism is operative with more than one energy level in the ion-ion system. Unfortunately, Moe and Petsch apparently did not extend their work to the collisions of protons and hydrogen molecules.

3.1.2 *Collisions of Protons with Hydrogen Atoms*

From the foregoing discussion it is seen that even in the simple $p + H_2$ collisions, interpretation of experiments in terms of the elementary processes which can occur is quite complex. A much simpler collision, although more difficult with which to experiment is the collision between protons and hydrogen atoms. Since there is only a single electron involved and only two nuclei, the only question about charge-rearranging collisions is whether the electron is transferred to the incident proton or whether the electron is left free. Electron production is identical to simple ionization. The entire collision is thus specified by determining only two cross sections and by measuring any two products of the collision.

Because of the extensive theoretical background on $p + H$ collisions (see Chapter 14) it was only natural that these be the first ion-neutral collisions to be studied experimentally following the development of modulated-atomic-beam techniques. Fite *et al.* (1958) first measured the cross section for slow proton production below 10 kev, but inadequate signal-to-noise ratio prevented separation of charge transfer and ionization processes. In a later improved experiment (Fite *et al.*, 1960) both ionization and charge transfer up to ion energies of 40 kev were studied.

In both experiments, the approach was similar to that used in studying the ionization of atomic hydrogen on electron impact (see Fig. 2). The only differences were that a proton beam replaced the electron beam and, in addition to detection of slow ions with a mass spectrometer, a second detector which did not discriminate against particle mass was used. This second detector consisted of a pair of plates on either side of and parallel to the plane defined by the two intersecting beams. In concept the method is identical to that used by Keene (1949) in his studies of charge transfer and ionization between ions and chemically stable gases. With a reversible electric field between the plates, either all slow positive ions or all slow electrons could be drawn to one of the two plates where a preamplifier separated the ac currents due to the two intersecting beams from the dc background ion currents. A magnetic field parallel to the electric field (similar to that of Gilbody and Hasted, 1957) was used in the later experiments to confine the slow particles until the imposed electric field could sweep them to the current detector. Comparison of the electron currents with the slow ion currents yielded the ratio of the ionization cross section $Q_1(e)$ to the sum of the charge transfer and ionization cross sections.

In order to determine absolute values in these experiments, a procedure similar to that employed in assigning absolute values to the electron-impact ionization cross section of H was used. By varying the furnace temperature with a fixed mass flow in the neutral beam, the dissociation fraction of the neutral beam could be varied arbitrarily, and from the slow ion signals (using either slow ion detector) ratios of atomic to molecular cross sections could be determined.

The atomic cross section Q_1 is the sum of the charge transfer and ionization cross sections. The molecular cross section is somewhat more complicated. When the parallel plate detector was used, the cross section for the production of all slow ions, irrespective of e/m, was appropriate. The eight processes yielding slow ions in $p + H_2$ collisions group themselves so that the cross section of interest, $Q_2(H^+ + H_2^+) = \sigma_{1,0} + Q_2(e) + Q(1, 1, -1) + \sigma_{1,-1}$ where $\sigma_{1,0}$ is the electron capture cross section, $Q_2(e)$ is the total cross section for production of free electrons, and the two last cross sections are for the processes $p + H_2 \rightarrow p + p + H^-$ and $p + H_2 \rightarrow H^- + p + p$, respectively. Since no slow negative ion signals could be detected in the mass spectrometer, $Q(1, 1, -1)$ could be neglected at the energies of these experiments, and measurements of Stier and Barnett (1956) had shown that $\sigma_{1,-1}$ has values of only about 0.01 $\sigma_{1,0}$ in the energy range of interest, so it too could be neglected. When a purely molecular beam was run, data could be obtained from which $Q_2(H^+ + H_2^+)$ could be related to $\sigma_{1,0}$ for $p + H_2$ collisions. Thus, the experiment measured the ratio $Q_1/\sigma_{1,0}(H_2)$. Absolute values for Q_1 were then obtained by using the values of the molecular electron capture cross section which was determined absolutely by Stier and Barnett (1956) and by Curran *et al.* (1959).

From the ratio $Q_1(e)/Q_1$ obtained by comparing slow particle signals upon reversal of the collection field and absolute values for Q_1, the total ion production cross section, absolute cross sections for ionization of the hydrogen atom could be assigned. Figure 24 shows the results of this cross section measurement and the Born approximation prediction made by Bates and Griffing (1953). For comparison, the charge transfer cross section is also shown in this figure.

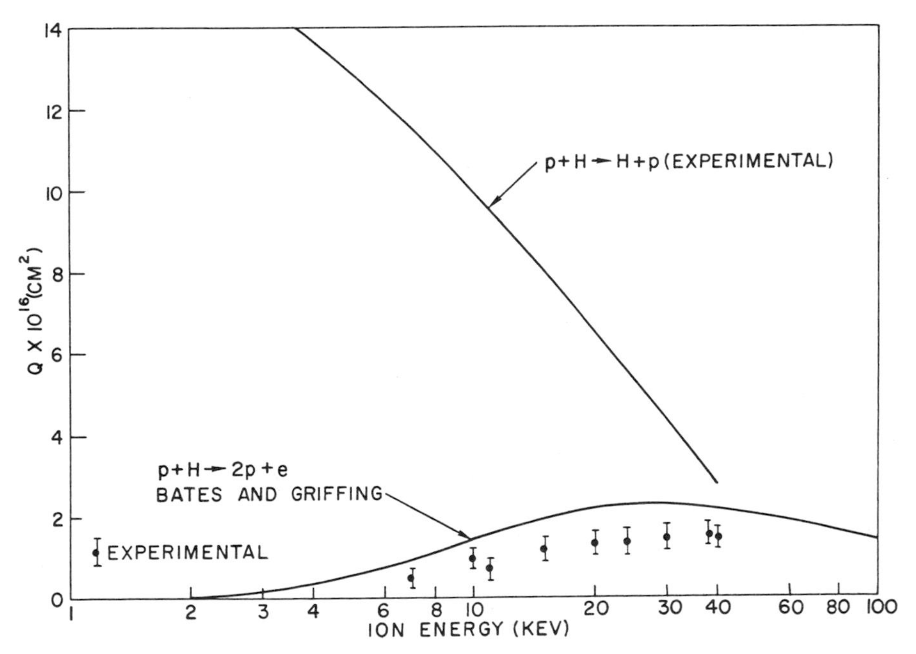

FIG. 24. Cross section for the process $p + H \rightarrow 2p + e$ (Fite *et al.*, 1960) compared with the Born approximation prediction (Bates and Griffing, 1953) and the charge transfer cross section.

It is of some interest to compare the predictions of electron-impact with proton-impact ionization of atomic hydrogen, and this is shown in Fig. 25, where the abscissa is the relative velocity between the colliding systems. The discrepancies between the Born approximation appear quite similar in both cases.

In the same series of experiments the charge transfer and electron production cross sections for $H^- + H$ collisions were also carried out (Hummer *et al.*, 1960) in the same manner as the $p + H$ collision experiments. The only important difference in experimental technique

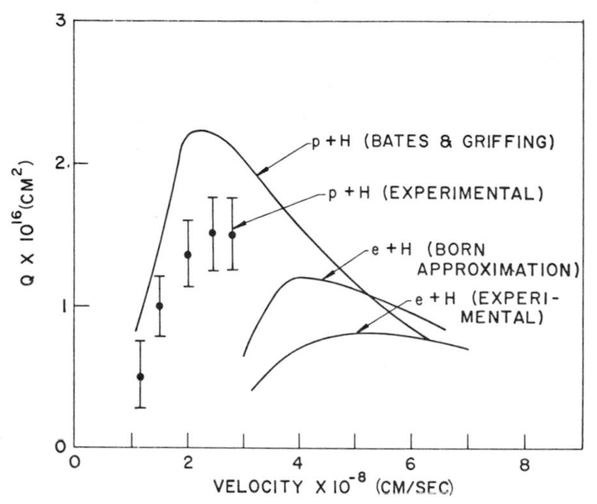

FIG. 25. Comparison of ionization of atomic hydrogen on proton impact (Fite *et al.*, 1960) and electron impact (Fite and Brackmann, 1958a).

was that here a weak magnetic field could be superposed parallel to the neutral beam. With this field off, both electrons and slow negative ions were indiscriminately detected together, but when the field was turned on electrons were prevented from reaching the collecting plate and only the slow negative ion current was measured. Thus, charge transfer and electron production processes could be separated in spite of both ions and electrons having negative charges.

3.1.3 *Ionization in Other Ion-Neutral Collisions*

It is appropriate to re-emphasize that the case of proton-hydrogen collisions was chosen in this discussion as an experimental problem with which to illustrate the various basic techniques employed in the laboratory study of charge-rearranging collisions. These problems have in fact constituted only a very small fraction of the work which has been done in ion-neutral collision processes.

Because it is the intention in this chapter to concentrate on measuring techniques rather than on results, detailed discussion of results on other ionization processes involving ions and neutrals will be omitted. The reader is referred to a recent excellent review by Fedorenko (1959) concerning ionization in collisions between positive ions and atoms generally. Negative ion stripping is treated in Chapter 18.

3.2 Excitation

The area of ion-impact excitation and charge transfer into excited states is very unexploited, compared to those processes which can be detected by the appearance of a charged particle. While several experiments on excitation of spectra have been carried out, very few have been done with the care required for a cross-section measurement. Two sets of experiments are particularly pertinent to discuss, however.

The first have been the experiments of Fan (1956), Carleton (1957) and Carleton and Lawrence (1958) on excitation of N_2 on proton impact, a process of particular importance in understanding the excitation of aurora by protons of solar origin. Carleton's apparatus was, very simply, an analyzed proton beam of diameter about 1/16 in., and energy of a few kev, which passed through nitrogen gas at pressures from 0.5 to 50 μ Hg. The light excited by the narrow beam along its length was the source for a liquid prism spectrograph. Primary or single collision processes were distinguished from multiple collision processes on the basis of the dependence of the signal on N_2 pressure. Spectrograms were taken in Carleton's first experiment which showed the first negative bands and the Meinel bands of N_2^+ which were produced in single collision processes. The first positive bands of neutral N_2 were seen also and were found to have a quadratic pressure dependence indicating that a product of the $p + H_2$ collision interacts with a second N_2 molecule. The product is identified on the basis of convincing experimental argument as a fast H atom formed by charge transfer. Carleton's spectra also show the Balmer lines formed in both a single and a double collision process. The single process is presumably electron capture into excited states, and the double process is interpreted as capture of the electron by the proton to form ground state H, with a subsequent neutral-neutral collision exciting the newly formed hydrogen atom.

Carleton and Lawrence (1958) went on to place absolute values on the excitation cross sections for $p + N_2$ excitation collisions. In these experiments, the spectrograph was replaced by a set of interference filters and detection was made with a photomultiplier rather than with a photographic plate. *In situ* calibration of the multiplier was provided. Also, these experiments included a simple charge transfer chamber of a

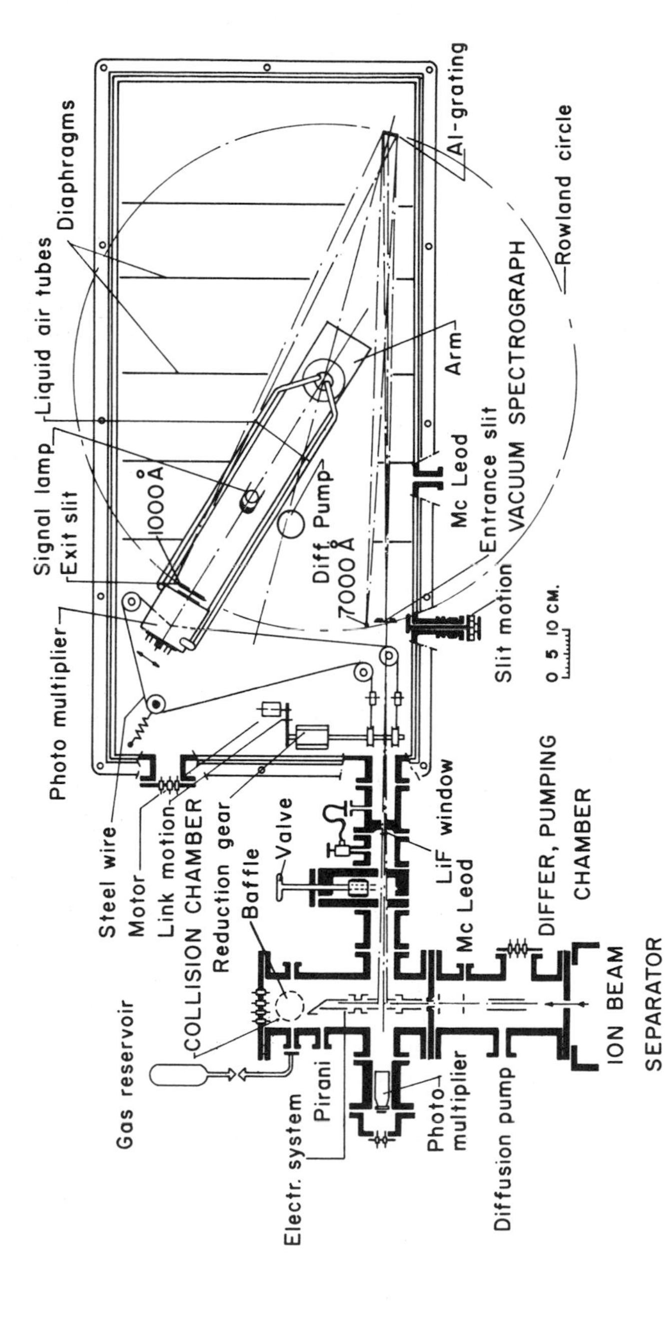

FIG. 26. The apparatus of Sluyters and Kistemaker (1959) for studying both charge rearranging collisions and excitation in ion-atom collisions.

kind similar to that used by Keene (1949) so that a comparison of charge transfer, ionization, and excitation could be made simultaneously.

Fan's experiments on the $p + N_2$ excitation were similar to Carleton's first experiment except in two regards. First, he used a much wider ion energy range, extending up to 350 kev, and second, his detector was a grating spectrograph.

Probably the most comprehensive work on excitation processes other than those related to auroral excitation phenomena has been that carried out by Sluyters and Kistemaker (1959) who studied the excitation processes in collisions between singly charged argon ions and various inert gases. Their experimental plan is shown in Fig. 26. The energy range was 5-24 kev and their detector was a vacuum grating monochromator, permitting a wavelength range from 1000 to 6500 A, and a photomultiplier which was calibrated *in situ* using a Philips tungsten standard light source. At the shorter wavelengths a sodium salicylate film was used to convert ultraviolet photons into visible light which could pass through the multiplier envelope.

These experiments also coupled electrical measurements of charge transfer and electron production with the optical measurements, as in the experiment of Carleton and Lawrence. The necessity of making these concurrent measurements arose from the fact that the gas in the collision chamber was at sufficiently high pressure that the incident ion beam became altered along its path length by charge changing collisions. Measuring slow ion production along the beam may be used to determine the actual ion and fast atom currents at the portion of the beam length which is the effective source of the radiation. These separate current components must be known for the measurement of absolute excitation cross sections.

TABLE III

SPECTRAL LINES DETECTED BY SLUYTERS AND KISTEMAKER (1959) IN THEIR STUDIES OF ION-NEUTRAL COLLISIONS

Reaction	Spectra[a]					
Ar^+-He	Ar II (52)	He I (6)				
Ar^+-Ne	Ar II (30)	Ne I (5)	Ne II (24)			
Ar^+-Ar	Ar I (19)	Ar II (120)	Ar III (10)			
Ar^+-Kr	Ar I (12)	Ar II (60)	Ar III (8)	Kr I (13)	Kr II (64)	Kr III(24
Ar^+-Xe	Ar I (12)	Ar II (82)	Xe I (15)	Xe II (90)	Xe III (14)	

[a] The numbers in parentheses are the numbers of identified spectral lines.

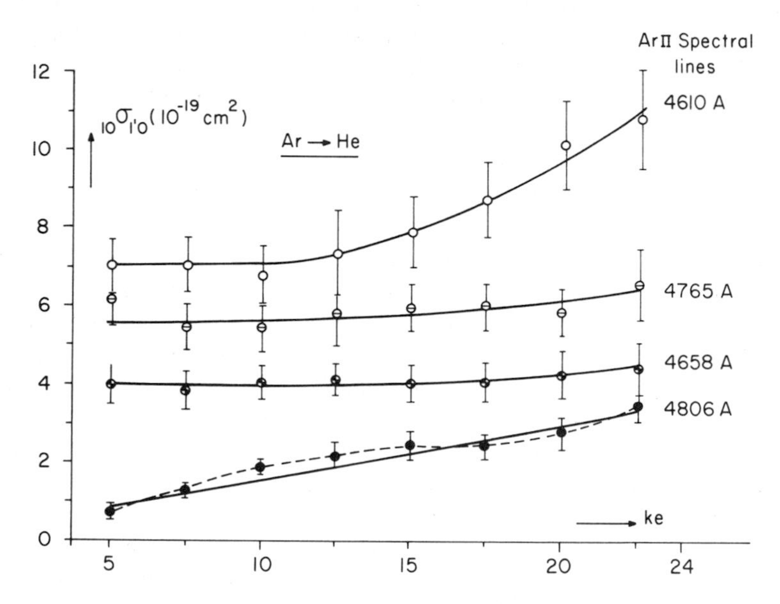

FIG. 27. Cross sections for excitation of several lines of singly ionized argon in collisions between Ar^+ and He (Sluyters and Kistemaker, 1959).

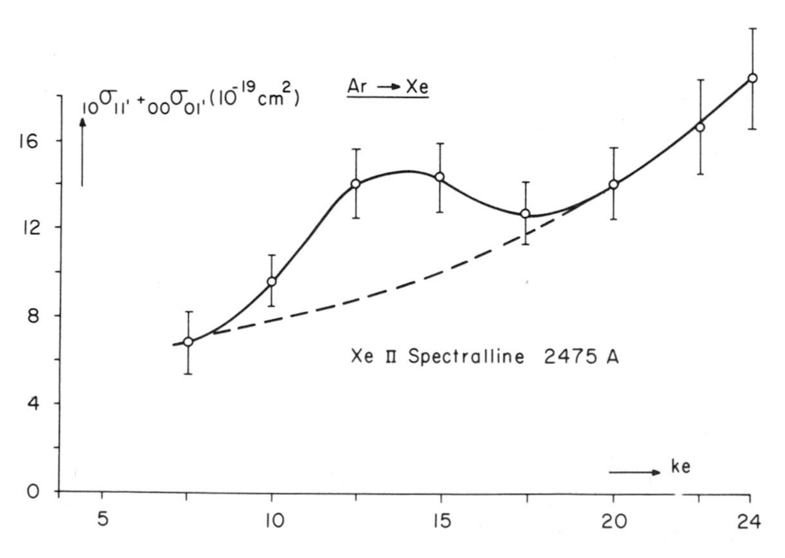

FIG. 28. Cross section for excitation of the 2475 A line in singly ionized xenon in collisions between Ar^+ and Xe (Sluyters and Kistemaker, 1959).

Table III lists the arc and spark spectra detected by Sluyters and Kistemaker in the different collisions.

Most spectra were observed in the spectral range 3000-5800 A and fall within the following categories: (1) excitation of the incident ion by the neutral, (2) excitation of the neutral, (3) single and double ionization of the neutral to excited ion states, and (4) charge transfer to an excited state of the neutralized incident ion. A typical result of these experiments is shown in Fig. 27, which presents cross sections for excitation of an incident argon ion on passage through helium gas. A less typical result, displaying a remarkable and unexplained maximum et 13 kev, is shown in Fig. 28. This is the cross section for ionization and excitation of xenon on impact by argon ions and atoms.

4 Neutral Impact Studies

No discussion of ionization and excitation cross section measurement would be complete without at least a brief reference to work collisions between two neutral particles. Massey and Burhop (1952) present a discussion of work prior to the year of publication of their book and outline the basic experimental methods employed in subsequent experiments. The experimental technique which is commonly used is to first produce a beam of ions and pass this beam through a collision chamber where charge transfer collisions neutralize a small fraction of the ion beam. The mixed beam then leaves the collision chamber and passes through a transverse electric field which removes charged particles from the beam. The fast neutral beam remaining then enters a second collision chamber.

In studying ionization, one of the various devices for detection of free electrons is situated in the second collision chamber. From knowledge of the initial fast ion current, the geometry and pressure in the first collision chamber and the charge transfer cross section for the ion and the gas in the first collision chamber, the fast neutral current is calculated, and it is this which is the primary current used to determine the neutral-neutral electron production cross section in experiments carried out in the second collision chamber. It is assumed that those fast neutrals reaching the second collision chamber have the energy of the primary fast ion beam.

With the exception of the introduction of a charge transfer or first collision chamber, the experimental techniques are identical to those used in ion-neutral collisions and thus warrant no further discussion here. Particularly interesting recent experiments on ionization in neutral-neutral collisions are those of Sluyters and associates (1959) on argon-inert gas collisions, and experiments of Bydin and Bukhteev (1960) on ionization produced in collisions of the alkalis with stable gas molecules, and the reader is referred to these original articles.

5 Electron-Ion Collisions

This chapter has, at numerous points, discussed novel experimental methods for application to what are basically "old" problems, i.e., problems on which considerable prior experimental work had been done. A very recent advance in atomic collision research has been the initiation of experimental research on excitation and ionization in collisions between charged particles, principally between electrons and ions. In this area, for which there is no experimental precedent, nothing has been published as yet, but the importance of this type of experiment for interpretation of thermonuclear fusion experiments and astrophysical phenomena warrants here a brief mention of research in progress.

The basic problem in the experimental study of collisions between electrons and ions is the small target particle (ion) density which can be achieved in most experiments. As a case in point, a helium ion beam of 5-kev energy and a current density of 10^{-5} amp/cm^2 has a number density of the order of 10^6 ions/cc. When it is realized that this number density is 6 orders of magnitude less than the number density of target molecules in ordinary collision experiments and 2 orders of magnitude less than the atom density in modulated-atomic-beam experiments on atomic hydrogen, the difficulties of these experiments may be appreciated. Nonetheless, the sensitive current detection instruments now available make it possible to perform experimentation in crossed-beam configurations on electron-ion ionization phenomena.

The first such experiment to near completion is that of Dolder *et al.* (1961) who have studied the simple ionization problem $e + He^+ \rightarrow 2e + He^{++}$ which is closely analogous to electron-impact ionization of atomic hydrogen. Their apparatus, which is illustrated diagrammatically in Fig. 29, consists of two ion-analyzing magnets with an intermediate electron gun. The first magnet selects a beam of He^+, and the second magnet separates the He^{++} from the He^+ primary beam.

The principal source of noise in this experiment is the production of He^{++} by stripping collisions of He^+ with residual gas in the vacuum. To isolate the doubly charged ions which are produced by this cause from those produced by electron impact, a type of modulation technique is used. The analyzed He^+ beam which enters the electron gun region is square-pulsed with a 50% duty cycle at an audio frequency, the pulsing

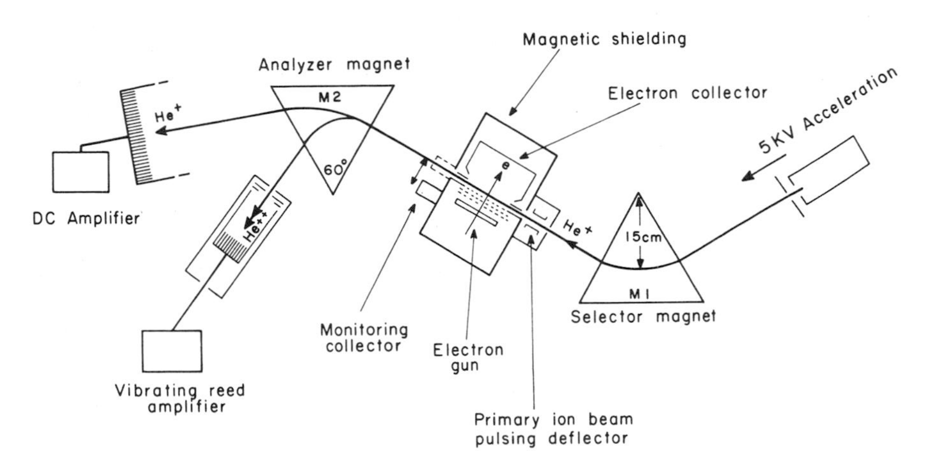

FIG. 29. Experimental plan of Harrison *et al.* (1961) for the study of the cross section for the process $e + He^+ \rightarrow 2e + He^{++}$ (crossed beam apparatus).

being produced by electric field deflection of the ion beam between the first magnet and the gun entrance slit. The electron gun current is also pulsed at the same frequency as the ion beam, but the duration of the electron pulses is less than that of the ion pulses. By first noting the He^{++} signal when phasing the electron pulses to lie completely within the ion pulses, and then subtracting the signal when the electron pulses are phased to lie completely outside the ion pulses, an indication of the He^{++} current component due to electron-impact ionization is obtained. The signal detectors are dc devices which permit simple and accurate calibration, and time-averaged currents under the two conditions are the measured quantities.

Experimental variations of this approach are obvious, and the success of the experiment of Dolder *et al.* implies considerable promise that other ion-electron collision cross section measurements may be successfully carried out. Particularly promising are experiments on $e + H_2^+ \rightarrow 2e + 2H^+$ which is under study at the Oak Ridge National Laboratory and on $e + H^- \rightarrow 2e + H$, the cross section for which has been calculated by Geltman (1960) and which is being examined experimentally at the National Bureau of Standards in the United States.

While the above-mentioned experiments are ionization experiments, the prognosis that excitation cross sections in electron-ion collisions can be measured also appears good. While severe signal-to-noise difficulties in such experiments are expected (since an ion source itself is such a strong source of the radiation to be detected in an exciting collision between an electron and an ion), the fact that the ion can be given considerable velocity prior to its interacting with a crossed-electron beam permits one to Doppler-shift the excitation radiation away from noise radiation from the ion source. Indeed, Branscomb (1960) has designed an experiment employing this scheme to study the excitation of the H and K lines in Ca^+.

Other experimental approaches to the study of electron-ion collision phenomena involve increasing the ion density by passing beams through plasmas. The thermal plasmas producible with cesium are particularly appealing in studying properties of the Cs^+ ion, and an even more interesting experiment is the study of excitation of O^{5+} on electron impact, which appears possible using magnetically confined plasmas of the type which have been developed in controlled thermonuclear research.

It seems likely that the study of high energy interactions between electrons and ions, and then between ions and ions, will constitute a major area for atomic collision cross section measurement in the future.

References

Afrosimov, V. V., and Fedorenko, N. V. (1957) *Zh. tekh. Fiz.* **27**, 2557; translation (1957) *Soviet Phys.—Tech. Phys.* **2**, 2378.

Afrosimov, V. V., Il'in, R. N., and Fedorenko, N. V. (1958) *Zh. eksper. teor. Fiz.* **34**, 1983; translation (1958) *Soviet Phys.—JETP* **7**, 968.

Allison, S. K. (1958) *Rev. mod. Phys.* **30**, 1137.

Bainbridge, K. T., and Jordan, E. B. (1936) *Phys. Rev.* **50**, 282.

Bates, D. R., and Griffing, G. W. (1953) *Proc. Phys. Soc.* **A66**, 961.

Bethe, H. A., and Salpeter, E. E. (1957) *In* Handbuch der Physik (S. Flügge, ed.), Vol. 35. Springer, Berlin.

Boyd, R. L. F., and Boksenberg, A. (1960) *In* "Proceedings of the Fourth International Conference on Ionization Phenomena in Gases, Uppsala, 1959" (N. R. Nilsson, ed.), Vol. 1, p. 529. North-Holland, Amsterdam.

Boyd, R. L. F., and Green, G. W. (1958) *Proc. Phys. Soc.* **A71**, 351.

Brackmann, R. T., Fite, W. L., and Hagen, K. E. (1958) *Rev. sci. Instrum.* **29**, 125.

Branscomb, L. M. (1960) Private communication.

Brode, R. B. (1929) *Phys. Rev.* **34**, 673.

Burhop, E. H. S. (1952) "The Auger Effect." Cambridge Univ. Press, London and New York.

Bydin, F. F., and Bukhteev, A. M. (1960) *Zh. tekh. Fiz.* **30**, 546; translation (1960) *Soviet Phys.—Tech. Phys.* **5**, 512.

Carleton, N. P. (1957) *Phys. Rev.* **107**, 110.

Carleton, N. P., and Lawrence, T. R. (1958) *Phys. Rev.* **109**, 1159.
Chubb, T. A., and Friedman, H. (1955) *Rev. sci. Instrum.* **26**, 493.
Clarke, E. M. (1954) *Canad. J. Phys.* **32**, 764.
Corrigan, S. J. B., and von Engel, A. (1958) *Proc. Roy. Soc.* **A245**, 335.
Craggs, J. D., and Massey, H. S. W. (1959) *In* "Handbuch der Physik" (S. Flügge, ed.), Vol. 37. Springer, Berlin.
Curran, R. K., Donahue, T. M., and Kasner, W. H. (1959) *Phys. Rev.* **114**, 490.
de Heer, F. J., Huizinga, W., and Kistemaker, J. (1957) *Physica* **23**, 181.
Dehmelt, H. G. (1957) *Phys. Rev.* **105**, 1487.
Dehmelt, H. G. (1958) *Phys. Rev.* **109**, 381.
Dibeler, V. H., and Reese, R. M. (1959) *J. chem. Phys.* **31**, 282.
Dolder, K. T., Harrison, M. F. A., and Thonemann, P. C., *Proc. Roy. Soc.* **A264**, 367, (1961)
Dorman, F. H., Morrison, J. D., and Nicholson, A. J. C. (1959) *J. chem. Phys*, **31**, 1335.
Dorrestein, R. (1942) *Physica* **9**, 447.
Fan, C. Y. (1956) *Phys. Rev.* **103**, 632.
Fedorenko, N. V. (1959) *Uspekhi fiz. Nauk* **68**, 481; translation (1959) *Soviet Phys.—Uspekhi* **2**, 526).
Fedorenko, N. V., Afrosimov, V. V., and Kaminker, D. M. (1956) *Zh. tekh. Fiz.* **26**, 1929; translation (1957) *Soviet Phys.—Tech. Phys.* **1**, 8161.
Fite, W. L., and Brackmann, R. T. (1958a) *Phys. Rev.* **113**, 1141.
Fite, W. L., and Brackmann, R. T. (1958b) *Phys. Rev.* **113**, 1151.
Fite, W. L., and Brackmann, R. T. (1959) *Phys. Rev.* **113**, 815.
Fite, W. L., Brackmann, R. T., and Snow, W. R. (1958) *Phys. Rev.* **112**, 1161.
Fite, W. L., Stebbings, R. F., and Brackmann, R. T. (1959) *Phys. Rev.* **116**, 356.
Fite, W. L., Stebbings, R. F., Hummer, D. G., and Brackmann, R. T. (1960) *Phys. Rev.* **119**, 663.
Fogel', Ya. M., Krupnik, L. I., and Safronov, B. G. (1955) *Zh. eksper. teor. Fiz.* **28**, 589.
Foner, S. N., and Nall, B. H. (1961) *Phys. Rev.* **122**, 512.
Fox, R. E. (1959) *In* "Advances in Mass Spectrometry" (J. D. Waldron, ed.), pp. 397-412. Pergamon, New York.
Fox, R. E., Hickam, W. M., Kjeldaas, T., and Grove, D. J. (1951) *Phys. Rev.* **84**, 859.
Fox, R. E., Hickam, W. M., and Kjeldaas, T. (1953) *Phys. Rev.* **89**, 555.
Franken, P., Sands, R., and Hobart, J. (1958) *Phys. Rev. Letters* **1**, 52.
Geltman, S. (1956) *Phys. Rev.* **102**, 171.
Geltman, S. (1960) Private communication.
Gilbody, H. B., and Hasted, J. B. (1957) *Proc. Roy. Soc.* **A240**, 382.
Gustafsson, E., and Lindholm, E. (1960) *Ark. Fys.* **18**, 219.
Haas, R. (1957) *Z. Phys.* **148**, 177.
Harrower, G. A. (1955) *Rev. sci. Instrum.* **26**, 850.
Hendrie, J. M. (1954) *J. chem. Phys.* **22**, 1503.
Hinteregger, H. E., and Watanabe, K. (1953) *J. Opt. Soc. Amer.* **43**, 604.
Hooper, J. W., McDaniel, E. W., Martin, D. W., and Harmer, D. S. (1961) *Phys. Rev.* **121**, 1123.
Hughes, A. L., and McMillen, J. H. (1932) *Phys. Rev.* **39**, 585.
Hughes, A. L., and Rojansky, V. (1929) *Phys. Rev.* **34**, 284.
Hummer, D. G., Stebbings, R. F., Fite, W. L., and Branscomb, L. M. (1960) *Phys. Rev.* **119**, 668.
Johnson, F. S., Watanabe, K., and Tousey, R. (1951) *J. Opt. Soc. Amer.* **41**, 702.
Keene, J. P. (1949) *Phil. Mag.* **40**, 369.

Khashaba, S., and Massey, H. S. W. (1958) *Proc. Phys. Soc.* **A71**, 574.
Krauss, M., Reese, R. M., and Dibeler, V. H. (1959) *J. Res. Natl. Bur. Stand.* **63A**, 201.
Lamb, W. E., and Retherford, R. C. (1950) *Phys. Rev.* **79**, 549.
Lamb, W. E., and Retherford, R. C. (1951) *Phys. Rev.* **81**, 222.
Lampe, F. W., Franklin, J. L., and Field, F. H. (1957) *J. Amer. Chem. Soc.* **79**, 6129.
Lawrence, E. O. (1926) *Phys. Rev.* **28**, 947.
Lichten, W. (1961) *Phys. Rev. Letters* **6**, 12.
Lichten, W., and Schultz, S. (1959) *Phys. Rev.* **116**, 1132.
Maier-Liebnitz, H. (1936) *Z. Phys.* **95**, 499.
Marmet, P., and Kerwin, L. (1960) *Canad. J. Phys.* **38**, 787.
Massey, H. S. W. (1949) *Rep. Progr. Phys.* **12**, 249.
Massey, H. S. W. (1956) *In* "Handbuch der Physik" (S. Flügge, ed.), Vol. 36. Springer, Berlin.
Massey, H. S. W., and Burhop, E. H. S. (1952) "Electronic and Ionic Impact Phenomena." Oxford Univ. Press (Clarendon), London and New York.
Moe, D. E., and Petsch, O. H. (1958) *Phys. Rev.* **110**, 1358.
Mohr, C. B. O., and Nicoll, F. H. (1932) *Proc. Roy. Soc.* **A138**, 229, 469.
Mohr, C. B. O., and Nicoll, F. H. (1933) *Proc. Roy. Soc.* **A142**, 320, 647.
Morrison, J. D. (1954) *J. chem. Phys.* **22**, 1219.
Morrison, J. D., and Nicholson, A. J. C. (1959) *J. chem. Phys.* **31**, 1320.
Nottingham, W. B. (1939) *Phys. Rev.* **55**, 203.
Novick, R., Lipworth, E., and Yergin, P. (1955) *Phys. Rev.* **100**, 1153.
Phelps, A. V. (1955) *Phys. Rev.* **99**, 1307.
Phelps, A. V. (1959) *Phys. Rev.* **114**, 1011.
Phelps, A. V., and Molnar, J. P. (1953) *Phys. Rev.* **89**, 1203.
Phelps, A. V., and Pack, J. L. (1955) *Rev. sci. Instrum.* **26**, 45.
Ramsey, N. F. (1956) "Molecular Beams." Oxford Univ. Press, London and New York.
Rubin, K., Perel, J., and Bederson, B. (1959) *Bull. Amer. Phys. Soc.* [2] **4**, 234.
Russek, A., and Thomas, M. T. (1958) *Phys. Rev.* **109**, 2015.
Russek, A., and Thomas, M. T. (1959) *Phys. Rev.* **114**, 1538.
Schulz, G. J. (1959) *Phys. Rev.* **116**, 1141.
Schulz, G. J., and Fox, R. E. (1957) *Phys. Rev.* **106**, 1179.
Seaton, M. J. (1959) *Phys. Rev.* **113**, 814.
Sluyters, T. J. M., and Kistemaker, J. (1959) *Physica* **25**, 1389.
Sluyters, T. J. M., de Haas, E., and Kistemaker, J. (1959) *Physica* **25**, 1376.
Stebbings, R. F. (1957) *Proc. Roy. Soc.* **A241**, 270.
Stebbings, R. F., Fite, W. L., Hummer, D. G., and Brackmann, R. T. (1960) *Phys. Rev.* **119**, 1939.
Stier, P. M., and Barnett, C. F. (1956) *Phys. Rev.* **103**, 896.
Tate, J. T., and Smith, P. T. (1932) *Phys. Rev.* **39**, 270.
von Koch, H., and Lindholm, E. (1961) *Ark. Fys.* **19**, 123.
Walker, W. C., Wainfan, N., and Weissler, G. L. (1955) *J. appl. Phys.* **26**, 1367.
Wannier, G. H. (1953) *Phys. Rev.* **90**, 873.
Watanabe, K., Zelikoff, M., and Inn, E. C. Y. (1953) *J. chem. Phys.* **21**, 1021.
White, H. E. (1931) *Phys. Rev.* **38**, 2016.
Wien, W. (1908) *Ann. Phys. (Leipzig)* **27**, 1025.
Wigner, E. P. (1948) *Phys. Rev.* **73**, 1002.
Woudenberg, J. P., and Milatz, J. M. (1941) *Physica* **8**, 871.
Yarnold, G. D., and Bolton, H. C. (1949) *J. sci. Instrum.* **26**, 38.

13.

Spectral Line Broadening in Plasmas

M. Baranger

1 *Introduction*

The study of the optical spectrum was, until fairly recently, the only tool at man's disposal in investigating the properties of bodies at extremely high temperatures. Even now that large scale plasmas can be controlled on earth, it retains an important role in one's knowledge of this state of matter. Some of its applications have been reviewed in Chapter 5 of this book. The present contribution is a survey of some recent developments in the theory of line shapes in plasmas. At sufficiently high densities, these shapes are due to interactions between the light-emitting atom and electrons or ions of the plasma. This is the justification for including the topic in this book. But, whereas most other chapters of the book are concerned with atoms in or near their ground state, line broadening affords a way of studying scattering or interaction of electrons and ions with atoms in highly excited states.

One important application is as a diagnostic tool in determining ion densities (cf. Kolb and Griem, Chapter 5, § 7.4). The line shapes are usually rather insensitive to the temperature, unless Doppler broadening dominates, so that temperature has to be determined in other ways. On the other hand, this means that it is not essential to have thermal equilibrium before one can apply the theory. The accuracy expected in the determination of the density is of the order of 20%. Another application is to opacity calculations. Those are often done by including only the continuous spectrum, but lines can actually increase the opacity by a sizable factor, most of the effect coming from the merging together of lines at high quantum numbers.

It is not our purpose here to provide a complete review of the theory of pressure broadening. This has already been done in several recent articles (Unsöld, 1955; Ch'en and Takeo, 1957; Breene, 1957; Traving, 1960). Rather, we want to restrict ourselves to broadening in a plasma because it is both the latest and the most accurate application of the general theory. We shall say nothing about molecular lines except this: while the same methods are used in calculating broadening in molecular spectra and in plasmas, the plasma calculations are more reliable because the forces there are simpler and have longer range. As a result of the long range, most of the broadening comes from weak collisions which are amenable to some kind of perturbation treatment.

The subject of spectral line broadening in plasmas has recently been reviewed by Margenau and Lewis (1959). We feel that the present endeavor is worthwhile, however, first because their article was written just before the first realistic calculations were done, second because we must take exception with their claim that the classical path approximation is not valid for electrons. We hope that the following will convince the reader that the classical path picture is good and that, moreover, a completely quantum mechanical calculation can be organized along lines very similar to the classical one, so that the validity of the latter is obvious at every step.

Considerations of space have prevented us from doing full justice to the broadening effects of both electrons and ions. We decided to stress the electrons because this is the newer development and also, now, the best known. The ion effect if considered only briefly in § 6.

2 *Basic Considerations*

2.1 Causes of Line Broadening in Ionized Gases

If the line originates in optically thick material, as happens often in astrophysics, its shape is strongly influenced by the successive emission and absorption processes that take place before the light escapes. This self-absorption effect is not the subject of the present review. In principle, it can be calculated if one knows the line shapes for optically thin samples and the conditions inside the thick material.

Other broadening effects that might be expected to compete with pressure broadening are Doppler broadening and natural broadening. The latter is due to the finite lifetime of the atomic excited state, arising from light emission itself. In all practical cases, it gives rise to widths that are orders of magnitude smaller than the observed ones. Hence it can be safely neglected, i.e., one can proceed as if an isolated excited atom went on emitting light forever.

As for Doppler broadening, it can be quite important, especially at high temperatures or low densities. For instance, for temperatures in the volt range, Doppler widths will often exceed pressure widths at electron densities of the order of $10^{15}\ \mathrm{cm}^{-3}$ or smaller. However, the theory of the Doppler effect is quite simple (Born, 1933). If M is the mass of the light-emitting atom or ion, ω the angular frequency, and ω_0 the unperturbed frequency, the line shape is

$$\left(\frac{Mc^2}{2\pi kT\omega_0^2}\right)^{1/2} \exp\left[-\frac{Mc^2}{2kT}\left(\frac{\omega-\omega_0}{\omega_0}\right)^2\right]. \tag{1}$$

After the effects of pressure broadening on the spectrum have been completely computed, one needs only to fold the above line shape into the result in order to incorporate Doppler broadening in the calculations.

Thus, the major effect to be considered is pressure broadening, which is caused by the perturbation of the light-emitting atom by other particles of the gas. The perturbers may be either neutral or charged, but the effect of charged ones is so much greater that neutral perturbers can be neglected as soon as there is any appreciable ionization. Hence there are two main broadening agents, ions and electrons. It was believed at one time that the effect of the electrons averaged out because of their fast motion and that only ions needed to be considered. This is definitely not true: for most atoms, electrons have a larger effect than ions; and for hydrogenic atoms, although the general aspect of the line can be obtained with ions alone, electrons constitute an important source of additional broadening and are essential if more than order-of-magnitude estimates are desired.

The difference between ions and electron comes only in their speed, not in the way they interact with the atom. The latter is the same for both, namely, the Coulomb interaction between the perturbing charge and the charged components of the atom. This may be approximated because most of the broadening is usually due to fairly distant perturbers. Hence it is usually sufficient to keep only the first term in a multipole expansion of the interaction, i.e., to write it

$$V = -\mathbf{d} \cdot \mathbf{E} \tag{2}$$

where $\mathbf{d}$ is the operator dipole moment of the atom and $\mathbf{E}$ the electric field produced by the perturbers at the center of the atom. This is for a neutral atom; if it is an ion which emits the light, one has to include also the zeroth order term of the expansion, the Coulomb potential $Z_1 Z_2 e^2/r$, r being the distance between the perturber and the center of the ion, and Z_1 and Z_2 the two charge numbers.

Since this is also the basic interaction in Stark effect, pressure broadening by ions and electrons is also referred to as Stark broadening. In the past this appellation has usually implied a point of view, namely, that the electric field is nearly static or that the motion of the perturbers is slow. On the other hand, the term "collision broadening" has been used also with a point of view, namely, that the motion is fast. The two effects are one and the same, of course. This is why we have used "pressure broadening," which is neutral and includes them both. The simplifications that arise in the theory in the limits of slow or fast motion will be considered shortly, in § 2.4.

2.2 The Spectrum of a Quantum Mechanical System

The system considered consists of a single light-emitting atom (which may also be an ion) immersed in a gas of perturbers. Perhaps the simplest point of view to take (Weisskopf, 1932; Jablónski, 1945) is to consider the whole system as a single quantum mechanical object, a gigantic molecule, with certain energy levels and stationary states. Then, the power radiated in the transition from initial state i to final state f, integrated over direction and summed over polarization, is given by the well-known formula† (Schiff, 1955, p. 261)

$$(4\omega_{if}^4/3c^3) \mid \langle f \mid \mathbf{d} \mid i \rangle \mid^2 \tag{3}$$

where $\omega_{if} = \omega_i - \omega_f$ is the frequency and $\mathbf{d}$ is the dipole moment of the whole system. The latter can be replaced by the dipole moment of the atom because we are interested in line shapes, not in that part of the spectrum which involves radiation by the perturbers and which forms a continuum.

To get the complete spectrum, one must sum over all possible final states and average over initial states, each of which occurs in the statistical ensemble with a certain probability, say ρ_i. Hence, the power radiated per unit frequency interval, $P(\omega)$, is given by

$$P(\omega) = (4\omega^4/3c^3)\, F(\omega) \tag{4}$$

with

$$F(\omega) = \sum_{if} \delta(\omega - \omega_{if}) \mid \langle f \mid \mathbf{d} \mid i \rangle \mid^2 \rho_i . \tag{5}$$

The double summation runs over all states of the system, but it is only for positive ω that $P(\omega)$ or $F(\omega)$ describe spontaneous emission. For negative ω, $P(\omega)$ is the energy absorbed out of a beam containing one photon per cross-sectional area $\pi^2 \lambda\!\!\!^{-2}$, where $\lambda\!\!\!^- = c/\mid \omega \mid$. For simplicity, we refer to $F(\omega)$ as "the spectrum," or "the line shape," since it is the more useful of the two functions. Its Fourier transform is more useful yet,

$$\Phi(s) = \int_{-\infty}^{+\infty} e^{-i\omega s} F(\omega)\, d\omega \tag{6}$$

$$= \sum_{if} e^{-i\omega_{if} s} \mid \langle f \mid \mathbf{d} \mid i \rangle \mid^2 \rho_i . \tag{7}$$

It satisfies+

$$\Phi(-s) = [\Phi(s)]^* \tag{8}$$

and F can be calculated back from

$$F(\omega) = \pi^{-1} \mathscr{R} \int_0^{\infty} e^{i\omega s} \Phi(s)\, ds \tag{9}$$

where $\mathscr{R}$ means the real part.

† A sum over the three components is implied in all equations involving the square of the dipole moment.

+ We use * for complex conjugate, † for Hermitian conjugate.

Since the system considered is macroscopic, the summations in (5) and (7) are really integrations and the functions $F(\omega)$ and $\Phi(s)$ are continuous, the latter tending to zero for large $|s|$. There is a very simple interpretation for $\Phi(s)$: it is the autocorrelation function of the light amplitude. To see this, call $A(t)$ the amplitude of the light train as a function of time. It may be written as a sum of monochromatic waves thus

$$A(t) = \int e^{-i\omega t}\, a(\omega)\, d\omega. \tag{10}$$

The autocorrelation function is the statistical average

$$[A(t+s)\, A^*(t)]_{\text{Av}} = \iint d\omega\, d\omega'\, e^{i\omega' t - i\omega(t+s)}\, [a(\omega)\, a^*(\omega')]_{\text{Av}}. \tag{11}$$

But there is no phase relation between components of different frequencies, hence the average on the right-hand side vanishes when $\omega \neq \omega'$. If we omit from consideration the slowly varying factor $4\omega^4/3c^3$ in (4), we can write

$$[a(\omega)\, a^*(\omega')]_{\text{Av}} = \delta(\omega - \omega')\, F(\omega). \tag{12}$$

Substituting in (11) and using (6) gives

$$[A(t+s)\, A^*(t)]_{\text{Av}} = \Phi(s). \tag{13}$$

Finally, one can rewrite (7) by introducing the time-evolution operator

$$T(s) = \exp\,(-\, iHs/\hbar) \tag{14}$$

which, when applied to an eigenstate $|\, a >$ of the Hamiltonian H, multiplies it by $e^{-i\omega_a s}$. Then (7) can be written as a trace, without explicit reference to initial and final states,

$$\Phi(s) = \text{Tr}\,[\mathbf{d}T^{\dagger}(s)\, \mathbf{d}T(s)\, \rho]. \tag{15}$$

2.3 The Classical Path Assumption

It is possible to develop the calculation of $\Phi(s)$ in a purely quantum mechanical way, and this will be done in § 3.6. However, it is most important to realize that, in all applications that have been made of the theory so far, it has turned out to be permissible to consider the perturbers as classical particles following prescribed trajectories. First, we shall see how the classical path assumption is incorporated in the calculation of (15), then we shall discuss its validity.

It is assumed that the time-dependent Schrödinger equation of the system has solutions of the form

$$\psi(t) = \chi(t)\,\varphi(t), \tag{16}$$

i.e., the product of a function $\chi(t)$ involving only the atomic coordinates by a function $\varphi(t)$ involving only the coordinates of the perturbers. It is further assumed that the function $\varphi(t)$ consists of wave packets and that it obeys a Schrödinger equation which is independent of the state of the atom

$$i\hbar\, d\varphi/dt = K\varphi. \tag{17}$$

The total Hamiltonian is the sum of the Hamiltonian of the atom, H_{A}, the perturbers' Hamiltonian K, and the interaction V,

$$H = H_{\mathrm{A}} + K + V. \tag{18}$$

If the light is emitted by a neutral atom, K is the sum of the kinetic energies of the perturbers and possibly also their mutual interactions. If the light comes from an ion, one includes in K the Coulomb potential $Z_1Z_2e^2/r$ between a perturber and the center of the ion; this is indeed independent of the state of the ion. In the latter case, the perturber wave packets describe hyperbolic trajectories, while in the neutral case they follow straight lines (at least if their mutual interactions are neglected).

Next, one finds the Schrödinger equation which is obeyed by $\chi(t)$ in this approximation. According to (16) $\chi(t)$ can be written

$$\chi(t) = \int \varphi^*(t)\,\psi(t)\,dx_p, \tag{19}$$

the integration extending over all perturber coordinates. Taking the time derivative of both sides, using the complex conjugate of (17) and the full Schrödinger equation for $\psi(t)$, one finds

$$i\hbar d\chi/dt = \int \varphi^*(t)\,(H_{\mathrm{A}} + V)\,\psi(t)\,dx_p. \tag{20}$$

Again using approximation (16) for $\psi(t)$ results in

$$i\hbar d\chi/dt = [H_{\mathrm{A}} + v(t)]\,\chi \tag{21}$$

with

$$v(t) = \int \varphi^*(t)\, V\varphi(t)\, dx_p. \tag{22}$$

In other words, $\chi(t)$ obeys a Schrödinger equation with a time-dependent potential $v(t)$ which is obtained from (2) by averaging over the perturber wave packets. If the packets are sufficiently small and do not spread much in time, the picture is that of classical perturbers giving rise to a time-dependent electric field which disturbs the quantum mechanical atom.

It is now possible to calculate the trace (15) in terms of these wave packet states, instead of the stationary states previously used. One must assume that the density matrix ρ is still diagonal with these states, i.e., that the energy spread in the packets is small compared to kT. This last assumption is actually slightly inconsistent. With this picture of prescribed classical paths, the perturbers transfer energy to the atom but the atom is not allowed to react back on the motion of the perturbers. The logical conclusion is that the atom would eventually reach infinite temperature, with all its states equally likely. One must make sure that the values of s for which one needs to calculate (15) are not so large that this has time to happen. Then the assumption is a reasonable one (see also Bloom and Margenau, 1953).

If one uses states (16) to calculate $\Phi(s)$, it can be written

$$\Phi(s) = \mathrm{Tr}\,[\mathbf{d}t^\dagger(s)\,\mathbf{d}t(s)\,\rho]_{\mathrm{Av}}. \tag{23}$$

Now, only states of the atom enter in the summation, ρ is the density matrix for the atom only, $t(s)$ is the evolution operator for the wave function χ and also obeys Schrödinger equation (21). But one must, in addition, perform a statistical average over all modes of motion of the perturbers. Equation (23) is the point of departure for line shape calculations which treat the perturbers classically (Anderson, 1949; Baranger, 1958b; Kolb and Griem, 1958).

For this classical path approximation to be valid, it must first be possible to use wave packets. A wave packet of size a has a momentum spread $\Delta p = \hbar/a$. If the wave packet is to hold together for any appreciable time, Δp must be small compared to p, the average momentum, i.e., a must be large compared to $\lambda\!\!\!^{-}$, the de Broglie wavelength divided by 2π. But, if it is to create the illusion of a classical particle, the wave packet must be small compared to its distance from the atom; therefore, the latter must be many times $\lambda\!\!\!^{-}$. The orbital angular momentum quantum number l is just the impact parameter measured in units of $\lambda\!\!\!^{-}$. Hence the l values for those perturbers that make most of the contribution to the broadening must be large. There will always be some perturbers with small l, but the approximation will be good if they do not contribute much to the broadening.

Another approximation that was made is that a perturber trajectory is fixed and does not depend on whatever interactions take place between the perturber and the atom. This will be good if most of the energy exchanges between the perturber and the atom involve an amount of energy $\Delta\epsilon$ small compared to the energy ϵ of the perturber. Here again, there will always be some collisions where a large amount of energy is exchanged; the approximation is valid if such collisions do not contribute a large fraction of the broadening. If one identifies ϵ with kT, the same condition ensures that the inconsistency mentioned earlier in the treatment of the density matrix does not produce any serious trouble.

In practice, the foregoing conditions are always met for electron perturbers, and *a fortiori* for ion perturbers. The condition on l follows from the fact that the charge-dipole interaction (2) is long range and, therefore, most of the scattering comes from distant perturbers undergoing weak collisions. The condition on $\Delta\epsilon$ follows then because, from considerations of adiabaticity, $\Delta\epsilon$ cannot be much larger than $\hbar/\tau$, where $\tau = \rho/v$ is the collision time, ρ being the impact parameter and v the velocity†; but $\hbar/\tau$ is just $2\epsilon/l$.

To see that l is large for a typical electron perturber, one can write $l = mv\rho/\hbar$. A typical impact parameter is of order $n^{-1/3}$, where n is the electron density, since the Debye length is typically a few times $n^{-1/3}$. As for v, it is of order $(kT/m)^{1/2}$. Hence one finds that l^3 is of the order of the "statistical factor"

$$\Gamma = \frac{2(2\pi mkT)^{3/2}}{nh^3}, \tag{24}$$

i.e., the number of quantum mechanical states available to each electron. Small Γ means that the electron gas is degenerate; Γ large compared to unity means that it is still subject to Maxwell-Boltzmann statistics. In typical cases, of course, Γ is very large. For instance, if the temperature is 1 ev and the density 10^{15} cm^{-3}, Γ equals 6×10^6.

But it is not enough to have shown that l is large for a typical electron. One must still be sure that most of the broadening is not due to those electrons that are very close. This will be ascertained in detail in § 4 [see discussion accompanying eqs. (104), (105), and (112)]. The reason

† We have been unable to avoid a certain amount of duplication of notations, which we do not expect to produce any confusion. We use ρ for the density matrix and also for the impact parameter. Here v is the velocity, but in (22) it is an interaction energy.

is simply this: there is an upper limit to the amount of broadening a close electron can produce. This can be compared with the contribution from distant electrons, and the latter is found to dominate because of the long-range nature of the interaction.

2.4 Two Limiting Approximations

It was mentioned earlier [Eq. (13)] that $\Phi(s)$ is essentially the autocorrelation function for the light amplitude. If the atom were alone in a vacuum, this would just be proportional to $e^{-i\omega_0 s}$ or a sum of such terms, each representing a single sharp line. In the presence of interactions, however, $\Phi(s)$ is perturbed and eventually goes to zero for large s: after a certain time, the light loses all memory of its original phase. The question then arises: how does the time that it takes the light to lose its memory compare with the collision time (the duration of a collision)? If it is either much larger or much smaller, the problem of calculating $\Phi(s)$ can be greatly simplified.

Let us first consider the case where $\Phi(s)$ goes to zero long before a single collision is completed. Then, clearly, the motion of the perturbers may be disregarded in the calculation of $\Phi(s)$. One may first assume all perturbers at fixed positions and obtain a spectrum of sharp lines; then one performs a statistical average over positions, which gives a broadened spectrum. We shall call this the "quasi-static approximation." The name "statistical approximation" has often been used in the literature, but it is not appropriate since statistics are just as important in other cases.

The other limit is that where the time that it takes $\Phi(s)$ to differ appreciably from its unperturbed value is very long compared to a collision time. Then, in an interval of a few collision times, $\Phi(s)$ suffers but a small change. Since the collisions are statistically independent, this change depends only on the present value and $\Phi(s)$ obeys a linear differential equation with constant coefficients, whose solution is a damped monochromatic wave. The archetype of such a theory is the Lorentz impact theory, in which it is assumed that collisions are instantaneous and occur with frequency ν, each collision interrupting the light train completely. Then, the autocorrelation function is obviously $e^{-i\omega_0 s-\nu s}$ and the line shape is proportional to $[(\omega - \omega_0)^2 + \nu^2]^{-1}$, which has since become known as a Lorentz shape. The same Lorentz shape arises whenever this fast collision approximation, or "impact approximation," is valid, but there may also be a shift in addition to the width ν. Moreover, if one is dealing with a group of overlapping lines, the differential equation is obeyed by a matrix instead of a single function and things get a little more involved (see § 3.3).

The regions of validity of the two approximations may be delimited roughly as follows. The time that it takes the light to lose memory of its original phase is also the inverse of the linewidth. Thus, the quasi-static approximation is good if the linewidth is large compared to the inverse of a typical collision time, the impact approximation is good if it is small. More details are given in later sections. One can also say that the impact approximation is valid if strong collisions occur at intervals much longer than a collision time. A strong collision is one which disturbs $\Phi(s)$ by a large amount. On the other hand, the quasi-static approximation is good if strong collisions are always going on.

The importance of the impact approximation is due to the fact that, in practice, it is always valid for the treatment of electron perturbers. To see this, consider a typical electron with velocity v, impact parameter ρ, colliding with an excited atom consisting of an electron in an orbit of principal quantum number ν around a core of charge Ze. A typical dipole transition matrix element in the atom is of order $e\nu^2 a_0/Z$, hence interaction (2) is of order $e^2\nu^2 a_0/Z\rho^2$. For the impact approximation to be good, a typical collision must be weak, i.e., the product of the interaction by the collision time must be small compared to $\hbar$; this ensures that the collision can be treated by perturbation theory and does not produce much disturbance. Hence, one must have

$$\frac{e^2\nu^2 a_0}{Z\rho^2}\,\frac{\rho}{v} \ll \hbar \tag{25}$$

or

$$\frac{v^3\rho^3\hbar^3 Z^3}{e^6\nu^6 a_0^3} \gg 1. \tag{26}$$

Estimating ρ by $4\pi\rho^3/3 = n^{-1}$, v as $(kT/m)^{1/2}$, and omitting some numerical factors, one finds that (26) reduces to

$$\Gamma Z^3/\nu^6 \gg 1 \tag{27}$$

where Γ is again the statistical factor (24). The latter being always quite large, condition (27) is satisfied even for fairly large values of the quantum number ν.

On the other hand, the quasi-static approximation may or may not be valid for the treatment of ion perturbers. If it does not hold, it may happen that the impact approximation holds, which makes calculations very easy since, then, ions and electrons can be treated together. But, more usually, neither limiting approximation is valid. This possibility is considered in § 6, but it may be said right now that there exists no good treatment of this intermediate region; this is one of the most serious gaps in the present theory. However, the situation is not as bad as one might think because, in many cases, electron broadening is the more important one and an error in the ion contribution does not affect the total much. Also, even if the quasi-static approximation breaks down in the line center, it remains valid in the wings (see § 6) which are of great interest, for instance, in astrophysical applications.

When the quasi-static approximation does hold, the problem of calculating the effect of the ions splits into two parts. First, since the interaction is given by (2), one must calculate the effect of a static electric field on the line; this is the well-known Stark effect. Second, one must calculate the probability distribution of the electric field vector and average the Stark pattern over all possible fields.

To summarize, the solution of a problem of pressure broadening in a plasma will usually proceed as follows. First, the atom is assumed to lie in a fixed electric field created by the ions, with attending Stark splitting. The effect of the electrons on the corresponding sharp lines is computed with the impact approximation. Then one averages the electron broadened spectrum with the probability distribution of the ionic field. Finally, one investigates possible corrections due to motion of the ions.

3 *The Impact Approximation*

3.1 Introduction

Simplified pictures of the broadening effects of collisions on a spectral line have existed for a long time (Lorentz, 1906; Weisskopf, 1932; Lenz, 1933; Burkhardt, 1940; Lindholm, 1941, 1942, 1945; Foley, 1946; Baranger, 1958a). Most of them involve the adiabatic assumption, i.e., they replace the full interaction between the perturber and the atom by a potential depending only on the position of the perturber. However, in the problem of electron broadening, there are a number of complications which are essential if any kind of accuracy is demanded of the theory. First there is the fact that electrons can, and often do, give *inelastic collisions*; this makes the adiabatic assumption invalid. A related difficulty is the degeneracy of the atomic states, i.e., a collision can be elastic and still change the state of the atom, again making the adiabatic assumption invalid (Spitzer, 1940). Another complication arises because, as was just seen, one has to calculate the broadening by electrons of spectral lines which are already closely split by the ionic field. Since electron and ion effects are of the same order of magnitude, one is faced with a problem of *overlapping lines*. The first realistic treatment of inelastic collisions is due to Anderson (1949); that of overlapping lines to Kolb (1957), Kolb and Griem (1958), and Baranger (1958b).

It was the impact approximation that made these new developments possible, and therefore a large fraction of this review is devoted to its study. Without it, the problem of electron broadening would be hopelessly difficult. Essentially, it says that the average collision is weak, so that it usually takes many collision times to make the light train lose memory of its phase. This does not mean that all collisions must be weak; the impact approximation is not to be confused with the Born approximation. Those collisions that are strong are treated as such. But since, in general, it takes many statistically independent collisions to disturb the atom, it is only some kind of average effect of the collisions that matters. All happens as if some time-independent perturbation $\mathscr{H}$ had been added to the atomic Hamiltonian H_{A}. The times involved in the calculation of $\mathscr{H}$ are much larger than collision times, so that $\mathscr{H}$ depends only on the net result of a collision, not on its detailed development. In other words, $\mathscr{H}$ can be expressed in terms of S-matrix elements, or scattering amplitudes. The light is the same as if it were emitted by an isolated atom with Hamiltonian $H_{\text{A}} + \mathscr{H}$ and, since $\mathscr{H}$ is not Hermitian, the energy levels have an imaginary part and the lines have a width.

3.2 Treatment of the Interaction in the Lower State

We shall take the classical path point of view from here until § 3.6.

It often happens that the interaction of the perturbers with the atom in the lower state of a given line is much weaker than their interaction with the upper state. This is so because the atom is more tightly bound, or less polarizable, in its lower state. Hence it is often a good approximation to neglect the interaction with the lower state completely, which simplifies things considerably.

This simplification is easily carried out in the equations of § 2.3. In (23), one can replace $t^{\dagger}(s)$ by $e^{i\omega_f s}$, where $\hbar\omega_f$ is the single energy of the lower level for the line or group of lines considered. In fact, one can take ω_f as origin of frequencies and omit this term entirely. Then, one may introduce an operator D acting only between upper atomic states and defined by

$$\langle a \mid D \mid b\rangle = \sum_{\alpha} \langle a \mid \mathbf{d} \mid \alpha\rangle \langle \alpha \mid \mathbf{d} \mid b\rangle. \tag{28}$$

Here, we have used Latin letters to designate upper states and Greek letters for the lower ones; the lower level may be degenerate, hence the sum on α. Finally, it turns out that one condition for the validity of the impact approximation (§ 5.1) is that the linewidth be small compared to the energy of the perturbers, or kT. Hence the density matrix ρ in (23) can be considered constant and we shall disregard it. With these simplifications, (23) becomes

$$\Phi(s) = \mathrm{Tr}\,[D t_{\mathrm{Av}}(s)] \tag{29}$$

where, now, only upper states are involved.

However, there are some lines for which the interaction in the lower state cannot be neglected. This happens often for hydrogenic atoms, because linear Stark effect does not increase with quantum number as fast as quadratic Stark effect. Then, perhaps the best point of view to take is to try to reduce the calculation of (23) again to something like (29). This can be done (Baranger, 1958b) by considering a "doubled atom," each state of which corresponds to two states of the original atom, i.e., to a line in its spectrum. More precisely, a state of the original atom is associated in a direct product with the complex conjugate of another state, so that the energy levels of the doubled atom will be the difference between the energy of an upper state and that of a lower state. We use the notation $\mid a\alpha^*\rangle\rangle$ to designate a state of the doubled atom and we define two operators

$$\langle \alpha \mid t^\dagger(s) \mid \beta\rangle \langle b \mid t(s) \mid a\rangle = \langle\langle b\beta^* \mid \Theta(s) \mid a\alpha^*\rangle\rangle, \tag{30}$$

$$\langle a \mid \mathbf{d} \mid \alpha\rangle \langle \beta \mid \mathbf{d} \mid b\rangle = \langle\langle a\alpha^* \mid \Delta \mid b\beta^*\rangle\rangle. \tag{31}$$

One can also consider $\Theta(s)$ as the direct product of $t(s)$ with the transpose of $t^\dagger(s)$. Then, if ρ is disregarded again as above, (23) becomes

$$\begin{aligned}\Phi(s) &= \sum_{ab\alpha\beta} [\langle a \mid \mathbf{d} \mid \alpha\rangle \langle \alpha \mid t^\dagger(s) \mid \beta\rangle \langle \beta \mid \mathbf{d} \mid b\rangle \langle b \mid t(s) \mid a\rangle]_{\mathrm{Av}} \\ &= \sum_{ab\alpha\beta} [\langle\langle a\alpha^* \mid \Delta \mid b\beta^*\rangle\rangle \langle\langle b\beta^* \mid \Theta(s) \mid a\alpha^*\rangle\rangle]_{\mathrm{Av}} \\ &= \mathrm{Tr}\,[\Delta\, \Theta_{\mathrm{Av}}(s)],\end{aligned} \tag{32}$$

which is the same as (29), but for the doubled system. Therefore, the considerations that follow are confined mostly to the case of no lower state interaction.

3.3 Evaluation of the Effective Perturbation $\mathscr{H}$

The next task is the evaluation of $t_{\mathrm{Av}}(s)$. It is actually more convenient to deal with the interaction representation evolution operator,

$$u(s) = \exp(iH_{\mathrm{A}}s/\hbar)\, t(s). \tag{33}$$

The change occurring in $u_{\mathrm{Av}}(s)$ when s increases by Δs can be written

$$\begin{aligned} \Delta u_{\mathrm{Av}}(s) &= u_{\mathrm{Av}}(s + \Delta s) - u_{\mathrm{Av}}(s) \\ &= [\{u(s + \Delta s, s) - 1\}\, u(s)]_{\mathrm{Av}}. \end{aligned} \tag{34}$$

Here, $u(s + \Delta s, s)$ is the interaction representation evolution operator between times s and $s + \Delta s$. It has the well-known perturbation expansion

$$\begin{aligned} u(s + \Delta s, s) - 1 = &- i\hbar^{-1} \int_s^{s+\Delta s} dx\, v'(x) \\ &- \hbar^{-2} \int_s^{s+\Delta s} dx \int_s^x dy\, v'(x)\, v'(y) + \ldots, \end{aligned} \tag{35}$$

with

$$v'(x) = \exp(iH_{\mathrm{A}}x/\hbar)\, v(x) \exp(-\, iH_{\mathrm{A}}x/\hbar). \tag{36}$$

The last average in (34) is over a product of two factors. The impact approximation is valid if Δs can be found such that: (1) Δs is so large that the two factors are actually statistically independent and may be averaged separately; (2) Δs is so small that the average of the first factor is small compared to unity, in which case it will be shown that it can be written

$$-\, i\hbar^{-1} \exp(iH_{\mathrm{A}}s/\hbar)\, \mathscr{H} \exp(-\, iH_{\mathrm{A}}s/\hbar)\, \Delta s, \tag{37}$$

$\mathscr{H}$ being a time-independent operator which will be calculated. Then $u_{\mathrm{Av}}(s)$ obeys the differential equation

$$i\hbar\, du_{\mathrm{Av}}(s)/ds = \exp(iH_{\mathrm{A}}s/\hbar)\, \mathscr{H} \exp(-\, iH_{\mathrm{A}}s/\hbar)\, u_{\mathrm{Av}}(s) \tag{38}$$

whose solution, expressed in terms of $t(s)$, is

$$t_{\text{Av}}(s) = \exp\left[-i(H_{\text{A}} + \mathscr{H})\, s/\hbar\right]. \tag{39}$$

All happens as though the constant operator $\mathscr{H}$ had been added to H_A.

The interaction v occurring in (35) is the total interaction of the atom with all the perturbers, which are assumed statistically independent of each other. The crux of the present argument is that, when the impact approximation is valid, one can calculate $u_{\text{Av}}(s + \Delta s, s) - 1$ assuming a single perturber to be present, and then multiply by N, the number of perturbers. To see this, consider two possibilities. First, there may occur only weak collisions during time interval Δs. Then, one can evaluate $u(s + \Delta s, s)$ by perturbation theory, i.e., keep only the lowest nonvanishing term in (35). In this case, the contributions of the various perturbers are just additive, and the procedure advocated is correct. The second possibility is that, in time Δs, there occurs a strong collision. This is a rare event, otherwise the second condition postulated above on Δs could not hold and the impact approximation would not be valid. When it occurs, not much error is committed by disregarding for that time interval all other weak collisions that may be going on together with the strong one, i.e., considering a single perturber again.

Hence, the problem is reduced to evaluating $u_{\text{Av}}(s + \Delta s, s)$ for a single perturber. In order to satisfy the first condition on Δs, one must make it much larger than an average collision time. Then it is a good approximation to calculate the average by including in it all collisions whose time of closest approach s_0 falls inside the interval $(s + \Delta s, s)$, and excluding those for which s_0 falls outside. The error thus made is but a small end correction. Moreover, for a collision that is included, one might as well replace $u(s + \Delta s, s)$ by $u(+\infty, -\infty)$. The latter can be written

$$u(+\infty, -\infty) = \exp(iH_{\text{A}}s_0/\hbar)\, S \exp(-iH_{\text{A}}s_0/\hbar), \tag{40}$$

where S is the S-matrix calculated for a similar collision with time of closest approach 0,

$$S = 1 - i\hbar^{-1} \int_{-\infty}^{+\infty} dx\, v'(x) - \hbar^{-2} \int_{-\infty}^{+\infty} dx \int_{-\infty}^{x} dy\, v'(x)\, v'(y) + \dots . \tag{41}$$

It is permissible to replace s_0 in (40) by s. This is because off-diagonal elements of u are needed only for two states which may give rise to overlapping lines, i.e., whose energy difference is of the order of magnitude of the width. But the product of the width by $\Delta s/\hbar$ is small as the result of condition (2) imposed on Δs. Since s_0 differs from s by less than Δs, very little error is involved in the replacement.

Finally, after multiplying this one-perturber average by N, one sees that $\{u_{\mathrm{Av}}(s + \Delta s, s) - 1\}$ takes form (37) with $\mathscr{H}$ given by

$$\mathscr{H} = -i\hbar \int (1 - S)\, d\nu \tag{42}$$

where the integral runs over all possible types of collision, each with its frequency $d\nu$. In practice, $\int d\nu$ consists of

$$n \int_0^\infty f(v)\, v dv \int_0^\infty 2\pi\rho d\rho \quad \text{(average over angles)} \tag{43}$$

where n is the perturber density, $f(v)$ the velocity distribution, and ρ the impact parameter.

A similar argument is applicable to cases where the interaction in the lower state is taken into account. One finds that $\Theta_{\mathrm{Av}}(s)$ is obtained by adding to the Hamiltonian of the doubled atom a constant operator given by

$$\mathscr{H} = -i\hbar \int (1 - S_i S_f^*)\, d\nu \tag{44}$$

where S_i acts on the upper part, S_f^* the lower part of a line.

3.4 Effect of Electron Correlations

When the foregoing theory is applied to electron broadening of hydrogenic levels, the integrals are found to diverge logarithmically on the weak collision side, i.e., large ρ. One way to eliminate this divergence (Lewis, 1961) is to note that the impact approximation is never valid for very weak collisions, since they have very long collision times. But a much more important cause of error is the fact that, so far, the perturbers have been assumed to move completely independently of each other. This is notoriously untrue of charged perturbers, whose correlations give rise to such important effects as Debye shielding and plasma oscillations.

Since the Debye length

$$r_{\mathrm{D}} = (kT/4\pi ne^2)^{1/2} \tag{45}$$

is usually somewhat larger than the nearest-neighbor distance, strong collisions are not affected by electron correlations. To examine their influence on weak collisions, it is sufficient to look at the first non-vanishing term in perturbation expansion (35). This is the second order term; the average of the first order term vanishes because the average electric field vector is zero. Hence, one requires $[\mathbf{E}(x)\,\mathbf{E}(y)]_{\text{Av}}$, $\mathbf{E}$ being the total electric field. In the previous section, this has been replaced by $N[\mathbf{E}_1(x)\,\mathbf{E}_1(y)]_{\text{Av}}$, $\mathbf{E}_1$ being the field of a single perturber, electron 1. The procedure is incorrect because correlations are present. But it is always correct to replace one of the two fields by $N\mathbf{E}_1$, thus $N[\mathbf{E}_1(x)\mathbf{E}(y)]_{\text{Av}}$; the second field must remain the total field. In performing the average, one can average over the positions and velocities of all other electrons before those of electron 1. Therefore, the problem is reduced to the following: Knowing that electron 1 is at $\mathbf{r}$, with velocity $\mathbf{v}$, at time x, find the average total electric field at the origin at time y. The answer is some sort of shielded field which exists only in the vicinity of electron 1. As a first approximation, one may assume that the electron moves along a straight trajectory at constant speed and use the Debye shielded field. This is correct for particles which do not have too high a velocity. For the high velocity particles, the Debye sphere is distorted. Since it turns out that slow electrons broaden lines more effectively than fast ones, the Debye approximation is probably sufficient.

Thus, the simplest way to take into account electron correlations is to shield one of the two fields in the second order term of the perturbation expansion, otherwise treating the electrons as independent. Instead of shielding the field, one might also use two unshielded fields but cutoff the ρ integration for large impact parameters. It can be shown (Griem *et al.*, 1962a) then that the cutoff should come at $1.1 r_{\text{D}}$.

3.5 The Line Shape

Going back to (39), the general result of the impact approximation, one can proceed to derive the line shape, using (29) and (9). The integral over s converges, as (42) shows that the anti-Hermitian part of $\mathscr{H}$ is negative definite. One finds

$$\pi\hbar^{-1} F(\omega) = -\mathscr{I}\,\mathrm{Tr}\,[D(\hbar\omega - H_{\text{A}} - \mathscr{H})^{-1}] \tag{46}$$

where $\mathscr{I}$ means the imaginary part. One needs to include in the trace only those upper states that contribute to the group of overlapping lines being considered. In general, one will have to invert a matrix whose number of lines and columns is the number of overlapping lines.

In the particular case of an isolated line, arising from upper state $|\, i\rangle$, the matrix is one by one and the shape is Lorentzian,

$$F(\omega) = \frac{1}{\pi}\frac{w_i \langle i \,|\, D \,|\, i\rangle}{(\omega - \omega_i - d_i)^2 + w_i^2}, \tag{47}$$

with a shift d_i and a width w_i given by

$$\hbar d_i = \mathscr{R} \langle i \mid \mathscr{H} \mid i \rangle, \tag{48}$$

$$\hbar w_i = -\mathscr{I} \langle i \mid \mathscr{H} \mid i \rangle. \tag{49}$$

$\mathscr{H}$ can itself be written in terms of the diagonal element of the S-matrix by (42). One can write

$$\langle i \mid S \mid i \rangle = e^{-\eta - i\varphi} \tag{50}$$

where φ is a real phase shift and η a real positive number. Then, the width and shift become

$$d_i = \int e^{-\eta} \sin \varphi \, d\nu, \tag{51}$$

$$w_i = \int (1 - e^{-\eta} \cos \varphi) \, d\nu. \tag{52}$$

Special cases of these expressions are: (1) If η is infinite, every collision is inelastic, there is no shift and the width is just the collision frequency; this is the Lorentz theory. (2) If η is zero, all collisions are elastic, the result is that of the adiabatic theory (Lindholm, 1941; Foley, 1946).

In the case where interaction is also present in the final state, the derivation of the line shape is quite similar to that of (46), but now all operators act on states of the doubled atom. We recall that the Hamiltonian of the latter is $H_{\mathrm{A}i} - H^*_{\mathrm{A}f}$, with $H_{\mathrm{A}i}$ acting on the upper part of a line, $H^*_{\mathrm{A}f}$ on the lower part. One finds

$$\pi \hbar^{-1} F(\omega) = -\mathscr{I} \operatorname{Tr} [\Delta(\hbar\omega - H_{\mathrm{A}i} + H^*_{\mathrm{A}f} - \mathscr{H})^{-1}], \tag{53}$$

$\mathscr{H}$ being given now by (44). The number of lines and columns of the matrix to be inverted is equal to the number of overlapping lines. For an isolated line $\mid a\alpha^* \rangle\rangle$, one gets a Lorentz shape with shift and width

$$\hbar d = \mathscr{R} \langle\langle a\alpha^* \mid \mathscr{H} \mid a\alpha^* \rangle\rangle, \tag{54}$$

$$\hbar w = -\mathscr{I} \langle\langle a\alpha^* \mid \mathscr{H} \mid a\alpha^* \rangle\rangle. \tag{55}$$

If the diagonal elements of the S-matrix are written again in the form

$$\langle a \mid S \mid a \rangle = e^{-\eta_i - i\varphi_i}, \tag{56}$$

$$\langle \alpha \mid S \mid \alpha \rangle = e^{-\eta_f - i\varphi_f}, \tag{57}$$

the shift and width become

$$d = \int e^{-\eta_i - \eta_f} \sin (\varphi_i - \varphi_f)\, dv, \tag{58}$$

$$w = \int [1 - e^{-\eta_i - \eta_f} \cos (\varphi_i - \varphi_f)]\, dv. \tag{59}$$

Great simplifications are brought into the calculation of the line shape in spherically symmetric situations, i.e., when the Hamiltonian H_A of the atom is invariant under rotations and the collisions themselves happen isotropically. Then the sum over magnetic quantum numbers in the traces of (46) or (53) can be eliminated, which greatly reduces the size of the matrices to be inverted. For instance, a line between two levels of angular momentum quantum numbers j_i and j_f can be treated as isolated, even if the j's are nonzero. We refer to Anderson (1949) and Baranger (1958b) for the details and confine ourselves to a few simple remarks.

First, such a spherical situation is the exception rather than the rule in plasmas, since one has to compute the electron broadening of atomic lines in the constant field of the ions, and the latter destroys spherical symmetry. However, it is usually allowable to forget about this electric field in the calculation of $\mathscr{H}$, though not in the calculation of the line shape. Then, $\mathscr{H}$ is a spherically symmetric operator. In case of an isolated line with no lower state interaction but nonvanishing j in the upper state, this means that $\mathscr{H}$ is a multiple of the unit matrix as far as the magnetic quantum number m is concerned. Since D has the same property, it is clear how the matrices in (46) are reduced to one by one. Actually, one can use this *a priori* knowledge of spherical symmetry to simplify the calculation of the average (43) occurring in the calculation of $\mathscr{H}$. For the simple case just referred to, it is not necessary to average over all possible orientations of a collision. It is sufficient to calculate $\langle m \mid \mathscr{H} \mid m \rangle$ assuming a definite orientation and then average over m; this effectively averages over orientations.

3.6 Quantum Mechanical Theory

We shall now sketch the impact theory for quantum mechanical perturbers. Although classical paths are sufficient for most applications, a full quantum mechanical treatment may serve to dispel any remaining doubts about the validity of the classical theory. It will also clarify the relation between the effective perturbation $\mathscr{H}$ and the scattering amplitude, thus tying our subject with the rest of this book, which is largely about quantum mechanical processes.

We shall endeavor to make the quantum theory as similar as possible to the classical path theory developed above. This will be achieved by keeping the representation of the perturbers as wave packets. A more rigorous presentation can be given (Baranger, 1958c), at the cost of losing some of the physical insight.

The starting point is (15), the Hamiltonian being given by (18). At first, we shall disregard the interaction in the lower state. If the lower state energy is taken as origin, (15) becomes

$$\Phi(s) = \mathrm{Tr}\,[D \exp(iKs/\hbar) \exp(-\,iHs/\hbar)\,\rho]. \tag{60}$$

The trace involves upper atomic states and the perturbers' states. We shall perform the sum over perturber states alone, denoted by Tr_p, and show that $\Phi(s)$ is then reduced to the same form as in the classical case, (29) and (39). This trace replaces the average over all perturber motions of the classical theory. As for ρ, which should be proportional to $e^{-H/kT}$, it will be replaced by the product of ρ_p and ρ_A, proportional to $e^{-K/kT}$ and $e^{-H_A/kT}$, respectively. This will be justified later.

The perturbers' states are taken to be wave packets but, now, no effort is made to make the packets appear like classical particles. On the contrary, the packets are taken much larger than the region in which the interaction V is appreciable. As in the classical case, one introduces the interaction representation evolution operator,

$$U(s) = \exp[i(H_A + K)\,s/\hbar] \exp(-\,iHs/\hbar). \tag{61}$$

Note that $U(s)$ does not change as long as a wave packet does not interact with the atom. One studies the time variation of

$$U_{Av}(s) = \mathrm{Tr}_p\,[U(s)\,\rho_p] \tag{62}$$

by writing

$$\begin{aligned} \Delta U_{Av}(s) &= U_{Av}(s + \Delta s) - U_{Av}(s) \\ &= [\{U(s + \Delta s, s) - 1\}\, U(s)]_{Av} \end{aligned} \tag{63}$$

with

$$\begin{aligned} U(s + \Delta s, s) &= U(s + \Delta s)\, U^{-1}(s) \\ &= \exp[i(H_A + K)(s + \Delta s)/\hbar] \exp(-\,iH\Delta s/\hbar) \exp(-\,i(H_A + K)\,s/\hbar]. \end{aligned} \tag{64}$$

Again, the impact approximation is valid if Δs can be found such that: (1) Δs is so large that the trace in the right-hand side of (63) splits into a product, i.e., the two factors involve different wave packets; (2) Δs is so small that the average of the first factor is small compared to unity, in which case it can be written in the form (37). From there on, the classical argument can be taken over in the quantum mechanical case. In particular, the line shape of § 3.5 is unchanged. We shall now examine the conditions under which there exists a Δs satisfying the requirements above and calculate $\mathscr{H}$.

If the collision time of a wave packet is defined as the time during which it overlaps with the interaction region, the first condition will be satisfied if Δs is taken much larger than a collision time. Hence, there is no sense in making the wave packets unnecessarily large. However, one does want them appreciably larger than the interaction region, for reasons that will become clear. Then, the various wave packets overlap with each other in general. As for the second condition on Δs, it can also be satisfied if each collision disturbs the atom very little. This does not mean that the collision can be calculated by perturbation theory. The disturbance may be small because the amplitude of the wave packet is small, not because the interaction is weak. It does mean, as in the classical case, that the effects of the various perturbers on the atom are additive. Each wave packet sees an isolated atom, without interference from the other wave packets. To calculate the first average in the right-hand side of (63), one just adds the contributions from all packets that arrive during time Δs.

We proceed to calculate the contribution of such a packet. If $\varphi(t)$ is its wave function, one needs

$$\langle b\varphi(0) \mid U(s + \Delta s, s) \mid a\varphi(0)\rangle, \tag{65}$$

a and b being two atomic states, but this is equal to

$$e^{i\omega_b s} \langle b\varphi(s) \mid \exp [i(H_{\mathrm{A}} + K) \Delta s/\hbar] \exp (-iH\Delta s/\hbar) \mid a\varphi(s)\rangle\, e^{-i\omega_a s} \tag{66}$$

This is the overlap of two wave functions. One function is $\exp (-iH\Delta s/\hbar) \mid a\varphi(s)\rangle$, i.e., the wave packet propagated to time $s + \Delta s$, together with the scattered wave arising from its interaction with the atom. The other function is $\exp [-i(H_{\mathrm{A}} + K) \Delta s/\hbar] \mid b\varphi(s)\rangle$, i.e., the wave packet propagated freely without interaction. In the case of elastic scattering, $a \equiv b$, and provided Δs be much larger than the collision time and the wave packet much larger than the region of interaction, such an overlap is obviously of the form

$$1 + xf(0), \tag{67}$$

$f(0)$ being the forward scattering amplitude and x a coefficient. A direct calculation of x by integration in coordinate space is possible, but does not seem to have appeared in the literature. It is not difficult if one takes care to smooth out the edges of the packet to avoid diffraction effects. However, an indirect calculation of x can be made if one uses (67) to derive the well-known optical theorem (Schiff, 1955, p. 105). In either case, one finds

$$x = i\lambda q \tag{68}$$

where λ is the wavelength and q the number of particles per unit cross section in the incident packet.

In the case of inelastic scattering, $a \neq b$, it is not clear at first sight that (66) has anything to do with a scattering amplitude. A true scattering process would involve an energy change in the perturber by $\hbar(\omega_a - \omega_b)$, while (66) has the same wave packet before and after scattering. However, one must keep in mind that such a matrix element is needed only if $\omega_a - \omega_b$ is of the order of magnitude of the width, which is itself much smaller than Δs^{-1}. Hence $(\omega_a - \omega_b)\,\Delta s$ is small and there is not enough time for any appreciable dephasing to appear due to the slight change of perturber energy in scattering. Then, the argument given in the elastic case is still valid, the scattering amplitude being taken at either energy. If the general inelastic scattering amplitude is denoted by $\langle b \mid f(\theta, \varphi) \mid a\rangle$, expression (65) is equal to

$$\delta_{ab} + i\lambda q e^{i\omega_b s} \langle b \mid f(0) \mid a\rangle\, e^{-i\omega_a s}. \tag{69}$$

There remains only to add the contributions from all the collisions occurring in time interval Δs. This is achieved by replacing q by the total number of perturbers incident per unit cross section, which is $nv\Delta s = hn\Delta s/m\lambda$, and averaging the scattering amplitude over perturber energies and directions. Then, $\{U(s + \Delta s, s) - 1\}_{\text{Av}}$ is seen to have form (37) with

$$\langle b \mid \mathscr{H} \mid a\rangle = -\,(2\pi\hbar^2 n/m)\,[\langle b \mid f(0) \mid a\rangle]_{\text{Av}}. \tag{70}$$

It is time to justify the approximation made on the density matrix. This is easy since, out of all the wave packets that contribute to $U(s + \Delta s, s)$, only a small fraction are interacting with the atom at time s. For those that do not interact, the exact density matrix and the free density matrix (i.e., without V) are identical.

A similar argument can be given if the interaction of the perturbers with the lower state cannot be neglected. The perturber part of the trace in (15) is

$$\mathrm{Tr}_p\,[(e^{iHs/\hbar})_f\,(e^{-iHs/\hbar})_i\,\rho_p], \tag{71}$$

subscripts i and f designating upper and lower atomic states, respectively.

We consider instead

$$W_{Av}(s) = \mathrm{Tr}_p\,[(e^{iHs/\hbar}\,e^{-iH_A s/\hbar})_f\,(e^{iH_A s/\hbar}\,e^{-iHs/\hbar})_i\,\rho_p]. \tag{72}$$

At time $(s + \Delta s)$ this can be written

$$W_{Av}(s + \Delta s) = \mathrm{Tr}_p\,[(e^{iHs/\hbar}\,e^{-iH_A s/\hbar})_f\,(e^{iH_A s/\hbar}\,e^{iH\Delta s/\hbar}\,e^{-iH_A(s+\Delta s)/\hbar})_f$$
$$\times\,(e^{iH_A(s+\Delta s)/\hbar}\,e^{-iH\Delta s/\hbar}\,e^{-iH_A s/\hbar})_i\,(e^{iH_A s/\hbar}\,e^{-iHs/\hbar})_i\,\rho_p]. \tag{73}$$

One argues that the operator appearing in the middle of this expression, namely,

$$\begin{aligned} W(s + \Delta s, s) = {} & (e^{iH_A s/\hbar}\,e^{iH\Delta s/\hbar}\,e^{-iH_A(s+\Delta s)/\hbar})_f \\ & \times\,(e^{iH_A(s+\Delta s)/\hbar}\,e^{-iH\Delta s/\hbar}\,e^{-iH_A s/\hbar})_i, \end{aligned} \tag{74}$$

can be averaged independently of the other operators, because different wave packets are involved. One argues also that this average is very close to unity and can be obtained by adding the contributions of the various packets. For a single packet,

$$\langle b\alpha\varphi(s)\,|\,(e^{iH\Delta s/\hbar}\,e^{-iH_A\Delta s/\hbar})_f\,(e^{iH_A\Delta s/\hbar}\,e^{-iH\Delta s/\hbar})_i\,|\,a\beta\varphi(s)\rangle \tag{75}$$

is just the overlap between two scattering wave functions. In one function, the packet is scattered by the atom in an upper state; in the other, by the atom in a lower state. The overlap has four terms: the overlap of the incident packets, which is unity; two overlaps between the incident part of one wave function and the scattered part of the other, which are proportional to the forward scattering amplitudes, as previously; and finally, the overlap between the two scattered waves, whose evaluation presents no problem. Hence, (75) is equal to

$$\begin{aligned} & \delta_{ab}\delta_{\alpha\beta} + i\lambda q\,[\delta_{\alpha\beta}\,\langle b\,|\,f(0)\,|\,a\rangle - \delta_{ab}\,\langle\beta\,|\,f(0)\,|\,\alpha\rangle^*] \\ & \quad + q\iint\langle\beta\,|\,f(\theta,\varphi)\,|\,\alpha\rangle^*\,\langle b\,|\,f(\theta,\varphi)\,|\,a\rangle\,\sin\theta d\theta d\varphi. \end{aligned} \tag{76}$$

At this point, it becomes expedient to transpose all operators involving the lower atomic states, as in § 3.2. We shall use $\bar{W}_{Av}$ for the transpose of W_{Av} and $\Theta_{Av}(s)$ for the transpose of (71). Then, $\bar{W}_{Av}(s)$ is seen to satisfy the differential equation

$$\begin{aligned} i\hbar d\bar{W}_{Av}(s)/ds = {} & \exp\,(iH_{Ai}s/\hbar - iH^*_{Af}\,s/\hbar) \\ & \times\,\mathscr{H}\,\exp\,(-\,iH_{Ai}s/\hbar + iH^*_{Af}s/\hbar)\,\bar{W}_{Av}(s) \end{aligned} \tag{77}$$

with

$$\langle\langle b\beta^* \mid \mathscr{H} \mid a\alpha^*\rangle\rangle = [-(2\pi\hbar^2 n/m)(\delta_{\alpha\beta}\langle b \mid f(0) \mid a\rangle - \delta_{ab}\langle\beta \mid f(0) \mid \alpha\rangle^*) + i\hbar nv \iint \langle\beta \mid f(\theta,\varphi) \mid \alpha\rangle^* \langle b \mid f(\theta,\varphi) \mid a\rangle \sin\theta d\theta d\varphi]_{\mathrm{Av}}. \tag{78}$$

The solution can be written

$$\Theta_{\mathrm{Av}}(s) = \exp[-i(H_{\mathrm{A}i} - H_{\mathrm{A}f}^* + \mathscr{H})s/\hbar]. \tag{79}$$

The line shape is given by (53).

For an isolated line, the shift and width are given solely in terms of elastic scattering amplitudes, by (48), (49), (70) or (54), (55), and (78). In case of no lower state interaction, one finds

$$d = -(2\pi\hbar n/m)\mathscr{R}[f(0)]_{\mathrm{Av}} \tag{80a}$$

$$w = (2\pi\hbar n/m)\mathscr{I}[f(0)]_{\mathrm{Av}} \tag{80b}$$

$$= (\tfrac{1}{2}nv\sigma)_{\mathrm{Av}}. \tag{81}$$

In the last expression, use has been made of the optical theorem to write the imaginary part of the forward scattering amplitude in term of the total cross section σ. The analogy with the classical expressions (51), (52), and (43) is brought out by writing the forward scattering amplitude as a sum over partial waves

$$f(0) = (\hbar/2imv)\,\Sigma_l\,(2l+1)(e^{2i\delta_l} - 1), \tag{82}$$

δ_l being a complex phase shift with positive imaginary part. The classical result if obtained when δ_l is computed by the WKB method and the sum over l replaced by an integral. It holds whenever many values contribute to the answer and strongly inelastic collisions do not predominate.

For an isolated line with lower state interaction, the width can be given an interesting form if one transforms the forward amplitudes appearing in (78) by the optical theorem and separates the cross sections into their elastic and inelastic parts. There comes

$$w = [\tfrac{1}{2}nv\{\sigma_{i\ \mathrm{in}} + \sigma_{f\ \mathrm{in}} + \iint \sin\theta d\theta d\varphi \mid f_i(\theta,\varphi) - f_f(\theta,\varphi)\mid^2\}]_{\mathrm{Av}}. \tag{83}$$

The cross sections appearing in this formula are inelastic, while the amplitudes are elastic. Thus, the broadening effects of the inelastic scattering on the initial and final states add incoherently, while those of the elastic scattering subtract coherently.

4 Applications of the Impact Approximation to Broadening by Electrons

4.1 Basic Approximations

It has just been seen that the calculation of the effective perturbation $\mathscr{H}$ is equivalent to the calculation of amplitudes or cross sections for the scattering of electrons by atoms in excited states. Much of this book is devoted to calculations of this nature, but usually for ground states. The fact that we are dealing with highly excited states and moderately fast electrons (say, upward of 1 ev) greatly simplifies matters. Whereas the scattering of slow electrons by a complicated atom in its ground state can often be quite untractable, our problem can be resolved by a set of fairly simple approximations whose expected accuracy if of the order of 20%.

The first of these approximations is the classical path assumption, of which enough has been said already. The second, mentioned in § 3.5, consists in neglecting, while calculating $\mathscr{H}$, the splitting of the atomic energy levels by the ionic field. The justification is that the product of the electron collision time and the interaction between the atom and the ions is small compared to $\hbar$. In fact, this is exactly condition (25) for the validity of the impact approximation. Hence, the electron goes by too quickly to notice any splitting.

Even then, the problem of calculating the collision matrix S of § 3.3 is untractable. It is tempting here to use perturbation theory. But this is correct only for sufficiently large impact parameters, those corresponding to "weak" collisions. Close collisions are "strong." Now, it is easy to calculate an upper limit to the contribution of close collisions. For instance, in (52), the worst that could happen would be that $1 - e^{-\eta} \cos \varphi$ is replaced by 2. When such limits have been set, it is found that most of the contribution (say, over 70%) to $\mathscr{H}$ actually comes from weak collisions, those for which perturbation theory is valid. It is therefore permissible to evaluate the contribution from strong collisions only crudely. This is usually done by saying that all coherence is lost between the light emitted before and that emitted after a strong collision, i.e., the S-matrix is replaced by zero. In other words, strong collisions are treated by the Lorentz theory. This is the same schematic treatment of strong collisions that was originated by Anderson (1949) for molecular broadening problems. But it is worth pointing out that the approximation is much better in plasma problems.

Here, the long range nature of the forces ensures the dominance of weak collisions; with molecules, strong collisions are often preponderant and the approximation becomes very questionable.

We shall consider separately four classes of calculations, depending on whether the light is emitted by a neutral atom or an ion, and whether its energy levels are hydrogenic, i.e. degenerate in l, or not. Note that complete degeneracy is not necessary for the applicability of the $\mathscr{H}$ computed for hydrogenic atoms; it is sufficient, as seen earlier, that the splitting between states of same principal quantum number be small compared to $\hbar/\tau$, τ being a typical collision time.

4.2 Electron Broadening in Nonhydrogenic Neutral Atoms

This case has been worked out by Griem *et al.* (1962a). A theory very similar in spirit was proposed by Vainshtein and Sobel'man (1959) but their quite unjustified lack of confidence in the classical path picture led them to make several unwarranted approximations, which invalidate all their numerical results.

The model for the atom is that of a single valence electron moving in a central potential. Fine structure is neglected. A typical atom to which this is applicable is helium. Cases involving more complicated coupling of angular momenta can be handled equally well, if the type of coupling is known. In this section, we shall consider isolated lines only. It is usually permissible to neglect the interaction in the lower state. Then, the shift and width are given by (48), (49), (42), and (43) and are independent of the magnetic quantum number of the upper state.

The first order term in the perturbation expansion of $\langle i \mid 1 - S \mid i \rangle$ has a vanishing angular average. The second order term is

$$\hbar^{-2} e^2 \sum_{\sigma \nu k} \langle i \mid r_\sigma \mid k \rangle \langle k \mid r_\nu \mid i \rangle \int_{-\infty}^{+\infty} du_1 \int_{-\infty}^{u_1} du_2 \, e^{i\omega_{ik}(u_1 - u_2)} \, E_{1\sigma}(u_1) \, E_{1\nu}(u_2) \qquad (84)$$

where $\mathbf{r}$ is the coordinate of the atomic electron and

$$\mathbf{E}_1 = e\mathbf{r}_1 / r_1^3 \qquad (85)$$

is the field produced at the origin by the perturber, whose coordinate is $\mathbf{r}_1$. Assuming an undisturbed straight trajectory, one can write

$$\mathbf{r}_1(u) = \boldsymbol{\rho} + \mathbf{v}u, \qquad (86)$$

$\boldsymbol{\rho}$ being the impact parameter and $\mathbf{v}$ the velocity.

The average of $E_{1\sigma}(u_1)\,E_{1\nu}(u_2)$ over all orientations of the collision is

$$\tfrac{1}{3}\,\delta_{\sigma\nu}\,e^2(\rho^2 + v^2 u_1 u_2)\,r_1^{-3}(u_1)\,r_1^{-3}(u_2). \tag{87}$$

It is convenient to introduce the dimensionless variables

$$x = vu/\rho, \qquad z_{ik} = \omega_{ik}\rho/v. \tag{88}$$

Then, the angular average of (84) takes the form

$$\tfrac{2}{3}\,(e^2/\hbar\rho v)^2 \sum_{\sigma k} \langle i \mid r_\sigma \mid k\rangle \langle k \mid r_\sigma \mid i\rangle\, C(z_{ik}) \tag{89}$$

with

$$\begin{aligned} C(z_{ik}) &= A(z_{ik}) + iB(z_{ik}) \\ &= \tfrac{1}{2}\int_{-\infty}^{+\infty} dx_1 \int_{-\infty}^{x_1} dx_2\, e^{iz_{ik}(x_1 - x_2)}\,(1 + x_1x_2)\,(1 + x_1^2)^{-3/2}\,(1 + x_2^2)^{-3/2}. \end{aligned} \tag{90}$$

The real part A of C can be expressed in terms of Bessel functions,

$$A(z) = z^2[K_0^2(|\,z\,|) + K_1^2(|\,z\,|)], \tag{91}$$

and the imaginary part B as a principal value integral involving A

$$B(z) = \pi^{-1}\,\mathrm{pv}\int_{-\infty}^{+\infty} (z - z')^{-1}\,A(z')\,dz'. \tag{92}$$

Both functions are plotted in Fig. 1; A is even and B is odd. For $z \gg 1$, $A(z)$ vanishes exponentially while $B(z) \approx \pi/4z$.

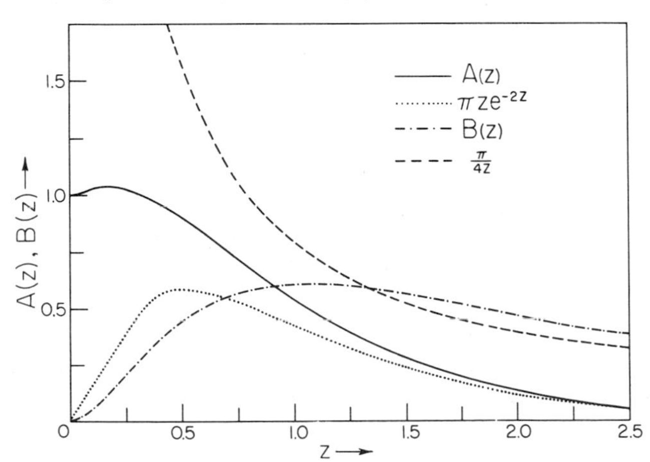

FIG. 1. The functions $A(z)$, $B(z)$, and their asymptotic limits (Griem *et al.*, 1962a).

The intermediate states in (89) must have an orbital angular momentum quantum number l_k differing from l_i by one unit. The most important states are those that are close to $| i \rangle$ in energy, otherwise z_{ik} takes on very large values and both A and B are small. One can perform the sum over σ and m_k and express (89) in terms of a radial integral,

$$(89) = \tfrac{2}{3} \lambda\!\!\!^{-2} \rho^{-2} \sum_{\nu_k, l_k} \mathscr{R}_{ik} C(z_{ik}), \tag{93}$$

with

$$\lambda\!\!\!^{-} = \hbar / mv, \tag{94}$$

$$\begin{aligned} \mathscr{R}_{ik} &= \sum_{\sigma, m_k} | \langle i | r_\sigma | k \rangle |^2 / a_0^2 \\ &= [\max(l_i, l_k)/(2l_i + 1)]\, a_0^{-2} \left[\int_0^\infty R_i(r)\, R_k(r)\, r^3 dr \right]^2. \end{aligned} \tag{95}$$

Here, ν_k is the principal quantum number of state k, $\max(l_i, l_k)$ is the larger of l_i and l_k, and it is understood that the sum over l_k in (93) has only two terms, $l_i - 1$ and $l_i + 1$. Hence, what comes in is the standard radial integral for the oscillator strength, but usually for a very low frequency transition.

If there were only weak collisions, the shift and width would therefore be given by

$$w_i + id_i = \int_0^\infty f(v)\, dv \frac{4\pi}{3} nv \lambda\!\!\!^{-2} \int_0^\infty \frac{d\rho}{\rho} \sum_{\nu_k l_k} \mathscr{R}_{ik} C(z_{ik}). \tag{96}$$

The real part of this formula can also be considered as giving an expression for the total inelastic scattering cross section. The elastic part is not included and would come out of higher order terms. It is true that it is usually, though not always, smaller. The problem now is to do something about the strong collisions. This is made imperative by the fact that the ρ integral in (96) diverges at the lower limit. One must cut it off at some $\rho_{\min}$, an impact parameter where $(1 - S)$ starts to be of order unity, and use the Lorentz theory for collisions closer than $\rho_{\min}$. Clearly, there is a lot of arbitrariness in the choice of $\rho_{\min}$, but there are two criteria to help us.

The first criterion is that, at least if $l_i = 0$, the theory should be correct in the adiabatic limit. This is the case where the great majority of collision times are large compared to all ω_{ik}^{-1}, so that all z_{ik} are large compared to unity. Then $A(z_{ik})$ is negligible and all collisions are elastic. At any one time, the state of the atom is the same as if it were immersed in a constant electric field and one can calculate a nonperturbative expression for S using the adiabatic approximation and Stark effect theory (Lindholm, 1941; Foley, 1946). This is strictly correct only if i is an s state, but it is often approximately true in other cases too because the Stark shift is often rather weakly dependent on the magnetic quantum number. The second criterion in choosing a cutoff was given by Griem and Shen (1961b). They pointed out that the true width and shift satisfy a certain dispersion relation and proposed that the approximate expressions also be made to satisfy this relation.

A cutoff which obeys both criteria is that of Vainshtein and Sobel'man (1959). They replace (96) by

$$w_i + id_i = n \int_0^\infty f(v)\, v dv \int_0^\infty 2\pi\rho d\rho (1 - e^{-\eta - i\varphi}) \tag{97}$$

with

$$\eta + i\varphi = \tfrac{2}{3}\, \lambda\!\!\!^{-2} \rho^{-2} \sum_{v_k l_k} \mathscr{R}_{ik} C(z_{ik}). \tag{98}$$

For weak collisions, this is the same as (96) and, for strong collisions for which $\eta \gg 1$, it gives the Lorentz theory. In the adiabatic limit, when η is negligible but φ is not, it gives the exact expression. Therefore, it includes in w_i the elastic scattering by the polarization potential as well as the inelastic cross section. With these various desirable properties, the Vainshtein-Sobel'man cutoff is probably the best one to use, although it is not in any sense exact and must remain a guess. For reasons of computational simplicity, the calculations of Griem *et al.* (1962a) use a different prescription, which also gives the correct adiabatic limit though not the dispersion relation. Any cutoff which has the right adiabatic limit will be correct for small electron velocities, while for high velocities one can check that strong collisions do not usually contribute more than 25% of the total width. Hence, the problem of strong collisions is not a critical one and accuracy of the order of 20% should be obtainable with any reasonable cutoff.

According to § 3.4, the integration in (96) or (97) should also be cut off on the high ρ side at $1.1 r_D$. Here, however, the integral would be convergent in any case. This modification may yield an appreciable reduction in the shift. For the width, it is usually not very important because $A(z)$ itself cuts off sharply for $z > 1$. The value of z corresponding to the Debye length cutoff, z_D, is of order ω_{ik}/ω_P, ω_P being the plasma frequency

$$\omega_P = (4\pi n e^2/m)^{1/2}. \tag{99}$$

As the Debye length is usually larger than the nearest-neighbor distance, ω_P is usually smaller than a typical $\hbar/\tau$. But (see § 4.1) the nonhydrogenic case is that where ω_{ik} is larger than $\hbar/\tau$, hence z_D is usually larger than unity.

It is convenient for comparison with later sections to write the weak collision part of the width as

$$w_i = \int_0^\infty f(v)\,dv\,\frac{4\pi}{3}\,nv\lambda\!\!\!^{-2}\sum_{\nu_k l_k}\mathscr{R}_{ik}\log\frac{\rho_{\max}}{\rho_{\min}} \tag{100}$$

where the upper cutoff $\rho_{\max}$ is function of k and takes into account the fact that $A(z)$ is 1 for small z but cuts off for $z > 1$. One can show that, for small z,

$$\int_z^\infty z'^{-1}\,dz' A(z') \approx \log(2/\gamma z) \tag{101}$$

with

$$\gamma = 1.781, \tag{102}$$

and therefore $\rho_{\max}$ should be taken equal to

$$\rho_{\max} = (2/\gamma)\,v/|\,\omega_{ik}\,| = 1.123v/|\,\omega_{ik}\,|. \tag{103}$$

This cutoff occurs because, for larger values of ρ, the time dependence of the interaction is too slow to induce transitions and the atom follows the change of electric field adiabatically. The Debye cutoff on the width is necessary only when $|\,\omega_{ik}\,|$ is smaller than ω_P, in which case $|\,\omega_{ik}\,|$ should be replaced by ω_P in estimating $\rho_{\max}$.

Finally, we would like to check the validity of the classical path approximation. For this, we make an estimate of $\rho_{\min}$, the strong collision cutoff. If this comes out larger than $\lambda\!\!\!^{-}$, it means that most of the broadening arises from high perturber angular momenta and therefore the classical theory holds. One can estimate $\rho_{\min}$ by writing that (93) is of order unity. If ν is the principal quantum number, $\mathscr{R}_{ik}$ is typically of order ν^4. For C, we will make two estimates and take the smaller value of $\rho_{\min}$. One estimate is $C \simeq 1$ (valid for small z_{ik}), the other is $C \approx z^{-1}$ (large z_{ik}). The two expressions for $\rho_{\min}$ are

$$\rho_{\min}^2 \approx \lambda\!\!\!^{-2}\nu^4 \tag{104}$$

$$\rho_{\min}^3 \approx \lambda\!\!\!^{-3}\nu^4(2\epsilon/\hbar\omega_{ik}) \tag{105}$$

where ϵ is the perturber energy, which is always larger than $\hbar\omega_{ik}$. As ν is typically 3 or 4, $\rho_{\min}$ is always larger than $\lambda\!\!\!^{-}$.

Had this not turned out to be the case, the classical theory would have been in error since, in quantum mechanics, ρ is quantized and varies by steps of $\lambda\!\!\!^{-}$. The proper procedure, then, would have been to use $\lambda\!\!\!^{-}$ (actually, $\lambda\!\!\!^{-}/\gamma$) as the lower cutoff in integral (96) and omit any strong collision term. If $\epsilon \gg \hbar\omega_{ik}$, the width so obtained is identical to that of Kivel (1955), which can also be obtained by substituting in (81) the inelastic cross section calculated in Born approximation. However, as shown previously, the Born approximation is not valid so that Kivel's widths are too large, though not by an order of magnitude. Similarly, Kivel's shift is identical with that in (96) (here, the integral converges for small ρ) and can also be obtained from standard, second order, time-independent perturbation theory. It is correct in the high-velocity limit ($z_{ik\ \min} \ll 1$), but not in the adiabatic limit. It can be shown that†

$$\int_0^\infty z^{-1}\, dz B(z) = \pi/2. \tag{106}$$

4.3 Electron Broadening of Hydrogen Lines

The case of hydrogen differs from the previous one in two main respects. First, because of the special degeneracy, one can omit the exponentials in an expression such as (84), which simplifies the integrations considerably; the function $C(z)$ is then replaced by unity. At least, this is a good approximation if most of the contribution comes from intermediate states with the same principal quantum number as the level studied, which is usually the case. The second new feature is that, again because of the degeneracy, one has to use the complete formula (46) or (53) appropriate to overlapping lines, for each value of the ionic field. The analysis was carried through by Griem *et al.* (1959) and Mozer (1960). The principal difference between the two theories is that Griem *et al.* neglect the interaction in the lower state, while Mozer shows that this is not very good for the low Balmer lines, as the following will also make clear.

The starting point is (44). Both S-matrices are expanded according to (41), but $v'(x)$ may be replaced by $v(x)$. For weak collisions, only terms of second order in the electric field are retained. There are three such terms: the second order term in S_i times the zeroth order term in S_f^*, the zeroth order term in S_i times the second order term in S_f^*,

† Both (101) and (106) can be obtained by integrating over z in (90) before performing the (x_1, x_2) integrations.

and the product of the two first order terms. Hence, including weak collisions only, one can write

$$\hbar^2 e^{-2} \langle\langle a\alpha^* \mid 1 - S_i S_f^* \mid b\beta^* \rangle\rangle$$

$$\begin{aligned} &= \delta_{\alpha\beta} \sum_{\sigma\nu c} \langle a \mid r_\sigma \mid c\rangle \langle c \mid r_\nu \mid b\rangle \int_{-\infty}^{+\infty} du_1 \int_{-\infty}^{u_1} du_2\, E_{1\sigma}(u_1)\, E_{1\nu}(u_2) \\ &\quad + \delta_{ab} \sum_{\sigma\nu\gamma} \langle \alpha \mid r_\sigma \mid \gamma\rangle^* \langle \gamma \mid r_\nu \mid \beta\rangle^* \int_{-\infty}^{+\infty} du_1 \int_{-\infty}^{u_1} du_2 E_{1\sigma}(u_1)\, E_{1\nu}(u_2) \\ &\quad - \sum_{\sigma\nu} \langle a \mid r_\sigma \mid b\rangle \langle \alpha \mid r_\nu \mid \beta\rangle^* \int_{-\infty}^{+\infty} du_1 \int_{-\infty}^{+\infty} du_2 E_{1\sigma}(u_1) E_{1\nu}(u_2). \end{aligned} \tag{107}$$

The average of $E_{1\sigma}\, E_{1\nu}$ over all orientations is given by (87). The integrals over u_1 and u_2 are elementary and one finds

$$\langle\langle a\alpha^* \mid \mathscr{H} \mid b\beta^* \rangle\rangle_{\text{weak collisions}}$$
$$= -\, i\hbar\, \langle\langle a\alpha^* \mid Q \mid b\beta^* \rangle\rangle \int_0^\infty f(v)\, dv (4\pi/3)\, nv\lambda\!\!\!^{-}{}^2 \int_0^\infty \rho^{-1} d\rho, \tag{108}$$

$$\langle\langle a\alpha^* \mid Q \mid b\beta^* \rangle\rangle = \sum_\sigma \Big(\sum_c \langle a \mid r_\sigma \mid c\rangle \langle c \mid r_\sigma \mid b\rangle$$
$$+ \sum_\gamma \langle a \mid r_\sigma \mid \gamma\rangle^* \langle \gamma \mid r_\sigma \mid \beta\rangle^* - 2\langle a \mid r_\sigma \mid b\rangle \langle \alpha \mid r_\sigma \mid \beta\rangle^* \Big) \Big/ a_0^2. \tag{109}$$

If ν_i and ν_f are the two principal quantum numbers, matrix elements of $\mathbf{r}$ in the initial and final states are of order $\nu_i^2 a_0$ and $\nu_f^2 a_0$, respectively. The cross terms in (109) can therefore be appreciable for the H_α and H_β lines, in spite of the fact that their number is smaller than that of the direct terms. Their effect, actually, is to reduce the broadening.

The ρ-integral in (108) diverges at both ends. Here, the upper cutoff at $1.1 r_D$ is essential, of course. The lower cutoff is still more arbitrary than in the last section, but again its exact value is not very critical. Then, a Lorentz term is added to take care of strong collisions.

The actual job of evaluating the various matrix elements of Q, those of Δ in (31), and inverting a matrix in order to calculate the line shape (53), involves a fair amount of work in the case of Balmer lines. For instance, for H_β, the matrices are really 64×64 and further approximations must be made. Note, that, at this stage, the splitting due to the ion field must be taken into account in H_{Ai} and H_{Af}^*. As the representation $\mid a\alpha^* \rangle\rangle$ has not yet been specified, one can choose, for instance, that in which H_{Ai} and H_{Af}^* are diagonal. In that case, Mozer (1960) found that $\mathscr{H}$ was mostly diagonal too and he performed the matrix inversion by perturbation theory, considering the off-diagonal elements as small. The high Balmer lines lend themselves to many further approximations and were studied by Griem (1960). Once the line shape (53) has been calculated, one must still integrate it with the probability distribution of the ionic field (see § 6) and fold in the Doppler effect if it is important.

The comparison of these calculations with experiments is deferred to § 7. The same effective perturbation $\mathscr{H}$ (usually without lower state interaction) may be used to study electron broadening in nonhydrogenic atoms, such as helium, when levels of same principal quantum number are nearly degenerate, i.e., when the forbidden line in the neighborhood of an allowed line is appreciably excited. This has been done (Griem, 1959; Griem *et al.*, 1962a) with considerable success.

4.4 Electron Broadening in Nonhydrogenic Ions

This case differs from that of nonhydrogenic neutrals by the fact that the electrons follow hyperbolic trajectories instead of straight lines. As a result, the integrations are slightly more difficult and the function $C(z)$ is changed. But a more interesting consequence is that the Coulomb attraction provides an automatic cutoff at the lower end of the integration in (96). This unexpected result arises because, when the hyperbola is strongly curved, the axial component of the electric field of the perturber changes sign twice during the collision, so that its first order effects tend to cancel. Hence, the cutoff will occur for an impact parameter such that the Coulomb scattering angle is of order 1 radian. If this is larger than the strong collision cutoff $\rho_{\min}$ of § 4.2, the addition of Coulomb attraction between the perturbers and the atom actually decreases the broadening! If it is smaller, the straight path result is not appreciably changed.

It is of interest to compare this new cutoff with $\lambda\!\!\!^{-}$ in order to check once more into the validity of the classical path approximation. If L is the angular momentum in units of $\hbar$ and η the usual Coulomb parameter (Z' is the charge of the ion)

$$\eta = Z'e^2/\hbar v, \tag{110}$$

the relationship between L and the scattering angle θ is

$$L = \eta \cot(\theta/2). \tag{111}$$

The cutoff comes when $\cot(\theta/2)$ is of order unity, i.e., for $L \approx \eta$. But η is always larger than unity, hence the cutoff always comes before quantum corrections set in. It is well known also in the theory of Coulomb scattering that large η corresponds to the classical limit. To show that η is always large, one can start by writing

$$\eta^2 = Z'^2 \,\mathrm{Ry}/\epsilon \approx Z'^2 \,\mathrm{Ry}/kT \tag{112}$$

where Ry is the Rydberg energy. But Z'^2 Ry is always larger than I, the ionization potential of the last ionized electron, at least if one excludes the K shell; and I is of order $kT \log \Gamma$ by Saha's equation, Γ being the parameter (24). Thus η^2 is larger than $\log \Gamma$, which is always appreciably larger than unity.

Study of this case was made by Baranger and Stewart (1962). Most applications arise in opacity calculations, where one is concerned with the width more than the shift. The foregoing discussions have implied neglect of the interaction in the lower state. Of course, one is not allowed to neglect the scattering of the electron by the Coulomb field when the ion is in its lower state. But (83) shows that this cancels against the identical scattering in the upper state. Other sources of elastic scattering are usually less important than inelastic scattering, hence, for an isolated line, the problem is reduced once more to calculating $\sigma_{i\ \mathrm{in}}$. In most opacity applications, perturbation theory is valid all the way to close collisions, i.e. the Coulomb cutoff occurs before the strong collision cutoff.† Then the calculation of $\sigma_{i\ \mathrm{in}}$ becomes identical to the perturbation treatment of the cross section for Coulomb excitation of a nucleus, a problem which has received a lot of attention and for which there exists an exhaustive review article (Alder *et al.*, 1956) including a treasury of formulae and curves relevant to the present study. As this review is completely quantum mechanical, it becomes convenient to discuss our problem quantum mechanically too in spite of the fact that we are in the classical limit.

Instead of (96), one finds the following width

$$w_i = \int_0^\infty f(v)\, dv\, \frac{4\pi}{3}\, nv\lambda\!\!\!^{-2} \sum_{\nu_k l_k} \mathscr{R}_{ik} \frac{9}{32\pi^2} f_{E1}(\eta_i, \eta_k). \tag{113}$$

Here f_{E1} is the function defined by Alder *et al.* with one difference: whereas they are interested in the case where the Coulomb interaction is repulsive, ours is attractive and the relation between the two cases is (see their Eq. II.B.59)

$$(f_{E1})_{\mathrm{att}} = e^{2\pi|\xi|}(f_{E1})_{\mathrm{rep}}. \tag{114}$$

† If the strong collision cutoff is the one that is effective, one can use the results of § 4.2 without change.

ξ is defined by

$$\xi = \eta_k - \eta_i \tag{115}$$

and η_i and η_k are the Coulomb parameters of the perturber before and after the inelastic collision, respectively. In our case, ξ is always small compared to η so that one can write

$$\xi = -\eta\hbar\omega_{ik}/2\epsilon. \tag{116}$$

In the classical limit, f_{E1} depends only on ξ and $(f_{E1})_{\text{rep}}$ is graphed by Alder *et al.* (Fig. II.4). It is an even function of ξ. For small ξ, one has approximately

$$(9/32\pi^2) f_{E1}(\xi) \approx \log(2/\gamma \mid \xi \mid), \tag{117}$$

i.e., one can write (100) again, with the same $\rho_{\max}$ but with a lower cutoff

$$\rho_{\min} = \mid \xi \mid v / \mid \omega_{ik} \mid = Z' e^2 / 2\epsilon \tag{118}$$

provided by the Coulomb interaction. This is also the quantity usually called a, half the major axis of the hyperbola. In terms of angular momentum, it corresponds exactly to

$$L_{\min} = \eta, \tag{119}$$

the relation surmised earlier.

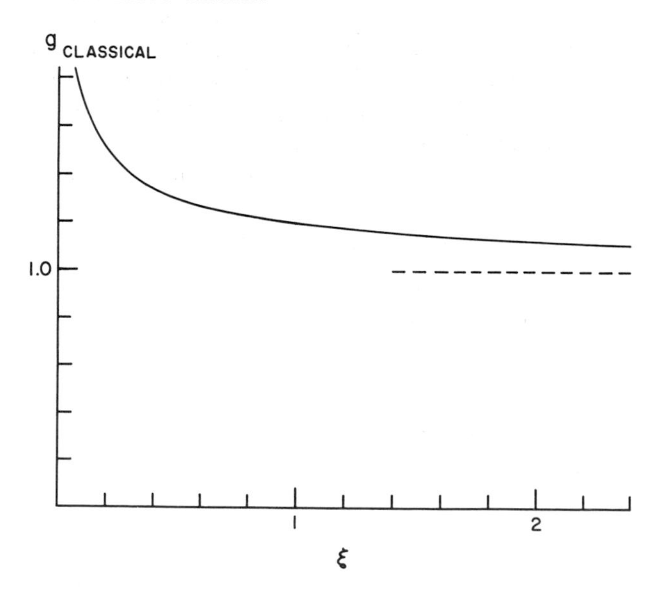

FIG. 2. The Gaunt factor in the classical limit, i.e., large η.

The function f_{E1} can also be expressed in terms of the Gaunt factor g (Gaunt, 1930)

$$(9/32\pi^2) f_{E1}(\eta_i, \eta_k) = (\pi/\sqrt{3})\, g(\eta_i, \eta_k). \tag{120}$$

The Gaunt factor equals unity for large ξ and, even at $\xi = 1$ (and η large), it is still only 1.2. A rough graph is given in Fig. 2. By comparing (103) and (118), one sees that the limit of large ξ or unit Gaunt factor is that where the lower, Coulomb cutoff starts to become larger than the upper, adiabatic cutoff. Then, all the inelastic cross section is contributed by strongly curved hyperbolae. This is also known as the Kramers limit (Kramers, 1923), but it does not usually arise in width calculations.

4.5 Electron Broadening in Hydrogenic Ions

Electron broadening in hydrogenic ions was investigated by Griem and Shen (1961a). The discussion is similar to that of § 4.3 except that the time integrals have to be performed for hyperbolic trajectories. They show very simply that, for each impact parameter, this reduces matrix element (107) by the factor $(1 + \eta^2/L^2)$, and the same result comes out of the general formulae of Alder *et al.* Of course, the dipole matrix elements appropriate to an ion of charge Z' must be used in (107) or (109). If the Coulomb cutoff (118) is smaller than the strong collision cutoff, the results of § 4.3 are not appreciably changed. This is the more common situation for lines produced in shock tubes. Uncertainties associated with the strong collision contribution are not critical as this contribution is small. On the other hand, if the Coulomb cutoff is the larger one, the ρ integral in (108) becomes $\log(\rho_{\max}/\rho_{\min})$ with $\rho_{\max} = 1.1 r_D$ and $\rho_{\min}$ given by (118). Then, the broadening is actually decreased.

5 Corrections to the Impact Approximation

5.1 Validity of the Impact Approximation

As mentioned previously in § 2.4 and throughout § 3, the impact approximation is valid when the average collision is weak, so that the time necessary to disturb the atom appreciably is large compared to τ, a representative collision time. Then, the question that arises is, how far down in the wings can one expect the shape of a line to be correctly given by the impact theory. The answer is that it holds for frequencies small compared to τ^{-1}. It is clear that the expression derived in § 3.3 for $\Phi(s)$ holds only for $s \gg \tau$. Hence its Fourier transform $F(\omega)$ is correct only for $\omega \ll \tau^{-1}$, with frequencies counted from the unperturbed line. If the width and shift are both small compared to τ^{-1}, then the central core of a line is well represented by a Lorentz shape. The Lorentz shape may also stop holding in regions of the spectrum where several distant lines make approximately equal contributions; but, if the frequencies involved are small compared to τ^{-1}, the correct answer can still be obtained by using overlapping line theory.

Perhaps it is worth noting at this point that under no circumstances can the collision time be smaller than $\hbar/\epsilon$, ϵ being the electron energy, which is of order kT. This lower limit on τ is due to the finite size of the wave packets, which are necessarily large compared to $\lambda\!\!\!^{-}$. Hence τ^{-1} is always small compared to $kT/\hbar$.

5.2 The One-Electron Approximation

Sometimes, one is interested in finding the spectrum far away in the wings, where the impact approximation does not hold. An instance of this arises in the opacity problem. To calculate the Rosseland mean opacity (Chandrasekhar, 1957, p. 212), one integrates the product of a certain weight function by the inverse of the photon absorption coefficient. Thus the result is mostly sensitive to the form of the latter in the frequency regions where it is small, i.e., in the windows, far from the line centers. The cores of strong lines may be replaced by regions of infinite absorption, but the shape of the windows must be known accurately (Baranger and Stewart, 1962).

Fortunately, there exists a simple prescription to calculate these remote parts of the spectrum. If the impact approximation is valid in the line centers, then the wings and the windows can be calculated with the one-electron approximation. This consists in calculating the spectrum with a single perturbing electron present, then multiplying by the total number of electrons. To understand why it is so, one has only to realize that, for frequencies very different from the unperturbed line frequencies, $F(\omega)$ arises from the behavior of $\Phi(s)$ at very short times. But, even for relatively long times such as the Δs of § 3.3 or § 3.6, we saw that the various electrons simply add their effects if the impact approximation is valid. *A fortiori*, they will add for times shorter than that.

It is clear, for instance, that the one-electron approximation is capable of explaining how the far wings of a line might eventually turn into the quasi-static shape (see § 6). This would arise from values of s smaller than the collision time in a close collision. Then the electron motion becomes negligible, provided quantum mechanics does not have to be taken into account. It is also clear that the one-electron approximation is never valid in the line cores. It actually gives a divergent result there. The rounded top of the Lorentz curve is always a many-perturber effect. However, when the impact approximation is good, there is a large overlap between its region of validity and that of the one-electron approximation. The latter holds for $\omega \gg w$ and the former for $\omega \ll \tau^{-1}$.

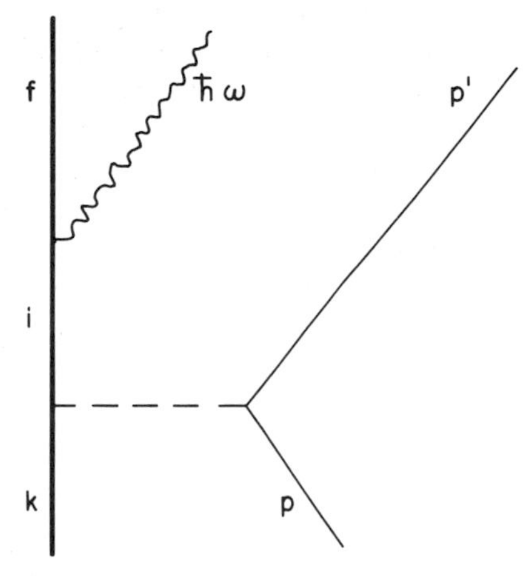

FIG. 3. Feynman diagram for the window calculation. The heavy line represents states of the ion, the lighter solid lines represent the perturbing electron in Coulomb scattering states. The wavy line is the photon and the broken line the interaction between the ion and the electron (exclusive of its Coulomb part). Another diagram, where the electron interacts with the ion after emission of the photon, has been neglected.

The argument for validity of the one-electron approximation may also be given a slightly more quantum mechanical form. Consider an excited atom being perturbed by electrons and consequently in a time-dependent state. It enters state i which is susceptible of emitting light with simultaneous decay of the atom to state f. But we are interested in a light frequency ω which is very different from ω_{if}. Hence, the original state i is very energy nonconserving and, by the uncertainty principle, cannot last a time much longer than $(\omega - \omega_{if})^{-1}$. Of course, the energy deficiency is made up by the perturbing electrons and the over-all process is energy conserving. We want to look at the wings and the windows, where $(\omega - \omega_{if})^{-1} \ll w^{-1}$. Again we appeal to § 3.3 or § 3.6 to argue that, when the impact approximation is good for the line cores, the perturbing effects of the various electrons are additive for times smaller than w^{-1}. In other words, w^{-1} is essentially the average interval between collisions and, if the excess energy must be carried away in a time short compared to w^{-1}, it has to occur through a single collision.

We shall sketch briefly the calculation of the window spectrum for the case of light emitted by an ion (Baranger and Stewart, 1962). In the upper state, the ion interacts with a single electron. Coulomb functions must be used as unperturbed electron wave functions. The perturbation in the lower state will be neglected, which simplifies the algebra considerably. The perturbation in the upper state will be treated in first order. This is usually good because Coulomb effects cutoff the contribution of strong collisions. Then the process can be represented by the Feynman diagram of Fig. 3. The ion starts in state k and interaction with the electron induces transition to state i, while the electron jumps from state p (energy ϵ) to state p' (energy ϵ'). Then, the ion emits the photon and decays to state f. The interaction is given by (2) and the field by (85). The spectrum is given by (5), but what was called i there is now a scattering state built upon the unperturbed states k and p of the ion and the electron. Hence

$$F(\omega) = \sum_{kfp'p\mu} \rho_k \rho_p \delta(\omega_k + \epsilon/\hbar - \omega_f - \omega - \epsilon'/\hbar)$$

$$\times \left| \sum_i \frac{\langle f \mid d_\mu \mid i \rangle \langle i\varphi_{p'}^- \mid \mathbf{d} \cdot \mathbf{E}_1 \mid k\varphi_p^+ \rangle}{\hbar(\omega + \omega_f - \omega_i)} \right|^2 \tag{121}$$

The δ-function takes care of over-all energy conservation. As for φ_p^+ and $\varphi_{p'}^-$, they are Coulomb wave functions for the electron in the field of the ion, with outgoing and ingoing scattered waves, respectively.

The result can again be expressed in terms of the functions f_{E1} or g of § 4.4. The Coulomb wave functions are expanded in spherical harmonics. The normalization of the incoming plane wave in φ_p^+ is taken to be n per unit volume in order to add the effects of all the electrons. There is a little bit of algebra with Clebsch-Gordan coefficients and one finds

$$F(\omega) = \int_0^\infty f(v)\, dv \sum_{kfii'} \rho_k \frac{4}{3} \frac{e^6}{\hbar^2} \frac{n}{v} \delta(l_i, l_{i'})$$

$$\times \frac{\max(l_i, l_f) \max(l_i, l_k)}{2l_i + 1} \frac{R_{if} R_{i'f} R_{ik} R_{i'k}}{(\omega + \omega_f - \omega_i)(\omega + \omega_f - \omega_{i'})} \frac{\pi}{\sqrt{3}} g(\epsilon, \epsilon') \tag{122}$$

with

$$R_{if} = \int_0^\infty R_i(r)\, R_f(r)\, r^3 dr. \tag{123}$$

The summations in (122) do not include any magnetic quantum numbers and, besides, one must have $l_k = l_i \pm 1$, $l_f = l_i \pm 1$. The final electron energy ϵ' still appearing as argument of the Gaunt factor is given by energy conservation,

$$\epsilon - \epsilon' = \hbar(\omega + \omega_f - \omega_k). \tag{124}$$

As the classical limit is always valid, g depends only on the variable ξ, which can usually be approximated by

$$\xi = \eta\hbar(\omega + \omega_f - \omega_k)/2\epsilon. \tag{125}$$

One sees that ξ vanishes for $\omega = \omega_{kf}$, which is the frequency of a forbidden line. The corresponding divergence in $F(\omega)$ will be discussed in the next section. If one stays away from these forbidden lines, the conditions of the Kramers approximation (ξ large) are often realized and one can replace the Gaunt factor by unity.

Expression (122) has been applied to the opacity problem by Pyatt and Stewart (1962). The continuous spectrum that it yields in the windows is much stronger than the light scattering given by the Kramers-Heisenberg (1925) formula and may be of the same order of magnitude as the contribution from bound-free transitions, especially in the vicinity of absorption edges. One has to reckon also with the quasi-static broadening arising in the windows from ion perturbers approaching very close to the light-emitting ion. But this is not as important as one might think, because the Coulomb repulsion between the two ions tends to keep them apart (see § 6).

To show the overlap between the one-electron approximation and the impact approximation, we shall use (122) to calculate the width of an isolated line $i \to f$. Only terms with $i' = i$ contribute. One can replace ρ_k by ρ_i as levels k and i are usually close. For unperturbed lines, the spectrum would be given by (5). After summation over magnetic quantum numbers, (5) becomes

$$F(\omega) = \sum_{if} \rho_i e^2 R_{if}^2 \max(l_i, l_f)\, \delta(\omega - \omega_{if}). \tag{126}$$

If a line is broadened, the δ-function is replaced by a Lorentz shape

$$\delta(\omega - \omega_{if}) \to \pi^{-1} w_i / [(\omega - \omega_{if} - d_i)^2 + w_i^2]. \tag{127}$$

Far from the line center, one can neglect the shift and the width in the denominator. Then, comparison with (122) yields

$$w_i = \int_0^\infty f(v)\, dv \frac{4\pi}{3} \frac{e^4 n}{\hbar^2 v} \sum_k \frac{\max(l_i, l_k)}{2l_i + 1} R_{ik}^2 \frac{\pi}{\sqrt{3}} g(\epsilon, \epsilon'), \tag{128}$$

which is identical with (113) with the understanding that, in (113), one has neglected the variation of the Gaunt factor with ω and taken it at the unperturbed line frequency. It is this neglect of the change in the Gaunt factor which constitutes the impact approximation.

5.3 Effect of Debye Shielding

As in the broadening of hydrogen lines, the logarithmic divergence arising at forbidden line frequencies is due to large impact parameters and disappears when shielding is taken into account. The recipe for taking into account shielding was already explained in § 4.2: if $|\omega + \omega_f - \omega_k|$ is smaller than ω_P, one should use ω_P in the calculation of ξ and the Gaunt factor. The recipe is only approximate as our treatment of Debye shielding was quite simplified and, in any case, the truth should certainly be a smooth function of ω. But $|\omega + \omega_f - \omega_k|$ enters only as the argument of a logarithm, so that this simple recipe is probably sufficient.

For example, consider the validity of the impact approximation applied to line $i \to f$, near which there is a forbidden line $k \to f$. If $\omega_P \gg |\omega_i - \omega_k|$, then the Debye cutoff is operative in the whole frequency region; the overlapping lines formulation of the impact approximation is valid everywhere and the effective perturbation $\mathscr{H}$ is the same as for hydrogenic levels. But if $\omega_P \ll |\omega_i - \omega_k|$, the impact approximation is good only in the immediate vicinity of the allowed line; to calculate the window between it and the forbidden line, the variation of the Gaunt factor with ω must be included. The same results hold for lines emitted by neutrals, with the logarithm of Eq. (100) replacing $\pi g/\sqrt{3}$.

It is worthwhile repeating this discussion from a somewhat more classical viewpoint. In § 5.1, we used the collision time τ as a measure of the time for which the electric field is correlated. Because of the long range correlations, we should really have used ω_P^{-1} which is usually larger. This is of special importance for the calculation of hydrogen lines (Lewis, 1961). Actually, the impact approximation stops being valid for frequencies of order ω_P which, for typical hydrogen lines, is not very far in the wing at all.

For a general discussion of the correlations, one can restrict oneself to weak collisions, since this is where the trouble arises. We shall use a general formula for $F(\omega)$ which does not make the impact approximation, but includes the perturbing field in second order only. It gives the contribution of the weak collisions to the wings and the windows, just like the one-electron approximation combined with perturbation theory in the previous section. We neglect interaction in the final state and take its frequency as origin. According to (29), the second order term in $\Phi(s)$ is

$$\Phi(s) = -\hbar^{-2} \sum_{ii'k\sigma\nu} \langle i' \mid D \mid i \rangle \langle i \mid d_\sigma \mid k \rangle \langle k \mid d_\nu \mid i' \rangle$$

$$\times \int_0^s dt \int_0^t dt' e^{-i\omega_i(s-t)} e^{-i\omega_k(t-t')} e^{-i\omega_{i'}t'} [E_\sigma(t) E_\nu(t')]_{\text{Av}} \tag{129}$$

and the corresponding $F(\omega)$ is given by (9). The average of the product of two electric fields depends only on the difference $t - t'$. If one changes to the variables $s - t = u$, $t - t' = \tau$, $t' = v$, all of which vary from 0 to $+\infty$, the integrals over u and v are trivial and one finds

$$F(\omega) = \pi^{-1} \mathscr{R} \sum_{ii'k\sigma\nu} (\omega - \omega_i)^{-1} (\omega - \omega_{i'})^{-1}$$

$$\times \langle i' \mid D \mid i \rangle \langle i \mid d_\sigma \mid k \rangle \langle k \mid d_\nu \mid i' \rangle \int_0^\infty d\tau e^{i(\omega-\omega_k)\tau} [E_\sigma(\tau) E_\nu(0)]_{\text{Av}}. \tag{130}$$

For an isolated line, ω_i and $\omega_{i'}$ are the same. The impact approximation consists in replacing ω by ω_i in the τ integral; then, one gets back the wing of a Lorentz line, with the width of § 4.2. Therefore, the proper way to correct the impact theory is this: wherever ω_{ik} occurs, one must replace it by $\omega - \omega_k$ or, if the unperturbed line is now taken as origin, $\omega + \omega_{ik}$. In particular, the cutoff $\rho_{\max}$ to use in (100) is either $1.1 r_D$ or $1.1 v / |\, \omega + \omega_{ik} \,|$, whichever comes first.

In the hydrogen case where $\omega_{ik} = 0$, this means that, for $|\, \omega \,| > \omega_P$, the weak collision logarithm must be reduced by an amount $\log (|\, \omega \,| / \omega_P)$. Since the original logarithm is usually between 2 and 5, this means that the reduction in the electron contribution to wing broadening may be appreciable for frequencies that are several times the plasma frequency. Frequencies of astrophysical interest are actually in that range. This correction was not included in published work on wing broadening of hydrogen lines (Griem *et al.*, 1959; Griem, 1960); however, it is partly compensated by the fact that the contribution from strong collisions was not included either.

6 *Broadening by Ions*

If the quasi-static approximation is going to be made, one must first calculate the Stark splitting of the line considered. This is standard (Bethe and Salpeter, 1957). In hydrogenic atoms, one has linear Stark effect; in other atoms, it is quadratic for small fields and linear for large ones. The same radial integrals (95) or (123) enter in this calculation as already occurred in the electron contribution.

Next one must know the electric field distribution. This was calculated long ago by Holtsmark (1919) for independent ions. Actually, the ions are correlated within themselves and also with the electrons. The importance of these correlations is measured by the quantity r_0/r_D, with $4\pi r_0^3/3 = n^{-1}$. In all shock tube or arc experiments, r_0/r_D is in the neighborhood of $\frac{1}{2}$ so that the correction is important. Ecker and Müller (1958) proposed to include it by replacing the Coulomb field (85) used by Holtsmark by the Debye field

$$(e\mathbf{r}_1/r_1^3)\,(1 + r_1/r'_D)\exp\,(-\,r_1/r'_D). \tag{131}$$

They want r'_D to represent shielding of the field of an ion both by the other ions and the electrons, hence they take

$$r'_D = (kT/8\pi ne^2)^{1/2} \tag{132}$$

if the ions are singly charged. This procedure is questionable when it comes to represent the shielding by the other ions, since the Debye shielded field is a long-time average while one desires the distribution of the instantaneous field. Mozer and Baranger (1960) showed that one could take the ion-ion correlations into account by means of a cluster expansion and obtained ionic field distributions both at a neutral atom (Fig. 4) and at a singly charged ion. Actually, their results are not very different from those of Ecker and Müller.

To investigate the validity of the quasi-static assumption (Spitzer, 1940; Holstein, 1950), one remarks that the motion of an ion introduces into the line shape Fourier components of order τ^{-1}, where τ is the ion collision time. If δ represents the splitting caused by the field of this ion, motion can be neglected if

$$\delta \gg \tau^{-1}. \tag{133}$$

For a very close ion, τ^{-1} is large, but δ increases faster than τ^{-1} so that the quasi-static picture is better for close ions. If the width of a line is large compared to a typical τ^{-1}, the quasi-static approximation is valid for the whole line. Even if this is not the case, the quasi-static approximation is still valid in the wings of a line, as soon as condition (133) becomes satisfied. This is very useful in astrophysical applications, where one is often interested only in the wings.

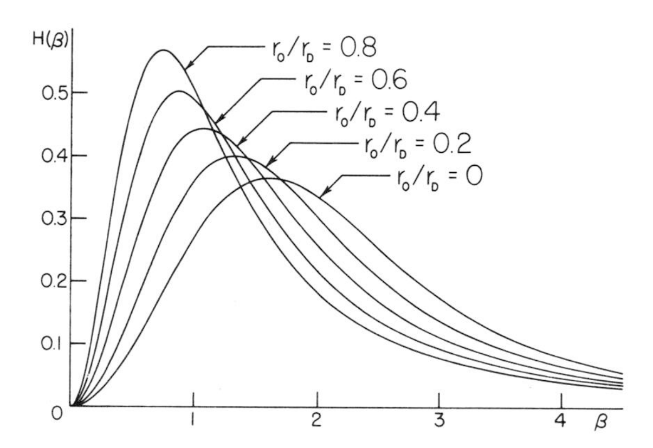

FIG. 4. Probability distribution of the ionic field at a neutral point, for various values of the parameter r_0/r_D. For notations, see text following (134). The case $r_0/r_D = 0$ is the Holtsmark distribution (after Mozer and Baranger, 1960).

The large electric fields associated with the far wings are due to a single ion approaching very close. Questions of correlation are then irrelevant and the field distribution tends toward the asymptotic form of the Holtsmark distribution,

$$H(\beta) \approx 1.5\beta^{-3/2}. \tag{134}$$

Here, β is the electric field in units of e/r_0^2 and the normalization is

$$\int_0^\infty H(\beta)\, d\beta = 1. \tag{135}$$

However, if the line is emitted by an ion, the repulsion between the two ions must be taken into account and one finds for singly charged ions the asymptotic form (Lewis and Margenau, 1958)

$$H(\beta) \approx 1.5\beta^{-3/2} \exp\left[-\tfrac{1}{3}(r_0/r_D)^2 \beta^{1/2}\right]. \tag{136}$$

The wings of ion lines are thus cut down by the Coulomb repulsion. In a window calculation such as that of § 5.2, the contribution of ion perturbers is then small compared to that of the electrons.

In nonhydrogenic atoms, for which the Stark splitting δ is smaller than in hydrogen, one often happens to be in the case where neither the quasi-static nor the impact limit is valid for the ions, as far as the line center is concerned. Fortunately, this is the case where ion broadening is smaller than electron broadening so that some pretty crude approximations can be made. Griem *et al.* (1962a) have taken ion motion into account by assuming that ion interactions with the atom are scalarly additive. This is not strictly correct; one must add the electric fields, not the interactions themselves. But it is true in the impact limit, as we have seen repeatedly. It is also true in the wings in the quasi-static limit, as discussed above. Hence it can be expected to be a reasonable compromise when the quasi-static approximation does not hold in the line center, especially since the over-all effect is small. The above authors neglected also the rotational degeneracy, treating all states like s states. Then, the adiabatic approximation can be used, since ions are much slower than electrons, and the line shape is given by the phase integral formula (Anderson, 1952)

$$\Phi(s) = e^{-ids-ws} \left[\left\{ \exp\left(- i\hbar^{-1} \int_0^s v_1(t)\, dt \right) \right\}_{\mathrm{Av}} \right]^N \tag{137}$$

Here, $v_1(t)$ is the interaction of the atom in its upper state with a single ion and the average is over all types of collisions. The first factor in $\Phi(s)$ takes into account the width and shift due to electrons, as calculated in § 4. Griem *et al.* (1962a) actually integrated and Fourier transformed expression (137) numerically, assuming straight trajectories for the ions. Debye shielding was not taken into account since this is only a small correction.

7 *Comparison with Experiment*†

The last few years have seen the simultaneous completion of realistic line shape calculations, based on the theory developed in earlier sections, and of clean and reproducible experiments against which this theory can be checked. In this section, we shall give a few examples of the agreement between the two. Further details can also be found in Chapter 5 § 6.1, § 6.2, and especially § 7.4, where the results of some recent calculations on H_β and the He I lines are given.

† The suggestions of Dr. H. R. Griem concerning this section are gratefully acknowledged.

Figure 5 shows a comparison with theory of an H_β profile arising in a water-stabilized arc (Bogen, 1957). The temperature $T = 10{,}400°K$ and the electron density $n = 2.2 \times 10^{16}\ cm^{-3}$ were determined by Bogen from measured absolute line intensities, assuming thermal

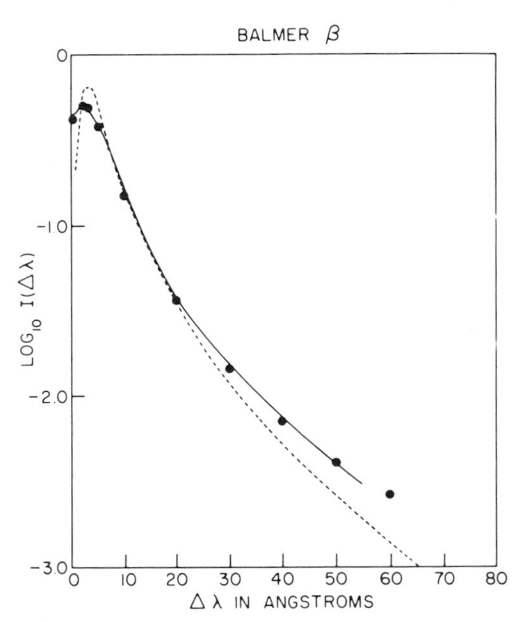

FIG. 5. The solid line is the H_β profile of Bogen (1957). The black dots are theoretical points from Mozer (1960). The broken line is the Holtsmark theory.

equilibrium. The theoretical points are calculated for this density; only the total intensity is adjusted. Although Bogen's profile might be subject to large errors in the far wings, the agreement is seen to be good. On the other hand, a calculation done with the Holtsmark theory, i.e., without including any electron contribution and without shielding the ion field, is in definite disagreement with experiment.

Figures 6 and 7 refer to neutral helium lines. The experiment was performed with a pulsed arc by Wulff (1958). The temperature is about 30,000°K, but he cannot really determine the electron density independently. However, since he has many lines, one can try to fit a theoretical profile to each line and see if the various values of the density so obtained are consistent. They are indeed, the deviations from the mean value of $2.5 \times 10^{16}\ cm^{-3}$ being at most 10% in all cases. This agreement is not trivial, as can be seen by repeating the fit with the adiabatic theory (Lindholm, 1941; Foley, 1946; Unsöld, 1955), in which case deviations from the mean value by factors of two are found. In the figures, the theoretical profiles have been computed for this mean density, $n = 2.5 \times 10^{16}$ cm^{-3}. Figure 6 is an isolated line, while in Fig. 7 a small forbidden component is beginning to show up and is rather well reproduced by theory.

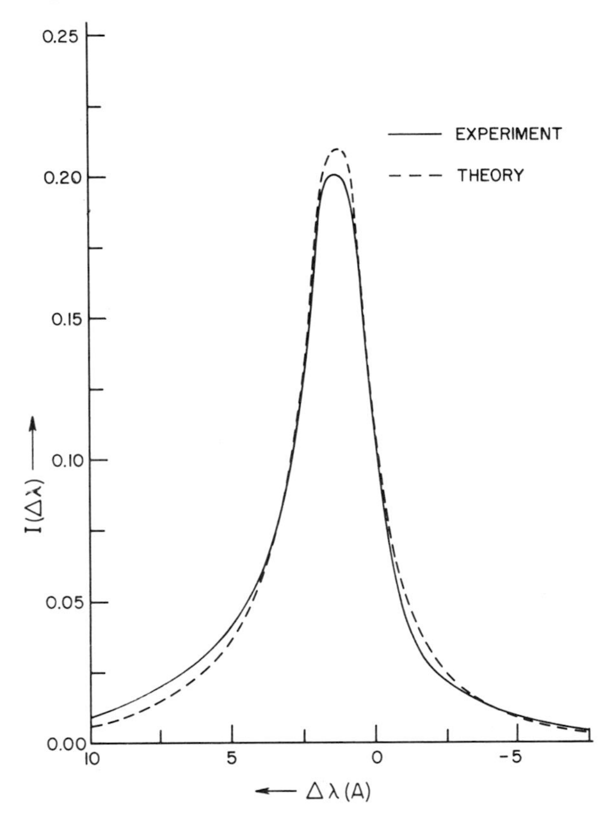

FIG. 6. Experimental and theoretical profiles of He I 4713 A (Griem *et al.*, 1962a).

The H_γ profile in Fig. 8 was obtained by Wiese *et al.* (1962) in a wall-stabilized arc. They fitted the theoretical profile to half-width and maximum intensity, from which an electron density $n = 8.45 \times 10^{16}$ cm^{-3} was deduced. However, there is an independent measurement of n: a temperature of 12,530°K is deduced from the absolute intensity of H_γ and n follows by assuming thermal equilibrium. The accuracy of this method is believed to be better than 10%, and it yields $n = 8.30 \times 10^{16}$ cm^{-3}. Figure 9 is an H_β profile by Shumaker and Wiese (1961). Here again, they sought the density that gives the best fit with the theore-

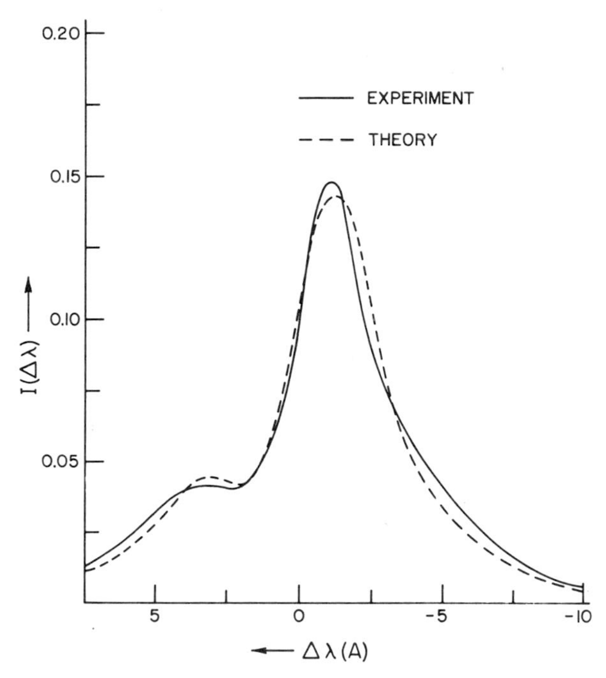

FIG. 7. Experimental and theoretical profiles of He I 3965 A (Griem *et al.*, 1962a).

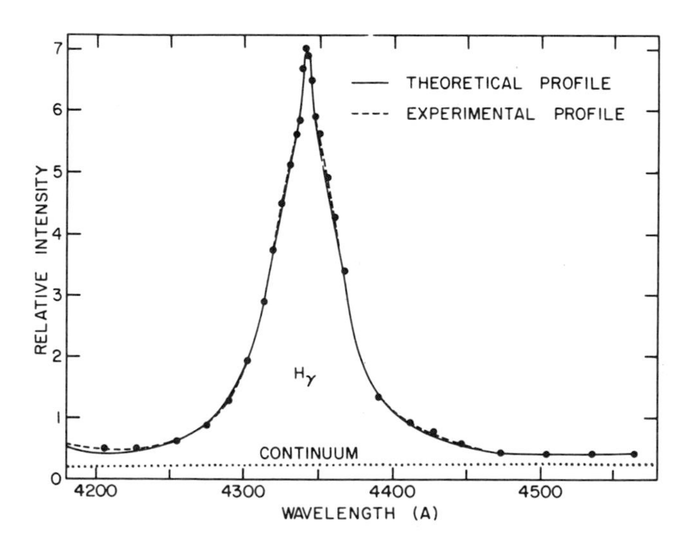

FIG. 8. Experimental and theoretical profiles of H_γ (Wiese *et al.*, 1962), communicated before publication by Dr. Wiese).

tical profile and found $n = 9.2 \times 10^{16}$ cm^{-3}. The point of the experiment was to show that the measurement of line profiles or half-widths can be used as a thermometer. The H_β line arises from a trace of hydrogen introduced in an oxygen arc. From the density n, they deduce a temperature of 12,830°K by assuming thermal equilibrium. This compares

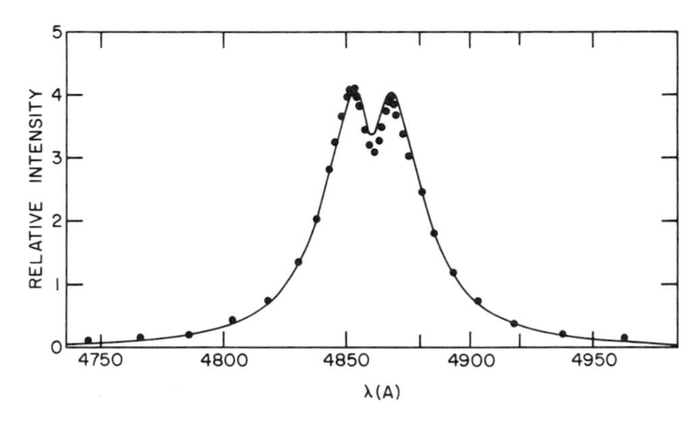

FIG. 9. Experimental points and theoretical line for H_β (Shumaker and Wiese (1961), communicated before publication by Dr. Wiese).

very well with the value of 12,710°K obtained from absolute line intensities. Generally, an error of 10% in the electron density will correspond to only 2% in the temperature. Similar work, but in shock tubes, has been done by Doherty (1961). In Fig. 9, one can also notice the asymmetry between the blue and the red wing of the line. This is understood as arising from several small effects: quadratic Stark effect, the factor ω^4 in (4), the changeover from frequency to wavelength scale (Griem, 1954).

In the past, arc experiments have tended to be more accurate than those done with shock tubes, but there are indications that this may not continue to be the case. Figures 10-12 show Balmer line profiles obtained in shock tubes by Berg *et al.* (1962). Electron densities were determined from absolute continuum intensities and temperatures from relative line and continuum intensities. These density and temperature were used to calculate the theoretical profiles; only the absolute intensity was adjusted. The profiles were gotten by scanning with monochromators from shot to shot and there results a certain scattering in the points, but the half-widths seem to be quite reproducible. The same authors have obtained similar profiles for neutral and ionized helium lines. Table I gives a summary of their findings. The errors accompanying the calculated widths and shifts are only a reflection of the error in the

TABLE I

COMPARISON OF MEASURED AND CALCULATED WIDTHS AND SHIFTS FOR VARIOUS HYDROGEN AND HELIUM LINES[a]

Line	Transition	Density (10^{17} cm^{-3})	Temperature (10^3 °K)	Full half-width (A)		Ratio of meas. and calc. widths	Shift of maximum (A)	
				Measured	Calculated		Measured	Calculated
H_α	2-3	0.92 ± 0.05	25 ± 2	7.8 ± 1.0	8.1 ± 0.3	0.97 ± 0.15	—	—
H_β	2-4	0.80 ± 0.04	14 ± 1	40.7 ± 2.0	41.3 ± 1.2	0.99 ± 0.08	—	—
H_γ	2-5	0.94 ± 0.05	14 ± 1	49.2 ± 2.5	50.4 ± 1.5	0.98 ± 0.08	—	—
He I 5876	2^3P-3^3D	1.59 ± 0.15	49 ± 1	5.5 ± 0.6	6.2 ± 0.6	0.89 ± 0.20	+ 0.7 ± 1.0	− 0.6 ± 0.1
He I 5876	2^3P-3^3D	1.26 ± 0.12	44 ± 1	4.9 ± 0.5	5.2 ± 0.5	0.94 ± 0.20	0.0 ± 1.0	− 0.5 ± 0.1
He I 5016	2^1S-3^1P	1.65 ± 0.08	24 ± 2	13.0 ± 1.3	14.5 ± 0.7	0.90 ± 0.15	− 4.8 ± 1.0	− 6.0 ± 0.3
He I 4713	2^3P-4^3S	1.30 ± 0.07	20 ± 2	14.0 ± 1.4	15.2 ± 0.7	0.92 ± 0.15	+ 6.0 ± 1.0	+ 8.9 ± 0.4
He I 4471	2^3P-4^3D	1.30 ± 0.07	20 ± 2	45.0 ± 4.5	46.6 ± 1.5	0.97 ± 0.15	− 4.5 ± 1.0	—
He I 3889	2^3S-3^3P	1.50 ± 0.08	26 ± 2	4.5 ± 0.5	4.2 ± 0.2	1.07 ± 0.15	+ 1.2 ± 1.0	+ 1.1 ± 0.1
He I 3188	2^3S-4^3P	1.50 ± 0.08	29 ± 2	13.4 ± 1.3	12.6 ± 0.7	1.06 ± 0.15	+ 4.1 ± 1.0	+ 3.8 ± 0.2
He II 4686	3-4	1.59 ± 0.15	49 ± 1	4.9 ± 0.5	5.1 ± 0.5	0.96 ± 0.20	− 2.2 ± 1.0	− 0.9 ± 0.1
He II 4686	3-4	1.26 ± 0.12	43 ± 1	4.2 ± 0.4	4.4 ± 0.5	0.97 ± 0.20	− 1.0 ± 1.0	− 0.7 ± 0.1
He II 3203	3-5	1.22 ± 0.12	47 ± 1	14.0 ± 1.4	13.8 ± 1.4	1.01 ± 0.20	—	—

[a] Table from Berg *et al.* (1962), communicated before publication by Dr. Griem.

independent measurement of n and do not include any estimate of errors in the theory. It is seen that one can expect agreement between the measured and calculated half-widths within 5% for hydrogen and 10% for helium. The agreement between the shifts is not so good. For those

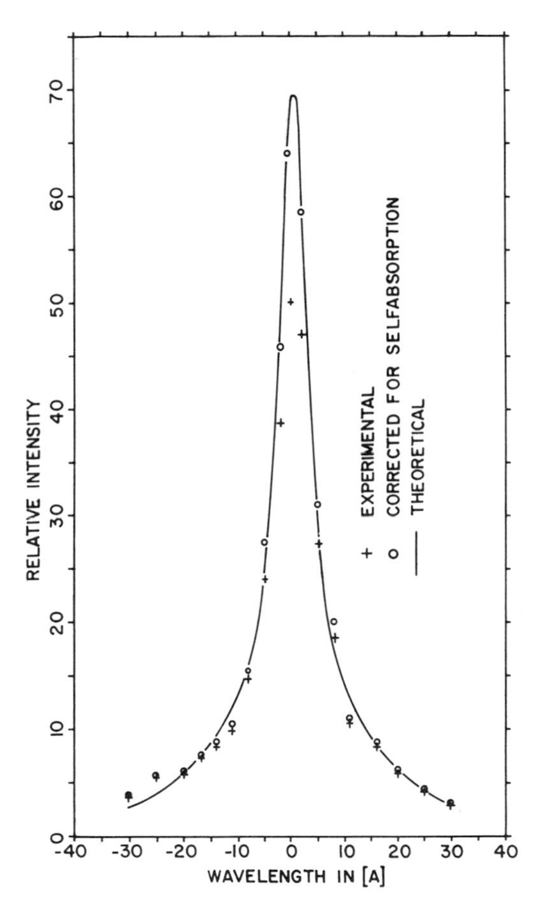

FIG. 10. Experimental and theoretical profiles of H_α (Berg *et al.*, 1962), communicated before publication by Dr. Griem).

few shifts that are significant, the calculated value seems to be a little too large. The reason for this is not clear at the moment, but it might be pointed out that the shift of neutral helium lines receives a large contribution from the ions, the calculation of whose effects, outlined at the end of § 6, is extremely crude. Fortunately, the width is very insensitive to the way in which the ions are treated.

In conclusion, one can say that, in the range of temperatures and densities covered by the above experiments, the measurement of line widths (not line shifts) appears to be the simplest way to determine the electron density, its accuracy being comparable to that of the other

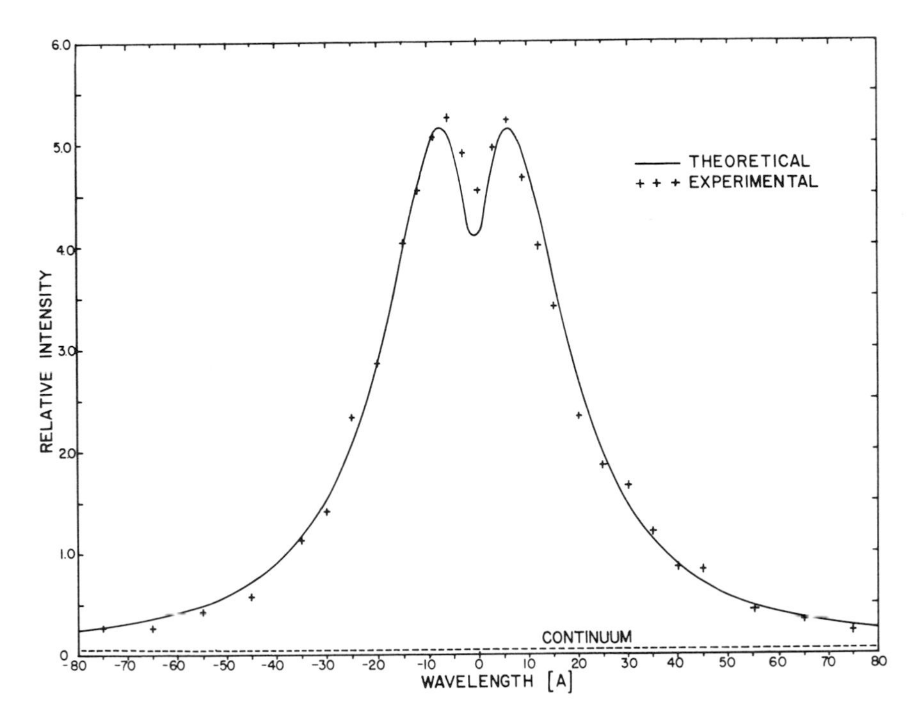

FIG. 11. Experimental and theoretical profiles of H_β (Berg *et al.*, 1962), communicated before publication by Dr. Griem).

methods. Moreover, in conjunction with the equations of thermal equilibrium if they hold, and if the pressure and chemical composition are known, it is also one of the best thermometers.

We end by giving references to the most complete calculations of theoretical profiles. They are: for H_β, Griem *et al.* (1962b); for other hydrogen lines, Griem *et al.* (1960); for neutral helium, Griem *et al.* (1962a); for ionized helium, Griem *et al.* (1962c). No detailed calculations have yet been done for other atoms, although in most cases this is possible along the lines of § 4.2 or § 4.4. As pointed out there, such calculations require an estimate of the oscillator strength between two close excited levels. This can be obtained, for instance, by the method of Bates and Damgaard (1949).

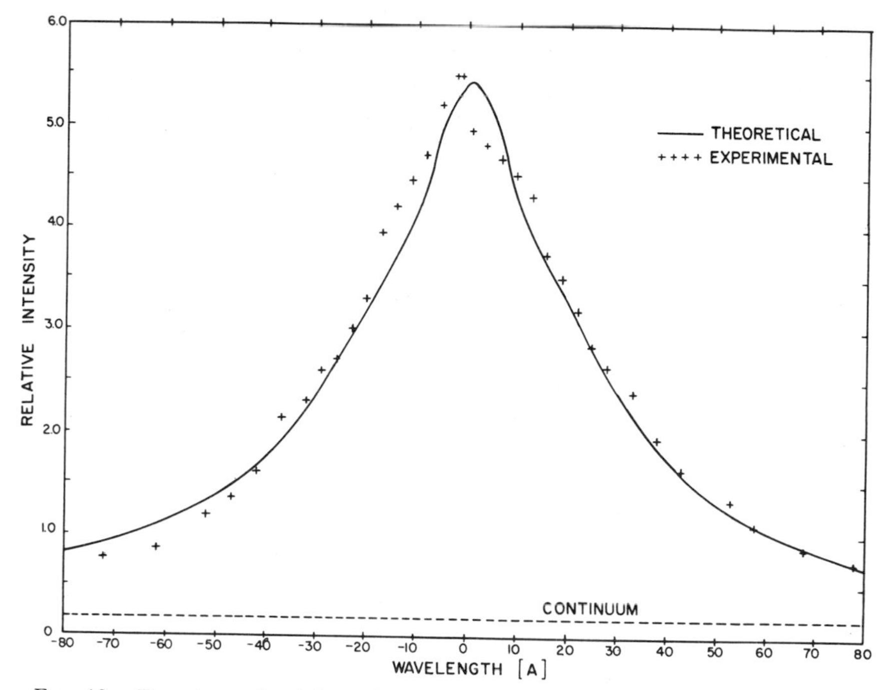

FIG. 12. Experimental and theoretical profiles of H_γ (Berg *et al.*, 1962), communicated before publication by Dr. Griem).

REFERENCES

Alder, K., Bohr, A., Huus, T., Mottelson, B., and Winther, A. (1956) *Rev. mod. Phys.* **28**, 432.

Anderson, P. W. (1949) *Phys. Rev.* **76**, 647.

Anderson P. W. (1952) *Phys. Rev.* **86**, 809.

Baranger, M. (1958a) *Phys. Rev.* **111**, 481.

Baranger, M. (1958b) *Phys. Rev.* **111**, 494.

Baranger, M. (1958c) *Phys. Rev.* **112**, 855.

Baranger, M., and Stewart, J. C. (1962) To be published.

Bates, D. R., and Damgaard, A. (1949) *Phil. Trans.* **A242**, 101.

Berg, H. F., Ali, A. W., Lincke, R., and Griem, H. R. (1962) Measurement of stark profiles of neutral and ionized helium and hydrogen lines from shock-heated plasmas in electromagnetic T-tubes. *Phys. Rev.* Jan.

Bethe, H. A., and Salpeter, E. E. (1957) "Quantum Mechanics of One- and Two-Electron Atoms." Academic Press, New York.

Bloom, S., and Margenau, H. (1953) *Phys. Rev.* **90**, 791.

Bogen, P. (1957) *Z. Phys.* **149**, 62.

Born, M. (1933) "Optik," p. 432. Springer, Berlin.

Breene, R. G., Jr. (1957) *Rev. mod. Phys.* **29**, 94.

Burkhardt, G. (1940) *Z. Phys.* **115**, 592.
Chandrasekhar, S. (1957) "An Introduction to the Study of Stellar Structure." Dover, New York.
Ch'en, S., and Takeo, M. (1957) *Rev. mod. Phys.* **29**, 20.
Doherty, L. R. (1961) Ph. D. thesis, University of Michigan, Ann Arbor, Michigan.
Ecker, G., and Müller, K. G. (1958) *Z. Phys.* **153**, 317.
Foley, H. M. (1946) *Phys. Rev.* **69**, 616.
Gaunt, J. A. (1930) *Phil. Trans.* **A229**, 163.
Griem, H. R. (1954) *Z. Phys.* **137**, 280.
Griem, H. R. (1959) *Bull. Amer. Phys. Soc.* [II] **4**, 236.
Griem, H. R. (1960) *Astrophys. J.* **132**, 883.
Griem, H. R., and Shen, K. Y. (1961a) *Phys. Rev.* **122**, 1490.
Griem, H. R., and Shen, C. S. (1961b) Application of a dispersion relation to the electron impact widths and shifts of isolated spectral lines from neutral atoms (to be published).
Griem, H. R., Kolb, A. C., and Shen, K. Y. (1959) *Phys. Rev.* **116**, 4.
Griem, H. R., Kolb, A. C., and Shen, K. Y. (1960) *U. S. Naval Research Lab. Rep. NRL* **5455**.
Griem, H. R., Baranger, M., Kolb, A. C., and Oertel, G. (1962a) Stark broadening of neutral helium lines in a plasma. *Phys. Rev.* Jan.
Griem, H. R., Kolb, A. C., and Shen, K. Y. (1962b) Stark profile calculations for the H_β line of hydrogen. *Astrophys. J.* Jan.
Griem, H. R., Kolb, A. C., and Shen, K. Y. (1962c) *U. S. Naval Research Lab. Rep.* (to be published).
Holstein, T. (1950) *Phys. Rev.* **79**, 744.
Holtsmark, J. (1919) *Ann. Phys. (Leipzig)* **58**, 577.
Jablónski, A. (1945) *Phys. Rev.* **68**, 78.
Kivel, B. (1955) *Phys. Rev.* **98**, 1055.
Kolb, A. C. (1957) Ph. D. thesis, University of Michigan, Ann Arbor, Michigan.
Kolb, A. C., and Griem, H. R. (1958) *Phys. Rev.* **111**, 514.
Kramers, H. A. (1923) *Phil. Mag.* **46**, 836.
Kramers, H. A., and Heisenberg, W. (1925) *Z. Phys.* **31**, 681.
Lenz, W. (1933) *Z. Phys.* **80**, 423.
Lewis, M. (1961) *Phys. Rev.* **121**, 501.
Lewis, M., and Margenau, H. (1958) *Phys. Rev.* **109**, 842.
Lindholm, E. (1941) *Ark. Mat. Astron. Fys.* **28B** (3).
Lindholm, E. (1942) Inaugural-Dissertation, Stockholm.
Lindholm, E. (1945) *Ark. Mat. Astron. Fys.* **32A** (17).
Lorentz, H. A. (1906) *Proc. Roy. Acad. Sci. (Amsterdam)* **8**, 591.
Margenau, H., and Lewis, M. (1959) *Rev. mod. Phys.* **31**, 569.
Mozer, B. (1960) Ph. D. thesis, Carnegie Institute of Technology, Pittsburgh, Pennsylvania.
Mozer, B., and Baranger, M. (1960) *Phys. Rev.* **118**, 626.
Pyatt, K. D., Jr., and Stewart, J. C. (1962) To be published.
Schiff, L. I. (1955) "Quantum Mechanics," 2nd ed. McGraw-Hill, New York.
Shumaker, J. B., and Wiese, W. L. (1961) *In* "Proceedings of the 3rd Symposium on Temperature, Its Measurement and Control in Science and Industry. Reinhold, New York.
Spitzer, L. (1940) *Phys. Rev.* **58**, 348.
Traving, G. (1960) "Über die Theorie der Druckverbreiterung von Spektrallinien." G. Braun, Karlsruhe.

Unsöld, A. (1955) "Physik der Sternatmosphären." Springer, Berlin.
Vainshtein, L. A., and Sobel'man, I. I. (1959) *Optika i Spektrosk.* **6**, 440.
Weisskopf, V. F. (1932) *Z. Phys.* **75**, 287.
Wiese, W. L., Paquette, D. R., and Solarski, J. E. (1962) *In* "Proceedings of the Fifth International Conference on Ionization Phenomena in Gases, Munich, 1961." North-Holland, Amsterdam (to be published in 2 volumes).
Wulff, H. (1958) *Z. Phys.* **150**, 614.

14.

Theoretical Treatment of Collisions between Atomic Systems

D. R. Bates

Collisions between atomic systems differ markedly from collisions involving free electrons. They are more complicated in that a wider variety of reaction path may be followed: for example, the projectile as well as the target may have structure and hence may undergo excitation or ionization; and again, in addition to the possibility of electron exchange, the possibility of charge transfer arises. Further, owing to nuclei being much more massive than an electron, the energy of relative motion remains above the inelastic thresholds down to much lower velocities of relative motion.† This is of great importance since the velocity, rather than the energy, is the parameter determining whether many of the approximations which might be suggested may properly be used; and in many cases the lower its value the more refined must be an approximation if it is to be acceptable. Fortunately, the massive nature of nuclei also enables some simplification to be effected. Because of it,

† A proton with the same velocity as an electron which is just able to ionize a hydrogen atom has an energy of about 25 kev.

the nuclei of the colliding systems may be regarded as classical particles except at very low velocities of relative motion; indeed, in many cases they may even be regarded as classical particles of infinite mass.

1 First Born Approximation

When the velocity of relative motion of the colliding systems is high, the cross section for a given process may be obtained by the use of the first Born approximation. In this approximation it is assumed that the incident and scattered waves associated with the relative motion are plane, or, alternatively, that the nuclei are classical particles of infinite mass; it is assumed that all the transition matrix elements are weak; and it is further assumed that there is no exchange of electrons between the systems. The simplified form of the first Born approximation known as the Bethe approximation (cf. Mott and Massey, 1949; Seaton, 1955) will not be discussed.

1.1 Wave Treatment

The wave treatment is formally the same as in the case of an encounter with a free electron. As is well known, it gives the cross section for a collision in which there is a transition from an initial state p to a final state q to be

$$Q(p, q) = \frac{M^2 v_q}{2\pi\hbar^4 v_p} \int_{-1}^{1} |\mathcal{N}|^2 \, d(\cos\theta) \tag{1}$$

where

$$\mathcal{N} = \int \Psi_p^* V \Psi_q \, d\tau \tag{2}$$

$$\cos\theta = \hat{\mathbf{v}}_p \cdot \hat{\mathbf{v}}_q, \tag{3}$$

M is the reduced mass of the colliding systems, $\mathbf{v}_p$ and $\mathbf{v}_q$ are the velocities of relative motion when the states indicated by the subscripts are occupied, Ψ_p and Ψ_q are the corresponding wave functions suitably normalized, and V is the interaction potential.

1.1.1 *Excitation and Ionization of Atoms by Bare Nuclei*

Taking the change of momentum instead of the change in the direction of the relative velocity vector to be the variable in the integral, (1) becomes

$$Q(p, q) = \frac{1}{2\pi\hbar^2 v_p^2} \int_{K_{\min}}^{K_{\max}} |\mathcal{N}|^2 K \, dK \tag{4}$$

with

$$\mathbf{K} = \boldsymbol{\varkappa}_p - \boldsymbol{\varkappa}_q \tag{5}$$

$$\boldsymbol{\varkappa}_p = M\mathbf{v}_p/\hbar, \qquad \boldsymbol{\varkappa}_q = M\mathbf{v}_q/\hbar. \tag{6}$$

It is seen that

$$K_{\min} = \kappa_p - \kappa_q \tag{7}$$

$$\simeq \frac{\Delta E}{\hbar v_p}\left(1 + \frac{\Delta E}{2Mv_p^2}\right), \tag{8}$$

ΔE being the threshold energy of the process; and that

$$K_{\max} = \kappa_p + \kappa_q \tag{9}$$

which is large enough to be taken as infinite. The values of the limits differ from those in the electron case where $K_{\min}$ is seriously underestimated by (8) and $K_{\max}$ is not effectively infinite unless the velocity of relative motion is high. An obvious consequence of the difference is that the cross section ${}^pQ$ for an inelastic proton-atom collision approaches the cross section ${}^eQ$ for the corresponding electron-atom collision from *above* as the velocity is increased. The cross sections are related quantitatively. Using (4) it may readily be shown that ${}^eQ$ at velocity v may be expressed in terms of ${}^pQ$ at velocities v_1 and v_2 such that

$$v_1 = v/(1 + \delta) \tag{10}$$

and

$$v_2 = \delta v_1 \tag{11}$$

where

$$\delta = \Delta E/2mv_1^2 \qquad (\leqslant 1), \tag{12}$$

m being the electronic mass; thus,

$${}^eQ(v) = \frac{1}{(1 + \delta)^2}\{{}^pQ(v_1) - \delta^2\, {}^pQ(v_2)\} \tag{13}$$

(Bates and Griffing, 1953).

In the simplest group of processes of the type under consideration, the target A is a nucleus of charge $Z_a e$ with a single orbital electron and the projectile B is a bare nucleus of change $Z_b e$ so that

$$\Psi_p = \chi_p(\mathbf{r}_a) \exp(i\boldsymbol{\varkappa}_p \cdot \mathbf{R}), \qquad \Psi_q = \chi_q(\mathbf{r}_a) \exp(i\boldsymbol{\varkappa}_q \cdot \mathbf{R}) \tag{14}$$

$$V = e^2 \left\{\frac{Z_a Z_b}{R} - \frac{Z_b}{|\mathbf{R} - \mathbf{r}_a|}\right\} \tag{15}$$

where $\mathbf{r}_a$ and $\mathbf{R}$ are the position vectors of the electron and of nucleus B relative to nucleus A and where χ_p and χ_q are the relevant atomic eigenfunctions normalized to unity as usual. Substitution in (4) yields

$$Q(p, q) = \left[\frac{8Z_b^2}{s^2} \int_{t_{\min}}^{\infty} |\mathscr{I}(p, q)|^2 t^{-3} \, dt\right] \pi a_0^2 \tag{16}$$

with

$$\mathscr{I}(p, q) = \int \chi_p^*(\mathbf{r}_a) \chi_q(\mathbf{r}_a) \exp(-i\mathbf{t} \cdot \mathbf{r}_a) \, d^3\mathbf{r}_a \tag{17}$$

$$s^2 = \tfrac{1}{2} m v_p^2 / I_{\mathrm{H}} \tag{18}$$

$$t_{\min} = \frac{\Delta E(p, q)}{2s} \left\{1 + \frac{m \Delta E(p, q)}{4Ms^2}\right\}, \tag{19}$$

r_a being here measured in units of a_0, the radius of the first Bohr orbit of hydrogen and $\Delta E(p, q)$ in units of I_{H}, the ionization potential of hydrogen.

Bates and Griffing (1953) have carried out calculations on transitions from the $1s$ state to the $2s$, $2p$, $3s$, $3p$, $3d$, and κ (continuum) states. The $\mathscr{I}$'s may be expressed in terms of

$$\tau = t/Z_a; \tag{20}$$

thus,

$$\begin{aligned}
|\mathscr{I}(1s, 2s)|^2 &= 2^{17}\tau^4/(4\tau^2 + 9)^6 \\
|\mathscr{I}(1s, 2p)|^2 &= 2^{15} \times 3^2\tau^2/(4\tau^2 + 9)^6 \\
|\mathscr{I}(1s, 3s)|^2 &= 2^8 \times 3^7(27\tau^2 + 16)^2 \tau^4/(9\tau^2 + 16)^8 \\
|\mathscr{I}(1s, 3p)|^2 &= 2^{11} \times 3^6(27\tau^2 + 16)^2 \tau^2/(9\tau^2 + 16)^8 \\
|\mathscr{I}(1s, 3d)|^2 &= 2^{17} \times 3^7\tau^4/(9\tau^2 + 16)^8
\end{aligned} \tag{21a}$$

and

$$|\mathscr{I}(1s, \kappa)|^2 = \frac{2^8\kappa\tau^2(1 + 3\tau^2 + \kappa^2) \exp\left[-(2/\kappa)\tan^{-1}\{2\kappa/(1 + \tau^2 - \kappa^2)\}\right]}{3\{1 + (\tau - \kappa)^2\}^3 \{1 + (\tau + \kappa)^2\}^3 \{1 - \exp(-2\pi/\kappa)\}} \tag{21b}$$

in which κ is such that $\kappa^2 I_{\mathrm{H}}$ is the energy of the ejected electron (Massey and Mohr, 1933). Integrals (16) for the Q's are best treated by numerical methods.

In the low-energy region defined by

$$\Delta E \ll E \ll 100 M \Delta E \tag{22}$$

where the energy of relative motion E and excitation energy ΔE are in electron volts and the reduced mass M is on the chemical (^{16}O) scale, the derived cross sections for collisions giving $1s \rightarrow 2s$ and $1s \rightarrow 2p$ transitions may be represented by

$$Q(1s, 2s) = [1.3 \times 10^{-6} Z_b^2 E^4 / M^4 \Delta E(1s, 2s)^6] \, \pi a_0^2 \tag{23}$$

and

$$Q(1s, 2p) = [7.4 \times 10^{-9} Z_b^2 E^5 / M^5 \Delta E(1s, 2p)^7] \, \pi a_0^2. \tag{24}$$

It is seen that the cross sections are small, are rapidly increasing functions of the velocity of relative motion, and, for a given class of transition, are rapidly decreasing functions of the excitation energy. According to the first Born approximation these features are a general characteristic of collisions between atomic systems. Many such collisions do indeed exhibit them; but many do not. Use of the first Born approximation in the low-energy region concerned is, of course, unjustified.

The cross sections may also be expressed in simple algebraic form at high energies. Let $\mathscr{E}$ denote the projectile's impact energy in kev (the target being taken to be stationary). Introducing

$$x = \mathscr{E} / M_b Z_a^2 \tag{25}$$

it is found that if the numerical value of this parameter is greater than about 100, then

$$Q(1s, 2s) = 11.1 \frac{Z_b^2}{x Z_a^4} \left(1 - \frac{7.8}{x}\right) \pi a_0^2 \tag{26}$$

and

$$Q(1s, 2p) = 128 \frac{Z_b^2}{x Z_a^4} \left(\log x - 1.185 + \frac{4.1}{x}\right) \pi a_0^2. \tag{27}$$

Results in the intermediate energy region not covered by (23) and (24) or by (26) and (27) are presented in Fig. 1 for the case of

$$\mathrm{H}(1s) + \mathrm{H}^+ \rightarrow \mathrm{H}(2s \text{ or } 2p) + \mathrm{H}^+. \tag{28}$$

The corresponding results for other processes of the family such as

$$\mathrm{Li}^{2+}(1s) + \mathrm{He}^{2+} \rightarrow \mathrm{Li}^{2+}(2s \text{ or } 2p) + \mathrm{He}^{2+} \tag{29}$$

may be obtained from Fig. 1 by scaling, since to the first Born approximation

$$Q(p, q \mid Z_a, Z_b, M_b, \mathscr{E}) = \frac{Z_b^2}{Z_a^4} Q\left(p, q \,\middle|\, 1, 1, 1, \frac{\mathscr{E}}{M_b Z_a^2}\right) \tag{30}$$

(Bates and Griffing, 1953). It may be shown that the existence of a

Coulomb repulsion between the systems for processes like (29) does not appreciably influence the inelastic collision cross sections unless the impact energy is very low (cf. Bates and Boyd, 1962a).

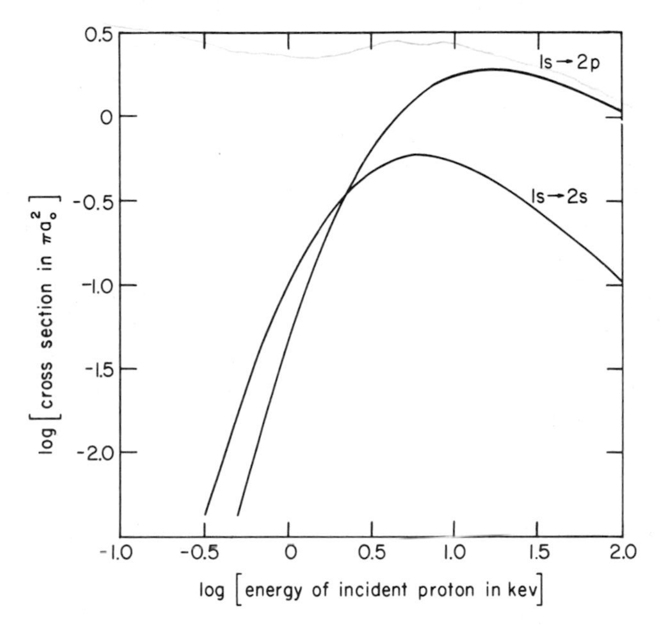

FIG. 1. First Born approximation to cross section—impact energy curves for $H(1s) + H^+ \rightarrow H(2s) + H^+$ and $H(1s) + H^+ \rightarrow H(2p) + H^+$ (Bates and Griffing, 1953).

Other excitation processes which have been treated by the first Born approximation are

$$\mathrm{He}\,(1s^2) + \mathrm{H}^+ \rightarrow \mathrm{He}\,(1s\,2p\,{}^1P) + \mathrm{H}^+, \tag{31}$$

(Moiseiwitsch and Stewart, 1954; Bell, 1961)

$$\mathrm{He}\,(1s^2) + \mathrm{H}^+ \rightarrow \mathrm{He}\,(1s\,3p\,{}^1P) + \mathrm{H}^+, \tag{32}$$

(Bell, 1961) and

$$\mathrm{Na}\,(3s) + \mathrm{H}^+ \rightarrow \mathrm{Na}\,(3p) + \mathrm{H}^+ \tag{33}$$

(Bell and Skinner, 1962). Results for (31) are presented later (Fig. 9).

Figure 2 gives the calculated cross sections for

$$\mathrm{H}\,(1s) + \mathrm{H}^+ \rightarrow \mathrm{H}^+ + e + \mathrm{H}^+ \tag{34}$$

(Bates and Griffing, 1953). Fite and associates (1960) have obtained rather smaller values in the laboratory (cf. Chapter 12, Fig. 26), but the measurements only cover the region below 40 kev.

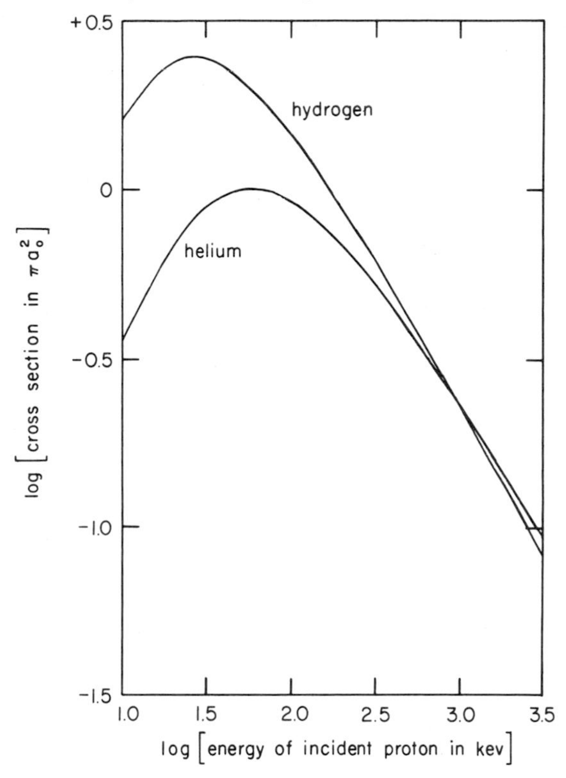

FIG. 2. First Born approximation to cross section—impact energy curves for $H(1s) + H^+ \rightarrow H^+ + e + H$ (Bates and Griffing, 1953) and $He(1s)^2 + H^+ \rightarrow He^+(1s) + e + H^+$ (Mapleton, 1958).

Erskine (1954) and Mapleton (1958) have investigated

$$He\,(1s^2) + X^{n+} \rightarrow He^+\,(1s) + e + X^{n+} \tag{35}$$

the former taking the projectile X^{n+} to be an alpha particle and the latter taking it to be a proton. On the first Born approximation the two cases are, of course, simply related, the cross section presented to an alpha particle being 4 times the cross section presented to a proton of the same velocity. Both described the ground state of the helium atom by the customary rather crude one-parameter variational wave function $\phi(r_1)\,\phi(r_2)$ with

$$\phi(r) = \left(\frac{\lambda^3}{\pi a_0^3}\right)^{1/2} \exp\,(-\,\lambda r/a_0), \qquad \lambda = 1.6875. \tag{36}$$

The main contribution to the ionization cross section comes from optically allowed transitions to p states of the continuum. It is therefore important that these states should be especially well represented. Erskine used Hartree wave functions. In order to take advantage of formula (21b) he was content to adopt Coulomb wave functions appropriate to a nucleus of charge λ† for the other states of the continuum. Mapleton adopted Coulomb wave functions for all states of the continuum, but he took the associated charge to be λ only for the s states and took it to be unity for the other states. The two sets of results are in very close agreement. Figure 2 shows those of Mapleton (which extend to lower energies than those of Erskine). There is fair accord with the experimental work of Fedorenko *et al.* (1960) at energies above about 50 kev. The significance of this is uncertain in view of the rather poor accuracy of the wave functions used.

The ionization of lithium by protons (McDowell and Peach, 1961) appears to be the only other case to have been treated in detail.

In some applications it is necessary to know the distribution in energy of the electrons which are ejected. This distribution is naturally obtained during the course of the calculation of the ionization cross section. A convenient means of displaying its main features is by a function $f(\mathscr{E} \mid \epsilon)$ giving the fraction of the electrons ejected by protons of energy $\mathscr{E}$ kev which have energy in excess of ϵI_H where I_H is as usual the ionization potential of hydrogen. Figure 3 shows $f(\mathscr{E} \mid \epsilon)$ for process (34) and for

$$Ne(2p)^6 + H^+ \rightarrow Ne^+(2p)^5 + e + H^+ \qquad (37)$$

(Bates *et al.*, 1957). It is apparent that $f(\mathscr{E} \mid \epsilon)$ is a more rapidly decreasing function of ϵ (at fixed $\mathscr{E}$) for the former than for the latter. This difference is as would be expected from the difference between the photoionization absorption continua of hydrogen and neon (which are determined by transition matrix elements related to those arising in impact ionization).

Inner-shell ionization. Ionization of the inner shells of atoms by protons or alpha particles is of considerable interest in that it leads to the emission of X-radiation. The perturbation arising from an encounter is weak since the charge on the projectile is small relative to the effective charge of the nucleus of the target. Consequently, the first Born approximation remains valid down to almost the threshold energy. A number

† This value of the charge was chosen to get orthogonality to the wave function of the ground state.

of comparisons showing the good agreement that exists between theory and experiment are given by Merzbacher and Lewis (1958) in their valuable survey of the field. Because of the lack of the necessary data, no comparison was made in the region near and beyond the maxima of the cross section versus energy curves. A systematic discrepancy may arise in this region due to simplifications introduced in the calculations.

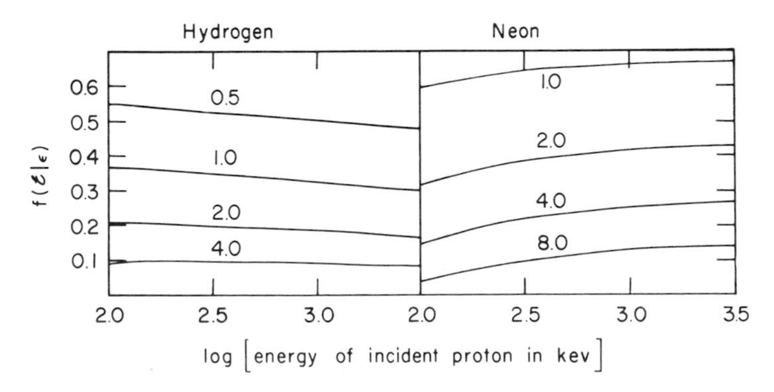

FIG. 3. Fraction of ejected electrons which have more kinetic energy than ϵ rydbergs where ϵ is as indicated on each curve. The set on the left refer to $H(1s) + H^+ \rightarrow H^+ + e + H^+$ (Bates and Griffing, 1953) and the set on the right to $Ne(2p)^6 + H^+ \rightarrow Ne^+(2p)^5 + e + H^+$ (Bates *et al.*, 1957).

The initial and final states were described by hydrogenic wave functions, the effective nuclear charge $Z_n e$ for the shell concerned being obtained from Slater's rules. If n is the principal quantum number of this shell, the associated ionization energy is $(Z_n/n)^2$ rydbergs, whereas the actual ionization energy is

$$I_n = \theta_n(Z_n/n)^2 \text{ rydbergs} \tag{38}$$

in which θ_n is a number less than unity. Some adjustment is needed to make the threshold of the calculated ionization curve occur at the correct energy. Bethe's procedure (cf. Henneberg, 1933) B, is to regard any inelastic collision in which the kinetic energy lost exceeds I_n as an ionization collision though the wave function for the final state is bound in character unless the kinetic energy lost exceeds $(Z_n/n)^2$. An alternative procedure due to Stobbe (1930) and to Massey and Mohr (1933), SMM, is to employ the actual ionization energy throughout the work despite its being inconsistent with the wave function adopted for the initial state. The calculations reported by Merzbacher and Lewis are based on the B procedure but most other calculations (including all those mentioned in this chapter) are based on the SMM procedure.

If the wave functions used are as indicated, the cross section Q_n^i describing the ionization of shell n by a nucleus of mass M_b, charge $Z_b e$, and energy E_b rydbergs may be expressed in the form

$$Q_n^i = \frac{Z_b^2}{Z_n^4} \mathscr{S}_n(\eta_n, \theta_n) \tag{39}$$

in which

$$\eta_n = E_b m / Z_n^2 M \tag{40}$$

and $\mathscr{S}_n(\eta_n, \theta_n)$ depends only on the variables indicated. Taking the K shell as an example, Arthurs (1959) has computed $\mathscr{S}_1$ as a function of η_1 for selected values of θ_1 using both the B and SMM procedures. In general, there is only close agreement in the region where the cross section is a rapidly increasing function of the energy (Fig. 4). Arthurs has adduced some evidence suggesting that the B procedure is inferior to the SMM procedure.

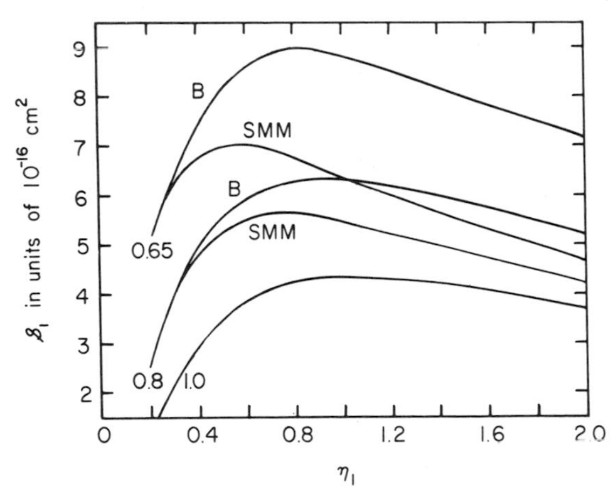

FIG. 4. Plots of $\mathscr{S}_1$ against η_1 for the values of θ_1 indicated on the curves: B, Bethe procedure; SMM, Stobbe-Massey-Mohr procedure (Arthurs, 1959).

Simultaneous excitation and ionization. A collision between a nucleus and a heavy atom may give an excited ion through inner shell ionization or through inner shell excitation followed by autoionization. There is also the possibility of a double transition. This possibility exists even for light atoms. It does not differ fundamentally from ordinary excitation or ionization.

Using the first Born approximation, Dalgarno and McDowell (1955) have investigated the simultaneous excitation and ionization of helium by fast protons

$$\mathrm{He}(1s^2) + \mathrm{H}^+ \rightarrow \mathrm{He}^+(nl) + e + \mathrm{H}^+. \tag{41}$$

They confined their attention to cases in which the azimuthal quantum number of the electron remaining bound is 1 or greater, such cases being relatively simple because the final wave function is automatically orthogonal to the initial wave function and because only spherically symmetrical states of the ejected electron need be taken into account,

The cross section for (41) depends on a transition integral like (17) and the overlap integral

$$S = \int \phi_{1s}(r) \ell_{\epsilon s}(r)\, d^3\mathbf{r} \tag{42}$$

in which ϕ_{1s} represents the orbital of one of the electrons of the normal helium atom and $\ell_{\epsilon s}$ represents the orbital of the ejected electron. Dalgarno and McDowell took ϕ_{1s} to be as in (36) and $\ell_{\epsilon s}$ to be the solution of the appropriate Hartree equation. The value of S is sensitive to the forms of these functions. In consequence the reliability of the derived cross sections (Table I) is poor. According to Dalgarno and McDowell an absolute error of a factor of 5 may well occur.

TABLE I

Cross Sections $Q(\pi a_0^2)$ for the Simultaneous Excitation and Ionization of Helium by Proton Impact[a] with, for Comparison, Cross Sections for Ordinary Excitation[b] and Ionization[c]

Final state of He$^+$ ion or He atom	Proton impact energy (kev)	250	500	750	1000	1500
Double transitions						
$2p$	$10^4 \times Q$	5.4	5.3	4.8	4.3	3.8
$3p$	$10^4 \times Q$	1.2	1.1	1.0	0.8_7	0.7_1
$4p$	$10^5 \times Q$	4.5	4.1	3.5	3.1	2.4
$3d$	$10^5 \times Q$	2.2	2.0	1.6	1.4	1.0
$4d$	$10^5 \times Q$	1.0	0.9_1	0.7_5	0.6_3	0.4_6
Single transitions						
(excitation) $1s\,2p\,^1P$	$10^2 \times Q$	9.2	6.0	4.5	3.8	2.8
(ionization) $1s$	$10 \times Q$	6.2	3.9	2.9	2.3	1.6_5

[a] Dalgarno and McDowell (1955).
[b] Moiseiwitsch and Stewart (1954).
[c] Erskine (1954).

1.1.2 *Excitation and Ionization of Atoms by Other Atoms*

Consider an encounter between a target hydrogen atom and a projectile hydrogen atom distinguished by the letters A and B, respectively. In addition to the single-transition processes,

$$H(1s \mid A) + H(1s \mid B) \rightarrow H(nl \mid A) + H(1s \mid B) \tag{43}$$

$$H(1s \mid A) + H(1s \mid B) \rightarrow H(1s \mid A) + H(n'l' \mid B), \tag{44}$$

it is necessary to take account of the double-transition processes

$$H(1s \mid A) + H(1s \mid B) \rightarrow H(nl \mid A) + H(n'l' \mid B). \tag{45}$$

The problem has been treated by Bates and Griffing (1953, 1954, 1955).

If $Q(1s - nl;\ 1s - n'l')$ is the cross section associated with the reaction path indicated the total cross section for raising A to the nl state is

$$Q(1s - nl;\ 1s - \Sigma) = \sum_{n'l'} Q(1s - nl;\ 1s - n'l'), \tag{46}$$

$\Sigma_{n'l'}$ signifying a summation over the discrete states and an integration over the continuum; that for raising B to the nl state is of course the same; that for raising either or both atoms to this state is

$$\mathcal{Q}((1s)^2 - (nl, \Sigma)) = 2Q(1s - nl;\ 1s - \Sigma) - Q(1s - nl;\ 1s - nl); \tag{47}$$

that for collisions in which a specified atom is excited or ionized is

$$Q(1s - \Sigma';\ 1s - \Sigma) = \sum_{nl \neq 1s}{}' Q(1s - nl;\ 1s - \Sigma); \tag{48}$$

and that for all inelastic collisions is

$$\mathcal{Q}((1s)^2 - (\Sigma, \Sigma')) - \sum_{nl \neq 1s}{}' Q(1s - 1s;\ 1s - nl) + \sum_{nl \neq 1s}{}' Q(1s - nl;\ 1s - \Sigma). \tag{49}$$

There is no difficulty in calculating the cross sections associated with the different reaction paths. Using (1) and noting that the initial and final wave functions are

$$\Psi_p = \chi_{1s}(\mathbf{r}_a)\, \chi_{1s}(\mathbf{r}_b) \exp\,(i\varkappa_{1s,1s} \cdot \mathbf{R}) \tag{50}$$

and

$$\Psi_q = \chi_{nl}(\mathbf{r}_a)\, \chi_{n'l'}(\mathbf{r}_b) \exp\,(i\varkappa_{nl,n'l'} \cdot \mathbf{R}) \tag{51}$$

and that the interaction potential is

$$V = e^2 \left\{ \frac{1}{|\mathbf{R} + \mathbf{r}_b - \mathbf{r}_a|} - \frac{1}{|\mathbf{R} + \mathbf{r}_b|} - \frac{1}{|\mathbf{R} - \mathbf{r}_a|} + \frac{1}{R} \right\}, \tag{52}$$

it may be shown that

$$Q(1s - nl;\, 1s - 1s) = \frac{8}{s^2} \int_{t_{\min}}^{\infty} |\mathscr{I}(1s, nl)|^2\, t^{-3} \left\{ 1 - \frac{16}{(4 + t^2)^2} \right\}^2 dt\, \pi a_0^2 \tag{53}$$

and

$$\underset{n'l' \neq 1s}{Q(1s - nl;\, 1s - n'l')} = \frac{8}{s^2} \int_{t_{\min}}^{\infty} |\mathscr{I}(1s, nl)\mathscr{I}(1s, n'l')|^2\, t^{-3}\, dt\, \pi a_0^2 \tag{54}$$

in which the notation is as in § 1.1.1. Two forms arise because in the case of (45) all but the last term of (52) contribute to the cross section, whereas in the case of (43) or (44) only the first term contributes (owing to orthogonality effects).

Closed expressions for asymptotes to the various total cross section versus impact energy curves may also be obtained (Bates and Griffing, 1954). It is apparent from (19) that at sufficiently high impact energies $t_{\min}$ may be taken to be zero so that

$$Q(1s - nl;\, 1s - \Sigma) \sim Q(1s - nl;\, 1s - 1s) + \left[\frac{8}{s^2} \int_0^{\infty} |\mathscr{I}(1s - nl)|^2 \sum_{n'l' \neq 1s}{}' |\mathscr{I}(1s - n'l')|^2\, t^{-3}\, dt \right] \pi a_0^2. \tag{55}$$

Noting that

$$\sum_{n'l'} |\mathscr{I}(1s - n'l')|^2 = 1 \tag{56}$$

(Bethe, 1930) and that

$$|\mathscr{I}(1s - 1s)|^2 = 256/(4 + t^2)^4, \tag{57}$$

it may be shown that

$$Q(1s - nl;\, 1s - \Sigma) \sim \left[\frac{16}{s^2} \int_0^{\infty} |\mathscr{I}(1s - nl)|^2 \left\{ 1 - \frac{16}{(4 + t^2)^2} \right\} t^{-3}\, dt \right] \pi a_0^2. \tag{58}$$

This may readily be evaluated in particular cases. For example, substituting from (21) and (22) followed by integration yields

$$Q(1s - 2s;\, 1s - \Sigma) \sim 4.3\mathscr{E}^{-1}\, \pi a_0^2, \tag{59}$$

$$Q(1s - 2p;\, 1s - \Sigma) \sim 21\mathscr{E}^{-1}\, \pi a_0^2, \tag{60}$$

and

$$Q(1s - C;\ 1s - \Sigma) \sim 128\mathscr{E}^{-1}\,\pi a_0^2, \tag{61}$$

C representing the entire continuum and $\mathscr{E}$ being the impact energy in kev. Collisions which do not change the state of the projectile contribute only

$$Q(1s - 2s;\ 1s - 1s) \sim 0.72\mathscr{E}^{-1}\,\pi a_0^2, \tag{62}$$

$$Q(1s - 2p;\ 1s - 1s) \sim 1.8\mathscr{E}^{-1}\,\pi a_0^2, \tag{63}$$

and

$$Q(1s - C;\ 1s - 1s) \sim 36\mathscr{E}^{-1}\,\pi a_0^2. \tag{64}$$

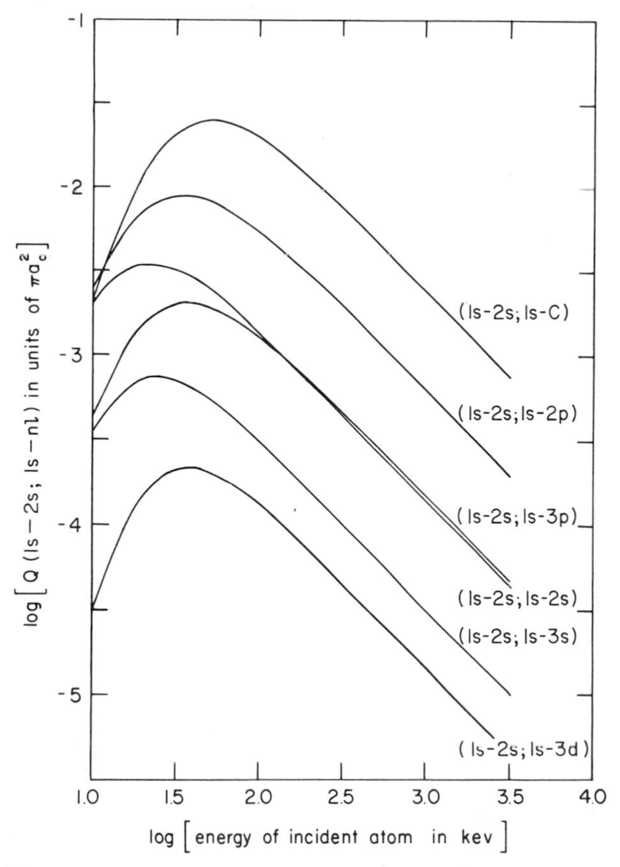

FIG. 5. First Born approximation to cross section—impact energy curves for H(1s) + H(1s) → H(2s) + H(nl) where nl represents 2s, 2p, 3s, 3p, 3d, or C (the continuum) (Bates and Griffing, 1954).

From (48) it may similarly be shown that

$$Q(1s - \Sigma'; 1s - \Sigma) \sim 168\mathscr{E}^{-1}\,\pi a_0^2 \tag{65}$$

and from (49) that

$$\mathscr{Q}((1s)^2 - (\Sigma, \Sigma')) \sim 210\mathscr{E}^{-1}\,\pi a_0^2. \tag{66}$$

A major contribution to both these comes from collisions giving rise to double ionization.

The double-transition cross sections $Q(1s - 2s;\ 1s - 2s, 2p, 3s, 3p, 3d,$ or $C)$ and $Q(1s - 2p;\ 1s - 2s, 2p, 3s, 3p, 3d,$ or $C)$ are given in Figs. 5 and 6. Combined with the single-transition cross sections $Q(1s - 2s;\ 1s - 1s)$ and $Q(1s - 2p;\ 1s - 1s)$ given in Fig. 7, they enable the total

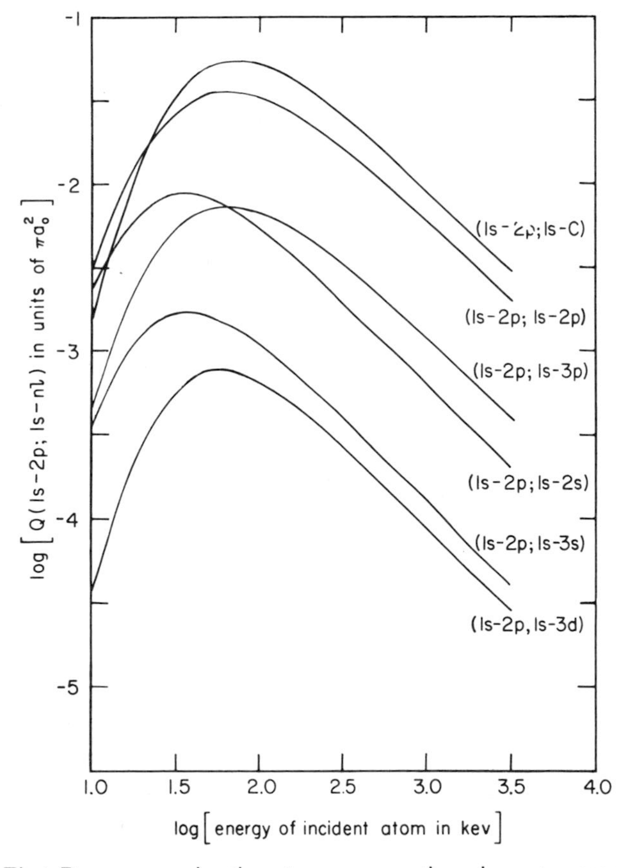

FIG. 6. First Born approximation to cross section—impact energy curves for $H(1s) + H(1s) \rightarrow H(2p) + H(nl)$ where nl represents $2s$, $2p$, $3s$, $3p$, $3d$, or C (the continuum) (Bates and Griffing, 1954).

cross sections $Q(1s - 2s;\ 1s - \Sigma)$ and $Q(1s - 2p;\ 1s - \Sigma)$ to be computed since $Q(1s - 2s;\ 1s - nl)$ and $Q(1s - 2p;\ 1s - nl)$ may be neglected without significant error when n is 4 and greater. These total cross sections are also given in Fig. 7. It is apparent that whereas single-transitions predominate at low impact energies, double transitions predominate at high impact energies.

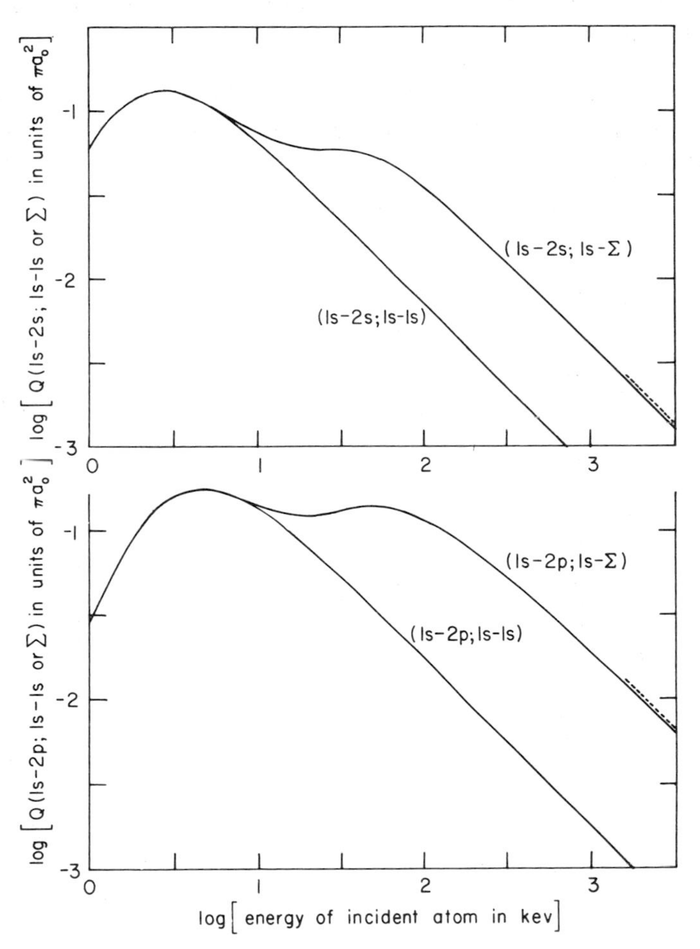

FIG. 7. First Born approximation to cross section—impact energy curves for H(1s)—H(1s) collisions in which the stationary hydrogen atom is excited to the 2s state (upper set) or to the 2p state (lower set) and the incident hydrogen atom is unaffected [curves marked $(1s - 2s;\ 1s - 1s)$ or $(1s - 2p;\ 1s - 1s)$] or is left in any state [curves marked $(1s - 2s;\ 1s - \Sigma)$ or $(1s - 2p;\ 1s - \Sigma)$]. The broken lines represent the asymptotic behavior of $Q(1s - 2s;\ 1s - \Sigma)$ and $Q(1s - 2p;\ 1s - \Sigma)$ as given by (59) and (60) of text (Bates and Griffing, 1954).

Moiseiwitsch and Stewart (1954) have carried out calculations on

$$\mathrm{He}(1s)^2 + \mathrm{H}(1s) \rightarrow \mathrm{He}(1s\,2p\,{}^1P) + \mathrm{H}(\Sigma) \tag{67}$$

transitions to discrete states for which n exceeds 3 being again neglected in the summation. Their results (Fig. 8) are similar to those just described. Moiseiwitsch and Stewart have also treated

$$\mathrm{He}(1s)^2 + \mathrm{He}^+(1s) \rightarrow \mathrm{He}(1s\,2p\,{}^1P) + \mathrm{He}^+(\Sigma). \tag{68}$$

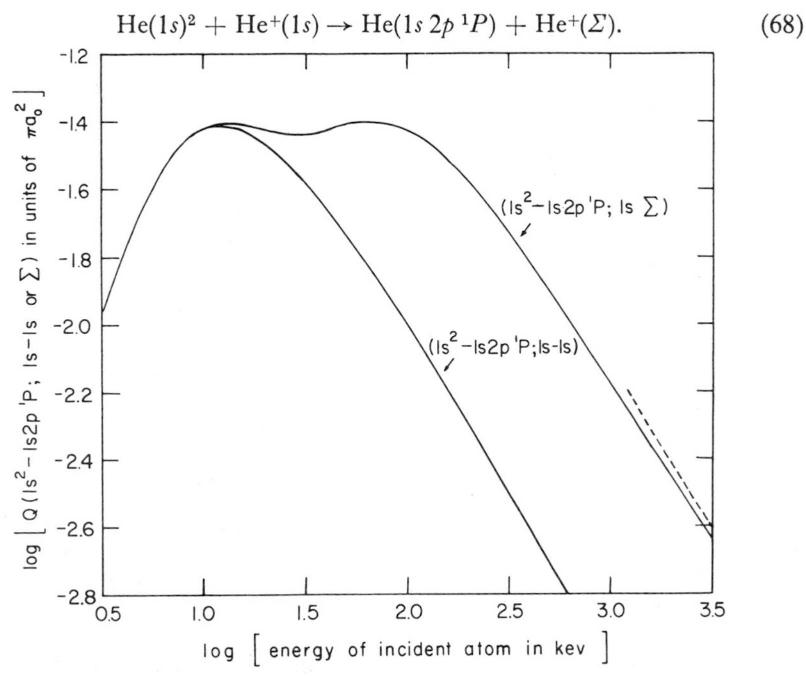

FIG. 8. First Born approximation to cross section—impact energy curves for $\mathrm{He}(1s)^2 - \mathrm{H}(1s)$ collisions in which the stationary helium atom is raised to the $1s\,2p\,{}^1P$ level and the incident hydrogen atom is unaffected [curve marked $(1s^2 - 1s2p\,{}^1P;\ 1s - 1s)$] or is left in any state [curve marked $(1s^2 - 1s2p\,{}^1P;\ 1s - \Sigma)$]. The broken line represents the asymptotic behavior of $Q(1s^2 - 1s2p\,{}^1P;\ 1s - \Sigma)$ (Moisewitsch and Stewart, 1954).

Double transitions are here quite unimportant (Fig. 9). This is as would be expected for a collision in which the particle responsible for the excitation is a positive ion. The comparison with

$$\mathrm{He}(1s)^2 + \mathrm{H}^+ \rightarrow \mathrm{He}(1s\,2p\,{}^1P) + \mathrm{H}^+ \tag{69}$$

for which results are included in Fig. 9, is instructive. It indicates that the screening of the helium nucleus by the orbital electron is slight at low impact energies but is so effective at moderate and high impact energies that the He^+ ion acts almost like a singly charged structureless particle.

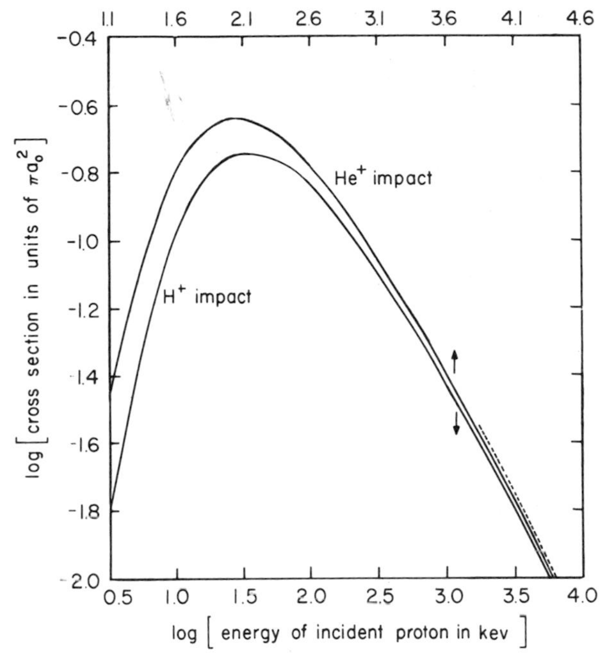

FIG. 9. Comparison of the cross section—energy curves for $He(1s)^2 - H^+$ and $He(1s)^2 - He^+(1s)$ collisions in which the helium atom is raised to the $1s2p\ ^1P$ level. The full lines are for single transition collisions, the broken line represents the asymptotic behavior of $Q(1s^2 - 1s2p\ ^1P;\ 1s - \Sigma)$ (Moiseiwitsch and Stewart, 1954).

The only other excitation processes which have been investigated are

$$H(1s) + He(2s\ ^3S) \rightarrow H(\Sigma) + He(2p \text{ or } 3p\ ^3P) \tag{70}$$

and

$$Ne(^1S) + He(2s\ ^3S) \rightarrow Ne(\Sigma) + He(2p \text{ or } 3p\ ^3P) \tag{71}$$

(Adler and Moiseiwitsch, 1957). When a many-electron atom is involved, as in (71), the task of summing and integrating over all the final states is formidable. It may be avoided without serious error being introduced by making use of a simple modification of the asymptotic formula (55) discussed earlier. In this modification $t_{\min}$ is given the value corresponding to the lowest double-transition which has to be taken into account. An alternative procedure is to carry out calculations on one of the double transitions. The contribution from the others may then be estimated because the cross section versus impact energy curves concerned may be taken to have a common shape (which is determined by the calculations just mentioned) and because in addition a formula corresponding to (58) may be derived for the asymptote to the total. Adler and Moiseiwitsch (1957) found that the two procedures give much the same results.

Unfortunately, the labor involved in the treatment of ionization is in general very heavy since the evaluation of the contribution of double-ionization collisions to the total cross section entails triple numerical integrations. Detailed calculations on complex systems are scarcely feasible. Some empirical methods for making predictions have been proposed (Russek and Thomas, 1958, 1959; Firsov, 1959; Bulman and Russek, 1961). The development of such methods is important but shall not be discussed here.

Figure 10 gives the results of the calculations of Bates and Griffing (1955) on the cross section $Q(1s - C;\ 1s - 1s)$ for

$$H(1s) + H(1s) \rightarrow H^+ + e + H(1s) \tag{72}$$

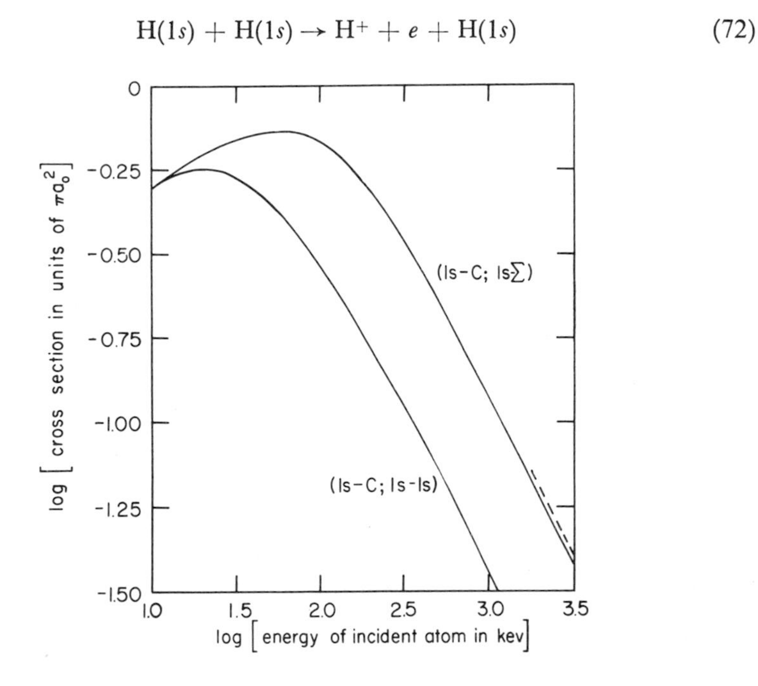

FIG. 10. First Born approximation to cross section—energy curves for $H(1s) - H(1s)$ collisions in which the stationary hydrogen atom is ionized and the incident hydrogen atom is unaffected [curve marked $(1s - C;\ 1s - 1s)$] or is left in any state [curve marked $(1s - C;\ 1s - \Sigma)$]. The broken line represents the asymptotic behavior of $Q(1s - C;\ 1s - \Sigma)$ (Bates and Griffing, 1955).

and the cross section $Q(1s - C;\ 1s - \Sigma)$ for

$$\mathrm{H}(1s) + \mathrm{H}(1s) \rightarrow \mathrm{H}^+ + e + \mathrm{H}(\Sigma) \tag{73}$$

(transitions to discrete states with n greater than 3 being ignored). It is seen that as the impact energy is increased the contribution from double-transitions grows relative to that from single-transitions and ultimately becomes the greater. Boyd *et al.* (1957) have shown that this also happens in the case of

$$\mathrm{He}^+(1s) + \mathrm{H}(1s) \rightarrow \mathrm{He}^{2+} + e + \mathrm{H}(\Sigma) \tag{74}$$

(Fig. 11). They have further shown that double-transitions are, however,

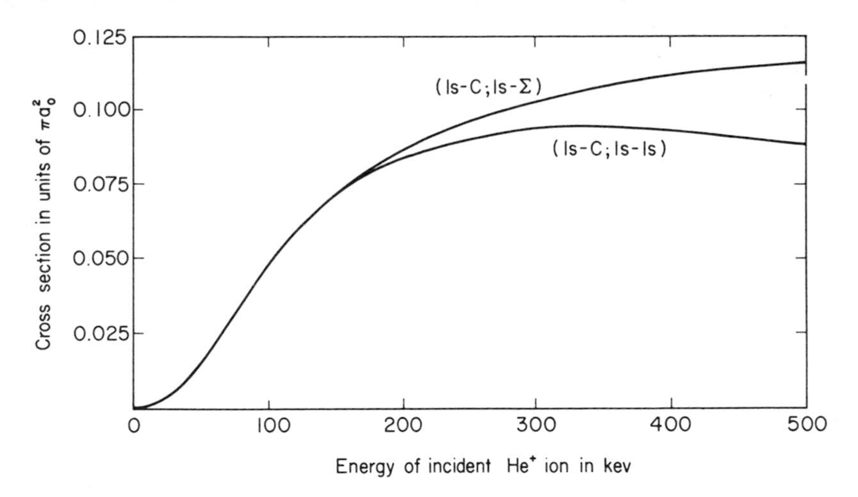

FIG. 11. First Born approximation to cross section-energy curves for $\mathrm{H}(1s) - \mathrm{He}^+(1s)$ collisions in which the stationary hydrogen atom is unaffected [curve marked $(1s - C;\ 1s - 1s)$] or is left in any state [curve marked $(1s - C;\ 1s - \Sigma)$] and in which the incident helium ion is stripped (Boyd *et al.*, 1957).

of little importance if the positive ion, instead of *undergoing* a specific change as in (74), *brings about* a specific change as in

$$\mathrm{H}(1s) + \mathrm{He}^+(1s) \rightarrow \mathrm{H}^+ + e + \mathrm{He}^+(\Sigma). \tag{75}$$

The situation is thus the same as for excitation.

At impact energies sufficiently high for charge transfer to be neglected, the cross section for the ionization of the projectile is equivalent to the cross section for electron loss from the projectile.

Figure 12 compares the calculated cross section for

$$H(1s) + He(1s)^2 \to H^+ + e + He(\Sigma) \tag{76}$$

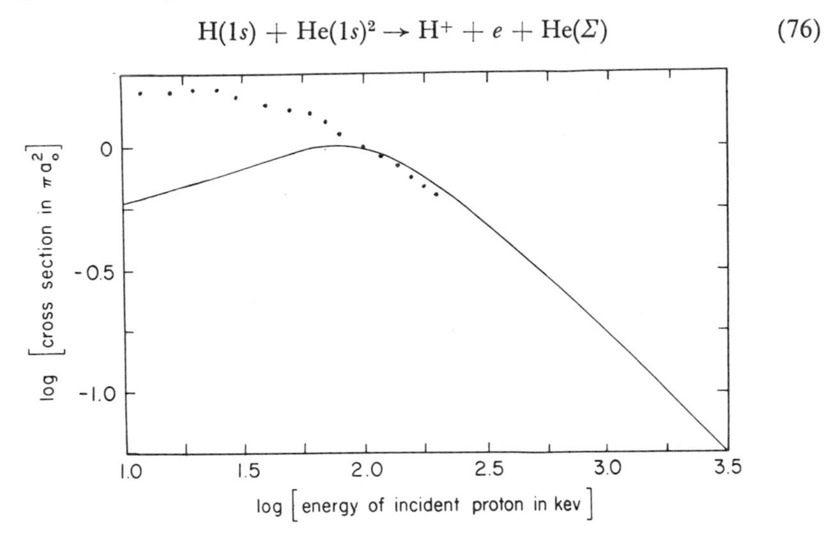

FIG. 12. Electron loss from hydrogen atoms passing through helium gas. The full line gives the first Born approximation to the cross section—energy curve (Bates and Williams, 1957); and the points give the measurements (Stier and Barnett, 1956).

(Bates and Williams, 1957) with the measured cross section for electron loss from hydrogen atoms passing through helium gas (Stier and Barnett, 1956). The agreement is quite good at impact energies above about 100 kev. Caution must be exercised in drawing conclusions since the one-parameter variational wave function defined in (36) was used to describe the ground state of helium. Simplifying assumptions made in theoretical work of Sida (1955) on

$$H^-(1s)^2 + He(1s)^2 \to H(1s) + e + He(1s)^2 \tag{77}$$

and of McDowell and Peach (1959) on

$$H^-(1s)^2 + H(1s) \to H(1s) + e + H(\Sigma) \tag{78}$$

prevent information on range of validity of the first Born approximation from being derived from a comparison of their results with the laboratory data of Hasted (1952) and of Hummer *et al.* (1960).

The computed energy distribution of the electrons ejected in fast $H(1s) - H(1s)$ single- and double-transition collisions is shown in Fig. 13. Because of the relatively high importance of close encounters they are broader than the corresponding distribution for fast $H(1s) - H^+$ collisions (which is shown for comparison).

The energy concerned in Fig. 13 is the energy associated with the motion of the electron relative to its parent nucleus. In the case of an electron originating from the projectile what is actually of most interest

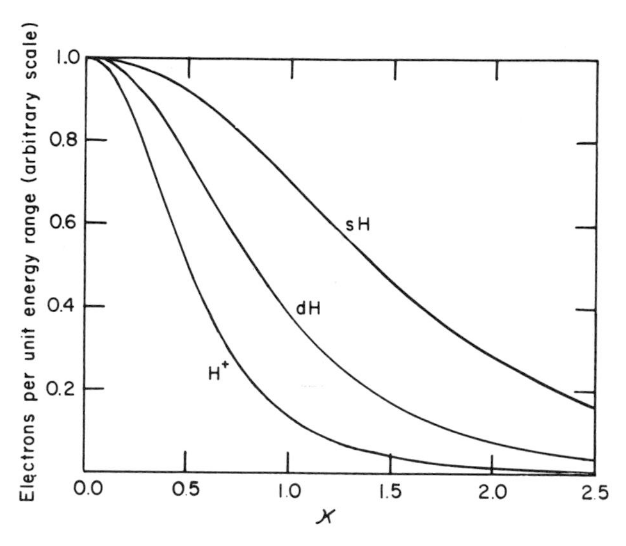

FIG. 13. Energy distribution of electrons ejected from stationary hydrogen atom. Curves sH and dH, respectively, refer to single- and double-transition collisions with other hydrogen atoms; curve H^+ to collisions with protons. In all cases the impact energy is 1000 kev. The units of κ are such that κ^2 is the kinetic energy of the ejected electron in rydbergs (Bates and Griffing, 1955).

is the energy associated with the motion relative to the target. There may be a considerable difference: for example, an electron ejected with zero energy relative to a 1-Mev projectile hydrogen atom has an energy of over 500 ev relative to the target. In order to convert the energy distribution from one frame of reference to the other, it is necessary to know the angular distribution. Calculations on this have been carried out by Dalgarno and Griffing (1958). There is a sharp peak in the direction that would be expected from classical considerations (Fig. 14).

1.1.3 *Charge Transfer*

In applying the first Born approximation to the charge transfer process

$$(\textit{nucleus } \mathrm{A} + \textit{electron } e)_p + \textit{nucleus } \mathrm{B} \rightarrow \textit{nucleus } \mathrm{A} + (\textit{nucleus } \mathrm{B} + \textit{electron } e)_q$$

full account must be taken of the change in the translational motion of the electron that accompanies a jump from state p around one nucleus to state q around the other.

Denote the position vector of the center of mass of (A + e) relative to B by $\boldsymbol{\rho}$ that of the center of mass of (B + e) relative to A by $\boldsymbol{\sigma}$ and the masses of A, B, and e by M_a, M_b, and m, respectively, but otherwise

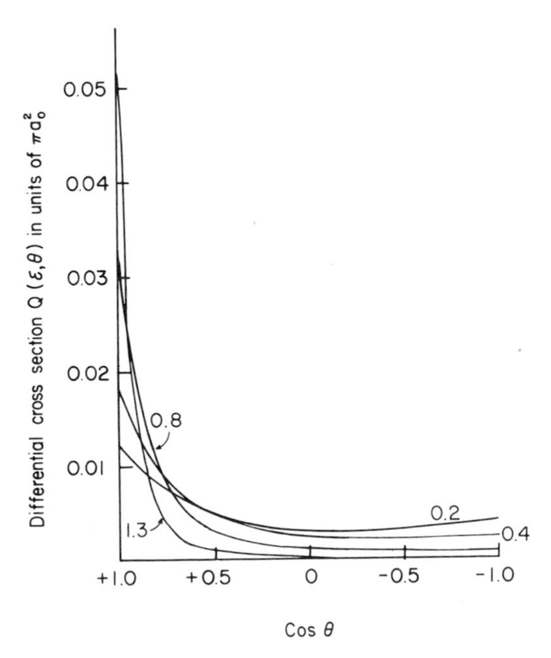

FIG. 14. Angular distribution of electrons ejected in the ionization of stationary hydrogen atoms by incident 1000-kev hydrogen atoms. The number on each curve is the kinetic energy ϵ of the ejected electron in rydbergs; θ is the angle the direction of ejection makes with the direction of impact; $Q(\epsilon, \theta)$ is the differential cross section normalized so that $\int_0^{\epsilon_{\max}} \int_{-1}^{+1} Q(\epsilon, \theta)\, d(\cos\theta)\, d\epsilon$ is the ionization cross section (Dalgarno and Griffing, 1958).

use the notation of § 1.1, with each velocity taken to be that of the atomic system relative to the bare nucleus (so that if no deflection were suffered $\mathbf{v}_p$ and $\mathbf{v}_q$ would be antiparallel). The initial and final wave functions are then given by

$$\Psi_p = \chi_p(\mathbf{r}_a) \exp(i\boldsymbol{\varkappa}_p \cdot \boldsymbol{\rho}), \qquad \Psi_q = \chi_q(\mathbf{r}_b) \exp(i\boldsymbol{\varkappa}_q \cdot \boldsymbol{\sigma}) \tag{79}$$

where

$$\boldsymbol{\varkappa}_p = \frac{(M_a + m)\, M_b \mathbf{v}_p}{(M_a + M_b + m)\, \hbar}, \qquad \boldsymbol{\varkappa}_q = \frac{(M_b + m)\, M_a \mathbf{v}_q}{(M_a + M_b + m)\, \hbar}. \tag{80}$$

Substitution in (2) gives the relevant matrix element to be

$$\mathscr{N} = \iint \chi_p^*(\mathbf{r}_a)\, \chi_q(\mathbf{r}_b) \exp[-i(\boldsymbol{\varkappa}_p \cdot \boldsymbol{\rho} - \boldsymbol{\varkappa}_q \cdot \boldsymbol{\sigma})]\, V\, d^3\mathbf{r}_a\, d^3\mathbf{r}_b. \tag{81}$$

As pointed out by Brinkman and Kramers (1930), it is convenient to introduce the vectors

$$\boldsymbol{\alpha} = \boldsymbol{\varkappa}_q + \frac{M_a}{M_a + m} \boldsymbol{\varkappa}_p, \qquad \boldsymbol{\beta} = -\boldsymbol{\varkappa}_p - \frac{M_b}{M_b + m} \boldsymbol{\varkappa}_q \tag{82}$$

so that (81) may be written

$$\mathscr{N} = \iint \chi_p^*(\mathbf{r}_a)\, \chi_q(\mathbf{r}_b) \exp\left[-i\{\boldsymbol{\alpha} \cdot \mathbf{r}_a + \boldsymbol{\beta} \cdot \mathbf{r}_b\}\right] V\, d^3\mathbf{r}_a\, d^3\mathbf{r}_b. \tag{83}$$

The simplest charge transfer process is

$$\mathrm{H}(1s \mid \mathrm{A}) + \mathrm{H}^+(\mid \mathrm{B}) \rightarrow \mathrm{H}^+(\mid \mathrm{A}) + \mathrm{H}(1s \mid \mathrm{B}). \tag{84}$$

For the systems on the right of (84) the interaction potential is

$$e^2 \left\{ \frac{1}{|\mathbf{r}_a - \mathbf{r}_b|} - \frac{1}{r_a} \right\} \qquad \text{[post]} \tag{85}$$

and for the systems on the left it is

$$e^2 \left\{ \frac{1}{|\mathbf{r}_a - \mathbf{r}_b|} - \frac{1}{r_b} \right\} \qquad \text{[prior].} \tag{86}$$

The final result is independent of whether it is the post or the prior expression that is adopted.† In their early work Oppenheimer (1928) and Brinkman and Kramers (1930) took the interaction potential to be

$$-e^2/r_a \qquad \text{[post]} \tag{87}$$

(or $-e^2/r_b$ [prior]) omitting the nuclear-nuclear term because it is physically unrealistic that such a term should influence the probability of charge transfer. Saha and Basu (1945) and Takayanagi (1952) have followed the same procedure but Bates and Dalgarno (1952) and Jackson and Schiff (1953) have included the nuclear-nuclear term in the belief that its presence corrects partially for defects in the elementary treatment of the problem. These defects originate from the non-orthogonality of the initial and final atomic eigenfunctions. Explicit account of the nonorthogonality has been taken by Bates (1958b) who

† This independence is not peculiar to process (84), it is general, provided the atomic eigenfunctions used are exact (Schiff, 1949) or satisfy certain conditions (Bates *et al.*, 1950).

has shown that instead of using (85) or (87) for V in (83), it is necessary to use the effective interaction potential

$$\frac{e^2}{1-S^2}\left\{-\frac{1}{r_a}+\int\frac{\chi_{1s}^*(\mathbf{r}_b)\,\chi_{1s}(\mathbf{r}_b)}{r_a}\,d^3\mathbf{r}_b\right\} \qquad \text{[post]}\dagger \tag{88}$$

in which

$$S^2=\left|\int\chi_{1s}^*(\mathbf{r}_a)\,\chi_{1s}(\mathbf{r}_b)\exp\left\{\frac{im}{\hbar}\mathbf{v}_{1s}\cdot\mathbf{r}_a\right\}d^3\mathbf{r}_b\right|^2. \tag{89}$$

An expression differing from (88) only in that the factor $1/(1-S^2)$ is missing has been obtained independently by Bassel and Gerjuoy (1960).

With the aid of the principle of the conservation of energy it may be proved that to the lowest order in m/M

$$\alpha^2=\beta^2=w/a_0^2, \qquad \hat{\boldsymbol{\alpha}}\cdot\hat{\boldsymbol{\beta}}=\frac{s^2}{2w}-1 \tag{90}$$

where

$$w=\frac{2M^2s^2}{m^2}(1+\cos\theta)+\frac{s^2}{4}, \qquad s=mva_0/\hbar, \tag{91}$$

$\mathbf{v}$ being the velocity of relative motion (from which the subscript is omitted as unnecessary since the collision is elastic). It follows that the matrix element (83) may be expressed as a function of w and that the minimum value of this dimensionless parameter is $s^2/4$ and the maximum value is effectively infinite. Referring to (1) it may hence be seen that the charge transfer cross section for (84) is given by

$$Q^X(1s\mid \mathrm{A};\,1s\mid \mathrm{B})=\frac{1}{4\pi^2s^2}\int_{s^2/4}^{\infty}|N^X(w)|^2\,dw \qquad (\pi a_0^2) \tag{92}$$

with

$$N^X(w)=\frac{1}{\pi}\iint V^X\exp\left[i(\boldsymbol{\alpha}\cdot\mathbf{r}_a+\boldsymbol{\beta}\cdot\mathbf{r}_b)-r_a-r_b\right]d^3\mathbf{r}_a d^3\mathbf{r}_b \tag{93}$$

in which the superscript X is (ne), $(ne+nn)$, or *(ortho)*, according to whether potential (87), (85), or (88) is used, and in which all quantities are in atomic units.

When only the simple nuclear-electronic term is included, potential (87), the necessary integrations may readily be performed. It is found that

$$Q^{(ne)}(1s\mid \mathrm{A};\,1s\mid \mathrm{B})=\left[\frac{2^{18}}{5s^2(s^2+4)^5}\right]\pi a_0^2 \tag{94}$$

† The analogous prior potential is equivalent.

if the atom is taken to be at rest, equals $24.9_7\, s^2$, this formula may be written approximately as

$$Q^{(ne)}(1s \mid \mathrm{A};\, 1s \mid \mathrm{B}) = \frac{1.28 \times 10^{13}}{\mathscr{E}(\mathscr{E} + 100)^5}\, \pi a_0^2. \tag{95}$$

In the low-energy region (where the treatment is, of course, invalid) the cross section is large tending to infinity as $\mathscr{E}$ tends to zero. This behavior is not characteristic of charge transfer collisions but is associated with the fact that (84) is elastic. Because of the change in the translational motion of the electron, the cross section ultimately falls off extremely rapidly as $\mathscr{E}$ is increased. Should allowance not be made for the change, the lower limit in the integration over w in (92) would be not $s^2/4$ but zero, and the charge transfer cross section would be given not by (95) but by

$$Q^{(ne)}(1s \mid \mathrm{A};\, 1s \mid \mathrm{B}) = \frac{1.28 \times 10^{3}}{\mathscr{E}}\, \pi a_0^2. \tag{96}$$

Comparing (95) and (96) it is seen that the effect of the change when $\mathscr{E}$ is 25, 50, 100, and 200 kev is to reduce the cross section by factors of about 3, 8, 32, and 243, respectively.

The nuclear-nuclear term of potential (85) causes considerable trouble. Bates and Dalgarno (1952) have carried out the spatial integrations over this term using a series expansion; Jackson and Schiff (1953) have carried them out with the aid of Fourier transforms and have obtained a closed expression. The algebraic results are complicated and will not be presented. Once again great simplification ensues if account is not taken of the change in the translational motion of the electron. In this unrealistic circumstance the formula for the charge transfer cross section reduces to

$$Q^{(ne+nn)}(1s \mid \mathrm{A};\, 1s \mid \mathrm{B}) = \frac{1.50 \times 10^{2}}{\mathscr{E}}\, \pi a_0^2 \tag{97}$$

(Bates and Dalgarno, 1952). The factors corresponding to those listed at the end of the preceding paragraph are about 3, 6, 20, and 120.

As would be expected, the corrected potential (88) is extremely difficult to handle. McCarroll (1961) has, however, evaluated $Q^{(ortho)}(1s \mid \mathrm{A};\, 1s \mid \mathrm{B})$ with the help of a digital computer.†

† He actually used not the formula derived from the wave treatment but instead the equivalent formula derived from the impact parameter treatment (§ 1.2.2).

Figure 15 depicts the cross sections associated with the three potentials. At very high energies $Q^{(ne)}$ $(1s \mid A;\ 1s \mid B)$ is the closer to $Q^{(ortho)}$ $(1s \mid A;\ 1s \mid B)$ (to which indeed it tends asymptotically); but elsewhere $Q^{(ne+nn)}$ $(1s \mid A;\ 1s \mid B)$ is the closer.

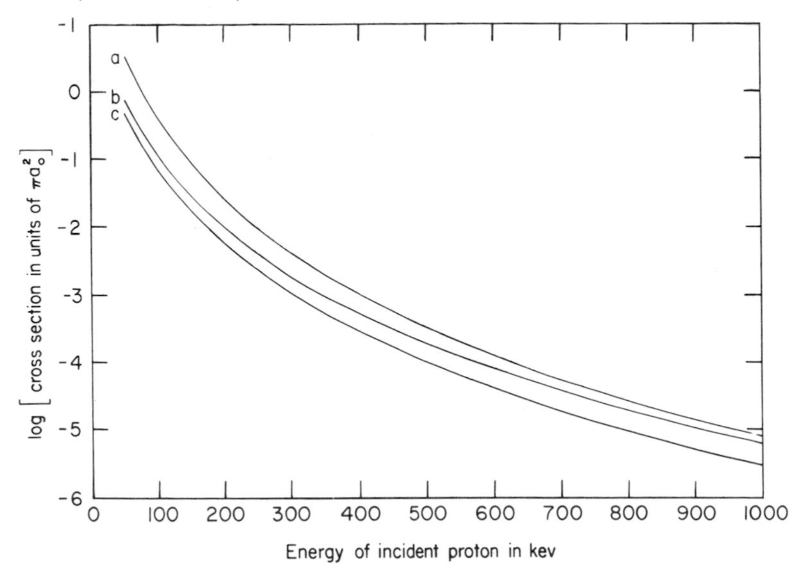

FIG. 15. Cross section—impact energy curves for $H(1s) + H^+ \rightarrow H^+ + H(1s)$. Curve a is $Q^{(ne)}$, curve b is $Q^{(ortho)}$, and curve c is $Q^{(ne+nn)}$ (McCarroll, 1961).

Collisions involving a single electron have naturally received most attention from theorists. Using a potential which, like (87), contains only the simple nuclear-electronic term, expressions have been derived for the cross sections describing capture from the $1s$ state around a nucleus of charge $Z_a e$ into the nl state around a nucleus of charge $Z_b e$ with nl either $1s$, $2s$, $2p$, $3s$, $3p$, $3d$, $4s$, $4p$, $4d$, or $4f$ (Brinkman and Kramers, 1930; Saha and Basu, 1945; Bates and Dalgarno, 1953). Table II gives numerical results for when both nuclei are protons. Leaving the special $1s - 1s$ case aside, the cross sections are initially rapidly increasing functions of the impact energy, pass through maxima at impact energies between 10 and 20 kev and finally become rapidly decreasing functions of the impact energy. For fixed l they fall off as n is increased and for fixed n they fall off as l is increased except in the region between about 0.5 and 100 kev where the $s - p$ cross sections are the greatest. Oppenheimer (1928) has shown that the cross section is proportional to $1/n^3$ at very high energies where capture is mainly into s states.

TABLE II

Cross Sections $Q^{(ne)}(1s \mid A; nl \mid B)$ for $H(1s) + H^+ \rightarrow H^+ + H(nl)$ Calculated Using Potential (87) of Text[a]

log s^2[b] \ Final state:	1s	2s	3s	4s	2p	3p	4p	3d	4d	4f
	log (capture cross section in units of πa_0^2)									
−1.0	1.73	−0.88	−1.86	−2.39	−1.12	−2.18	−2.73	−3.43	−3.90	−5.67
−0.9	1.62	−0.69	−1.62	−2.14	−0.84	−1.85	−2.39	−3.02	−3.47	−5.16
−0.8	1.51	−0.54	−1.42	−1.92	−0.60	−1.55	−2.08	−2.65	−3.08	−4.69
−0.7	1.39	−0.42	−1.25	−1.73	−0.40	−1.30	−1.80	−2.31	−2.73	−4.27
−0.6	1.27	−0.34	−1.11	−1.58	−0.24	−1.08	−1.57	−2.03	−2.43	−3.90
−0.5	1.14	−0.29	−1.02	−1.47	−0.13	−0.91	−1.38	−1.80	−2.17	−3.59
−0.4	1.00	−0.28	−0.96	−1.39	−0.07	−0.79	−1.24	−1.63	−1.98	−3.35
−0.3	0.86	−0.30	−0.94	−1.36	−0.05	−0.72	−1.15	−1.52	−1.85	−3.18
−0.2	0.71	−0.35	−0.96	−1.36	−0.08	−0.70	−1.11	−1.48	−1.79	−3.09
−0.1	0.54	−0.43	−1.01	−1.40	−0.15	−0.74	−1.13	−1.50	−1.79	−3.08
0.0	0.36	−0.54	−1.10	−1.48	−0.28	−0.82	−1.21	−1.58	−1.86	−3.14
+0.1	0.17	−0.68	−1.22	−1.60	−0.45	−0.97	−1.34	−1.74	−2.00	−3.30
0.2	−0.05	−0.86	−1.38	−1.75	−0.67	−1.16	−1.52	−1.96	−2.21	−3.53
0.3	−0.28	−1.06	−1.57	−1.94	−0.93	−1.40	−1.76	−2.25	−2.49	−3.84
0.4	−0.54	−1.30	−1.80	−2.17	−1.24	−1.70	−2.05	−2.59	−2.82	−4.22
0.5	−0.82	−1.57	−2.07	−2.44	−1.59	−2.04	−2.39	−2.99	−3.22	−4.67
0.6	−1.13	−1.88	−2.38	−2.74	−1.98	−2.43	−2.77	−3.44	−3.67	−5.18
0.7	−1.46	−2.21	−2.72	−3.08	−2.41	−2.85	−3.20	−3.94	−4.17	−5.75
0.8	−1.82	−2.59	−3.09	−3.45	−2.88	−3.31	−3.66	−4.49	−4.71	−6.36
0.9	−2.21	−2.99	−3.49	−3.86	−3.37	−3.81	−4.16	−5.07	−5.29	−7.02
1.0	−2.62	−3.41	−3.92	−4.29	−3.90	−4.33	−4.68	−5.68	−5.90	−7.72
1.1	−3.06	−3.86	−4.37	−4.74	−4.44	−4.88	−5.23	−6.32	−6.54	−8.44
1.2	−3.52	−4.33	−4.85	−5.22	−5.01	−5.46	−5.80	−6.98	−7.20	−9.20
1.3	−3.99	−4.83	−5.34	−5.71	−5.60	−6.05	−6.40	−7.67	−7.89	−9.98
1.4	−4.49	−5.33	−5.85	−6.22	−6.21	−6.66	−7.00	−8.37	−8.59	−10.77
1.5	−5.00	−5.85	−6.37	−6.74	−6.83	−7.28	−7.63	−9.09	−9.31	−11.58
1.6	−5.52	−6.38	−6.90	−7.28	−7.46	−7.91	−8.26	−9.82	−10.04	−12.41
1.7	−6.06	−6.93	−7.45	−7.82	−8.10	−8.55	−8.90	−10.56	−10.78	−13.25
1.8	−6.60	−7.48	−8.00	−8.37	−8.75	−9.20	−9.55	−11.31	−11.53	−14.09
1.9	−7.15	−8.03	−8.56	−8.93	−9.41	−9.86	−10.21	−12.06	−12.28	−14.95
2.0	−7.71	−8.60	−9.12	−9.50	−10.08	−10.53	−10.88	−12.83	−13.05	−15.81
2.1	−8.28	−9.17	−9.69	−10.07	−10.74	−11.20	−11.55	−13.60	−13.82	−16.68
2.2	−8.85	−9.74	−10.27	−10.64	−11.42	−11.87	−12.22	−14.37	−14.59	−17.55
2.3	−9.42	−10.32	−10.85	−11.22	−12.10	−12.55	−12.90	−15.15	−15.37	−18.43
2.4	−10.00	−10.90	−11.43	−11.80	−12.78	−13.23	−13.58	−15.93	−16.15	−19.31
2.5	−10.58	−11.48	−12.01	−12.38	−13.46	−13.91	−14.26	−16.71	−16.93	−20.19

[a] Bates and Dalgarno (1953).

[b] The energy of the incident proton is $24.97\ s^2$ kev.

Though the nuclear-nuclear term makes the work very lengthy, Jackson and Schiff (1953) have used potential (85) in calculations on

$$H(1s \mid A) + H^+(\mid B) \to H^+(\mid A) + H(2s \text{ or } 2p \mid B). \tag{98}$$

They found that to a close approximation

$$Q^{(ne+nn)}(1s \mid A;\, 2s \mid B)/Q^{(ne+nn)}(1s \mid A;\, 1s \mid B) = Q^{(ne)}(1s \mid A;\, 2s \mid B)/Q^{(ne)}(1s \mid A;\, 1s \mid B) \tag{99}$$

and

$$Q^{(ne+nn)}(1s \mid A;\, 2p \mid B)/Q^{(ne+nn)}(1s \mid A;\, 1s \mid B) = Q^{(ne)}(1s \mid A;\, 2p \mid B)/Q^{(ne)}(1s \mid A;\, 1s \mid B). \tag{100}$$

Taking this as justification for assuming that Oppenheimer's $1/n^3$ law is valid for potential (85) and using the law even at moderate energies, they computed the total charge transfer cross section

$$Q^{(ne+nn)}(1s \mid A;\, \Sigma \mid B) = \sum_{\text{discrete } nl} Q^{(ne+nn)}(1s \mid A;\, nl \mid B) \tag{101}$$

from the approximate formula

$$Q^{(ne+nn)}(1s \mid A;\, \Sigma \mid B) = Q^{(ne+nn)}(1s \mid A;\, 1s \mid B) + c\{Q^{(ne+nn)}(1s \mid A;\, 2s \mid B) + Q^{(ne+nn)}(1s \mid A;\, 2p \mid B)\} \tag{102}$$

with

$$c = 8\sum_{n=2}^{\infty} \frac{1}{n^3} = 1.616. \tag{103}$$

Comparison has been made with laboratory data obtained from investigations on the passage of protons through hydrogen gas, taking one hydrogen molecule to be approximately equivalent to two hydrogen atoms. Tuan and Gerjuoy (1960) have, however, shown that such comparison is, for fundamental reasons, improper.

Using the relevant potential of the same type as (85), Schiff (1954) has calculated the cross sections for

$$H(1s) + He^{2+} \to H^+ + He^+(1s, 2s, \text{ or } 2p). \tag{104}$$

Because of the exact energy balance the $1s - 2s$ or $2p$ transitions are found by Schiff to be more probable than the $1s - 1s$ transitions except at extremely high velocities of relative motion. Though his results lend little support for the supposition that the empirical relations (99) and (100) are general, Schiff used a formula similar to (102) to compute the total charge transfer cross section. He found that an alpha particle of energy between about 100 and 600 kev is some 6 times as effective at capturing an electron from atomic hydrogen as is a proton of the same velocity.

Using both a potential like (87) and a potential like (85) and describing the helium atom by the one-parameter variational wave function (36) Bransden *et al.* (1954) have evaluated the cross section for

$$\mathrm{He}(1s^2) + \mathrm{H}^+ \rightarrow \mathrm{He}^+(1s) + \mathrm{H}(1s). \tag{105}$$

From a comparison with the laboratory data available, they concluded that the use of the nuclear-electronic term alone is unsatisfactory, $Q^{(ne)}$ being much too large at least for H^+ energies up to 150 kev. The calculated $Q^{(ne+nn)}$ is in fair accord with the measurements, but undue significance cannot be attached to this since some rather uncertain mathematical simplications were made to reduce the great complications arising from the nuclear-nuclear term. Mapleton (1961) attacked the problem with great vigor some years later. He adopted the same helium wave function as did Bransden *et al.* but avoided further approximation in the evaluation of $Q^{(ne+nn)}(1s;\,1s)$ having at his disposal powerful high-speed computing facilities. Mapleton did not confine his attention to the main reaction path—he also calculated $Q^{(ne+nn)}(nl;\,n'l')$ with $(nl;\,n'l')$ taken as $(1s;\,2s)$, $(1s;\,2p)$, $(1s;\,3s)$, $(1s;\,3p)$, $(1s;\,3d)$, $(2s;\,1s)$, $(2p;\,1s)$, $(2s;\,2s)$, $(2s;\,2p)$, or $(2p;\,2s)$; and he investigated the magnitude of the post-prior discrepancy (which in spite of the rather crude helium wave function employed he found to be not more than 20 % in the single-transition cases). His results (average of the post and the prior) are displayed in Table III. It is seen that simultaneous capture and excitation has quite a considerable cross section. The estimated total charge transfer cross section is in good agreement with the measured values of Stier and Barnett (1956) and Barnett and Reynolds (1958). Further calculations with a more refined helium wave function would be of interest. They should yield larger cross sections at high velocities of relative motion since wave function (36) underestimates the extreme tail of momentum distribution of the $\mathrm{He}(1s)^2$ electrons.

1.2 Impact Parameter Treatment

An account of the derivation of the impact parameter version of the first Born approximation will be given since it is rather less well known than the derivation of the wave version.

TABLE III

CROSS SECTIONS $Q^{(ne+nn)}(nl; n'l')$ FOR $He(1s)^2 + H^+ \rightarrow He^+(nl) + H(n'l')$[a]

$\mathscr{E}$[b]:	7.03	12.5	22.2	39.5	70.3	125	222	395	703	1000
					Cross sections[c] in units of πa_0^2					
$(1s; 1s)$	12.85	7.16	3.75	1.64	5.61^{-1}	1.38^{-1}	2.33^{-2}	2.68^{-3}	2.16^{-4}	4.01^{-5}
$(1s; 2s)$	5.62^{-1}	3.38^{-1}	2.33^{-1}	1.43^{-1}	6.20^{-2}	1.75^{-2}	3.15^{-3}	3.66^{-4}	2.90^{-5}	5.29^{-6}
$(1s; 2p)$	8.18^{-1}	5.63^{-1}	3.00^{-1}	1.33^{-1}	4.35^{-2}	8.93^{-3}	1.07^{-3}	7.62^{-5}	3.52^{-6}	4.54^{-7}
$(1s; 3s)$	1.4^{-1}	8.37^{-2}	5.94^{-2}	3.91^{-2}	1.79^{-2}	5.20^{-3}	9.46^{-4}	1.10^{-4}	8.68^{-6}	1.58^{-6}
$(1s; 3p)$	2.15^{-1}	1.60^{-1}	8.96^{-2}	4.12^{-2}	1.43^{-2}	3.05^{-3}	3.74^{-4}	2.69^{-5}	1.2^{-6}	
$(1s; 3d)$	9.25^{-3}	9.50^{-3}	6.17^{-3}	2.83^{-3}	9.14^{-4}	1.66^{-4}	1.52^{-5}	7.37^{-7}	2.1^{-8}	
$(2s; 1s)$	1.02^{-1}	1.06^{-1}	8.35^{-2}	4.32^{-2}	1.37^{-2}	2.70^{-3}	3.82^{-4}	3.67^{-5}	2.76^{-6}	5.05^{-7}
$(2p; 1s)$	3.65^{-3}	1.26^{-2}	2.22^{-2}	2.0^{-2}	9.49^{-3}	2.44^{-3}	3.55^{-4}	3.11^{-5}	1.61^{-6}	1.98^{-7}
$(2s; 2s)$		5.0^{-3}	5.4^{-3}	3.88^{-3}	1.6^{-3}	3.8^{-4}	5.4^{-5}	5.0^{-6}	3.65^{-7}	6.6^{-8}
$(2s; 2p)$		4.8^{-3}	6.0^{-3}	4.4^{-3}	1.75^{-3}	3.15^{-4}	3.0^{-5}	1.55^{-6}	5.4^{-8}	
$(2p; 2s)$		4.4^{-4}	1.25^{-3}	1.65^{-3}	1.09^{-3}	2.85^{-4}	5.4^{-5}	4.65^{-6}	2.35^{-7}	
$(\Sigma; \Sigma)$[d]					7.47^{-1}	1.84^{-1}	3.08^{-2}	3.45^{-3}	2.73^{-4}	5.0^{-5}

[a] Mapleton, 1961.

[b] $\mathscr{E}$ is the energy of the incident proton in kev.

[c] The superscripts give the power of 10 by which the entries must be multiplied.

[d] The row (Σ, Σ) gives the estimated total charge transfer cross section.

1.2.1 *Excitation and Ionization*

Let the nucleus A of the target be located at the fixed origin of the coordinate system and let the nucleus B of the projectile have position vector **R** and move with constant velocity **v** in the positive sense along a line parallel to and distant ρ from the Z-axis. The electronic wave function may be represented by the expansion

$$X_p(\mathbf{r}_a, t) = \sum_n a_{pn}(t)\, \chi_n(\mathbf{r}_a) \exp\left[-i\epsilon_n t/\hbar\right] \tag{106}$$

where the subscript p indicates the state occupied initially and where $\chi_n(r_a)$ and ϵ_n are the eigenfunctions and eigenenergies in the absence of the interaction potential $V^{\mathrm{B}}(r_b)$ due to the projectile so that

$$\left\{-\frac{\hbar^2}{2m}\nabla^2 + V^{\mathrm{A}}(r_a)\right\}\chi_n(\mathbf{r}_a) = \epsilon_n \chi_n(\mathbf{r}_a), \tag{107}$$

$V^{\mathrm{A}}(r_a)$ being the potential field of the nucleus and core of the target. Reminders of the dependence of the expansion coefficients $a_{pn}(t)$ on the impact parameter ρ and azimuthal angle Φ are not included. The Schrödinger equation for X is

$$\left\{-\frac{\hbar^2}{2m}\nabla^2 + V^{\mathrm{A}} + V^{\mathrm{B}}\right\} X = i\hbar\frac{\partial X}{\partial t}. \tag{108}$$

Substituting from (106) and using (107) it is seen that if the origin of time is chosen so that

$$Z = vt, \tag{109}$$

then

$$i\hbar \sum_n \frac{\partial a_{pn}(Z)}{\partial Z}\chi_n(\mathbf{r}_a)\exp\left[\frac{-i\epsilon_n Z}{\hbar v}\right] = \frac{V^{\mathrm{B}}}{v}\sum_n a_{pn}(Z)\,\chi_n(\mathbf{r}_a)\exp\left[\frac{-i\epsilon_n Z}{\hbar v}\right], \tag{110}$$

the expansion coefficients now being taken to be functions of Z instead of t for convenience. Multiplication by

$$\chi_q^*(\mathbf{r}_a)\exp\left[\frac{i\epsilon_q Z}{\hbar v}\right]$$

followed by integration over $\mathbf{r}_a$ space yields

$$i\hbar\frac{\partial a_{pq}(Z)}{\partial Z} = \frac{1}{v}\sum_n a_{pn}(Z)\mathscr{V}_{qn}(\mathbf{R})\exp\left[\frac{-i}{\hbar v}(\epsilon_n - \epsilon_q)\, Z\right] \tag{111}$$

where

$$\mathscr{V}_{qn}(\mathbf{R}) = \int \chi_q^*(\mathbf{r}_a)\, V^{\mathrm{B}} \chi_n(\mathbf{r}_a)\, d^3\mathbf{r}_a. \tag{112}$$

The boundary conditions are

$$a_{pn}(-\infty) = \delta_{pn}, \tag{113}$$

δ_{pn} being the Kronecker delta function indicated. Substitution of the zero order approximation to the solution

$$a_{pn}(Z) = \delta_{pn} \tag{114}$$

in the right of (111) and integration with respect to Z gives for the first order approximation

$$a_{pp}(Z) = 1 - \frac{i}{\hbar v} \int_{-\infty}^{Z} \mathscr{V}_{pp}(\mathbf{R})\, dZ \tag{115}$$

$$a_{pq}(Z) = \frac{-i}{\hbar v} \int_{-\infty}^{Z} \mathscr{V}_{qp}(\mathbf{R}) \exp\left[\frac{-i}{\hbar v}(\epsilon_p - \epsilon_q)\, Z\right] dZ. \tag{116}$$

The probability that state q is occupied after the encounter is given by

$$\mathscr{P}_{pq} = |\, a_{pq}(\infty)\, |^2 \tag{117}$$

and hence the cross section associated with $p \to q$ transitions is given by

$$Q(p, q) = \int_0^{\infty} \int_0^{2\pi} \rho\, \mathscr{P}_{pq}\, d\rho\, d\Phi \tag{118}$$

which may usually be written

$$Q(p, q) = 2\pi \int_0^{\infty} \rho\, \mathscr{P}_{pq}\, d\rho \tag{119}$$

(Gaunt, 1927). From the physical assumptions made it is apparent that the combination of (116), (117), and (119) is equivalent to the first Born approximation of the wave treatment. This equivalence has been verified mathematically by Arthurs (1961).

The impact parameter version of the first Born approximation does not readily lead to closed expressions for the various cross sections as does the wave version; but it is nevertheless of considerable value, the picture it gives of collisions being very instructive.

When v the velocity of relative motion is sufficiently high, the exponential in the basic formula

$$\mathscr{P}_{pq} = \frac{1}{\hbar^2 v^2} \left| \int_{-\infty}^{\infty} \mathscr{V}_{pq}(\mathbf{R}) \exp \left[\frac{-i(\epsilon_p - \epsilon_q) Z}{\hbar v} \right] dZ \right|^2 \tag{120}$$

may be replaced by unity. It follows† that $\mathscr{P}_{pq}$ ultimately falls off as v^{-2} if $\mathscr{V}_{pq}$ is independent of v, which it is for simple excitation or ionization.

Many oscillations of the exponential occur within the effective range a of $\mathscr{V}_{pq}$ when v is so low that

$$\frac{(\epsilon_p - \epsilon_q)\, a}{\hbar v} \ll 2\pi. \tag{121}$$

In this region $\mathscr{P}_{pq}$ is hence predicted to be a rapidly increasing function of v and to be extremely small. The recognition that many collisions are nearly adiabatic when condition (121) is satisfied is due to Massey (1949). For reasons which will be apparent later exceptions to the rule are not uncommon. Considerations initiated by (121) have led to useful semiempirical formulae (Chapter 18, § 5).

Detailed calculations have been performed for the processes

$$\mathrm{H}(1s) + \mathrm{H}^+ \rightarrow \mathrm{H}(2s \text{ or } 2p) + \mathrm{H}^+ \tag{122}$$

and

$$\mathrm{H}(1s) + \mathrm{H}(1s) \rightarrow \mathrm{H}(2s \text{ or } 2p) + \mathrm{H}(1s) \tag{123}$$

(Bates, 1958a). The asymptotic values of the expansion coefficients concerned may readily be expressed in terms of the modified Bessel functions of the third kind. Representative results are depicted in Fig. 16. As would be expected, distant collisions are relatively more important for high velocities of relative motion than for low velocities of relative motion, for the optically allowed *s-p* transitions than for the optically forbidden *s-s* transitions and for ion-atom collisions than for atom-atom collisions.

Bell (1961), and Bell and Skinner (1962) have carried out a similar study of

$$\mathrm{He}(1s)^2 + \mathrm{H}^+ \rightarrow \mathrm{He}(1s, 2p, \text{ or } 3p\ {}^1P) + \mathrm{H}^+ \tag{124}$$

and

$$\mathrm{Na}(3s) + \mathrm{H}^+ \rightarrow \mathrm{Na}(3p) + \mathrm{H}^+. \tag{125}$$

† The position is otherwise in the case of electron exchange or of charge transfer where the fall off of $\mathscr{P}_{pq}$ with is very rapid.

As they realized, (125) is too strong a transition to be successfully treated by the first Born approximation† (cf. p. 594).

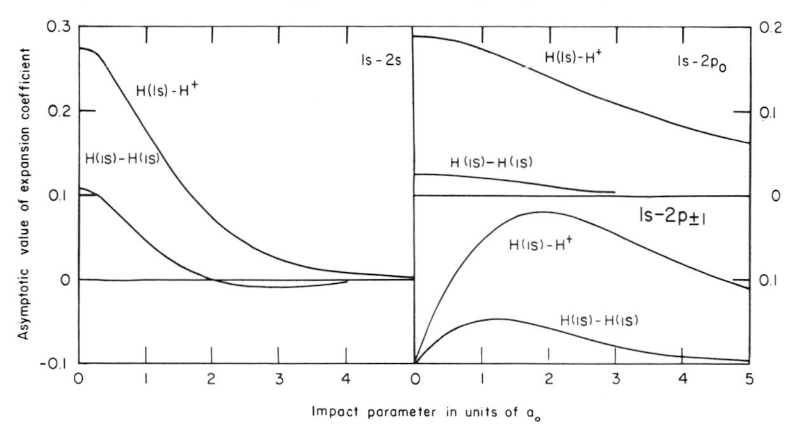

FIG. 16. First Born approximation to $a_{1s,nl}(\infty) - \rho$ curves with $nl = 2s$, $2p_0$, or $2p_{\pm 1}$ for H(1s) — H$^+$ collisions and H(1s) — H(1s) collisions at 100-kev impact energy. (The subscript 0 is used to indicate that the angular part of the eigenfunctions is cos θ and the subscript ± 1 that it is sin θ cos ϕ).

Information on the range of validity of the first Born approximation may be obtained from the impact parameter version. Necessary general conditions are (i) that the nuclei move with constant relative velocity vector and (ii) that the total population in states other than the initial state remains small so that if Z_M is the value of Z at which

$$P(Z) = \sum_{q \neq p} | a_{pq}(Z) |^2 \tag{126}$$

is greatest for any given ρ, then

$$P(Z_M) \ll 1. \tag{127}$$

With the aid of classical orbit theory it may easily be shown that unless very strong Coulomb forces are involved condition (i) is satisfied down to low velocities of relative motion (Bates and Boyd, 1962a).

Condition (ii) may cause quite severe restriction. It is of course satisfied at high velocities of relative motion; but it may be violated at moderate and low velocities of relative motion and indeed the calculated value of $P(Z_M)$ may even exceed unity. The restriction imposed has as yet only been investigated for collisions of the H(1s) — H$^+$ family.

† In the case of head-on collisions the calculated probability of excitation is unity at an impact energy of just over 10 kev.

Even in this case the results available are not sufficient to determine $P(Z_M)$ properly because (a) they relate to the values of the expansion coefficients after the encounter is over, and (b) they concern only the excitation of the 2-quantum level. However, as regards (a) it may be demonstrated that $| a_{1s,nl}(Z_M) |^2$ may be taken to equal $| a_{1s,nl}(\infty) |^2$, at least beyond the maximum of the cross section versus energy curve; and as regards (b), undue error can scarcely be introduced by putting

$$\sum_{nl \neq 1s}' | a_{1s,nl}(\infty) |^2 = r\{| a_{1s,2s}(\infty) |^2 + | a_{1s,2p}(\infty) |^2\} \tag{128}$$

where r is the known ratio of the total inelastic cross section to the excitation cross section for the 2-quantum level. Figure 17 shows P'/Z_b^2

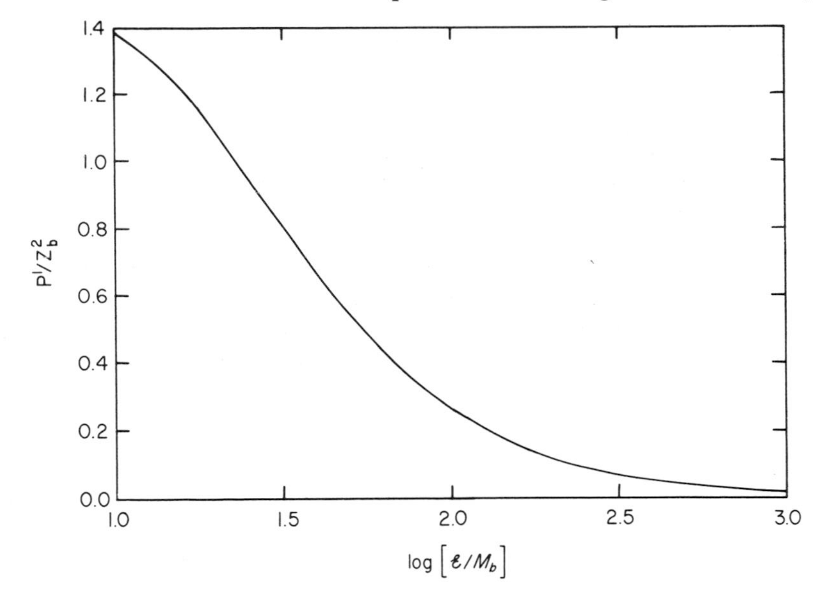

FIG. 17. Estimated first Born approximation to the probability that a nucleus of charge $Z_b e$ mass M_b (on chemical scale) and energy $\mathscr{E}$ (in kev) causes excitation or ionization of a normal hydrogen atom in a head-on collision.

plotted against log $(\mathscr{E}/M_b)$ where P' is the estimated value of $P(Z_M)$ for head-on collisions (which are the most critical) and where $Z_b e$, M_b, and $\mathscr{E}$ are the charge, mass (chemical ^{16}O scale), and energy (kev) of the projectile. If $\mathscr{E}_X$ is the value of $\mathscr{E}$ at which P' falls to the value X, it is reasonable to regard the inequality (127) as being seriously violated when

$$\mathscr{E} < \mathscr{E}_{0.5} \tag{129}$$

and as being satisfied when

$$\mathscr{E} > \mathscr{E}_{0.1}. \tag{130}$$

It may be seen from Fig. 17 that if the projectile is a proton $\mathscr{E}_{0.5}$ and $\mathscr{E}_{0.1}$ are about 50 and 200 kev, respectively. In the case of an alpha particle the corresponding energies are about 16 times greater, that is, about 800 and 3200 kev.

If the probabilities derived from the first Born approximation (or any other approximation) are unacceptably high for some range of impact parameters, it may be advantageous when computing the cross section from (119) to replace them by smaller values even though these be empirical. The contrivance has been used in several cases: for example, by Purcell (1952) and Seaton (1955) in their investigations on the excitation of the *2s-2p* transition of hydrogen by proton impact.

A further necessary general condition is that electron exchange should be unimportant. Little can be said about this at present. Exchange effects naturally fall off rapidly as the velocity of relative motion is increased. They do not, of course, enter at all if the projectile is a bare nucleus.

In considering the range of validity of the first Born approximation, account should ideally be taken of relevant characteristics of the particular transitions of interest. For a given pair of colliding systems a velocity of relative motion which is within the range for one transition may be below the limit to the range for another transition. We shall return to this later (§ 2.2).

1.2.2 *Charge Transfer*

Letting $\varphi_m(\mathbf{r}_b)$ be the eigenfunctions and η_m the eigenenergies in state m around the projectile B it is seen that as an alternative to (106) the electronic wave function may be written in the form

$$X_p(\mathbf{r}_a, t) = \sum_m b_{pm}(t)\,\varphi_m(\mathbf{r}_b)\exp\left[\frac{imvz_a}{\hbar} - \frac{i}{\hbar}(\eta_m + \tfrac{1}{2}mv^2)\,t\right] \tag{131}$$

with

$$z_a = \mathbf{r}_a \cdot \hat{\mathbf{v}}. \tag{132}$$

Each term of this expansion is an exact solution of the Schrödinger equation (108) in the limit of infinite nuclear separation.

Substituting (131) in (108) and proceeding as in § 1.2.1 it may be shown that

$$b_{pq}(\infty) = \frac{-i}{\hbar v}\int_{-\infty}^{\infty} \mathscr{B}_{pq}\,dZ \tag{133}$$

where

$$\mathscr{B}_{pq} = \left[\int \varphi_q^*(\mathbf{r}_b) \exp\left[\frac{-imvz_a}{\hbar}\right] V^A(r_a)\, X_p(\mathbf{r}_a, t)\, d^3\mathbf{r}_a\right] \exp\left[\frac{i}{\hbar v}(\eta_q + \tfrac{1}{2}mv^2)Z\right] \tag{134}$$

If X_p is taken to be $\chi_p(\mathbf{r}_a) \exp[-i\epsilon_p t/\hbar]$ as usual, (134) becomes

$$\mathscr{B}_{pq} = \int \varphi_q^*(\mathbf{r}_b)\, V^A(r_a)\, \chi_p(\mathbf{r}_a) \exp\left[\frac{-imvz_a}{\hbar}\right] d^3\mathbf{r}_a$$

$$\exp\left[\frac{-i}{\hbar v}(\epsilon_p - \eta_q - \tfrac{1}{2}mv^2)\, Z\right] \tag{135}$$

which on putting

$$z = z_a - \tfrac{1}{2} Z \tag{136}$$

reduces to

$$\mathscr{B}_{pq} = g_{pq} \exp\left[\frac{-i}{\hbar v}(\epsilon_p - \eta_q)\, Z\right] \tag{137}$$

where

$$g_{pq} = \int \varphi_q^*(\mathbf{r}_b)\, V^A(r_a)\, \chi_p(\mathbf{r}_a) \exp\left[\frac{-imvz}{\hbar}\right] d^3\mathbf{r}_a. \tag{138}$$

The position regarding the post and prior interactions is of course the same as in the wave treatment (cf. § 1.1.3).

Since the nuclei are assumed to move with *constant* relative velocity vector, it is clear that the force between them cannot affect the probability of charge transfer. The impact parameter treatment might therefore be judged to require that the interaction potential contain only the nuclear-electronic term. However, $V^A(r_a)$, introduced in (107) and appearing in (138), is not in fact uniquely defined. Formally, any function of **R**, including the controversial nuclear-nuclear term, may be added. As in § 1.1.3 the arbitrariness in the results is due to the nonorthogonality of the eigenfunctions (Takayanagi, 1955; Bates, 1958b; Sil, 1960). Taking proper cognisance of the nonorthogonality, it may be shown that g_{qp} in (137) should be replaced by

$$\frac{h_{qp} - S_{qp}h_{pp}}{1 - |S_{qp}|^2} \tag{139}$$

where

$$h_{qp} = \int \varphi_q^*(\mathbf{r}_b)\, V^B(r_b)\, \chi_p(\mathbf{r}_a) \exp\left[\frac{-imvz}{\hbar}\right] d^3\mathbf{r}_a \tag{140}$$

$$h_{pp} = \int \chi_p^*(\mathbf{r}_a)\, V^B(r_b)\, \chi_p(\mathbf{r}_a)\, d^3\mathbf{r}_a \tag{141}$$

and

$$S_{qp} = \int \varphi_q^*(\mathbf{r}_b)\, \chi_p\, (\mathbf{r}_a) \exp\left[\frac{-imvz}{\hbar}\right] d^3\mathbf{r}_a. \tag{142}$$

It is observed that (139) is not affected by the addition of a function of $\mathbf{R}$ to $V^{\mathrm{B}}(r_b)$.

Using (133), (137), and (138) with only the nuclear-electronic term in the interaction potential, Brinkman and Kramers (1930) have obtained an expression for the probability of charge transfer from the $1s$ state around a nucleus of charge $Z_a e$ to the $1s$ state around a nucleus of charge $Z_b e$. In the important case of charge transfer from hydrogen atoms to protons their expression for the probability reduces to†

$$\mathscr{P}^{(ne)}_{1s\mathrm{A},1s\mathrm{B}} = \frac{2^6 \rho^4}{s^2(4+s^2)^2}\{K_2(\tfrac{1}{2}\rho[4+s^2]^{1/2})\}^2 \tag{143}$$

in which the impact parameter ρ is in atomic units, s denotes $mv/\hbar$ and K_2 is the modified Bessel function. It is apparent that an increase in v, when v is high, leads to an increase in the *relative* contribution of close encounters to the charge transfer cross section; and that in the limit the only encounters which are effective are those for which ρ is vanishingly small. This is just what would be expected physically, bearing in mind the difficulty of giving much momentum to an electron in an encounter (cf. Bates and McCarroll, 1962).

The inclusion of the nuclear-nuclear term in the interaction potential is not easy mathematically, but a rather complicated expression for $\mathscr{P}^{(ne+nn)}_{1s\mathrm{A},1s\mathrm{B}}$ has been obtained by Schiff (1954). The evaluation of $\mathscr{P}^{(ortho)}_{1s\mathrm{A},1s\mathrm{B}}$ is best carried out by numerical methods (Bassel and Gerjuoy, 1960; McCarroll, 1961). A closed formula may be obtained if the denominator of (139) is replaced by unity (Murakaver, 1961).

Even when the energy $\mathscr{E}$ is moderate, $\mathscr{P}^{(ne)}_{1s\mathrm{A},1s\mathrm{B}}$ exceeds unity over a range of ρ giving a major contribution to the calculated charge transfer cross section and $\mathscr{P}^{(ne+nn)}_{1s\mathrm{A},1s\mathrm{B}}$ exceeds unity for small ρ (tending logarithmically to infinity as ρ tends to zero); but unless $\mathscr{E}$ is low, $\mathscr{P}^{(ortho)}_{1s\mathrm{A},1s\mathrm{B}}$ behaves properly, which is satisfactory since it is based on what is believed to be the correct treatment. The fact that the calculated values of $\mathscr{P}^{(ne)}_{1s\mathrm{A},1s\mathrm{B}}$, $\mathscr{P}^{(ne+nn)}_{1s\mathrm{A},1s\mathrm{B}}$, and $\mathscr{P}^{(ortho)}_{1s\mathrm{A},1s\mathrm{B}}$ are not all less than unity is not in itself evidence that the transition matrix elements involved are basically incorrect—it merely means that back-coupling to the initial state should be taken into account.

† The superscripts (ne), $(ne + nn)$, and $(ortho)$ have the significance assigned to them on page 573.

Similar calculations on

$$\mathrm{H}(1s) + \mathrm{He}^{2+} \rightarrow \mathrm{H}^{+} + \mathrm{He}^{+}(1s,\ 2s,\ \text{or}\ 2p) \tag{144}$$

have been carried out by McCarroll and McElroy (1962) and McElroy (1962). However, the effect of distortion (§ 2.1) on the cross sections of these is very great (cf. page 592).

2 Higher Approximations

There are two complementary approaches to the problem of developing satisfactory higher approximations: to begin by taking either (i) full account of what are judged to be the more important matrix elements neglected in the first Born approximation or (ii) partial account of all the matrix elements. Approach (i) is followed in the distortion approximation and approach (ii) is attempted in the second Born approximation.

Inelastic collisions between pairs of ions are usually no more difficult to treat by the higher approximation than are inelastic collisions between ions and neutral atoms because the Coulomb force may be ignored unless the velocity of relative motion v is very low (Bates and Boyd, 1962a). The excitation of hydrogenic ions by bare nuclei is especially simple. It is found that $\{Z_a^4/Z_b^2\}Q$ is a function only of Z_a/Z_b and of v/Z_a where Q is the excitation cross section and $Z_a e$ and $Z_b e$ are the nuclear charges of the target and projectile, respectively (Bates, 1961b). Theoretical (or indeed experimental) results obtained on hydrogen atoms may hence be scaled to apply to hydrogenic ions.

2.1 Distortion and Other Special Approximations

Since it is simpler than the wave treatment, the impact parameter treatment will be used.

The diagonal matrix elements on the right of (111) are not multiplied by oscillatory exponential factors as are, in general, the other matrix elements. They are therefore likely to be particularly important. Allowance for them may easily be made. Putting

$$a_{pq}(Z) = c_{pq}(Z) \exp\left[\frac{-i}{\hbar v}\int_0^Z \mathscr{V}_{qq}\, dZ\right], \tag{145}$$

(111) for $\partial a_{pq}(Z)/\partial Z$ becomes

$$i\hbar \frac{\partial c_{pq}(Z)}{\partial Z} = \frac{1}{v} \sum_{n \neq q}' c_{pn}(Z) \mathscr{V}_{qn}(\mathbf{R}) \exp\left[\frac{-i}{\hbar v} \gamma_{nq}(Z)\right] \tag{146}$$

with

$$\gamma_{nq}(Z) = (\epsilon_n - \epsilon_q) Z + \int_0^Z \{\mathscr{V}_{nn}(\mathbf{R}) - \mathscr{V}_{qq}(\mathbf{R})\} dZ. \tag{147}$$

The first order approximation to the solution may be obtained by substituting the zero order approximation

$$c_{pn}(Z) = \delta_{pn} \tag{148}$$

in the right and then integrating. It gives the probability of a transition from state p to state q to be

$$\mathscr{P}_{pq} = \frac{1}{\hbar^2 v^2} \left| \int_{-\infty}^{\infty} \mathscr{V}_{qp}(\mathbf{R}) \exp\left[\frac{-i}{\hbar v} \gamma_{pq}(Z)\right] dZ \right|^2. \tag{149}$$

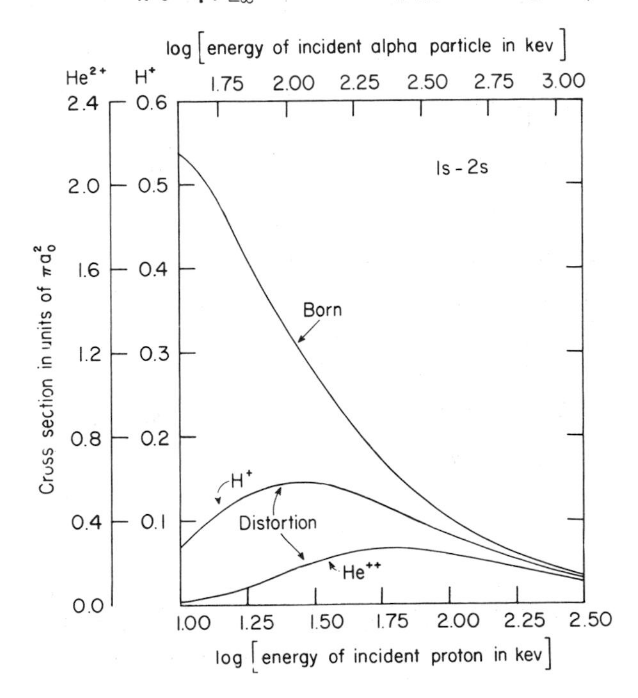

FIG. 18. Comparison of first Born and distortion approximations to the cross section—energy curves for $H(1s) + H^+$ (or He^{2+}) $\rightarrow H(2s) + H^+$(or He^{2+}). The lower and inner scales refer to H^+ impact; and the upper and outer scales to He^{2+} impact.

This is the distortion approximation. The simple derivation presented above is that given originally by Bates (1959). Another derivation has been given by Mittleman (1961). Clearly (149) differs from (120) of the first Born approximation only in that perturbed values of the eigen-energies replace the exact values.

Figures 18 and 19 compare the cross sections for

$$\mathrm{H}(1s) + \mathrm{H}^{+}(\text{or } \mathrm{He}^{2+}) \rightarrow \mathrm{H}(2s) + \mathrm{H}^{+}(\text{or } \mathrm{He}^{2+}) \tag{150}$$

and

$$\mathrm{H}(1s) + \mathrm{H}^{+}(\text{or } \mathrm{He}^{2+}) \rightarrow \mathrm{H}(2p) + \mathrm{H}^{+}(\text{or } \mathrm{He}^{2+}) \tag{151}$$

obtained by the distortion approximation with those obtained by the first Born approximation (Bates, 1959, 1961a). As would be expected, the greater the charge on the projectile the greater the effect of distortion. It is seen that distortion is much more important for (150) than for (151) and further that its influence is not even the same qualitatively, the cross sections for (150) being decreased at all energies but the cross section for (151) being decreased only at low energies and being increased

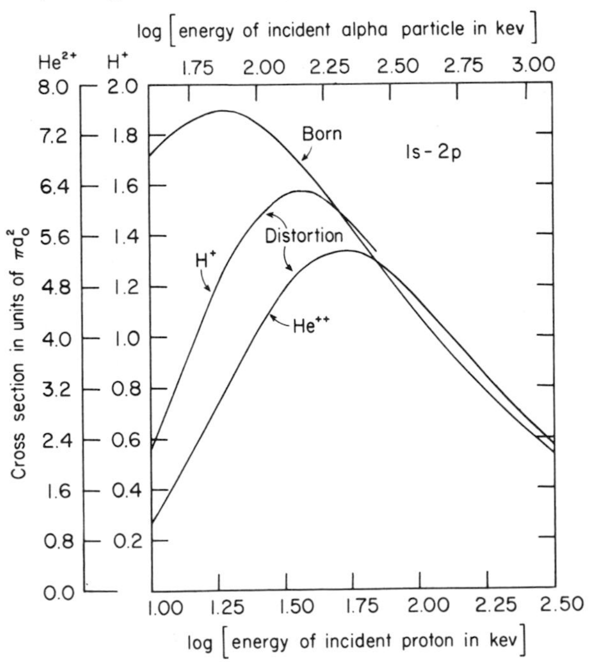

FIG. 19. Comparison of first Born and distortion approximations to the cross section—energy curves for $\mathrm{H}(1s) + \mathrm{H}^{+}$ (or He^{2+}) $\rightarrow \mathrm{H}(2p) + \mathrm{H}^{+}$ (or He^{2+}). The lower and inner scales refer to H^{+} impact; and the upper and outer scales to He^{2+} impact.

at high energies. The difference arises from two causes. Firstly, close encounters contribute relatively more to the $1s$-$2s$ transitions than to the $1s$-$2p$ transitions and such encounters are more affected by distortion than are distant encounters. Secondly, $\mathscr{V}_{2s,1s}$ and $\mathscr{V}_{2p_{\pm 1},1s}$ are symmetrical with respect to Z, but $\mathscr{V}_{2p_0,1s}$ is antisymmetrical so that (149) gives

$$\mathscr{P}_{1s,2s} = \frac{4}{\hbar^2 v^2} \left| \int_0^\infty \mathscr{V}_{2s,1s}(\mathbf{R}) \cos\left\{\frac{\gamma_{1s,2s}(Z)}{\hbar v}\right\} dZ \right|^2 \tag{152}$$

and

$$\mathscr{P}_{1s,2p_{\pm 1}} = \frac{4}{\hbar^2 v^2} \left| \int_0^\infty \mathscr{V}_{2p_{\pm 1},1s}(\mathbf{R}) \cos\left\{\frac{\gamma_{1s,2p_{\pm 1}}(Z)}{\hbar v}\right\} dZ \right|^2 \tag{153}$$

with the exponential replaced by a *cosine* function in both cases, but gives

$$\mathscr{P}_{1s,2p_0} = \frac{4}{\hbar^2 v^2} \left| \int_0^\infty \mathscr{V}_{2p_0,1s}(\mathbf{R}) \sin\left\{\frac{\gamma_{1s,2p_0}(Z)}{\hbar v}\right\} dZ \right|^2 \tag{154}$$

with the exponential replaced by a *sine* function. The interaction makes the initial and final potential energy curves separate as the colliding systems approach and thus makes

$$-\gamma_{1s,2s}(Z),\ -\gamma_{1s,2p_{\pm 1}}(Z) \text{ and } -\gamma_{1s,2p_0}(Z) > (\epsilon_{2s \text{ or } 2p} - \epsilon_{1s})\, Z. \tag{155}$$

It follows that when the energy is high, distortion *decreases* the calculated values of $Q(1s, 2s)$ and $Q(1s, 2p_{\pm 1})$ but *increases* that of $Q(1s, 2p_0)$. In the region where the effects of distortion on $Q(1s, 2p_0)$ and $Q(1s, 2p_{\pm 1})$ are in the opposite sense, the effect on

$$Q(1s, 2p) \equiv Q(1s, 2p_0) + Q(1s, 2p_{\pm 1}) \tag{156}$$

is, of course, reduced by cancellation. Owing to the difference in the relative importance of close and distant collisions in the two cases, the effect of distortion on $Q(1s, 2p_{\pm 1})$ is less than on $Q(1s, 2p_0)$ and hence the tendency is for $Q(1s, 2p)$ to be increased.

It should not be assumed that accurate cross sections for processes (150) and (151) are obtained merely by taking distortion into account. Other refinements are needed.

Processes (150) and (151) lead to states of the same energy.† The cross section $Q(1s, 2s)$ for (150) is small. Its value would be expected to be affected markedly by the existence of (151) for which the cross

† The degeneracy gives rise to secular terms in (111) in addition to those taken into account in the distortion approximation.

section $Q(1s, 2p)$ is quite large. As will be seen in § 2.2, this expectation is fulfilled. The influence of (150) on (151) is probably much less important than the inverse, but it requires investigation. Using (146) it is a trivial task to write down the relevant coupled differential equations, but to solve these equations entails much labor.

If the projectile is a proton the values of $Q(1s, 2s)$ and $Q(1s, 2p)$ are likely to be affected by the resonance charge transfer processes

$$H(1s) + H^+ \rightarrow H^+ + H(1s) \tag{157}$$

and

$$H(2s, 2p) + H^+ \rightarrow H^+ + H(2s, 2p) \tag{158}$$

(which have very large cross sections at low energies). It should be noted too (cf. § 3) that in close encounters there is very strong coupling between the $2p\sigma$ and $2p\pi$ states of the quasi-H_2^+ molecule [which correspond in the separated atoms limit to $H(1s) + H^+$ and to $H(2p_x) + H^+$, respectively]. Clearly H^+ and $He^+(1s)$ ions of the same speed must differ more in their effectiveness in exciting (or ionizing) atomic hydrogen than might be expected from the slight extent (cf. Fig. 9) they differ in their potential fields.

Bell (1961) has investigated the excitation of normal helium atoms to the $1s2p\ ^1P$ and $1s3p\ ^1P$ levels by proton and alpha particle impact. He found that distortion is more important for the higher level than the lower. The effect on the cross section versus impact energy curves is qualitatively as depicted in Fig. 19.

Distortion similarly modifies the first Born approximation for the probability of charge transfer, the eigenenergies of the isolated systems in (137) being replaced by the values of these eigenenergies when perturbed by the interaction (Bates, 1958b). In the case of symmetrical resonance charge transfer the corrections to the eigenenergies corresponding to the initial and final states are naturally the same so that the effect of distortion vanishes because of cancellation. McCarroll and McElroy (1962) have investigated

$$H(1s) + He^{2+} \rightarrow H^+ + He^+(1s) \tag{159}$$

for which the distortion is unusually great, since though there is no Coulomb interaction between the initial systems there is a Coulomb interaction between the final systems. The potential energy surfaces involved are much closer together at small nuclear separations. In consequence, the inclusion of allowance for distortion leads to an increase in the cross section (Fig. 20). McElroy (1962) has also investigated the asymmetric or (accidental) resonance (§ 3.1.2) processes,

$$H(1s) + He^{2+} \rightarrow H^{+} + He^{+}(2s \text{ or } 2p). \tag{160}$$

The most striking effect of distortion on these is to make the cross sections fall very rapidly as the impact energy is decreased towards zero. Distortion has little influence on (98) which differs from (160) in that a Coulomb interaction is not involved (McElroy, 1962).

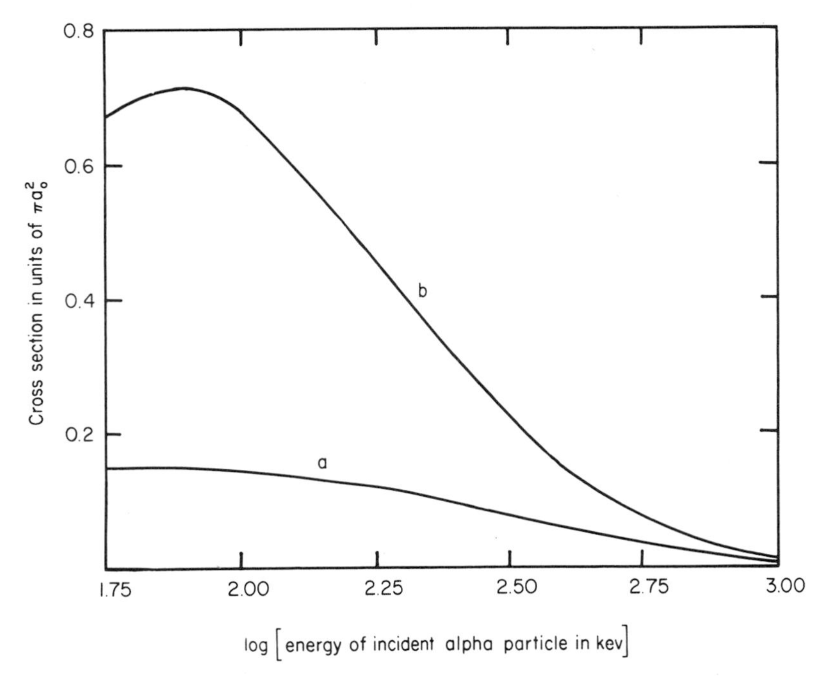

FIG. 20. Effect of distortion on cross section—impact energy curve for $H(1s) + He^{2+} \rightarrow H^{+} + He^{+}(1s)$. Curve a is $Q^{(ortho)}$ ignoring distortion; curve b is $Q^{(ortho)}$ with allowance for distortion (McCarroll and McElroy, 1962).

Matrix elements other than those associated with distortion may readily be taken into account in principle, but they are rather troublesome in practice, since they give rise to coupled differential equations. Little detailed work has yet been done.

Skinner (1962) has refined the distortion calculations on (151) by including the matrix element joining the $2p_0$ and $2p_{\pm 1}$ states. The coupling influences the components $1s \rightarrow 2p_0$ and $1s \rightarrow 2p_{\pm 1}$ markedly but in the opposite sense; it influences the complete transition $1s \rightarrow 2p$ only slightly (cf. Table IV). Bell and Skinner (1962) later investigated the effect of back-coupling to the $1s$ state finding it to be small. Back-coupling is, however, very important in the excitation of the resonance line of sodium by proton or alpha-particle impact or in other processes involving a strong transition. Taking account of it naturally ensures that the calculated probability of excitation is less than unity.

TABLE IV

PROCESS $H(1s) + H^+ \rightarrow H(2p) + H^+$ [a,b]

Transition	$1s \rightarrow 2p_0$			$1s \rightarrow 2p_{\pm 1}$			Total $1s \rightarrow 2p$		
	Cross section in units of πa_0^2								
$\log \mathscr{E}$	1B	D	DC ($p_0 - p_{\pm 1}$)	1B	D	DC ($p_0 - p_{\pm 1}$)	1B	D	DC ($p_0 - p_{\pm 1}$)
1.00	1.072	0.287	0.520	0.650	0.274	0.120	1.722	0.561	0.640
1.25	1.054	0.685	1.002	0.837	0.516	0.341	1.891	1.201	1.343
1.50	0.859	0.862	1.103	0.896	0.697	0.560	1.755	1.559	1.663
1.75	0.620	0.724	0.875	0.815	0.722	0.653	1.435	1.446	1.528
2.00	0.411	0.498	0.577	0.680	0.636	0.607	1.091	1.135	1.184
2.25	0.257	0.310	0.344	0.524	0.507	0.496	0.781	0.817	0.840
2.50	0.155	0.179	0.195	0.380	0.375	0.371	0.535	0.554	0.566

[a] Skinner, 1962.

[b] Column 1B refers to the first Born approximation, column D to the distortion approximation and column DC ($p_0 - p_{\pm 1}$) to the approximation in which account is taken of distortion and of the coupling between the $2p_0$ and $2p_{\pm 1}$ states; $\mathscr{E}$ is the energy of the incident proton in kev.

2.2 SECOND BORN APPROXIMATION

Substitution of the first order approximation (115) and (116) in the right of the set of differential equations (111) for the expansion coefficients followed by integration with respect to Z yields the second order approximation

$$a_{pq}(\infty) = \frac{-i}{\hbar v}\left[\int_{-\infty}^{\infty} \mathscr{V}_{qp}(\mathbf{R}) \exp\left\{\frac{-i}{\hbar v}(\epsilon_p - \epsilon_q) Z\right\} dZ + \Delta_{pq}\right] \quad (161)$$

in which the correction term is given by

$$\Delta_{pq} = \frac{-i}{\hbar v}\sum_n \int_{-\infty}^{\infty}\left\{\int_{-\infty}^{Z} \mathscr{V}_{np}(\mathbf{R}') \exp\left\{\frac{-i}{\hbar v}(\epsilon_p - \epsilon_q) Z'\right\} dZ'\right\} \mathscr{V}_{qn}(\mathbf{R}) \exp\left\{\frac{-i}{\hbar v}(\epsilon_n - \epsilon_q) Z\right\} dZ \quad (162)$$

where $\mathbf{R}'$ has coordinates (Z', ρ, Φ). The second Born approximation to the probability that the transition occurs is then

$$\mathscr{P}_{pq} = |\, a_{pq}(\infty) \,|^2_{X4\text{th}} \tag{163}$$

the postscript X4th being a reminder that *no* terms of the fourth order in the interaction potential are to be included in evaluating $\mathscr{P}_{pq}$ from (161) and (163) since *all* such terms are not known. Failure to heed the reminder may lead to serious error (cf. Kingston *et al.*, 1960).

The second Born correction term Δ_{pq} may be regarded as allowing for the possibility that the $p \to q$ transition may take place through the $p \to n \to q$ sequence of virtual transitions. Distortion (as understood in § 2.1) corresponds to the $p \to p \to q$ and $p \to q \to q$ sequences. Integration by parts shows that its effective contribution to Δ_{pq} is

$$\Delta'_{pq}(\text{dist}) = \frac{-i}{\hbar v} \int_{-\infty}^{\infty} \left[\int_{-\infty}^{Z} \left\{ \mathscr{V}_{pp}(\mathbf{R}') - \mathscr{V}_{qq}(\mathbf{R}') \right\} dZ' \right.$$

$$\left. \mathscr{V}_{qp}(\mathbf{R}) \exp \left\{ \frac{-i}{\hbar v} (\epsilon_p - \epsilon_q) Z \right\} dZ. \right] \tag{164}$$

Similarities between the structure of the integral in (149) and the structure of (161) with Δ_{pq} replaced by Δ'_{pq} (dist) are immediately apparent. The matrix element $\mathscr{V}_{qp}$ is contained in the right of (164) as in the leading term of (161) so that the importance of the partial correction term Δ'_{pq} (dist) does not depend on whether the transition is weak or strong. The full second Born correction Δ_{pq} has different properties. It is especially important if $p \to q$ is a weak transition, and if there is a state n such that $p \to n$ and $n \to q$ are strong transitions: for example, the excitation of the $1s \to 2s$ transition of atomic hydrogen by proton impact is markedly influenced by the $1s \to 2p \to 2s$ sequence (Kingston *et al.*, 1960). In contrast, if $p \to q$ is a strong transition Δ_{pq} is likely to be unimportant (excepting perhaps for the part due to distortion). However, on proceeding to the third Born approximation, a further correction is introduced which includes a contribution (corresponding to what is called *back-coupling*) from the $p \to q \to p \to q$ sequence. For strong transition this contribution may be considerable; thus in the excitation of sodium by protons the $3s - 3p$ transition is greatly influenced by the $3s \to 3p \to 3s \to 3p$ sequence (Bell and Skinner, 1962). The existence of sequences enabling the final state to be reached indirectly through intermediate states does not of course necessarily make the cross section greater than it would otherwise be. Because of interference the cross section may be diminished. In general, the tendency of sequences is to strengthen weak transitions and weaken strong transitions. For excitation and ionization the effects associated with the $(N+1)$th Born approximation fall off with one higher power of the velocity of relative motion than do the effects associated with the Nth Born approximation; but for charge transfer this is not necessarily the case (Bates and McCarroll, 1962).

Attempts at carrying out the summation in (162) analytically have not proven successful. It has as yet only been possible to take a few of the terms into account.

Kingston *et al.* (1960) have used the second Born approximation in an investigation of the excitation of the $1s \rightarrow 2s$ transition of atomic hydrogen by proton impact. They followed the wave treatment since they were also interested in excitation by electrons. They confined their attention to (a) $1s \rightarrow 1s \rightarrow 2s$ and $1s \rightarrow 2s \rightarrow 2s$ (which correspond to distortion) and (b) $1s \rightarrow 2p \rightarrow 2s$ (which is responsible for the main part of what may be regarded as the effect of the polarization of the atom by the incident proton). With (a) alone they obtained excellent agreement with the results of the distortion approximation; with (b) in addition they found that the calculated cross section increased to close to that given by the first Born approximation, the effects of distortion and polarization almost canceling (Table V).

TABLE V

PROCESS $H(1s) + H^+ \rightarrow H(2s) + H^+$ [a]

Case[b]	$k = 1.0$	1.5	2.0	3.0[c]
	Cross section in units of πa_0^2			
(i)	0.290	0.148	0.0877	0.0404
(ii)	0.148	0.110	0.0743	0.0376
(iii)	0.161	0.119	0.0777	0.0383
(iv)	0.588	0.243	0.125	0.0495

[a] Kingston *et al.*, 1960.

[b] (i) First Born approximation; (ii) distortion approximation; (iii) second Born approximation including only the direct transition and the $1s \rightarrow 1s \rightarrow 2s$ and $1s \rightarrow 2s \rightarrow 2s$ transitions (distortion effects); (iv) second Born approximation including also the $1s \rightarrow 2p \rightarrow 2s$ transition (polarization effects).

[c] k ($\equiv Mv/\hbar$) is such that $29.6k^2$ is the energy of the incident proton in kev.

3 *Slow Collisions*

The range of velocities of relative motion over which reliable calculations on collision cross sections may be carried out can in practice be extended only moderately by going from the first Born approximation to the distortion, second Born or similar approximations since the number of matrix elements which are important generally increases rapidly as the velocity is decreased. It is clear that the expansion in terms of the eigenfunctions of the isolated target is unsuitable for the study of slow encounters. Mott (1931) suggested that the expansion should be in terms of the eigenfunctions which would describe the quasi-molecule formed by the target and the projectile if their relative position vector **R** were momentarily fixed. This expansion forms the basis of what is called the *perturbed stationary state* or *pss* approximation (cf. Mott and Massey, 1949; Bates *et al.*, 1953).

In discussing the *pss* approximation, we shall in the first instance exclude the special case of symmetrical resonance charge transfer.

It was originally assumed that a transition from one state of the quasi-molecule to another is very unlikely in a slow encounter and therefore that the only matrix element which need be taken into account is that joining the states of the quasi-molecule which tend in the separated atoms limit to the initial and final states of the target.† On this false premise the impact parameter treatment gives that the probability $\mathscr{P}_{pq}$ of the target being excited from state p to state q is $|a_{pq}(\infty)|^2$ where if $\psi_p(\mathbf{r} \mid \mathbf{R})$ and $\psi_q(\mathbf{r} \mid \mathbf{R})$ are the eigenfunctions and $E_p(R)$ and $E_q(R)$ are the eigenenergies of the appropriate states of the quasi-molecule, then

$$a_{pq}(\infty) = -\int_{-\infty}^{\infty} \mathscr{M}(p, q \mid \mathbf{R}) \exp\left\{\frac{-i}{\hbar v}\int^{Z} [E_p(R') - E_q(R')]\, dZ'\right\} dZ \tag{165}$$

with

$$\mathscr{M}(p, q \mid \mathbf{R}) = \frac{Z}{R}\int \psi_q^*(\mathbf{r} \mid \mathbf{R}) \frac{\partial}{\partial R} \psi_p(\mathbf{r} \mid \mathbf{R})\, d^3\mathbf{r} + \frac{\rho}{R^2}\int \psi_q^*(\mathbf{r} \mid \mathbf{R})\, i\mathscr{L}_x \psi_p(\mathbf{r} \mid \mathbf{R})\, d^3\mathbf{r} \tag{166}$$

† To avoid unnecessary complication it is supposed throughout this chapter that there is a one-one correlation between the molecular and the atomic states.

and

$$i\mathscr{L}_x = -\sin\phi\,\frac{\partial}{\partial\theta} - \cot\theta\cos\phi\,\frac{\partial}{\partial\phi}, \tag{167}$$

θ and ϕ being the angular coordinates of $\mathbf{r}$ referred to a rotating system (with z polar axis along the internuclear line).

Until a few years ago it was widely but incorrectly believed that the *pss* approximation, formulated as just indicated, tends to the first Born approximation in the limit of weak interactions and high velocities of relative motion. Apparent proofs of this have been advanced. They are invalid because they fail to make proper allowance for the rotation of the internuclear line during the encounter.

An understanding of the *pss* approximation may most readily be gained by examining its true limit the *perturbed rotating atom* or *pra* approximation. In this (Bates, 1957), the eigenfunctions of the quasi-molecule are replaced by the eigenfunctions of the target perturbed by the projectile, the same rotating frame of reference being used; and the eigenenergies of the quasi-molecule are replaced by the eigenenergies of the isolated target—for consistency they should strictly be replaced by the eigenenergies of the target perturbed by the projectile but the difference is unimportant at high velocities of relative motion. With the replacements in (166) which have just been indicated, Bates (1958a) carried out calculations on the excitation of the $1s \rightarrow 2s$ and $1s \rightarrow 2p$ transitions of hydrogen by protons and by other hydrogen atoms. For the $1s \rightarrow 2s$ transition the *pra* approximation naturally gives precisely the same results as does the first Born approximation (the atomic eigenfunctions involved being spherically symmetrical). The position with regard to the $1s \rightarrow 2p$ transition is far otherwise. Taking the case of proton impact, the *pra* approximation gives

$$a_{1s,2p_0}(\infty) = \frac{16\sqrt{2}\alpha\rho}{9}\left[\frac{128}{81}\int_0^{3/2}\frac{t^2K_1(\rho[\alpha^2+t^2]^{1/2})}{(\alpha^2+t^2)^{1/2}}\,dt - \frac{3\rho}{2A^2}K_0(\rho A) - \frac{2}{A}\left(1+\frac{3}{2A^2}\right)K_1(\rho A)\right] \tag{168}$$

and

$$a_{1s,2p_{\pm1}}(\infty) = \frac{8\sqrt{2}\rho}{9}\left[\frac{128}{81}\int_0^{3/2} t^2K_0(\rho[\alpha^2+t^2]^{1/2})\,dt - \tfrac{4}{3}K_0(\rho A)\right] \tag{169}$$

whereas the first Born approximation gives

$$a_{1s,2p_0}(\infty) = \frac{8\sqrt{2}\alpha^2}{9}\left[\frac{256}{81}K_0(\rho\alpha) - \left\{\frac{256}{81}+\frac{2\rho^2}{A^2}\right\}K_0(\rho A) - \frac{\rho}{A}\left\{\frac{32}{9}+\frac{4}{A^2}\right\}K_1(\rho A)\right] \tag{170}$$

and

$$a_{1s,2p_{\pm 1}}(\infty) = \frac{8\sqrt{2}\alpha}{9}\left[\frac{256\alpha}{81} K_1(\rho\alpha) - \frac{32\rho}{9} K_0(\rho A) - A\left\{\frac{256}{81} + \frac{2\rho^2}{A^2}\right\} K_1(\rho A)\right] \tag{171}$$

in which

$$\alpha = \frac{3}{8v}, \quad A = \frac{3}{2}\left(1 + \frac{1}{16v^2}\right)^{1/2}, \tag{172}$$

v is the velocity of relative motion in atomic units, and the K's are the modified Bessel functions. The results clearly differ. In particular, for large impact parameters ρ combination of (168) and (169) yields

$$\mathscr{P}_{1s,2p} \sim \frac{2^{11}\pi\rho}{3^7} \exp(-3\rho) \tag{173}$$

as the velocity of relative motion v tends to infinity, which is absurd, while combination of (170) and (171) yields

$$\mathscr{P}_{1s,2p} \sim \frac{2^6\pi\rho^3}{3^5 v^2} \exp(-3\rho), \tag{174}$$

which is correct. The difference for slow encounters is also striking. As may be seen most easily from the wave treatment, the cross section given by the *pra* approximation again rises above the cross section given by the first Born approximation at low v and indeed passes through a subsidiary maximum (Bates, 1957). Close encounters are here responsible.

At any given v the assumption that the interaction of the projectile causes the eigenfunctions to follow the rotation of the internuclear line is invalid if ρ is less than some value ρ_1 since the rotation is then too rapid and is invalid also if ρ is more than some value ρ_2 since the interaction is then too weak. As v is decreased ρ_1 becomes smaller and ρ_2 becomes greater. The assumption regarding the eigenfunctions is clearly completely invalid when v is high. From what has just been said concerning ρ_1 and ρ_2 it might be thought that if v were decreased enough the assumption would become justified over a sufficient range of ρ to ensure that it could not lead to serious error in the excitation cross section. This is not the case. The region $\rho < \rho_1$ remains of major importance (Bates, 1958a).

Allowance must be made in the *pra* approximatuon for the reluctance of the eigenfunctions to follow the rotation of the internuclear line. This reluctance is acknowledged mathematically by the existence of coupling between states differing only in magnetic quantum number. The coupling is strong. When account is taken of it the modified *pra* approximation tends to the first Born approximation in the weak interaction and high velocity limits (and indeed is equivalent to the first Born approximation if unperturbed eigenenergies are used).

In the case of the $1s \rightarrow 2p$ transition the modified equations for the expansion coefficients are

$$\frac{\partial a_{1s,2p_0}}{\partial Z} = \left\{ \frac{\partial}{\partial Z} \mathscr{I}^0_{1s,2p}(R) \right\} \exp\left[\frac{-i}{\hbar v} \beta_{1s,2p_0}\right] + \underline{a_{1s,2p_{\pm 1}} \frac{\rho}{R^2} \exp\left[\frac{-i}{\hbar v} \beta_{2p_{\pm 1},2p_0}\right]} \tag{175}$$

and

$$\frac{\partial a_{1s,2p_{\pm 1}}}{\partial Z} = -\frac{\rho}{R^2} \mathscr{I}^0_{1s,2p}(R) \exp\left[\frac{-i}{\hbar v} \beta_{1s,2p_{\pm 1}}\right] \underline{- a_{1s,2p_0} \frac{\rho}{R^2} \exp\left[\frac{-i}{\hbar v} \beta_{2p_0,2p_{\pm 1}}\right]} \tag{176}$$

with

$$\mathscr{I}^0_{1s,2p} = \frac{1}{\eta_{1s}(\infty) - \eta_{2p}(\infty)} \int \chi^*_{1s}(\mathbf{r})\, V(\mathbf{r}, \mathbf{R})\, \chi_{2p_0}(\mathbf{r})\, d^3\mathbf{r} \tag{177}$$

$$\beta_{n,m} = \int_{-\infty}^{Z} \{\eta_n(R) - \eta_m(R)\}\, dZ \tag{178}$$

the χ's being the unperturbed eigenfunctions of the target (with z polar axis along the internuclear line), the η's being the perturbed eigenenergies and $V(\mathbf{r}, \mathbf{R})$ being the interaction potential. As usual, the subscript 0 denotes the state for which the angular part of the eigenfunction is $\cos\theta$ and the subscript ± 1 denotes the state for which the angular part of the eigenfunction is $\sin\theta \cos\phi$. The coupling terms are underlined. They are particularly important if the impact parameter is small (since the maximum of the modulus is then large) or if the velocity is high (since they are then effectively secular).

Coupling terms must similarly be retained in the basic equations of the *pss* approximation. Bearing in mind their physical significance it is apparent that special attention must be paid to the coupling terms which connect the initial or final states with states which differ only in magnetic quantum number either in the united atom limit or in the separed atoms limit.† The equations may readily be written down but form a com-

† The need for these coupling terms has only recently been realized. Because of this all published applications of the *pss* approximation to inelastic collisions are incorrect.

plicated set. Furthermore, in addition to the coupling associated with rotation, it is necessary to take into account coupling with any states having potential energy surfaces which approach close to the potential energy surfaces of the states directly concerned. The general prospect for detailed calculations is not promising.

3.1 Charge Transfer

3.1.1 *Symmetrical Resonance*

At low velocities of relative motion the probability of symmetrical resonance charge transfer

$$X + X^+ \rightarrow X^+ + X \tag{179}$$

is, as will be seen later, high up to quite large values of the impact parameter. Close encounters give only a small contribution to the cross section. Consequently, the neglect, in the original version of the *pss* approximation, of the coupling of the states directly concerned to states differing only in magnetic quantum number in the united atom limit cannot cause significant error in the cross section. The neglect of the corresponding coupling to states differing only in magnetic quantum number in the separated atoms limit is justified at sufficiently low velocities. Moreover, this coupling does not arise for most of the processes studied because the initial and final atomic states happen to be spherically symmetrical.

The *pss* approximation was first applied to symmetrical charge transfer by Massey and Smith (1933). Bates *et al.* (1953) discussed the theory further and established the relationship between the wave and impact parameter treatments. Later Bates and McCarroll (1958, 1962) removed inconsistencies revealed by the work of Bates, Massey, and Stewart and took account of the change in the translational motion of the active electron.

We shall first give a general description of the mechanism by which charge transfer is effected.

Before the encounter the active electron is in the neighborhood of a particular nucleus. It is not described by one of the eigenfunctions of the quasi-molecule; instead, it is described by the superposition of a pair of these eigenfunctions, one gerade and the other ungerade. If the velocity of relative motion is slow the coefficients associated with these eigenfunctions are not altered appreciably by the encounter. However, the relative phase of the eigenfunctions is changed because of the difference between the eigenenergies when the nuclear separation is finite. Hence, the superposition does not remain the same. This corresponds to the occurrence of charge transfer. It is to be noted that an electronic transition is *not* involved. In consequence, the cross section does not behave in the customary manner as the velocity tends to zero. Instead, if it were not for the effect of the interaction† on the motion of the nuclei it would tend logarithmically to infinity.

If the effect of the change in the translational motion of the active electron is ignored, it is found from the *pss* approximation that the probability that symmetrical resonance charge transfer occurs in an encounter is $\sin^2 \zeta$ where

$$\zeta = \frac{1}{\hbar} \int_{R_c}^{\infty} \frac{\epsilon^+(R) - \epsilon^-(R)}{v(R)} \left\{ 1 - \frac{v(\infty)^2 \rho^2}{v(R)^2 R^2} \right\}^{-1/2} dR \tag{180}$$

$\epsilon^+(R)$ and $\epsilon^-(R)$ being the eigenenergies of the relevant gerade and ungerade states of the quasi-molecule, $v(R)$ being the velocity of relative motion and R_c being the greatest value of R making the integrand vanish (Massey and Smith, 1933). Unless the encounter is very slow or Coulomb forces are involved $v(R)$ may be put equal to $v(\infty)$ without significant error. Denoting it by v (as usual), (180) becomes

$$\zeta = \frac{1}{\hbar v} \int_{\rho}^{\infty} \frac{R\{\epsilon^+(R) - \epsilon^-(R)\}}{(R^2 - \rho^2)^{1/2}} dR. \tag{181}$$

Unless v is high, $\sin^2 \zeta$ oscillates rapidly in the region of small ρ and may there be replaced by its average value. If ρ^* is the largest value of ρ for which ζ is $\pi/2$, the formula for the charge transfer cross section may to a sufficient approximation be written

$$Q = \tfrac{1}{2} \pi \rho^{*2} + 2\pi \int_{\rho^*}^{\infty} \rho \sin^2 \zeta \, d\rho. \tag{182}$$

The procedure for computing the cross section has been discussed by Firsov (1951), Demkov (1952), Holstein (1952), and Dalgarno and McDowell (1956).

A formula essentially similar to (181) may be obtained using atomic eigenfunctions as in § 1.2.2 allowing for their nonorthogonality and allowing also for back-coupling between the final and initial states (Gurnee and Magee, 1957). The only difference is that the LCAO ap-

† The effect of the interaction is to increase the cross section if one system is an ion and the other a neutral atom, and to decrease the cross section if both systems are ions (cf. Bates and Boyd, 1962b).

proximation to $\epsilon^+(R)$ and $\epsilon^-(R)$ appear in place of the exact eigenenergies (Bates and Lynn, 1959). On this model symmetrical resonance charge transfer entails an electronic transition. The behavior in the low velocity limit might therefore seem anomalous. There is, however, no anomaly since *in addition* to the energy balance being exact at infinite nuclear separation no distortion effects arise (cf. p. 592).

Calculations with exact or quite accurate eigenenergies have been carried out on the following processes:

$$H(1s) + H^+ \rightarrow H^+ + H(1s) \tag{183}$$

(Dalgarno and Yadav, 1953; Ferguson, 1961);

$$H(2s, 2p) + H^+ \rightarrow H^+ + H(2s, 2p) \tag{184}$$

(Boyd and Dalgarno, 1958);

$$H(1s) + H^-(1s)^2 \rightarrow H^-(1s)^2 + H(1s) \tag{185}$$

(Dalgarno and McDowell, 1956);

$$He(1s)^2 + He^+(1s) \rightarrow He^+ + He(1s)^2 \tag{186}$$

(Massey and Smith, 1933; Firsov, 1951; Demkov, 1952; Jackson, 1954; Moiseiwitsch, 1956);

$$He(1s)^2 + He^{2+} \rightarrow He^{2+} + He(1s)^2 \tag{187}$$

(Ferguson and Moiseiwitsch, 1959). Some results are given in Figs. 21 and 22.

A number of more complicated symmetrical resonance charge transfer processes have been treated, taking the atomic wave functions to be nodeless and spherically symmetrical, which is a rather poor approximation (Firsov, 1951; Demkov, 1952; Gurnee and Magee, 1957; Karmohapatro and Das, 1958; Karmohapatro, 1959, 1960). Moderate accord with experiment is achieved (cf. Chapter 18, § 4).

As would be expected, the smaller the energy of the orbital concerned the greater in general is the cross section at a given velocity of relative motion (cf. Firsov, 1951; Demkov, 1952; Iovitsu and Ionescu-Pallas, 1960).

The refinement of the *pss* approximation which takes into account the change in the translational motion of the active electron (Bates and McCarroll, 1958) has been applied to process (183) by Ferguson (1961).

In a complementary approach McCarroll (1961) has made allowance for this change in the formulation due to Gurnee and Magee (1957)—or, what is equivalent, has extended the range of validity of the high-energy approximation (139) by introducing back-coupling. The results of

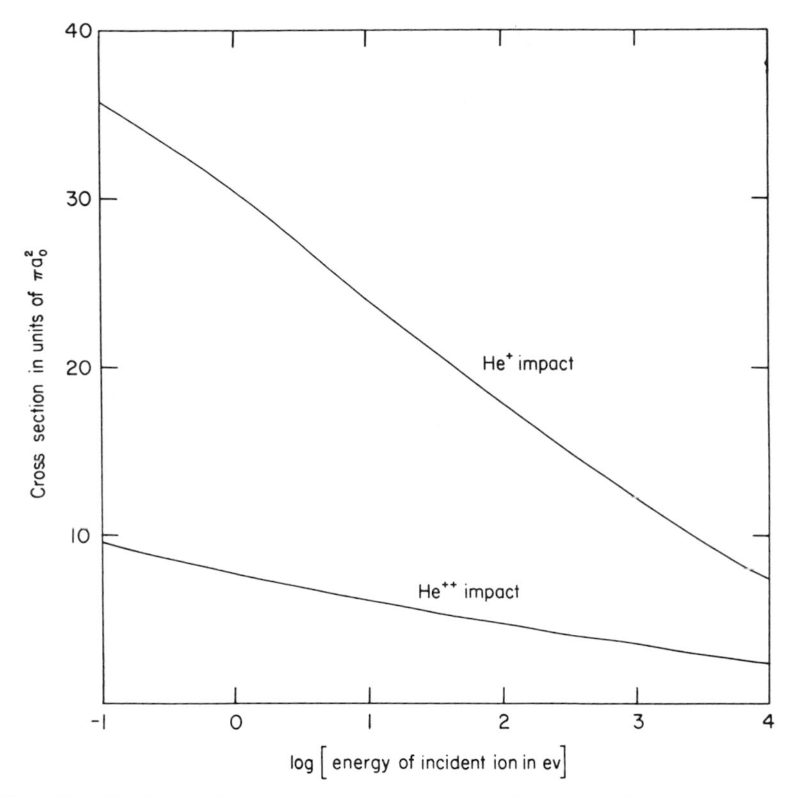

FIG. 21. Single and double symmetrical resonance charge transfer: *pss* approximation to the cross section—energy curves for $He(1s)^2 + He^+(1s) \rightarrow He^+(1s) + He(1s)^2$ (Moiseiwitsch, 1956) and $He(1s)^2 + He^{2+} \rightarrow He^{2+} + He(1s)^2$ (Ferguson and Moiseiwitsch, 1959).

Ferguson (based on the molecular eigenfunction expansion) and McCarroll (based on the atomic eigenfunction expansion) are included in Fig. 22, together with the results obtained by Fite and associates (1960) in the laboratory. Slight oscillations occur in the theoretical curves because of the presence of the sine function in the integral giving the cross section.

Comparing curves M_1 and M_2 or curves A_1 and A_2, it is seen that the effect of the change in the translational motion of the active electron becomes appreciable at an impact energy of only a few kev.

In the low-energy region M_2 should be the most accurate of the theoretical curves and in the high-energy region A_2 should be the most accurate. The agreement with experiment is satisfactory in the low-energy region, a possible error of 15% being indicated by Fite *et al*; but is poor in the high-energy region.

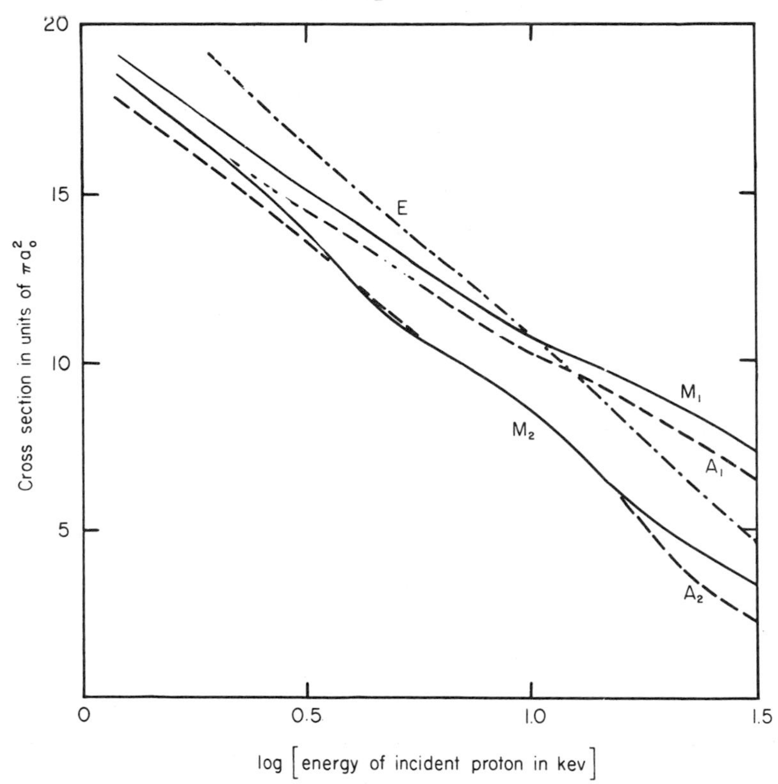

FIG. 22. Comparison of cross section—impact energy curves for $H(1s) + H^+ \rightarrow H^+ + H(1s)$. Curves A_1 and A_2 are based on an expansion in *atomic* eigenfunctions, the first ignoring and the second taking account of the effect of the change in the translational motion of the electron (McCarroll, 1961); curves M_1 and M_2 are correspondingly based on a expansion in *molecular* eigenfunctions (Ferguson, 1961); curve E is experimental (Fite *et al.*, 1960) for capture into *any* state.

3.1.2 *Asymmetric (or Accidental) Resonance*

In asymmetric (or accidental) resonance charge transfer the system which loses the electron is not the same species as the system which gains the electron but nevertheless the energy balance is exact (or very close): examples are

$$N_2(X\,{}^1\Sigma_g^+\; v = 0) + O^+({}^2D) \rightarrow N_2^+(A\,{}^2\Pi_u\; v = 1) + O({}^3P) + 0.0 \text{ ev.} \tag{188}$$

and

$$O(^3P_j) + H^+ \rightarrow H(1s) + O^+(^4S) \begin{cases} +0.01 \text{ ev } (j = 0) \\ +0.00 \text{ ev } (j = 1) \\ -0.02 \text{ ev } (j = 2) \end{cases} \quad (189)$$

The mechanism by which the charge transfer takes place is fundamentally different from the mechanism involved in the symmetrical resonance case (§ 3.1.1).

When the separation between the nuclei is great there is a pair of degenerate (or almost degenerate) eigenfunctions of the quasi-molecule, one describing the active electron in an orbital of the system of which it is initially part, the other describing it in an orbital of the system of which it becomes part if transfer occurs. Transfer involves an electronic transition. In spite of the energy balance its probability is low when the nuclei are far apart because the overlap between the eigenfunctions is then slight. As the nuclei approach each other this overlap increases; but so also, in general does the difference between the associated eigenenergies which tends to inhibit a transition if the velocity of relative motion is low. In consequence, if the effect of the interaction on the motion of the nuclei were ignored† the cross section would tend to zero as the velocity of relative motion is decreased indefinitely§ in marked contrast to the symmetrical resonance case where, as already noted, it would rise logarithmically to infinity (Bates and Lynn, 1959).

If atomic eigenfunctions are used in the expansion and the procedure indicated on page 602 is again followed the formula obtained for the cross section for asymmetric resonance charge transfer is in the first order approximation the same as the corresponding formula for symmetrical resonance charge transfer (Gurnee and Magee, 1957). It might therefore seem strange that the behavior in the two cases is so different at low velocities. The explanation is simple. It is that the exact cancellation which prevents the cross section in the symmetrical case from being affected by distortion does not occur in the asymmetrical case.

Provided the initial and final atomic states are *s* states, the formula mentioned in the preceding paragraph is valid throughout the region in which the velocity is high enough for the effect of distortion to be neglected, and is low enough for the effect of the change in the translational motion of the active electron to be also neglected. The formula shows that the cross section for asymmetric resonance charge transfer may be very large.

† See footnote on p. 602.

§ This behavior exemplified by the results obtained by McElroy (1962) on process (160) —see p. 593.

If other than s states are involved account should be taken of the coupling to states differing only in magnetic quantum number. This is not done in the derivation of the formula. Gurnee and Magee (1957) and Rapp and Ortenburger (1961) have used it in such cases ignoring the angular dependence of the eigenfunctions. In spite of the crudeness of this procedure the results obtained are in fair agreement with laboratory data.

If the energy balance is not exact the coupled differential equations for the expansion coefficients associated with the atomic eigenfunctions cannot be solved analytically even for s-s transitions unless the interaction potential has a special form. Rosen and Zener (1932) and Gurnee and Magee (1957) have made conjectures regarding the asymptotic amplitudes of the expansion coefficients (which are what determine the cross section). Neither conjecture is correct (Bates and Lynn, 1959; Skinner, 1961).

3.2 Spin-Change Collisions

The simplest example of a spin-change collision is

$$\mathrm{H}(1s)\,[F = 1] + \mathrm{H}(1s) \rightarrow \mathrm{H}(1s)\,[F = 0] + \mathrm{H}(1s) \tag{190}$$

in which the hydrogen atom in the excited hyperfine level is deactivated through the exchange of electron spin with the hydrogen atom in an unspecified hyperfine level. This process was first discussed by Purcell and Field (1956).

A quantal treatment of spin-change collisions for the case when the change of internal energy is small compared with the energy of relative motion, has been given by Dalgarno (1961). It is then basically similar to the treatment of resonance charge transfer.

For simplicity Dalgarno took the colliding atoms A and B to be in doublet spin states and supposed the spin-orbit coupling to be negligible. He showed that the cross section for the process in which the spin eigenfunction of atom A changes from X_{A}^{i} to X_{A}^{f} and that of B changes from X_{B}^{i} to X_{B}^{f} is

$$Q = |\langle \mathrm{X}_{\mathrm{A}}^{f}\,\mathrm{X}_{\mathrm{B}}^{f} \mid \mathbf{S}_{\mathrm{A}} \cdot \mathbf{S}_{\mathrm{B}} \mid \mathrm{X}_{\mathrm{A}}^{i}\,\mathrm{X}_{\mathrm{B}}^{i} \rangle|^{2} Q_{st} \tag{191}$$

where $\mathbf{S}_{\mathrm{A}}$ and $\mathbf{S}_{\mathrm{B}}$ are the total electron spin operators for A and B and

$$Q_{st} = 8\pi \int_{0}^{\infty} \rho \sin^{2} \Gamma \, d\rho \tag{192}$$

with

$$\Gamma = \frac{1}{\hbar v} \int_{\rho}^{\infty} \frac{R\{\epsilon_t(R) - \epsilon_s(R)\}}{(R^2 - \rho^2)^{1/2}} dR, \tag{193}$$

$\epsilon_t(R)$ and $\epsilon_s(R)$ being the eigenenergies of the triplet and singlet states of the quasi-molecule formed by the colliding systems. The similarity of (193) to (181) is immediately apparent.

Dalgarno evaluated (191) for process (190) at velocities of relative motion corresponding to temperatures between 1°K and 10,000°K. He found that in this range of temperatures the cross section falls slowly from 9.4 to 4.1 πa_0^2. The cross section for the reverse process is 3 times as great.

3.3 Pseudocrossing of Potential Energy Surfaces

Once again let $\psi_p(\mathbf{r} \mid \mathbf{R})$ and $\psi_q(\mathbf{r} \mid \mathbf{R})$ be exact eigenfunctions of the quasi-molecule formed by the colliding systems, the first corresponding at infinite nuclear separation to the active electron in state p of system A and the second to the active electron in state q of system B; and let $E_p(R)$ and $E_q(R)$ be the associated eigenenergies. It may happen that at some nuclear separation R_c there is a pseudocrossing of the potential energy surfaces with related change in the character of the eigenfunctions so that whereas for $R \gg R_c$ $\psi_p(\mathbf{r} \mid \mathbf{R})$ and $\psi_q(\mathbf{r} \mid \mathbf{R})$ describe $(\mathrm{A} + e)_p + \mathrm{B}$ and $\mathrm{A} + (\mathrm{B} + e)_q$, respectively, for $R \ll R_c$ they describe $\mathrm{A} + (\mathrm{B} + e)_q$ and $(\mathrm{A} + e)_p + \mathrm{B}$, respectively (Fig. 23). In this circumstance it is

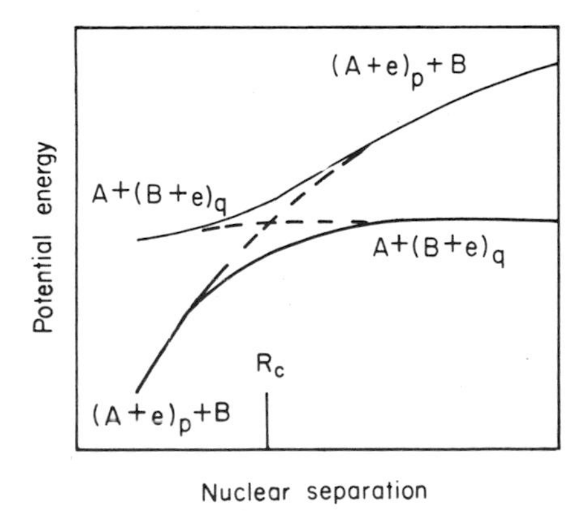

Fig. 23. Pseudocrossing of potential energy surfaces. The full curves represent $E_p(R)$ and $E_q(R)$; the broken curves represent $h_{pp}(R)$ and $h_{qq}(R)$. In the region to the right of pseudocrossing $E_p(R)$ [and $h_{pp}(R)$] lie above $E_q(R)$ [and $h_{qq}(R)$].

convenient when treating the collision problem to introduce linear combinations of $\psi_p(\mathbf{r} \mid \mathbf{R})$ and $\psi_q(\mathbf{r} \mid \mathbf{R})$, one of which $\phi_p(\mathbf{r} \mid \mathbf{R})$ describes $(\mathrm{A} + e)_p + \mathrm{B}$ for *any* internuclear distance and the other of which $\phi_q(\mathbf{r} \mid \mathbf{R})$ describes $\mathrm{A} + (\mathrm{B} + e)_q$ for *any* internuclear distance.† These combinations satisfy the equations

$$\mathscr{H}\phi_p(\mathbf{r} \mid \mathbf{R}) = h_{pp}(R)\,\phi_p(\mathbf{r} \mid \mathbf{R}) + h_{qp}(R)\,\phi_q(\mathbf{r} \mid \mathbf{R}) \tag{194}$$

and

$$\mathscr{H}\phi_q(\mathbf{r} \mid \mathbf{R}) = h_{pq}(R)\,\phi_p(\mathbf{r} \mid \mathbf{R}) + h_{qq}(R)\,\phi_q(\mathbf{r} \mid \mathbf{R}) \tag{195}$$

where $\mathscr{H}$ is the Hamiltonian operator and

$$h_{st} = \int \phi_s^*(\mathbf{r} \mid \mathbf{R})\, \mathscr{H}\phi_t(\mathbf{r} \mid \mathbf{R})\, d^3\mathbf{r}. \tag{196}$$

By definition R_c is such that

$$h_{pp}(R_c) = h_{qq}(R_c). \tag{197}$$

If it is assumed (i) that the effect of the change in the translational motion of the active electron may be ignored, (ii) that terms involving the time derivations $(\partial/\partial t_{\mathbf{r}_1})\,\phi_p$ and $(\partial/\partial t_{\mathbf{r}_2})\,\phi_q$ are small enough to be neglected, (iii) that only the initial and final states p and q need be taken into account, (iv) that the energy of relative motion at the crossing is very much greater than the difference between the exact eigenenergies of the quasi-molecule there the equations for the expansion coefficients c_{pp} and c_{pq} associated with states p and q, respectively, reduce to

$$i\hbar v \frac{\partial c_{pp}}{\partial Z} = h_{pq} c_{pq} \exp\left\{\frac{-i}{\hbar v} \int^Z (h_{qq} - h_{pp})\, dZ\right\} \tag{198}$$

$$i\hbar v \frac{\partial c_{pq}}{\partial Z} = h_{qp} c_{pp} \exp\left\{\frac{-i}{\hbar v} \int^Z (h_{pp} - h_{qq})\, dZ\right\}. \tag{199}$$

A simple formula for the probability $\mathscr{P}_{pq}$ may be obtained from (198) and (199) if it is further assumed (v) that transitions only occur to an appreciable extent in a very narrow zone around the crossing, so that it is permissible to take

$$h_{pp} - h_{qq} = (Z - Z_c)\,\alpha \tag{200}$$

and

$$h_{pq} = h_{qp} = \beta \tag{201}$$

† Unless R_c is small it is a sufficient approximation to take ϕ_p and ϕ_q to be the same as the atomic eigenfunctions (except at high velocities). It is necessary to take explicit account, as in (131) of the translational motion of the nuclei (Bates, 1960). If this is not done and if the mathematics are then developed rigorously false terms arise (cf. Mordvinov and Firsov, 1960).

where

$$Z_c = \mathbf{R}_c \cdot \hat{\mathbf{v}} \tag{202}$$

and where α and β are constants. Accepting (200) and (201) the asymp totic amplitudes of the required solutions of (198) and (199) may be shown to be given by

$$| c_{pp}(\infty) |^2 = P \tag{203}$$

$$| c_{pq}(\infty) |^2 = 1 - P \tag{204}$$

with

$$P = \exp(-\eta/v) \tag{205}$$

$$\eta = \frac{2\pi}{\hbar} h_{pq}^2 \Big/ (h'_{pp} - h'_{qq}), \tag{206}$$

the primes indicating differentiation with respect to Z and all quantities being evaluated at the crossing (Landau, 1932; Zener, 1932). Clearly P is the probability that the system is described by ϕ_p and $1 - P$ is the probability that it is described by ϕ_q after the crossing is traversed once. Noting that in an actual encounter the crossing is traversed twice it is apparent that the probability of a transition from state p to state q is

$$\mathscr{P}_{pq} = 2P(1 - P). \tag{207}$$

This, with P as in (205), is the Landau-Zener formula. Figure 24 depicts the form of the $\mathscr{P}_{pq}$ versus v curve.

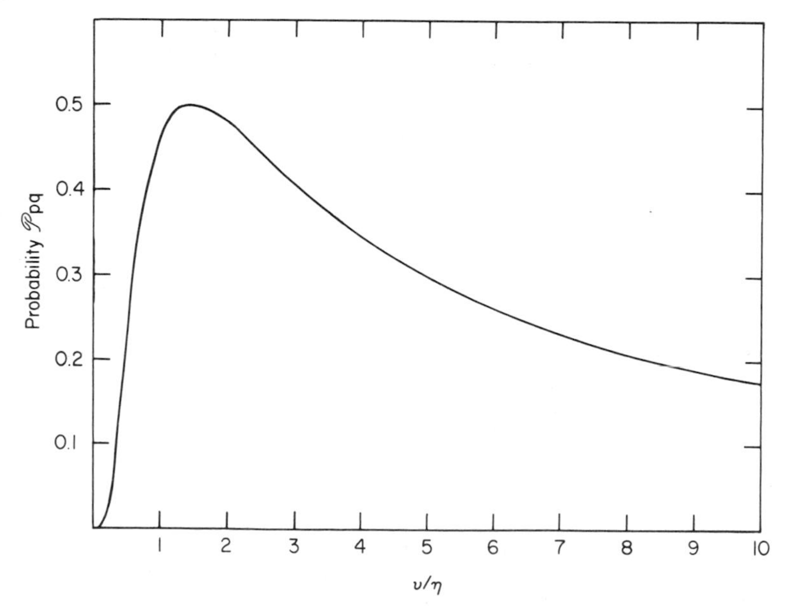

FIG. 24. Landau-Zener probability—velocity curve.

The Landau-Zener formula shows that charge transfer or excitation† may have a high probability even when the velocity of relative motion is so low that the behavior might be expected from condition (121) to be nearly adiabatic. This is of great importance.

A slightly or moderately exothermic process of type

$$A^- + B^+ \rightarrow A + B \tag{208}$$

or of type

$$A + B^{n+} \rightarrow A^+ + B^{(n-1)+} \tag{209}$$

has a pseudocrossing at a nuclear separation which exceeds the mean atomic diameter. The evaluation of the Landau-Zener formula in the case of such a process is not difficult. A number of computations have been performed (Magee, 1940, 1952; Bates and Massey, 1943; Bates and Moiseiwitsch, 1954; Dalgarno, 1954; Bates and Lewis, 1955; Bates and Boyd, 1956; Boyd and Moiseiwitsch, 1957). It has, however, recently been pointed out (Bates, 1960) that the Landau-Zener formula has very serious limitations which must affect some of the results. The limitations arise in part from assumptions (i), (ii), (iii), and (iv) made in obtaining the simplified equations (198) and (199); and they arise in part also from assumption (v) made in solving these equations. We shall consider the assumptions in turn.

Assumption (i) is justified only if the velocity of relative motion is so low that $\hbar/mv$ exceeds the linear extent l of the more compact of the two atomic wave functions since the cancellation within the adopted matrix elements is otherwise not as great as within the true matrix elements. An equivalent statement of this condition is that the energy of relative motion $\mathscr{E}$ must be less than

$$\mathscr{E}_t = (2.5 \times 10^4\, M/l^2)\ \text{ev} \tag{210}$$

where the reduced mass M is on the chemical scale and l is in atomic units. As $\mathscr{E}$ is increased above $\mathscr{E}_t$, cancellation effects tend to make the true probability fall below the probability given by the Landau-Zener formula.

The limitation imposed by assumptions (ii) and (iii), which are linked, may be much more severe. Suppose that either of the atomic states concerned is not an s state. The time derivative of its eigenfunction $(\partial/\partial t_{\mathbf{r}_1})\,\phi_p$ or $(\partial/\partial t_{\mathbf{r}_2})\,\phi_q$ then contains a considerable component of the eigenfunction of states differing only in magnetic quantum number. In

† Excitation, like charge transfer, may occur through the pseudocrossing of potential energy surfaces.

consequence, there is a strong coupling to these states and account must be taken of them in the expansion used for the wave function describing the colliding systems. Instead of it being necessary to solve two coupled equations, like (198) and (199), for the expansion coefficients, it is necessary to solve three or more such equations.† This has not yet been done. However, it is not difficult to see how the results may be affected qualitatively. Figure 25 illustrates the position for *s*-*p* transi-

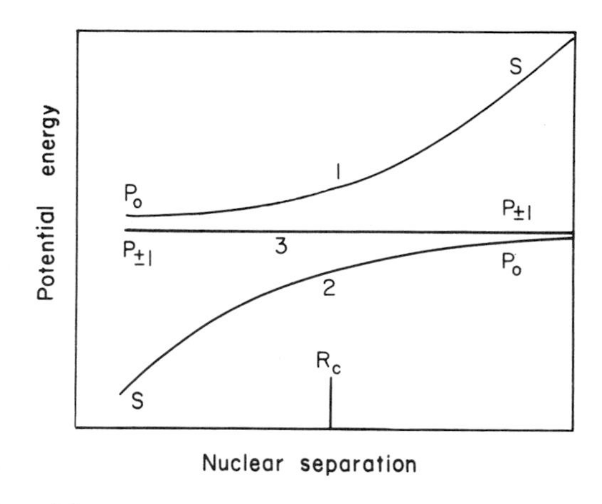

FIG. 25. Potential energy curves associated with s, p_0, and $p_{\pm 1}$ atomic states with pseudocrossing at R_c.

tions. It shows the exact potential energy curves of the two molecular states allowed for in the Landau-Zener formula, that is, the molecular states associated with the s and p_0 atomic states; and it shows in addition the corresponding curve of the molecular state associated with the $p_{\pm 1}$ atomic state. Suppose that the s state is occupied initially. If the systems approach and recede sufficiently slowly then according to the Landau-Zener formula there is very little chance of a jump from curve 1 to curve 2 (Fig. 25) and thus very little chance of the p_0 state being occupied finally. However, because of the strong coupling between the p_0 and $p_{\pm 1}$ states there may actually be a considerable chance of a jump from curve 1 to curve 3 when the representative point in Fig. 25 is to the left of the pseudocrossing, and also a considerable chance of a jump from curve 3 to curve 2 when it is to the right of the pseudocrossing. The neglect in the Landau-Zener formula of the coupling arising from the rotation of the polar axis may this lead to a serious underestimate of the probability of a transition at low velocities of relative

† An example of the equations which arise has been given by Bates (1960).

motion. If only s states are involved such coupling does not arise. However, even in this case the time derivatives $(\partial/\partial t_{\mathbf{r}_1})\,\phi_p$ and $(\partial/\partial t_{\mathbf{r}_2})\,\phi_q$ must eventually become important as the velocity is increased.

In general, assumption (iv) is fully justified over the energy range of most interest. It causes relatively little concern.

Because of assumption (v) the Landau-Zener formula may be incorrect even for encounters for which (198) and (199) are valid. The most obvious failure is in the high velocity limit; thus, the Landau-Zener formula predicts the rate at which the probability of a transition diminishes is as v^{-1} whereas the true rate is as v^{-2} (the effect of the change in the translational motion of the active electron being ignored in both instances). It may be shown that assumption (v) is responsible (Bates, 1960). Transitions are *not* confined to a narrow zone around the crossing. This makes (200) and (201), which are of course correct only over a narrow zone, unacceptable. Owing to their adoption in the derivation of the Landau-Zener formula the width of the ill-defined zone over which transitions take place in the mathematical model increases *indefinitely* as $v^{1/2}$ which is physically absurd.

Fallacious results are obtained not only when v is very high. Even when v is but moderate in magnitude, the transition zone may be too wide for it to be permissible to take $|\,h_{pq}\,|$ to be constant as in assumption (v). The rapid variation of $|\,h_{pq}\,|$ with R may profoundly alter the form of the $\mathscr{P}_{pq}$ versus v curve. In particular, it may cause $\mathscr{P}_{pq}$ to rise to a second maximum on the high v side of the maximum corresponding to that given by the Landau-Zener formula (cf. Bates, 1960).

It is unreasonable to expect it to be possible to remove the defects of the Landau-Zener formula by a simple modification. The course of an encounter is not in general determined by the matrix elements associated with two states of the quasi-molecule at the nuclear separation of the pseudocrossing. Reliable quantitative predictions can scarcely be made without solving the relevant set of coupled equations.

3.4 Ionization and Detachment

Intersecting potential energy surfaces may facilitate transitions into the continuum making it possible for ionization and detachment to occur readily even in very slow encounters. The mechanisms have been discussed by Bates and Massey (1954).

Suppose that at some nuclear separation R_c the potential energy surface on which two atoms A and B approach intersects a potential energy surface of the complex consisting of A, B^+, and a free electron of zero energy (Fig. 26). If the nuclear separation is less than R_c a radiationless transition resulting in autoionization

$$A + B \rightarrow A + B^+ + e \tag{211}$$

may take place. It is apparent that the probability that the reaction path indicated in (211) is followed in a slow encounter may be high; thus,

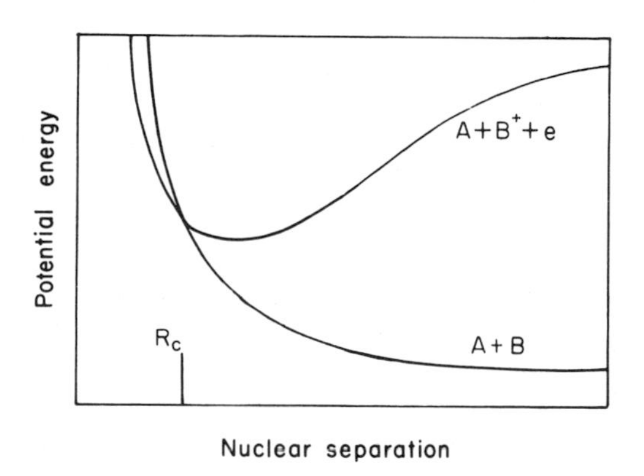

FIG. 26. Collisional ionization through intersecting potential energy surfaces.

if τ_{pq} is the life towards autoionization, ρ is the impact parameter and v is the velocity of relative motion (taken to be constant), the probability is given by

$$\mathscr{P}_{pq} = 1 - \exp\left[-2(R_c^2 - \rho^2)^{1/2}/v\tau_{pq}\right]. \tag{212}$$

This formula is not valid unless the nuclear separation remains less than R_c for much longer than the time associated with the motion of the electron.

As in the case of excitation and charge transfer, even qualitative predictions about a specific process are in general prevented by lack of information on the relevant potential energy surfaces. However, it may be said that the existence of suitable intersections is not favored by the characters and relative positions of the potential energy surfaces of the systems on either side of

$$A + B^+ \rightarrow A^+ + B^+ + e \tag{213}$$

or on either side of

$$A + B^+ \rightarrow A + B^{2+} + e. \tag{214}$$

Hence the probabilities of such processes would not be expected to be high at low velocities.

Returning to (211) it should be noted that the colliding systems may become bound

$$A + B \rightarrow AB^+ + e. \tag{215}$$

This process, *associative ionization*, is of the same family as *dissociative recombination*

$$AB^+ + e \rightarrow A + B \tag{216}$$

(cf. Chapter 7). Attempts have been made to use the principle of detailed balancing to determine the cross section for associative ionization from measurements on the cross section (or rather the rate coefficient) for dissociative recombination. They are unjustified since the systems on the left of (215) and (216) are in general in their ground states whereas those on the right may be in excited states (and in the case of the molecular ion may have considerable vibrational and rotational energy).

Detachment may naturally occur by the same mechanism as ionization. An important difference is that associative detachment is in most instances possible even in thermal encounters: examples are

$$H + H^- \rightarrow H_2 + e \tag{217}$$

and

$$O + O^- \rightarrow O_2 + e. \tag{218}$$

Estimates based on (212) indicate that in a favorable case the rate coefficient may be as high as about 10^{-10} cm³/sec.

There is a subtly different mechanism for detachment. As the atom and negative ion approach, the affinity may decrease and at some internuclear separation R_0 may reach zero. At separations less than R_0 the excess electron is not bound to the system. In slow encounters a detachment cross section of order πR_0^2 is to be expected.

3.5 Resonance Effects

In this section we shall attempt to provide answers to such questions as "Under what circumstances does a very close energy balance in a collision process lead to a cross section far exceeding the gas kinetic, and under what circumstances is the closeness of the energy balance an irrelevance as far as the cross section is concerned?" Certain aspects of the subject have been discussed in earlier sections of this chapter.* Brief mention of these will be made for the sake of completeness.

Symmetrical resonance charge transfer (179) is the most striking example of the association of exact energy balance with an extremely large cross section at low velocities of relative motion. The peculiar

* See also Massey and Burhop (1952).

way in which the process takes place has already been described (§ 3.1.1). As will be recalled an electronic transition is not involved. Symmetrical resonance excitation transfer* and spin-change collisions (§ 3.2) are essentially similar. An electronic transition is also not involved in the case of a simple chemical reaction like

$$AB + C \rightarrow A + BC \tag{219}$$

or

$$AB + CD \rightarrow AC + BD. \tag{220}$$

For such reactions, however, resonance effects do not arise: the course of an encounter is determined by the motions of a set of interacting classical particles and the cross section is not critically dependent on the closeness of the energy balance.

An appreciation of the position regarding processes requiring an electronic transition (and not proceeding through a pseudocrossing of potential energy surfaces) may be obtained from formula (149) of the distortion approximation.

Let M be the reduced mass of the colliding systems; let E be their energy of relative motion and for simplicity take it to be constant; let $\mathscr{V}_{st}(\mathbf{R})$ be a matrix element of the interaction energy; and let

$$\Delta\epsilon_{pq}(\mathbf{R}) = \epsilon_p - \epsilon_q + \mathscr{V}_{pp}(\mathbf{R}) - \mathscr{V}_{qq}(\mathbf{R}) \tag{221}$$

where ϵ_p and ϵ_q are the atomic eigenenergies indicated. Measure M on the chemical scale; measure E, $\mathscr{V}_{qp}(\mathbf{R})$, and $\Delta\epsilon_{pq}(\mathbf{R})$ in electron volts; and measure Z in atomic units of length. According to (149), the probability that a transition from state p to state q occurs in an encounter is then

$$\mathscr{P}_{pq} = \frac{34M}{E} \left| \int_{-\infty}^{\infty} \mathscr{V}_{qp}(\mathbf{R}) \exp\left[- i(5.8 \frac{M^{1/2}}{E^{1/2}} \int_0^Z \Delta\epsilon_{pq}(\mathbf{R})\, dZ \right] dZ \right|^2. \tag{222}$$

The value of $\mathscr{P}_{pq}$ is greatly influenced by the cancellation due to the oscillatory exponential factor. This cancellation is severe if

$$5.8 \frac{M^{1/2}\, \overline{\Delta\epsilon_{pq}}\, \delta Z}{E^{1/2}} \gg 1 \tag{223}$$

where $\overline{\Delta\epsilon_{pq}}$ is the average value of $\Delta\epsilon_{pq}(\mathbf{R})$ in the region giving the main contribution to $\mathscr{P}_{pq}$ and δZ is the width of the region. It is apparent that the closeness of the energy balance is here of great importance. This is in accord with observation (cf. Mitchell and Zemansky, 1934; Massey, 1949).

* The transfer cross section may be small at low velocities due to the colliding atoms being kept apart by repulsive forces (cf. Buckingham and Dalgarno, 1952).

For an excitation transfer process

$$\mathrm{A} + \mathrm{B}' \rightarrow \mathrm{A}' + \mathrm{B} \tag{224}$$

in which the transitions A′ — A and B′ — B are optically allowed, $\mathscr{V}_{qp}(\mathbf{R})$ is a slowly decreasing function of R and may be considerable in magnitude. If $\epsilon_p - \epsilon_q$ is small*, $\mathscr{P}_{pq}$ may therefore be high even when the impact parameter ρ is large. Hence, the excitation transfer cross section may be much greater than the gas kinetic cross section. In contrast, if an optically forbidden (particularly spin charge) transition is involved $\mathscr{V}_{qp}(\mathbf{R})$ falls off rapidly as R is increased. Consequently, $\mathscr{P}_{pq}$ also falls off rapidly. Even if $\epsilon_p - \epsilon_q$ is zero, the excitation cross section cannot therefore be very much greater than the gas kinetic cross section; and it may indeed be smaller since for the close encounters which are important in such processes $\overline{\Delta\epsilon_{pq}}$ may be large enough for (223) to be satisfied at thermal energies† (cf. Bates, 1962).

In a collision with a molecule an excited atom may be destroyed in several ways other than by giving up its energy to excite the molecule electronically

$$\mathrm{A}' + \mathrm{BC} \rightarrow \mathrm{A} + \mathrm{BC}'; \tag{225}$$

thus, it may take part in atom-atom interchange

$$\mathrm{A}' + \mathrm{BC} \rightarrow \mathrm{B} + \mathrm{AC} \tag{226}$$

or it may excite the molecule vibrationally

$$\mathrm{A}' + \mathrm{BC} \rightarrow \mathrm{A} + \mathrm{BC}^*. \tag{227}$$

The closeness of the energy balance is of critical importance for (225) as for (224) but, as will be seen, it matters little for (226) and (227).

Clearly (226) is simply a chemical reaction like (219) so that the remarks on page 616 apply. As an example of the class, mention may be made of the quenching of the resonance radiation of cadmium by molecular hydrogen which occurs through

$$\mathrm{Cd}(^3P_1) + \mathrm{H}_2 \rightarrow \mathrm{CdH} + \mathrm{H} \tag{228}$$

with a cross section of $2.1 \times 10^{-16}\ \mathrm{cm}^2$ (Lipson and Mitchell, 1935). Electronically excited molecules also may be destroyed by atom-atom interchange

$$\mathrm{AC}' + \mathrm{B} \rightarrow \mathrm{A} + \mathrm{BC}. \tag{229}$$

* The resonance is extremely sharp.

† In the case of symmetrical resonance $\overline{\Delta\epsilon_{pq}}$ is of course zero.

It has been suggested that such destruction is important in connection with the nightglow (Bates, 1955).

Certain processes of type (227) are known to be very effective; thus, the cross section for

$$Na(^2P) + H_2 \rightarrow Na(^2S) + H_2^* \tag{230}$$

is as great as 2.3×10^{-15} cm² (Norrish and Smith, 1940). This might at first seem surprising since it is difficult to convert electronic energy directly into vibrational energy. The explanation is that the potential energy hypersurface associated with the systems on the left and also that associated with the systems on the right have quasi-crossings with a third potential energy hypersurface via which the process proceeds (Magee and Ri, 1941; Laidler, 1942). The mechanism can, perhaps, be most easily understood by considering the three potential energy surfaces involved in the analogous but simpler process

$$H(3s, 3p, \text{ or } 3d) + H(1s) \rightarrow [H^- + H^+] \rightarrow H(2s \text{ or } 2p) + H(1s) \tag{231}$$

in which electronic energy is converted into translational energy (Bates and Lewis, 1955).

The position regarding charge transfer

$$A + B^+ \rightarrow A^+ + B \tag{232}$$

is rather similar to that for excitation transfer in the optically forbidden case. However, account must be taken of the long-range attraction between the neutral atom and the positive ion which not only increases the number of close encounters but increases the energy of relative motion in such encounters. A useful formula for the rate coefficient κ_c describing close encounters (all of which do not of course lead to charge transfer) is

$$\kappa_t = 2.3 \times 10^3 (\alpha/M)^{1/2} \text{ cm}^3/\text{sec} \tag{233}$$

where α is the polarizability of the neutral atom in cgs units and M is the reduced mass on the chemical scale (Gioumousis and Stevenson, 1958).

If a pseudocrossing of the potential energy surfaces exists, $\Delta\epsilon_{pq}$ is there a minimum. At low velocities of relative motion the transitions occur mainly near the pseudocrossing (§ 3.1.3). Formula (222) is useless here since it does not take into account the effect of the back-coupling which is very pronounced. Guidance may be obtained from the Landau-Zener formula (207) provided its limitations are borne in mind. In the important cases (208) and (209) of exothermic recombination between a positive ion A^+ and a negative ion B^- and of exothermic electron capture from an atom A by an ion $B^{n(+)}$ of multiple charge, the nuclear separation at the pseudocrossing is given approximately by

$$R_c = 27/\sigma(\epsilon_p - \epsilon_q) \begin{cases} \sigma = 1, \text{ ionic recombination} \\ \sigma = n - 1, \text{ electron capture} \end{cases}$$

where R_c is in atomic units and $(\epsilon_p - \epsilon_q)$ is in electron volts. If $(\epsilon_p - \epsilon_q)$ is small, R_c is clearly very large. The matrix element $\mathscr{V}_{qp}(R_c)$ is then minute and hence so also is the probability $\mathscr{P}_{pq}$ of recombination or capture. The greatest collision cross sections arise not when the energy balance is exact, but when an energy of a few electron volts is liberated (cf. Bates and Massey, 1954).

References

Adler, J., and Moiseiwitsch, B. L. (1957) *Proc. Phys. Soc.* **A70**, 117.
Arthurs, A. M. (1959) *Proc. Phys. Soc.* **73**, 681.
Arthurs, A. M. (1961) *Proc. Cambridge Phil. Soc.* **57**, 904.
Barnett, C. F., and Reynolds, H. K. (1958) *Phys. Rev.* **109**, 355.
Bassel, R. H., and Gerjuoy, E. (1960) *Phys. Rev.* **117**, 749.
Bates, D. R. (1955) *J. atmos. terrest. Phys.* **6**, 171.
Bates, D. R. (1957) *Proc. Roy. Soc.* **A243**, 15.
Bates, D. R. (1958a) *Proc. Roy. Soc.* **A245**, 299.
Bates, D. R. (1958b) *Proc. Roy. Soc.* **A247**, 294.
Bates, D. R. (1959) *Proc. Phys. Soc.* **73**, 227.
Bates, D. R. (1960) *Proc. Roy. Soc.* **A257**, 22.
Bates, D. R. (1961a) *Proc. Phys. Soc.* **A77**, 59.
Bates, D. R. (1961b) *Proc. Phys. Soc.* **78**, 1080.
Bates, D. R. (1962), *Faraday Soc. Discussions No.* **33**.
Bates, D. R., and Boyd, A. H. (1962a) *Proc. Phys. Soc.* (in press).
Bates, D. R. and Boyd, A. H. (1962b) *Proc. Phys. Soc.* (in press).
Bates, D. R., and Boyd, T. J. M. (1956) *Proc. Phys. Soc.* **A69**, 910.
Bates, D. R., and Dalgarno, A. (1952) *Proc. Phys. Soc.* **A65**, 919.
Bates, D. R., and Dalgarno, A. (1953) *Proc. Phys. Soc.* **A66**, 972.
Bates, D. R., and Griffing, G. W. (1953) *Proc. Phys. Soc.* **A66**, 961.
Bates, D. R., and Griffing, G. W. (1954) *Proc. Phys. Soc.* **A67**, 663.
Bates, D. R., and Griffing, G. W. (1955) *Proc. Phys. Soc.* **A68**, 90.
Bates, D. R., and Lewis, J. T. (1955) *Proc. Phys, Soc.* **A68**, 173.
Bates, D. R., and Lynn, N. (1959) *Proc. Roy. Soc.* **A253**, 141.
Bates, D. R., and McCarroll, R. (1958) *Proc. Roy. Soc.* **A245**, 175.
Bates, D. R., and McCarroll, R. (1962) *Advances in Phys.* (in press).
Bates, D. R., and Massey, H. S. W. (1943) *Phil. Trans.* **A239**, 269.
Bates, D. R., and Massey, H. S. W. (1954) *Phil. Mag.* **45**, 111.
Bates, D. R., and Moiseiwitsch, B. L. (1954) *Proc. Phys. Soc.* **A67**, 805.
Bates, D. R., and Williams, A. (1957) *Proc. Phys. Soc.* **A70**, 306.
Bates, D. R., Fundaminsky, A., and Massey, H. S. W. (1950) *Phil. Trans.* **A243**, 93.
Bates, D. R., Massey, H. S. W., and Stewart, A. L. (1953) *Proc. Roy. Soc.* **A216**, 437.

Bates, D. R., McDowell, M. R. C., and Omholt, A. (1957) *J. atmos. terrest. Phys.* **10**, 51.
Bell, R. J. (1961) *Proc. Phys. Soc.* **78**, 903.
Bell, R. J., and Skinner, B. G. (1962) *Proc. Phys. Soc.* (in press).
Bethe, H. A. (1930) *Ann. Phys.* (*Leipzig*) **5**, 325.
Boyd, T. J. M., and Dalgarno, A. (1958) *Proc. Phys. Soc.* **72**, 694.
Boyd, T. J. M., and Moiseiwitsch, B. L. (1957) *Proc. Phys. Soc.* **A70**, 809.
Boyd, T. J. M., Moiseiwitsch, B. L., and Stewart, A. L. (1957) *Proc. Phys. Soc.* **A70**, 110.
Bransden, B. H., Dalgarno, A., and King, N. M. (1954) *Proc. Phys. Soc.* **A67**, 1075.
Brinkman, H. C., and Kramers, H. A. (1930) *Proc. Acad. Sci. Amsterdam* **33**, 973.
Buckingham, R. A., and Dalgarno, A. (1952) *Proc. Roy. Soc.* **A213**, 506.
Bulman, J. B., and Russek, A. (1961) *Phys. Rev.* **122**, 506.
Dalgarno, A. (1954) *Proc. Phys. Soc.* **A67**, 1010.
Dalgarno, A. (1961) *Proc. Roy. Soc.* **A262**, 132.
Dalgarno, A., and Griffing, G. W. (1958) *Proc. Roy. Soc.* **A248**, 415.
Dalgarno, A., and McDowell, M. R. C. (1955) In "The Airglow and the Aurorae" (A. Dalgarno and E. B. Armstrong, eds.), p. 340. Pergamon, New York.
Dalgarno, A., and McDowell, M. R. C. (1956) *Proc. Phys. Soc.* **A69**, 615.
Dalgarno, A., and Yadav, H. N. (1953) *Proc. Phys. Soc.* **A66**, 173.
Demkov, U. H. (1952) *Uchenye Zapiski Leningrad. Gosudarst. Univ. im. A. A. Zhdanova Ser. Fiz. Nauk No. 8* **146**, 74.
Erskine, G. A. (1954) *Proc. Roy. Soc.* **A224**, 362.
Fedorenko, N. V., Afrosimov, V. V., Il'in, R. N., and Solov'ev, E. S. (1960) *In* "Proceedings of the Fourth International Conference on Ionization of Gases, Uppsala, 1959" (N. R. Nilsson, ed.), p. 47. North-Holland, Amsterdam.
Ferguson, A. F. (1961) *Proc. Roy. Soc.* **A 246**, 540.
Ferguson, A. F., and Moiseiwitsch, B. L. (1959) *Proc. Phys, Soc.* **74**, 457.
Firsov, O. B. (1951) *Zh. eksper. teor. Fiz.* **21**, 1001.
Firsov, O. B. (1959) *Zh. eksper. teor. Fiz.* **36**, 1517; translation (1959) *Soviet Phys.—JETP* **9**, 1076.
Fite, W. L., Stebbing, R. F., Hummer, D. G., and Brackmann, R. T. (1960) *Phys. Rev.* **119**, 663.
Gaunt, J. A. (1927) *Proc. Cambridge Phil. Soc.* **23**, 732.
Gioumousis, G., and Stevenson, D. P. (1958) *J. chem. Phys.* **29**, 294.
Gurnee, E. F., and Magee, J. L. (1957) *J. chem. Phys.* **26**, 1237.
Hasted, J. B. (1952) *Proc. Roy. Soc.* **A212**, 235.
Henneberg, W. (1933) *Z. Phys.* **86**, 592.
Holstein, T. (1952) *J. phys. Chem.* **56**, 832.
Hummer, D. G., Stebbings, R. F., Fite, W. L., and Branscomb, L. M. (1960) *Phys. Rev.* **119**, 668.
Iovitsu, I. P., and Ionescu-Pallas, N. (1960) *Soviet Phys.—Tech. Phys.* **4**, 781.
Jackson, J. D. (1954) *Canad. J. Phys.* **32**, 60.
Jackson, J. D., and Schiff, H. (1953) *Phys. Rev.* **89**, 359.
Karmohapatro, S. B. (1959) *J. chem. Phys.* **30**, 538.
Karmohapatro, S. B. (1961) *Proc. Phys. Soc.* **77**, 416.
Karmohapatro, S. B., and Das, T. P. (1958) *J. chem. Phys.* **29**, 240.
Kingston, A. E., Moiseiwitsch, B. L., and Skinner, B. G. (1960) *Proc. Roy. Soc.* **A258**, 237.
Laidler, K. J. (1942) *J. chem. Phys.* **10**, 34, 43.
Landau, L. D. (1932) *Phys. Z. Sowjetunion* **2**, 46.
Lipson, H. C., and Mitchell, A. C. G. (1935) *Phys. Rev.* **48**, 625.
McCarroll, R. (1961) *Proc. Roy. Soc.* **A246**, 547.

McCarroll, R., and McElroy, M. (1962) *Proc. Roy. Soc.* **A.** (in press).
McDowell, M. R. C., and Peach, G. (1959) *Proc. Phys. Soc.* **74**, 463.
McDowell, M. R. C., and Peach, G. (1961) *Phys. Rev.* **121**, 1383.
McElroy, M. (1962) *Proc. Roy. Soc.* **A.** (in press).
Magee, J. L. (1940) *J. chem. Phys.* **8**, 687.
Magee, J. L. (1952) *Disc. Faraday Soc.* **12**, 33.
Magee, J. L., and Ri, T. (1941) *J. chem. Phys.* **9**, 638.
Mapleton, R. A. (1958) *Phys. Rev.* **109**, 1166.
Mapleton, R. A. (1961) *Phys. Rev.* **122**, 528.
Massey, H. S. W. (1949) *Rep. Progr. Phys.* **12**, 248.
Massey, H. S. W., and Burhop, E. H. S. (1952) "Electronic and Ionic Impact Phenomena." Oxford Univ. Press, London and New York.
Massey, H. S. W., and Mohr, C. B. O. (1933) *Proc. Roy. Soc.* **A140**, 613.
Massey, H. S. W., and Smith, R. A. (1933) *Proc. Roy. Soc.* **A142**, 142.
Merzbacker, E., and Lewis, H. W. (1958) *In* "Handbuch der Physik" (S. Flügge, ed.), Vol. 34, p. 166. Springer, Berlin.
Mitchell, A. C. G., and Zemansky, M. W. (1934) "Resonance Radiation and Excited Atoms." Cambridge Univ. Press, London and New York.
Mittleman, M. H. (1961) *Phys. Rev.* **122**, 499.
Moiseiwitsch, B. L. (1956) *Proc. Phys. Soc.* **A69**, 653.
Moiseiwitsch, B. L., and Stewart, A. L. (1954) *Proc. Phys. Soc.* **A67**, 1069.
Mordvinov, Y. P., and Firsov, O. B. (1960) *Zh. eksper. teor. Fiz.* **39**, 427; translation (1960) *Soviet Phys.—JETP* **12**, 301.
Mott, N. F. (1931) *Proc. Cambridge Phil. Soc* **27**, 523.
Mott, N. F., and Massey, H. S. W. (1949) "The Theory of Atomic Collisions," 2nd ed. Oxford Univ. Press, London and New York.
Murakaver, Yu. E. (1961) *Zh. eksper. teor. Fiz.* **40**, 1080; translation (1961) *Soviet Phys.—JETP* **13**, 762.
Norrish, R. G. W., and Smith, W. M. (1940) *Proc. Roy. Soc.* **A176**, 295.
Oppenheimer, J. R. (1928) *Phys. Rev.* **31**, 349.
Purcell, E. M. (1952) *Astrophys. J.* **116**, 457.
Purcell, E. M., and Field, G. B. (1956) *Astrophys. J.* **124**, 542.
Rapp, D., and Ortenburger, I. B. (1960) *J. chem. Phys.* **33**, 1230.
Rosen, N., and Zener, C. (1932) *Phys. Rev.* **40**, 502.
Russek, A., and Thomas, M. T. (1958) *Phys. Rev.* **109**, 2015.
Russek, A., and Thomas, M. T. (1959) *Phys. Rev.* **114**, 1538.
Saha, M. N., and Basu, D. (1945) *Indian J. Phys.* **19**, 121.
Schiff, H. (1954) *Canad. J. Phys.* **32**, 393.
Schiff, L. I. (1949) "Quantum Mechanics." McGraw-Hill, New York
Seaton, M. J. (1955) *Proc. Phys. Soc.* **A68**, 457.
Sida, D. W. (1955) *Proc. Phys. Soc.* **A68**, 240.
Sil, N. C. (1960) *Proc. Phys. Soc.* **75**, 194.
Skinner, B. G. (1961) *Proc. Phys. Soc.* **77**, 551.
Skinner, B. G. (1962) *Proc. Phys. Soc.* **79** (in press).
Stier, P. M., and Barnett, C. F. (1956) *Phys. Rev.* **103**, 896.
Stobbe, M. (1930) *Ann. Phys. (Leipzig)*, **7**, 661.
Takayanagi, K. (1952) *Sci. Rep. Saitama Unvi.* **1**, 9.
Takayanagi, K. (1955) *Sci. Rep. Saitama Univ.* **2**, 33.
Tuan, T. F., and Gerjuoy, E. (1960) *Phys. Rev.* **117**, 756.
Zener, C. (1932) *Proc. Roy. Soc.* **A137**, 696.

15.

Range and Energy loss

A. Dalgarno

Except at high energies where bremsstrahlung is the dominant energy loss process, electrons passing through a material lose energy almost entirely through excitation and ionization of the material particles. When the velocity of the electrons is reduced to the point at which excitation and ionization are no longer energetically possible, the slowing down continues through the transfer of momentum in elastic collisions until thermal equilibrium with the material is attained. For heavy charged particles passing through a material, the behavior is more complicated in that the incident particles may also be excited and ionized and they may capture electrons from the atoms of the material. However, for heavy particles bremsstrahlung is never an important loss mechanism.

1 Energy Loss of Charged Particles

Experimental data (Whaling, 1958) on the energy loss of charged particles are conveniently analyzed within the framework of an approximate theoretical treatment due to Bethe (1930), which is valid at high energies where electron capture may be ignored.

1.1 Bethe's Theory for Heavy Particles

Suppose that the ground state of a neutral atom of mass M and charge Z is described by a wave function $\psi_0(\mathbf{r})$, $\mathbf{r}$ denoting collectively the position vectors $\mathbf{r}_i$ of the atomic electrons. Then the differential cross section for excitation to a state described by a wave function $\psi_t(\mathbf{r})$ by the impact of a structureless particle of charge Z', mass M', and velocity v is given by the Born approximation (Mott and Massey, 1949) as

$$I_t(q)\, dq = \frac{8\pi}{\hbar^4 k^2} Z'^2 \mathcal{M}^2 e^4 \frac{dq}{q^3} \mid X_t \mid^2 \tag{1}$$

where $\mathcal{M}$ is the reduced mass of the colliding particles defined by

$$\mathcal{M} = (MM')/(M + M'), \tag{2}$$

$\mathbf{q}$ is the difference between the initial relative momentum $\mathbf{k}$ and the final relative momentum $\mathbf{k}_t$, and X_t is the matrix element

$$X_t = \sum_{i=1}^{Z} (\psi_0, \exp(i\mathbf{q} \cdot \mathbf{r}_i)\, \psi_t). \tag{3}$$

The total excitation cross section is given by

$$Q_t = \int_{q_{\min}}^{q_{\max}} I_t(q)\, dq \tag{4}$$

where $q_{\max}$ is the maximum value of the momentum change and $q_{\min}$

is the minimum value. If E_0 and E_t are the energies of the ground and excited states, respectively,

$$k^2 - k_t^2 = \frac{2\mathscr{M}}{\hbar^2}(E_t - E_0) \tag{5}$$

and for fast collisions we may use the approximation

$$q_{\min} = k - k_t \sim \frac{\mathscr{M}}{k\hbar^2}(E_t - E_0). \tag{6}$$

The value of $q_{\max}$ is determined from the conservation of momentum in the collision of a heavy particle of mass $\mathscr{M}$ and an atomic electron of mass m and is therefore given by

$$q_{\max} = \frac{2km}{m + \mathscr{M}} \sim \frac{2km}{\mathscr{M}}. \tag{7}$$

The efficiency of the transition in causing a loss of kinetic energy E of the heavy particle may be described by a *stopping cross section*

$$(E_t - E_0)\,Q_t = \int_{q_{\min}}^{q_{\max}} (E_t - E_0)\, I_t(q)\, dq \tag{8}$$

and the energy loss in passing through unit distance in a gas of number density n cm^{-3}, the *stopping power*, is given by

$$-\frac{dE}{dx} = n\,\mathbf{S}_t(E_t - E_0)\,Q_t \tag{9}$$

where the summation $\mathbf{S}_t$ includes an integration over the continuum and so allows for energy loss through ionization.

Bethe introduces a generalized oscillator strength

$$F_t(q) = \frac{2m}{k^2 q^2}(E_t - E_0)\,|\,X_t\,|^2 \tag{10}$$

and proves the sum rule

$$\mathbf{S}_t F_t(q) = Z. \tag{11}$$

In terms of $F_t(q)$, (9) may be written

$$-\frac{dE}{dx} = n\,\frac{4\pi Z'^2 e^4}{mv^2}\,\mathbf{S}_t \int_{q_{\min}}^{q_{\max}} \frac{F_t(q)}{q}\, dq. \tag{12}$$

The integral is now separated into two contributions at $q = q_0$ where

$$q_0^2 = \frac{2\mathscr{M} \mid E_0 \mid}{\hbar^2}, \tag{13}$$

the two contributions to the energy loss corresponding roughly to small and large momentum changes. The energy loss $-dE''/dx$ arising from large momentum changes follows immediately from (12) and (11) in the form

$$-\frac{dE''}{dx} = n\frac{4\pi Z'^2 e^4}{mv^2} Z \ln(2km/\mathscr{M}q_0), \tag{14}$$

but because $q_{\min}$ depends upon the index t, a further approximation is needed to deal with the energy loss $-dE'/dx$ arising from small momentum changes. If $\exp(i\mathbf{q} \cdot \mathbf{r}_i)$ is expanded in powers of q and only the first nonvanishing term is retained, $F_t(q)$ becomes equal to the optical oscillator strength

$$f_t = \frac{2m}{\hbar^2}(E_t - E_0)\left|\sum_{i=1}^{Z}(\psi_0, Z_i\psi_t)\right|^2 \tag{15}$$

which satisfy the Thomas-Kuhn sum rule

$$\mathbf{S}_t f_t = Z. \tag{16}$$

Then from (12) and (6),

$$-\frac{dE'}{dx} = n\frac{4\pi Z'^2 e^4}{mv^2} Z\left\{\ln\frac{k\hbar^2 q_0}{\mathscr{M}} - \mathbf{S}_t f_t \frac{\ln(E_t - E_0)}{Z}\right\}. \tag{17}$$

On adding the two contributions, it follows that

$$-\frac{dE}{dx} = n\frac{4\pi Z'^2 e^4}{mv^2} Z \ln\frac{2mv^2}{I} \tag{18}$$

where I is an average excitation energy defined by

$$\ln I = \frac{\mathbf{S}_t f_t \ln(E_t - E_0)}{\mathbf{S}_t f_t}. \tag{19}$$

Equation (18) is frequently written in the form

$$-\frac{dE}{dx} = n\frac{4\pi Z'^2 e^4}{mv^2} B \tag{20}$$

where

$$B = Z \ln \frac{2mv^2}{I} \tag{21}$$

is a dimensionless quantity called the *stopping number*.

It follows from (18) that the stopping power should depend on the velocity of the incident particle but not on its mass. This conclusion, which is also obtained in more accurate theoretical treatments, is supported by many experiments (cf. Phillips, 1953; Brolley and Ribe, 1955). It also follows from (18) that $-(1/Z'^2)(dE/dx)$ should be independent of the charge of the incident particle. This conclusion is a consequence of the Born approximation and is not expected to hold at low energies, even if it were permissible to ignore the effect of capture and loss processes (§ 1.9).

1.2 The Stopping Power of Atomic Hydrogen

An indication of the accuracy of Bethe's theory can be obtained for the case of the stopping power of atomic hydrogen arising from excitation and ionization by proton impact, which has also been computed using the Born approximation for the cross sections but with no additional approximation (Dalgarno and Griffing, 1955). A comparison is effected in Fig. 1, (18) being computed with a value of 15.0 ev for I (Walske,

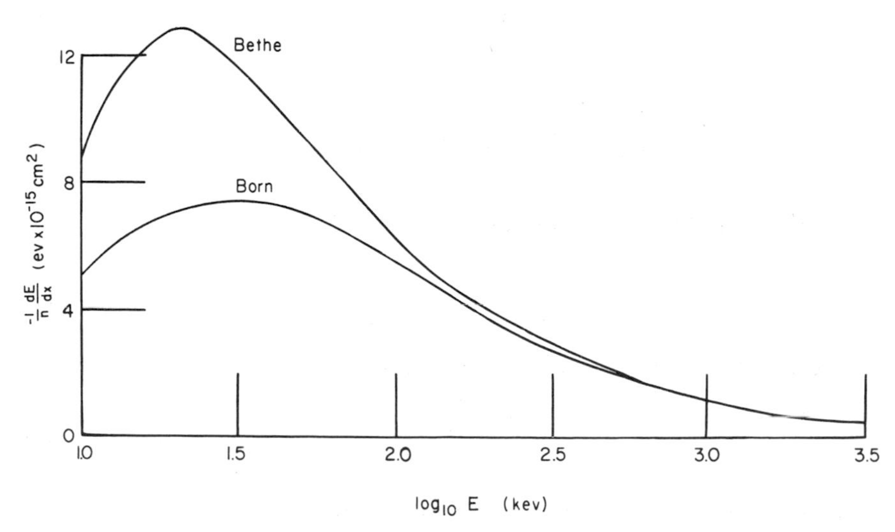

FIG. 1. The Born approximation and the Bethe approximation for the stopping power of atomic hydrogen towards a beam of protons, taking account of excitation and ionizaton only.

1952). The Bethe theory consistently overestimates the stopping power (because it includes transitions to states which are not energetically accessible), but the error does not become serious until the impact energy is reduced to below 100 kev and is there less than that caused by ignoring capture and loss processes. For heavier elements the error will be larger and it is necessary to introduce corrections.

1.3 THE STOPPING POWER OF HELIUM

The only gas for which I has been calculated directly and for which experimental data exist is helium, the value obtained for I being 41.5 ev (Dalgarno and Lynn, 1957).† The predictions of (18) are compared in Fig. 2 with the measurements of Phillips (1953), Reynolds *et al.* (1953),

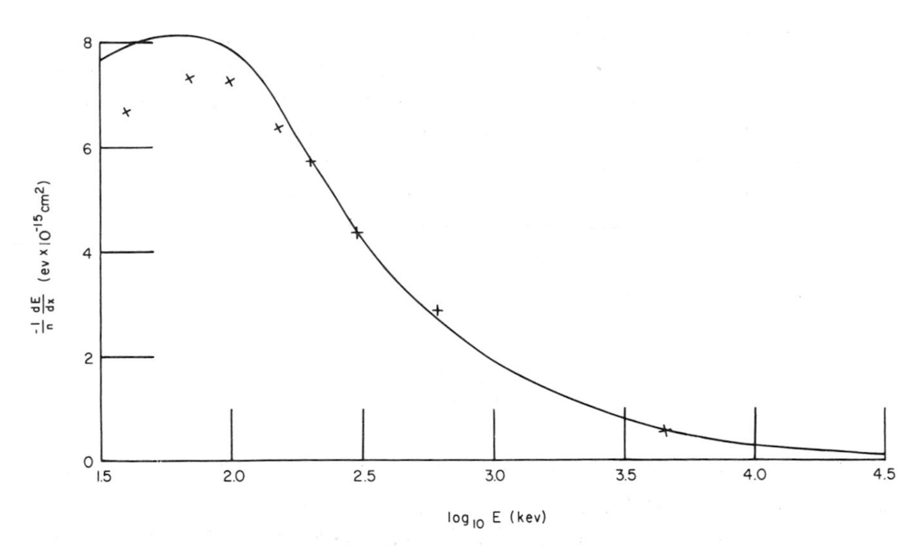

FIG. 2. The stopping power of helium towards a beam of protons computed using the Bethe approximation. The + points are the experimental measurements.

Weyl (1953), and Brolley and Ribe (1955), the incident particles being protons, deutrons, or tritons. The agreement is satisfactory and it is probable that much of the discrepancy at lower energies would be removed by using the Born approximation in place of the Bethe formula.§

† Dalgarno and Lynn actually give I = 40.2 ev. The value given here is the result of a similar but improved calculation.

§ Because of the approximate atomic functions used, the Born approximation calculations of Erskine (1954) (suitably scaled) do not give a reliable indication.

1.4 Inner Shell Corrections

Born's approximation is satisfactory for incident velocities v such that

$$\frac{Z'e^2}{\hbar v} \ll 1, \tag{22}$$

but Bethe's approximation requires in addition that

$$E \gg \frac{M'}{m} E_i \tag{23}$$

where E_i is the ionization potential of the atomic electrons. This latter condition is a serious restriction on the use of (18), the ionization potentials of the inner K-shell electrons being several hundred electron volts even for light elements.

Provided the charge of the incident particle Z' is small compared to the nuclear charge Z of the atom, so that distortion of the wave functions of the atomic electrons is small, the Born approximation gives a satisfactory description of the excitation and ionization of the inner shell electrons by heavy particles with velocities much lower than that of the atomic electrons. The severity of the restriction (23) may be lessened by using the Born approximation for the inner shell electrons, which can be adequately represented by hydrogenic wave functions, and retaining (18) for the outer shell electrons.

The stopping number can be written

$$B = Z \ln \frac{2mv^2}{I} - \sum_i C_i \tag{24}$$

where C_i is the correction for the ith shell. The binding corrections have been computed for the K-shell electrons (Brown, 1950; Walske, 1952) and for the L-shell electrons (Walske, 1956) and there is general agreement with experimental data (cf. Brandt, 1958).

1.5 Values of the Average Excitation Potential

The average excitation potential I has been calculated directly only for atomic hydrogen and helium and it must in general be determined from experimental data.† Some representative values (cf. Sternheimer, 1959) are listed in Table I.

† An indirect method of calculating I has been suggested by Dalgarno (1960).

TABLE I

VALUES OF AVERAGE EXCITATION ENERGY I (EV)

Be	C	Al	Cu	Pb	Air
64	78	166	371	1070	94

By using statistical models of the atom, Bloch (1933) and Lindhard and Scharff (1953) have shown that

$$I \sim KZ \tag{25}$$

and the experimental data suggest that the Bloch constant K is about 13 ev.

1.6 THE STOPPING NUMBERS OF COMPOUNDS

Measurements by Reynolds *et al.* (1953) for a number of substances show that for proton energies above 150 kev, the stopping number of a compound XY is equal, to within the experimental error of 2%, to the sum of the stopping numbers of the individual atoms X and Y,

$$B(\mathrm{XY}) = B(\mathrm{X}) + B(\mathrm{Y}), \tag{26}$$

except for nitric oxide, $B(\mathrm{NO})$ being about 4% greater than $\frac{1}{2}\{B(\mathrm{N_2}) + B(\mathrm{O_2})\}$. Slight deviations of this order are to be expected for compounds containing C, N, O, and F (Platzmann, 1952). For energies below 150 kev, the additive rule (26) breaks down.

1.7 RELATIVISTIC EFFECTS

For relativistic velocities of the incident heavy particle, the stopping number must be modified to the form

$$B = Z\left\{\ln\frac{2mv^2}{I} - \ln(1-\beta^2) - \beta^2\right\} - \sum_i C_i \tag{27}$$

where $\beta = v/c$ (Møller, 1932; Bethe, 1938). Thus, at velocities close to the velocity of light, the energy loss increases with increasing energy. The increase is partly due to the increase in the maximum energy that can be transferred to the atomic electrons and partly due to the Lorentz contraction of the Coulomb field of the heavy particle which increases the efficiency of energy transfer in distant collisions. A similar effect occurs in the classical theory (Bohr, 1915). The increase is limited by the density effect.

1.8 The Density Effect

At high velocities, it is no longer possible to regard the atoms of the material as isolated. There is a decrease in the effective interaction between the incident particles and the material due to the polarization of the material. A general discussion has been given by A. Bohr (1948) and many explicit calculations have been made (cf. Sternheimer, 1956).

1.9 Capture and Loss

As the incident particles are slowed down to velocities of the same order as those of the atomic electrons, the charge Z' is altered by the increasing efficiency of electron capture and formula (35) becomes invalid. A semiclassical theory has been developed by Bell (1953), extending earlier work by N. Bohr (1948), which appears to be successful in predicting the average charge Z'_{eff} of fission fragments. It is then assumed that the associated energy loss may be obtained by replacing Z' by Z'_{eff}.

A detailed discussion of the energy loss for protons moving in a gas of atomic hydrogen has been given by Dalgarno and Griffing (1955), using collision cross sections obtained with the Born approximation. The incident protons capture electrons to form neutral atoms according to the process

$$H^+ + H \rightarrow \sum_t H_t + H^+ \tag{28}$$

where the subscript t denotes the tth excited state. The neutral atoms may be transformed back into protons by the loss process

$$H + H \rightarrow H^+ + e + \mathbf{S}_t H_t. \tag{29}$$

Under the usual experimental conditions, an equilibrium between the charged and neutral components of the beam is rapidly attained such that the number of captures per second equals the number of losses. The fractional charged content of the beam is given by

$$f(H^+) = (1 + \sigma)^{-1} \tag{30}$$

and the neutral content by

$$f(H) = (1 + \sigma^{-1})^{-1} \tag{31}$$

where σ is the ratio of the total capture cross section to the total loss cross section. The computed values are given in Table II.

TABLE II

ANALYSIS OF ENERGY LOSS CONTRIBUTIONS[a]

$\log_{10} E$ (kev)	1.00	1.25	1.50	1.75	2.00	2.25	2.50	2.75	3.00	3.25	3.50
Proton impact											
Ionization	2.62	4.29	5.15	5.07	4.18	3.04	2.07	1.39	0.887	0.538	0.347
Excitation	2.60	2.64	2.32	1.82	1.37	0.959	0.648	0.426	0.274	0.174	0.109
Capture excitation	1.87	1.81	0.939	0.258	4.11^{-2}	4.07^{-3}	2.86^{-4}	1.57^{-5}	7.18^{-7}	2.93^{-8}	1.10^{-9}
Capture momentum loss	5.74	5.17	3.67	1.76	0.555	0.117	1.77^{-2}	1.99^{-3}	1.80^{-4}	1.39^{-5}	9.70^{-7}
Total proton stopping power	12.83	13.90	12.08	8.92	6.15	4.12	2.73	1.82	1.16	0.712	0.456
Neutral atom impact											
Single ionization	1.79	2.52	3.09	3.33	3.11	2.56	1.89	1.32	0.780	0.570	0.326
Double ionization	3.81^{-3}	4.65^{-2}	0.293	0.902	1.58	1.80	1.55	1.12	0.709	0.429	0.262
Single excitation	0.518	0.332	0.198	0.113	6.43^{-2}	3.67^{-2}	2.04^{-2}	1.14^{-2}	6.50^{-3}	3.61^{-3}	2.05^{-3}
Double excitation	2.32^{-2}	8.35^{-2}	0.144	0.161	0.136	9.52^{-2}	6.03^{-2}	3.83^{-2}	2.21^{-2}	1.28^{-2}	8.03^{-3}
Simultaneous excitation and ionization	2.04^{-2}	0.127	0.380	0.603	0.584	0.430	0.278	0.177	9.65^{-2}	6.07^{-2}	3.28^{-2}
Capture excitation	0.493	0.366	0.252	0.090	0.026	0.00	—	—	—	—	—
Total neutral stopping power	2.84	3.48	4.35	5.19	5.50	4.93	3.80	2.68	1.62	1.07	0.623
Actual proton stopping power	0.51	1.26	2.66	4.74	5.27	4.01	2.72	1.82	1.16	0.712	0.456
Actual neutral stopping power	2.73	3.16	3.39	2.43	0.79	0.13	0.01	0.00	—	—	—
Stopping power of beam[b]	3.29	4.47	6.09	7.19	6.06	4.14	2.73	1.82	1.16	0.712	0.456

[a] Units of (ev × 10^{-15} cm^2) ($1.23^{-2} = 1.23 \times 10^{-2}$).

[b] These figures include a contribution from H^-.

The energy loss associated with capture consists of the excitation energy and the energy required for the transfer of the electron from its parent nucleus to the capturing proton. The energy loss produced by the neutral component of the beam arises not only from excitation and ionization of the gas but also from excitation and ionization of the beam particle. Double processes of the kind

$$H + H \rightarrow H^+ + e + H_t \tag{32}$$

must be included.

The detailed contributions to stopping power of the various processes are listed in Table II. At the lower energies the major loss for protons is associated with the transfer of electron momentum during capture. At high energies only ionization and excitation are important, ionization being about 3 times as effective as excitation. For neutral atom impact, single ionization is the most important process at all energies and only at low energies does excitation make an effective contribution. At energies greater than about 150 kev the neutral component of the beam is negligible and all the energy loss occurs through proton impacts, while at energies below 45 kev most of the loss is due to neutral atom impacts.

1.10 Nuclear Scattering Losses

In the slowing down of fission fragments near the end of their path, an important contribution comes from elastic collisions with the nuclei of the stopping material. It is given by

$$-\frac{1}{n}\frac{dE}{dx} = \frac{4\pi e^4}{Mv^2} Z'^2 Z^2 \ln\left(\frac{\mathscr{M} v^2 \rho}{Z' Z e^2}\right) \tag{33}$$

(A. Bohr, 1948) where ρ is the impact parameter beyond which the energy loss is zero due to the screening of the nucleus by the atomic electrons. To (33), there should be added a small amount due to energy loss in ordinary gas kinetic collisions, but this has a negligible effect on the range.

2 *Range of Charged Particles*

The mean distance through which a particle initially with energy E travels in a material before it is stopped is related to the stopping power by

$$R(E) = \int_0^E dE\Big/\left(-\frac{dE}{dx}\right) = \frac{M'}{Z'^2}\frac{m}{4\pi e^4 n}\int_0^v \frac{v^3 dv}{B(v)}. \tag{34}$$

The extensive data have been summarized recently by Whaling (1958).

2.1 Relationships between Ranges of Different Particles

The mean distance through which a particle of mass M' and charge Z' travels in decreasing its velocity from v_1 to v_2 is given by

$$R(v_1) - R(v_2) = \frac{M'}{Z'^2} \{f(v_1) - f(v_2)\} \tag{35}$$

where the function $f(v)$ is independent of the mass and charge, provided v_2 is not so small that charge neutralization is significant. A range-energy relationship for α-particles can therefore be obtained from that for protons,

$$R_{\mathrm{H}}(v) = \frac{M'_{\mathrm{H}}}{M'_{\alpha}} \left(\frac{Z'_{\alpha}}{Z'_{\mathrm{H}}}\right)^2 R_{\alpha}(v) - c, \tag{36}$$

in an obvious notation, c being a small constant which takes account of the effect of capture and loss at low energies and which must, in general, be determined experimentally. For air at 15°C and 760 mm Hg, Bethe and Ashkin (1953) give

$$R_{\mathrm{H}}(v) = 1.007 R_{\alpha}(v) - 0.20 \text{ cm}. \tag{37}$$

Capture and loss are the same for all isotopes so that for them the constant c vanishes. In particular, the range of a μ-meson with a mass $210m$ is

$$R_{\mu}(E) = 0.114 R_{\mathrm{H}}(8.8E). \tag{38}$$

A typical range-energy relationship is illustrated in Fig. 3. It refers to the passage of protons in molecular nitrogen (Whaling, 1958).

2.2 Straggling

Because the energy loss does not occur continuously but consists of a large number of discrete processes, the ranges of the particles of an initially monoenergetic group will be distributed about the mean range. Similarly, the energies lost in traveling the same path length will be distributed about a mean energy loss. These fluctuation effects are termed "straggling."

The mean-square fluctuation of the energy of the particles after traveling a path length X is

$$[(E^2)_{\mathrm{av}} - (E_{\mathrm{av}})^2]_X = n \int_0^X dx\, \mathbf{S}_t (E_t - E_0)^2 Q_t \tag{39}$$

which may be evaluated using the Bethe theory as in § 1.1 (Livingston and Bethe, 1937).

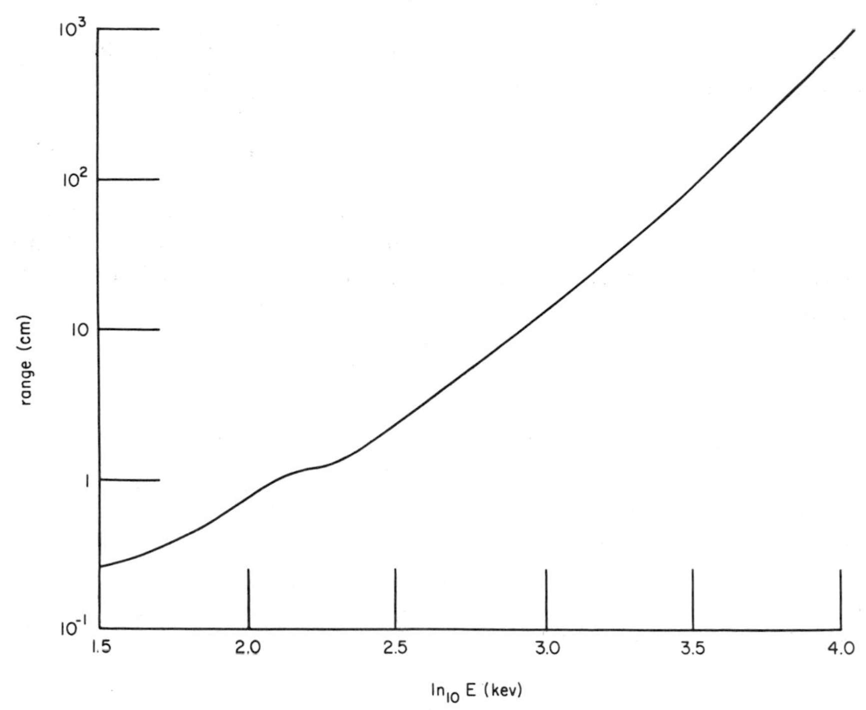

FIG. 3. The range-energy relation for protons moving through molecular nitrogen at 15°C and 760 mm Hg.

Similarly, the mean-square fluctuation of the distance traveled by particles which have lost the energy E is given by

$$\frac{d}{dE}[(X^2)_{\text{av}} - (X_{\text{av}})^2] = \frac{d}{dX}[(E^2_{\text{av}}) - (E_{\text{av}})^2]\left(\frac{dE}{dx}\right)^{-3} \tag{40}$$

which may be evaluated after substitution of (39). Integrating (40) from the initial energy E_0 to zero gives the mean-square fluctuation of the particle range

$$\begin{aligned}(R - R_0)^2_{\text{av}} &= [(R^2)_{\text{av}} - R_0^2] \\ &= n \int_0^{E_0} \frac{\mathbf{S}_t(E_t - E_0)^2 Q_t}{(dE/dx)^3}\, dE.\end{aligned} \tag{41}$$

These formulae are in satisfactory agreement with observations (Bloembergen and Van Heerden, 1951; Mather and Segre, 1951) apart from some deviations due probably to multiple scattering (Bichsel, 1960) which contributes to the experimental range.

Formula (41) takes account only of losses due to excitation and ionization. For fission products, the nuclear scattering loss is more important near the end of their range and it causes a large straggling amounting to about 10% of the range (N. Bohr, 1948).

3 *Ionization by Charged Particles*

The ionization produced by a heavy charged particle consists of the *primary* ionization produced by direct ionization by heavy particle impact and the secondary ionization produced by the impact of the electrons ejected during the primary ionization. At low energies the secondary ionization is negligible, but at high energies it may be as large as the primary ionization.

3.1 Mean Energy Per Ion Pair

It is convenient to characterize the ionization produced by the *mean energy per ion pair*

$$W = E/J(E) \tag{42}$$

where $J(E)$ is the total number of ion pairs produced by the absorption of the heavy particle in the material. For a large number of substances, the remarkable fact is observed that W is nearly constant over a wide range of energies. The production of ions per centimeter path is then obtained very easily as

$$p = -\frac{dE}{dx}\bigg/W \tag{43}$$

and the shape of the curve of ionization as a function of residual range (the Bragg curve) is the same as the shape of the curve of energy loss. This has useful experimental applications (cf. Bethe and Ashkin, 1953).

There is no general theoretical explanation of the near-constancy of W, but Fano (1946) has given a qualitative explanation. Fano considers the total energy available for ionization, which may be possessed by the heavy particle or by the secondary electrons. In a collision in which the atom is excited this available energy is reduced by the excitation energy W_e. In a collision in which the atom is ionized and the kinetic energy of the ejected electron is insufficient to cause further ionization, the ionization energy W_i is lost from the available energy. If the kinetic energy is sufficient to cause further ionization, only the energy equal to the ionization potential W_I is lost. The average amount of energy spent per ion produced is then

$$\omega = \frac{Q_e W_e + Q_1 W_1 + Q_I W_I}{Q_1 + Q_I} \tag{44}$$

where the Q's are the corresponding cross sections. The *ratios* of the cross sections are insensitive to energy and ω should therefore be nearly constant.

A detailed analysis has been presented by Dalgarno and Griffing (1958) for protons in atomic hydrogen. They calculate not W but

$$\omega(E) = 1/(dJ/dE) \tag{45}$$

which is related to W by

$$\frac{1}{\omega} = \frac{d}{dE}\left(\frac{E}{W}\right). \tag{46}$$

If either is constant, $\omega = W$.

The results are shown in Fig. 4. There is a spread of only 2.5 ev about a mean of 36 ev for energies increasing from 10 kev to over

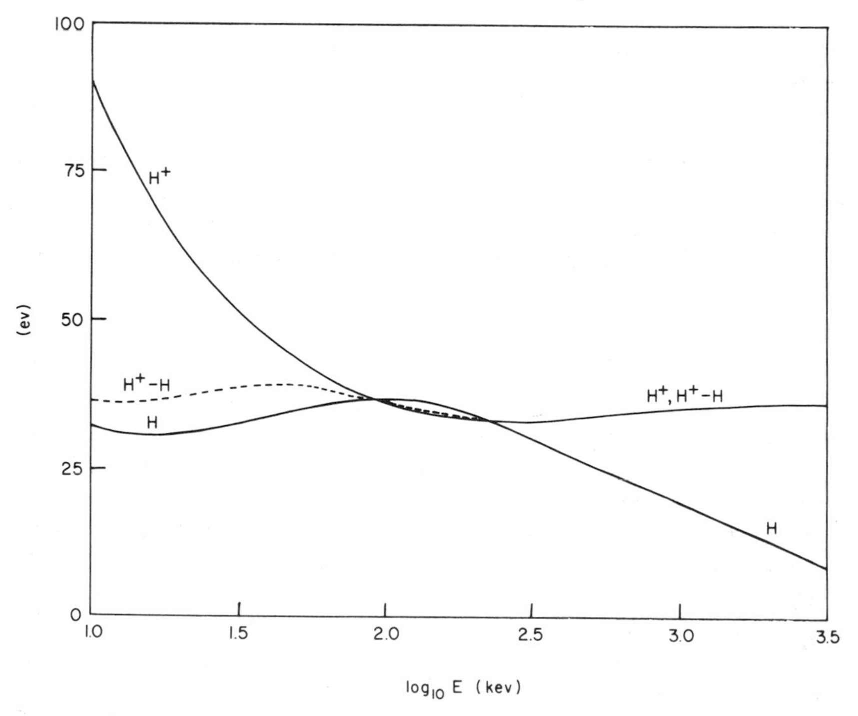

FIG. 4. The mean specific energy per ion pair for a proton beam, a neutral hydrogen beam, and an H^+-H beam in charge equilibrium moving through atomic hydrogen.

1 Mev. Of particular interest is the fact that the progressive neutralization of the beam through capture and loss processes extends considerably the range of energy over which ω is nearly constant.

A similar near-constancy of ω has been found by Erskine (1954) for α-particles from 1 to 6 Mev in helium, capture and loss effects being unimportant. Erskine calculated that ω varies by less than 1.5% about a mean of 40.9 ev.

Exceptions to the general constancy of ω are to be expected, but the experimental data are not definitive. There are indications that ω (and W) eventually increase as the energy decreases (cf. Valentine and Curran, 1958) as indeed, must happen, since the fraction of energy expended on excitation and on momentum transfer in elastic collisions must increase as the ionization threshold is approached.

A list of values of W measured for several gases is given in Table III. They range from 22 ev for Xe to 46 ev for He.

TABLE III

VALUES OF THE MEAN SPECIFIC ENERGY PER ION PAIR

Gas	He	Ne	Ar	Kr	Xe	H_2	N_2	O_2	Air
(ev)	46	37	26	24	22	36	37	33	36

3.2 Fluctuations of Ionization

For a given incident energy E, the number of ions produced will fluctuate about a mean. The theoretical development is similar to that for range straggling (Fano, 1947).

4 *Energy Loss of Electrons*

For high-energy electrons, the energy loss through excitation and ionization can be analyzed using the Bethe theory of § 1.1 with some slight modifications. We shall ignore loss through bremsstrahlung except to remark that the ratio of radiative loss to collision loss is approximately

$$\frac{(dE/dx)_{\rm rad}}{(dE/dx)_{\rm coll}} \sim \frac{EZ}{800} \tag{47}$$

(Bethe and Heitler, 1934), where E is measured in Mev.

4.1 Bethe's Theory for Electrons

From (7), the maximum value of the momentum change for electron impact is k so that (18) is replaced by

$$-\frac{dE}{dx} = n\frac{4\pi e^4}{mv^2} Z \ln \frac{mv^2}{I} \tag{48}$$

which differs from (18) by a factor of 2 in the logarithm. A further modification is necessary because of the indistinguishability of the two electrons in an ionizing collision. Bethe (1930) defines the electron with higher energy as the primary electron so that the maximum energy loss in any collision is $\frac{1}{4}mv^2$ not $\frac{1}{2}mv^2$. On using now the formula of Mott (1930), Bethe obtains

$$-\frac{dE}{dx} = n\frac{4\pi e^4}{mv^2} Z \ln \frac{mv^2}{2I}\sqrt{\frac{e}{2}}. \tag{49}$$

The difference between (18) and (49) consists of a small factor in the logarithm and protons and electrons of the same (nonrelativistic) velocity lose energy at much the same rate.

4.2 Relativistic Effects

For the relativistic stopping power, Bethe (1933) obtains

$$\begin{aligned}-\frac{dE}{dx} = n\frac{2\pi e^4}{mv^2} Z \Big\{ &\ln \frac{mv^2E}{2I^2(1-\beta^2)} - (2\sqrt{1-\beta^2} - 1 + \beta^2)\ln 2 \\ &+ 1 - \beta^2 + \tfrac{1}{8}(1-\sqrt{1-\beta^2})^2 \Big\}\end{aligned} \tag{50}$$

where E is the total energy of the electron minus the rest energy. When β is small, (50) goes over into (49). The relativistic increase has been observed in many experiments (cf. Birkhoff, 1958).

4.3 Density Effect

The density effect noted in § 1.8 is more easily observed for electrons since it depends on the particle velocity. The predicted size of the effect for molecular hydrogen at 1-atm pressure is illustrated in Fig. 5, which shows also the relativistic increase.

There is an interesting connection between the density effect and Cerenkov radiation, which has been discussed by Budini (1953).

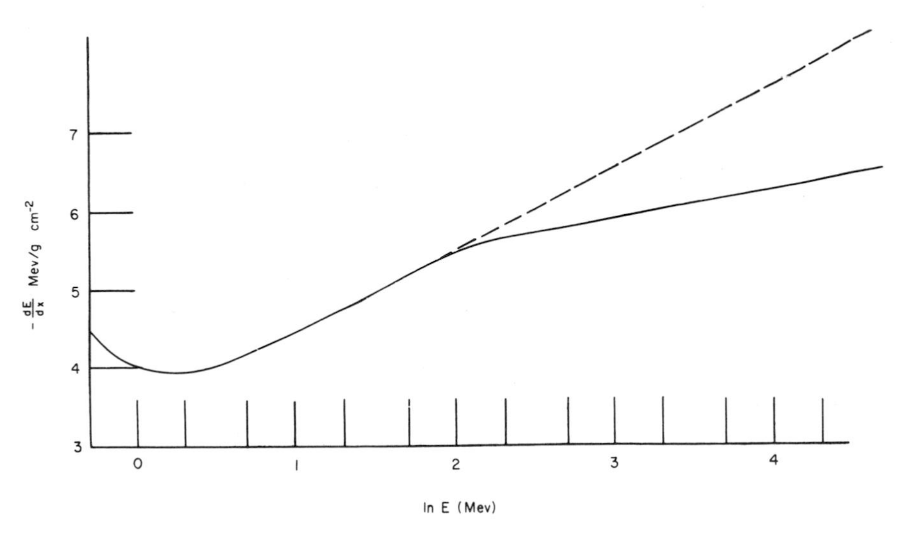

FIG. 5. Energy loss of electrons in molecular hydrogen at 15°C and 760 mm Hg. The dotted curve is obtained when the density effect is neglected.

5 *Range of Electrons*

Because of the small mass of the electron, the number of scattering processes associated with a given energy loss is much larger than that for a heavy particle of the same energy. Electrons will follow different paths in passing through a gas and the range is not a useful parameter.

5.1 The Probable Energy Loss

The straggling of the energy loss is increased over that associated with multiple scattering by the possibility that the energy loss in a single collision may be a large fraction of the energy of the incident electron. The theory of the straggling of electrons has been developed by Landau (1944) and Blunck and Leisegang (1950) who show that the most probable rate of energy loss is given by

$$-\left(\frac{dE}{dx}\right)_{\text{prob}} = n\,\frac{2\pi e^4}{mv^2}\,Z\left\{\ln\left[\frac{4\pi n e^4 x Z}{I^2(1-\beta^2)}\right] - \beta^2 + 0.37\right\} \tag{51}$$

which is, in general, less than the average energy loss.

The full width of the straggling distribution at half-maximum height is about

$$T = n\frac{8\pi e^4 xZ}{mv^2} \tag{52}$$

so that the straggling amounts to about 25% of the most probable energy loss.

These calculations refer to the energy loss along the path of the particle, but do not predict the path of the particle.

5.2 Multiple Scattering

The distribution of path length and the average increase in path length has been discussed by Yang (1951). It cannot be observed directly and further calculation is necessary to determine its effect on the straggling distribution. The mathematical description is complicated. The most refined treatment appears to be that of Spenser (1955) who gives curves showing the distribution of energy deposition for electrons of various energies. However, the theory does not take proper account of the secondary electrons.

6 *Ionization by Electrons*

Just as for heavy particles, it seems that the mean energy per ion pair produced by the absorption of an electron beam is a constant over a large energy range and for a large number of substances (cf. Valentine and Curran, 1958). The values of W are not greatly different from those for heavy particles (Table III).

As the energy decreases, ω will eventually increase rapidly. In Fig. 6, the calculations of Dalgarno and Griffing (1958) for atomic hydrogen are reproduced, the increase in ω setting in at an impact energy of 60 ev. Dalgarno and Griffing calculate ω by solving numerically the integral equation for the number of ion pairs

$$Q_\Sigma J(E) = Q_c(E) + \sum_t Q_t(E)\, J(E - E_0 + E_t) \\ + \int_0^{E-E_c} Q_c(E, \epsilon)\; J(E - \epsilon - E_c) + J(\epsilon)\} \, d\epsilon \tag{53}$$

where $Q_c(E, \epsilon)$ is the cross section for the ejection of an electron with energy ϵ, Q_c is the total ionization cross section, and E_c is the ionization potential (Knipp *et al.*, 1953; Erskine, 1954). A similar calculation has been performed by Erskine (1954) for helium.

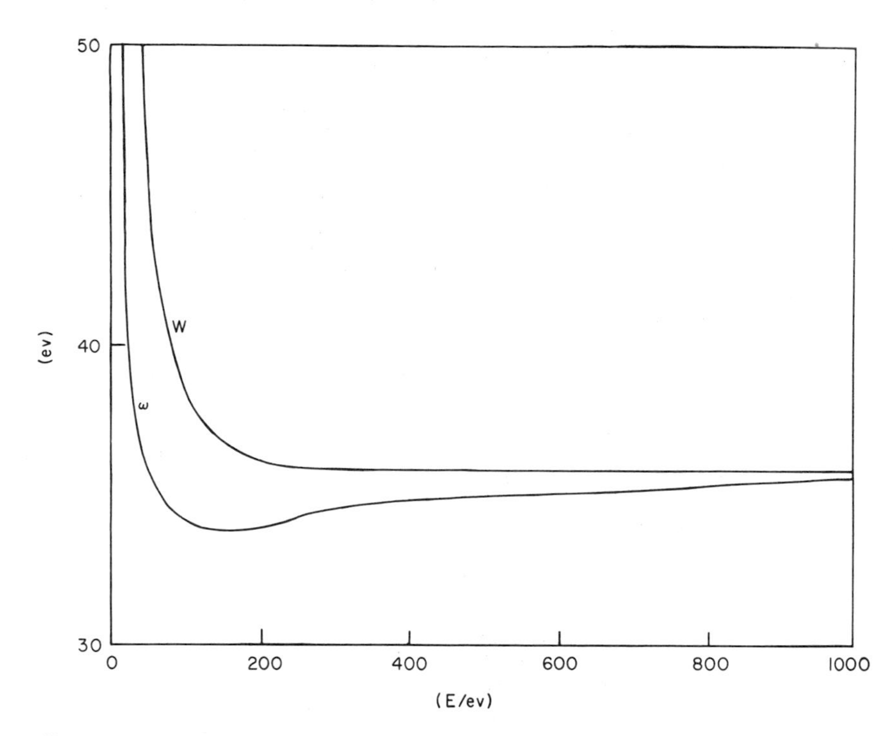

FIG. 6. The mean specific energy ω per ion pair and the average energy W per ion pair for electrons in atomic hydrogen.

REFERENCES

Bell, G. I. (1953) *Phys. Rev.* **90**, 548.

Bethe, H. A. (1930) *Ann. Phys. (Leipzig)* **5**, 325.

Bethe, H. A. (1933) *In* "Handbuch der Physik" Vol. **24** (2), p. 273. Springer, Berlin.

Bethe, H. A. (1938) *Z. Phys.* **76**, 293.

Bethe, H. A., and Ashkin, J. (1953) "Experimental Nuclear Physics." Wiley, New York.

Bethe, H. A., and Heitler, W. (1934) *Proc. Roy. Soc.* **A146**, 83.

Bichsel, H. (1960) *Phys. Rev.* **120**, 1012.

Birkhoff, R. D. (1958) *In* "Encyclopaedia of Physics" (S. Flügge, ed.), Vol. 34, p. 53. Springer, Berlin.

Bloch, F. (1933) *Z. Phys.* **81**, 363.

Bloembergen, N., and Van Heerden, J. (1951) *Phys. Rev.* **83**, 561.

Blunck, I., and Leisegang, S. (1950) *Z. Phys.* **128**, 500.

Bohr, A. (1948) *K. Danske Vidensk. Selsk. mat. fys. Medd.* **24**, 191.

Bohr, N. (1915) *Phil. Mag.* **30**, 581.
Bohr, N. (1948) *K. Danske Vidensk. Selsk. mat. fys. Medd.* **18**, 8.
Brandt, W. (1958) *Phys. Rev.* **112**, 1624.
Brolley, J. E., and Ribe, F. L. (1955) *Phys. Rev.* **98**, 112.
Brown, L. M. (1950) *Phys. Rev.* **79**, 297.
Budini, P. (1953) *Nuovo Cimento* **10**, 236.
Dalgarno, A. (1960) *Proc. Phys. Soc.* **76**, 422.
Dalgarno, A., and Griffing, G. W. (1955) *Proc. Roy. Soc.* **A232**, 423.
Dalgarno, A., and Griffing, G. W. (1958) *Proc. Roy. Soc.* **A248**, 415.
Dalgarno, A., and Lynn, N. (1957) *Proc. Phys. Soc.* **A70**, 802.
Erskine, G. A. (1954) *Proc. Roy. Soc.* **A224**, 361.
Fano, U. (1946) *Phys. Rev.* **70**, 44.
Fano, U. (1947) *Phys. Rev.* **72**, 26.
Knipp, J. K., Eguchi, T., Ohta, M., and Nagata, S. (1953) *Progr. theor. Phys.* **10**, 24.
Landau, L. D. (1944) *J. Phys. USSR* **8**, 201.
Lindhard, J., and Scharff, M. (1953) *K. Danske Vidensk. Selsk. mat. fys. Medd.* **27**, (15).
Livingston, M. S., and Bethe, H. A. (1937) *Rev. mod. Phys.* **9**, 245.
Mather, R. L., and Segré, E. (1951) *Phys. Rev.* **84**, 191.
Møller, C. V. (1932) *Ann. Phys. (Leipzig)* **14**, 531.
Mott, N. F. (1930) *Proc. Roy. Soc.* **A126**, 259.
Mott, N. F., and Massey, H. S. W. (1949) "The Theory of Atomic Collisions," 2nd ed. Clarendon Press, (Oxford), London and New York.
Phillips, J. A. (1953) *Phys. Rev.* **90**, 532.
Platzman, R. L. (1952) "Symposium on Radiobiology" (J. L. Nickson, ed.). Wiley, New York.
Reynolds, H. K., Dunbar, D. N. F., Wengel, W. A., and Whaling, W. (1953) *Phys. Rev.* **92**, 742.
Spenser, L. V. (1955) *Phys. Rev.* **98**, 1597.
Sternheimer, R. M. (1956) *Phys. Rev.* **115**, 137.
Sternheimer, R. M. (1959) *Phys. Rev.* **115**, 137.
Valentine, J. M., and Curran, S. C. (1958) *Rep. Progr. Phys.* **21**, 1.
Walske, M. C. (1952) *Phys. Rev.* **88**, 1283.
Walske, M. C. (1956) *Phys. Rev.* **101**, 940.
Weyl, P. K. (1953) *Phys. Rev.* **91**, 289.
Whaling, W. (1958) *In* "Encyclopaedia of Physics" (S. Flügge, ed.), Vol. 34, p. 193. Springer, Berlin.
Yang, C. N. (1951) *Phys. Rev.* **84**, 599.

16.

Diffusion and Mobilities

A. Dalgarno

Introduction

The phenomenon of *diffusion* is the transfer of mass from one region of space to another that occurs because of a gradient in the concentration of the material. The diffusive flow is in the direction opposite to that of the gradient and the flux is proportional to the magnitude of the gradient.

Thus if $\mathbf{j}$ is the particle flux and n the particle concentration,

$$\mathbf{j} = - \mathscr{D} \operatorname{grad} n \tag{1}$$

where $\mathscr{D}$ is a constant called the *diffusion coefficient*.

A different phenomenon occurs if some of the particles of a gas are charged and a small electric field is applied. The charged particles acquire a drift velocity $\mathbf{v}_d$ which is proportional to the magnitude E of the electric field. Thus

$$v_d = \mathscr{K} E \tag{2}$$

where $\mathscr{K}$ is a constant called the *ionic mobility*.

There is a close similarity between the mathematical descriptions of the two phenomena and $\mathscr{D}$ and $\mathscr{K}$ are related according to

$$\mathscr{K} = e\mathscr{D}/kT \tag{3}$$

where T is the gas temperature. If $\mathscr{D}$ is measured in $\text{cm}^2\ \text{sec}^{-1}$ and $\mathscr{K}$ in $\text{cm}^2\ \text{volt}^{-1}\ \text{sec}^{-1}$, (3) is

$$\mathscr{K} = 1.17 \times 10^4\ \mathscr{D}/T. \tag{4}$$

1 Diffusion

1.1 The Diffusion Coefficient

The kinetic theory of spherically symmetric (monatomic) gases has been developed rigorously by Enskog and Chapman (cf. Chapman and Cowling, 1939). For the diffusion coefficient in a binary mixture of gases consisting of particles of mass M_1 and concentration n_1 and of particles of mass M_2 and concentration n_2, it yields the formula

$$\mathscr{D} = \frac{3}{16(n_1 + n_2)} \left(\frac{2\pi kT}{\mu}\right)^{1/2} \frac{1 + \epsilon_0}{\bar{Q}_d} \tag{5}$$

where μ is the reduced mass $M_1M_2/(M_1 + M_2)$, ϵ_0 is a small correction factor, and $\bar{Q}_d$ is an averaged diffusion (or momentum transfer) cross section

$$\bar{Q}_d = \tfrac{1}{2} \int_0^\infty x^2 Q_d(x) \exp(-x)\, dx, \tag{6}$$

x being related to the energy $\mathscr{E}$ of relative motion of the two particles by

$$x = \frac{\mathscr{E}}{kT} = \tfrac{1}{2}\mu v^2/kT \tag{7}$$

and v being the relative velocity.

If Q_d is independent of x, $\overline{Q}_d = Q_d$, and $\mathscr{D}$ is proportional to $(T/\mu)^{1/2}$, a result which is also obtained by simple mean free path considerations (cf. Hirschfelder *et al.*, 1954).

1.2 The Diffusion Cross Section

The diffusion cross section is given by

$$Q_d = 2\pi \int_0^\pi I(\theta)\,(1 - \cos\theta)\sin\theta\, d\theta \tag{8}$$

where $I(\theta)$ is the differential cross section for elastic scattering through an angle θ (Mott and Massey, 1949) by the interaction potential $V(R)$ between the two particles, R being their separation. According to the quantal theory of elastic scattering (Chapter 9, § 1.3), the differential cross section is given by

$$I(\theta) = \frac{1}{k^2}\left|\sum_{l=0}^{\infty}(2l+1)\,e^{i\eta_l}\sin\eta_l P_l(\cos\theta)\right|^2 \tag{9}$$

where $k = \mu v/\hbar$ and the phase shifts η_l are such that the regular solution of

$$\frac{d^2\phi_l}{dR^2} + \left\{k^2 - \frac{2\mu}{\hbar^2}V(R) - \frac{l(l+1)}{R^2}\right\}\phi_l = 0 \tag{10}$$

behaves asymptotically as

$$\phi_l \sim k^{-1}\sin(kr - \tfrac{1}{2}l\pi + \eta_l). \tag{11}$$

It follows after substituting (9) into (8) that

$$Q_d = \frac{4\pi}{k^2}\sum_{l=0}^{\infty}(l+1)\sin^2(\eta_l - \eta_{l+1}). \tag{12}$$

Except at very low temperatures, Jeffreys' formula

$$\eta_l = \int_{R_0}^{\infty}\left\{k^2 - \frac{2\mu V}{\hbar^2} - \frac{l(l+1)}{R^2}\right\}^{1/2} dR - \int_{R_0'}^{\infty}\left\{k^2 - \frac{l(l+1)}{R^2}\right\}^{1/2} dR, \tag{13}$$

the lower limits being the outermost zeros of the integrands, provides an adequate approximation to the phase shifts. On introducing the impact parameter p defined by

$$p = \sqrt{l(l+1)}\,k, \tag{14}$$

(13) may be written

$$\eta(p) = k\int_{R_0}^{\infty}\left\{1 - \frac{V}{\mathscr{E}} - \frac{p^2}{R^2}\right\}^{1/2} dR - k\int_{R_0'}^{\infty}\left\{1 - \frac{p^2}{R^2}\right\}^{1/2} dR. \tag{15}$$

In the case of collisions between heavy particles, the number of phase shifts contributing to (12) is very large and we may replace the summation over l by an integration over p. Consistent with this replacement,

$$\eta_l - \eta_{l+1} = -\,k^{-1}\frac{\partial\eta}{\partial p} \tag{16}$$

so that (12) becomes

$$Q_d = 4\pi\int_0^{\infty} p\cos^2\alpha(p)\,dp \tag{17}$$

where

$$\alpha(p) = \int_{R_0}^{\infty} dR/R\Phi(R) \tag{18}$$

$$\Phi(R) = \left\{\frac{R^2}{p^2} - 1 - \frac{R^2}{p^2}\frac{V(R)}{\mathscr{E}}\right\}^{1/2}. \tag{19}$$

Formula (17) is just that obtained from a classical description of the scattering and a quantal description is required only in those cases where more than one possible interaction potential occurs (§ 1.9). The success of the classical theory stems from the factor of $1 - \cos\theta$ in the integrand of (8) which suppresses the contribution from the scattering at small angles.

1.3 Temperature Dependence of $\mathscr{D}$

The variation of $\mathscr{D}$ with temperature depends upon the explicit form of the interaction potential. For a rigid sphere model, the diffusion coefficient at constant density is proportional to $T^{1/2}$ and the diffusion coefficient at constant pressure is proportional to $T^{3/2}$. For more realistic models, it will vary rather more rapidly. Thus, for a potential directly proportional to R^{-n}, simple dimensional considerations show that Q_d varies as $\mathscr{E}^{-2/n}$ and $\mathscr{D}$ at constant density as $T^{2/n}\sqrt{T}$.

At low temperatures, the scattering is dominated by the long-range interaction which for atoms in their ground states falls off as R^{-6}. The diffusion coefficient at constant density therefore varies as $T^{5/6}$ at low temperatures (but not so low that quantum effects are important). At high temperatures, the scattering is dominated by the short-range interaction which for many systems varies roughly as R^{-12}. The diffusion coefficient at constant density therefore increases roughly as $T^{2/3}$ at high temperatures. For the diffusion in a He-N_2 mixture, for example, Walker and Westenberg (1958) find experimentally that $\mathscr{D}$ varies as $T^{0.691}$.

1.4 Calculations of $\mathscr{D}$

Extensive calculations of $\mathscr{D}$ have been carried out based on the Lennard-Jones potential

$$V(R) = 4\epsilon\left\{\left(\frac{\sigma}{R}\right)^{12} - \left(\frac{\sigma}{R}\right)^{6}\right\} \tag{20}$$

(Kihara and Kotani, 1943; de Boer and van Kranendonk, 1948; Hirschfelder *et al.*, 1948, 1949; Rowlinson, 1949) and on the modified Buckingham potential

$$V(R) = \frac{\epsilon}{1 - 6/\alpha}\left\{\frac{6}{\alpha}\exp\left[\alpha(1 - R/\sigma)\right] - \left(\frac{\sigma}{R}\right)^{6}\right\} \tag{21}$$

(Mason, 1954) for a wide range of the parameters ϵ, σ, and α. The forms adopted are useful representations of the interactions between atoms in their ground states.

The parameters may be chosen either from theoretical considerations or so that the available experimental data on equilibrium properties and transport phenomena (frequently viscosity data) are reproduced as closely as possible. The calculations then yield values of $\mathscr{D}$ (and of other transport coefficients) for a range of physical conditions more extensive than those obtaining in the measurements.

The diffusion coefficient in an argon-helium mixture: From viscosity data on pure helium and pure argon, the values of the parameters ϵ

and σ in (20) may be obtained. Then using the combination rules

$$\sigma_{ij} = \frac{1}{2}(\sigma_{ii} + \sigma_{jj}) \tag{22}$$

$$\epsilon_{ij} = (\epsilon_{ii}\epsilon_{jj})^{1/2} \tag{23}$$

in an obvious notation, Hirschfelder *et al.* (1954) find that for a helium-argon mixture, $\sigma = 3.059$ A and $\epsilon = 21.3k$ °K. The corresponding value of $\mathscr{D}$ at a temperature of 273°K and a pressure of 760 mm Hg is 0.653 cm^2 sec^{-1} compared with the measured value of 0.641 cm^2 sec^{-1}. A more extended comparison has been given by Weissman *et al.* (1960).

The range of separations R which effectively control the values of the transport coefficients depends on the temperature, higher temperatures involving smaller separations. The accuracy of the theoretical predictions may be much reduced if the temperatures at which predictions are made are significantly different from those of the measurements from which the interaction potential is derived. This limitation may be overcome by using an increased number of experimental data and more flexible representations of $V(R)$ for their analysis (Buckingham, 1961).

1.5 The Correction Factor for $\mathscr{D}$

Formulae for ϵ_0, the correction factor in (5), have been given by Chapman and Cowling (1939) and by Kihara (1953). They involve collision integrals similar to $\bar{Q}_d$ and may be evaluated straightforwardly once the interaction potential has been selected. The correction factor depends slightly on the relative concentrations of the (inter)-diffusing particles.

Mason (1957) has prepared tables of ϵ_0 corresponding to the modified Buckingham potential. He shows that in practical circumstances ϵ_0 never exceeds 0.05 and is usually much less.

The experimental values tend to be somewhat larger than the predicted values (Nettley, 1954).

1.6 Diffusion in Mixtures

Hirschfelder *et al.* (1954) have given a formula for the diffusion in a multicomponent gas which can be evaluated without difficulty when all the possible interaction potentials have been specified. As yet, little application of it has been made.

In the case when the concentration of one of the components is very small compared with those of the others, the formula simplifies and the diffusion coefficient of the rare component is given by

$$\mathscr{D}_j^{-1} = \sum_i f_i \, \mathscr{D}_{ij}^{-1} \tag{24}$$

where f_i is the fraction of the ith particles and $\mathscr{D}_{ij}$ is the binary diffusion coefficient for the $i - j$ mixture.

1.7 Experimental Data

The available experimental data have been reviewed by Hirschfelder *et al.* (1954) and by Westenberg (1957). Subsequent measurements have been carried out by Walker and Westenberg (1958, 1959, 1960), Amdur and Schatzki (1957), Srivastava and Srivastava (1959), Srivastava and Barua (1959), Walker *et al.* (1960), Wise (1960) and others. Binary diffusion coefficients have been measured for mixtures of most of the common laboratory gases using the Loschmidt or the Stefan technique (cf. Jost, 1952) at and below room temperature and the point source technique (Walker and Westenberg, 1958) at higher temperatures. Wise and Ablow (1958) have suggested a method for measuring the diffusion coefficients of labile species.

A representative sample of the data is presented in Table I.

TABLE I

Binary Diffusion Coefficients at 300°K and 760 mm Hg

Mixture	$\mathscr{D}$ (cm² sec⁻¹)
CO_2-N_2	0.17
He-N_2	0.74
He-Ar	0.76
Ne-Ar	0.33
Ne-Kr	0.26
Ar-Kr	0.14

1.8 Diffusion Coefficients of Excited Atoms

The calculation of the diffusion coefficients of excited atoms is complicated by a possible multiplicity of potential energies of approach.

Provided there are no symmetry effects (§ 1.9), (5) is still valid with the modification that the cross section Q_d must be replaced by a weighted mean of cross sections, each associated with one of the interaction potentials (Mason *et al.*, 1959). A more serious difficulty is introduced by the possibility that many of the collisions may be nonadiabatic. This aspect has received little attention.

Despite the lack of detailed calculations, it is clear on general grounds that the interaction potentials will usually be large in magnitude and with long ranges with the consequence that the values of $\mathscr{D}$ will be smaller (and may be much smaller) than those for atoms in ground states. Thus, Phelps (1959) finds that $\mathscr{D}$ for metastable ^3P neon atoms in helium is 0.82 cm^2 sec^{-1} at 300°K and 760 mm Hg compared with the value of 1.06 cm^2 sec^{-1} for ordinary neon in helium (Srivastava and Barua, 1959).

1.9 Symmetry Effects

The classical description fails for the diffusion of atoms through a gas of similar atoms when more than one interaction potential occurs. This situation always arises in the diffusion of excited atoms through a gas of normal atoms and in the self-diffusion of an atom which does not have a closed shell configuration.

Two normal hydrogen atoms, for example, may approach each other in either the $^1\Sigma_g$ or the $^3\Sigma_u$ state of the molecule H_2, the former of which is symmetrical with respect to interchange of the nuclei and the latter antisymmetrical. Equations (5) and (12) still apply, but because of the quantal symmetry requirements (Massey and Mohr, 1934) the phase shifts η_l are determined by the $^1\Sigma_g$ potential when l is odd and by the $^3\Sigma_u$ potential when l is even. There is in consequence no relation between the quantal and classical cross sections. It may be shown that the quantal diffusion cross section is approximately twice the cross section for the process in which the two electrons are interchanged. For the diffusion of, say, Na (^2P) in Na (^2S) the electron exchange process may be represented by

$$\mathrm{Na}\,(^2P) + \mathrm{Na}\,(^2S) \rightarrow \mathrm{Na}\,(^2S) + \mathrm{Na}\,(^2P), \tag{25}$$

and is often called the excitation transfer process.

The effect of the symmetry requirements is usually to produce very large cross sections and very small diffusion coefficients. Thus, the self-diffusion coefficient of atomic hydrogen at 273°K and 760 mm Hg is about 0.3 cm^2 sec^{-1}, nearly an order of magnitude smaller than classical considerations suggest.

There is some direct experimental support for the importance of the symmetry requirements, Phelps (1959) having measured a value of 0.20 cm^2 sec^{-1} for the diffusion coefficient of metastable ^3P neon atoms in neon gas at 300°K and 760 mm Hg while the self-diffusion coefficient for neon is about 0.48 cm^2 sec^{-1}.

Calculations by Buckingham and Dalgarno (1952) of $\mathscr{D}$ for metastable helium suggest that the effect of symmetry requirements may be unusually small (though significant) in the case of the metastable inert gases. Certainly for excited atoms which may combine optically with the ground state a much larger effect is to be expected, the two possible interaction potentials having a long range (King and Van Vleck, 1939; Mulliken, 1960).

A larger body of experimental data confirming the significance of the quantal symmetry requirements is given by measurements of the mobility of ions in their parent gases (§ 2.7).

1.10 Diffusion at Low Temperatures

At low temperatures, the quantal expression (12)† must be used for Q_d. Qualitatively, this results in oscillations of Q_d but the effect on $\mathscr{D}$ is negligibly small except for the light elements H, H_2, and He. Even for these, the alteration in the diffusion coefficients is less than 0.5% at room temperature.+

At low temperatures and high densities, quantum symmetry requirements must be satisfied in the structure of the Boltzman equation and not merely in the description of the collisions (Uehling and Uehling, 1933; Uehling, 1934; Hellund and Uehling, 1939). Some relevant calculations on He at 1°K have been carried out by de Boer (1943).

2 Mobilities

2.1 Ionic Mobility at Low Field Strengths

At low field strengths, the ionic mobility $\mathscr{K}$ is given by§

$$\mathscr{K} = e\mathscr{D}/kT \tag{26}$$

† It may be necessary to include the effect of nuclear spin (cf. Hirschfelder *et al.*, 1954).

+ See Buckingham *et al.* (1958) for some detailed numerical comparisons of classical and quantal calculations.

§ Equation (26) is not exactly correct but is a very good approximation (cf. p. 242).

where e is the charge on the ion. The calculation of $\mathcal{K}$ is therefore entirely similar to the calculation of $\mathcal{D}$ for a neutral particle except that the interaction potential between an ion and a gas atom has a much longer range and at large distances depends upon properties of the gas only. This allows the derivation of a general formula of wide applicability.

2.2 Variation of Mobility with Temperature

At large separations,

$$V(R) = -\tfrac{1}{2}\alpha e^2/R^4 \tag{27}$$

where α is the polarizability of the gas atoms.

The corresponding diffusion cross section is

$$Q_d = 2.210\pi(\alpha e^2/2\mathscr{E})^{1/2} \tag{28}$$

and the mobility is given by

$$\mathcal{K} = 35.9/\sqrt{\alpha\mu}\ \text{cm}^2\ \text{volt}^{-1}\ \text{sec}^{-1} \tag{29}$$

(cf. Dalgarno *et al.*, 1958) where α is measured in atomic units (a_0^3), the reduced mass μ is measured in units of the proton mass, and $\mathcal{K}$ is referred to a constant gas density of 2.69×10^{19} cm^{-3}. Formula (29) may be appropriately called the Langevin formula.

The linear dependence on $1/\sqrt{\mu}$ is predicted by classical theory for all interactions, but its constancy with temperature is true only for a pure R^{-4} interaction. Formula (29) also demonstrates that for a pure R^{-4} interaction, the mobility is independent of the charge state of the ion.

Since α is a property of the gas, it follows from (29) that $\mathcal{K}\sqrt{\mu}$ is independent of the nature of the ion. This is useful in assessing the accuracy of experimental data for the independence of $\mathcal{K}\sqrt{\mu}$ on the particular ion must be accompanied by its constancy with temperature.

Since the scattering at low temperatures is determined by the forces at large separations, all mobilities must ultimately conform with the predicted behavior provided sufficiently low temperatures can be reached before quantum modifications become important. The effect of the short-range forces persists down to quite low temperatures and observed mobilities may not be precisely constant but may vary slowly with temperature.

At high temperatures, the scattering depends on the short-range forces. The position is similar to that of neutral particles and a mobility falling off roughly as $T^{-1/3}$ is to be expected. In the intermediate region, some cancellation of the repulsive short-range forces and the attractive long-range forces will occur and the mobility will pass through a maximum.

Strictly, the long range interaction potential contains further attractive terms, the most important of which varies as R^{-6}, and Dalgarno *et al.* (1958) have pointed out that as a consequence the mobility may pass through a slight minimum. Detailed calculations by Mason and Schamp (1958) confirm this suggestion.

2.3 Experimental Data on Ions in Unlike Gases

Values of $\mathscr{K}\sqrt{\mu}$, resulting from various measurements at about 290°K (cf. Tyndall, 1938) are given in Table II. In the cases of xenon, krypton, and argon, $\mathscr{K}\sqrt{\mu}$ is essentially constant in harmony with (29),

TABLE II

$\mathscr{K}\sqrt{\mu}$ for Various Ions in the Inert Gases at 290°K

Ion \ Gas	He	Ne	Ar	Kr	Xe
Li	38.6	25.2	11.4	9.4	7.3
Na	41.9	26.8	11.5	9.3	7.5
K	41.0	27.4	11.7	9.6	7.4
Rb	39.3	27.2	11.7	9.5	7.4
Cs	36.3	25.5	11.5	9.5	7.4

but for helium and neon, gases of low polarizabilities, $\mathscr{K}\sqrt{\mu}$ depends upon the particular ion. We expect therefore that $\mathscr{K}$ should vary with temperature in helium and neon but not in the other three inert gases.

The observed variations with temperature (Pearce, 1936; Hoselitz, 1941) are reproduced in Table III. For xenon and krypton $\mathscr{K}$ is virtually independent of temperature (if we ignore the measurement in krypton at 90°K), but for argon it seems necessary to reject the measurements at and below 195°K. For helium, the variations are qualitatively in harmony with the theory.

TABLE III

VARIATION OF $\mathscr{K}$ WITH TEMPERATURE[a]

T°K	Li^+-He	K^+-Ar	Rb^+-Kr	Cs^+-Xe
20.5	20.0	—	—	—
78	21.8	1.30	—	—
90	22.2	1.52	1.15	—
195	23.9	2.34	1.57	1.02
273	—	—	1.575	1.005
291	25.8	2.81	1.58	1.01
370	—	—	1.59	1.01
389	27.8	—	—	—
400	—	3.07	—	—
460	—	2.95	1.64	1.03
483	29.2	—	—	—

[a] Units: cm^2 $volt^{-1}$ sec^{-1}.

2.4 CLUSTERING

It is probable that the anomously small mobilities observed in krypton at 90°K and in argon at and below 195°K are due to clustering, which leads to an increase in the mass of the ion. Clustering has been studied by Munson and Tyndall (1939) for polar gases and by Munson and Hoselitz (1939) for the inert gases. In the case of water vapor it appears that the chance of attaching the first water molecule is low, but once one attachment has occurred further attachment occurs efficiently and the final cluster is built rapidly (Massey and Burhop, 1952). For the rare gases, Munson and Hoselitz (1939) were able to analyze their observations to yield dissociation energies for clusters consisting of Li^+ and inert gas atoms. For Li^+-He, they derive a dissociation energy of about 0.07 ev, a value which is in harmony with that obtained from other more direct arguments (§ 2.5).

The theory of clustering does not provide a complete explanation of the observations even in its most refined form (Bloom and Margenau, 1952) and the phenomenon is worthy of further attention.

2.5 CALCULATIONS OF MOBILITIES

The problem of calculating the low field mobility of an ion was solved by Langevin (1905) giving results essentially equivalent to the statement (26) in which the correction factor ϵ_0 in (5) is put equal to zero and it is assumed that the number density of the ions is much smaller than of the gas particles. Langevin also carried out detailed calculations (see also Hassé, 1926) using an interaction potential

$$\begin{aligned} V(R) &= \infty & R &< \rho \\ &= -\tfrac{1}{2}\alpha e^2/R^4 & R &> \rho \end{aligned} \tag{30}$$

which revealed the essential features of the variation of mobility with temperature and which are in harmony with (29) in the limit of vanishing temperature. Similar calculations using the representation

$$V(R) = \epsilon \left\{ \left(\frac{\sigma}{R}\right)^8 - \left(\frac{\sigma}{R}\right)^4 \right\} \tag{31}$$

have been carried out by Hassé and Cook (1931), but neither (30) nor (31) provides an adequate representation of the short-range forces. An important set of tables of collision integrals corresponding to the more realistic model

$$V(R) = \frac{\epsilon}{2}\left[(1+\gamma)\left(\frac{\sigma}{r}\right)^{12} - 4\gamma\left(\frac{\sigma}{r}\right)^6 - 3(1-\gamma)\left(\frac{\sigma}{r}\right)^4\right] \tag{32}$$

has been prepared by Mason and Schamp (1958) and they have been able to reproduce most of the data, any failure being attributable to clustering. In particular, Mason and Schamp obtain a value of 0.05 to 0.06 ev for the dissociation energy of Li^+He, a result obtained with somewhat lower precision also by Dalgarno *et al.* (1958).

The derived interaction potentials are sensitive to the details of the mobility data and it may be misleading to use them to extend the measurements to higher temperatures or to higher electric field strengths. It would be desirable to supplement the mobility data by measurements of direct elastic scattering, thereby allowing a more accurate determination of the short-range forces.

Corrections to the mobility formula: Mason and Schamp (1958) have discussed in detail the corrections to (26) which is only a first approximation. Their discussion is based upon the kinetic theory development of Kihara (1953). For a pure R^{-4} interaction, the small quantity ϵ_0, appearing in (5), vanishes and (26) is exact. For other interactions, (26) is no longer exact but even in the most unfavorable cases it is unlikely to be in error by more than 5%.

It is interesting to note that (26) is exact (to the second order of approximation) whenever the mobility passes through a maximum or minimum as the temperature is changed.

2.6 Mobility at High Field Strengths

For a pure R^{-4} interaction, the mobility is a constant, but for other interactions it varies with the electric field. According to the theory of Kihara (1953), the mobility can be expressed as a power series in E^2:

$$\mathscr{K} = \mathscr{K}_0 + \mathscr{K}_2 E^2 + \mathscr{K}_4 E^4 + \cdots \tag{33}$$

where $\mathscr{K}_m$ are complicated functions of collision integrals, which have been written out by Mason and Schamp (1958). Mason and Schamp have discussed the convergence of (33) and have shown that the greatest value of E for which it may be used varies within wide limits, depending upon the force law, the temperature, and the ionic mass. In the common experimental cases, (33) should be useful for electric fields E and pressures p for E/p up to about 10 volts/cm · mm Hg.

There are some interesting features in the variation of $\mathscr{K}$ with E. At a temperature $T_{\min}$ or $T_{\max}$ corresponding to a minimum or maximum in $\mathscr{K}$, the coefficient $\mathscr{K}_2$ effectively vanishes and $\mathscr{K}$ is constant over a wide range of E. For $T < T_{\max}$, $\mathscr{K}$ initially increases with increasing field strength while for $T > T_{\max}$, $\mathscr{K}$ initially decreases. For $T < T_{\min}$, $\mathscr{K}$ initially decreases and for $T > T_{\min}$, $\mathscr{K}$ initially increases (Kihara, 1953; Mason and Schamp, 1958).

A detailed comparison of these predictions with observations is prevented by the absence of measurements in suitable systems.

No completely satisfactory theory exists for high values of E, although Wannier (1953) has succeeded in treating the case in which the particles are represented as rigid spheres. (This is actually quite a good approximation to the cases of ions moving in their parent gases discussed in the following section.) Wannier finds that for high E, the drift velocity v_d varies as $(E/p)^{1/2}$ and is independent of temperature. Wannier's theory is supported by several experimental studies (Hornbeck, 1951; Varney, 1952, 1953).

2.7 Mobilities of Ions in Their Parent Gases

The symmetry effects described in § 1.9 always arise in the description of the mobilities of atomic ions in their parent gases and lead to anomalously small values. The theory was first given by Massey and Mohr (1934) in a calculation of the mobility of He^+ in He at room temperature. Massey and Mohr obtained a value of 11 cm^2 volt^{-1} sec^{-1} whereas Tyndall and Powell (1932) had observed a mobility of about 20 cm^2

volt^{-1} sec^{-1}. The discrepancy remained unresolved until Meyerott (1944) suggested that the ion of high mobility observed by Tyndall and Powell might be the heavier ion He_2^+. Mass spectrometric investigations (cf. Loeb, 1955) support this interpretation which has now been confirmed by the observation of two distinct mobilities in helium.

By using a semiclassical version of the quantal theory (Holstein, 1952; Dalgarno, 1958), it may be shown that the diffusion cross section for an ion X^+ moving in its parent atomic gas X is twice the cross section for the resonance charge transfer process

$$X^+ + X \rightarrow X + X^+ \tag{34}$$

except at very low temperatures where the contribution of the long-range attractive forces cannot be ignored. Resonance charge transfer cross sections decrease slowly in the low-energy region as the impact energy increases (cf. Chapter 14, § 3.1.1, and Chapter 18, § 4) with the result that the mobility of an ion in its parent gas decreases steadily as the gas temperature is increased, a variation different from that occurring for an ion in an unlike gas. This difference can be of considerable assistance in identifying the ion under observation. Thus, the variation observed by Tyndall and Pearce (1935) in helium shows clearly that the ion cannot be He^+, as was originally supposed.

The experimental data have been reviewed by Loeb (1955) and by Dalgarno (1958). As a typical sample, we reproduce in Fig. 1 the

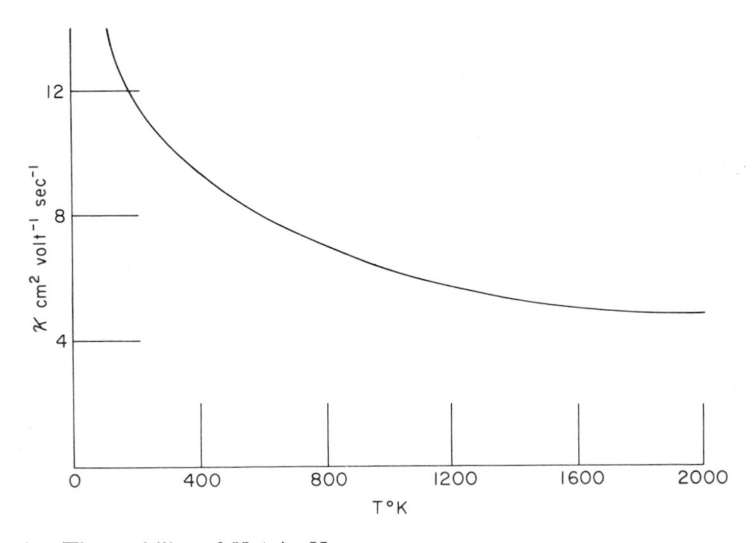

FIG. 1. The mobility of He^+ in He.

measurements on He^+ in He carried out by Biondi and Chanin (1957) and extended to higher temperatures by Dalgarno (1958) using data on charge transfer cross sections. The values agree well with the explicit quantal calculations of Lynn and Moiseiwitsch (1957).

Charge transfer also affects the mobility of doubly charged ions X^{++} in their parent gas X. The long-range polarization forces are several times larger and the cross section for (double) charge transfer several times smaller than in the singly charged case, and it is probable that doubly charged ions behave more like ions moving in an unlike gas. Explicit calculations have been made by Ferguson and Moiseiwitsch (1959) for He^{++} in He.

2.7.1 *Diatomic Ions in Their Parent Atomic Gases*

The mobilities of diatomic ions X_2^+ in X have no special features and ions behave like ions moving in an unlike gas.

2.7.2 *Diatomic Ions in Their Parent Molecular Gases*

Because of difficulties over the identity of the ions involved, there are few definitive data on the low field mobilities of ions such as X_2^+ in a gas of diatomic molecules X_2. At high field strengths the identification problems are fewer because high electric fields cause the breakup of any clusters and complexes which may have formed.

Varney (1960) observes only a single species in O_2 with an extrapolated zero field mobility of 2.25 cm^2 volt^{-1} sec^{-1}. This is so small that the identification of the ion as O_2^+ seems unambiguous and the value suggests that charge transfer plays an important role in such cases, though the magnitude of the effect is less than for the atomic ions.

The measurements in N_2 reveal a curious variation of the observed mobility with electric field strength which Varney (1953) and Kovar *et al.* (1957) interpret as referring to an ion which changes its structure so that at low fields it is N_4^+ and at high fields it is N_2^+. At intermediate fields the ion changes from one to the other many times during its time of flight. Similar results are obtained in carbon monoxide (Varney, 1953).

Many identification problems remain to be solved before a complete understanding can be reached, an interesting example being provided by argon. Three distinct mobilities are observed in Ar with low field values of 1.5, 1.8, and 2.6 cm^2 volt^{-1} sec^{-1} (Beaty, 1961). It seems certain that the lowest value refers to Ar^+ and that one of the others refers to Ar_2^+, probably the largest one. Perhaps the intermediate ion is Ar_4^+.

2.8 Mobilities in Molecular Gases

The interaction between an ion and a molecule contains terms depending upon the orientation of the molecule, the most important of which falls off as R^{-2} for a heteronuclear molecule and as R^{-3} for a homonuclear molecule. Their coefficients depend upon properties of the molecular gas only and at low temperatures, $\mathcal{K}\sqrt{\mu}$ should again be independent of the ion. Table IV reproduces some measurements in N_2 which confirm this independence.

TABLE IV

$\mathcal{K}\sqrt{\mu}$ for Various Ions in N_2 at about 290°K

Ion	Li	Na	K	Rb	Cs	Al	Ga	Kr
$\mathcal{K}\sqrt{\mu}$	9.3	10.1	10.2	10.3	10.3	10.1	10.2	10.3

Arthurs and Dalgarno (1960) have presented a theory of scattering by molecules which takes proper account of the orientation dependence of the interaction potential. One consequence of the presence of a term in R^{-3} is that the mobility of an ion in a molecular gas should not be independent of temperature at low temperatures but should decrease with decreasing temperature passing through a minimum at some very low temperature. The available experimental data do decrease with decreasing temperature but this may be due to clustering.

2.9 Mobilities in Gas Mixtures

It follows from the theory of Chapman and Cowling (1939) that the mobility of an ion in a multicomponent gas is given by

$$\mathcal{K}^{-1} = \sum_i f_i/\mathcal{K}_i. \tag{35}$$

This is known as Blanc's law. Deviations from it may occur if the nature of the ion is changing during its time of flight (Overhauser, 1949). The law is in harmony with observation (McDaniel and Crane, 1957).

2.10 Experimental Methods

Many methods have been used for the measurement of mobilities and a detailed description of them has been given by Loeb (1955).

Those most frequently used include the "four-guage" method of Tyndall and Powell (1932) and the Bradbury-Nielsen shutter method (cf. Crompton and Elford, 1959) for ions moving in various gases and two more recent methods due to Hornbeck (1951) and to Biondi and Chanin (1954) for ions moving in their parent gases.

Mobilities may also be determined from measurements of ionic diffusion.

2.11 Ambipolar Diffusion

The ions and electrons in a plasma move under the action of concentration gradients and an electric field. Then if a plus and minus superscript indicate, respectively, an ion and an electron property, the equations of motion of the ions and electrons are

$$\mathbf{j}^- = -\mathscr{D}^- \operatorname{grad} n^- - \mathscr{K}^- \mathbf{E} n^- \tag{36}$$

$$\mathbf{j}^+ = -\mathscr{D}^+ \operatorname{grad} n^+ + \mathscr{K}^+ \mathbf{E} n^+ \tag{37}$$

and the currents $\mathbf{j}^-$ and $\mathbf{j}^+$ will be equal provided

$$|n^+ - n^-| \ll n^-. \tag{38}$$

Eliminating $\mathbf{E}$ from (36) and (37), it follows that

$$\mathbf{j} = -\mathscr{D}_a \operatorname{grad} n \tag{39}$$

where $\mathscr{D}_a$, the ambipolar diffusion coefficient, is given by

$$\mathscr{D}_a = \frac{(\mathscr{D}^+\mathscr{K}^- + \mathscr{D}^-\mathscr{K}^+)}{\mathscr{K}^+ + \mathscr{K}^-} \tag{40}$$

and superscripts are no longer necessary in (39).

If the ions and electrons are in equilibrium with the gas at a temperature T,

$$\frac{\mathscr{D}^+}{\mathscr{K}^+} = \frac{\mathscr{D}^-}{\mathscr{K}^-} = \frac{kT}{e} \tag{41}$$

and

$$\mathscr{D}_a = 2\mathscr{D}^+\mathscr{K}^-/(\mathscr{K}^+ + \mathscr{K}^-)$$

$$\sim 2\mathscr{D}^+. \tag{42}$$

The mobility of an ion is therefore related to the ambipolar diffusion coefficient by

$$\mathcal{K} = \tfrac{1}{2} e \mathcal{D}_a / kT. \quad (43)$$

If the number density of the ions falls below about 10^8 cm^{-3}, the diffusion is no longer ambipolar. The transition from ambipolar to free diffusion has been investigated by Allis and Rose (1954).

Measurements of $\mathcal{D}_a$ have been made for a number of gases by Biondi and Brown (1949), Biondi (1950, 1951), Persson and Brown (1955), and Faire and Champion (1959).

REFERENCES

Allis, W. P., and Rose, D. J. (1954) *Phys. Rev.* **93**, 84.
Amdur, I., and Schatzki, T. F. (1957) *J. chem. Phys.* **29**, 1049.
Arthurs, A. M., and Dalgarno, A. (1960) *Proc. Roy. Soc.* **A256**, 540, 552.
Beaty, E. C. (1961) *In* "Proceedings of the Fifth International Conference on Ionization Phenomena in Gases, Munich, 1961." (H. Maecker ed.) Vol. I, p. 183. North-Holland, Amsterdam.
Biondi, M. A. (1950) *Phys. Rev.* **79**, 733.
Biondi, M. A. (1951) *Phys. Rev.* **83**, 1078.
Biondi, M. A., and Brown, S. C. (1949) *Phys. Rev.* **75**, 1700; **76**, 302.
Biondi, M. A., and Chanin, L. M. (1954) *Phys. Rev.* **94**, 910.
Biondi, M. A., and Chanin, L. M. (1957) *Phys. Rev.* **106**, 473.
Bloom, S., and Margenau, H. (1952) *Phys. Rev.* **85**, 670.
Buckingham, R. A. (1961) *Planet. Space Sci.* **3**, 205.
Buckingham, R. A., and Dalgarno, A. (1952) *Proc. Roy. Soc.* **A213**, 327, 506.
Buckingham, R. A., Davies, A. E., and Davies, A. R. (1958) *Proc. Conf. Thermodynamic and Transport Properties of Fluids, London*, 1957.
Chapman, S., and Cowling, T. G. (1939) "The Mathematical Theory of Non-uniform Gases." Cambridge Univ. Press, London and New York.
Crompton, R. W., and Elford, M. T. (1959) *Proc. Phys. Soc.* **74**, 497.
Dalgarno, A. (1958) *Phil. Trans.* **A250**, 428.
Dalgarno, A., McDowell, M. R. C., and Williams, A. (1958) *Phil. Trans.* **A250**, 411.
de Boer, J. (1943) *Physica* **10**, 348.
de Boer, J., and van Kranendonk, J. (1948) *Physica* **14**, 442.
Faire, A. C., and Champion, K. S. W. (1959) *Phys. Rev.* **113**, 1.
Ferguson, A. F., and Moiseiwitsch, B. L. (1959) *Proc. Phys. Soc.* **74**, 457.
Hassé, H. R. (1926) *Phil. Mag.* **1**, 139.
Hassé, H. R., and Cook, W. R. (1931) *Phil. Mag.* **12**, 554.
Hellund, E. J., and Uehling, E. A. (1939) *Phys. Rev.* **56**, 818.
Hirschfelder, J. O., Bird, R. B., and Spotz, E. L. (1948) *J. chem. Phys.* **16**, 968.
Hirschfelder, J. O., Bird, R. B., and Spotz, E. L. (1949) *J. chem. Phys.* **17**, 1343.
Hirschfelder, J. O., Curtiss, C. F., and Bird, R. B. (1954) "Molecular Theory of Gases and Liquids." Wiley, New York.
Holstein, T. (1952) *J. phys. Chem.* **56**, 832.
Hornbeck, J. A. (1951) *Phys. Rev.* **83**, 374; **84**, 621.

Hoselitz, K. (1941) *Proc. Roy. Soc.* **A177**, 200.
Jost, W. (1952) "Diffusion in Solids, Liquids, and Gases." Academic Press, New York.
Kihara, T. (1953) *Rev. mod. Phys.* **15**, 831.
Kihara, T., and Kotani, M. (1943) *Proc. phys.-math. Soc. Japan* **25**, 602.
King, G. W., and Van Vleck, J. H. (1939) *Phys. Rev.* **55**, 1165.
Kovar, F. R., Beaty, E. C., and Varney, R. W. (1957) *Phys. Rev.* **107**, 1490.
Langevin, P. (1905) *Ann. Chem. Phys.* **8**, 245.
Loeb, L. B. (1955) "Basic Processes of Gaseous Electronics." Univ. Calif. Press, Berkeley California.
Lynn, N., and Moiseiwitsch, B. L. (1957) *Proc. Phys. Soc.* **A70**, 574.
McDaniel, E. W., and Crane, H. R. (1957) *Rev. sci. Instrum.* **28**, 684.
Mason, E. A. (1954) *J. chem. Phys.* **22**, 169.
Mason, E. A. (1957) *J. chem. Phys.* **27**, 782.
Mason, E. A., and Schamp, H. W. (1958) *Ann. Phys. (New York)* **4**, 233.
Mason, E. A., Vanderslice, J. T., and Yos, J. M. (1959) *Phys. of Fluids* **2**, 688.
Massey, H. S. W., and Burhop, E. H. S. (1952) "Electronic and Ionic Impact Phenomena." Clarendon Press, London and New York.
Massey, H. S. W., and Mohr, C. B. O. (1934) *Proc. Roy. Soc.* **A144**, 188.
Meyerott, R. E. (1944) *Phys. Rev.* **66**, 242.
Mott, N. F., and Massey, H. S. W. (1949) "The Theory of Atomic Collisions," 2nd ed. Clarendon Press, London and New York.
Mulliken, R. S. (1960) *Phys. Rev.* **120**, 1674.
Munson, R. J., and Hoselitz, K. (1939) *Proc. Roy. Soc.* **A172**, 43.
Munson, R. J., and Tyndall, A. M. (1939) *Proc. Roy. Soc.* **A172**, 28.
Nettley, P. T. (1954) *Proc. Phys. Soc.* **B67**, 753.
Overhauser, A. W. (1949) *Phys. Rev.* **76**, 250.
Pearce, A. F. (1936) *Proc. Roy. Soc.* **A155**, 490.
Persson, K. B., and Brown, S. C. (1955) *Phys. Rev.* **100**, 729.
Phelps, A. V. (1959) *Phys. Rev.* **114**, 1011.
Rowlinson, J. S. (1949) *J. chem. Phys.* **17**, 101.
Srivastava, B. N., and Srivastava, K. P. (1959) *J. chem. Phys.* **30**, 984.
Srivastava, K. P., and Barua, A. K. (1959) *Indian J. Phys.* **33**, 229.
Tyndall, A. M. (1938) "The Mobility of Positive Ions in Gases." Cambridge Univ. Press, London and New York.
Tyndall, A. M., and Pearce, A. F. (1935) *Proc. Roy. Soc.* **A149**, 426.
Tyndall, A. M., and Powell, C. F. (1932) *Proc. Roy. Soc.* **A136**, 145.
Uehling, E. A. (1934) *Phys. Rev.* **46**, 917.
Uehling, E. A., and Uehling, E. E. (1933) *Phys. Rev.* **43**, 552.
Varney, R. N. (1952) *Phys. Rev.* **88**, 362.
Varney, R. N. (1953) *Phys. Rev.* **89**, 708.
Varney, R. N. (1960) *J. chem. Phys.* **33**, 1709.
Walker, R. E., and Westernberg, A. A. (1958) *J. chem. Phys.* **29**, 1139, 1147.
Walker, R. E., and Westenberg, A. A. (1959) *J. chem. Phys.* **31**, 519.
Walker, R. E., and Westenberg, A. A. (1960) *J. chem. Phys.* **32**, 436.
Walker, R. E., de Haas, N., and Westenberg, A. A. (1960) *J. chem. Phys.* **32**, 1314.
Wannier, G. H. (1953) *Bell Syst. techn. J.* **32**, 170.
Weissman, S., Saxena, S. C., and Mason, E. A. (1960) *Phys. of Fluids* **3**, 510.
Westenberg, A. A. (1957) *Combustion and Flame* **1**, 217.
Wise, H. (1960) *J. chem. Phys.* **31**, 1414.
Wise, H., and Ablow, C. M. (1958) *J. chem. Phys.* **29**, 634.

17.

High-Energy Elastic Scattering of Atoms, Molecules, and Ions

Edward A. Mason and Joseph T. Vanderslice

1 *Introduction*

The elastic scattering of beams of high-energy atoms, molecules, and ions by gases leads to valuable information on the forces between the interacting particles. In principle, this is one of the most direct ways of obtaining such information. However, even though the method is straightforward in principle, the analysis of the actual experiments to obtain meaningful force law information can often be quite difficult.

There are two main reasons for using molecular beams to obtain information about intermolecular forces. Other methods of obtaining such information, such as the analysis of gaseous transport and equation of state data, involve averaging over the thermal velocity distribution of the molecules, which introduces some ambiguity into the final results. Beam experiments seek to avoid this difficulty by the use of monoenergetic beams. This advantage is somewhat offset, however, by the necessity of averaging over apparatus geometry. The second reason for using beams is that it is possible to obtain intermolecular force information at close distances of molecular approach by working with high-energy beams. This range of molecular interactions is not susceptible to investigation at the present time by the more conventional means of gaseous transport property and equation of state measurements, inasmuch as the temperatures required are so high (of the order of 10^4 °K). Such short-range intermolecular forces are of fundamental interest for what light they can throw on basic questions concerning the electronic structure of molecules, and molecular beam experiments have stimulated, and been stimulated by, a number of quantum-mechanical calculations on such electronically simple systems as H-H_2, H-He, He^+-He, and He-He (discussed further in § 9). The molecular beam results for more complicated systems have also furnished valuable tests of more approximate theoretical and semiempirical calculation schemes (Mason and Vanderslice, 1958b; Gáspár, 1960).

An obvious direct application of the short-range intermolecular force information obtainable from scattering experiments is the calculation of the properties of gases at very high temperatures by well-established methods of kinetic theory and statistical mechanics (Amdur, 1953; Mason and Vanderslice, 1958d; Amdur and Mason, 1958; Amdur and Ross, 1958). Indeed, high-energy scattering experiments can be considered as an indirect method of determining high-temperature properties without the use of high temperatures, and this feature undoubtedly accounts for the recent upsurge of interest in such experiments. Some applications of scattering results have also been made to the calculation of ion mobilities (Mason and Vanderslice, 1959b), and the estimation of steric effects in molecules (Mason and Kreevoy, 1955; Westheimer, 1956; Kreevoy and Mason, 1957). We may look forward to the time when such applications can be extended to the calculation of more complex phenomena like vibrational relaxation times and reaction rate constants.

2 *Scope of Present Survey*

The principal discussions in this chapter will concern the major experimental work in high-energy beams, i.e., beams with kinetic energies from about 5 to 5000 ev in the laboratory coordinate system. Clearly "high energy" in this connection means only high with respect to ordinary thermal energies ($\sim$0.025 ev). In this energy range experimental techniques are all basically similar, and it is not too difficult to obtain fairly monoenergetic collimated beams, which makes the analysis of results more clear cut. Other simplifications which result from working in this range are that the scattering gas molecules can be considered stationary before collision, and that the intermolecular forces are often of a rather simple type, since the long-range dispersion and induction forces are usually almost negligible. The low-energy limit of the range is determined by the difficulty of preventing straggling and spreading of slow ion beams (even neutral atom beams always start out as ion beams which are later neutralized), and the high-energy limit is determined by the onset of inelastic collisions involving electronic excitation and ionization.

Although the kinetic energy of the beams is high, the potential energy of interaction between the beam and scattering particles is in the range of about 0.1 to 10 ev because the experiments usually involve only small-angle scattering. Thus, the only events "seen" by the experiments involve very glancing collisions. This limitation to small-angle scattering is both a blessing and a curse. It is a blessing in that one can use beams of conveniently high kinetic energy, and the mathematical analysis of small-angle scattering is comparatively simple. It is a curse in that the experimental results are quite sensitive to apparatus geometry, beam width, and the intensity distribution across the beam, all of which strongly influence the effective angular aperture of the apparatus, i.e., the minimum angle through which a beam particle must be deflected in order to be counted as scattered from the beam. These factors must be taken into account in order to interpret the results properly. Much of the early work on elastic scattering cannot be interpreted quantitatively because these factors were largely ignored, and the present survey will therefore not include work earlier than about 1940. This earlier work has been adequately summarized elsewhere (Massey and Burhop, 1952).

We omit from this survey all discussion of so-called "thermal beams," which use rather different experimental techniques and which would require another chapter as long as the present one to survey them properly. Reviews of early work on thermal beams are available (Fraser, 1931; Massey and Burhop, 1952; Smith, 1955; Ramsey, 1956), but there has been a revival of interest in the past 5 years and considerable experimental and theoretical work has been reported in the literature. Particularly active workers have been Pauly (1957, 1959, 1960, and previous papers) and Bernstein (Schumacher *et al.*, 1960; Bernstein, 1960; Hostettler and Bernstein, 1960; and previous papers).

We also omit discussion of results above 5 kev because no observations of strictly elastic scattering have been made in this region, the apparatus angular aperture necessary to avoid "seeing" the inelastic collisions being almost impossibly small. It should be mentioned, however, that it is quite accurate to calculate the scattering of the beam as if the collisions were truly elastic, since the electronic excitation and ionization energies involved are so small compared to the beam kinetic energy (Everhart *et al.*, 1955; Sida, 1957; Firsov, 1958; Lane and Everhart, 1960a).

Note added in proof: Since the above was written, Lane and Everhart (1960b) have reported potential energy functions between several ions and atoms in the range 1-60 kev (for the potential energy), corresponding to ion-atom separations of the order of 10^{-2} A. Although such collisions are inelastic, the inelastic energies involved are probably but a negligible fraction of the kinetic and potential energies involved in the collisions. Lane and Everhart studied differential cross sections for He^+ in He, Ne, Ar; Ne^+ in Ne, Ar; and Ar^+ in Ar; at beam energies of 25, 50, and 100 kev. Their work is particularly significant because they measured a sufficient angular distribution to be able to calculate $V(r)$ directly from their data without any assumption as to form other than that $V(r)$ be a monotonically decreasing function (Firsov, 1953; Keller *et al.*, 1956). The results could all be represented by a screened Coulomb potential of the form $V(r) = (Z_1 Z_2 e^2/r)\,\chi(r)$, where $Z_1 e$ and $Z_2 e$ are the nuclear charges and $\chi(r)$ is a screening function. A Thomas-Fermi model calculation of $\chi(r)$ gave slightly better agreement with the experimental $V(r)$ curves than an exponential form for $\chi(r)$. Lane and Everhart's work furnishes a beautiful practical example of how much more information can be obtained when the differential cross section rather than the total cross section is measured. One general word of caution is in order, however: for many potentials the quantum differential cross sections do not converge uniformly to the classical limit, so that no matter how small the de Broglie wavelength is there will be an angle (not necessarily near $\theta = 0$) for which the quantum and classical results differ by a large amount (Ford and Wheeler, 1959).

Limitations of space prevent discussion of collisions involving resonant charge or excitation exchange, although these are elastic processes, strictly speaking. Such phenomena are reviewed elsewhere in this volume.

3 *Classical Scattering Approximation*

The angular distribution of particles scattered from a beam is described both quantum mechanically and classically by a differential scattering cross section $\sigma(\theta)$, defined by the statement that the number of particles scattered through angles between θ and $\theta + d\theta$ is $2\pi J\sigma(\theta) \sin \theta \, d\theta$, where J is the flux density of the incident unscattered beam in particles per unit area per unit time, and $\sigma(\theta)$ has the dimensions of area. Because of axial symmetry of the beam, there is no dependence on azimuth angle, only on the polar deflection angle θ. In most beam scattering experiments it is not the angular distribution of scattered particles which is measured, but the total fraction of the beam scattered into all angles greater than some given angle θ_0, the angular aperture of the apparatus. This is usually called (not strictly correctly) the total scattering cross section $S(\theta_0)$, given by

$$S(\theta_0) = 2\pi \int_{\theta_0}^{\pi} \sigma(\theta) \sin \theta \, d\theta. \tag{1}$$

The true total cross section is the limiting value as $\theta_0 \rightarrow 0$. Since $\sigma(\theta)$ is peaked very sharply around $\theta = 0$, and the sharpness of the peak increases rapidly with increasing energy, measured values of $S(\theta_0)$ will be only a small fraction of the true total cross section $S(0)$ unless low-energy (thermal) beams and detectors of very high angular resolution are used. Good approximations to $S(0)$ have been obtained experimentally with thermal beams and very narrow detectors, but the resolution required to observe true total cross sections with high-energy beams is so high that there is scant prospect of their being measured. McDowell (1958) has performed some interesting calculations of $S(0)$ for ions of energies up to 10^4 ev, but there is little likelihood of ever relating such calculations to experimental measurements.

There is, in fact, no reason to attempt the measurement of $S(0)$, since measurements of $S(\theta_0)$ will do just as well for the determination of intermolecular forces. There is even an advantage to the use of $S(\theta_0)$ instead of $S(0)$, since $S(\theta_0)$ can be treated entirely classically (with a considerable resultant saving in mathematical labor) if θ_0 is larger than a "critical" angle θ_c. Below θ_c the classical approximation rapidly fails, but above θ_c the classical approximation is quite accurate. The classical approximation always fails at sufficiently small angles, with the result that the classical expression for $S(0)$ always diverges for a force law without a finite cutoff distance. According to Massey and Mohr (1933, 1934), the value of this critical angle is approximately

$$\theta_c \approx \lambda/2r_0 = \pi\hbar/(\mu v r_0), \tag{2}$$

where λ is the de Broglie wavelength of the colliding pair of particles, $\mu = m_1 m_2/(m_1 + m_2)$ is their reduced mass, v is their initial relative velocity, and r_0 is a characteristic molecular dimension, usually taken as the distance of closest approach. The critical angle for a thermal beam of helium atoms is of the order of 10 deg, for a 10-ev helium beam is of the order of 1/2 deg, and for a 1000-ev helium beam is of the order of 1/20 deg. It is correspondingly smaller for beams of heavier particles.

It appears that a completely classical analysis of practically all the available high-energy scattering measurements is valid, and they will be discussed in classical terms from here on. For experiments involving very light atoms and very small angular apertures, the classical analysis has been questioned (Wu, 1958). This point will be discussed further in § 7.3.

It should be mentioned that the mass and energy dependence of the true total quantum-mechanical cross section $S(0)$ is quite different from that of the classical "total" cross section $S(\theta_0)$, $(\theta_0 > \theta_c)$. For a potential energy of interaction of the form $V(r) = \pm K/r^s$, $S(0)$ varies as $(K/\hbar v)^{2/(s-1)}$ (Massey and Mohr, 1934; Landau and Lifshitz, 1958), whereas $S(\theta_0)$ varies as $(K/E\theta_0)^{2/s}$, where $E = \frac{1}{2}\mu v^2$ (Kennard, 1938; Gordon, 1957). Thus, the energy dependence of $S(\theta_0)$ is much stronger than that of $S(0)$, and the fraction of the true total cross section measured by an apparatus of fixed θ_0, i.e., the ratio $S(\theta_0)/S(0)$, rapidly decreases with increasing beam energy. It should also be mentioned that $S(\theta_0)$ is influenced largely by more energetic collisions than is $S(0)$ for a given beam energy, and so $S(\theta_0)$ gives more information on the short-range forces than does $S(0)$.

4 *Elementary Description of Scattering Measurements*

A straightforward method of performing a scattering experiment is shown schematically in Fig. 1. A collimated beam of monoenergetic particles passes through a scattering chamber SC containing scattering gas of known density. Some of the beam particles are deflected and miss hitting the detector D. If the detector has a width of $2a$ and the beam width is much smaller than $2a$, then the aperture of the apparatus in the laboratory coordinate system is $\theta_{0a} \approx \tan\theta_{0a} = a/l$, provided $\Delta x \ll l$. All of the scattering is supposed to occur in the length Δx. The beam intensity is measured with (I) and without (I_0) scattering gas in the scattering chamber, and the density of the scattering gas is determined by measuring its pressure and temperature. The cross section S is then calculated from the relation

$$I/I_0 = \exp(-nS\Delta x), \tag{3}$$

where $n = p/kT$ is the gas number density. This relation holds as long as n and Δx are small enough that multiple scattering does not occur. The experimental test for single scattering is that S be independent of variations of n. This procedure gives S as a function of the beam energy.

The foregoing description is clearly an idealization, and the actual geometry of the experiments is complicated by the following factors: (1) Δx is not entirely negligible compared to l. (2) The definition of Δx is complicated by the presence of small clouds of scattering gas effusing from the entrance and exit holes of the scattering chamber. Both these effects are usually small and, consequently, corrections can be easily calculated. More serious effects are: (3) The beam is slightly divergent and its width is at least as large as the detector width (infinitesimally narrow beams are nice in theory, but one must have enough intensity to measure). (4) Since the exit hole of SC must be small to reduce the effusion of the scattering gas, and the beam must have a finite width to give enough intensity, the edges of the exit hole may partly determine θ_0 for beam particles near the edge of the beam. These effects make the definition of θ_0 less clear than the ideal, and a rather complicated averaging procedure is necessary to find the effective angular aperture to use in interpreting the results in terms of intermolecular forces.

An apparatus of this sort can be essentially one-dimensional, with slits in the scattering chamber and the detector in the form of a long strip (perpendicular to the plane of the paper in Fig. 1), and such an apparatus geometry has been employed by Amdur and his co-workers in their experiments on the scattering of beams of neutral atoms with energies from 500 to 2100 ev by gases (Amdur *et al.*, 1948; Amdur and Harkness, 1954, and many subsequent papers). They have also used an apparatus with circular holes in SC and a circular detector of radius

a in experiments on the scattering of a He beam by He gas in the energy range from 150 to 1500 ev (Amdur *et al.*, 1961).

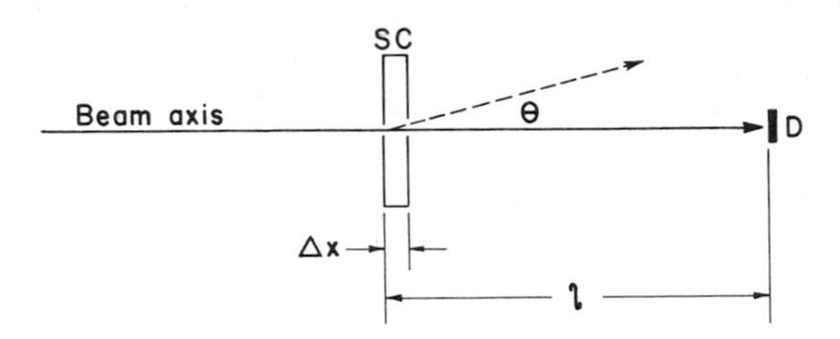

FIG. 1. Schematic diagram of idealized scattering experiment.

Another popular geometry is obtained by making $\Delta x = l$ and using circular beams and detectors. This geometry has been employed by Amdur and co-workers in their first experiments with neutral beams of energies from 300 to 1100 ev (Amdur and Pearlman, 1940, 1941, and several subsequent papers), and by Simons and co-workers in their experiments with ion beams of energies from about 4 up to 400 ev (Simons *et al.*, 1942; Muschlitz *et al.*, 1956; Cramer and Simons, 1957, and many other papers). In this arrangement there is no difficulty about the definition of the length of the scattering path, because $\Delta x = l$ and the gas cloud at the entrance hole of SC is negligible because the scattering path l is long. A difficulty occurs for long scattering paths, however—the apparatus aperture varies from arc tan (a/l) at the beginning of the path up to $\pi/2$ at the end. Clearly the cross sections calculated from (3) for such a geometry are averages of some sort over the length of the scattering path. They are also averages over a finite beam width, since infinitesimally narrow beams have only infinitesimal intensity.

Still a third geometry is obtained by arranging that the detector can be moved off the beam axis to measure the angular distribution of scattered particles. Essentially this procedure has been employed by Berry (1949, 1955) in his experiments on the scattering of neutral beams of neon and argon atoms with energies from 300 to 3500 ev. These experiments are much more difficult because the scattered intensity per unit solid angle is so small. Furthermore, at large angles the detector undoubtedly collects a lot of unknown excited atomic and ionic debris resulting from inelastic collisions, which is avoided when the detector is kept on the beam axis. Experiments of this sort are consequently sometimes difficult to interpret accurately. In compensation, much more detailed information is obtained since the angular distribution is measured, and not just its integral.

We might summarize this section by the statement that elastic scattering measurements are easy in principle but more difficult in practice, both in execution and in interpretation. In the next section we consider some of the details of the actual experiments, and in the following sections will consider the theoretical interpretation of scattering measurements, including the corrections needed to permit a quantitative interpretation in terms of intermolecular forces.

5 *Experimental Methods*

In this section we review briefly the experimental methods used in high-energy elastic scattering measurements. These have been employed chiefly by Amdur, Simons, Berry, and their students and co-workers. For details of apparatus and technique, reference should be made to the original papers.

5.1 Amdur's Apparatus

A schematic diagram of one of Amdur's apparatuses is shown in Fig. 2 (Amdur *et al.*, 1948; Amdur and Harkness, 1954). It employs an ion source based on the design of Lamar and Luhr (1934). A low voltage arc is formed between the filament F and the anode A. Positive ions are drawn off from this arc by the enclosing cathode C, and some pass through the exit grid E. These ions are accelerated to the desired energy by the ion gun G, and some of the ions are neutralized by resonant charge exchange with residual neutral gas molecules in the vicinity of the hole in G. A mixed beam of fast ions and neutrals emerges from G; the ion component is removed by the condenser H, and the neutral component passes through the scattering chamber SC to the detector D. Thermal detectors are used, either a thermopile or a fine butt-welded thermocouple. There is probably some energy spread in the neutral beam, caused by comparatively high gas pressures (0.1-

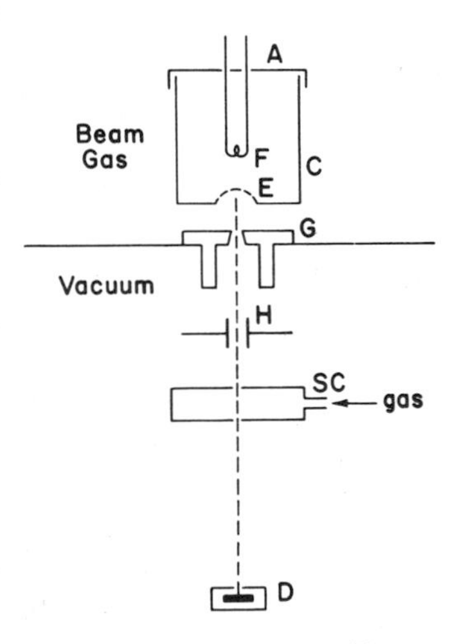

Fig. 2. Schematic diagram of Amdur and co-workers' neutral beam apparatus.

0.2 mm Hg) and voltage drops in the arc, and by neutralization of some ions between E and G before being accelerated to the full ion gun voltage. This type of apparatus is thus not suitable for the study of neutral beams with energies below about 100 ev. The detection of lower energy beams is also much more difficult. With this apparatus have been studied the scattering of rare gas atoms in rare gases and in nitrogen, and of H atoms in helium. An example of the data obtained is given in Fig. 6.

An earlier apparatus (Amdur and Pearlman, 1940) employed a much larger scattering path, extending from G nearly to D. With this apparatus have been studied the systems H-H_2, D-D_2, He-He, and Ar-Ar. This apparatus has recently been rebuilt and modified, including a limited scattering path as illustrated in Fig. 2, and used to study the He-He system further (Amdur *et al.*, 1961). Because of different average angular apertures, the three apparatuses give intermolecular potentials valid in different ranges, and tend to complement rather than duplicate each other.

More recently a new apparatus has been constructed with an improved ion source. Magnetic selection of the ions before neutralization and many other refinements of experimental technique have been incorporated, but results have not yet been published (Amdur and Jordan, private communication).

5.2 Simons' Apparatus

Several different apparatuses have been used by Simons and his co-workers in their studies of ion beams, and Fig. 3 is intended to represent the main features of all the apparatuses only in very schematic fashion. Ions are produced in the ion source and formed into a beam by the focusing elements FE of the fore chamber. Momentum selection occurs in the magnet chamber M, the emerging beam is again focused

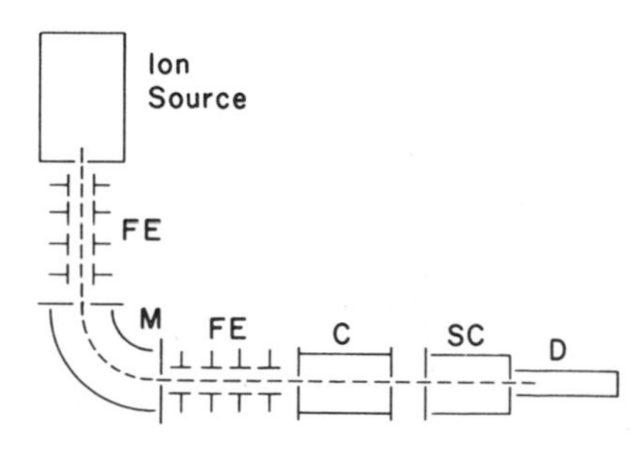

Fig. 3. Schematic diagram of Simons and co-workers' ion beam apparatus.

by the elements FE of the post chamber, collimated by the chamber C, and finally enters the scattering chamber SC. The ions arriving in the detector chamber D are collected and measured with a dc amplifier. Fast differential pumping is used to keep the gas pressures suitably low in the different sections of the apparatus. The scattering section also contains various elements, not shown in Fig. 3, for measurement of ionization and of electron exchange and detachment.

The first apparatuses (Russell *et al.*, 1941; Simons *et al.*, 1942) employed a Lamar and Luhr (1934) ion source, and the scattering of H^+, H_2^+, and H_3^+ ions in a wide variety of gases was investigated over a period of years. An example of the data obtained is given in Fig. 7. More recently an improved apparatus with a mass spectrometer ion source has been constructed (Cramer and Simons, 1957), with which the scattering of rare gas ions in rare gases, and of D^+ and D_2^+ ions in D_2 have been investigated. An example of the data obtained from this apparatus is given in Fig. 6.

The method is also suitable for negative ions, which have been produced by electron bombardment of a jet of gas introduced into the ion source through a nozzle (Muschlitz *et al.*, 1956; Muschlitz, 1957). This apparatus has been used to investigate the scattering of H^- ions in H_2, He, Ne, and Ar, and of O^- and O_2^- ions in O_2.

5.3 Berry's Apparatus

Berry has constructed an apparatus for the measurement of the angular distribution of scattered neutral atoms, with which he has studied the scattering of Ar in Ar (Berry, 1949) and of Ne in Ne (Berry, 1955). A schematic diagram of the apparatus is shown in Fig. 4. The

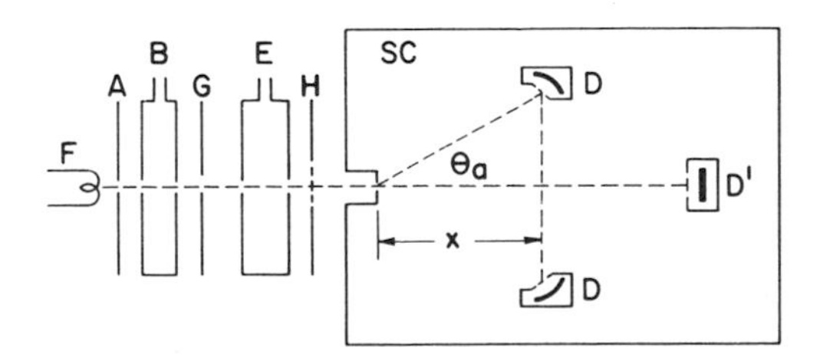

Fig. 4. Schematic diagram of Berry's neutral beam apparatus.

ion source is based on the design of an efficient ion source operating at low gas pressures ($< 10^{-3}$ mm Hg) developed by Finkelstein (1940). Electrons accelerated from the filament F ionize some beam gas atoms in the field-free region B. The resulting ions are accelerated to the desired energy by G, and charge exchange with parent gas at a few microns pressure occurs in E. Unneutralized ions are rejected by H, and the neutral beam enters the scattering chamber SC, which contains the scattering gas and two detectors, both of which can be moved back and forth along the beam axis. Detector D measures the scattered atoms directly and is in the form of a ring to receive the full azimuthal angle. Detector D′ can be pushed up to the entrance hole of SC to measure the unscattered beam intensity and to serve as a beam stop to measure the background recorded by D. Detection is accomplished by measuring the secondary electron ejection from tantalum sheet by the impinging neutral atoms. The minimum deflection angle observed in this apparatus is about 15 deg, which is not a very small angle.

If the detector is moved from position x to $x + dx$, atoms scattered along a path length x are still received, but their intensity is slightly decreased because the primary beam must pass through an extra thickness dx of the scattering gas. This effect decreases the detector current by the amount $nISdx$, where I is the beam current along the path and S is the cross section for beam attenuation. The detector current is much more greatly increased, however, by now receiving atoms scattered through an apparatus angle θ_a from the region dx at the beginning of the scattering path. This increases the detector current by the amount $\alpha n I_0[2\pi\sigma'(\theta_a) \sin \theta_a \, d\theta_a] \, dx$, where I_0 is the initial beam current (measured by the detector D′), α is a correction factor to allow for different secondary electron ejection coefficients at D′ (which receives unscattered atoms) and at D (which receives atoms scattered at an angle θ_a), and $\sigma'(\theta_a)$ is the differential cross section for both beam and scattering atoms,

$$\sigma'(\theta_a) = \sigma(\theta_a) + \sigma(\tfrac{1}{2}\pi - \theta_a), \tag{4}$$

since the detector cannot distinguish between scattered and recoil atoms (Bohm, 1951). The subscript a means that the laboratory or apparatus coordinate system is being used. The differential cross section can thus be found from the equation

$$dI = \alpha n I_0 \, [2\pi\sigma'(\theta_a) \sin \theta_a d\theta_a] \, dx - nISdx. \tag{5}$$

An experiment measures n, I_0, and I as a function of x. The value of $d\theta_a$ as a function of x is determined by the apparatus design. The quantities S and α have to be determined in separate experiments, and then (5) can be solved for $\sigma'(\theta_a) \sin \theta_a$. The term involving S in (5) amounts to at most a total correction of 15%, but α is obviously more important.

It is difficult to assess the reliability of Berry's measurements without more experimental detail than is supplied in his papers. The experimental difficulties are greater than those encountered by Amdur and by Simons, who do not measure angular distributions, but only total scattering cross sections. However, the extra information given by the angular distribution permits the calculation of numerical values of $V(r)$ without the necessity of assuming an algebraic form (Hoyt, 1939; Berry, 1949). Berry's results for $V(r)$ of Ne-Ne and Ar-Ar systems do not fall in the same range of r as do Amdur's results for the same systems, but if a comparison is made by extrapolation (which is always a risky procedure for potentials), only order of magnitude agreement is obtained. This is probably satisfactory under the circumstances (Amdur *et al.*, 1950a). Certainly it cannot be said that the two sets of results are in disagreement. No other reliable independent checks on Berry's results appear to be available.

6 Elementary Classical Theory of Scattering

The measured cross sections must now be related to molecular properties, which is done by consideration of individual molecular collisions. In this section we give the theory for an experiment of idealized geometry. Most of the results are well known, but certain aspects of their application to high-energy scattering experiments have often been overlooked. The following section considers the corrections needed in actual experiments.

6.1 Impact Parameter Formulation

In a classical description the collision trajectories are well defined, and for a given relative energy ($E = \frac{1}{2}\mu v^2$) each value of the impact parameter b results in a definite value of the scattering angle. Thus, the number of particles falling on a molecular target area between b and $b + db$ will be scattered into angles between θ and $\theta + d\theta$, so that we can write $2\pi b db = 2\pi\sigma(\theta) \sin \theta d\theta$, from which we obtain

$$S(\theta_0) = 2\pi \int_{b(\pi)}^{b(\theta_0)} b db = \pi b^2(\theta_0), \tag{6}$$

since $b(\pi) = 0$ (head-on collision). In the impact parameter formulation $S(\theta_0)$ thus has a simple physical interpretation: it is equal to the target area per molecule which a beam particle must hit in order to be scattered through an angle greater than θ_0 and so miss the detector. In a scattering volume of area A and length dx there are $nAdx$ scattering molecules presenting a total target area of $nS(\theta_0)\,Adx$. The fraction of A thus "blocked" for beam particles is thus $nS(\theta_0)\,dx$, which must be equal to the fractional loss of intensity of the beam passing through dx, so that

$$dI/I = -\,nS(\theta_0)\,dx. \tag{7}$$

On integration this becomes the same as (3) from which the experimental cross section is calculated. The experiment thus measures $\pi b^2(\theta_0)$.

The connection between the cross section and the intermolecular forces appears through the relation between scattering angle and impact parameter (Kennard, 1938; Bohm, 1951),

$$\theta(b, E) = \pi - 2b \int_{r_c}^{\infty} [1 - b^2/r^2 - V(r)/E]^{-1/2}\, r^{-2}\, dr, \tag{8}$$

where r_c is the distance of closest approach in the collision, given by

$$1 - b^2/r_c^2 - V(r_c)/E = 0. \tag{9}$$

If $V(r)$ is known, $\theta(b, E)$ can be calculated and in principle solved to give $b(\theta, E)$, from which we obtain $S(\theta_0, E)$ by (6). The reverse process of calculating $V(r)$ if $S(\theta_0, E)$ is known is more difficult, and usually involves first the selection of a suitable model for $V(r)$, i.e., the assumption of an algebraic form for $V(r)$ containing a few disposable parameters. The calculation of $S(\theta_0, E)$ is then carried through with these parameters, and the values of the parameters determined by comparing the calculations with experimental results. Since many different models for $V(r)$ can reproduce a given set of measured cross sections, it is necessary to have some *a priori* knowledge of the nature of $V(r)$ if the results are to have physical significance. Once a physically realistic model is chosen, however, the potential parameters can be determined with some precision. Much of the difficulty of interpreting many scattering experiments, especially those involving complex systems, arises from a lack of *a priori* knowledge of the nature of $V(r)$.

The situation is a little more complicated if both attractive and repulsive intermolecular forces are significant in producing the observed scattering. At the high beam energies considered in the present article, the only important attractive forces are those usually thought of in connection with chemical binding, the familiar second order induction and dispersion forces being too weak to be more than a small perturbation. A curve of scattering angle vs. impact parameter is shown in Fig. 5 for a case where both attractive and repulsive forces are important.

Positive angles correspond to net repulsion and negative angles to net attraction, although of course the scattering experiment cannot distinguish between the two. It is apparent that there may be three values

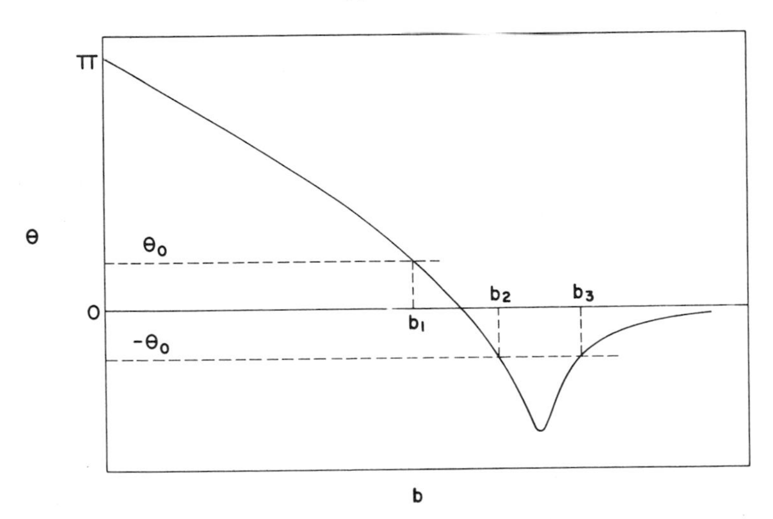

FIG. 5. Schematic representation of the scattering angle for a force law involving both attraction and repulsion.

of b corresponding to a given aperture θ_0. Scattering through angles greater than θ_0 will occur only if the impact parameter is less than b_1 or is between b_2 and b_3, shown in Fig. 5, so that (6) must be modified to read (Myers, 1955, 1956; Mason, 1955)

$$S(\theta_0) = \pi b_1^2(\theta_0) + \pi[b_3^2(\theta_0) - b_2^2(\theta_0)]. \tag{10}$$

The target area per molecule for scattering thus looks like ◉. As E is increased, the minimum in the $\theta(b)$ curve rises and eventually the outer ring of the target area vanishes and there is no longer any contribution to the scattering from the attractive forces. If the attractive energy is very great compared to E, the minimum is lower (approaching $-\infty$ in the limit), b_3 is very large, and b_1 and b_2 are nearly equal, so that the cross section can be taken as πb_3^2 with little error.

6.2 Relative and Apparatus Coordinates

The preceding discussion has been given in terms of the relative or center-of-mass coordinate system, rather than in terms of an apparatus or laboratory coordinate system fixed in space. In apparatus coordinates the energy W is given by $W = \frac{1}{2} m_1 v^2 = (m_1/\mu)E$, where m_1 is the mass of a beam particle. The relation between the relative angle θ and the apparatus angle θ_a is (Kennard, 1938; Bohm, 1951)

$$\tan \theta_a = \frac{m_2 \sin \theta}{m_1 + m_2 \cos \theta}, \tag{11}$$

where m_2 is the mass of a scattering molecule. The total cross section $S(\theta_0)$ is unaffected by the transformation to apparatus coordinates, although $\sigma(\theta)$ is affected, except that the relative aperture θ_0 is different from the apparatus aperture θ_{0a}. For a given apparatus, it is θ_{0a} which is fixed, not θ_0. The foregoing relations are derived on the assumption that the scattering molecule is initially at rest, an assumption which introduces no error in the present cases.

6.3 Small-Angle Approximation

The mathematical analysis is greatly simplified if the relative scattering angle θ is small. Equation (11) then becomes $\theta_a = \theta/[1 + (m_1/m_2)]$, from which we find, in conjunction with $W = (m_1/\mu)E$, that $W\theta_a = E\theta$. The greatest simplification comes in the relation between θ_0 and S. Equation (9) is substituted into (8) to eliminate b, after which the square root is expanded. The first term is (Kennard, 1938)

$$\theta = (r_c/E) \int_{r_c}^{\infty} [V(r_c) - V(r)] (r^2 - r_c^2)^{-3/2} \, r dr. \tag{12}$$

From (9) we have, without approximation,

$$S = \pi b^2 = \pi r_c^2 [1 - V(r_c)/E]. \tag{13}$$

If (12) is a good approximation, then the last term on the right of (13) is small, although not always completely negligible (Kells, 1948).

It is instructive to examine the results of (12) and (13) for some simple potential forms. For $V(r) = \pm K/r^s$, where K is positive, (12) yields

$$\theta = \pm C_s K/(E r_c^s) = C_s V(r_c)/E, \tag{14}$$

where $C_s = \pi^{1/2} \Gamma(\frac{1}{2} s + \frac{1}{2})/\Gamma(\frac{1}{2} s)$, from which we obtain with (13)

$$S(\theta_0) = \pi [C_s K/(E\theta_0)]^{2/s} [1 \mp (\theta_0/C_s)], \tag{15}$$

the negative sign in the last term corresponding to a repulsive potential and the positive sign to an attractive potential. From (15) it is apparent that a plot of log S vs. log E or log W will give a straight line of slope $-2/s$ if an inverse power potential is a good model. The potential parameter K can be calculated from the intercept. This result holds true even with nonideal geometry, only the interpretation of the aperture θ_0 being changed. The result for this particular potential is actually independent of the small angle approximation (12), as can be proved by changing variables in (8) to show that θ is a function of the single variable Eb^s (Hirschfelder *et al.*, 1954).

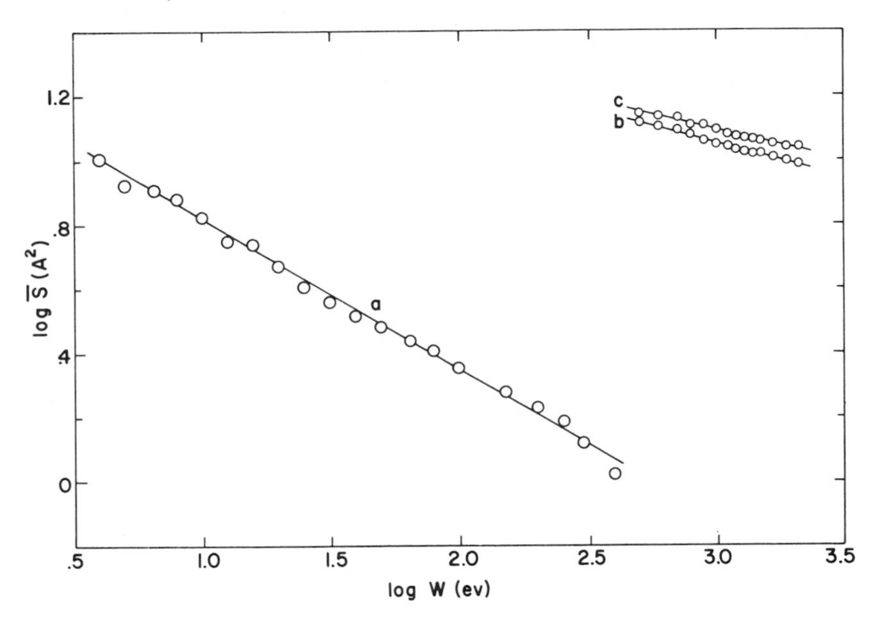

FIG. 6. Logarithmic plots illustrating the applicability of an inverse power potential. Curve a, Ne^+ in He (Cramer, 1958); curves b and c, Ne in Ne for two different size detectors (Amdur and Mason, 1955a).

Figure 6 shows some typical experimental results which can be interpreted in terms of an inverse power potential, the ion scattering results of Cramer (1958) on a Ne^+ ion beam scattered in He, and the atom scattering results of Amdur and Mason (1955a) on a Ne beam in Ne. The latter are shown for two different detectors corresponding to two different values of θ_0, showing that the value of S is dependent on θ_0 as required by (15).

Another useful potential form is $V(r) = \pm \epsilon \exp(-r^s/\rho^s)$, where ϵ, ρ, and s are positive constants. Equation (12) can be manipulated by the method of Amdur and Pearlman (1941) to yield an asymptotic series giving θ as a function of r_c (Mason and Vanderslice, 1957a). This series cannot be solved explicitly for r_c as a function of θ and then substituted into (13) to give $S(\theta_0)$, but must be treated parametrically with respect to r_c. However, the strongest dependence of θ on r_c in the series is as $\exp(-r_c^s)$, and since $S \approx \pi r_c^2$, it turns out that the results of the calculations can be represented accurately over a wide range as a linear relation between $S^{s/2}$ and $\log E$ (or $\log W$). This is also found to hold as well for nonideal geometry, which in a sense only changes the interpretation of θ_0. As an illustration, a plot of $S^{1/2}$ vs. $\log W$ is shown in Fig. 7 for the experimental measurements of Simons *et al.* (1943a) on H_3^+ in He, showing that a simple exponential can interpret the results (Mason and Vanderslice, 1957a).

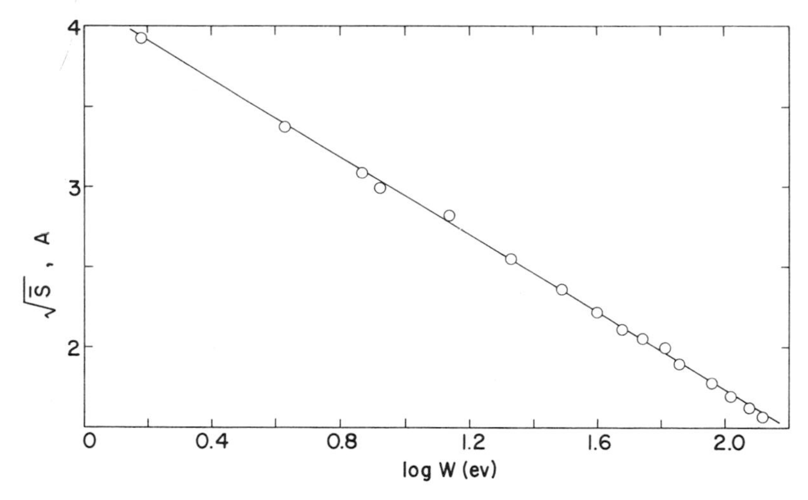

FIG. 7. Semilogarithmic plot illustrating the applicability of an exponential potential for H_3^+ in He measurements (Simons *et al.*, 1943a).

If the potential is more complicated and must be represented by a sum of several inverse-power or exponential terms, the expression for θ calculated from (12) becomes a corresponding sum of terms, and one may have the complication of several impact parameters giving the same angle, as in (10). Except in special cases it is not possible to eliminate r_c between $\theta(r_c)$ and $S(r_c)$ to obtain $S(\theta_0)$, but it is still possible to calculate $S(\theta_0)$ numerically. Explicit solutions have been worked out for several two-term inverse-power potentials exhibiting both attraction and repulsion (Myers, 1955, 1956; Mason, 1955; Mason and Vanderslice, 1957a), and numerical solutions in tabular form for the Morse two-term exponential function (Mason and Vanderslice, 1958e). The dependence of S on W is more complicated than in cases where the potential is monatomic. Nevertheless, this more complicated dependence does show up in the experimental results and good agreement between theory and experiment can be obtained.

If the relative scattering angle is not small, so that the approximate (12) cannot be used, recourse must be made to the accurate (8), which almost invariably means extensive numerical integration. This situation can arise even if the apparatus aperture θ_{0a} is small, since the relative aperture θ_0 can still be large if the mass of a beam particle is much larger than the mass of a scattering particle. This is shown by (11), which also shows that the maximum apparatus aperture allowable if any scattering at all is to be observed is

$$\max \theta_{0a} = \arc\tan [(m_1/m_2)^2 - 1]^{-1/2}. \tag{16}$$

For apertures greater than this maximum no scattering at all will be observed no matter how large the cross section is, since the heavy beam particles cannot be deflected enough by the light scattering particles to miss the detector. Even if the aperture is small enough to detect scattering, the target area per scatterer always has a hole in the center ◉, corresponding to the fact that head-on collisions and nearly head-on collisions do not deflect the beam particle appreciably but only slow it down, so that it still reaches the detector. The analysis of such experiments is quite involved mathematically unless θ_{0a} is very small, and has never been carried through. The recent results on Ne^+ in He and Ar^+ in Ne reported by Cramer (1958, 1959) are seriously influenced by these effects. Earlier experiments on Ar in He (Amdur *et al.*, 1954) and on Ar in Ne (Amdur and Mason, 1956c) escaped this difficulty because of a much smaller apparatus angular aperture.

6.4 Range of Validity of Intermolecular Forces Determined from Cross-Section Measurements

Given a potential $V(r)$ obtained from cross-section measurements, it is important to inquire as to its range of validity. Reference to (8) shows that in principle the angle of deflection is determined by all intermolecular separations between r_c and ∞, but in fact the integrand is so strongly peaked around r_c that for all practical purposes it is only the interaction around r_c that matters. Thus, it is essentially $V(r_c)$ that is determined, and not an average over all values of r out to infinity. For small-angle scattering we further have $S \approx \pi r_c^2$, so that the range of validity is easily obtained directly from the minimum and maximum values of the measured cross sections:

$$(S_{\min}/\pi)^{1/2} \leq r \leq (S_{\max}/\pi)^{1/2}. \tag{17}$$

For a given apparatus, i.e., a given θ_{0a} and range of W, the values of r calculated from (17) of course vary widely as molecular systems of different "sizes" are studied. However, in terms of the interaction energies themselves the range is almost independent of the particular molecular system studied. This can be most easily illustrated for an inverse power potential. On rearranging (14) and remembering that $W\theta_{0a} \approx E\theta_0$, we obtain

$$W_{\min}\,\theta_{0a}/C_s \leq V(r) \leq W_{\max}\,\theta_{0a}/C_s, \tag{18}$$

which depends only weakly on the particular form of $V(r)$ through the constant C_s. This result also illustrates how interaction energies of the order of 1 ev are determined from scattering measurements with beams of the order of 1000-ev kinetic energy—the aperture is of the order of 10^{-3} radian.

The preceding results are of course not so clear cut for nonideal apparatus geometry, but they are usually interpreted as being approximately correct in an average sense.

7 *Analysis of Experiments*

In this section we take up the difficult problem of accounting for various nonideal aspects of the experiments, especially nonideal geometry.

7.1 Finite Beam Width and Scattering Path

It is impossible in practice to obtain the infinitesimal beams and scattering paths which would lead to an unambiguous value of the apparatus aperture, and a real experiment involves some sort of average aperture which must be evaluated if the results are to be interpreted quantitatively. The proper quantity to average is not the angular aperture itself, but the cross section. The usual procedure is to average the cross section first over the length of the scattering path (including nonuniformities of scattering gas density along the path), and then to average again over the width of the unscattered beam, including a factor for the intensity distribution across the width. This is not a strictly correct procedure, but the formal exhibition of the approximation involved does not seem to have been published before, so we present it here.

For concreteness we assume cylindrical symmetry with the x-axis lying along the beam direction, although this is not essential to the argument. The beam intensity at any point is $i(x, r)$, where r is the radial distance from the x-axis, and the total intensity intercepted by a detector of radius a would be $I(x) = 2\pi \int_0^a i(x, r) r dr$. The cross section $S(\theta_0)$ depends on x and r because θ_0 does. Scattering reduces the intensity of each infinitesimal beam at r by an amount

$$di(x, r) = -n(x)\, i(x, r)\, S(x, r)\, dx, \tag{19}$$

where possible nonuniformities of scattering gas are allowed for by making n a function of x. On integrating over the area intercepted by a detector, we obtain

$$dI(x) = -n(x)\, I(x)\, \langle S(x)\rangle\, dx, \tag{20}$$

where

$$\langle S(x)\rangle = \int_0^a i(x, r)\, S(x, r)\, r dr \Big/ \int_0^a i(x, r)\, r dr, \tag{21}$$

and another integration over x yields the usual relation given previously by (3), namely, $I = I_0 \exp(-\bar{n}\, \overline{\langle S\rangle}\, \Delta x)$, where

$$\overline{\langle S\rangle} = (\bar{n}\, \Delta x)^{-1} \int_0^{\Delta x} n(x)\, \langle S(x)\rangle\, dx,$$

$$\bar{n} = (\Delta x)^{-1} \int_0^{\Delta x} n(x)\, dx. \tag{22}$$

The correct procedure is therefore to average first over beam width and second over scattering path. Usually the reverse procedure is carried out, largely because $i(x, r)$ appearing in (21) is never known experimentally, only the intensity distribution of the unscattered beam, $i_0(r)$. We may expect that the value of $\langle S(x)\rangle$ calculated using $i_0(r)$ in place of $i(x, r)$ will not be greatly in error, since this weighting factor occurs in both numerator and denominator of (21). On replacing $i(x, r)$ by $i_0(r)$, we can interchange the orders of integration in (22) and obtain the usual result

$$\overline{\langle S\rangle} \approx \langle \bar{S}\rangle = \int_0^a i_0(r)\, \bar{S}(r)\, r dr \Big/ \int_0^a i_0(r)\, r dr, \tag{23}$$

where

$$\bar{S}(r) \equiv (\bar{n}\Delta x)^{-1} \int_0^{\Delta x} n(x)\, S(x, r)\, dx. \tag{24}$$

Equations (23) and (24) are the basis for the usual corrections for nonideal geometry. For details of the often complicated computations necessary in particular cases, and for a discussion of the special case where the unscattered beam is wider than the detector, reference should be made to the original papers (Amdur and Harkness, 1954; Mason and Vanderslice, 1957a, 1958a).

For an inverse power potential it is clear from (15) that averaging S is the same, within a small correction term, as calculating an average aperture by averaging $\theta_0^{-2/s}$ rather than θ_0 itself. This is the procedure used by Amdur and Harkness (1954), and they showed the importance of carrying out such averaging. Measurements were made with two detectors of very different geometric apertures, 13.4 and 0.96 minutes. For the He-He system the proper average apertures were calculated to be 6.99 and 3.84 minutes, respectively. This gives agreement for the potential functions calculated from the two sets of measured cross

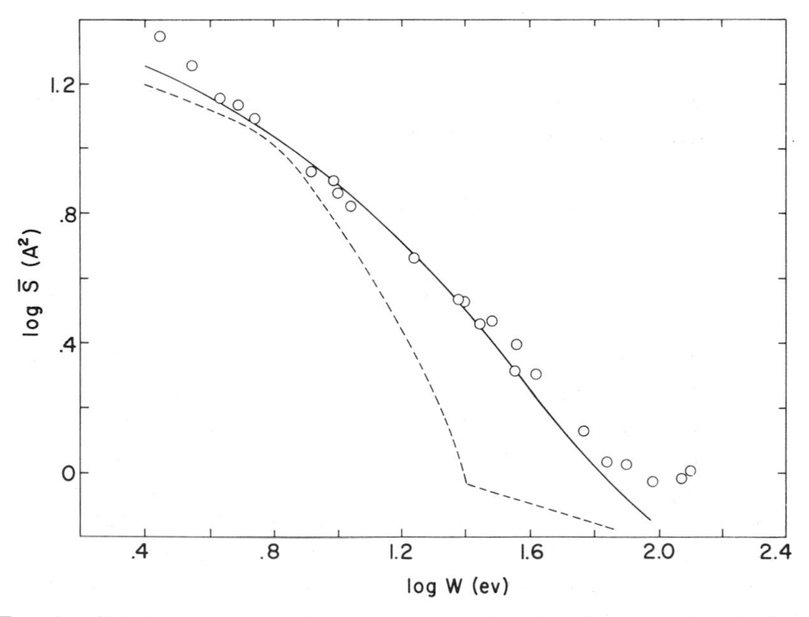

Fig. 8. Calculated and experimental cross sections for the scattering of H^+ ions in He, illustrating the importance of the beam width correction. The dashed curve is calculated for an infinitesimally narrow beam, and the solid curve for a beam of finite width.

sections within a factor of 1.3, compared with a factor of 0.17 if geometric apertures are used directly—nearly order of magnitude difference, showing the importance of the correction.

Another example of favorable agreement between theory and experiment being obtained only when the width of the ion beam is taken into account is shown in Fig. 8, which gives the cross sections for the elastic scattering of H^+ ions in He gas. The curves were calculated from the H^+-He potential obtained quantum mechanically, the dashed curve being calculated ignoring the width of the ion beam and the solid curve assuming the beam width to be the same as the detector width, but with an intensity that falls off linearly from a maximum at the center to zero at the edge (Mason and Vanderslice, 1957a). The circles are the experimental points of Simons *et al.* (1943a). There are no disposable parameters in the calculation, only universal constants and apparatus geometry. Lack of knowledge concerning beam width and intensity distribution across the beam prevents a more precise determination of the potential parameters in the assumed potential for many systems (Mason and Vanderslice, 1958a).

7.2 *A Priori* Knowledge of Potential Function

We have already mentioned in § 6.1 that some *a priori* knowledge of $V(r)$ is usually necessary before any physical significance can be attached to the potential parameters obtained from scattering experiments. Sometimes it is necessary to carry out approximate quantum-mechanical calculations to establish the nature of the potential (Mason and Vanderslice, 1957a, 1958a). One must know from other sources at least whether the potential is attractive or repulsive in the region of interest. The case of He^+ ions scattered in He gas furnishes a rather extreme example, in that two equally probable potential curves govern the scattering, one attractive and one repulsive (Mason and Vanderslice, 1957b, 1958e). The scattering measurements of Cramer and Simons (1957) have been analyzed for this case with the potential curves so determined shown in Fig. 9. The solid curves are the potentials determined from the scattering measurements and the points represent the quantum-mechanical calculations of Weinbaum (1935), Moiseiwitsch (1956), and Csavinszky (1959). The agreement is excellent. If one assumes a simple inverse-power attraction to represent the true potential, then the scattering measurements can be analyzed to give the dashed curve in Fig. 9 (Cramer and Simons, 1957). It is apparent that little physical significance can be attached to such a potential. Even though one could use it to recalculate the measured cross sections, it could probably not be used to predict or interpret other properties of the system.

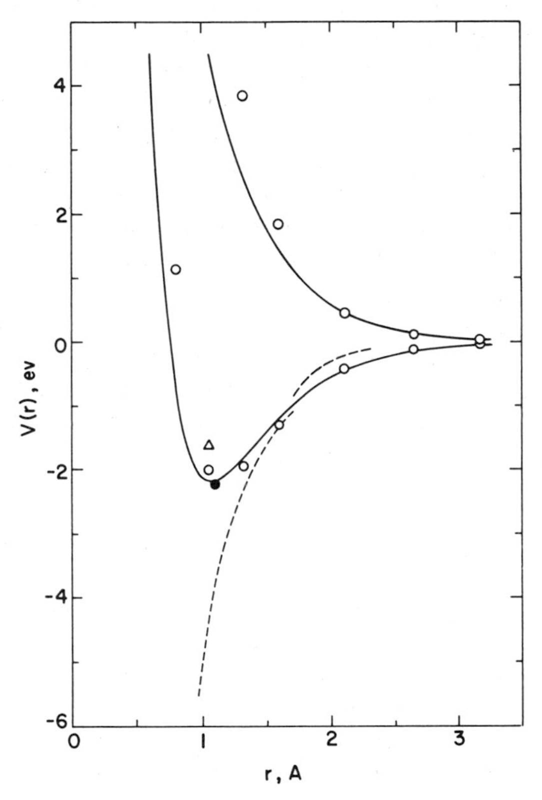

FIG. 9. Interaction energy between He^+ and He. The solid lines represent the scattering results. The points represent the following quantum-mechanical calculations: ●, Weinbaum (1935), variational calculation; ○, Moiseiwitsch (1956), effective nuclear charge of 2 for He^+ and 27/16 for He; △, Csavinszky (1959), "open shell" calculation. The dashed lines represent the scattering results on the assumption that the interaction is a simple inverse-power attraction.

7.3 QUANTUM CORRECTIONS

Wu (1958) has questioned the classical analysis of the scattering results for the He-He system (Amdur and Harkness, 1954). The usual criterion for the applicability of classical mechanics is that the uncertainty in the momentum be small compared to the momentum transferred in a collision.† The momentum uncertainty is usually taken to be about $\hbar/(2r_0)$, where r_0 is the distance of closest approach, and the momentum transferred is about $V(r_0)/v$. The criterion thus becomes $\hbar v/[2r_0 V(r_0)] \ll 1$ (Bohm, 1951). For Amdur and Harkness' results, this ratio is in the range 0.1-0.2, not extremely small compared to unity, but hardly large enough to be certain that the classical approximation is invalid (Amdur, 1958). For lack of a more quantitative criterion of validity, the results must therefore be judged by their agreement with independent determinations of the He-He intermolecular potential, and Amdur (1958) cites a good deal of independent evidence that the results of the classical approximation are indeed valid. This indirect evidence is rather convincing, but a definite theoretical answer to the question must await the quantum calculation of the angular distribution $\sigma(\theta)$ at small angles for the system in question, to establish definitely the region of validity of the classical approximation. In terms of scattering angles we have $V(r_0)/v \approx \mu v \theta_0$ for small angles, so that the previous criterion for classical scattering can be stated as $\theta_0 \gg \theta_c$, where θ_c is given by (2). Put in these terms the criterion seems too stringent, but further theoretical work is needed.

8 Summary of Experimental Results

Measurements on high-energy elastic scattering to obtain intermolecular potential energy information have not been extensive, and a great deal of work is still necessary in this largely unexplored field. Only relatively recently has such scattering work produced useful information, and this has been mainly due to the persistent and pioneering efforts of Amdur, Simons, and their co-workers. We have collected the available results into three groups, which are given in Tables I-III. Table I consists of those results which appear to be the most reliable, in the sense that account has been taken of the beam width corrections and of the form of the potential (e.g., attractive or repulsive) in the region of interest. Part A of Table I contains only those systems for which direct measurements have been made, and Part B contains three systems for which indirect results have been obtained from combination rules. These three systems have also been checked in other ways, however. By combination rules we mean empirical or semiempirical formulas by which the potential function of one system can be calculated from the

† Another criterion on the de Broglie wavelength is well satisfied.

potential functions of two or more related systems. For example, the He-Ar potential can be accurately calculated as the geometrical mean of the He-He and Ar-Ar potentials, which has been confirmed by direct measurement (Amdur *et al.*, 1954).

Table II consists of those results which we feel are less reliable, either because of greater experimental difficulties or because some of the corrections have not been made. Table III consists of those results which we consider to be essentially uninterpreted at present, although empirical potential functions which will reproduce the measured cross sections have in most cases been calculated. In many of these cases it is lack of suitable *a priori* knowledge of the potential function that prevents interpretation, even some of the apparently simpler systems probably having rather complicated forces. For instance, the forces in the systems $H^+ - H_2O$ and $H^+ - CH_4$ are possibly chemical valence forces, since H_3O^+ and CH_5^+ are quite stable species. The experiments themselves are not at fault, and are usually masterpieces of design and execution.

TABLE I

MOST RELIABLE POTENTIAL ENERGY FUNCTIONS DERIVED FROM SCATTERING MEASUREMENTS[a]

System	Potential function (ev)	Range (A)	Reference
		Part A	
He-He	$3.47/r^{5.03}$	0.97-1.48	Amdur *et al.* (1961)
	$4.71/r^{5.94}$	1.27-1.59	Amdur and Harkness (1954)
Ne-Ne	$312/r^{9.99}$	1.76-2.13	Amdur and Mason (1955a)
Ar-Ar	$849/r^{8.33}$	2.18-2.69	Amdur and Mason (1954)
Kr-Kr	$159/r^{5.42}$	2.42-3.14	Amdur and Mason (1955b)
Xe-Xe	$7.05 \times 10^3/r^{7.97}$	3.01-3.60	Amdur and Mason (1956*a*)
H-He	$2.34/r^{3.29}$	1.16-1.71	Amdur and Mason (1956b)
He-Ar	$62.1/r^{7.25}$	1.64-2.27	Amdur *et al.* (1954)
Ne-Ar	$630/r^{9.18}$	1.91-2.44	Amdur and Mason (1956c)
$He-N_2$	$74.3/r^{7.06}$	1.79-2.29	Amdur *et al.* (1957)
$Ar-N_2$	$755/r^{7.78}$	2.28-2.83	Amdur *et al.* (1957)
H^+-He	$1.90[(0.76/r)^4 - 2(0.76/r)^2]$	0.63-1.59	Simons *et al.* (1943a);
	$- 6.72/r^5$	1.59-2.65	Mason and Vanderslice (1957a)
H_2^+-He	$8.31/r^{7.00}$ (linear)	0.74-1.59	Simons *et al.* (1943a);
	$1.48/r^{4.00}$ (triangular)	0.74-1.59	Mason and Vanderslice (1957a)
H_3^+-He	$177 \exp(-r/0.301)$	0.89-2.22	Simons *et al.* (1943a);
			Mason and Vanderslice (1957a)
H^--He	$18.1 \exp(-r/0.491)$	0.77-2.0	Bailey *et al.* (1957);
			Mason and Vanderslice (1958a)
H^--H_2	$24.6 \exp(-2.18r)$	0.87-2.0	Muschlitz *et al.* (1956, 1957);
			Mason and Vanderslice (1958c)

TABLE I *(continued)*

System	Potential function (ev)	Range (A)	Reference
H^--Ne	$34.9 \exp(-2.20r)$	0.73-1.84	Bailey *et al.* (1957); Mason and Vanderslice (1958c)
H^--Ar	$60.1 \exp(-2.23r)$	0.77-2.15	Bailey *et al.* (1957); Mason and Vanderslice (1958c)
H^+-H_2	$2.7\{\exp[6(1 - r/1.5)] - 2 \exp 3(1 - r/1.5)]\}$	1.5-3.7	Simons *et al.* (1943c); Mason and Vanderslice (1959b)
H_3^+-H_2	$99.8 \exp(-r/0.376)$	1.48-2.45	Simons *et al.* (1943c); Mason and Vanderslice (1959b)
He^+-He	$2.16\{\exp[4.66(1 - r/1.080)] - 2 \exp[2.33(1 - r/1.080)]\}$	0.9-3.8 (attraction)	Cramer and Simons (1957); Mason and Vanderslice (1958e)
	$4.32 \exp[2.33(1 - r/1.080)]$	0.9-3.8 (repulsion)	
Ne^+-Ne	$0.71\{\exp[8(1 - r/1.7)] - 2 \exp[4(1 - r/1.7)]\}$;	1.8-4.2 (attraction)	Cramer (1958); Mason and Vanderslice (1958a); Pauling (1960)
	$1.42 \exp[4(1 - r/1.7)]$	1.8-4.2 (repulsion	
	Part B		
N_2-N_2	$595/r^{7.27}$	2.43-3.07	Amdur *et al.* (1957)
H-Ne	$26.3 \exp(-2.02r)$	0.78-2.5	Mason and Vanderslice (1958c)
H-Ar	$17.6 \exp(-1.42r)$	1.26-3.0	Mason and Vanderlisce (1958c)

[a] The potential is in electron volts when r is in angstroms. Part A, direct determinations; Part B, indirect determinations.

TABLE II

OTHER POTENTIAL ENERGY FUNCTIONS DERIVED FROM SCATTERING MEASUREMENTS[a]

System	Potential function (ev)	Range (A)	Reference
He-He	$2.88/r^{1.79}$	0.52-1.02	Amdur and Pearlman (1941); Amdur (1949)
Ne-Ne	$6.49 \times 10^3 \exp(-4.25r)$	0.4-1.0	Berry (1955)
Ar-Ar	$1.37 \times 10^4 \exp(-4.14r)$	0.6-1.2	Berry (1949, 1955)
Ar-Ar	$28.8/r^{4.33}$	1.37-1.84	Amdur *et al.* (1950*a*)
H-H_2	$6.72 \exp(-1.52r)$	0.4-1.2	Amdur *et al.* (1950b); Mason and Vanderslice (1958c)
D-D_2	$28.7 \exp(-5.17r^2)$	0.29-0.56	Amdur *et al.* (1950b)

[a] The potential is in electron volts when r is in angstroms.

TABLE III

OTHER SYSTEMS INVESTIGATED

System	Beam energy (ev)	Reference
H_2^+-H_2	8.3-75	Simons *et al.* (1943b)
H^+-H_2O		
H_2^+-H_2O	3-130	Simons *et al.* (1943d)
H_3^+-H_2O		
H^+-CH_4	1.1-156	
H_2^+-CH_4	3-125	Simons and Fryburg (1945)
H_3^+-CH_4	1.5-152	
H^+-C_2H_6	6.5-125	
H_2^+-C_2H_6	5-130	
H_3^+-C_2H_6	5-127	Simons and McAllister (1952)
H^+-C_3H_8	5.8-125	
H_2^+-C_3H_8	5.6-132	
H_3^+-C_3H_8	5.8-125	
H^+-nC_4H_{10}	1.9-117	
H_2^+-nC_4H_{10}	1.4-100	
H_3^+-nC_4H_{10}	2-100	Simons and Cramer (1950)
H^+-iC_4H_{10}	3.2-100	
H_2^+-iC_4H_{10}	3.8-100	
H_3^+-iC_4H_{10}	3.1-100	
H^+-C_2H_4	2.4-100.5	
H_2^+-C_2H_4	3.15-116	
H_3^+-C_2H_4	3-138	Simons and Unger (1945)
H^+-C_3H_6	3.7-118	
H_2^+-C_3H_6	4-114	
H_3^+-C_3H_6	4.5-116	
H^+-CF_4	5-130	
H_2^+-CF_4	5.2-131	
H_3^+-CF_4	5.7-124	Simons and Garber (1953)
H^+-C_2F_6	5-140	
H_2^+-C_2F_6	5.4-125	
H_3^+-C_2F_6	5-125	
He^+-Ne	4-400	Cramer (1958)
Ne^+-He		
Ne^+-Ar		
Ar^+-Ne	4-400	Cramer (1959)
Ar^+-Ar		
D^+-D_2	3.2-400	Cramer and Marcus (1960)
D_2^+-D_2	4-400	
O^--O_2	3-350	Muschlitz (1960)
O_2^--O_2	5-350	

9 Comparison of Scattering Results with Other Data

It is of interest to compare the results of the elastic scattering experiments with information available from other sources. Space limitations prevent our doing more than exhibiting a few examples and citing others. We have already seen that the scattering results for H^+ in He are in good agreement with quantum-mechanical calculations of the interaction energy (Fig. 8), and similarly for He^+ in He (Fig. 9). Scattering measurements and quantum-mechanical calculations have also been carried out for the system H-He, and found to be in good agreement (Amdur and Mason, 1956b; Mason *et al.*, 1956). Quantum-mechanical calculations of the repulsive interaction of two He atoms (Rosen, 1950; Griffing and Wehner, 1955; Sakamoto and Ishiguro, 1956; Lynn, 1958; Hashino and Huzinaga, 1958; Moore, 1960) are also in good agreement with the results of Amdur and Harkness (1954), which are valid between 1.27 and 1.59 A, and in reasonable agreement with the results of Amdur *et al.* (1961), which are valid between 0.97 and 1.48 A. Agreement is not so good for the earlier experiments giving results valid between 0.52 and 1.02 A (Amdur and Pearlman, 1941; Amdur, 1949). Here the experiments themselves are less certain, and the theoretical calculations are questionable because of their use of approximate wave functions which have an erroneous united atom limit (Buckingham, 1958). Somewhat similar remarks apply to the experimental and theoretical results for the H-H_2 system (Amdur and Pearlman, 1940; Amdur, 1943, 1949; Margenau, 1944; Aroeste and Jameson, 1959; Jameson and Aroeste, 1960).

It should be emphasized that the results for the rare gas and nitrogen interactions listed in Table IA and B are entirely consistent with potentials valid at larger distances, derived from gas and crystal properties. The comparisons are reported in the original scattering papers.

As was mentioned in the Introduction, scattering results can be used to calculate gas properties at high temperatures. In a few cases measurements of the gas properties are also available at temperatures high enough for a direct comparison to be made. In cases examined so far, the agreement has been satisfactory (Mason and Vanderslice, 1958d; Amdur and Mason, 1958). Two particularly striking recent examples of agreement come from the measurements of Walker and Westenberg (1958, 1959) of the diffusion coefficients of He-N_2 and He-Ar at unusually high temperatures. The agreement is remarkably good.

It has also been shown (Mason and Vanderslice, 1958a, c) that the elastic scattering results for H-H_2, H-He, H-Ne, and H-Ar derived from Amdur's measurements are consistent with the elastic scattering and electron detachment results for H^--H_2, H^--He, H^--Ne, and H^--Ar obtained with a Simons-type apparatus (Muschlitz *et al.*, 1956, 1957; Bailey *et al.*, 1957).

10 Summary

Although the experimental techniques are exacting and the theoretical interpretation often complicated, high-energy elastic scattering studies can supply information which is often unobtainable in any other way. Although the method now seems well established, much remains to be done. Further work at independent laboratories would be particularly valuable to check key results already obtained and to obtain further information. Few studies have yet been made on systems of interest in such fields as astrophysics, upper atmosphere physics and chemistry, and the physics and chemistry of combustion, detonations, and shock waves. Many interesting and important systems would require study by crossed beam techniques, which are not sufficiently developed as yet.

Acknowledgment

The authors wish to thank the U. S. National Aeronautics and Space Administration for support of their work.

References

Amdur, I. (1943) *J. chem. Phys.* **11**, 157.
Amdur, I. (1949) *J. chem. Phys.* **17**, 844.
Amdur, I. (1953) *Science* **118**, 567.
Amdur, I. (1958) *J. chem. Phys.* **28**, 987.
Amdur, I., and Harkness, A. L. (1954) *J. chem. Phys.* **22**, 664.
Amdur, I., and Mason, E. A. (1954) *J. chem. Phys.* **22**, 670.
Amdur, I., and Mason, E. A. (1955a) *J. chem. Phys.* **23**, 415.
Amdur, I., and Mason, E. A. (1955b) *J. chem. Phys.* **23**, 2268.
Amdur, I., and Mason, E. A. (1956a) *J. chem. Phys.* **25**, 624.
Amdur, I., and Mason, E. A. (1956b) *J. chem. Phys.* **25**, 630.
Amdur, I., and Mason, E. A. (1956c) *J. chem. Phys.* **25**, 632.

Amdur, I., and Mason, E. A. (1958) *Phys. of Fluids* **1**, 370.
Amdur, I., and Pearlman, H. (1940) *J. chem. Phys.* **8**, 7.
Amdur, I., and Pearlman, H. (1941) *J. chem. Phys.* **9**, 503.
Amdur, I., and Ross, J. (1958) *Combustion and Flame* **2**, 412.
Amdur, I., Glick, C. F., and Pearlman, H. (1948) *Proc. Amer. Acad. Arts Sci.* **76**, 101.
Amdur, I., Davenport, D. E., and Kells, M. C. (1950a) *J. chem. Phys.* **18**, 525.
Amdur, I., Kells, M. C., and Davenport, D. E. (1950b) *J. chem. Phys.* **18**, 1676.
Amdur, I., Mason, E. A., and Harkness, A. L. (1954) *J. chem. Phys.* **22**, 1071.
Amdur, I., Mason, E. A., and Jordan, J. E. (1957) *J. chem. Phys.* **27**, 527.
Amdur, I., Jordan, J. E., and Colgate, S. O. (1961) *J. chem. Phys.* **34**, 1525.
Aroeste, H., and Jameson, W. J. (1959) *J. chem. Phys.* **30**, 372.
Bailey, T. L., May, C. J., and Muschlitz, E. E. (1957) *J. chem. Phys.* **26**, 1446.
Bernstein, R. B. (1960) *J. chem. Phys.* **33**, 795.
Berry, H. W. (1949) *Phys. Rev.* **75**, 913.
Berry, H. W. (1955) *Phys. Rev.* **99**, 553.
Bohm, D. (1951) "Quantum Theory," Chapt. 21. Prentice-Hall, Englewood Cliffs, New Jersey.
Buckingham, R. A. (1958) *Trans. Faraday Soc.* **54**, 453.
Cramer, W. H. (1958) *J. chem. Phys.* **28**, 688.
Cramer, W. H. (1959) *J. chem. Phys.* **30**, 641.
Cramer, W. H., and Marcus, A. B. (1960) *J. chem. Phys.* **32**, 186.
Cramer, W. H., and Simons, J. H. (1957) *J. chem. Phys.* **26**, 1272.
Csavinszky, P. (1959) *J. chem. Phys.* **31**, 178.
Everhart, E., Stone, G., and Carbone, R. (1955) *Phys. Rev.* **99**, 1287.
Finkelstein, A. T. (1940) *Rev. sci. Instrum.* **11**, 94.
Firsov, O. B. (1953) *Zh. eksper. teor. Fiz.* **24**, 279.
Firsov, O. B. (1958) *Zh. eksper. teor. Fiz.* **34**, 447; translation, *Soviet Phys.—JETP* **7**, 308.
Ford, K. W. and Wheeler, J. A. (1959) *Ann. Phys. (New York)* **7**, 259.
Fraser, R. G. J. (1931) "Molecular Rays." Cambridge Univ. Press, London and New York.
Gáspár, R. (1960) *Acta phys. Hungar.* **11**, 71.
Gordon, M. M. (1957) *Amer. J. Phys.* **25**, 32.
Griffing, V., and Wehner, J. F. (1955) *J. chem. Phys.* **23**, 1024.
Hashino, T., and Huzinaga, S. (1958) *Progr. theor. Phys.* **20**, 631.
Hirschfelder, J. O., Curtiss, C. F., and Bird, R. B. (1954) "Molecular Theory of Gases and Liquids," p. 546. Wiley, New York.
Hostettler, H. U., and Bernstein, R. B. (1960) *Phys. Rev. Letters* **5**, 318.
Hoyt, F. C. (1939) *Phys. Rev.* **55**, 664.
Jameson, W. J., and Aroeste, H. (1960) *J. chem. Phys.* **32**, 374.
Keller, J. B., Kay, I., and Shmoys, J. (1956) *Phys. Rev.* **102**, 557.
Kells, M. C. (1948) *J. chem. Phys.* **16**, 1174.
Kennard, E. H. (1938) "Kinetic Theory of Gases," pp. 115-134. McGraw-Hill, New York.
Kreevoy, M. M., and Mason, E. A. (1957) *J. Amer. Chem. Soc.* **79**, 4851.
Lamar, E. S., and Luhr, O. (1934) *Phys. Rev.* **46**, 87.
Landau, L. D., and Lifshitz, E. M. (1958) "Quantum Mechanics," p. 416. Pergamon, New York.
Lane, G. H., and Everhart, E. (1960a) *Phys. Rev.* **117**, 920.
Lane, G. H., and Everhart, E. (1960b) *Phys. Rev.* **120**, 2064.
Lynn, N. (1958) *Proc. Phys. Soc.* **72**, 201.
Margenau, H. (1944) *Phys. Rev.* **66**, 303.
Mason, E. A. (1955) *J. chem. Phys.* **23**, 2457.

Mason, E. A., and Kreevoy, M. M. (1955) *J. Amer. Chem. Soc.* **77**, 5808.
Mason, E. A., and Vanderslice, J. T. (1957a) *J. chem. Phys.* **27**, 917.
Mason, E. A., and Vanderslice, J. T. (1957b) *Phys. Rev.* **108**, 293.
Mason, E. A., and Vanderslice, J. T. (1958a) *J. chem. Phys.* **28**, 253.
Mason, E. A., and Vanderslice, J. T. (1958b) *J. chem. Phys.* **28**, 432.
Mason, E. A., and Vanderslice, J. T. (1958c) *J. chem. Phys.* **28**, 1070.
Mason, E. A., and Vanderslice, J. T. (1958d) *Industr. engng. Chem.* **50**, 1033.
Mason, E. A., and Vanderslice, J. T. (1958e) *J. chem. Phys.* **29**, 361.
Mason, E. A., and Vanderslice, J. T. (1959a) *J. chem. Phys.* **30**, 599.
Mason, E. A., and Vanderslice, J. T. (1959b) *Phys. Rev.* **114**, 497.
Mason, E. A., Ross, J., and Schatz, P. N. (1956) *J. chem. Phys.* **25**, 626.
Massey, H. S. W., and Burhop, E. H. S. (1952) "Electronic and Ionic Impact Phenomena", Chapts. 7 and 8. Oxford Univ. Press (Clarendon), London and New York.
Massey, H. S. W., and Mohr, C. B. O. (1933) *Proc. Roy. Soc.* **A141**, 434.
Massey, H. S. W., and Mohr, C. B. O. (1934) *Proc. Roy. Soc.* **A144**, 188.
McDowell, M. R. C. (1958) *Proc. Phys. Soc.* **72**, 1087.
Moiseiwitsch, B. L. (1956) *Proc. Phys. Soc.* **A69**, 653.
Moore, N. (1960) *J. chem. Phys.* **33**, 471.
Muschlitz, E. E. (1957) *J. appl. Phys.* **28**, 1414.
Muschlitz, E. E. (1960) *In* "Proceedings of the Fourth International Conference on Ionization Phenomena in Gases, Uppsala, 1959" (N.R. Nilsson, ed.), Vol. 1A, p. 52. North-Holland, Amsterdam.
Muschlitz, E. E., Bailey, T. L. and Simons, J. H. (1956) *J. chem. Phys.* **24**, 1202.
Muschlitz, E. E., Bailey, T. L., and Simons, J. H. (1957) *J. chem. Phys.* **26**, 711.
Myers, V. W. (1955) *J. chem. Phys.* **23**, 755.
Myers, V. W. (1956) *J. chem. Phys.* **25**, 1284.
Pauling, L. (1960) "The Nature of the Chemical Bond," 3rd ed., p. 356. Cornell Univ. Press, Ithaca, New York.
Pauly, H. (1957) *Z. angew. Phys.* **9**, 600.
Pauly, H. (1959) *Z. Phys.* **157**, 54.
Pauly, H. (1960) *Z. Naturforsch.* **15**a, 277.
Ramsey, N. F. (1956) "Molecular Beams," Chapts. 1 and 2. Oxford Univ. Press, London and New York.
Rosen, P. (1950) *J. chem. Phys.* **18**, 1182.
Russell, A. S., Fontana, C. M., and Simons, J. H. (1941) *J. chem. Phys.* **9**, 381.
Sakamoto, M., and Ishiguro, E. (1956) *Progr. theor. Phys.* **15**, 37.
Schumacher, H., Bernstein, R. B., and Rothe, E. W. (1960) *J. chem. Phys.* **33**, 584.
Sida, D. W. (1957) *Phil. Mag.* **2**, 761.
Simons, J. H., and Cramer, W. H. (1950) *J. chem. Phys.* **18**, 473.
Simons, J. H., and Fryburg, G. C. (1945) *J. chem. Phys.* **13**, 216.
Simons, J. H., and Garber, C. S. (1953) *J. chem. Phys.* **21**, 689.
Simons, J. H., and McAllister, S. A. (1952) *J. chem. Phys.* **20**, 1431.
Simons, J. H., and Unger, L. G. (1945) *J. chem. Phys.* **13**, 221.
Simons, J. H., Francis, H. T., Fontana, C. M., and Jackson, S. R. (1942) *Rev. sci. Instrum.* **13**, 419.
Simons, J. H., Muschlitz, E. E., and Unger, L. G. (1943a) *J. chem. Phys.* **11**, 322.
Simons, J. H., Fontana, C. M., Francis, H. T., and Unger, L. G. (1943b) *J. chem. Phys.* **11**, 312.
Simons, J. H., Fontana, C. M., Muschlitz, E. E., and Jackson, S. R. (1943c) *J. chem. Phys.* **11**, 307.

Simons, J. H., Francis, H. T., Muschlitz, E. E., and Fryburg, G. C. (1943d) *J. chem. Phys.* **11**, 316.
Smith, K. F. (1955) "Molecular Beams." Methuen, London.
Walker, R. E., and Westenberg, A. A. (1958) *J. chem. Phys.* **29**, 1147.
Walker, R. E., and Westenberg, A. A. (1959) *J. chem. Phys.* **31**, 519.
Weinbaum, S. (1935) *J. chem. Phys.* **3**, 547.
Westheimer, F. H. (1956) "Steric Effects in Organic Chemistry" (M. S. Newman, ed.), Chapt. 12. Wiley, New York.
Wu, T.-Y. (1958) *J. chem. Phys.* **28**, 986.

18.

Charge Transfer and Collisional Detachment

John B. Hasted

1 Introduction

A complete understanding of a collision process is not possible without a satisfactory solution of the relevant quantal equations. The detailed study of collisions between the simplest systems is therefore of central importance. It is, however, also necessary to develop semiempirical theories which may be applied to collisions between complicated systems.

As the impact energy is increased from the threshold, the collision cross section for the various types of charge transfer and for detachment in general rises, passes through a maximum, and then falls off mono-

TABLE I

Process	Equation	Designation
Single charge transfer	$A^+ + B \to A + B^+$	10/01
Transfer ionization	$A^+ + B \to A + B^{2+} + e$	10/12e
Stripping	$A^+ + B \to A^{2+} + e + B$	10/2e0
Charge transfer to	$A^+ + B \to A' + B^+$	10/0′1
an excited level	$A^+ + B \to A + B^{+\prime}$	10/01′
Capture by neutral atom	$A + B \to A^- + B^+$	00/$\bar{1}$1
Collisional detachment	$A^- + B \to A + e + B$	$\bar{1}$0/0e0

tonically. In some cases evidence for fine structure has been reported (Donahue and Hushfar, 1959, 1960; Curran and Donahue, 1960), but this has not yet been confirmed.

It is convenient to refer to a collision by $p_i t_i / p_f t_f$ where p_i and p_f are the initial and final charges on the projectile in units of the charge on the proton, and where t_i and t_f are the corresponding charges on the target. A prime is inserted if one of the systems is excited and the symbol e if an electron is ejected (see Table I).

2 *Collision Chamber Techniques*

Slow charged particles are produced in charge transfer processes in which the projectile gains or loses one or more electrons, and in collisional detachment. The collection of these particles is the basis of the method of cross section measurement which is of primary importance at moderate energies.

Consider an ion beam passing through a collision chamber containing gas at temperature T and pressure P low enough to ensure that an ion does not make several collisions. If I_b is the beam current, the current of slow charged particles produced in path length l is given approximatively by

$$I_s = I_b \mathcal{N} \sigma \tag{1}$$

where

$$\mathcal{N} = lP/kT \tag{2}$$

is the number of atoms or molecules of the target gas per square centimeter traversed by the beam and σ is the collision cross section concerned. Hence, σ may be determined experimentally by collecting all the slow collision products formed in an accurately known path length.

Measurements were first made with the aid of a cage electrode in the collision chamber, the cage having holes through which the beam could pass (Goldmann, 1931; Rostagni, 1935). Elastic scattering of the beam ions takes place through such small angles that it may be ignored. The effects of the positive ions and electrons resulting from ionizing collisions, 10/11e, cancel. At high impact energies, transfer ionization, 10/02e, contributes to the positive charge observed and stripping, 10/2e0, to the negative charge. Charge transfer, 10/01, is however predominant at energies below a few kev.

In addition to charge transfer, $\bar{1}0/0\bar{1}$, a negative ion beam suffers collisional detachment $\bar{1}0/0e0$. All the electrons must be collected if a true detachment cross section is to be measured.

Since the slow collision products are very much less energetic than the primary ions it is possible to separate them by fields. This has been done with uniform electric fields transverse to the ion beam (Keene, 1949; Gilbody and Hasted, 1956; Donahue and Hushfar, 1959, 1960); with uniform electric and magnetic fields, parallel and transverse to the ion beam (Wolf, 1936-1939; Hasted, 1951-1952); with a uniform transverse magnetic field alone; (Hasted and co-workers, unpublished experiments); and with crossed electric and magnetic fields, to separate electrons from ions (Varney, 1936; Bailey, 1960; Moe, 1956; Stebbings *et al.*, 1960).

In the transverse electric field, with no magnetic field, the slow charged particles follow parabolic paths to the metal plates. If the potential difference is only comparable with their initial energy, the particles will not be collected at points transversely corresponding to their points of formation; and under certain circumstances the path length of collection will not correspond to the length of the collecting plates. However, if the potential difference is great enough the paths are almost exactly transverse. The achievement of saturation conditions is an indication of transverseness and hence of the accuracy of the collision path length.

It is sometimes the practice (Curran *et al.*, 1959) to measure the currents collected at successive condenser electrodes. This does not offer very great advantages. The errors in physical measurement and electrical alignment of the electrode system are far smaller than those arising from other sources, such as pressure measurement, and are likely to be increased by the use of several condenser electrodes. The statistical advantage may in any case be achieved by pressure variation. A check is, however, provided on the invariance of the beam composition as it passes through the collision chamber; and also on the absence of interference from end effects, and of pronounced forward or back scattering effects, arising at the collision chamber entrance of Faraday collection cage.

It is important that a high degree of uniformity of the transverse electric field should be maintained in the collision region. In the condenser electrode systems mentioned above there is the possibility of field penetration from the collision chamber walls. To avoid this, resistive spacers have been used by some workers (Hasted, 1951). It is not easy to obtain uniformity and stability with organic graphite colloid painted on insulator.

The secondary electrons due to collisions of ions with metal surfaces must be taken into consideration. Those from the primary ion collecting electrodes must be prevented from reaching the slow ion collectors. A transverse electric field, with or without parallel magnetic field, will confine them to the primary beam collector. Secondaries from the slow ion collectors, which are not nearly so numerous, are confined to the transverse lines of force. Where it is necessary, as in the measurement of ionization $10/11e$, to separate the slow positive and negative charge components, the secondaries arising at the negative collector may be suppressed by covering the collector with an even more negative grid. Further secondaries arise from collisions of the primary beam with the edges of the entrance slit of the collision chamber. They may be reflected back by a negatively charged electrode behind the entrance slit (Keene, 1949).

When the ions at the edges of the primary beam collide with the entrance slit edges they may be partially neutralized, perhaps with the production of metastable systems. Further neutralization may occur by collisions with the gas in the collision chamber. These effects are not normally important (Gilbody, 1956), but since neutrals may also undergo electron capture, $00/\bar{1}1$, their significance must be estimated.

The tightest control that can be kept upon the paths of charged particles is the control by means of parallel electric and magnetic lines of force. Whatever the initial path direction or energy of the particle, a sufficiently large magnetic field will confine it to a helical path of calculable pitch radius around the lines of force. It is important not only that the magnetic and electric fields be correctly aligned, but that the helical pitch radii be small compared with l the path length. If these conditions are satisfied there should be no difference between cross sections measured with or without magnetic field. This is a good test of the quality of an electrode system.

It is also possible to collect slow charged products on a cylinder with transverse magnetic field, but no electric field. The data on angular distribution of collision product ions obtained by Fedorenko (1959) make the measurement of total charge transfer cross sections much easier. As is expected, the slow ions, typically, are produced within a scattering cone of 80°-90° to the primary beam. The definition of a path length by the collection of ions on a simple cylinder, (or a series of simple cylinders) should be adequate, even though the slow particles follow straight line paths which are not perpendicular to the primary beam. This does not necessarily apply to the electrons with unknown angular distributions formed in ionizing or stripping collisions. They can, however, be confined to transverse helices by magnetic fields which are too weak to have much effect upon the ion paths. An advantage of this electrode system is that it requires no electric field. The high electric potentials ($\sim$ 2kv) necessary to separate completely the ions and fast electrons produced in ion-atom collisions at a few kev may be undesirable. Measurement of small currents at high potentials is technically difficult (the entire electrometer must float). Moreover, the accelerated electrons in the collision chamber may produce further ionization.

A magnetic field perpendicular both to the primary ion beam and to the "transverse" electric field applied to the condenser electrodes has been used in the measurements of Stebbings *et al.* (1960). In such a field configuration charged particles will follow trochoidal paths along the magnetic lines of force, so that the separation of electrons from negative ions becomes possible. The electrons, traveling in paths of small radii, will not be collected on the condenser electrodes; but the ions, traveling in paths of large radii, will hardly be disturbed by the magnetic field in their passage to the electrodes.

Currents of particles are measured by the following means:

(i) *Electrometers* of various types [cf. Du Bridge and Brown (1933) and A.E.R.E. (Harwell) Electronics Division *Manual for Equipment* 1079c].

(ii) *Thermocouples*: Fast neutral particles may be detected by their heating effect (cf. Allison, 1958). The method is simple but relatively insensitive. It cannot be used if the current is less than about 10^{10} particles/sec.

(iii) *Particle multipliers*: These depend upon electron emission from a bombarded metal surface (cf. Allison, 1958). The electrons are multiplied by Malter effect dynodes until sufficiently numerous to be counted as current pulses or smoothed into direct current (cf. Pierce, 1954). A 15-stage silver-magnesium or beryllium-copper multiplier may have a gain in the 10^6 to 10^7 region. Calibration is necessary. At impact energies below about 1000 ev the electron emission coefficient is so small that the multiplier is inefficient and difficult to use. As the impact energy is increased the coefficient rises to a flat maximum. It is independent of the charge state of the bombarding particles.

(iv) *Molecular beam detectors* may be used for neutral particles which are of too low energy for multipliers to be efficient: Thus, neutral particles may be ionized on the surface of a filament and then detected by their reduction of the space charge limitation on a diode (Taylor, 1929). Ionization by electron impact is now commonly used (Weiss, 1961).

(v) *Alternating current methods*: A beam of particles may be converted into a low-frequency (30-500 cps) square-wave either electrically or mechanically. Measurements may then be made with the aid of a phase-sensitive narrow-band amplifier, the phase discrimination being controlled by the chopper of the beam. The width of the band may be only 0.03 cps. Making the band narrow reduces radio noise and also molecular noise arising from unwanted particles in the collision chamber. Unfortunately, an absolute calibration of the amplifier may be difficult (Fite *et al.*, 1958).

(vi) *Nuclear physics techniques*: The scintillation counter (Sweetman, 1959) and photographic plate (Chalklin and Fremlin, 1960) have been used in ionic collision experiments. The former is sensitive to particle energy.

3 Mass Analysis Problems

Figure 1 shows the essentials of an experiment in which both target and beam products are mass analyzed. Each of the analyses has its own peculiar difficulties.

We shall first discuss the problems involved in producing mass analyzed ion beams of particularly low energies (down to ~ 1 ev) for use in

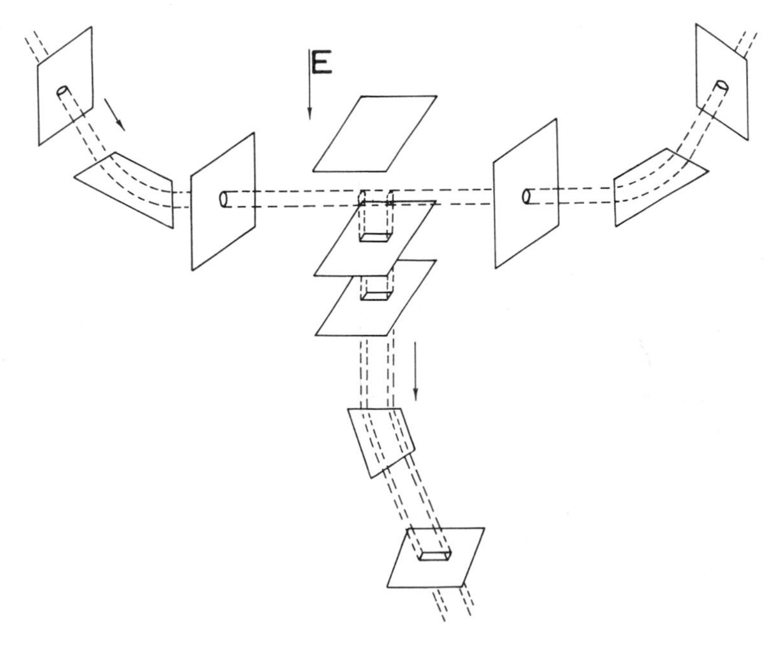

FIG. 1. Mass analysis of target and beam products.

collision studies. It is necessary to refer to a textbook on mass spectrometry (e.g., Barnard, 1952) and to the discussion of ion sources in Massey and Burhop (1952).

The ion densities in the standard ion sources due to Nier and Heyl may be obtained either from previous designs or by measuring currents to a probe immersed in the plasma. The ion current drawn from the source may be assumed to equal that drawn by a negative probe of the same surface area as the source aperture; for this current the expression given by Bohm *et al.* (1949) is

$$I_+ = 0.40 n_+ A \sqrt{\frac{2kT_e}{m_+}} \tag{3}$$

with A the probe area, n_+ and m_+ the ion density and mass, and T_e the electron temperature. The magnetic field is of major importance so far as the ion density in the source is concerned. With intense fields high densities can be obtained. Extraction is less efficient for intense fields owing to the ion path curvature. The optimum field for a particular design of source must be found by experiment. Space charge defocusing is avoided by the use of a strong electric field. It has been found (Willmore, 1955) that the current extracted from a source varies approximately as $V^{3/2}$ where V is the potential on the accelerating electrode immediately outside the source. Assuming that this is a variation with field, $E^{3/2}$, it might be imagined that a low-energy beam could be extracted more efficiently by reducing the source-to-accelerating-electrode distance while keeping V constant. This is limited by the effusion of gas from the source slit, the area of which must be designed according to (3). The limits are such that we must expect to have to retard the beam. This may be done either before or after mass analysis, or both, but there is a lower limit set on the energy at which mass analysis may be carried out, determined by the energy spread in the source.

In the case of a 180° mass spectrometer of radius of curvature ρ and potential V_a relative to the source, into which a homogeneous beam of ions of mass m_1, with angular divergence α, enters through an infinitely narrow slit, it may be shown that

$$\rho^2\alpha^2 \propto m_1 V_a. \tag{4}$$

If the energy spread in the source is δV_0, the beam will contain ions of energy from $V_a - \delta V_0$ to $V_a + \delta V_0$. Two ion species of masses m_1 and m_2 will be resolved if

$$m_1 V_a = m_2(V_a - \delta V_0) \tag{5}$$

so that the resolving power

$$\frac{m_2}{m_2 - m_1} = \frac{V_a}{\delta V_0}. \tag{6}$$

If, typically, δV_0 is 10 ev, then to achieve a resolving power of 20 (which is suitable for many purposes) it is necessary that V_a exceed 200 volts.

An efficient beam extraction may be obtained by an accelerating electrode A at a potential of $\sim$ 1200 volts, preceded by a focusing electrode F, and followed by an analysis chamber C at a potential of − 200 volts, both potentials relative to the source (Fig. 2). The focusing

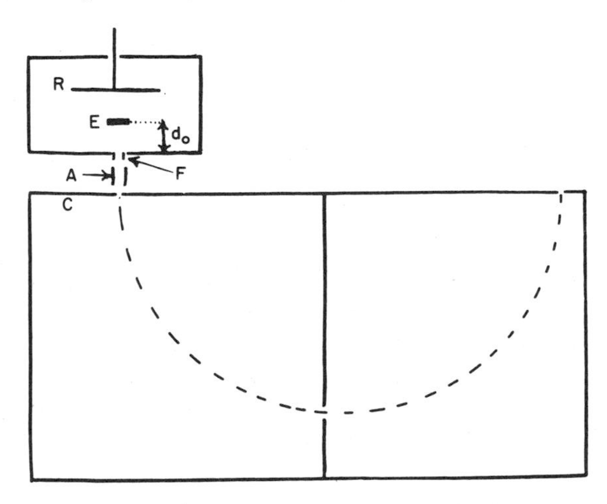

FIG. 2. Nier mass-spectrometer source with extraction and analysis sytem: E, electron beam perpendicular to plane of paper; R. repeller; F, optical focusing electrode; A, accelerating electrode; C, mass analysis chamber.

cylinders should be made of magnetic screening material. When the focusing electrode potential is adjusted for maximum current, the beam is parallel at both entrance and exit to the mass analyzer. By the use of a fine exit slit to the analyzer, the appropriate ion velocity is selected. For widths S of entrance and exit slits the emergent beam energy spread is

$$\delta V_a = \frac{2S}{\rho} V_a. \tag{7}$$

The divergence α of the beam must be such that

$$\alpha^2 \ll S/\rho. \tag{8}$$

Retardation experiments have shown that an energy spread δV_a of 0.5 ev can be obtained without difficulty.

The choice of mass-spectrometer geometry for the analysis of the beam is governed not only by normal focusing conditions but by the resolving power and the need for stability. At the entrance and exit to the magnetic field the beam is nearly parallel, and under these exceptional conditions the beam strength is proportional to slit width. In the 180° instrument not only is the resolving power greater but there is also more stability. Conditions in an ion source are likely to fluctuate, resulting in an alteration δV_a in the energy of the beam entering the spectrometer. In a 180° instrument this results in a displacement $\delta\rho$ given by

$$\frac{\delta\rho}{\rho} = \frac{\delta V_a}{2V_a}. \tag{9}$$

But in a 90° instrument the displacement may be several times larger, being given by

$$\delta\rho + l_1\theta \tag{10}$$

where l_1 is the distance between the collision chamber and the exit slit from the analysis chamber. In a small deflection instrument the displacement is less, the stability greater, but the resolution smaller.

For the lower energy beams V_a must be kept as small as possible, $\sim$ 100 volts, partly because of (7) and partly because serious overfocusing occurs in the retardation region.

In fact, Gustafsson and Lindholm (1958) have found that a retardation to an energy less than 0.03 of the initial energy is not practicable in a lens system.

The focal length of a single aperture at potential V with field strengths E_1 and E_2 in the object and image spaces is

$$f = 4V/(E_1 - E_2) \tag{11}$$

(Cosslett, 1944). If the potential V of the retarding aperture is 1 volt, its focal length is short enough to dominate the lens, and may, typically, be much less than 1 cm (convex) reducing the beam by perhaps 10^4 in about a 10-cm path. The aperture at potential V of 100 volts has a much longer focal length (concave), which is far from compensating for the second aperture.

It is possible to arrange a decelerating field such that the focusing effects of the two ends cancel, that is, such that

$$f_1 = |f_2|. \tag{12}$$

The field has a potential of the form

$$V(x) = a \exp(-bx) \tag{13}$$

where a and b are positive constants. Such a field may be produced approximately by a lens of many elements connected to a resistance chain. Detailed calculation by Willmore (1955) of the path followed by an ion confirms that the deviation is much reduced. Gustafsson and Lindholm (1958) have found, however, that an exponential field actually produces more deviation than certain empirically designed apertures, which are shown in Fig. 3.

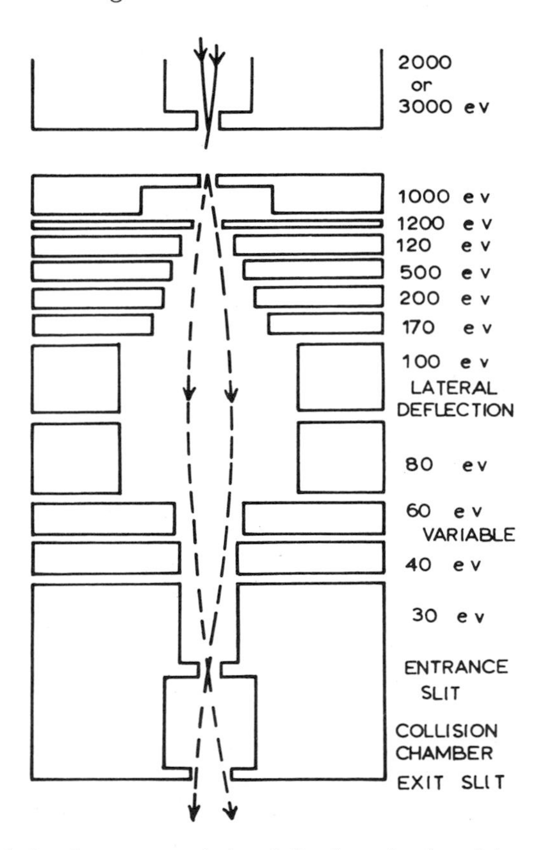

FIG. 3. Retardation lens system designed for the reduction of ion beam energy to 0.03 of its former energy by Gustafsson and Lindholm (1958).

We shall now discuss the extraction of slow collision products from a collision chamber into a mass analyzer. This is usually achieved with the aid of a transverse electric field (Fig. 1). The situation is simpler than in the extraction of ions from the Nier or Heyl mass-spectrometer source since there is no strong magnetic field in the collision chamber. The transverse electric field must be of sufficient strength to exceed by an order of magnitude the kinetic energy of the ions produced by collision. Owing to the angular distribution of these ions, they will diverge in their passage towards the extraction slit (E), and must be focused to form an image which itself forms the object of the (sector or 180°) magnetic focusing system. This is achieved by means of a second slit or aperture held at an accelerating potential. Ions formed with kinetic energy, as in the dissociation of molecules through anti-bonding states, may be unable to pass through such a system, but if they are able to do so, they will be velocity-analyzed according to (7) and (9). Retardation techniques may then be applied in the manner of Hagstrum (1951).

Two conditions must be shown experimentally to be satisfied. Firstly, the extracted beam strength must be independent of the transverse collision chamber field, just as in total charge collection experiments. Second, the analysis peaks must be found to be "flat-top," that is, independent of magnetic field over a certain range, when the exit slit is widened abnormally. Only then will the peak ion current with a narrow slit be directly proportional to the number of ions entering the mass spectrometer. The cross section for the production of a particular ion is only relative; calibration of the instrument against total charge collection measurements is necessary. Such calibration must be shown to be independent of secondary ion mass.

The 180° instrument is of greater resolution than the 90° or 60° sector analyzers, but of no better stability. However, the sector instruments are easier to position geometrically, especially when it is necessary to rotate them through small angles about the collision chamber for the measurement of differential cross sections (§ 5). In this type of experiment the transverse electric field is omitted.

We shall discuss briefly the mass analysis of the primary beam after collision (Fig. 3). It has been much employed by Flaks and co-workers for partial charge transfer 20/11 collisions and by Fogel and co-workers for double electron capture 10/12 collisions.

The ion beam passing through a collision chamber should be as nearly as possible parallel or should diverge slightly from a focus before the collision chamber. Such a beam will emerge nearly parallel from a magnetic analyzer, the focusing conditions being correct when a sector field is employed. The problems are therefore much the same as those encountered in the mass analysis of a primary ion beam emerging from an ion source and accelerated to an energy of the order of kilovolts.

The magnetic fields are required to be of the order of 10^3 to 10^4 gauss. This may be achieved with less magnetic material and fewer ampere turns by including the pole pieces in the vacuum chamber.

Continuous measurement and stabilization of the magnetic field can be arranged most conveniently by using the current from a rotating search coil mounted in the vacuum chamber. The most stable fields are those from permanent magnets. They may be varied using moveable shunts across the pole gap. The principle difficulty of design arises from the large mechanical forces to which such shunts are subject.

In the study of double capture $10/\bar{1}2$, the negative ions may be separated by mass analysis without separately designed sector instruments. Fogel and co-workers (1956-1959) have employed a collision chamber with transverse magnetic field intense enough for primary beam curvature. Similar techniques have been employed in much of the fast collision work described in Chapter 17.

Finally we shall discuss the Aston peak technique, an ingenious idea, used by Melton and Wells (1957) and by McGowan and Kerwin (1960), for avoiding the need for two mass analyzers, before and after collision, in the study of the impact of molecular ions upon gases.

Dissociation of ions by collision in the analysis region of the mass-spectrometer produces peaks at magnetic fields corresponding to non-integral m/e ratios smaller than those of the dissociated ion. Such peaks were first observed and discussed by Aston. They arise because an ion dissociating in collision with a gas molecule after acceleration will retain a velocity appropriate to the energy of the undissociated ion. In conventional mass spectrometers the peaks are usually broadened because the region in which collisions may occur is not equipotential. By separate pumping of the source and analysis chambers of a sector instrument, the two being connected only by a very fine slit and the source pressure being kept low, it may be arranged that collisions only occur in the (field-free) analysis chamber. Under these conditions the apparent ratio m^*/e^* is related to the primary molecular ion mass m_p and charge e_p and the fragment ion mass m_f and charge e_f, by the equation

$$m^*/e^* = m_f^2 e_p / m_p e_f^2. \tag{14}$$

In the dissociation of carbon monoxide, for example, apparent ratios are observed due to the following processes:

Process	Apparent ratio
$CO^+ + M \rightarrow C^+ + O + M$	5.14
$CO^+ + M \rightarrow C + O^+ + M$	9.14
$CO^{++} + M \rightarrow C^+ + O + M^+$	10.28

4 *Symmetrical Resonance Charge Transfer*

Because of the very large cross sections involved, symmetrical resonance charge transfer,

$$A^{n+} + A \rightarrow A + A^{n+} \tag{15}$$

is one of the simplest processes to study experimentally. The detailed quantal theory has been discussed in Chapter 14. We shall confine our attention to some approximate formulae of wide applicability.

Firsov (1951) has considered slow† encounters between singly charged ions and their parent atoms. He assumes that the passage of an electron from the atom to the ion is unlikely unless the internuclear distance becomes less than the value R_0 of ρ which makes ζ of (181) of Chapter 14 equal $1/\pi$, but that then the passage occurs readily. On this model the probability of transfer oscillates rapidly between 0 and 1 for impact parameters up to R_0, the mean probability in this range is 0.5, so that the charge transfer cross section is

$$\sigma = \tfrac{1}{2}\pi R_0^2. \tag{16}$$

Supposing the active electron to be in an s orbital, denoting the energy to remove it by I and taking the radial wave function to be of the form $r^\gamma \exp(-\alpha r)$ with

$$\left.\begin{aligned} \gamma &= (m_e e^4/2\hbar^2 I)^{1/2} && \text{[if electron in Coulomb field]} \\ &= 0 && \text{[otherwise]} \end{aligned}\right\} \tag{17}$$

$$\alpha = (2m_e I/\hbar^2)^{1/2}. \tag{18}$$

Firsov showed that R_0 satisfies the approximate equation

$$\alpha R_0 - (2\gamma - \tfrac{1}{2}) \ln \alpha R_0 - \frac{\gamma - \frac{1}{8}}{\alpha R_0} = \ln A + \ln \left[\frac{(2\pi^3)^{1/2}\alpha\hbar}{(2\gamma)!\, m_e v}\right] \tag{19}$$

where A is a normalizing factor (put equal to unity for singly charged ions) and v is the velocity of relative motion.

The cross section obtained for the H^+H reaction from Firsov's theory+ is compared with the laboratory data of Fite *et al.* (1958) in Fig. 4.

† *Slow* signifies that the velocity of relative motion is much less than the electron orbital velocity.

+ Acknowledgement is made to Mr. Antony Lee for carrying out the computations from (19) described in this section.

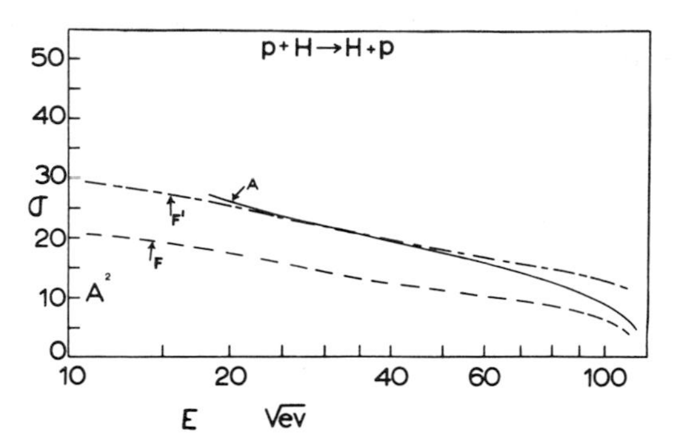

FIG. 4. H^+H resonance charge transfer: A (Fite *et al.*, 1958); F, Firsov formula (with A unity); F′, Firsov formula (with ln A unity).

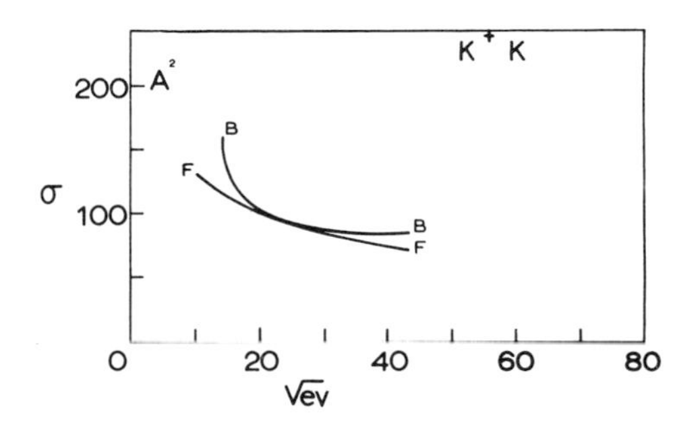

FIG. 5. K^+K resonance charge transfer: B (Bydin and Bukhteev, 1960); F, Firsov formula (with A unity).

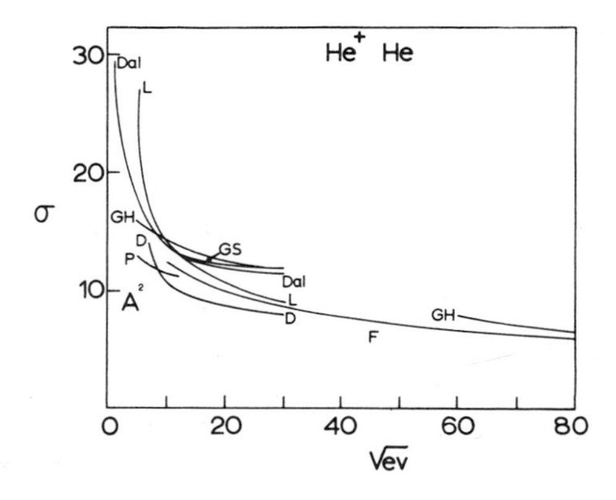

FIG. 6, He^+He resonance charge transfer: GH (Gilbody and Hasted, 1956); GS (Ghosh and Sheridan, 1957); D (Dillon *et al.*, 1955); Da (Dalgarno, 1958); P (Potter, 1954); L (Lindholm, 1960); F, Firsov formula (with A unity).

The agreement is not very good. It is improved if ln A rather than A is put equal to unity.

Figures 5-8 give the Firsov and experimental cross sections for the K^+K, He^+He, Ne^+Ne, and Ar^+Ar reactions. The last two do not involve s orbitals (as assumed in the theory), but this does not seem to make the accord any less satisfactory than it is in the other cases.

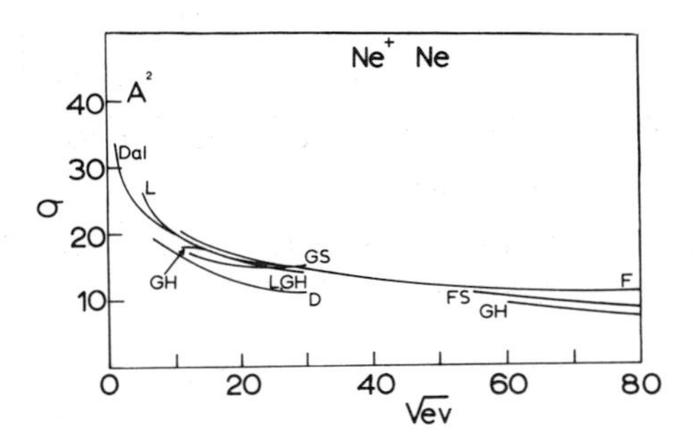

FIG. 7. Ne^+Ne resonance charge transfer: key as for Fig. 6 with FS (Flaks and Solov'ev, 1958).

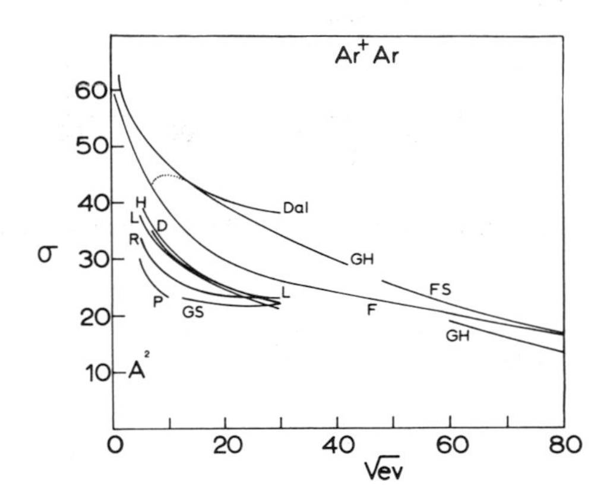

FIG. 8. Ar^+Ar resonance charge transfer: key as for Fig. 6 with R (Rostagni, 1935); H (Hasted, 1951).

Firsov's theory may be applied to negative ion charge transfer. For the H^-H reaction the agreement with the results of the detailed quantal calculation of Dalgarno and McDowell (1956) is essentially complete.

One of the important predictions of Firsov's theory is that the cross section σ is largely controlled by the ionization potential I provided the velocity of relative motion v is not extremely high. The prediction has strong experimental support even for double and triple charge transfer reactions (which the theory was not designed to cover). Arbitrarily taking v to be 10^7 cm/sec we have plotted σ vs. I over two decades in Fig. 9.

It is shown in Chapter 16 that the laboratory data on the mobility of an atomic ion moving in its parent gas can be analyzed to yield charge transfer cross sections. Table II compares some cross sections

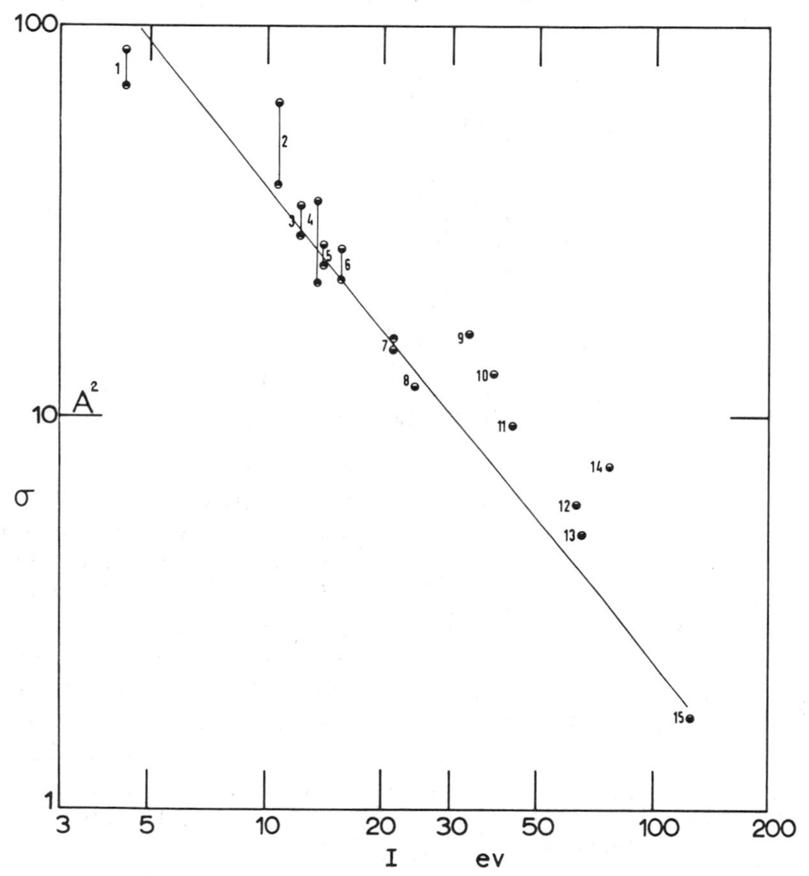

FIG. 9. Resonance charge transfer cross section at velocity of relative motion v of 10^7 cm/sec plotted against ionization energy I of atom: ⊖, experiment; ◓, Firsov's formula (with A unity). 1, K^+K (Bydin and Bukhteev, 1960); 2, Hg^+Hg (Palyukh and Sena, 1950) extrapolated; 3, Xe^+Xe (Flaks and Solov'ev, 1958); 4, H^+H, Fite *et al.* (1958); 5, Kr^+Kr (Flaks and Solov'ev, 1958); 6, Ar^+Ar (Flaks and Solov'ev, 1958; Gilbody and Hasted, 1956); 7, Ne^+Ne (Gilbody and Hasted, 1956); 8, He^+He (Gilbody and Hasted, 1956); 9, $Xe^{++}Xe$ (Flaks and Solov'ev, 1958) extrapolated; 10, $Kr^{++}Kr$ (Flaks and Solov'ev, 1958) extrapolated; 11, $Ar^{++}Ar$ (Flaks and Solov'ev, 1958) extrapolated; 13, $Xe^{3+}Xe$ (Hasted, unpublished data); 14, $Kr^{3+}Kr$ (Flaks and Filippenko, 1960) extrapolated; 15, $Ne^{3+}Ne$ (Flaks and Filippenko, 1960) extrapolated.

derived in this way by Dalgarno (1958) with the corresponding values given by Firsov's theory. Clearly Firsov's theory yields cross sections which are somewhat smaller at low energies than those suggested by the mobilities.

TABLE II

COMPARISON OF RESONANCE CHARGE TRANSFER CROSS SECTIONS σ CALCULATED FROM FIRSOV'S THEORY[a] WITH THOSE DEDUCED FROM MOBILITIES BY DALGARNO

	H	D	He	Ne	Ar	Kr	Xe
Firsov (10^{-16} cm^2)	53	57	30	41	65	78	97
Dalgarno (10^{-16} cm^2)	60	70	36	41	74	91	109

[a] With A taken to be unity. Impact energy 0.1 ev

Fetisov and Firsov (1960) have extended Firsov's theory to cover reactions involving doubly charged ions. They took the wave function to be of the form exp $[-\alpha(r_1 + r_2)]$ where

$$\alpha = \frac{1}{\hbar}\{2m_e(I_1 + I_2)\}^{1/2}, \tag{20}$$

I_1 and I_2 being the first and second ionization potentials. Substituting approximations to the corresponding *LCAO* energies in (181) of Chapter 14 they hence found the distance R_0 already defined. Figure 10 compares the calculated values of $\alpha^2\sigma$ with laboratory data on the inert gases obtained by Flaks and Solov'ev (1958) and by Hasted and his co-workers (unpublished). It is to be noted that the measured cross sections correspond to a combination of single and double charge transfer:

$$\sigma = {}_{20}\sigma_{02} + \tfrac{1}{2}\,{}_{20}\sigma_{11}. \tag{21}$$

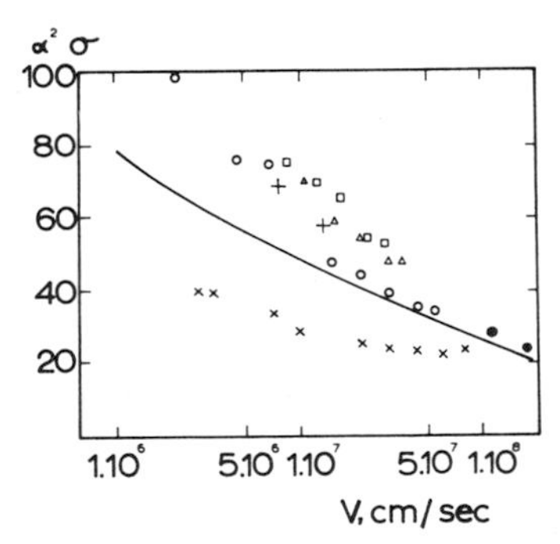

FIG. 10. Comparison of calculated results of Fetisov and Firsov (1960) on $\alpha^2\sigma$ for resonance 20/02 reactions with experimental data of Flaks and Solov'ev (1958) (×, Ne; ○, Ar; △, Kr; □, Xe) and of Hasted (unpublished) (+, He; ●, Ar).

Charge transfer reactions involving homonuclear diatomic molecules, $X_2^+X_2$ and $X_2^-X_2$, are only in energy resonance if the equilibrium nuclear separation is the same in the two charge states. Even in this case the higher vibrational levels do not correspond. Hence, the measured charge transfer cross section would not be as large as might at first be expected, should a considerable fraction of the molecular ions in the beam used be vibrationally excited (as is likely).

The equilibrium nuclear separations in the two charge states are in general different. Application of the Franck-Condon principle indicates that the energy defect of the reaction ranges from zero to E_4-E_1 if the ion is initially in the highest vibrational level which can be reached in this way (Fig. 11). The type of cross section function which arises may be seen from the experimental data on the $H_2^+H_2$, $O_2^+O_2$, and $O_2^-O_2$ reactions (Hasted and Smith, 1956; Gilbody and Hasted, 1956; Lindholm, 1960). Apparent discrepancies between sets of data may be due to the molecular ions studied being vibrationally excited to different degrees.

Using an approximate form of (181) of Chapter 14, Gurnee and Magee (1957) have carried out calculations on resonance charge transfer in the inert gases and in some diatomic gases. In general their cross sections are rather higher than those obtained from Firsov's theory.

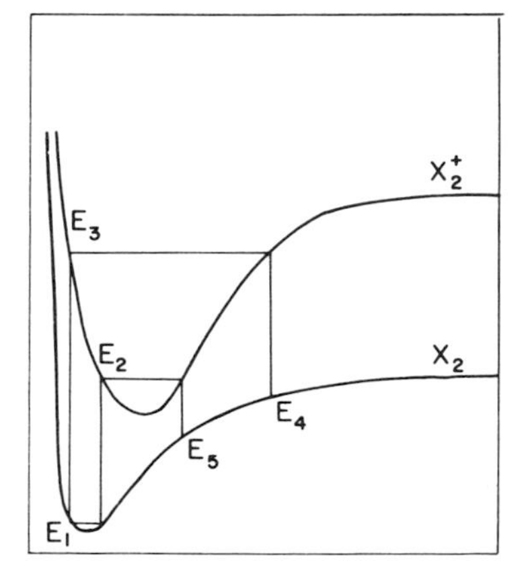

FIG. 11. Potential energy curves showing transitions between a diatomic molecule and its singly charged ion in the case where the equilibrium nuclear separations are different.

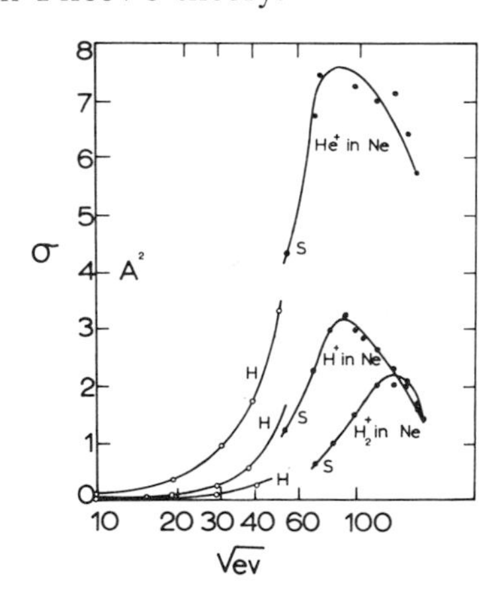

FIG. 12. Charge transfer cross section functions for H^+, H_2^+, and He^+ in neon: (H, Hasted; S, Stedeford) Stedeford and Hasted (1955).

5 *Charge Transfer Reactions between Unlike Systems*

Typical cross section functions for single charge transfer 10/01 between unlike systems are shown in Fig. 12.

The velocity of relative motion v_{m} at which a single charge transfer cross section function passes through its maximum is generally determined by the energy defect ΔE through the approximate relation

$$v_{\mathrm{m}}/a = |\Delta E|/h \tag{22}$$

where a is a length of the order 7×10^{-8} cm (Massey, 1949; Hasted, 1952, 1960). The region where $v \ll v_{\mathrm{m}}$ is termed the *adiabatic region*. In it the cross section function has the form

$$_{10}\sigma_{01} = C \exp(-a|\Delta E|/4hv), \tag{23}$$

C being a constant (dependent on the particular reaction).

Laboratory data may be complicated by the presence of metastable ions in the beam and by the formation of excited collision products.

An example of the anomalies which may arise from metastable ions is shown in Fig. 13, where a cross section function obtained with normal

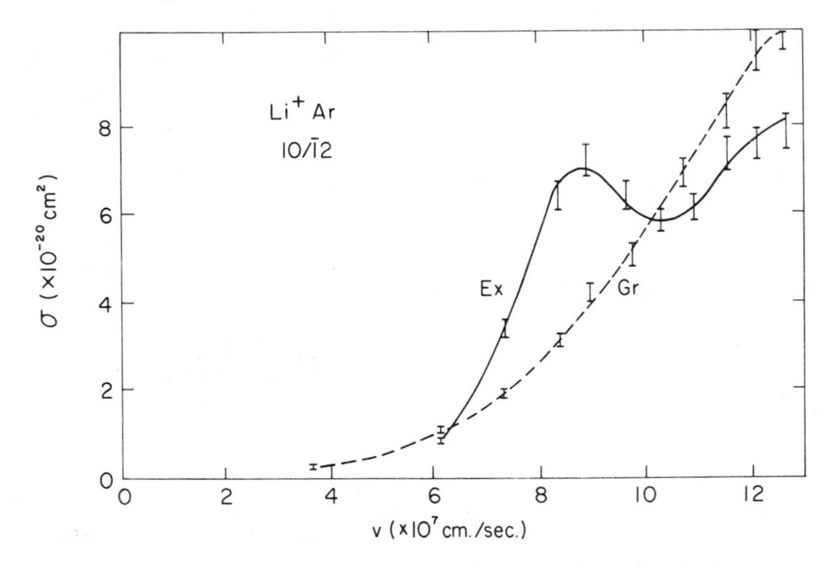

FIG. 13. Comparison of Li^+Ar $10/\bar{1}2$ cross section functions for ground state and partly excited Li^+ ion beams (Fogel *et al.*, 1959).

Li^+ ions from a surface ionization source is compared with the corresponding cross section function obtained with Li^+ ions from an electron impact source (Fogel *et al.*, 1959). The detection of metastable ions by retardation techniques (Hagstrum, 1960) may be of value in future studies.

Reactions in which the collision products are excited have been investigated using optical methods. It has long been known that slow collisions between He^+ ions and Cd atoms produce Cd II lines, the strongest being those resulting from 10/01′ reactions in close energy balance. The 10/0′1 reaction between 4-kev He^+ ions and Ne atoms leading to He ($1s3p\,^1P$) has been reported by Kondratiev (1960). Other 10/0′1 reactions have been reported by Carleton and Lawrence (1957, 1958) and by Sluyters *et al.* (1958, 1959).

Extensive studies of charge transfer between ions and molecules, including N_2, O_2, CO, H_2S, NH_3, and C_2H_5OH, have been carried out by Lindholm (1953, 1954, 1959, 1960).

If the equilibrium nuclear distances in the two charged states are not the same, ions formed in the source by the impact of electrons with neutral molecules will be vibrationally excited and the recombination energies which are released in vertical transitions, if the ions revert to molecules in charge transfer collisions, may have a considerable spread (Fig. 11). The experimental data which Stedeford and Hasted (1955) obtained in their studies with H_2^+ beams in inert gases are consistent with this picture.

Dissociative charge transfer may result from transitions to a bound state if the dissociation limit can be reached without serious violation of the Franck-Condon principle; or they may result from transitions to an unbound state. The observation of collision products with kinetic energy has not yet proved possible in the mass spectrometers employed in such experiments.

Fogel and co-workers (1956, 1957) have applied rule (22) to the $00/\bar{1}1$ and 10/12 reactions. They found the parameter a did not have the typical value (7×10^{-8} cm). The reactions are, however, in a special class. Owing to the long-range Coulomb forces which enter, the effective energy difference is not the energy difference $\Delta E(\infty)$ at infinite nuclear separation but is the energy difference $E(R_{\text{eff}})$ at a separation R_{eff} determined by the empirical equation

$$R_{\text{eff}} = 0.92\,(Z_1 + Z_2)^{1/3} \quad \text{(atomic units)} \tag{24}$$

in which Z_1 and Z_2 are the atomic numbers of the colliding systems (Hasted, 1961). For the $(n-1)0/\bar{1}n$ reaction, this energy difference is given by

$$\Delta E(R_{\text{eff}}) = \left\{ \Delta E(\infty) - \frac{27.2}{R_{\text{eff}}} \right\} \text{ ev.} \tag{25}$$

If the values of $\Delta E(R_{\text{eff}})$ obtained from (25) are used in (22) it is found that the parameter a becomes about 7×10^{-8} cm for 00/11 reactions (as it is for 10/01 reactions) and about 3.5×10^{-8} cm for $10/\bar{1}2$ reactions. Drukharev (1960) has shown that

$$a \simeq h/q_{\rm m} \tag{26}$$

where $q_{\rm m}$ is the momentum transferred. The fact that the value of a in the case of two-electron capture is about half the value of a in the case of one-electron capture is thus understandable.

On taking $R_{\rm eff}$ from (24) and using the experimental polarizabilities α, we can calculate the polarization energy

$$E_{\rm polar} = -\alpha e^2/2R_{\rm eff}^4. \tag{27}$$

If allowance for this is made, the conformity of single-charge transfer data to the maximum rule (22) and to the rate of rise rule (23) is improved (Hasted, 1961).

For a number of processes (e.g., Ne^+CO, Ar^+Kr, Ne^+Ar, C^+Kr, $Xe^+C_2H_4$) the cross sections in the adiabatic region are larger than would be expected. Some of these anomalies may be due to single or multiple crossovers (§ 6). In polyatomic collisions such as $Xe^+C_2H_4$ a complex may be formed which ultimately dissociates into components (Burton and Magee, 1952).

6 *Crossovers*

Some cross-section functions are seemingly anomalous in the low-energy region (cf. Fig. 14). This may be due to a pseudocrossing of potential energy curves. Such pseudocrossings or crossovers are common in cases where both the collision products or both the reactants are charged (but are not confined to these cases). In an $n0/(n-1)\,1$ reaction which is exothermic by ΔE ev the nuclear separation at the crossover is approximately

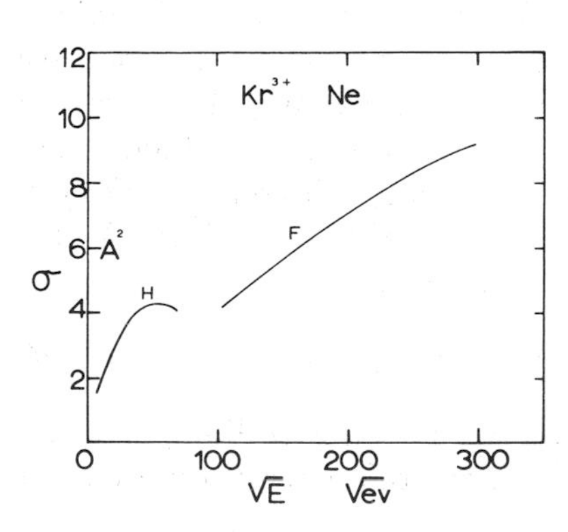

FIG. 14. $Kr^{3+}Ne$ 30/21 cross section function: H (Hasted *et al.*, 1960); F (Flaks and Filippenko, 1960).

$$R_x = [27.2(n-1)/\Delta E] \text{ atomic units.} \tag{28}$$

The Landau-Zener formula gives the maximum cross section to be

$$\sigma_{\rm max} = 1.4 p R_x^2 \tag{29}$$

where p is a certain statistical weighting factor. Furthermore, it expresses in terms of known quantities the parameter

$$\xi = \Delta U(R_x)^4/E_{\max} \tag{30}$$

where $\Delta U(R_x)$ is the energy separation at the crossing and $E_{\max}$ is the impact energy at which $\sigma_{\max}$ occurs (Boyd and Moiseiwitsch, 1957). The Landau-Zener formula is now believed to be seriously in error—especially if other than s orbitals are involved (Chapter 14, § 3.3). Nevertheless, it is to be noted that the measured maximum cross sections are within the limits of (29); and that the values of $\Delta U(R_x)$ obtained by substituting the observed $E_{\max}$ in (30) are of the order suggested by quantal calculations. The complications that arise from the presence of excited systems (Hasted *et al.* 1960) and the paucity of the data make further generalization unprofitable at present.

7 *Negative Ions and Collisional Detachment*

The detachment of electrons from negative ions in collisions with atoms or molecules

$$A^- + B \rightarrow A + e + B \ (\bar{1}0/0e0) \tag{31}$$

may be studied experimentally by the collection of the electrons. It is unnecessary to separate these from negative ions in the energy region below 2 kev if the system B does not readily form a negative ion. Hasted (1952, 1954) and Dukel'skii and Zandberg (1954) have measured detachment cross sections by the straightforward slow-charge collection technique.

Bailey (1960) has designed electrodes for cases in which charge transfer $\bar{1}0/0\bar{1}$ is possible. The primary beam is surrounded by two cylindrical grids, the inner one to screen it electrically, the outer one to accelerate the negative particles. Plates are arranged outside this as in the vanes of a paddle steamer. Potentials varying at a frequency of several megacycles are applied to alternate vanes to filter out the electrons. In their study of H^-H collisions Stebbings *et al.* (1960) separated the electrons from the negative ions with crossed electric and magnetic fields.

The total electron production cross section function for H^-H collisions at impact energies of a few kev and higher is presumably to be attributed to normal detachment $\bar{1}0/0e0$ with contributions from

$\bar{1}0/0e1e$ and $\bar{1}0/1ee0$. It lies above but is of the same form as the function obtained by McDowell and Peach (1959) using the Born approximation. At low (less than about 1 kev) energies the measured function is rather similar to a resonance charge transfer cross-section function (Fig. 15). A pseudocrossover is probably involved enhancing both $\bar{1}0/0e0$ and $\bar{1}0/\bar{1}1e$. Bates and Massey (1954) have discussed the mechanism (cf. Chapter 14, § 3.4).

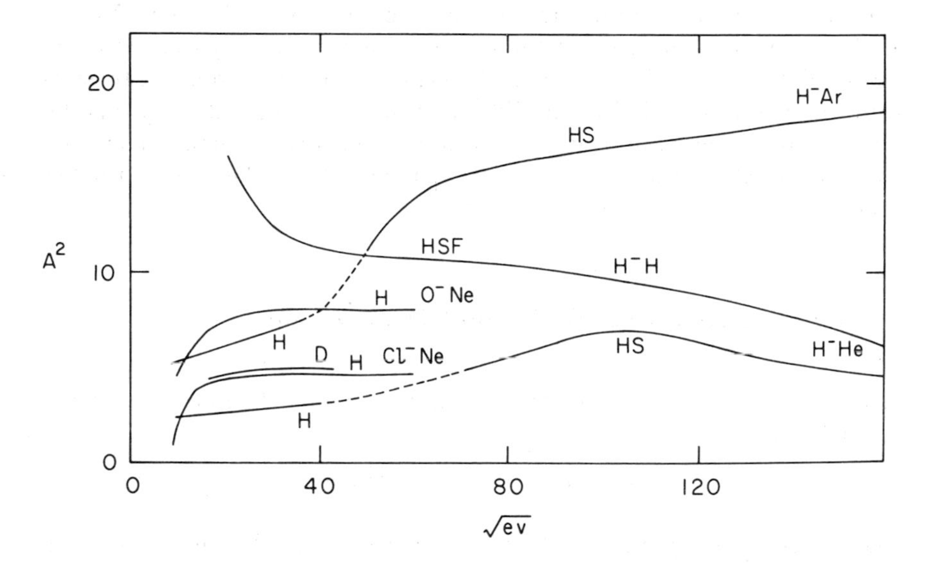

FIG. 15. Collisional detachment cross section functions: H (Hasted, 1952); D (Dillon *et al.*, 1955); HS (Stedeford and Hasted, 1955); HSF (Hummer *et al.*, 1960).

Detachment cross-section functions (Fig. 15) exhibit flat maxima at an impact energy of about 2 kev or rather lower (Hasted, 1952; Bydin and Dukel'skii, 1957). It has been observed (Gilbody and Hasted, 1958) that the maximum value of all heavy negative ion detachment cross sections at present known is proportional to the reduced mass of the colliding systems.

Fogel *et al.* (1957) have reported that the cross section for double detachment, $\bar{1}0/1ee0$ or $\bar{1}0/1e\bar{1}$, though much less than that for single detachment $\bar{1}0/0e0$ is not as small as might be expected.

Negative ion charge transfer has been studied by Muschlitz (1957). The $O_2^-O_2$ reaction behaves in the normal symmetrical resonance manner.

References

Allison, S. K. (1958) *Rev. mod. Phys.* **10**, 1137.
Bailey, T. L. (1960). Thirteenth Gaseous Electronics Conference, Monterey, California.
Barnard, G. P. (1952) "Modern Mass Spectrometry." Institute of Physics, London.
Bates, D. R., and Massey, H. S. W. (1954) *Phil. Mag.* [7] **45**, 173.
Bohm, D., Burhop, E. H. S., and Massey, H. S. W. (1949) *In* "Gaseous Discharges in Magnetic Fields" (Guthrie and Wakerling, eds.), Chapt. II. McGraw-Hill, New York.
Boyd, T. J. M., and Moiseiwitsch, B. L. (1957) *Proc. Phys. Soc.* **A60**, 809.
Burton, M., and Magee, J. L. (1952) *J. phys. Chem.* **56**, 842.
Bydin, Y. F., and Bukhteev, A. M. (1960) *Soviet Phys.—JETP* **4**, 10.
Bydin, Y. F., and Dukel'skii, V. M. (1957) *Zh. eksper. teor. Fiz.* **31**, 474.
Carleton, N. P., and Lawrence, T. R. (1957) *Phys. Rev.* **107**, 110.
Carleton, N. P., and Lawrence, T. R. (1958) *Phys. Rev.* **109**, 1159.
Chalklin, L. P., and Fremlin, J. H. (1960) *Proc. Phys. Soc.* **75**, 850.
Cosslett, V. E. (1944) "Electron Optics," p. 56. Oxford Univ. Press, London and New York.
Curran, R. K., and Donahue, T. M. (1960) *Phys. Rev.* **118**, 1233.
Curran, R. K., and Donahue, T. M., and Kasner, W. H. (1959) *Phys. Rev.* **114**, 490.
Dalgarno, A. (1958) *Phil. Trans.* **A250**, 426.
Dalgarno, A., and McDowell, M. R. C. (1956) *Proc. Phys. Soc.* **A69**, 615.
Dillon, J. A, Sheridan, W. F., Edwards, H. D., and Ghosh, S. N. (1955) *J. chem. Phys.* **26**, 776.
Donahue, T. M., and Hushfar, F. (1959) *Phys. Rev. Letters* **3**, 470.
Donahue, T. M., and Hushfar, F. (1960) *Nature* **186**, 1038.
Drukharev, G. F. (1960) *Soviet Phys.—JETP* **10**, 603.
Du Bridge, L. A., and Brown, H. (1933) *Rev. sci. Instrum.* **4**, 532.
Dukel'skii, V. M., and Zandberg, Y. (1954) *Soviet Phys.—Doklady* **99**, 947.
Fedorenko, N. V. (1959) *Soviet Phys.—Uspekhi* **68**, 481.
Fetisov, I. K., and Firsov, O. B. (1959) *Zh. eksper. teor. Fiz.* **37**, 95.
Fetisov, I. K., and Firsov, O. B. (1960) *Zh. eksper. teor. Fiz.* **37**, 67.
Firsov, O. B. (1951) *Zh. eksper. teor. Fiz.* **21**, 1001.
Fite, W. L., Brackmann, R. T., and Snow, W. R. (1958). *Phys. Rev.* **112**, 1161; **119**, 663.
Flaks, I. P., and Filippenko, L. G. (1960) *Soviet Phys.—JETP* **4**, 1005.
Flaks, I. P., and Solov'ev, E. S. (1958) *Soviet Phys.—JETP* **3**, 564, 577.
Fogel, Ya. M., Mitin, R. V., and Koval, A. G. (1956) *Zh. eksper. teor. Fiz.* **31**, 397.
Fogel, Ya. M., Ankudinov, V. A., and Slabospitskii, R. E. (1957) *Zh. eksper. teor. Fiz.* **32**, 453.
Fogel, Ya. M., Kozlov, V. F., Kalmykov, A. A., and Muratov, V. I. (1959) *Zh. eksper. teor. Fiz.* **36**(9), 929.
Ghosh, S. N., and Sheridan, W. F. (1957) *J. chem. Phys.* **26**, 480.
Gilbody, H. B. (1956) Ph. D. Thesis, University of London, London.
Gilbody, H. B., and Hasted, J. B. (1956) *Proc. Roy. Soc.* **A238**, 334.
Gilbody, H. B., and Hasted, J. B. (1958) *Proc. Phys. Soc.* **72**, 393.
Goldmann, F. (1931) *Ann. Phys. (Leipzig)* [5] **10**, 460.
Gurnee, E. F., and Magee, J. L. (1957) *J. chem. Phys.* **26**, 1237.
Gustafsson, E., and Lindholm, E. (1958) Private communication.
Hagstrum, H. D. (1951) *Rev. mod. Phys.* **23**, 185.
Hagstrum, H. D. (1960) *J. appl. Phys.* **31**, 897.
Hasted, J. B. (1951) *Proc. Roy. Soc.* **A205**, 421.

Hasted, J. B. (1952) *Proc. Roy. Soc.* **A212**, 235.
Hasted, J. B. (1954) *Proc. Roy. Soc.* **A222**, 74.
Hasted, J. B. (1959) *J. appl. Phys.* **30**, 25.
Hasted, J. B. (1960) *Advances in Electronics* **13**, 1.
Hasted, J. B. (1961) *Proc. Phys. Soc.* (in press).
Hasted, J. B., and Smith, R. A. (1956) *Proc. Roy. Soc.* **A235**, 349.
Hasted, J. B., Scott, J. T., and Chong, A. Y. J. (1960) *In* "Proceedings of the Fourth International Conference on Ionization Phenomena in Gases, Uppsala, 1959" (N. R. Nilsson, ed.). North-Holland, Amsterdam.
Hummer, D. G., Stebbings, R. F., and Fite, W. L. (1960) *Phys. Rev.* **119**, 668.
Keene, J. P. (1949) *Phil. Mag.* [7] **40**, 369.
Kondratiev, V. N. (1960) Second Humphrey Davy Lecture, Royal Institute of Chemistry, London.
Lindholm, E. (1953) *Proc. Phys. Soc.* **A66**, 1068.
Lindholm, E. (1954) *Ark. Fys.* **8**, 257, 433.
Lindholm, E. (1959) Private communication in respect of ethanol.
Lindholm, E. (1960) *Ark. Fys.* **18**, 219.
McDowell, M. R. C., and Peach, G. (1959) *Proc. Phys. Soc.* **74**, 463.
McGowan, W., and Kerwin, L. (1960) *Can. J. Phys.* **38**, 642.
Massey, H. S. W. (1949) *Rep. Prog. Phys.* **12**, 248.
Massey, H. S. W., and Burhop, E. H. S. (1952) "Electronic and Ionic Impact Phenomena." Oxford Univ. Press, London and New York.
Melton, C. E., and Wells, G. F. (1957) *J. chem. Phys.* **27**, 1132.
Moe, D. E. (1956) *Phys. Rev.* **104**, 694.
Muschlitz, E. E. (1957) *J. appl. Phys.* **28**, 1414.
Palyukh, B. M., and Sena, L. A. (1950) *Zh. eksper. teor. Fiz.* **20**, 481.
Pierce, J. R. (1954) "Theory and Design of Electron Beams." Van Nostrand, Princeton, New Jersey.
Potter, R. F. (1954) *J. chem. Phys.* **22**, 974.
Rostagni, A. (1935) *Nuovo Cimento* **12**, 134.
Sluyters, T. J. M., de Haas, E., and Kistemaker, J. (1958) *Rev. sci. Instrum.* **29**, 597.
Sluyters, T. J. M., de Haas, E., and Kistemaker, J. (1959) *Physica* **25**, 182.
Stebbings, R. F., Fite, W. L., and Hummer, D. G. (1960) *J. chem. Phys.* **33**, 1226.
Stedeford, J. B. H., and Hasted, J. B. (1955) *Proc. Roy. Soc.* **A227**, 466.
Sweetman, D. R. (1959) *Phys. Rev. Letters* **3**, 425.
Taylor, J. B. (1929) *Z. Phys.* **57**, 242.
Varney, R. N. (1936) *Phys. Rev.* **50**, 159.
Weiss, R. (1961) *Rev. sci Instrum.* **32**, 397.
Willmore, A. P. J. (1955) Thesis, University of London, London.
Wolf, F. (1936) *Ann. Phys.* (*Leipzig*) [5] **23**, 185, 627; **25**, 527, 737.
Wolf, F. (1937) *Ann. Phys.* (*Leipzig*) [5] **27**, 543; **28**, 361.
Wolf, F. (1938) *Ann. Phys.* (*Leipzig*) [5] **29**, 33; **30**, 313.
Wolf, F. (1939) *Ann. Phys.* (*Leipzig*) [5] **34**, 341.

19.

Electron Capture and Loss at High Energies *

Samuel K. Allison and M. Garcia-Munoz

* This work supported in part by The United States Atomic Energy Commission.

1 Introduction

The discussion in this chapter is confined to collisions in which the number of electrons bound to a high-energy† atomic projectile is changed. Investigations on the final state of the target are not considered.

2 General References and Previous Collections of Data

The first quantitative experimental data on the probability of charge changing collisions and on equilibrated charges were obtained by Henderson (1922) in connection with the singly charged component He^+ which exists among α-particles emitted from radioactive sources. This was followed by an important paper by Rutherford (1924), and further contributions by Kapitza (1924), Henderson (1925), Briggs (1927), and Jacobsen (1927). An account of this early work on α-particles ranging in energy from 406 to 7680 kev appears in the book "Radiations from Radioactive Substances" (Rutherford *et al.*, 1931).

The first quantitative work with artificially produced projectiles (protons) in our energy range seems to have been done by Bartels (1930) and reviews of this and other early work have been written by Rüchardt (1933) and by Geiger (1933). These articles appeared in the "Handbuch der Physik."

Several postwar reviews whose main emphasis is on experimental techniques and results have appeared. The earliest of these is in "Electronic and Ionic Impact Phenomena" (Massey and Burhop, 1952).

† The lower limit of the kinetic energy is rather arbitrarily taken to be 0.2 kev.

Charge changing collisions were also discussed in an article by Allison and Warshaw (1953). A complete summary of the then existing data on hydrogen and helium projectiles in our energy range has been prepared by Allison (1958c).

The theory and experiment of charge changing collisions have been discussed at several international conferences, such as those at Venice (1957), New York City (1958), Gatlinburg (1958), and Uppsala (1959).

3 *Mathematical Description of Charge Changing Probabilities; Notation*

There is a certain probability that a projectile having an electric charge of ie where i is a positive or negative integer or zero, and e is the negative of the electronic charge, will scatter into an element of solid angle $d\Omega$, having suffered a charge change from ie to fe. We can regard this probability as depending on the angle θ, which the scattered particle makes with its previous forward direction and we can express it by the differential cross section

$$d\sigma_{if}(\theta)/d\Omega. \tag{1}$$

The total cross section is then given by

$$\sigma_{if} = 2\pi \int_0^{\pi} [d\sigma_{if}(\theta)/d\Omega] \sin \theta d\theta. \tag{2}$$

Hasted designates by $_{ii'}\sigma_{ff'}$ the cross section for the process

$$P(ie) + T(i'e) \rightarrow P(fe) + T(f'e) \tag{3}$$

in which the charges on the projectile and target change from ie and $i'e$ to fe and $f'e$, respectively. Since this chapter is concerned with events in which i' and f' are not measured, the simpler form $_i\sigma_f$ or σ_{if} seems sufficient.

For a large category of charge changing events, namely, those in which the number of electrons lost or gained is small, essentially the entire contribution to the integral of (2) comes from a very small angular range in θ, near zero. A rough classical picture of a case in which one electron is stripped off the projectile indicates that the maximum angular deviation would be approximately $2m/M$ where m is the electronic, M an atomic, mass. Thus, the entire contribution to (2) would be from values of θ less than 0.001. We shall see an experimental confirmation of this in that $d\sigma_{12}/d\Omega$ for He^+ has been measured as a function of angle by Everhart and co-workers. Computation of σ_{12} by (2) yields values of the total cross section which agree with those measured by beam attenuation methods in which the defining apertures limited the beam to an angular spread comparable to this classical limit. A similar agreement was found in the capture cross section σ_{10} measured in both ways.

It has unfortunately become customary in the case of the diatomic elementary gases as target particles, to express cross sections per atom of gas by dividing by 2 the measured cross sections per molecule. This practice began with the early postwar work with hydrogen target gas, when it was wrongly anticipated that the proton projectile cross sections would be compared with theoretical calculations of protons capturing electrons from hydrogen atoms. The system cannot reasonably be extended to target gases with molecules such as CO, CH_4, NH_3. In this chapter we shall give cross sections per atom for the noble gases and the elementary gases with diatomic molecules, but cross sections per molecule for other cases.

A frequent experimental situation in work on charge changing collisions is schematically illustrated in Fig. 1. A unidirectional (at least

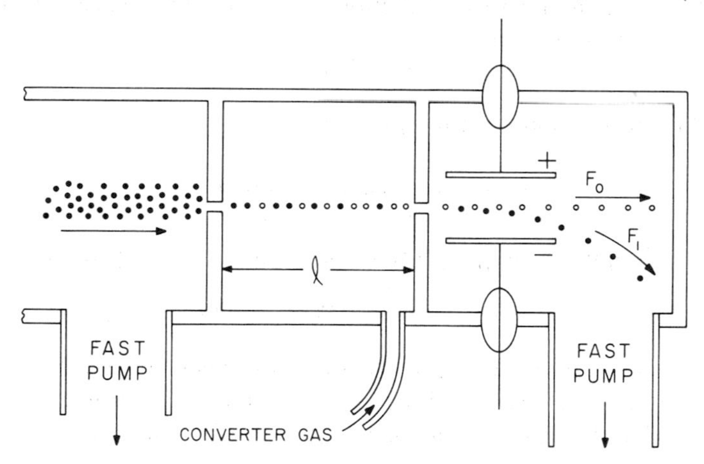

FIG. 1. Schematic experimental equipment for charge equilibration of an ionic beam.

within the limits mentioned above) monoenergetic ionic or atomic beam has been prepared and is moving in high vacuum† toward a

† In this context, "high vacuum" means a pressure low enough so that the mean free path for a charge changing collision is large (by a factor of 10, say) compared to the path length in vacuum which is limited by the dimensions of the apparatus. A cross section of 200×10^{-17} cm^2 is large for charge changing. If the dimensions of the apparatus are comparable to 1 meter, the condition is met by a pressure of $\lesssim 5 \times 10^{-6}$ mm Hg.

compartment into which a target gas may be introduced. The "prepared" beam has a known composition; in most cases it consists entirely of atoms or of ions all of the same charge. The beam passes through the compartment, called the converter cell or the equilibration chamber, and emerges into high vacuum, where it is deflected electrically or magnetically and thus analyzed into its charged components.

Notation

F_z Let the charge composition of the prepared beam be expressed by a set of fractions F_z where z is a positive or negative integer or zero, and each F_z corresponds to the fraction of particles in the beam with charge ze.

F_i Similarly, let the charge composition of the beam emerging from the converter cell be specified by a set of fractions F_i.

$F_{i\infty}$ As the pressure of gas in the converter cell is raised, each F_i will eventually reach a plateau, at a value $F_{i\infty}$ characteristic of the equilibrated beam.

$P(z, i)$ and $N(z, i)$ are certain functions of the cross sections which appear as coefficients of exponentials in the expression for F_i in the case of a beam in which all the particles had charge ze initially.

p Pressure in microns of mercury.

l Length of path in centimeters in the converter cell or in a compartment where cross sections are measured.

T Absolute temperature.

R Gas constant, 6.24×10^7 $\mu \times \text{cm}^3/\text{mol}\ ^\circ\text{C}$.

ξ Number of atoms per molecule for the diatomic elementary gases (otherwise unity).

π Number of atoms per cm² traversed by the beam in noble and diatomic elementary gases (otherwise, number of molecules per cm²).

We have

$$\pi = \frac{A\xi}{RT}(pl) \tag{4}$$

which at 20°C may be written

$$\pi = 3.30 \times 10^{13}\xi\,(pl). \tag{5}$$

The number of charged components in a projectile beam which play any significant role in the phenomena observed at a fixed particle velocity is limited. If the charge changing collisions of a prepared beam of protons ($F_z = F_1 = 1$) in the kinetic energy range 100-500 kev are under investigation, two components are sufficient since neutral atoms and protons are the observable constituents (i = 0, 1). The charge changing cross sections for a two component system in which the components differ by one electronic charge are characterized by σ_{if} with $f = i \pm 1$.

For a two-component system with $i = 0, 1$, consider the change dF_0 produced in the neutral fraction of the beam emerging from the converter cell (Fig. 1) as the pressure is raised by dp. The rise in pressure will cause a corresponding increase $d\pi$ in the number of target particles per cm^2; thus,

$$dF_0 = -\,dF_1 = (-\,F_0\sigma_{01} + F_1\sigma_{10})\,d\pi. \tag{6}$$

In a form which can conveniently be generalized to the three-component case discussed below, the solution may be summarized in the expression

$$F_i = F_{i\infty} + P(z, i)\exp\left[-\,\pi(\sigma_{01} + \sigma_{10})\right]; \qquad i = 0 \text{ or } 1. \tag{7}$$

The integration constants $F_{i\infty}$ are

$$F_{0\infty} = \sigma_{10}/(\sigma_{01} + \sigma_{10}); \qquad F_{1\infty} = \sigma_{01}/(\sigma_{01} + \sigma_{10}). \tag{8}$$

The two important cases (a) where the prepared beam is all atomic ($F_z = F_0 = 1$) and (b) where it is all ionic ($F_z = F_1 = 1$) lead to the values of $P(z, i)$ shown in Table I.

TABLE I

VALUES OF $P(z, i)$ IN EQ. (7) FOR TWO-COMPONENT SYSTEM

Condition of the initial beam	Values of $P(z, i)$	
	$i = 0$	$i = 1$
All neutral	$P(0, 0) = F_{1\infty}$	$P(0, 1) = -\,F_{1\infty}$
All charged	$P(1, 0) = -\,F_{0\infty}$	$P(1, 1) = F_{0\infty}$

At kinetic energies above 1000 kev, the charge changing phenomena in a prepared projectile beam of He^+ or He^{++} are describable in terms of these two components. The relevant cross sections are σ_{12} and σ_{21}. The two-component equations, (7) and (8), and the values of $P(z, i)$ easily transform to this case by increasing each numerical subscript or index by unity. Thus, the equations may be quickly modified for the discussion of any two-component system with $f = i \pm 1$.

The charge composition of a three-component system in which $i = 0$, 1, or 2 can be similarly studied in terms of the differential equations

$$dF_0/d\pi = -\,F_0(\sigma_{01} + \sigma_{02} + \sigma_{20}) + F_1(\sigma_{10} - \sigma_{20}) + \sigma_{20} \tag{9}$$

$$dF_1/d\pi = F_0(\sigma_{01} - \sigma_{21}) - F_1(\sigma_{10} + \sigma_{12} + \sigma_{21}) + \sigma_{21} \tag{10}$$

and the relation

$$F_2 = 1 - (F_0 + F_1). \tag{11}$$

It is convenient to introduce new constants which are linear combinations of the cross sections

$$a \equiv -\,(\sigma_{10} + \sigma_{12} + \sigma_{21}) \qquad f \equiv (\sigma_{10} - \sigma_{20})$$

$$b \equiv (\sigma_{01} - \sigma_{21}) \qquad g \equiv -\,(\sigma_{01} + \sigma_{02} + \sigma_{20}) \tag{12}$$

giving

$$dF_0/d\pi = gF_0 + fF_1 + \sigma_{20} \tag{13}$$

$$dF_1/d\pi = bF_0 + aF_1 + \sigma_{21}. \tag{14}$$

The solutions of physical interest are

$$F_i = F_{i\infty} + [P(z, i) \exp(\pi q) + N(z, i) \exp(-\pi q)] \exp(-\tfrac{1}{2}\pi \sum \sigma_{if}) \tag{15}$$

in which

$$q \equiv + \tfrac{1}{2}[(g - a)^2 + 4bf]^{1/2} \tag{16}$$

and

$$\sum \sigma_{if} = -(a + g). \tag{17}$$

From the form of (15) it is clear that the interpretation of $F_{i\infty}$ as the asymptotic limit of $F_i(\pi)$ with increasing π, is consistent with its definition as previously given, provided

$$\tfrac{1}{2} \sum \sigma_{if} > q. \tag{18}$$

This is a necessary condition for solutions of interest, since without it certain F_i's would either go negative or become infinite as π is increased. These equilibrated beam fractions can be expressed in terms of the fundamental cross sections by equating (9) and (10) to zero. They are

$$F_{0\infty} = (f\sigma_{21} - a\sigma_{20})/(ag - bf) \tag{19}$$

$$F_{1\infty} = (b\sigma_{20} - g\sigma_{21})/(ag - bf) \tag{20}$$

$$F_{2\infty} = [\sigma_{20}(a - b) + g(a + \sigma_{21}) - f(b + \sigma_{21})]/(ag - bf). \tag{21}$$

The coefficients $P(z, i)$ and $N(z, i)$ depend in value on the composition of the prepared, incident projectile beam (i.e., on index z) and on the component (i) in the converted beam under consideration. Table II gives their values. For brevity, the symbol

$$s \equiv \tfrac{1}{2}(g - a) \tag{22}$$

is used in the table.

The following useful relations between the P's, N's, and $F_{i\infty}$ values result from the condition that at $\pi = 0$ the beam emerges unchanged from the converter cell and consists entirely of one charge component:

$$F_{i\infty} = -[P(z, i) + N(z, i)]; \quad z \neq i, \tag{23}$$

$$1 - F_{i\infty} = [P(z, i) + N(z, i)]; \quad z = i. \tag{24}$$

The three-component solution outlined above is easily generalized to any such system characterized by values of i which are successive ntegers. It is only necessary to add or subtract an appropriate integer

TABLE II

VALUES OF $P(z, i)$ AND $N(z, i)$ IN EQ. (15) FOR A THREE-COMPONENT SYSTEM

Composition of the original beam	To calculate F_1, the fraction single charged in the converted beam	
100% neutrals; $z = 0$	$P(0, 1) = \frac{1}{2q} [bF_{2\infty} + (b + s - q) F_{1\infty}]$	$N(0, 1) = P(0, 1) - F_{1\infty}$
100% singly charged; $z = 1$	$P(1, 1) = -\frac{1}{2q} [(s - q)(1 - F_{1\infty}) + bF_{0\infty}]$	$N(1, 1) = 1 - F_{1\infty} - P(1, 1)$
100% doubly charged; $z = 2$	$P(2, 1) = \frac{1}{2q} [F_{1\infty} (s - q) - bF_{0\infty}]$	$N(2, 1) = P(2, 1) - F_{2\infty}$
	To calculate F_0, the neutral fraction in the converted beam	
100% of charge ze; $z = 0, 1, 2$	$P(z, 0) = P(z, 1)(s + q)/b$	$N(z, 0) = N(z, 1)(s - q)/b$
	To calculate F_2, the doubly charged fraction in the converted beam	
100% of charge ze; $z = 0, 1, 2$	$P(z, 2) = -P(z, 1)(b + s + q)/b$	$N(z, 2) = -N(z, 1)(b + s - q)/b$

to each numerical index or subscript. Thus, for the hydrogen system in the lower part of our energy range, where the charge states H^-, H^0, H^+, are observed, the integer to be subtracted is 1; for example, (19) becomes

$$F_{\bar{1}\infty} = (f\sigma_{10} - a\sigma_{1\bar{1}})/(ag - bf) \tag{25}$$

with

$$a \equiv -(\sigma_{0\bar{1}} + \sigma_{01} + \sigma_{10}), \tag{26}$$

etc. obtained from (12).

It is useful to consider charge changing cycles, at least in a two-component system where (in the absence of complicating factors such as capture into metastable states) they have a unique interpretation. The cycle consists in the following sequence of events. A projectile particle of charge ie changes charge to fe in a collision. Subsequently, in its path it undergoes the $f \rightarrow i$ type of collision and the cycle is completed. On the average, the distance for the completion of the cycle is the sum of the mean free paths for the two types of collisions. Thus, we may speak of a cross section for cycle completion

$$\sigma_c = \sigma_{if}\sigma_{fi}/(\sigma_{if} + \sigma_{fi}). \tag{27}$$

This concept is useful in discussing the contributions of charge changing collisions to the total stopping power losses incurred in traversing the medium.

4 *Experimental Equipment and Methods*

The experiments on the charge composition of ionic and atomic beams, and on the total cross sections for change of charge when one does not inquire into the changes in the target particles or the detailed electronic configurations of the projectiles, can be carried out with simple apparatus. The measurement of angular distributions of post-collision, charge changed projectiles involves some increase in complexity.

4.1 DETERMINATION OF RATIOS $F_{i\infty}$ AND F_i

4.1.1 *Equilibria in Gases*

In certain very simple cases it is possible to measure the charge composition of a beam without sorting it into its components by electric or magnetic deviation. For instance, if in a projectile beam only a neutral and a singly charged component exist, a change in the current collected by a Faraday cup will reveal a change in F_1, and the current at high vacuum is proportional to $F_1 + F_0$. The danger here is that admission of the target gas may scatter some of the beam out of the detector aperture, giving erroneously low values of F_1.

In the more general procedure, the beam emergent from the equilibrating gas chamber, or from a solid foil, is sorted into charged components by a variable, transverse, magnetic or electric field, and by changing the field each charged component is directed to a detector. If there are n components, a measurement of $(n - 1)$ independent ratios F_i will determine the F's, since $\Sigma F_i = 1$. In this case, an error may arise if a particular component F_i should be elastically scattered out of the beam more than some other charge (cf. Allison, 1958c).

4.1.2 F_i *Measurements from Single Scatterings into Large Angles*

The experiments previously mentioned concern highly collimated beams, from which projectiles scattering with angular deviations greater than $\sim 0.06°$ are excluded.

The scattered projectiles from collisions involving much greater scattering have been collected at various angles and their F_i values determined. These are not equilibrium distributions; on the contrary, they are the results of single collisions. The apparatus and methods, which are largely due to Everhart and his associates, are discussed in § 4.4.

To avoid confusion, it must be pointed out that the F_i values for the totality of beam particles scattered and collected at a fixed angle θ are not related to the charge changing cross sections through equations such as (6) or (9), (10). Thus, in the scattering of a proton beam into 5° the event whose probability is σ_{01} does not enter; the generating events are σ_{10} and σ_{11}, the latter a scattering without change of charge.

4.1.3 *Equilibria in Beams Emergent from Solids*

Simplification is possible if solids are used for charge equilibration, in that differential pumping is not necessary, for in most cases no gas need be admitted to the system. The early experiments with α-particles (Rutherford, 1924) were done by interposing mica or metal foils between radioactive source and scintillator. In the more recent researches on accelerated particles (Hall, 1950; Phillips, 1955; Reynolds *et al.*, 1955a; Teplova *et al.*, 1957; Nikolaev *et al.*, 1957; Northcliffe, 1960), the ion beam is passed directly through a foil, whose plane is perpendicular to the beam direction. The undeviated, emergent beam is analyzed electrically or magnetically for its charged components, the neutral component usually being obtained by subtraction.

Dissanaike (1953) in his research on the equilibration of helium beams in metal foils, used a different technique which is illustrated in Fig. 2. The beam of He^+ ions from a Van de Graaff accelerator entered

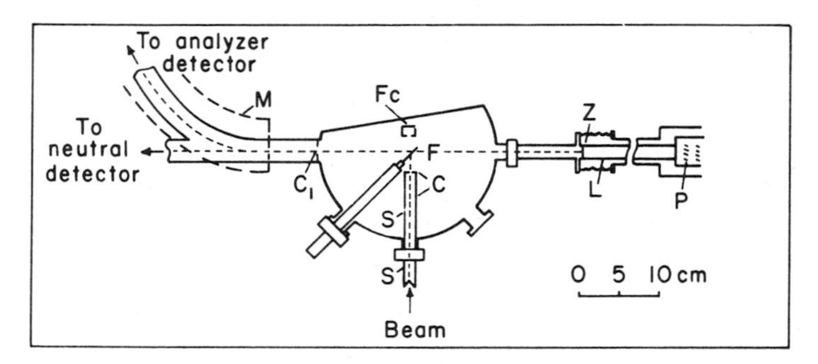

FIG. 2. Equipment used by Dissanaike (1953) for the study of the composition of helium beams emergent from solids.

the measuring equipment through the defining slits SS and impinged on the foil, which was placed in a plane at 45° to the beam direction. The foils were sufficiently thin (0.03-0.07 cm air equivalent) so that most of the beam was transmitted, and measured as a current in the Faraday cup Fc. The analysis of the charge components was performed on the particles which scattered through 90° and passed through the foil, toward the analyzing magnet M. The particles which scattered through 90° and emerged from the surface of the foil first encountered by the beam were used to monitor the intensity, and their scintillations on a ZnS screen were electrically amplified in a photomultiplier tube and counted. Since the kinetic energy lost in scattering at 90° is dependent on the mass of the target atom, the adjustment of the magnetic field for the analysis of the charged components could ensure that the particles had been scattered by the foil material (Be, Al, or Ag) and not by carbon or oxygen on the foil surface.

In the case of the neutral beam, the less energetic particles scattered from C and O could be distinguished from those deflected by Al and Ag by stopping them in foils placed in front of the detector. This scattering technique reduced beam intensities to the point where the analysis could be carried out by the method of counting single particles.

In planning an experiment, the question often arises as to what foil thickness is sufficient for charge equilibration of an ionic or atomic beam. We may make estimates as follows. Consider an estimate of the thickness of aluminum foil necessary to bring a beam of doubly charged alpha particles of 6780 kev to charge equilibrium in which, judging from the known ratios in air (Table XIV) $F_{2\infty} = 0.995$ and $F_{1\infty} = 0.005$. Since values of σ_{12} and σ_{21} in air are known (Table XIX), we can calculate the number of atoms per cm² which the He^{++} must traverse before $F_1/F_{1\infty} = 0.90$. By adding unity to each subscript and index in (7) and Table I we find this number to be

$$\pi_{\text{eff}} = (\log_e 10)/(\sigma_{12} + \sigma_{21}). \tag{28}$$

Since $\sigma_{12} = 1.7 \times 10^{-17}$ cm², $\sigma_{21} = 0.0084 \times 10^{-17}$ cm², we obtain $\pi_{\text{eff}} = 13.4 \times 10^{16}$ atoms of air per cm². Assuming the same number of atoms of Al are necessary, the equivalent Al foil has a thickness of 220 A. Little is known of the mechanism of charge changing in metals; furthermore, the estimate neglects any density effect (cf. § 16.1) which would decrease the necessary thickness. However, it would seem that any usable foils are amply thick for equilibration; the example was chosen as one which would give a relatively large value of π_{eff}. We may conclude that if the final 100 A or so traversed by the beam as it leaves the foil are in an environment (surface contamination, different density) different from the foil interior, the equilibrium will be characteristic of the surface rather than the material of which the foil is presumably composed. Experiments by Phillips (1955), to be discussed later, confirm this analysis.

4.2 Detectors

A deep Faraday cup which the beam enters through an opening subtending an angle $\lesssim 10^{-2}$ sterad at any point on the inner wall of the cup that may be struck by the direct beam is probably the most reliable detector of the current transported by the beam. If properly constructed, the current it registers should be independent of the sign and magnitude of its bias potential (region: $\pm$ 100 ev) with respect to other adjacent conductors. In experiments performed by the authors, a beam 0.2 cm in diameter entered a hollow, metallic, cylindrical cup through its open end of diameter 0.83 cm. The depth of the cup was 6.3 cm (probably excessive). Its response was independent of bias within the limits indicated.

4.2.1 *Secondary Electron Detectors*

In secondary electron detectors the ionic or atomic beam is measured by allowing it to strike a metal surface which can "see" other conductors in the vacuum, and which is biased electrically negative with respect to them. In our kinetic energy range, ionic or atomic particles will eject secondary electrons from the metal; these contribute a current which is equivalent to one which would result from additional positive ions impinging on the electrode. Thus, in the detection of a positive ion beam the true current is enhanced; for a neutral beam a positive current is provided, and the true current from a negative ion beam is diminished in magnitude and may even be reversed in direction. The ability to detect neutral beams is an advantage of these detectors, but there are also many possible sources of error in their use.

If the intensities of ionic beams of different charge composition are being compared, the number of secondary electrons per incident particle for each charge state of the projectiles must be known, or the beams must be charge equilibrated by passing through a foil or a gas converter cell before they are detected, common velocity being assumed.

If beams of different charge are being compared, and each is directed to the detector by the requisite electric or magnetic field, it must be established that the detectors' response is independent of the change in field from beam to beam. This is especially serious if the field is magnetic —there may well be a stray field at the detector which will affect the easily deviable secondary electrons.

The secondary electrons may ionize any gas near the detector, thus causing a spurious current to it because of its bias. If the gas pressure is changed in the detector compartment as in a cross section measurement, or because of gas effluent from the converter cell, the sensitivity to gas pressure of the detector must be investigated. A more detailed discussion of secondary electron detectors is available elsewhere (Allison, 1958a).

4.2.2 *Calorimetric Detectors*

Detectors which respond to the calories per second of heat developed when the ionic or atomic beam is stopped have been used by many investigators in this field: Rudnick (1931), Snitzer (1953), Fedorenko (1954a), Stier *et al.* (1954), Fogel' *et al.* (1957), Curran and Donahue (1959), Allison *et al.* (1960a). Their usefulness depends on the fact that it is almost always true that the values of (pl) or atoms/cm^2 through which ion beams are charge converted and equilibrated are so low that there is no appreciable diminution of their kinetic energy through the stopping process. Thus, at energies large compared to the energy release on electron capture, the response of a detector which measures the calories per second generated in stopping the beam should be independent of its charge state.

Absolute measurements of the power of an ionic or atomic beam are in most cases not necessary. The temperature attained in radiative equilibrium by a conductor of small mass (to avoid long time constants) heated by the beam is indicated by a suitable thermometric device. In the detector used by Stier *et al.* (1954), the beam was absorbed in a disc of platinum foil 0.95 cm in diameter and 0.0025 cm thick. A thermocouple junction of copper constantan wires each 0.0075 cm in diameter was welded to the disk on the side opposite to that struck by the beam.

A temperature indicator slightly less complicated than a thermocouple, and sufficiently accurate for these measurements is provided by the "thermistor,"† a highly temperature-sensitive resistance, obtainable in a bead about 0.04 cm in diameter. A typical unit has a room temperature resistance of 2500 ohms, decreasing 110 ohms for each 10°C temperature increase. In a design used by Allison *et al.* (1960b) such a thermistor was attached to the rear of a platinum disk similar to that mentioned above. The disk was held in place by quartz fibers inside a thick-walled housing consisting of 200 gm of copper. A second thermistor, inserted into the copper, read the "ambient" temperature around the detector; the significant reading was the temperature difference between the detector thermistor and the ambient thermistor. For a beam of about 6×10^{10} particles/sec moving with 20-kev kinetic energy, such a calorimeter will come to radiative equilibrium when the temperature of the platinum disk is from 5° to 6°C higher than the surrounding objects. This means a resistance change of from 500 to 1000 ohms.

4.2.3 *Particle-Counting Detectors*

Detectors which count each beam particle have certain advantages over continuous current detectors in that many of the correction factors for interpreting the relative intensities of beams of different charge do not apply. The early work of Henderson (1922) and of Rutherford (1924) was carried out using scintillations on a zinc sulfide screen. The only disadvantage of such detection methods as used at present is their electronic complexity and somewhat lesser stability and reliability.

At the lower end of the energy range of this chapter, where the particles cannot penetrate foils of thickness sufficient to maintain the pressure difference necessary for a gas filled ionization chamber, or proportional counter, scintillating crystals or secondary electron multipliers are indicated as detectors. Ziemba and associates (1960) have used a 10-stage Dumont type 6467 electron multiplier. The photosensitive surface and

† This may be purchased from the Victory Engineering Corporation, Union, New Jersey.

the glass envelope were removed to allow the beam particles to strike the first dynode directly. The receptacle for the tube was carefully positioned in the vacuum compartment before opening the tube. This operation took place in an atmosphere of dry nitrogen, and was carried out so that the time of exposure of the dynodes to atmospheric pressure was minimized. If too prolonged, the pulse-height response might be lowered by factors as large as 100. The useful life of such a detector seemed to be about 2 months.

4.3 Methods of Measuring Charge Changing Cross Sections

Measurements of charge distributions in equilibrated ionic and atomic beams cannot in themselves give information about individual cross sections since all F_i's are dimensionless, and auxiliary experiments are needed. There are approximately five different methods described in the literature for measuring σ_{if} cross sections. They will be briefly reviewed here.

4.3.1 *Cross Sections Through Measurement of $F_i(\pi)$ and $F_{i\infty}$.*

This method only applies to two-component systems whose components are known, either through the simplicity of the system (H^0 and H^+, for instance) or through auxiliary measurements. A beam of the desired kinetic energy is prepared by acceleration; its component fractions F_z must be known. For the following analysis one assumes that the beam consists entirely of one of the two possible components, and that they are characterized by $i = 0$ or 1. The necessary measurements could be carried out in an apparatus schematically as simple as that in Fig. 1. The initial beam enters from the left, consisting entirely of charge state 1. Gas is admitted to the converter cell to a measured pressure p, producing a π which is less than that required to equilibrate the beam. The value of F_1 in the emergent beam is measured, then more gas is admitted until the value of F_1 has further decreased to $F_{1\infty}$. From the measured values π, $F_1(\pi)$, and $F_{1\infty}$ the two cross sections can be determined from

$$\sigma_{01} = \frac{F_{1\infty}}{\pi} \log \left\{ \frac{1 - F_{1\infty}}{F_1 - F_{1\infty}} \right\} \tag{29}$$

$$\sigma_{10} = \frac{1 - F_{1\infty}}{\pi} \log \left\{ \frac{1 - F_{1\infty}}{F_1 - F_{1\infty}} \right\}. \tag{30}$$

4.3.2 *Measurement by Beam Attenuation in a Transverse Field*

A method which does not involve the measurement of the quantities $F_{i\infty}$ is to place either the original, accelerated beam, or the beam containing various charge states as it issues from the converter cell, into a measuring chamber in which there is either a transverse electric or magnetic field (cf. Fig. 3). The field holds each constituent of the equili-

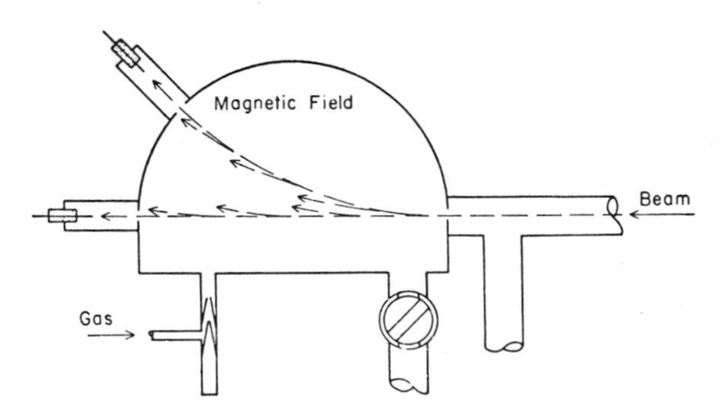

FIG. 3. Schematized apparatus for the measurement of charge changing cross sections by the method of attenuation in a transverse magnetic field.

brated beam, or the entire prepared beam, in an appropriate orbit. With $\pi = 0$ in the measuring chamber, values of the deflecting field and locations of the detectors are found such that the various beams enter their detectors through apertures which are larger, but of diameters comparable to that of the beam itself. When gas is admitted to the chamber, each beam particle whose charge is changed is lost to its beam because it can no longer continue in its old orbit, and a beam attenuation due to charge changing is observed.

If the system is one of only two components, the measurement of the attenuation can be uniquely interpreted; if the components are neutral and singly charged,

$$\sigma_{10} = -(1/\pi) \log_e R_1(\pi) \tag{31}$$

$$\sigma_{01} = -(1/\pi) \log_e R_0(\pi) \tag{32}$$

in which R_i is to be interpreted as the ratio of the detector response with gas in the measuring chamber to the response at high vacuum.

In this type of measurement, the effect of stray electric or magnetic fields on the detector need not be considered since the only change in the experimental conditions during a measurement cycle is in the gas pressure. It must be demonstrated that the admission of sufficient gas for a measurable beam attenuation does not affect the response of the detector (Montague, 1951; Ribe, 1951). Furthermore, it must be shown that the beam is not appreciably attenuated by elastic scattering when the gas is in the measuring chamber. This is usually demonstrated by turning off the field and allowing the beam to proceed in a straight line through the measuring chamber to a detector with the same entrance aperture, and so placed that the length of path is the same as that for the component beam under measurement. Gas is then admitted as if an attenuation experiment were in progress. It is essential to the validity of this method that under these circumstances the attenuation is negligibly small.

Another source of error may arise from the inadequacy of the differential pumping. If admission of gas into the measuring cell causes appreciable increases in pressure in the compartments through which the beam passes before entering the cell, the composition of the beam may be changed, particularly if the charge component that has been accelerated is under investigation, and no gas is deliberately introduced into the converter cell (Allison, 1958a).

The background reading of the detector must be investigated with and without gas in the measuring cell. The profile of a charged beam may easily be observed by varying the guide field slightly so that the beam moves across, and slightly to either side of, the detector aperture. In the case of measurements on a neutral beam this cannot be done: the profile is examined by a movable slit in front of the detector, or by a movable detector (Stier and Barnett, 1956; Allison, 1958b). If other beams than the one being measured are traversing the measuring cell, the ion spray from charge changing collisions of these beams may introduce a large error because the background will be raised with "gas in" (Allison *et al.*, 1956).

Figure 4 shows an experimental arrangement in which the transverse "guide" field is electric, but, since the experiment is designed to study electron loss from neutrals, the field does not actually guide, but only removes charged ions as fast as they are formed. The segmentation of the electric field is partially to enable a better measurement of the path length l, since through a series of measurements, using fractional parts of the total condenser length, a set of simultaneous equations may be found from which the additional Δl, the edge effect at the entrance and exit from the electric field, may be eliminated (Stier and Barnett, 1956; Kasner and Donahue, 1957).

In Fig. 4 some of the charged particles from the Cockcroft-Walton accelerator (top) are allowed to neutralize themselves in the converter, or "neutralizer" cell. Any remaining charged particles are removed in "electrostatic analyzer No. 1," so that a pure neutral beam enters the measuring chamber through a narrow cylindrical channel 1.2 cm long and 0.16 cm in diameter. The detector was sufficiently removed (1 meter) so that a slight electric field was sufficient to prevent any positives from entering its aperture.

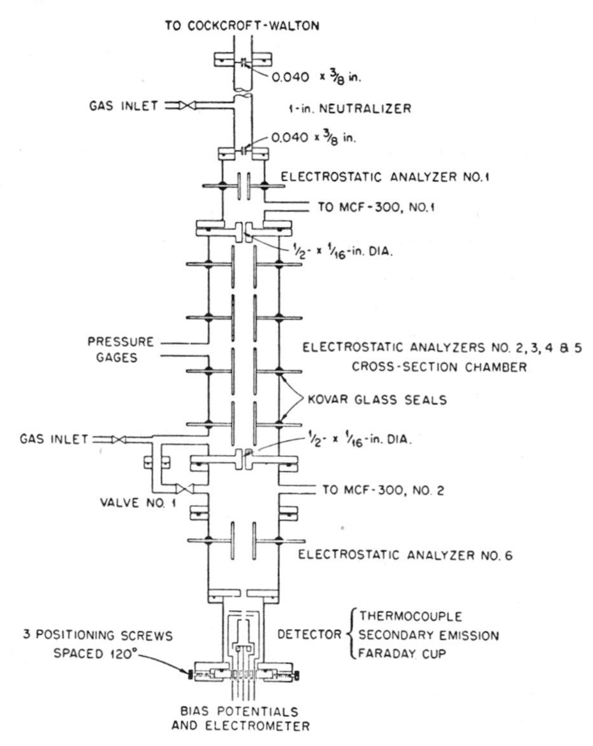

FIG. 4. Apparatus of Stier and Barnett (1956) for charge changing cross section measurements.

When this method is applied to a system of more than two components, the measurements indicated will not, without further data, give values of individual cross sections because the attenuation may be due to any event, (i.e., multiple or single electron loss or capture) which changes the charge of the moving ion or atom; thus, we actually measure

$$\Sigma_f \sigma_{if} = -(1/\pi) \log R_i \tag{33}$$

where the final charge fe has all possible values except ie. If the results thus obtained on $\Sigma_f \sigma_{if}$ are supplemented with data on the values of $F_{i\infty}$, however, it is usually possible to deduce some individual cross sections with meaningful accuracy. This procedure is illustrated in the measurement of cross sections for lithium projectiles to be discussed later (Allison *et al.*, 1960a).

4.3.3 *Measurement by Ion Collection*

This is sometimes referred to as the "condenser" method, and apparently was first devised and used by Goldmann (1931). It depends on the fact that electron capture by a fast moving ion traversing a gas must leave behind a positively charged target atom of relatively very low velocity, except in the rare case where the impact parameter of the collision is so small that a large momentum transfer is possible. Other positive and negative ions may be formed in the gas by ionization of the target atoms without a change in the electronic configuration of the projectile; these will, however, contribute equal numbers of positive and negative ions. Thus, if a net positive current is extracted by an electric field from a volume through which the beam is passing, this positive excess may be attributed to electron capture by the beam, and a cross section computed. Such a measurement could be carried out in the apparatus of Stier and Barnett shown in Fig. 4, if careful attention were paid to the currents collected by the condenser plates, rather than using the electric field generated by them to produce beam attenuation as in the previous method.

For a simple interpretation of a positive excess current it is essential that the only charge changing collision occurring in the volume is electron capture and that the number of electrons captured in a single collision be known. For a two-component system this means that the length along the beam from which slow ions are collected must be small compared to the mean free path for cycle completion (27). The beam, as it enters the region from which ions are collected, must be of known charge composition: preferably 100% in the singly charged state if the two possible components correspond to $i = 0, 1$. A principal technical difficulty in these experiments lies in electron emission from the collecting plates due to ultraviolet light or electron bombardment, and careful shielding is necessary. Also, the exact volume from which the collected ions have come must be known, and one must be sure that all of the ions have been swept out and have contributed to the observed current.

The interpretation of measurements taken by this ion-collection method becomes much more difficult for systems containing more than two components. In the three-component system He^0, He^+, He^{++}, such an experiment with He^+ as the incident beam would give $(\sigma_{10} - \sigma_{12})$; if He^{++} were the incident beam, it would give $(\sigma_{21} + 2\sigma_{20})$.

4.3.4 *Measurements Involving the Initial Behavior of the Growth Curve*

A prepared beam, all of whose particles have the same charge ze is passed through a gas converter cell, and the growth of the new charge component fractions F_i for $(i \neq z)$ is studied at very low pressures, where more than one collision in the cell is improbable. For each new component, the observed growth curve $F_i(\pi)$ is analytically expanded into a power series in π, so that

$$F_i(\pi) = \Sigma\, a_n \pi^n; \qquad n = 1, 2, ... \tag{34}$$

If in the series thus computed from the experimental data a_1 is finite, the probability that new charge component i can be formed in a single collision from the original component of charge ze is finite and is equal to a_1. The method is of special interest in determining the probability of capture or loss of two electrons in a single collision. It has been applied by Fogel' and his collaborators in several determinations (Fogel' and Mitin, 1956; Fogel' *et al.*, 1957, 1959) involving $\sigma_{1\bar{1}}$ for H^+, Li^+, etc; and $\sigma_{\bar{1}1}$ for H^-. These systems consist of three components. We have given the growth equation for a three-component system in which $i = 0, 1, 2$ in (15). Let us apply it to the growth of a neutral component from an accelerated beam which is all doubly charged. An example would be the growth of He^0 through double electron capture by He^{++}. With the proper subscripts and indices to apply to this case, (15) becomes

$$F_0 = F_{0\infty} + [P(2, 0) \exp (\pi q) + N(2, 0) \exp (-\pi q)] \exp \left(-\frac{\pi}{2} \Sigma\, \sigma_{if}\right). \tag{35}$$

Expanding in powers of π, collecting coefficients, and using Table II and (16)-(24) leads to

$$F_0 = \pi[F_{0\infty}(\tfrac{1}{2} \Sigma\, \sigma_{if} - s) - fF_{1\infty}] + \tfrac{1}{2}\pi^2[F_{0\infty}\{s\, \Sigma\, \sigma_{if} - q^2\} - \tfrac{1}{2}(\Sigma\, \sigma_{if})^2 + fF_{1\infty}\, \Sigma\, \sigma_{if}] + ... \tag{36}$$

Reduction of the coefficient of π shows that

$$F_{0\infty}(\tfrac{1}{2} \Sigma\, \sigma_{if} - s) - fF_{1\infty} = \sigma_{20} \tag{37}$$

so that if the growth curve of the neutral component has a finite slope at $\pi = 0$, there is a measurable cross section σ_{20}. The preceding equations can be applied to the initial growth of H^- from a proton beam by reducing all numerical subscripts and indices by unity. For this case Fogel' has given the result

$$F_{\bar{1}} = \pi\sigma_{1\bar{1}} + \tfrac{1}{2}\pi^2(\sigma_{10}\sigma_{0\bar{1}} + \sigma_{1\bar{1}}\sigma_{10} + \sigma_{1\bar{1}}^2 - \sigma_{11}\sigma_{\bar{1}0} - \sigma_{11} - \sigma_{\bar{1}1}) + ... \tag{38}$$

The method has also been used to measure the double capture of electrons by He^+ to form He^- (Dukel'skii *et al.*, 1956).

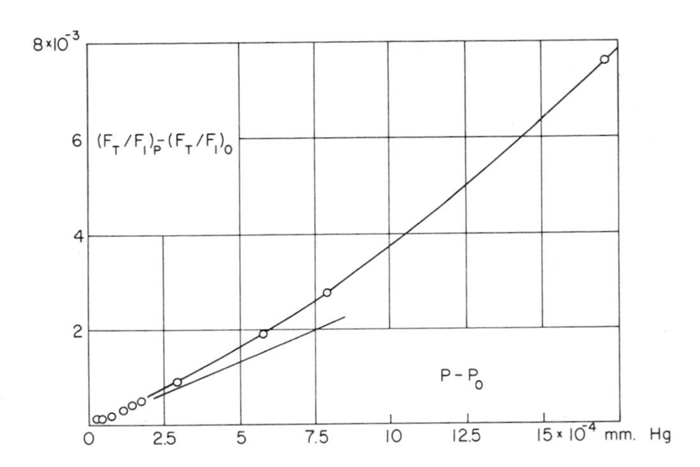

FIG. 5. Growth of H^- from a 29-kev H^+ beam traversing H_2 gas. After Fogel' and Mitin (1956).

Figure 5 shows the growth of H^- from H^+ as the proton beam is converted in H_2 gas in a cell 10 cm long according to Fogel' and Mitin (1956). The proton beam had 29-kev kinetic energy. The lowest possible pressure in the converter cell was 5×10^{-5} mm Hg, and the abscissae show excess of H_2 pressure over this base. The ordinates represent the increase in the H^- component as H_2 is admitted. The curve is a plot of the function

$$F_{\bar{1}}/F_1 = (F_{\bar{1}}/F_1)_0 + \gamma(p - p_0) + \delta(p - p_0)^2 \tag{39}$$

with $\gamma = 2.74$ and $\delta = 1.03 \times 10^3$. From the coefficient γ it results that $\sigma_{1\bar{1}}$ is 0.21×10^{17} cm² per atom of H_2 gas.

Allison (1958a) used a slight modification of this method in attempting to detect and measure double electron capture by He^{++} and double loss by He^0. Other information about the helium cross sections was already available, since Allison *et al.* (1956) had measured the sums $(\sigma_{20} + \sigma_{21})$ and $(\sigma_{01} + \sigma_{02})$ and the individual cross sections σ_{10} and σ_{12}. Further, the fractions $F_{i\infty}$ were known from the work of Snitzer (1953) and Barnett and Stier (1958).

In the work on σ_{20}, Allison measured the ratio F_0/F_1 as the components He^0 and He^+ grew from He^{++}, at very low converter cell pressures. The initial growth curve calculated for these components by (35) and the analogous equation for F_1 is not sensitive to a breakdown of the sum ($\sigma_{01} + \sigma_{02}$) into individual cross sections. Predicted curves showing F_0/F_1 as functions of π were calculated, each curve corresponding to an assumed value of ϵ in $\epsilon = \sigma_{20}/(\sigma_{20} + \sigma_{21})$. Comparison with experiment indicated that one of the choices of ϵ gave a curve fitting the data, and from this, and ($\sigma_{20} + \sigma_{21}$) previously measured by the attenuation method, individual values of σ_{20} and σ_{21} could be calculated.

A similar treatment of the growth of the fraction F_2/F_1 from a neutral He^0 beam permitted an evaluation of the individual cross sections σ_{02} and σ_{01}.

4.4 Measurements of $d\sigma_{zi}(\theta)/d\Omega$

Although the number of scatterings into unit solid angle at θ decreases roughly as $\csc^4(\theta/2)$, there are a detectable number of scattered projectiles with angular deviations of several degrees. Two types of study have been made on them; the charge fractions $F_i(\theta)$ have been determined, and the differential cross sections $d\sigma_{zi}(\theta)/d\Omega$, where the positive charge on the initial beam is ze. Our information concerning these phenomena is largely due to Fedorenko (1954a), and his report to the Venice Conference (Venice, 1957), Kaminker and Fedorenko (1955a), and several contributions from Everhart and his associates (i.e., Jones *et al.*, 1959). Figure 6 shows equipment used by the Everhart group to measure the angular dependence of charge changing cross sections. The equipment in the upper part of the figure is used for scattering angles above 1°. The electrostatic energy analyzer can be rotated about point *b*, using the flexible bellows to avoid breaking the vacuum. The beam at angle θ, as selected by apertures c and d, contains all the new charge components which have been produced in scattering events in the region of b, plus some incident beam particles which have been scattered without change of charge. The scattering volume (analogous to the converter cell in other techniques) is in this type of equipment determined by the geometry of the slit system, and must be computed. The electrostatic field separates the scattered beam and its components can, in turn, be directed to the detector aperture, behind which a Faraday cell detector is shown. The neutral beam must be measured by rotating the electron multiplier tube into position; also the electron multiplier is useful for very weak beams. A small Faraday cell "a" can be rotated into the beam for monitoring purposes. The gas pressures at b for sufficient scattering were on the order of 1 μ Hg; the base pressure at "b" corresponded to 5×10^{-6} mm. The sensitivity of detection was such that even at this low pressure scattered particles were measurable. The lower part of the figure shows the equipment adapted to the angular region below 1°, which is the range into which most particles having small values of $| z - i |$ are scattered. The design is such that all particles scattered by less than 1° enter the detectors, separated only according to charge.

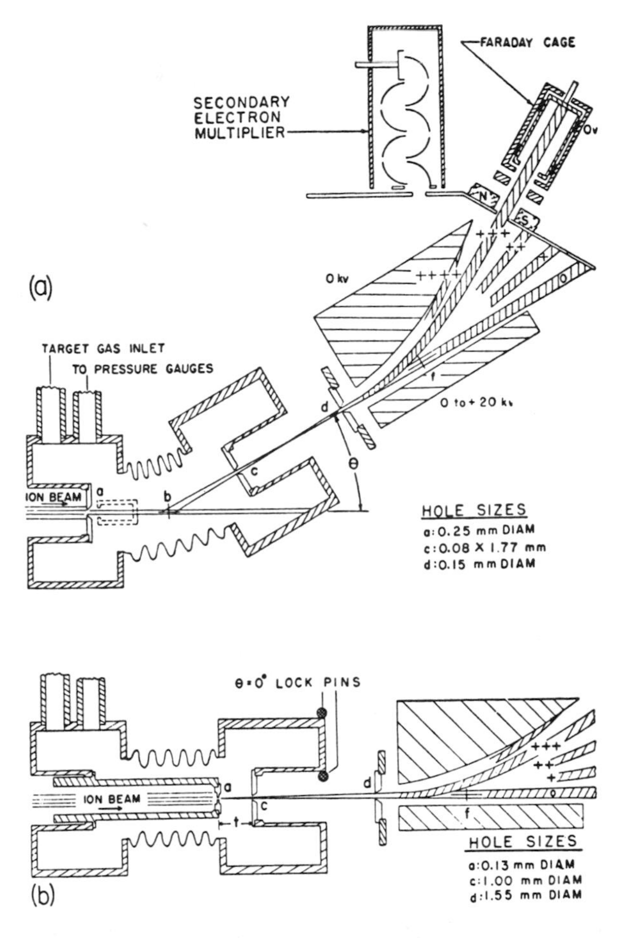

FIG. 6. Apparatus used by Everhart and his associates (Jones *et al.*, 1959) for measuring the differential cross sections $d\sigma_{zi}(\theta)/d\Omega$.

Figure 7 shows apparatus used by Fedorenko (Venice, 1957) which permits measurements of $d\sigma_{zi}/d\Omega$. Positive ions are extracted from the source S by a probe electrode E, and are accelerated to kinetic energies between 5 and 180 kev in the tube T. A pure beam, consisting 100% of a single charge component, is selected magnetically at M, and enters the collision or converter cell C_1. The current entering this cell is in the range 10^{-7}-10^{-6} amp. A mobile analyzer A_1 is attached to this chamber and can determine F_i values for beams of scattered particles which have undergone deviations from 77° to 92° from the beam direction.

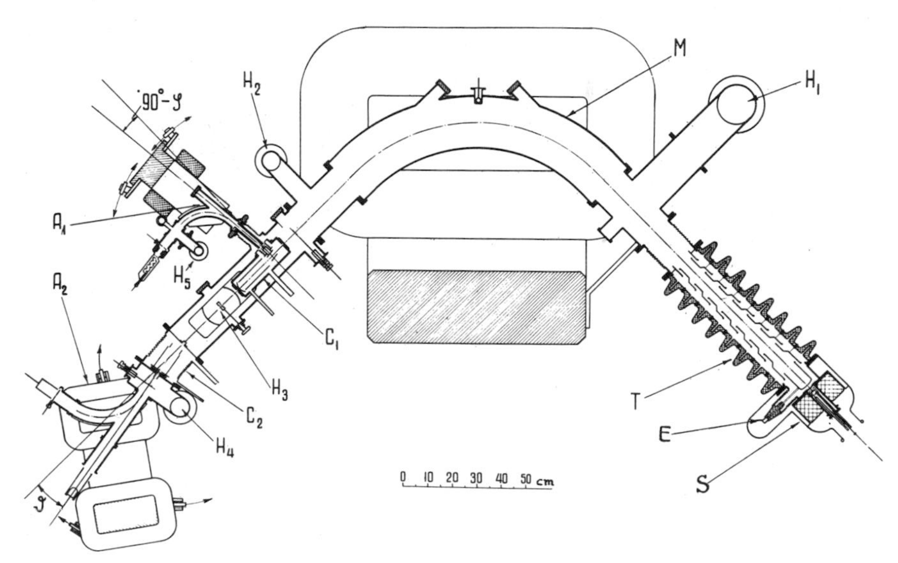

FIG. 7. Apparatus of Fedorenko for angular distribution measurements. (Venice, 1957).

For the investigation of smaller scattering angles, the beam is passed on into a second collision chamber C_2, from which the second movable magnetic analyzer A_2 can collect projectiles scattered through the range $\pm$ 17°. Provision is also made for the detection of fast neutrals formed in C_2.

5 *Results on Hydrogen Beams*†

5.1 CHARGE ANALYSIS OF SINGLE COLLISION SCATTERING INTO 5° IN THE LABORATORY SYSTEM

Ziemba *et al.* (1960), using the type of equipment shown in Fig. 6, have carried out the charge analysis of a proton beam scattered on various gases into an angle of 5°. This angle was fixed and was arbitrarily selected as representative of large angle charge changing scattering.

† A survey of the data existing on beams of the different elements, much more comprehensive than the data included in this chapter (up to December, 1960), has been compiled by the authors and can be obtained by writing to them.

The fraction F_0 of neutral atoms in the scattered beam plotted as a function of the proton kinetic energy show the existence of peaks and since the total scattering with and without change of charge, at a fixed angle, is a monotonic function of the kinetic energy, the maxima indicate the location of peaks in the electron capture (σ_{10}) process. The existence of this kind of resonant peaks in the single electron capture process was first discovered by Ziemba and Everhart (1959) using a helium beam in helium gas (cf. Fig. 10). The peaks in the curves persist at other angles comparable with 5°. When the peaks are plotted against a collision time interval (the range of interaction divided by the velocity), they appear equally spaced.

5.2 Composition of Equilibrated Hydrogen Beams in Various Gases

The values of $F_{\bar{1}}$, F_0, and F_1 in unidirectional hydrogen beams at charge equilibrium in various gases, reported in Tables III and IV are essentially from Stier and Barnett (1956) and Barnett and Reynolds (1958). In the energy interval 4-50 kev, the H^+ and H^- components issuing from the converter cell were measured in a Faraday cup. The H^0 component was measured by passing both the total equilibrated beam and then its neutral component through a second gas cell in order to re-establish equilibrium before entering the detector. At the higher energies, the fraction $F_{0\infty}$ was measured using a secondary electron detector, calibrated against a calorimetric detector. The $F_{1\infty}$ values were, in this higher range, obtained by subtraction from unity. Comparison of the $F_{0\infty}$ values in hydrogen and helium target gases is very instructive. At 25-kev kinetic energy the translational velocity of the proton through the target gas is equal to that of an electron in the first Bohr orbit in the hydrogen atom ($c/137 = 2.18 \times 10^8$ cm/sec). At velocities low compared with this, the fraction $F_{0\infty}$ is much greater in the chemically active target gases than in helium and to a lesser extent, neon (as can be seen in further data of Stier and Barnett on neon and argon). At higher velocities this difference between the chemically active gases and the lighter noble gases is not a dominating effect. Fogel' and Mitin (1956) and Whittier (1954) have reported some values of the ratio $F_{\bar{1}\infty}/F_{1\infty}$ of hydrogen beams in various gases. The essential features of the equilibrium fractions in a hydrogen target are illustrated in Fig. 8.

TABLE III

EQUILIBRIUM FRACTIONS OF A HYDROGEN BEAM IN HYDROGEN AND HELIUM[a]

Kinetic Energy (kev)	Hydrogen			Helium		
	$F_{\bar{1}\infty}$	$F_{0\infty}$	$F_{1\infty}$	$F_{\bar{1}\infty}$	$F_{0\infty}$	$F_{1\infty}$
3	0.008	0.897	0.095	—	—	—
4	0.010	0.895	0.095	—	0.125	0.875
5	0.014	0.891	0.095	—	0.185	0.815
7	0.019	0.886	0.095	—	0.280	0.720
9	0.020	0.885	0.095	—	0.400	0.600
11	0.020	0.880	0.100	0.003	0.450	0.547
13	0.020	0.875	0.105	0.007	0.530	0.463
15	0.020	0.855	0.125	0.008	0.550	0.442
20	0.018	0.797	0.185	0.009	0.600	0.391
25	0.016	0.764	0.220	0.009	0.600	0.391
30	0.014	0.706	0.280	0.010	0.599	0.391
35	—	0.680	0.320	—	0.575	0.425
40	—	0.620	0.380	—	0.535	0.465
45	—	0.580	0.420	—	0.515	0.485
50	—	0.525	0.475	—	0.490	0.510
60	—	0.447	0.553	—	0.447	0.553
70	—	0.361	0.639	—	0.369	0.631
80	—	0.276	0.724	—	0.329	0.671
90	—	0.237	0.763	—	0.292	0.708
100	—	0.183	0.817	—	0.254	0.746
150	—	0.063	0.937	—	0.118	0.882
200	—	0.024	0.976	—	0.060	0.940
250	—	0.075	0.993	—	0.032	0.968
300	—	0.0041	0.996	—	0.017	0.983
350	—	0.0024	0.998	—	0.011	0.989
400	—	0.0012	0.999	—	0.0064	0.994
450	—	0.0_357	—	—	0.0038	0.996
500	—	0.0_339	—	—	0.0028	0.997
600	—	0.0_316	—	—	0.0014	0.999
700	—	0.0_496	—	—	0.0_392	
800	—	0.0_454	—	—	0.0_352	
900	—	0.0_440	—	—	0.0_345	
1000	—	0.0_427	—	—	0.0_336	

[a] Stier and Barnett (1956), Barnett and Reynolds (1958).

TABLE IV

EQUILIBRIUM FRACTIONS OF A HYDROGEN BEAM IN NITROGEN AND OXYGEN[a]

Kinetic Energy (kev)	Nitrogen			Oxygen		
	$F_{\bar{1}\infty}$	$F_{0\infty}$	$F_{1\infty}$	$F_{\bar{1}\infty}$	$F_{0\infty}$	$F_{1\infty}$
3	—	—	—	0.015	0.880	0.105
4	0.004	0.871	0.125	0.016	0.849	0.135
5	0.006	0.854	0.140	0.018	0.827	0.155
7	0.011	0.815	0.175	0.018	0.782	0.200
9	0.012	0.778	0.210	0.016	0.729	0.255
11	0.014	0.756	0.230	0.015	0.710	0.275
13	0.013	0.712	0.275	0.014	0.681	0.305
15	0.012	0.688	0.300	0.013	0.661	0.320
20	0.010	0.640	0.350	0.010	0.610	0.380
25	0.008	0.587	0.405	0.009	0.581	0.410
30	0.006	0.544	0.450	0.008	0.517	0.475
35	—	0.500	0.500	—	0.500	0.500
40	—	0.480	0.520	—	0.460	0.540
45	—	0.430	0.560	—	0.440	0.560
50	—	0.410	0.590	—	0.405	0.595
60	—	0.384	0.616	—	0.364	0.636
70	—	0.314	0.686	—	0.318	0.682
80	—	0.240	0.760	—	0.281	0.719
90	—	0.208	0.792	—	0.241	0.759
100	—	0.172	0.828	—	0.212	0.788
150	—	0.095	0.905	—	0.122	0.878
200	—	0.038	0.962	—	0.051	0.942
250	—	0.019	0.981			
300	—	0.010	0.990			
350	—	0.0057	0.994			
400	—	0.0037	0.996			
450	—	0.0028	0.997			
500	—	0.0020	0.998			
600	—	0.0012	0.999			
700	—	$0.0_{3}77$				
800	—	$0.0_{3}50$				
900	—	$0.0_{3}39$				
1000	—	$0.0_{3}28$				

[a] Stier and Barnett (1956), Barnett and Reynolds (1958).

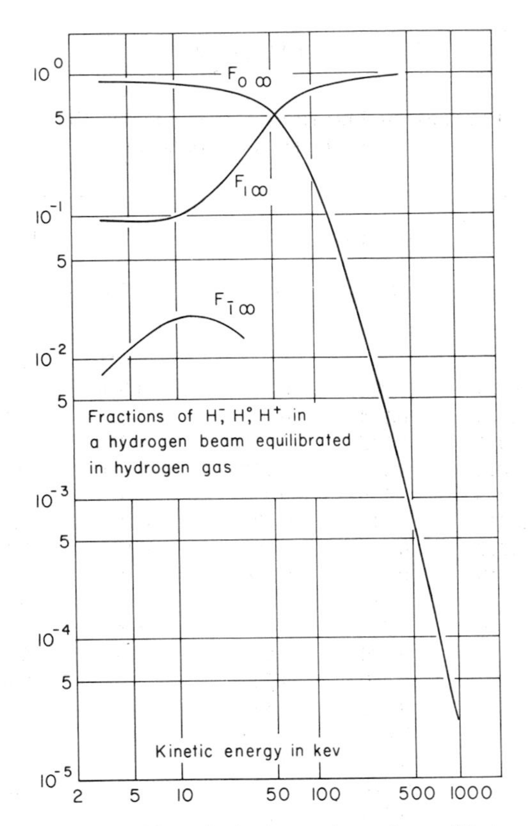

FIG. 8. The charge composition of a hydrogen beam in equilibrium in hydrogen gas

5.3 Composition of Hydrogen Beams Emergent from Solid Foils

Data on the equilibration of hydrogen beams in solid foils are presented in Table V. In the experiments of Phillips (1955) a special attempt was made to determine whether the charge composition of the beam is typical of the nature of the surface layers through which it left the foil, or is the same for foils of a pure material, irrespective of surface conditions. Phillips found that a layer of aluminum, freshly deposited, through which the beam emerged, imposed on it a charge composition which changed with time, presumably due to alteration of the surface layer in the highly imperfect vacuum used in this type of experiment. At a pressure of 2.4×10^{-5} cm Hg, a fresh surface changed to an "old" or time-independent condition in about 15 min; at a pressure of 4×10^{-6} mm the contamination time was much longer, and not measured. These values from "old" surfaces are approximately the same for all metals, and are listed in the right-hand columns of Table V; the other values are for freshly deposited surfaces. Hall (1950) observed little difference between the beam compositions emergent from Be, Al, and Ag foils; these were not freshly prepared and the uniformity rather supports the "old surface" idea; however, Hall found the beams emergent from gold to have larger values of $F_{0\infty}$ than from other metals. Fogel' *et al.* (1955b), Table V, found differences between the $F_{\bar{1}\infty}F/_{1\infty}$ ratios from Be and Al foils, and does not mention an ageing effect.

TABLE V

CHARGE DISTRIBUTION IN HYDROGEN BEAMS EMERGING FROM SOLID FOILS[a,b]

Kinetic	Beryllium				Aluminum				Old surfaces	
energy (kev)	$F_{\bar{1}\infty}$ (1)	$F_{0\infty}$ (1)	$F_{0\infty}$ (3)	$F_{\bar{1}\infty}/F_{1\infty}$ (2)	$F_{\bar{1}\infty}$ (1)	$F_{0\infty}$ (1)	$F_{0\infty}$ (3)	$F_{\bar{1}\infty}/F_{1\infty}$ (2)	$F_{\bar{1}\infty}$ (1)	$F_{0\infty}$ (1)
4	—	—	—	—	0.062	0.856	—	—	0.045	0.810
7	0.035	0.828	—	—	0.048	0.840	—	—	0.030	0.773
10	0.031	0.793	—	—	0.040	0.818	—	—	0.023	0.727
11.5	—	—	—	0.400						
14.3	—	—	—	0.323						
15	—	—	—	—	—	—	—	0.181		
20	0.019	0.671	0.44	—	0.024	0.725	0.44	—	0.012	0.612
20.2	—	—	—	—	—	—	—	0.109		
21.3	—	—	—	0.201						
25	0.014	0.611	—	—	0.017	0.643	—	—	0.011	0.56
27.4	—	—	—	0.134						
28.2	—	—	—	—	—	—	—	0.060		
50	0.013	0.341	0.32	—	0.0037	0.356	0.32	—	0.004	0.425
75	0.015	0.195	0.28	—	0.0015	0.20	0.28	—	0.001	0.30
100	—	0.134	0.18	—	0.0_36	0.120	0.18	—	0.0_38	0.20
125	—	0.08	0.14	—	—	0.08	0.14	—	0.0_32	0.13
150	—	0.048	0.091	—	0.0_34	0.045	0.091	—	—	0.088
175	—	0.03	0.061	—	—	0.03	0.061	—	—	0.05
200	—	0.01	0.046	—	—	0.02	0.046	—	—	0.03
250	—	—	0.022	—	—	—	0.022			
300	—	—	0.016	—	—	—	0.016			
350	—	—	0.011	—	—	—	0.011			
400	—	—	0.005	—	—	—	0.005			

[a] $F_{\bar{1}\infty} + F_{0\infty} + F_{1\infty} = 1$.

[b] Key to references: (1) Phillips (1955); (2) Fogel' *et al.* (1955b); (3) Hall (1950). (Phillips and Hall give further data for silicon monoxide, calcium, silver and gold foils.)

TABLE VI

SUMMARY OF CHARGE CHANGING CROSS SECTION INVESTIGATIONS WITH HYDROGEN BEAMS

Investigation	Quantity measured	kev range	Target gases	Method
Montague (1951)	σ_{01}	40-250	H_2	Attenuation in transverse magnetic field
Ribe (1951)	σ_{10}	35-150	H_2	Attenuation in transverse magnetic field
Kanner (1951)	σ_{01}, σ_{10}	30-300	Air	Attenuation in transverse magnetic field
Hasted (1952)	σ_{10}	0.025-4	Ar	Ion collection
Whittier (1954)	$\sigma_{\bar{1}0}$, σ_{10}^*	4-50	H_2	Attentuation in transverse magnetic field
Fogel' *et al.* (1955a)	σ_{10}	12.3-36.7	H_2	Ion collection
Stedeford (1955)	$\sigma_{\bar{1}0}$, σ_{10}	3-40	H_2, He, Ne, Ar, Kr, Xe	Ion collection
Hasted (1955)	$\sigma_{\bar{1}0}$, σ_{10}	0.1-4.0	H_2, He, Ne, Ar, Kr, Xe	Ion collection
Fogel' and Mitin (1956)	$\sigma_{1\bar{1}}$	10-30	O_2	$(dF_{\bar{1}}/d\pi)$ at low π-values
Gilbody and Hasted (1956)	σ_{10}	0.01-40	N_2, CO, NH_3	Ion collection
Stier and Barnett (1956)	$\sigma_{0\bar{1}}$, σ_{01} σ_{10}	3-30	H_2, He, N_2, O_2, Ne, Ar	Attenuation in an electric field
Fogel' *et al.* (1957)	$\sigma_{\bar{1}1}$	5-40	H_2, He, N_2, O_2, Ne, Ar, Kr, Xe	$dF_1/d\pi$ at low π-values
Afrosimov *et al.* (1958a)	σ_{10}	5-180	Ar	Ion collection and e/m analysis
Barnett and Reynolds (1958)	σ_{01}	250-1000	H_2, He, N_2, Ar	Attenuation in transverse electric field
Barnett and Reynolds (1958)	σ_{10}	250-1000	H_2, He, N_2, Ar	σ_{01}, plus $F_{1\infty}/F_{0\infty}$
Fogel' *et al.* (1958a)	$\sigma_{0\bar{1}}$, σ_{01}	5-40	H_2, He, Ne, N_2, O_2, Ar, Kr, Xe	$dF_{\bar{1}}/d\pi$; $dF_1/d\pi$ at low π-values
Fogel' *et al.* (1958b)	$\sigma_{1\bar{1}}$	3-65	H_2, He, N_2, Ne, Ar, Kr, Xe	$dF_{\bar{1}}/d\pi$ at low π-values
Rose *et al.* (1958)	$\sigma_{\bar{1}0}$	0.4-2000	H_2, O_2, Ar, CO_2	Attenuation in transverse field
Curran and Donahue (1959)	σ_{01}	5-40	H_2	Attenuation; ion collection
Il'in *et al* (1959)	σ_{10}	5-180	Air, N_2	Ion collection and e/m analysis
Fogel' *et al.* (1960)	σ_{01}	5-10	Kr	$dF_1/d\pi$ at low π-values

TABLE VII

ELECTRON CAPTURE CROSS SECTIONS FOR H^0 AND H^+ MOVING IN HYDROGEN GAS[a,b]

Kinetic energy (kev)	$\sigma_{0\bar{1}}$			$\sigma_{1\bar{1}}$	σ_{10}					
	(1)	(2)	(3)	(4)	(1)	(5)	(6)	(7)	(3)	(8)
0.2	—	—	—	—	—	—	—	3.8		
0.4	—	—	—	—	—	—	—	11.0		
0.6	—	—	—	—	—	—	—	18		
0.8	—	—	—	—	—	—	—	22		
1	—	—	—	—	—	—	—	26		
2	—	—	—	—	—	—	—	40		
3	—	—	—	—	—	—	—	44	38	
4	0.64	—	—	0.13	39	—	—	40	40	
5	0.83	—	0.7	—	40	—	—	40	40	41
7	1.1	—	1.9	0.28	40	—	—	42	41	44
9	1.2	1.0	2.0	—	40	—	—	41	40	43
11	1.2	1.0	1.8	0.41	40	44	—	41	39	41
13	1.2	0.95	1.7	—	39	39	—	37	37	38
15	1.1	0.90	1.5	0.78	35	35	—	32	34	35
20	0.95	0.80	1.4	1.00	29	28	—	31	28	30
25	0.81	0.70	1.1	0.88	25	22	—	22	23	25
30	0.70	0.60	1.1	0.55	19	18	—	—	19	21
35	—	0.50	1.0	0.30	16	14	15	—	16	18
40	—	0.40	0.9	0.12	14	11	12	13	14	14
45	—	—	0.7	0.08	11	—	9.0	—	12	11
50	—	—	0.6	0.06	9.0	—	7.0	—	9.9	8.0
60	—	—	—	0.05	5.9	—	4.9	—	—	5.0
70	—	—	—	—	3.9	—	2.9	—	—	3.0
80	—	—	—	—	2.6	—	1.8	—	—	1.5
90	—	—	—	—	1.9	—	1.3	—	—	1.2
100	—	—	—	—	1.3	—	0.95	—	—	0.5
150	—	—	—	—	0.29	—	0.22	—	—	
200	—	—	—	—	0.088					
250	—	—	—	—	0.022					
300	—	—	—	—	0.010					
350	—	—	—	—	0.0050					
400	—	—	—	—	0.0024					
450	—	—	—	—	0.0010					
500	—	—	—	—	$0.0_{3}62$					
600	—	—	—	—	$0.0_{3}24$					
700	—	—	—	—	$0.0_{3}11$					
800	—	—	—	—	$0.0_{4}57$					
900	—	—	—	—	$0.0_{4}38$					
1000	—	—	—	—	$0.0_{4}26$					

[a] Units of 10^{-17} cm^2 per hydrogen atom.

[b] Key to references: (1) 4-150 kev, Stier and Barnett (1956); 200-1000 kev, Barnett and Reynolds (1958); (2) Fogel' *et al.* (1958a); (3) Curran and Donahue (1959); (4) Fogel' *et al.* (1958b); (5) Fogel' *et al.* (1955a); (6) Ribe (1951); (7) Hasted (1955); Stedeford (1955); (8) Afrosimov *et al.* (1958b).

TABLE VIII

ELECTRON LOSS CROSS SECTIONS FOR H^- AND H^0 MOVING IN HYDROGEN GAS[a,b]

Kinetic energy (kev)	$\sigma_{\bar{1}0}$			$\sigma_{\bar{1}1}$	σ_{01}				
	(1)	(2)	(3)	(4)	(2)	(5)	(6)	(7)	(8)
4	62	51	—	—	4.2	—	—	—	—
5	62	51	—	4.2	4.2	—	—	—	0.8
7	62	51	—	4.4	4.2	—	—	—	2.5
9	68	52	—	4.4	4.2	—	—	—	3.0
11	61	51	—	4.4	4.6	—	—	3.0	2.3
13	60	49	—	4.3	5.0	—	—	3.5	3.3
15	59	46	—	4.3	5.4	—	—	4.0	3.9
20	48	42	—	4.3	6.4	—	—	5.5	4.6
25	45	40	—	4.3	7.3	—	—	5.9	5.1
30	41	38	—	4.3	8.0	—	—	6.0	5.4
35	38	—	—	4.3	8.0	—	—	6.6	5.7
40	35	—	—	4.3	8.0	—	6.8	6.8	5.3
45	33	—	—	—	8.0	—	6.7		
50	30	—	—	—	7.9	—	6.7		
60	—	—	—	—	7.4	—	6.1		
70	—	—	—	—	6.9	—	5.8		
80	—	—	—	—	6.5	—	5.5		
90	—	—	—	—	6.1	—	5.2		
100	—	—	—	—	5.8	—	4.7		
150	—	—	—	—	4.3	—	3.8		
175	—	—	—	—	4.0	—	3.5		
200	—	—	—	—	3.6	3.6	3.2		
250	—	—	—	—	—	2.9	2.8		
300	—	—	—	—	—	2.4	2.4		
350	—	—	—	—	—	2.0			
400	—	—	8.5	—	—	2.0			
450	—	—	—	—	—	1.8			
500	—	—	—	—	—	1.6			
600	—	—	6.7	—	—	1.4			
700	—	—	—	—	—	1.2			
800	—	—	5.2	—	—	1.0			
900	—	—	—	—	—	0.95			
1000	—	—	4.5	—	—	0.95			
1200	—	—	3.8						
1400	—	—	3.3						

[a] Units of 10^{-17} cm^2 per hydrogen atom.

[b] Key to references: (1) Whittier (1954); (2) Stier and Barnett (1956); (3) Rose *et al.* (1958); (4) Fogel' *et al.* (1957); (5) Barnett and Reynolds (1958); (6) Montague (1951); (7) Fogel' *et al.* (1958a); (8) Curran and Donahue (1959).

TABLE IX

ELECTRON LOSS AND CAPTURE CROSS SECTIONS FOR HYDROGEN IONS AND ATOMS MOVING IN HELIUM GAS[a, b]

Kinetic energy (kev)	$\sigma_{\bar{1}0}$		$\sigma_{\bar{1}1}$	σ_{01}		$\sigma_{0\bar{1}}$		$\sigma_{1\bar{1}}$	σ_{10}	
	(1)	(2)	(3)	(2)	(4)	(2)	(4)	(5)	(2)	(1)
0.2	26	—	—	—	—	—	—	—	—	0.24
0.4	30	—	—	—	—	—	—	—	—	0.22
0.6	34	—	—	—	—	—	—	—	—	0.25
0.8	36	—	—	—	—	—	—	—	—	0.30
1	37	—	—	—	—	—	—	—	—	0.50
2	40	—	—	—	—	—	—	—	—	1.0
3	45	—	—	—	—	—	—	—	—	1.5
4	50	57	—	11.5	—	—	—	—	—	2.5
5	55	56	4.5	13.3	8.0	—	0.25	—	2.6	3.3
7	65	56	4.5	15	—	—	—	0.008	5.6	6.0
9	72	51	4.5	15	9.7	0.32	0.38	—	8.0	9.2
11	75	51	4.4	15	—	0.56	—	0.013	10.5	11.5
13	70	50	4.3	15	—	0.60	—	—	14	12
15	68	49	4.2	14	11	0.62	—	0.020	18	16
20	58	44	3.5	14	10	0.68	0.60	0.032	19	18
25	50	41	3.2	13	11	0.67	0.65	0.056	19.5	18.4
30	48	—	3.0	13	10	0.60	0.63	0.069	19	17.5
35	—	—	2.7	13	—	—	—	0.072	17.5	15.7
40	—	—	2.5	12	9.7	—	0.50	0.069	16	14
45	—	—	—	12	—	—	—	0.063	14.5	
50	—	—	—	12	—	—	—	0.048	11.8	
60	—	—	—	12	—	—	—	0.016	9.7	
70	—	—	—	11	—	—	—	—	6.3	
80	—	—	—	10	—	—	—	—	4.9	
90	—	—	—	9.2	—	—	—	—	3.8	
100	—	—	—	8.8	—	—	—	—	3.0	
150	—	—	—	6.9	—	—	—	—	0.93	
200	—	—	—	5.6	—	—	—	—	0.36	

[a] Units of 10^{-17} cm^2 per helium atom (see Table XIII for σ_{01}, σ_{10} values at higher energies).

[b] Key to references: (1) Hasted (1952), Stedeford (1955); (2) Stier and Barnett (1956); (3) Fogel' *et al.* (1957); (4) Fogel' *et al.* (1958a); (5) Fogel' *et al.* (1958b).

TABLE X

CHARGE CHANGING COLLISION CROSS SECTIONS FOR HYDROGEN BEAMS IN NITROGEN, AIR, AND AMMONIA GASES[a, b]

Kinetic energy (kev)	Nitrogen				Air	Nitrogen					Air		NH_3
	$\sigma_{\bar{1}0}$	$\sigma_{\bar{1}1}$	σ_{01}		σ_{01}	$\sigma_{0\bar{1}}$		$\sigma_{1\bar{1}}$	σ_{10}		σ_{10}		σ_{10}
	(1)	(2)	(1)	(3)	(4)	(1)	(3)	(5)	(1)	(7)	(6)	(4)	(7)
0.2	—	—	—	—	—	—	—	—	—	5	—	—	47
0.4	—	—	—	—	—	—	—	—	—	13	—	—	67
0.6	—	—	—	—	—	—	—	—	—	24	—	—	79
0.8	—	—	—	—	—	—	—	—	—	38	—	—	87
1	—	—	—	—	—	—	—	—	—	47	—	—	91
2	—	—	—	—	—	—	—	—	—	70	—	—	92
4	88	—	10	—	—	0.52	—	—	59	—	—	—	—
5	90	10.4	12	3.2	—	0.80	0.75	—	58	80	—	—	102
7	88	10.5	14	8.0	—	1.20	1.00	0.25	55	77	—	—	97
9	81	10.8	16	10	—	1.20	1.15	—	58	70	—	—	82
11	78	11.1	16	12	—	1.50	1.20	0.47	50	62	54	—	73
13	74	11.2	18	12	—	1.60	1.15	—	48	58	—	—	68
15	70	11.5	19	13	—	1.60	1.10	0.53	43	52	48	—	60
20	68	12.5	20	15	—	1.10	0.90	0.49	39	42	41	—	46
25	68	13.5	22	16	—	0.92	0.70	0.36	34	35	—	—	35
30	66	14.9	22	17	—	0.75	0.56	0.33	28	28	32	21	26
35	—	16.1	23	18	—	—	0.50	—	26	23	—	19	19
40	—	17.5	23	18	23	—	0.50	—	22	—	25	17	
45	—	—	24	—	22.9	—	—	—	20	—	—	15	
50	—	—	24	—	22.8	—	—	—	18	—	20	13	

[a] Units of 10^{-17} cm^2 per target atom (per molecule in ammonia) (see Table XI for σ_{01}, σ_{10} values at higher energies).

[b] Key to references: (1) Stier and Barnett (1956); (2) Fogel' *et al.* (1957); (3) Fogel' *et al.* (1958a); (4) Kanner (1951); (5) Fogel' *et al.* (1958b); (6) Il'in *et al.* (1959); (7) Gilbody and Hasted (1956).

TABLE XI

CHARGE CHANGING CROSS SECTIONS FOR HYDROGEN BEAMS MOVING IN NITROGEN GAS AND IN AIR[a, b]

Kinetic energy (kev)	Nitrogen		Air		
	σ_{01}	σ_{10}	σ_{01}	σ_{10}	
	(1)	(1)	(2)	(2)	(3)
50	24.5	18	22.8	13	20
60	24	15	22.4	10.5	16
70	24	11	22.1	8.4	13
80	24	7.6	21.7	7.6	11
90	24	6.3	21.4	5.3	8.0
100	24	5.0	20.6	4.2	7.0
150	21	2.2	19.3	—	3.5
200	19	0.75	17.6		
250	18	0.34	15.8		
300	15	0.16	14.1		
350	14	0.08			
400	13	0.05			
450	12	0.032			
500	10	0.020			
600	9.6	0.011			
700	8.7	0.0067			
800	8.0	0.0040			
900	7.7	0.0030			
1000	7.0	0.0020			

[a] Units of 10^{-17} cm^2 per target gas atom.
[b] Key to references: (1) Stier and Barnett (1956), Barnett and Reynolds (1958); (2) Kanner (1951); (3) Il'in *et al.* (1959).

TABLE XII

CHARGE CHANGING CROSS SECTIONS FOR HYDROGEN BEAMS IN OXYGEN, CARBON MONOXIDE, AND CARBON DIOXIDE GASES[a, b]

Kinetic energy (kev)	Oxygen									CO	CO_2
	$\sigma_{\bar{1}0}$		$\sigma_{\bar{1}1}$	σ_{01}		$\sigma_{0\bar{1}}$		$\sigma_{1\bar{1}}$	σ_{10}	σ_{10}	$\sigma_{\bar{1}0}$
	(1)	(2)	(3)	(1)	(4)	(1)	(4)	(5)[c]	(1)	(6)	(2)
0.2	—	—	—	—	—	—	—	—	—	190	
0.4	—	—	—	—	—	—	—	—	—	260	
0.6	—	—	—	—	—	—	—	—	—	268	
0.8	—	—	—	—	—	—	—	—	—	269	
1	—	—	—	—	—	—	—	—	—	270	
2	—	—	—	—	—	—	—	—	—	270	
4	—	—	—	8.5	—	1.0	—	—	49	—	
5	50	—	9.1	9.5	3.2	1.0	0.70	—	47	260	
7	51	—	9.3	13.5	5.1	1.1	0.77	—	45	243	
9	53	—	9.5	14.5	7.0	1.2	0.80	0.57	43	222	
11	54	—	9.7	15.5	8.2	1.2	0.80	0.56	40	202	
13	56	—	10.0	16.0	9.1	1.2	0.79	0.53	37	190	
15	56	—	10.3	19	10	1.1	0.78	0.50	33	174	
20	58	—	11.3	19	11	0.95	0.71	0.44	29.2	140	
25	59	—	12.4	19	12	0.82	0.66	0.35	28.0	118	
30	59	—	13.9	20	13	0.72	0.57	0.27	23.0	103	
35	—	—	15.2	20	—	—	0.50	—	20.0	92	
40	—	—	16.7	20	13	—	0.40	—	18.7	86	
45	—	—	—	20	—	—	—	—	18.2		
50	—	—	—	20	—	—	—	—	14.0		
60	—	—	—	20.3	—	—	—	—	11.6		
70	—	—	—	20.4	—	—	—	—	9.5		
80	—	—	—	20.5	—	—	—	—	8.0		
90	—	—	—	20.5	—	—	—	—	6.5		
100	—	—	—	20.5	—	—	—	—	5.5		
150	—	—	—	19.5	—	—	—	—	2.7		
200	—	—	—	19.0	—	—	—	—	1.0		
400	—	34.4	—	—	—	—	—	—	—	—	82.5
500	—	—	—	—	—	—	—	—	—	—	74.0
600	—	27.6	—	—	—	—	—	—	—	—	61.7
700	—	—	—	—	—	—	—	—	—	—	—
800	—	23.5	—	—	—	—	—	—	—	—	54.5
1000	—	19.1	—	—	—	—	—	—	—	—	52.1
1500	—	—	—	—	—	—	—	—	—	—	44.0
1750	—	—	—	—	—	—	—	—	—	—	39.7

[a] Units—oxygen: 10^{-17} cm^2/atom; CO, CO_2: 10^{-17} cm^2/molecule.

[b] Key to references: (1) Stier and Barnett (1956); (2) Rose *et al.* (1958); (3) Fogel' *et al.* (1957); (4) Fogel' *et al.* (1958a); (5) Fogel' and Mitin (1956); (6) Gilbody and Hasted (1956).

[c] Values probably too large by a factor of 1.5 to 2. [see Fogel' *et al.* (1958b)].

5.4 Experimental Results in the Charge Changing Cross Sections of Hydrogen Beams

The experimental results on charge changing collision cross sections for H^-, H^0, and H^+ projectiles are given in Tables VI to XIII. The maximum observed value of $F_{\bar{1}\infty}$ in an equilibrated beam is 0.04 according to Stier and Barnett (1956) in argon gas at 4-kev kinetic energy. Thus, in a great many experiments hydrogen can be treated as a two-component system, and any of the experimental methods discussed in § 4.3 can be applied. Table VI gives a summary of the experimental methods which have been used.

TABLE XIII

Charge Changing Collision Cross Sections for Hydrogen Beams in Helium, Neon, and Argon Gases[a, b]

Kinetic energy (kev)	Helium		Neon		Argon			
	σ_{01} (1)	σ_{10} (1)	σ_{01} (1)	σ_{10} (1)	$\sigma_{\bar{1}0}$ (2)	σ_{01} (1)	σ_{10} (1)	σ_{10} (3)
50	12	12	20	11.5	—	48	33	45
60	12	9.7	20	9.8	—	48	22	36
70	11	6.3	20	8.5	—	48	20	—
80	10	4.9	20	7.5	—	48	16	26
90	9.2	3.8	19.5	6.2	—	48	12	—
100	8.8	3.0	—	5.8	—	45	9.8	18
150	7.0	0.93	19	2.8	—	40	3.8	6
200	5.6	0.36	18	1.4	—	37	1.1	
250	4.9	0.16	—	—	—	30	0.38	
300	4.0	0.70	—	—	—	29	0.15	
350	3.6	0.038	—	—	—	26	0.074	
400	3.1	0.020	—	—	62	22	0.050	
450	2.9	0.011	—	—	—	22	0.038	
500	2.4	0.0068	—	—	52	20	0.029	
600	2.2	0.0030	—	—	51	19	0.019	
700	2.0	0.0018	—	—	—	19	0.015	
800	1.6	0.0_385	—	—	43	18	0.011	
900	1.5	0.0_368	—	—	—	17	0.0088	
1000	1.4	0.0_352	—	—	42	16	0.0076	
1500	—	—	—	—	36			
1750	—	—	—	—	33			

[a] Units of 10^{-17} cm^2 per noble gas atom.

[b] Key to references: (1) Below 200 kev, Stier and Barnett (1956); above 200 kev, Barnett and Reynolds (1958); (2) Rose *et al.* (1958); (3) Afrosimov *et al.* (1958a).

Figure 9 shows the energy dependence of the cross sections of a hydrogen beam in hydrogen gas. The features displayed are not widely different in other chemically active target gases. The capture cross sections $\sigma_{0\bar{1}}$, $\sigma_{1\bar{1}}$, and σ_{10} fall off very rapidly at high energies whereas the loss cross sections $\sigma_{\bar{1}1}$, $\sigma_{\bar{1}0}$, and σ_{01} fall off quite slowly.

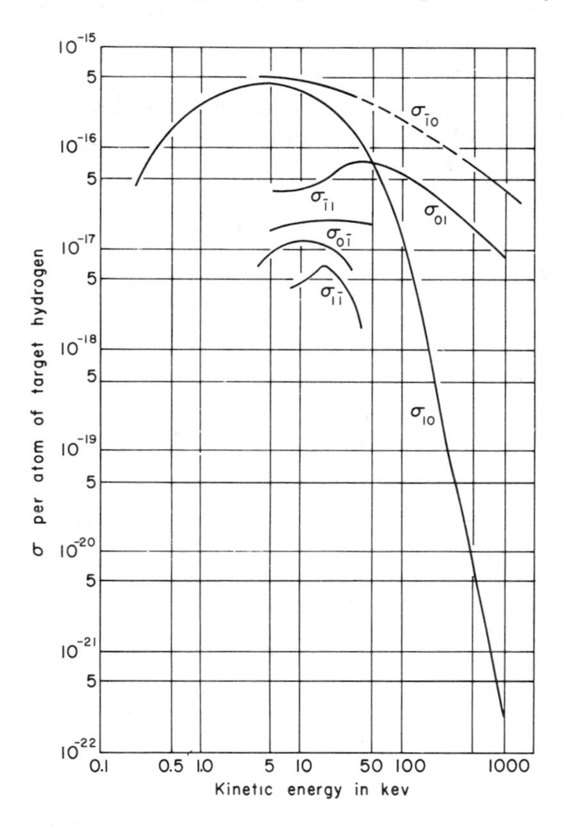

FIG. 9. The charge changing collision cross sections for hydrogen atoms and ions in hydrogen gas.

6 *Results on Helium Beams*

6.1 THE ANGULAR DISTRIBUTION OF PARTICLES SCATTERED FROM A HELIUM BEAM

Figure 10 shows the fraction F_0 in a helium beam scattered into a fixed angle 5° when it passes through helium gas, as a function of the projectile energy. It presents resonant peaks as in the case mentioned previously of scattering-with-capture from a hydrogen beam.

It is well known that quantum mechanics predicts a resonant exchange when an electron can be in either of two identical potential wells, as is

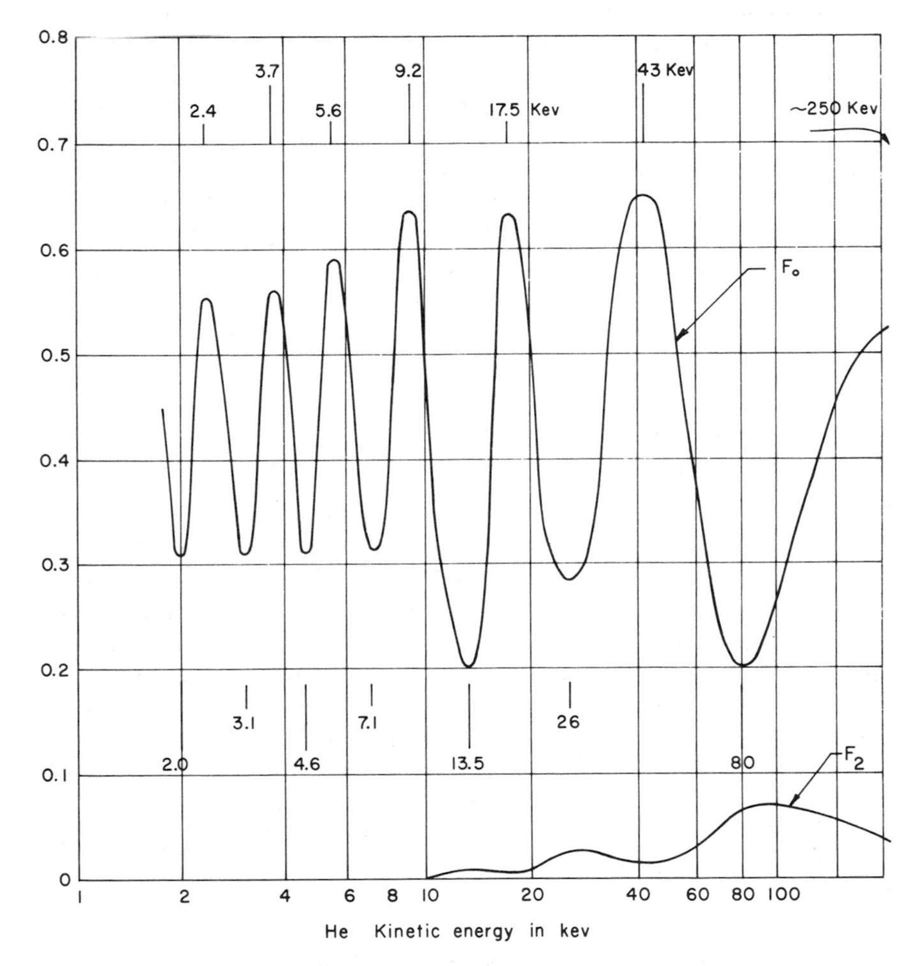

FIG. 10. Fractions of He^0 and He^{++} in the particles scattered at 5° (lab.) in single collisions of a He^+ beam traversing helium gas, versus the He^+ kinetic energy (Ziemba *et al.*, 1960).

the case in a close approach of a He^+ ion to a He^0 atom. The frequency of this exchange is determined by the difference in the energy of the eigenstates corresponding to the symmetrized and antisymmetrized eigenfunctions of the situation. The period of the resonance oscillation is Planck's constant divided by this difference. Transfer of the electron is most likely if the time of interaction, determined by the velocity and the range of atomic forces, is in the vicinity of an integral number of half-periods. If Fig. 10 is replotted so that the abscissas represent times proportional to the inverse of the velocity of approach of projectile to target, the peaks appear equally spaced, lending substance to the interpretation suggested above.

Fuls *et al.* (1957) have measured the charge composition of the particles scattered from a helium beam in helium, neon, and argon. These components are not to be interpreted as being in charge equilibrium in the sense commonly used in this chapter, since they are produced by single scatterings from a beam of He^+ ions, and the process σ_{11} is included. The oscillatory character of the process whose cross section is σ_{10} is clearly seen by fixing on the angle 4° in helium and noting that He^0 is 0.24 at 25 kev, 0.55 at 50 kev and 0.28 at 100 kev.

6.2 Charge Composition of an Equilibrated Helium Beam

The equilibrium fractions $F_{i\infty}$ in a rectilinear helium beam traversing various gases, and emerging from solids are given in Tables XIV and XV. Some of the data shown in Table XIV were computed from observed cross sections using (19)-(21). The fractions $F_{0\infty}$ are not listed, because they can easily be found, in each case, by subtracting from unity the sum of the $F_{i\infty}$'s which are given. It may be noted that in the region around 100 kev, the target gas neon is notably more effective than the other gases shown in producing the doubly charged constituent He^{++}.

6.3 The Differential Scattering Cross Section for Helium Projectiles

Jones *et al.* (1959) and Fuls *et al.* (1957) have measured the cross section in cm²/sterad for scattering (summation of all scatterings irrespective of charge) at various laboratory angles and energies in helium, neon, and argon. Thus, the absolute value of the differential cross section corresponding to a charge component may be found from the charge analysis of the scattered beam.

TABLE XIV

EQUILIBRIUM FRACTIONS OF A HELIUM BEAM IN VARIOUS TARGET GASES[a]

Kinetic energy (kev)	Hydrogen $F_{\bar{1}\infty}+F_{0\infty}+F_{1\infty}+F_{2\infty}=1$			Helium $F_{0\infty}+F_{1\infty}+F_{2\infty}=1$		Nitrogen $F_{0\infty}+F_{1\infty}+F_{2\infty}=1$		Air $F_{0\infty}+F_{1\infty}+F_{2\infty}=1$		Oxygen $F_{0\infty}+F_{1\infty}+F_{2\infty}=1$		Neon $F_{0\infty}+F_{1\infty}+F_{2\infty}=1$		Argon $F_{0\infty}+F_{1\infty}+F_{2\infty}=1$	
	$F_{\bar{1}\infty}$ (1)	$F_{1\infty}$ (2)	$F_{2\infty}$ (2)	$F_{1\infty}$ (2)	$F_{2\infty}$ (2)	$F_{1\infty}$ (2)	$F_{2\infty}$ (2)	$F_{1\infty}$ (3)	$F_{2\infty}$ (3)	$F_{1\infty}$ (2)	$F_{2\infty}$ (2)	$F_{1\infty}$ (2)	$F_{2\infty}$ (2)	$F_{1\infty}$ (4)	$F_{2\infty}$ (4)
4	—	—	—	—	—	0.020									
8	—	0.151	—	0.028	—	0.037	—	—	—	0.027	—	0.018	—	0.05	
12	—	0.161	—	0.045	—	0.054	—	—	—	0.054	—	0.030	—	0.016	
16	—	0.166	—	0.062	—	0.072	—	—	—	0.071	—	0.041	—	0.024	
20	—	0.166	—	0.075	—	0.094	—	—	—	0.097	—	0.060	—	0.028	
40	0.0_452	0.175	—	0.154	0.0023	0.190	—	—	—	0.220	—	0.164	0.0012	0.115	
60	0.0_310	0.195	—	0.204	0.0039	0.275	—	—	—	0.293	0.0021	0.251	0.0014	0.200	
80	0.0_316	0.240	—	0.248	0.0048	0.348	0.0023	—	—	0.356	0.0040	0.320	0.0078	0.279	0.0012
100	0.0_320	0.290	0.001	0.304	0.006	0.406	0.0041	0.358	0.004	0.408	0.0069	0.379	0.012	0.345	0.0026
125	0.0_323	0.374	0.002	0.357	0.009	0.474	0.0084	0.427	0.008	0.464	0.012	0.435	0.017	0.428	0.0052
150	0.0_323	0.458	0.004	0.410	0.011	0.530	0.015	0.506	0.012	0.511	0.018	0.482	0.023	0.500	0.0092
175	0.0_320	0.500	0.008	0.441	0.015	0.580	0.024	0.540	0.0175	0.556	0.029	0.521	0.029	0.560	0.016
200	0.0_317	0.543	0.012	0.473	0.019	0.625	0.035	0.578	0.024	0.588	0.034	0.551	0.035	0.604	0.024
250	—	0.623	0.029	0.586	0.038	—	—	0.619	0.050	—	—	—	—	0.76	0.047
300	—	0.684	0.067	0.623	0.062	—	—	0.630	0.077	—	—	—	—	0.77	0.085
350	—	0.744	0.104	0.660	0.085	—	—	0.647	0.127	—	—	—	—	0.74	0.14
400	—	0.723	0.164	0.679	0.120	—	—	0.640	0.180	—	—	—	—	0.71	0.22
450	—	0.702	0.223	0.698	0.154	—	—	0.645	0.238	—	—	—	—	0.65	0.29
480	—	—	—	0.68	0.20	—	—	0.66	0.29	—	—	—	—	0.62	0.33
646	—	—	—	—	—	—	—	0.50	0.50						
1696	—	—	—	—	—	—	—	0.12	0.88						
4440	—	—	—	—	—	—	—	0.015	0.985						
6780	—	—	—	—	—	—	—	0.005	0.995						

[a] Key to references: (1) Jorgensen (1958); (2) 4-200 kev, Barnett and Stier (1958); 200-450 kev, Allison *et al.* (1956), Allison (1958a); 480 kev, Snitzer (1953); (3) 100-450 kev, computed from cross sections, Allison (1958a); 480 kev, Snitzer (1953); 646-6780, Rutherford (1924); (4) 8-200 kev, Barnett and Stier (1958); 200-480 kev, Snitzer (1953).

TABLE XV

THE CHARGE COMPOSITION OF HELIUM BEAMS EQUILIBRATED IN SOLIDS[a]

Kinetic energy (kev)	Scattered and transmitted in Be, A1, and Ag foils			$F_{1\infty}$ emergent from		$F_{i\infty}$ emergent from celluloid		
	$F_{0\infty}$ (1)	$F_{1\infty}$ (1)	$F_{2\infty}$ (1)	Mica foil (2)	Au foil (2)	$F_{0\infty}$ (3)	$F_{1\infty}$ (3)	$F_{2\infty}$ (3)
130	0.58	0.40	0.017					
200	0.39	0.57	0.04					
300	0.24	0.65	0.11					
400	0.15	0.65	0.20					
406	—	—	—	0.68				
500	0.09	0.59	0.32	—				
590	—	—	—	0.60				
600	0.06	0.52	0.42	—				
640	—	—	—	0.50				
700	0.02	0.43	0.55	—				
750	—	—	—	—	—	0.017	0.368	0.615
787	—	—	—	0.35				
800	—	0.36	0.64					
900	—	0.29	0.71					
1000	—	0.24	0.76					
1100	—	0.20	0.80					
1260	—	—	—	0.16				
1330	—	—	—	—	—	0.002	0.145	0.853
1355	—	—	—	0.14	0.14			
1521	—	—	—	Ag foil	0.045			
1696	—	—	—	0.12	—			
1935	—	—	—	0.061	0.065			
2323	—	—	—	Al foil	0.045			
2408	—	—	—	0.038	0.038			
2719	—	—	—	0.032	0.036			
3406	—	—	—	—	0.024			
4262	—	—	—	Cu foil	0.015			
4440	—	—	—	0.015				
4915	—	—	—	0.012				
6028	—	—	—	0.0079	0.0088			
7000	—	—	—	—	—	0.0_31	0.005	0.995
7684	—	—	—	0.062	0.050			

[a] Key to references: (1) Dissanaike (1953); (2) Rutherford (1924), Henderson (1922); Nikolaev *et al.* (1957).

6.4 Total Charge Changing Cross Sections for Helium Beams

A summary of investigations leading to cross-section determinations for helium beams is given in Table XVI. Charge changing cross sections for helium beams through hydrogen, helium, nitrogen, oxygen, and air gases are given in Tables XVII-XIX.

In the region 150 to 700 kev, helium cannot even approximately be treated as less than a three-component system, since He^0, He^+, and

TABLE XVI

Summary of Charge Changing Cross Section Investigations with Helium Beams

Investigation	Quantity measured	kev range	Target gases	Method
Rutherford (1924)	σ_{21}	646-6780	Air	Attenuation in a magnetic field
Rudnick (1931)	σ_{01}, σ_{10}	5-20	He	Attenuation in a transverse electric field; $F_{0\infty}$
Hasted (1955)	σ_{10}	0.2-4	H_2, He, Ne, Ar, Kr	Ion collection
Stedeford (1955)	σ_{10}	4-40	H_2, He, Ne, Ar, Kr	Ion collection
Allison *et al.* (1956)	σ_{10}, σ_{12} ($\sigma_{20}+\sigma_{21}$)	100-450	H_2, He, Air	Attenuation in a magnetic field
Dukel'skii *et al.* (1956)	$\sigma_{1\bar{1}}$	60-180	He, Ne, Ar, Kr	Growth of $F_{\bar{i}}$ at small π-values
Fedorenko *et al.* (1956)	σ_{10}	20-100	He	Ion collection
Gilbody and Hasted (1956)	σ_{10}	0.010-0.3 3-40	He	Ion collection
de Heer (1956)	σ_{10}	10-20	H_2	Ion collection
Ghosh and Sheridan (1957)	σ_{10}	0.025-0.85	He	Ion collection
Allison (1958a)	σ_{20}, σ_{21}	100-450	H_2, He, air	F_2/F at low π plus $\sigma_{20}+\sigma_{21}$
Allison (1958b)	σ_{01}	100-450	H_2, He, air	Attenuation in a magnetic field
Barnett and Stier (1958)	σ_{01}, σ_{10}	20-200	H_2, He, N_2, O_2, Ne, Ar	Attenuation of He^0 in electric field, $F_{0\infty}$, $F_{1\infty}$
Jones *et al.* (1959)	σ_{10}, σ_{12}	25-100	He, Ne, Ar	Integration of angular distribution
Fogel' *et al.* (1960)	$\sigma_{0\bar{1}}$, σ_{01}	10-50	Kr, Xe	$dF_1/d\pi$ and $dF_{\bar{1}}/d\pi$ at low π-values

He^{++} all coexist, constituting beam fractions of more than 2%. None of the cross-section measuring methods gives simply and directly a single probability σ_{if}; additional information, such a $F_{i\infty}$ determinations, is needed to interpret an experimental result. In spite of this there is, in general, good agreement on those cross sections which have been measured by various investigators.

TABLE XVII

Charge Changing Collision Cross Sections for Helium Beams in Hydrogen[a, b]

Kinetic energy (kev)	σ_{01}	σ_{02}	σ_{10}			σ_{12}	σ_{20}	σ_{21}
	(1)	(1)	(2)	(3)	(4)	(5)	(6)	(6)
0.2	—	—	7.5					
0.4	—	—	7.0					
0.6	—	—	6.5					
0.8	—	—	6.0					
1	—	—	5.5					
2	—	—	4.2					
4	0.25	—	3.5					
6	—	—	4.2					
8	0.70	—	4.7	3.9				
12	1.0	—	5.7	5.4	6.4			
16	1.5	—	7.5	6.8	7.9			
20	1.8	—	7.8	8.0	10.1			
25	—	—	8.5	10.5				
30	2.0	—	10	—				
35	—	—	11	11				
40	2.4	—	12	—				
60	3.0	—	—	—				
80	3.9	—	—	12				
100	4.2	—	—	9.8	—	0.13	—	38
125	5.1	—	—	8.7	—	0.18	—	32
150	5.5	—	—	7.1	—	0.24	1.1	26
175	5.8	—	—	5.7	—	0.37	1.0	24
200	5.9	—	—	4.8	—	0.50	0.98	21
250	6.0	—	—	3.2	—	0.64	0.87	14
300	6.2	—	—	2.2	—	0.81	0.53	11
350	6.2	<0.1	—	1.2	—	0.98	0.20	6.8
400	6.3	—	—	0.95	—	0.89	0.16	4.9
450	6.7	<0.1		0.70	—	0.80	0.12	2.4

[a] Units of 10^{-17} cm^2 per hydrogen atom.

[b] Key to references: (1) 4-200 kev, Barnett and Stier (1958); 200-450 kev, Allison (1958b); (2) Hasted (1955), Stedeford (1955); (3) 8-200 kev, Barnett and Stier (1958); 200-450 kev Allison (1958a); (4) de Heer (1956); (5) Allison *et al.* (1956); (6) Allison (1958a).

TABLE XVIII

CHARGE CHANGING COLLISION CROSS SECTIONS FOR HELIUM BEAMS IN HELIUM GAS[a,b]

Kinetic energy (kev)	σ_{01} (1)	σ_{01} (2)	σ_{01} (3)	σ_{02} (2)	$\sigma_{1\bar{1}}$ (4)	σ_{10} (1)	σ_{10} (2)	σ_{10} (5)	σ_{10} (6)	σ_{10} (7)	σ_{10} (8)	σ_{10} (9)	σ_{12} (7)	σ_{12} (10)	σ_{20} (11)	σ_{21} (11)
0.2	—	—	—	—	—	—	—	108	—	—	145	124				
0.4	—	—	—	—	—	—	—	95	—	—	134	120				
0.6	—	—	—	—	—	—	—	92	—	—	128	120				
0.8	—	—	—	—	—	—	—	86	—	—	122	120				
1	—	—	—	—	—	—	—	82	—	—	117					
2	—	—	—	—	—	—	—	70-82								
4	—	0.46	—	—	—	—	72	74	—	—	71					
6	—	—	—	—	—	—	—	69	—	—	64					
8	1.8	1.8	1.8	—	—	34	60	63	—	—	58					
12	2.9	3.0	—	—	—	33	58	58	—	—	51					
16	3.8	4.0	2.6	—	—	31	53	55	—	—	49					
20	4.1	4.8	3.2	—	—	29	50	52	50	—	46					
25	—	—	3.8	—	—	—	—	48	44	38	43	—	0.093			
30	—	5.9	4.6	—	—	—	43	44	41	—	40					
35	—	—	5.7	—	—	—	—	39	37	—	40					
40	—	6.2	—	—	—	—	38	38	34	—	39					
50	—	—	—	—	—	—	—	—	—	30	—	—	0.26			
60	—	7.3	—	—	0.0_32	—	30	—	28							
80	—	8.0	—	—	0.0_32	—	24	—	25							
100	—	7.6	—	—	0.0_32	—	17	—	20	19	—	—	0.34	0.56	$\sigma_{20} + \sigma_{21} = 38$	
125	—	8.5	—	—	0.0_32	—	15	—	—	—	—	—	—	0.60	$\sigma_{20} + \sigma_{21} = 26$	
150	—	9.0	—	—	0.0_32	—	12	—	—	—	—	—	—	0.64	5.7	19
175	—	9.4	—	—	0.0_32	—	10	—	—	—	—	—	—	0.76	4.7	17
200	—	9.7	—	—	—	—	8.6	—	—	—	—	—	—	0.88	3.7	15
250	—	9.9	—	—	—	—	6.2	—	—	—	—	—	—	1.0	2.7	14
300	—	9.8	—	—	—	—	4.4	—	—	—	—	—	—	1.2	1.9	12
350	—	9.4	—	<0.2	—	—	3.1	—	—	—	—	—	—	1.5	1.1	11
400	—	8.9	—	—	—	—	2.2	—	—	—	—	—	—	1.9	1.1	9.2
450	—	8.2	—	<0.2	—	—	1.5	—	—	—	—	—	—	2.1	1.1	7.5

[a] Units of 10^{-17} cm^2 per helium atom.

[b] Key to references: (1) Rudnick (1931); (2) 4-200 kev, Barnett and Stier (1958); 200-450 kev, Allison (1958b); (3) Fogel' *et al.* (1960); (4) Dukel'skii *et al.* (1956); (5) Hasted (1955), Stedeford (1955); (6) Fedorenko *et al.* (1956); (7) Jones *et al.* (1959); (8) Gilbody and Hasted (1956); (9) Ghosh and Sheridan (1957); (10) Allison *et al.* (1956); (11) Allison (1958a).

TABLE XIX

CHARGE CHANGING CROSS SECTIONS FOR HELIUM BEAMS IN NITROGEN, OXYGEN, AND AIR[a, b]

Kinetic energy (kev)	Nitrogen		Oxygen		Air					
	σ_{01} (1)	σ_{10} (1)	σ_{01} (1)	σ_{10} (1)	σ_{01} (2)	σ_{02} (3)	σ_{10} (2)	σ_{12} (4)	σ_{20} (3)	σ_{21} (5)
4	—	—	0.27							
8	0.83	21.5	1.1	39						
12	1.7	27	2.2	39						
16	2.3	30	3.2	41						
20	3.2	32	4.5	41						
40	7.2	32	9.3	34						
60	11	29	14	30						
80	14	26	16	28						
100	16	24	17.5	24	16.4	—	29	0.79	$\sigma_{20} + \sigma_{21} = 79$	
125	19	20	19.5	20						
150	20	18	20	19	21.0	$0.2 + 0.2$	20	1.2	11.8	50
175	23	15	22	16						
200	24	13	23	15	21.8	—	16	2.2	8.5	42
250	—	—	—	—	22.7	0.5 ± 0.4	12	3.1	5.2	35
350	—	—	—	—	21.2	0.8 ± 0.5	7.3	5.1	2.1	25
450	—	—	—	—	23.2	1.3 ± 0.4	4.0	6.6	1.3	17
646	—	—	—	—	—	—	—	6.2	—	6.2
1696	—	—	—	—	—	—	—	3.7	—	0.50
4440	—	—	—	—	—	—	—	2.4	—	0.036
6780	—	—	—	—	—	—	—	1.7	—	0.0084

[a] Units of 10^{-17} cm^2 per target atom.

[b] Key to references: (1) Barnett and Stier (1958); (2) Allison (1958b); (3) Allison (1958a); (4) 100-450 kev, Allison *et al.* (1956); 646-6780 kev, Rutherford (1924); (5) 100-450 kev, Allison (1958a); 646-6780 kev, Rutherford (1924).

6.5 EFFECTS OF METASTABLE ATOMS AND IONS IN CHARGE CHANGING COLLISION EXPERIMENTS

An effect which may well be due to metastable states of helium and neon was discovered by Barnett and Stier (1958). In order to measure the loss cross section σ_{01} for He^0 and Ne^0 projectiles, beams containing these atoms were prepared by conversion of He^+ and Ne^+ beams in hydrogen. The charged components of the emergent, equilibrated beam were then deflected away, and σ_{01} for the neutral constituent measured by the attenuation in an electric field when gas was admitted. If the pressure p in the converter cell was low (2-8 μ) the value of σ_{01} was dependent on this pressure, and increased as p decreased. It is likely that the cross section for electron loss from an excited atom is greater than that from a normal atom; and it is also likely that the neutral beam coming from the converter cell contained a higher percentage of metastable atoms at low pressure than at high pressure since the probability of the destruction of such atoms is an increasing function of the pressure. A similar experiment with H^0 and N^0 did not show such an effect.

Barnett and Stier (1958) and Allison (1958b) found that there is a large pressure region in which, for a given converter cell, there is a constant value of σ_{01}. This probably means that at high pressures there is no significant fraction of metastable helium atoms among the emergent neutrals. A more conservative conclusion is that if there is such a metastable fraction, it is invariant with converter cell pressure over a considerable range.

Jorgensen (1958) investigated the production of He^- as the hydrogen pressure was increased in a converter cell 50 cm long through which an originally homogeneous He^+ beam was passing. The pressure of hydrogen was varied from 4 to 34 μ; the He^+ kinetic energy from 40 to 200 kev. As the pressure increased, the fractional output of He^- did not rise to a plateau, but had a rather broad maximum. The values of $F_{\bar{1}\infty}$ quoted from Jorgensen in Table XIV are actually these maxima, rather than the pressure-independent values indicated by the notation.

To explain this behavior, Jorgensen uses the predictions of Holøien and Midtdal (1955) that the third electron in He^- is actually bound to the metastable configuration $1s2s\,^3S$.

Fogel' *et al.* (1959) have obtained evidence that the cross section $\sigma_{1\bar{1}}$ for a lithium beam in hydrogen, argon, and krypton depends on whether the Li^+ ions are produced from a heated lithium mineral (spodumene or β-eucryptite) or from a radio-frequency ion source. It is suggested that the beam from the hot filament is free from ions in the metastable states whereas the radio-frequency produced beam contains ions in these states. Fogel' *et al.* also find that metastable ions may exist in B^+, O^+, and F^+ beams.

7 *Results on Lithium Beams*

7.1 Equilibrium Charge Compositions of Lithium Beams

With the exception of some values of the ratio $F_{2\infty}/F_{1\infty}$ in air (Leviant *et al.*, 1955), the experimental values of $F_{i\infty}$ for lithium beams in gases are due to Allison *et al.* (1960a). The quantities directly determined by experiment were $F_{\bar{1}\infty}/F_{1\infty}$, $F_{0\infty}$ (calorimetric detector), $F_{2\infty}/F_{1\infty}$, and $F_{3\infty}/F_{1\infty}$, provided these ionic ratios exceeded 5×10^{-6}. The main target gases were hydrogen, helium, nitrogen, nitrous oxide, and propane.

Table XX shows the equilibrium fractions of lithium beams in hydrogen and helium. The maximum yield of Li^- seems to be about $F_{\bar{1}\infty} = 3 \times 10^{-4}$, attainable in these two gases and (at much lower

TABLE XX

EQUILIBRIUM FRACTIONS OF A LITHIUM BEAM IN HYDROGEN AND IN HELIUM GASES[a]

Li^7 kinetic energy (kev)	Hydrogen					Helium				
	$F_{\bar{1}\infty}$	$F_{0\infty}$	$F_{1\infty}$	$F_{2\infty}$	$F_{3\infty}$	$F_{\bar{1}\infty}$	$F_{0\infty}$	$F_{1\infty}$	$F_{2\infty}$	$F_{3\infty}$
20	$0.0_{4}15$	0.18	0.82							
30	$0.0_{4}48$	0.17	0.83							
40	$0.0_{4}80$	0.18	0.82	—	—	—	<0.09	>0.90	$0.0_{2}70$	
50	$0.0_{3}15$	0.20	0.80	$0.0_{4}73$	—	—	<0.09	>0.90	$0.0_{2}83$	
75	$0.0_{3}18$	0.21	0.79	$0.0_{3}29$	—	$0.0_{4}9$	—	—	0.012	
100	$0.0_{3}20$	0.22	0.78	$0.0_{2}11$	—	$0.0_{3}17$	0.17	0.81	0.016	
150	$0.0_{3}24$	0.23	0.77	$0.0_{2}48$	—	$0.0_{3}28$	0.19	0.79	0.021	
200	$0.0_{3}25$	0.22	0.78	0.011	—	$0.0_{3}32$	0.18	0.79	0.026	
250	$0.0_{3}35$	0.19	0.81	0.020	—	$0.0_{3}35$	0.16	0.81	0.031	$0.0_{4}13$
300	$0.0_{3}31$	0.16	0.84	0.029	—	$0.0_{3}32$	0.14	0.82	0.039	$0.0_{4}39$
350	$0.0_{3}28$	0.11	0.85	0.042	$0.0_{4}16$	$0.0_{3}28$	0.10	0.85	0.049	$0.0_{4}75$
400	$0.0_{3}26$	0.09	0.86	0.051	$0.0_{4}54$	$0.0_{3}26$	0.076	0.86	0.060	$0.0_{3}20$
475	$0.0_{3}20$	0.06	0.87	0.070	$0.0_{3}31$	$0.0_{3}20$	0.051	0.85	0.10	$0.0_{3}56$

[a] Allison *et al.* (1960a).

energies) in nitrous oxide and propane. A remarkable feature is the relatively high $F_{2\infty}$ in helium at the lower energies; thus, for a 60-kev Li^7 beam, $F_{2\infty}$ is 1.3×10^{-4} and 4.7×10^{-4} in hydrogen and nitrogen but is 9.4×10^{-3} in helium (Allison *et al.*, 1960b).

Teplova *et al.* (1957) have studied the equilibrium charge composition of lithium beams emergent from celluloid foil.

7.2 TOTAL CHARGE CHANGING CROSS SECTIONS FOR LITHIUM BEAMS

Table XXI gives a summary of the investigations reported on lithium charge changing cross sections. Most of the numerical results are from experiments of Allison *et al.* (1960a) using the method of beam attenuation in a magnetic field. As discussed in § 4.3 this method actually measures $\Sigma_f \sigma_{if}$, and in the lithium system i and f may range through the values $-1, 0, 1, 2, 3$. Thus, simplifying assumptions and further

TABLE XXI

SUMMARY OF CHARGE CHANGING CROSS SECTION INVESTIGATIONS WITH LITHIUM BEAMS

Investigation	Quantity measured	kev range	Target gases	Method
Hasted (1952)	$(\sigma_{\bar{1}0} + \sigma_{\bar{1}1})$	0.2-3	Neon	Ion collection
Leviant *et al.* (1955)	σ_{12}	80-250	Air	$(dF_2/d\pi)$ at low π-values
Fogel' *et al.* (1959)	$\sigma_{1\bar{1}}$	5-60	H_2	$(dF_{\bar{1}}/d\pi)$ at los π-values
Van Eck and Kistemaker (1960)	σ_{01}, σ_{10}	5-22.5	H_2, He	Ion collection
Allison *et al.* (1960a)	$(\sigma_{\bar{1}0} + \sigma_{\bar{1}1})$ $(\sigma_{31} + \sigma_{32})$	20-480	H_2, He, N_2	Attenuation in a magnetic field
Allison *et al.* (1960a)	σ_{01}, σ_{10} σ_{21}, σ_{12}	10-480	H_2, He, N_2	Attenuation in a magnetic field plus $F_{i\infty}$ values

information concerning the $F_{i\infty}$ values are essential for the deduction of individual cross sections from the direct experimental results. The assumptions used by Allison were:

(1) Electron loss cross sections in which two or more electrons of widely different ionization potential are stripped off in one collision are (in the rather low velocity range 5×10^7 to 4×10^8 cm/sec covered by the experiments) small compared with πa_0^2 $(8.8 \times 10^{-17}$ cm$^2)$. Thus, $\sigma_{\bar{1}3}$, σ_{03}, σ_{13}, $\sigma_{\bar{1}2}$, and σ_{02} were neglected in additive combinations with other, more probable cross sections.

(2) Electron capture cross sections in which several electrons are caught, one of these having a binding energy to lithium less than it had in the target gas molecule, are small compared with πa_0^2. Such cross sections are, in this case, $\sigma_{3\bar{1}}$, $\sigma_{2\bar{1}}$, σ_{30}, σ_{20}.

(3) Any cross section sum $\Sigma \sigma_{if}$ may safely be assumed not to exceed a certain upper limit, even at such velocities that $F_{i\infty}$ is so small that it is not feasible to prepare a beam for the cross-section determination. This upper limit is approximately the maximum value of $\Sigma \sigma_{\bar{1}f}$.

By using these three assumptions, and the measured values of $F_{i\infty}$ together with the cross-section sums in the basic equations (19)-(21), it was possible to deduce individual values of σ_{01}, σ_{10}, σ_{21}, σ_{12}, and the sums $(\sigma_{\bar{1}0} + \sigma_{\bar{1}1})$, $(\sigma_{31} + \sigma_{32})$.

7.3 Properties of Isoelectronic Projectiles at a Common Velocity

The study of charge changing cross sections in the hydrogen, helium, and lithium projectile systems makes possible the comparison of projectiles having the same electron configuration, and at the same velocity. At a velocity of 3.62×10^8 cm/sec the kinetic energies are 68, 273, and 478 kev, respectively (Li^7). The cross sections σ_{10}, σ_{21}, $(\sigma_{32} + \sigma_{31})$ in hydrogen are 3.0, 12, and 30×10^{-17} cm^2 per H atom, respectively. These are in the order 1 : 4 : 9 or Z^2. Such a simple relation is not found in more complex target gases.

Another comparison of the behavior of isoelectronic projectiles may be made by looking at the electron capture by H^0, He^+, and Li^{++} traversing various target gases at the same velocity. The capture cross sections to be compared are $\sigma_{0\bar{1}}$, σ_{10}, σ_{21} for the three projectiles, respectively (Fig. 11). There is a monotonic increase in the capture probability in hydrogen and in nitrogen as the electron affinity of the projectile

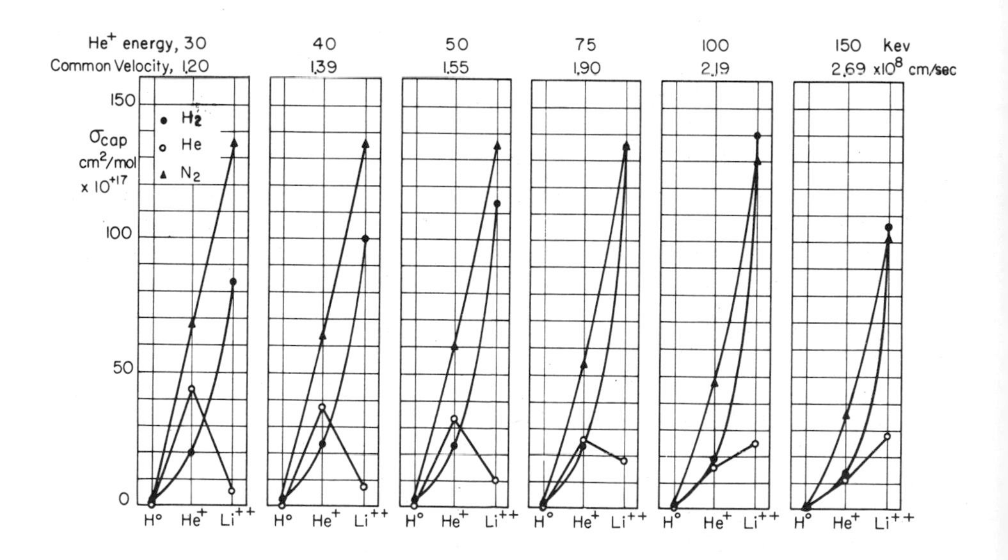

Fig. 11. Comparison of the charge changing collision probabilities of projectile atoms and ions of the same electronic structure and the same translational velocity through the gases H_2, He, and N_2. The isoelectronic structure is the nucleus plus one *s* electron. The anomalously large cross section for the resonant reaction He (H_e^+, He) He^+ at low velocities is clearly indicated.

increases (0.73, 24.5, and 75 ev, respectively). Owing to the resonance between He^+ and He^0, the behavior in helium is entirely different: at the lower velocities σ_{10} is much larger than would be predicted by interpolation between $\sigma_{0\bar{1}}$ and σ_{21}.

8 Results on Boron Atomic and Ionic Beams; the Average Ionic Charge

Nikolaev *et al.* (1957) have studied the equilibrium charge of boron beams traversing various gases and emerging from a celluloid. In the same investigation they found $F_{i\infty}$ values for carbon, nitrogen, oxygen, fluorine, and neon beams. The energies of the beams were in the range 0.2-6 Mev. It was found that at each energy the distribution around the average

$$\bar{\imath} = \sum_i iF_{i\infty} \tag{40}$$

is closely Gaussian so that

$$F_{(i-\bar{\imath}),\infty} = \frac{1}{\beta\sqrt{2\pi}} \exp \frac{-(i-\bar{\imath})^2}{2\beta^2} \tag{41}$$

where β is a constant. For boron, carbon, nitrogen, oxygen, and neon beams it is recommended that β be taken to be 0.65, 0.73, 0.80, 0.83, and 0.76, respectively.

Fogel' *et al.* (1960) give cross sections for electron capture and loss by boron atoms.

9 Results on Carbon Beams

Electron capture and loss cross sections for carbon atoms moving in various target gases are available from experiments by Fogel' *et al.* (1958c). Charge changing collision cross sections for C^- and C^+ ions have been measured by Hasted (1952), Fogel' *et al.* (1956, 1958b), and Gilbody and Hasted (1956).

10 Results on Nitrogen Beams

Reynolds *et al.* (1955b) reports a charge composition analysis of a prepared 26-Mev nitrogen beam, in three charged states; $F_z = 1$ for

$z = 5, 6, 7$, after passing through various zapon thicknesses. The thinner foils (10 μgm/cm^2) are not thick enough to equilibrate this high energy beam.

Stier *et al.* (1954) and Nikolaev *et al.* (1957) give values of $\bar{i}$, defined in (40) for nitrogen beams in hydrogen, air, and argon in the range 0.1 to 6 Mev. By using (41) an estimate of $F_{i\infty}$ values in the vicinity of i may hence be made.

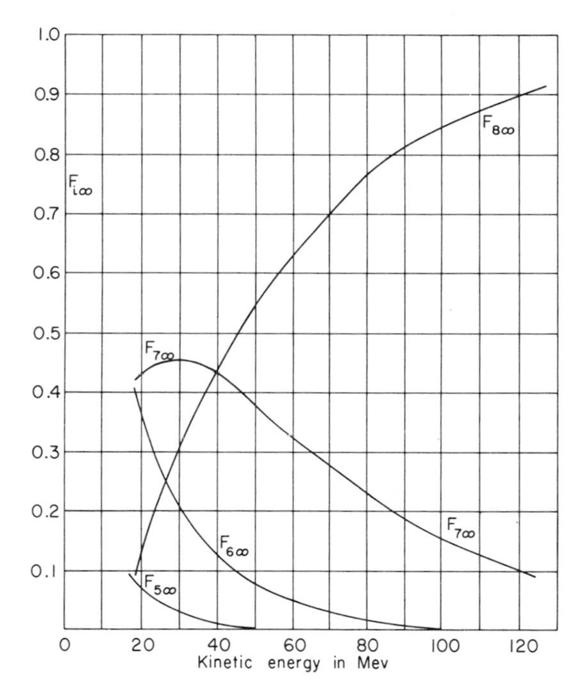

FIG. 12. Equilibrium fractions in an oxygen beam emerging from aluminum foil (Northcliffe, 1960).

Information concerning the composition of charge equilibrated nitrogen beams, of energies from 20 kev to 30 Mev, has been obtained by Stier *et al.* (1954), Reynolds *et al.* (1954, 1955a), and Korsunsky *et al.* (1955).

Data from various sources (Hasted, 1951, 1952; Fedorenko, 1954a, b; de Heer, 1956; Gilbody and Hasted, 1956) are available on the charge changing collision cross sections of N^0 and N^+ in the low-energy range from 0.2 to 60 kev. In the high-energy region Korsunsky *et al.* (1955) give electron loss cross sections from 0.485 to 1.180 Mev. At 26 Mev Reynolds *et al.* (1955b) has estimated electron capture cross sections in a solid for N^{6+} and N^{7+}, and loss cross sections for N^{5+} and N^{6+}. These are the only cases of cross sections given for collisions in the solid phase.

11 Results on Oxygen Beams

Fogel' *et al.* (1956) have obtained data on equilibrium charge fractions in oxygen beams at low energies (13-45 kev). The charge composition of oxygen beams from the Yale heavy ion accelerator in the energy range 20 to 120 Mev is shown in Fig. 12 (Northcliffe, 1960).

Cross sections for charge changing collisions of oxygen atoms and ions are reported by Hasted (1951, 1952, 1953) and Fogel' *et al.* (1956, 1958b, c).

12 Results on Atomic and Ionic Beams of the Halogens

There seem to be no direct determinations of equilibrium charge fractions $F_{i\infty}$ of halogen ionic and atomic beams. The halogens are distinguished by the ease with which they form negative ions, and considerable data on $\sigma_{\bar{1}0}$ and $\sigma_{0\bar{1}}$ exist (Hasted, 1952, 1953; Dukel'skii and Zandberg, 1952; Fedorenko, 1954a; de Heer, 1956; Fogel' *et al.*, 1958b, 1960).

13 Results on Neon Beams

13.1 Charge Composition of Scattered and Unidirectional Neon Beams

Following the discovery of the resonances which occur when the neutral fraction in the scattering is plotted against energy for fixed angle (5° lab.) scattering of a He^+ beam by He (Fig. 10), other similar cases were investigated. In Fig. 13 the solid line curve for F_0 might be expected to show such an effect, since we are dealing with electron

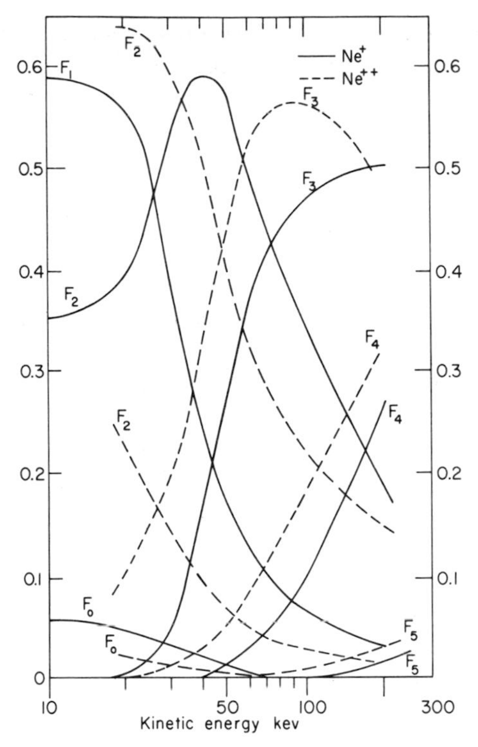

FIG. 13. Charge composition, in fractions of the total scattering into 5° (lab.) from single collisions of Ne^+ and Ne^{++} on Ne targets (Ziemba *et al.*, 1960).

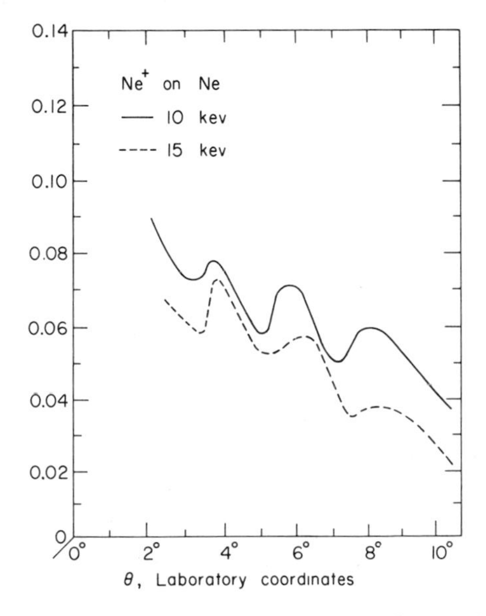

FIG. 14. Fraction of Ne^0 in the particles scattered from a Ne^+ beam at various laboratory angles. Neon gas target (Ziemba *et al.*, 1960).

capture by Ne^+ from neon atoms, but there is no evidence of it. However, in Fig. 14 we see evidence of resonance in the capture by Ne^+ from neon gas, as shown by peaks in the plot of the neutral fraction in the scattering, at constant energy, against scattering angle. The peaks are less prominent at 15 kev than at 10 kev (particularly at the larger scattering angles). At 25 kev Jones *et al.* (1959) and Fuls *et al.* (1957) failed to detect peaks.

Studies of the composition of charge-equilibrated unidirectional neon beams are given by Stier *et al.* (1954), Hubbard and Lauer (1955), and Nikolaev *et al.* (1957).

13.2 VALUES OF $d\sigma_{zi}(\theta)/d\Omega$ AND OF TOTAL CHARGE CHANGING CROSS SECTIONS, FOR NEON BEAMS

The works of Fuls *et al.* (1957) and Jones *et al.* (1959) give absolute values of $d\sigma(\theta)/d\Omega$ for the scattering (irrespective of ultimate charge state) of a Ne^+ ionic beam in neon and argon target gases. On using these values, the absolute differential cross section corresponding to a given charge component as being scattered from a Ne^+ beam may be found by multiplying by the corresponding charge fraction.

Figure 15, from the work of Fedorenko *et al.* (1960a) shows values of $d\sigma_{20}(\theta)/d\Omega$ and $d\sigma_{21}(\theta)/d\Omega$ for Ne^{++} ions moving in neon and krypton gases. The double electron capture process is essentially confined to angles very near the forward direction ($\theta < 0.5°$).

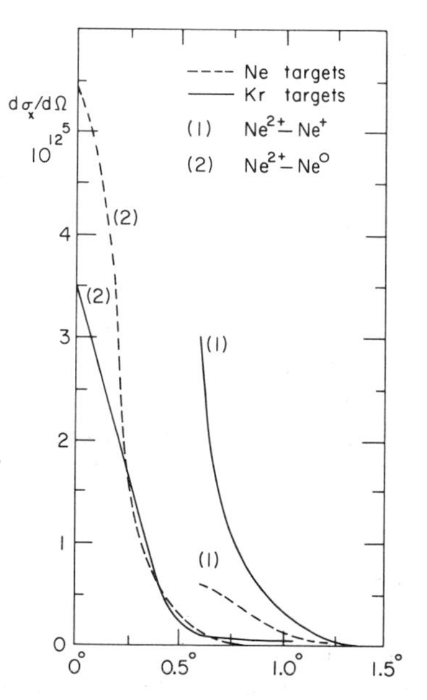

FIG. 15. Absolute differential electron capture cross sections for 33-kev neon ions from neon and krypton (Fedorenko *et al.*, 1960a).

Results of researches on the charge changing cross sections for neon projectiles in various target gases can be found in Hasted (1951), Fedorenko (1954a, b), Gilbody and Hasted (1956), de Heer (1956), Ghosh and Sheridan (1957), Flaks and Solov'ev (1958a, b), Flaks and Filippenko (1959), Jones *et al.* (1959), and Fedorenko *et al.* (1960a).

14 Results on Argon Beams

14.1 Charge Composition of Scattered and of Unidirectional Argon Beams

The information available on charge components in argon beams is similar in content to that for neon beams. In an experiment by Ziemba, Lockwood, Morgan, and Everhart similar to that which, when conducted with helium, showed characteristic resonance peaks (Fig. 10), the fraction F_0 of the scattered particles gives no indication of this effect, but as we have seen it is also not in evidence in neon (Fig. 13). Argon, however, exhibits an effect analogous to that seen in neon in Fig. 14. An earlier survey of the fraction of neutral Ar in the scattering by Fuls *et al.* (1957) and Jones *et al.* (1959) shows an indication of the maximum at 25 kev in the neighborhood of 4°.

Results on the equilibration of unidirectional argon beams in various gases are reported by Stier *et al.* (1954).

14.2 Values of $d\sigma_{zi}(\theta)/d\Omega$ and of Total Charge Changing Cross Sections, for Argon Beams

Fuls *et al.* (1957) and Jones *et al.* (1959) have measured the differential cross sections for all types of scattering with and without charge exchange, for Ar^+ in Ar. By using them in conjunction with the corresponding charge fractions, these may be converted into absolute units of $d\sigma_{zi}/d\Omega$.

Kaminker and Fedorenko (1955b) give detailed information on the scattering, with and without change of charge, from an Ar^+ beam traversing the noble gases. For a 75-kev Ar^+ beam, capture of an electron from argon is confined very closely to the forward direction; 90% of the projectiles, after the capture, appear within 0.3° (0.006 radian) of the direction of the original beam. The angular distribution of the events in which an electron is stripped from the argon is quite different; a significant probability for this extends to larger angles. The angular distribution of the probability for simultaneous stripping of 4 electrons is quite remarkable, showing a maximum at approximately 6° scattering angle (lab.) which is almost 100 times as high as a secondary maximum for this event, which occurs at 0°.

The charge changing collision cross sections σ_{if} for an argon beam have been measured by Hasted (1951), Fedorenko (1954a, b), Kaminker and Fedorenko (1955b), Fedorenko *et al.* (1956), de Heer (1956), Gilbody and Hasted (1956), Ghosh and Sheridan (1957), Flaks and Solov'ev (1958a), Sluyters *et al.* (1959), and Jones *et al.* (1959). There are large discrepancies between the results of various observers on σ_{10} in krypton gas.

15 Results on Krypton and Xenon Beams

Everhart and his associates have made extensive studies of the scattering of fast ion beams through various angles, and in particular, the scattering angle 5° in the laboratory system has been chosen as typical of large angle scattering. The work by Ziemba *et al.* (1960) shows the charge composition of the total scattered beam at this angle. This is to be compared with the analogous results for argon, neon (Fig. 13), and helium (Fig. 10). There is no certain evidence in the heavier noble gases for the resonances found in helium.

When, however, the fraction of neutral krypton atoms in the scattering of a 25-kev Kr^+ beam is studied at various scattering angles, there is faint evidence of a resonance peak at 7°.

Fedorenko *et al.* (1960b) gives some numerical values of differential cross sections (the scattering being separated out into charge components).

Data on charge changing collision cross sections for krypton atoms and ions can be found in the papers of de Heer (1956), Ghosh and Sheridan (1957), Flaks and Solovyev (1958a, b), Flaks and Filippenko (1959), and Fedorenko *et al.* (1960a).

Charge changing collision cross sections for xenon ions moving in the noble gases are given by Ghosh and Sheridan (1957), Flaks and Solov' ev (1958a, b). The resonant cross section σ_{10} for Xe^+ in xenon at low kinetic energies reaches the very large value of 515×10^{-17} cm^2.

16 Charge Changing Collisions of Fission Fragments

The heavy nuclei which undergo fission after the absorption of a thermal neutron divide assymmetrically into light and heavy fragments having velocities of the order 10^9 cm/sec. The light fragments having abundances above 0.1% range in atomic number Z from about 35 to 47, and have velocities of 5 to $6v_0$ ($v_0 = c/137 = 2.19 \times 10^8$ cm/sec). Heavy fragments of comparable abundance range in Z from 53 to 63, and in velocity between 3 and $4v_0$.

16.1 Equilibrium Charge of Fission Fragments

Lassen (1945, 1955) observed the $z_p e/M_p V_p$ spectrum of the fission fragments by placing the source in a magnetic field. The subscript p refers to the fission fragment, or projectile, z_p is the charge of the moving ion, M_p its mass, and V_p its velocity. From knowledge obtained otherwise concerning the average energy of the two groups, Lassen was able to compute an average charge which characterizes them as they emerge from a uranium metal foil. In later experiments gas was admitted to the space between the source of fragments and the detector, and an equilibrium charge distribution, characteristic of the gas and lower than that of the nascent fragments, was determined. These averages are shown in Table XXII. Since the fissions in the uranium were initiated by neutron capture the effective depth of production in the metal was probably sufficient to impose a charge equilibrium on the emergent fragments which was typical of the solid foil. It is significant that this is higher than the gas-equilibrated values. Bohr and Lindhard (1954) have ascribed the difference to the fact that in a solid, and by inference, even in a gas at high pressure, the time interval between successive impacts is so brief that a projectile system which has been excited is more likely to suffer an ionizing impact than to return to the ground state with the emission of radiation.

TABLE XXII

Equilibrium Charges of Fission Fragments[a]

Medium traversed	Light fragment		Heavy fragment	
	$v_1 \sim 6v_0$	$v_1' \sim 5v_0$	$v_2 \sim 4v_0$	$v_2' \sim 3v_0$
H_2	15.8	13.4	12.6	9.2
He	14.1	11.7	11.6	8.6
N_2	15.1	13.8	13.9	10.5
Ar	15.4	13.7	14.6	10.4
U	20.0		22.0	
Mica		19.4		18.0

[a] Lassen (1955).

16.2 Capture and Loss of Electrons by Fission Fragments at Charge Equilibrium

Bell (1953) and Bohr and Lindhard (1954) have shown that the capture and loss cross sections, σ_c and σ_l, of a highly ionized projectile, like Xe^{15+}, of atomic number Z_p and ionic charge z_p moving with velocity V_p in a gas like argon, of moderately high atomic number Z_t are given approximately by

$$\sigma_c = Z_t^{1/3} z_p^2 (V_0/V_p)^3 \pi a_0^2 \quad (42)$$

$$\sigma_l = Z_t^{2/3} Z_p^{4/3} z_p^{-3} (V_p/V_0)^2 \pi a_0^2 \quad (43)$$

where V_0 is $c/137$ as before.

When the fission fragment is at charge equilibrium as it moves through the target substance the capture and loss cross sections have a common value which may be found by elimination of z_p from (42) and (43). The resulting cross section is roughly

$$\sigma = \sqrt{Z_t Z_p} (v_0/V_p) \pi a_0^2. \quad (44)$$

For a typical light fragment with $z_p = 35$ and $V_p = 4v_0$ moving in argon, we predict about $4.4\pi a_0^2$; for a heavy fragment with $z_p = 54$ and $V_p = 3.5v_0$, the charge changing collision cross section should be $9.4\pi a_0^2$.

Table XXIII

Charge Changing Collision Cross Sections at Charge Equilibrium for Fission Fragments Moving in Various Gases[a–c]

Gas	Light fragments		Heavy fragments	
	$V_p \sim 6v_0$	$V_p \sim 5v_0$	$V_p \sim 4v_0$	$V_p \sim 3v_0$
H_2	0.06 ± 0.02	0.8 ± 0.3	1.6 ± 0.6	8 ± 2
He	0.7 ± 0.3	4 ± 2	9 ± 2	17 ± 5
N_2	1.4 ± 0.5	8 ± 4	15 ± 8	20 ± 6
Ar	2.5 ± 0.8	—	24 ± 8	—

[a] Units of 8.8×10^{-17} cm² ($= \pi a_0^2$) per target atom.

[b] Lassen (1955).

[c] The values corresponding to velocities $6v_0$, $5v_0$, $4v_0$, and $3v_0$ refer to ionic charges z_p of 20, 19.4, 22, and 18 electronic units, respectively.

Lassen was able to make estimates of collision cross sections at equilibrium by studying the magnetic spectrum of the fission fragments traversing a chamber in which the gas pressure could be raised from low values to those in the range where the charge was independent of pressure. His results are shown in Table XXIII. They are in qualitative agreement with (44) in that the light fragments show lower cross section values than the heavy, and within each group the particles of higher velocity have lower cross sections.

References

Afrosimov, V. V., Il'in, R. N., and Fedorenko, N. V. (1958a) *Zh. tekh. Fiz.* **28**, 2266; (1958) *Soviet Phys.—JETP* **3**, 2080.

Afrosimov, V. V., Il'in, R. N., and Fedorenko, N. V. (1958b) *Zh. eksper. teor. Fiz.* **34**, 1398; (1958) *Soviet Phys.—JETP* **7**, 968.

Allison, S. K. (1958a) *Phys. Rev.* **109**, 76.

Allison, S. K. (1958b) *Phys. Rev.* **110**, 670.

Allison, S. K. (1958c) *Rev. mod. Phys.* **30**, 1137.

Allison, S. K., and Warshaw, S. D. (1953) *Rev. mod. Phys.* **25**, 779.

Allison, S. K., Cuevas, J., and Murphy, P. G. (1956) *Phys. Rev.* **102**, 1041.

Allison, S. K., Cuevas, J., and Garcia-Munoz, M. (1960a) *Phys. Rev.* **120**, 1266.

Allison, S. K., Cuevas, J., and Garcia-Munoz, M. (1960b) *Rev. sci. Instrum.* **31**, 1193.

Barnett, C. F., and Reynolds, H. K. (1958) *Phys. Rev.* **109**, 355.

Barnett, C. F., and Stier, P. M. (1958) *Phys. Rev.* **109**, 385.

Bartels, H. (1930) *Ann. Phys. (Leipzig)* **6**, 957.

Bell, G. I. (1953) *Phys. Rev.* **90**, 548.

Bohr, N., and Lindhard, J. (1954) *K. Danske Vidensk. Selsk. mat. fys. Medd.* **28** (7).

Briggs, G. H. (1927) *Proc. Roy. Soc.* **A114**, 341.

Curran, R. K., and Donahue, T. M. (1959) Office of Naval Research Tech. Report ONR-8.

de Heer, F. J. (1956) Ph.D. thesis, University of Leyden, Holland.

Dissanaike, G. A. (1953) *Phil. Mag.* **7**, 44, 1051.

*Dukel'skii, V. M., and Zandberg, E. J. (1952) *Dokl. Akad. Nauk SSSR* **82**, 33.

Dukel'skii, V. M., Afrosimov, V. V., and Fedorenko, N. V. (1956) *Zh. eksper. teor. Fiz.* **30**, 792; (1956) *Soviet Phys.—JETP* **3**, 764.

Fedorenko, N. V. (1954a) *Zh. tekh. Fiz.* **24**, 769.

Fedorenko, N. V. (1954b) *Zh. tekh. Fiz.* **24**, 2113.

Fedorenko, N. V., Afrosimov, V. V., and Kaminker, D. M. (1956) *Zh. tekh. Fiz.* **26**, 1929; (1956) *Soviet Phys.—JTP* **1**, 1861.

Fedorenko, N. V., Flaks, I. P., and Filippenko, L. G. (1960a) *Zh. eksper. teor. Fiz.* **38**, 719; (1960) *Soviet Phys.—JETP* **11**, 519.

Fedorenko, N. V., Filippenko, L. G., and Flaks, I. P. (1960b) *Zh. tekh. Fiz.* **30**, 49.

Flaks, I. P., and Filippenko, L. G. (1959) *Zh. tekh. Fiz.* **29**, 1100; (1960) *Soviet Phys.—JTP* **4**, 1005.

Flaks, I. P., and Solov'ev, E. S. (1958a) *Zh. tekh. Fiz.* **28**, 599; (1958) *Soviet Phys.—JTP* **3**, 564.

* This article not available to authors.

Flaks, I. P., and Solove'v, E. S. (1958b) *Zh. tekh. Fiz.* **28**, 612; (1958) *Soviet Phys.—JTP* **3**, 577.
Fogel', Ya. M., and Mitin, R. V. (1956) *Zh. eksper. teor. Fiz.* **30**, 450; (1956) *Soviet Phys.—JETP* **3**, 334.
Fogel', Ya. M., Krupnik, L. I., and Safronov, B. G. (1955a) *Zh. eksper. teor. Fiz.* **28**, 589; (1955) *Soviet Phys.—JETP* **1**, 415.
Fogel', Ya. M., Safronov, B. G., and Krupnik, L. I. (1955b) *Zh. eksper. teor. Fiz.* **28**, 711; (1955) *Soviet Phys.—JETP* **1**, 546 .
Fogel', Ya. M., Mitin, R. V., and Koval, A. G. (1956) *Zh. eksper. teor. Fiz.* **31**, 397; (1957) *Soviet Phys.—JETP* **4**, 359.
Fogel', Ya. M., Ankudinov, V. A., and Slabospitskii, R. E. (1957) *Zh. eksper. teor. Fiz.* **32**, 453; (1957) *Soviet Phys.—JETP* **5**, 382.
Fogel', Ya. M., Ankudinov, V. A., Filippenko, D. V., and Topolia, N. V. (1958a). *Zh. eksper. teor. Fiz.* **34**, 579; (1958) *Soviet Phys.—JETP* **7**, 400.
Fogel', Ya. M., Mitin, R. V., Kozlov, V. F., and Romashko, N. D. (1958b) *Zh. eksper. teor. Fiz.* **35**, 565; (1959) *Soviet Phys.—JETP* **8**, 390.
Fogel', Ya. M., Ankudinov, V. A., and Filippenko, D. V. (1958c) *Zh. eksper. teor. Fiz.* **35**, 868; (1959) *Soviet Phys.—JETP* **8**, 601.
Fogel', Ya. M., Kozlov, V. F., Kalmykov, A. A., and Muratov, V. I. (1959) *Zh. eksper. teor. Fiz.* **36**, 1312; (1959) *Soviet Phys.—JETP* **9**, 929.
Fogel', Ya. M., Ankudinov, V. A., and Filippenko, D. V. (1960) *Zh. eksper. teor. Fiz.* **38**, 26; (1960) *Soviet Phys.—JETP* **11**, 18.
Fuls, E. N., Jones, P. R., Ziemba, F. P., and Everhart, E. (1957) *Phys. Rev.* **107**, 704.
Gatlinburg (1958) Proc. Conf. on the Penetration of Charged Particles in Matter, Gatlinburg, Tennessee. (Edwin A. Uehling, ed.), *in* Nat. Acad. Sci. Publ. No. 752 (1960).
Geiger, H. (1933) *In* "Handbuch der Physik" (S. Flügge, ed.), Vol. 22, p. 221. Springer, Berlin.
Ghosh, S. N., and Sheridan, W. F. (1957) *Indian J. Phys.* **31**, 337.
Gilbody, H. B., and Hasted, J. B. (1956) *Proc. Roy. Soc.* **A238**, 334.
Goldmann, F. (1931) *Ann. Phys. (Leipzig)* **10**, 460.
Hall, T. (1950) *Phys. Rev.* **79**, 504.
Hasted, J. B. (1951) *Proc. Roy. Soc.* **A205**, 421.
Hasted, J. B. (1952) *Proc. Roy. Soc.* **A212**, 235.
Hasted, J. B. (1953) *Proc. Roy. Soc.* **A222**, 74.
Hasted, J. B. (1955) *Proc. Roy. Soc.* **A227**, 466.
Henderson, G. H. (1922) *Proc. Roy. Soc.* **A102**, 496.
Henderson, G. H. (1925) *Proc. Roy. Soc.* **A109**, 157.
Holøien, E., and Midtdal, J. (1955) *Proc. Phys. Soc.* **A68**, 815.
Hubbard, E. L., and Lauer, E. J. (1955) *Phys. Rev.* **98**, 1814.
Il'in, R. N., Afrosimov, V. V., and Fedorenko, N. V. (1959) *Zh. eksper. teor. Fiz.* **36**, 41; (1959) *Soviet Phys.—JETP* **9**, 29.
Jacobsen, J. C. (1927) *Nature* **117**, 858.
Jones, P. R., Ziemba, F. P., Moses, H. A., and Everhart, E. (1959) *Phys. Rev.* **113**, 182.
Jorgensen, T. (1958) Proc. Gatlinburg Conf., Natl. Acad. Sci. Publ. No. 752, p. 72 (1960).
Kaminker, D. M., and Fedorenko, N. V. (1955a) *Zh. tekh. Fiz.* **25**, 1843.
Kaminker, D. M., and Fedorenko, N. V. (1955b) *Zh. tekh. Fiz.* **25**, 2239.
Kanner, H. (1951) *Phys. Rev.* **84**, 1211.
Kapitza, P. (1924) *Proc. Roy. Soc.* **A106**, 602.
Kasner, W. H., and Donahue, T. M. (1957) University of Pittsburgh O. N. R. Project Report No. 3.

Korsunsky, M. I., Pivovar, L. I., Markus, A. M., and Leviant, K. L. (1955) *Dokl. Akad. Nauk SSSR* **103**, 399.

Lassen, N. O. (1945) *Phys. Rev.* **68**, 142.

Lassen, N. O. (1955) *K. Danske Vidensk. Selsk. mat. fys. Medd.* **30** (8).

Leviant, K. L., Korsunsky, M. I., Pivovar, L. I., and Podgornyi, I. M. (1955) *Dokl. Akad. Nauk SSSR* **103**, 403.

Massey, H. S. W., and Burhop, E. H. S. (1952) "Electronic and Ionic Impact Phenomena" Oxford Univ. Press (Clarendon), London and New York.

Meyer, H. (1937) *Ann. Phys. (Leipzig)* **30**, 635.

Montague, J. H. (1951) *Phys. Rev.* **81**, 1026.

New York (1958) Conference on the physics of electronic and atomic collisions. Sponsored by the Office of Naval Research (USA), the Air Force Office of Scientific Research (USA), and New York University in New York.

Nikolaev, V. S., Dmitriev, I. S., Fateeva, L. N., and Teplova, Y. A. (1957) *Zh. eksper. teor. Fiz.* **33**, 1325; (1958) *Soviet Phys.—JETP* **6**, 1019.

Northcliffe, L. C. (1960) *Phys. Rev.* **120**, 1744.

Phillips, J. A. (1955) *Phys, Rev.* **97**, 404.

Reynolds, H. L., Scott, D. W., and Zucker, A. (1954) *Phys. Rev.* **95**, 671.

Reynolds, H. L., Wyly, L. D., and Zucker, A. (1955a) *Phys. Rev.* **98**, 474.

Reynolds, H. L., Wyly, L. D., and Zucker, A. (1955b) *Phys. Rev.* **98**, 1825.

Ribe, F. L. (1951) *Phys. Rev.* **83**, 1217.

Rose, P. H., Connor, R. J., and Bastide, R. P. (1958) *Bull. Amer. Phys. Soc.* [II] **3**, 40.

Rüchardt, E. (1933) *In* "Handbuch der Physik" (S. Flügge, ed.), Vol. 22, p. 103. Springer, Berlin.

Rudnick, P. (1931) *Phys. Rev.* **38**, 1342.

Rutherford, E. (1924) *Phil. Mag.* **47**, 277.

Rutherford, E., Chadwick, J., and Ellis, C. D. (1931) "Radiations from Radioactive Substances." Cambridge Univ. Press, London and New York.

Sluyters, T. J. M., de Haas, E., and Kistemaker, J. (1959) *Physica* **25**, 1376.

Snitzer, E. (1953) *Phys. Rev.* **89**, 1237.

Stedeford, J. B. H. (1955) *Proc. Roy. Soc.* **A227**, 466.

Stier, P. M., and Barnett, C. F. (1956) *Phys. Rev.* **103**, 896.

Stier, P. M., Barnett, C. F., and Evans, G. E. (1954) *Phys. Rev.* **96**, 973.

Teplova, Y. A., Dmitriev, I. S., Nikolaev, V. S., and Fateeva, L. N. (1957) *Zh. eksper. teor. Fiz.* **32**, 974; (1957) *Soviet Phys.—JETP* **5**, 797.

Uppsala (1959) "Proceedings of the Fourth International Conference on Ionization Phenomena in Gases, Uppsala, 1959" (N. R. Nilsson, ed.). North-Holland, Amsterdam.

Van Eck, J., and Kistemaker, J. (1960) *Physica* **26**, 553.

Venice (1957) "Proceedings of the Third International Conference on Ionization Phenomena in Gases, Venice, 1957." Italian Phys. Soc., Milan.

Whittier, A. C. (1954) *Canad. J. Phys.* **32**, 275.

Ziemba, F. P., and Everhart, E. (1959) *Phys. Rev. Letters* **2**, 299.

Ziemba, F. P., Lockwood, G. J., Morgan, G. H., and Everhart, E. (1960) *Phys. Rev.* **118**, 1552.

20.

Relaxation in Gases

J. D. Lambert

1 The Nature of the Relaxation Process

This chapter is concerned with the interchange of translational, vibrational, and rotational energy between molecules in collision. For all ordinary purposes *translational* energy may be regarded as freely interchanged in collisions, and a gas can take up or lose translational energy at a rate depending only on the intermolecular collision rate. The two forms of *internal* energy, vibrational and rotational, are both quantized, and are not freely interchanged with translational energy in collision. The probability of a transition between either vibrational or rotational energy and translational energy occurring in a collision is always less than unity, so that a molecule may, on the average, undergo a considerable number of collisions before it can gain or lose internal energy. This results in a *relaxation* of the internal energy when the gas undergoes any process involving very rapid energy change; the internal energy lags behind the translational energy so that, at a particular instant, the vibrational or rotational temperature of the gas differs from the translational temperature.

If P_{10} is the transition probability per collision for a change in internal quantum number from $1 \rightarrow 0$, the average number of collisions required for a molecule to lose one quantum will be $Z_{10} = 1/P_{10}$. If Z is the total number of collisions one molecule suffers per second, a *relaxation time*, β, may be defined by the equation

$$Z_{10} = Z\beta. \tag{1}$$

Z is proportional to the gas pressure, and, since Z_{10} is a constant for a particular transition, the actual value of β is inversely proportional to the pressure. For convenience, relaxation times are usually referred to a pressure of 1 atm. Equation (1) is an approximation, as the gas kinetic collision number, Z, requires modifying by a factor taking into account the equilibrium Boltzmann distribution of molecules between quantum states, e.g., for a two-state system involving an energy quantum, ϵ, the correct equation is

$$Z_{10} = Z\beta(1 + e^{-\epsilon/kT}) \tag{2}$$

which approximates closely to (1) for large values of ϵ (Herzfeld and Litovitz, 1959).

2 Phenomena Associated with Relaxation

2.1 Ultrasonic Dispersion and Absorption

The passage of a sound wave through a gas involves rapidly alternating adiabatic compression and rarefaction. The adiabatic compressibility of a gas is a function of γ, the ratio of the specific heats, and the classical expression for the velocity, V, of sound in a perfect gas is

$$V^2 = \gamma RT/M = \frac{RT}{M}\left(1 + \frac{R}{C_V}\right). \tag{3}$$

When the period of the compression-rarefaction cycle becomes comparable with the relaxation time, the internal temperature of the gas lags behind the translational temperature for the whole of the cycle, so that the effective values of C_V and of γ become frequency dependent. This results in a *dispersion zone*, over which V changes from a value corresponding to the static specific heat, C_{V_0} at lower frequencies, to a value corresponding to $C_\infty = (C_{V_0} - C_i)$, at higher frequencies, C_i

being the contribution due to the relaxing form of internal energy. The variation of sound velocity with cyclic frequency, ω, is given by

$$V^2 = \frac{RT}{M}\left(1 + \frac{R(C_{V_0} + C_\infty \omega^2 \beta^2)}{(C_{V_0}^2 + C_\infty^2 \omega^2 \beta^2)}\right). \tag{4}$$

The relaxation time, β, may thus be calculated from measurements of sound velocity over a range of frequencies covering the dispersion zone. Dispersion usually occurs at ultrasonic frequencies ranging from 50 kc sec^{-1} to 10 Mc sec^{-1}, and the velocities may be accurately measured by an acoustic interferometer employing a piezoelectric quartz oscillator. Since β is inversely proportional to the pressure, doubling the pressure is exactly equivalent to halving the frequency in the context of (4), and it is usually convenient to make measurements at constant frequency and varying pressure. Corrections for gas imperfection are essential, and require an accurate knowledge of the second virial coefficient of the gas over a range of temperatures. The dispersion process is accompanied by a nonclassical *absorption* of sound, rising to a maximum at a value of frequency/pressure in the center of the dispersion zone. The absorption coefficient, μ, is given by

$$\mu = \frac{2\pi\omega C_\infty \beta}{C_{V_0}}\left[\frac{V^2/V_0^2 - 1}{(V^2/V_0^2)\,(C_\infty \omega\beta/C_{V_0})^2 + 1}\right], \tag{5}$$

and measurements of absorption may also be used to evaluate β. Measurements of ultrasonic dispersion and absorption are the most convenient and accurate way of investigating relaxation times involving low-lying energy levels. The technique of acoustic interferometry can be applied only at temperatures below 300°C, as suitable transducers which will operate efficiently as ultrasonic generators at higher temperatures are not available. A novel acoustic method, employing measurement of reverberation time in a large resonant cavity, with an external transducer operating at frequencies below the relaxation frequency, has been used by Lukasik and Young (1957) at temperatures as high as 1186°K. Edmonds and Lamb (1958a, b) have used a similar method at 25°C and obtained relaxation times in excellent agreement with ultrasonic velocity measurements. An extensive and detailed theoretical and experimental treatise on ultrasonic dispersion and absorption has recently been published by Herzfeld and Litovitz (1959), and there are recent reviews by Richardson (1955) and McCoubrey and McGrath (1957).

2.2 AERODYNAMIC PHENOMENA

2.2.1 *Shock Waves*

When a shock wave travels through a gas, a thin layer behind the shock undergoes extremely rapid adiabatic compression, resulting in an increase of density and temperature. If there is relaxation of internal energy, the translational energy alone is affected directly behind the shock, resulting in an initial rise in temperature and density corresponding to an effective specific heat, $(C_p - C_i)$. As the energy subsequently leaks into the internal degrees of freedom, the translational temperature decreases and the density increases until they reach values corresponding to the static value of C_p a short distance behind the shock. The density profile behind the shock front thus depends on the relaxation time. The theory of this effect has been worked out by Bethe and Teller (1941), and measurements of density profiles may be made in shock tubes by various optical methods, thus enabling calculations of relaxation times. (cf. Smiley and Winkler, 1954; Green and Hornig, 1953). The results are less accurate than those obtained by ultrasonic methods, and there are obvious experimental difficulties in ensuring chemical purity of materials, but the method has the great advantage of enabling relaxation times to be obtained for temperatures up to 3000°K (at higher temperatures chemical decomposition sets in).

Application of spectroscopic techniques to shock tube experiments gives interesting confirmation of the reality of the relaxation effect. Windsor and associates (1957) investigated the infrared emission arising from vibrational excitation of carbon monoxide during passage of a shock wave, and obtained relaxation times in fair agreement with those calculated from density profile measurements. Clouston *et al.* (1958, 1959) have employed a sodium line reversal technique to investigate temperatures in shock waves. They find that, in nitrogen, the sodium atom excitation temperature is in equilibrium with the *vibrational* temperature of the gas rather than with the translational temperature, and rises from a relatively low value immediately behind the front at a rate consistent with the vibrational relaxation time of nitrogen as measured by conventional methods.

2.2.2 *The Impact Tube*

Rapid adiabatic compressions and rarefactions with accompanying temperature changes also occur in the flow of gases past obstacles. The time in which these changes take place is controlled by the dimensions of the obstacle and the velocity of flow. If the time intervals involved are comparable with the relaxation time of the gas, there will be a lag in the adjustment of the internal heat capacity, resulting in an increase of entropy and a dissipation of energy. Kantrowitz (1946) has developed a quantitative theory of this effect, and has calculated relaxation times for a number of gases by measuring the pressure developed in a small Pitot tube which forms an obstacle in a rapid gas stream (Huber and Kantrowitz, 1947). This method is not very accurate, and requires large quantities of gas whose purity is difficult to ensure, but it has enabled, for example, a relaxation time to be obtained for water vapor at temperatures between 400° and 700°K. This effect of relaxation on aerodynamic flow past obstacles is of technical importance where gases other than air are used in wind tunnel tests. The drag on an aerofoil in a relaxing gas, such as carbon dioxide, will be different from the drag in air at the same Mach number and Reynolds number. The influence of relaxation on rapid adiabatic expansion is also important in calculations involving nozzle flow in rocket motors (Penney and Aroeste, 1955).

2.3 Photochemical and Chemical Phenomena

Lipscomb *et al.* (1956) have produced oxygen molecules in a high state of vibrational excitation by the flash photolysis of chlorine dioxide or nitrogen dioxide in an excess of inert gas. The rate of decay of this vibrational excitation can be directly followed by spectroscopic techniques, and estimates made of the vibrational relaxation times corresponding to heteromolecular collisions of the excited oxygen with various other gases. This technique enables investigation of the energy transfer process from *higher* vibrational energy levels (up to 8 quanta in this case), in contrast to techniques previously discussed, which are mainly concerned with $0 \rightleftharpoons 1$ transfers. The estimated relaxation times show the same general trends as observed for lower energy transitions (§ 5). Vibrational deactivation of O_2 by N_2 is extremely inefficient, ($Z_{10} > 10^7$), while the more efficient deactivators are free radicals, which show incipient chemical reactivity, and molecules with near resonant vibrational levels.

Molecules with high vibrational excitation are also produced in exothermic *chemical* reactions. Garvin *et al.* (1960) have shown that OH radicals possessing up to 10 quanta of vibrational excitation are produced in the cold "atomic flame" which results on mixing atomic hydrogen with ozone. Interpretation of the emission spectrum shows that, while the *vibrational* energy of the OH radicals present rapidly approaches an equilibrium Boltzmann distribution, there is considerable time lag in reaching thermal equilibrium with rotation and translation (cf. § 5). Cashion and Polanyi (1960) have investigated the behavior of vibrationally excited HCl and DCl molecules, formed in the atomic reactions: H + Cl_2; H + HCl; H + DCl; and D + HCl, by following infrared emission, and reach conclusions which are similar. Harrington *et al.* (1960) produce vibrationally excited *sec*-butyl-d_1 radicals by addition of D atoms to *cis*-butene-2, and conclude from the chemical kinetics of the system that *easy* collisional deactivation occurs. This is in accord with expectation for a radical possessing internal rotation (see § 4.2). Frey and Kistiakowsky (1957) produced vibrationally "hot" methylene radicals by photolysis of ketene; this was allowed to react with ethylene to produce a vibrationally "hot" *cyclo*propane molecule, which subsequently isomerized to propylene. They infer that the vibrational excitation of the cyclopropane molecule is much more easily removed by collision than that of the methylene radical. As the vibrational frequencies of the methylene radical are much higher than those of *cyclo*propane; this is again in accordance with expectation (§ 4).

While the techniques described in this section are so far only sufficiently developed to yield semiquantitative information about relaxation processes, the general importance of relaxation phenomena for chemical kinetics is obvious. The accumulation of sufficient vibrational energy in a particular interatomic bond in the molecule is the main criterion for chemical reaction occurring, and an understanding of the way in which vibrational energy transfer takes place between molecules, and between different fundamental vibrational modes in a molecule, is basic to an understanding of reaction mechanisms. Relaxation phenomena may be especially important in very fast reactions, such as occur in high-temperature flames (Gaydon, 1948).

2.4 Spectroscopic Phenomena

A polar molecule may be vibrationally excited by absorption of intense infrared radiation of the appropriate wavelength. The energy absorbed will be subsequently degraded by intermolecular collisions to thermal (translational) energy at a rate depending on the relaxation time. If *pulsed* radiation is used, sound will be generated in the gas at an intensity depending on the relation between the pulse frequency and the relaxation time. This is the basis of the *spectrophone*, originally devised by Slobodskaya (1948), which is in principle an ideal method for following the relaxation of excitation of a specific intramolecular vibration. Unfortunately, the experimental difficulties are formidable, and so far only semiquantitative results have been obtained (Jacox and Bauer, 1957). Relaxation of vibrational and rotational energy following photoexcitation is of importance in the consideration of energy distribution among molecules in the upper atmosphere, where intermolecular collision rates are very low (cf. Heaps and Herzberg, 1952). Fluorescence phenomena also involve collisional deactivation of a vibrationally excited molecule in an upper electronic state; a review of this aspect of fluorescence in gases has been published by Stevens and Boudart (1957).

2.5 Gas Kinetic Effects

The classical kinetic theory of transport properties relates viscosity, η, and thermal conductivity, k, of a gas at moderate pressures by the relation

$$k = 2.5\eta C_V/M \tag{6}$$

(Chapman and Cowling, 1939). This based on the assumption of free interchange of translational and internal energy at every collision. Experimental values of the ratio, k/η, for polyatomic gases are consistently lower than given by (6), and Eucken (1913) proposed the view that, if there is negligible collisional interchange between translational and internal energy, the latter is transported by a simple diffusion process, giving the modified relation

$$k = (\eta/M)(2.5C_{\text{trans}} + C_{\text{vib}} + C_{\text{rot}}). \tag{7}$$

Recent systematic measurements of transport properties of polyatomic molecules (Schäfer, 1943; Craven and Lambert, 1951; Lambert *et al.*, 1955) have given rise to further modifications of this relation, but have confirmed the view that, for most molecules, there is negligible interchange of translational with vibrational (and, possibly, rotational) energy in the contest of (7). It appears that *very* rapid interchange ($Z_{10} \leqslant 5$) is required to render the Eucken assumption invalid, and that this *may* occur with long chain hydrocarbons, which are also shown by ultrasonic measurements (§ 4.2) to suffer easy energy transfer. Recent measurements of the pressure variation of thermal conductivity, as pressure falls towards the Knudsen domain (Waelbroeck and Zuckerbrodt, 1958), show evidence of rotational relaxation for hydrogen and oxygen, giving values; $Z_{10} = 300$ for H_2, and $Z_{10} = 20$ for O_2, which are in striking agreement with ultrasonic values (§ 3). Diffusion rates of polyatomic molecules under gas kinetic conditions also appear to be influenced by ease of energy transfer (Clarke and Ubbelohde, 1957), and a similar influence may operate in thermal diffusion (Danby *et al.*, 1957).

3 *Rotational Relaxation*

Rotational quanta are much smaller than vibrational quanta, and the probability of a translation-rotation energy transfer occurring in collision is correspondingly higher. Rotational relaxation times are therefore very small: for most gases the average number of intermolecular collisions needed for transfer is < 10, corresponding to a relaxation time of the order, 10^{-9} sec. Rotational relaxation phenomena thus play a comparatively unimportant role, and are also difficult to measure. This, coupled with the fact that rotational energy is of less *chemical* significance than vibrational energy, has led to rotational relaxation being given comparatively little attention.

The theoretical treatment of translation-rotation transfer is in principle less simple than that of translation-vibration. Molecules will be distributed among a variety of rotational levels under most experimental conditions; these levels differ in magnitude, and a number of different transition probabilities have to be considered (in contrast to vibrational relaxation, where the $1 \rightleftharpoons 0$ transition predominates). The problem is simplified for hydrogen, where the quantum is much larger, and only $0 \rightleftharpoons 2$ transitions for $p - H_2$ and $1 \rightleftharpoons 3$ transitions for $o - H_2$ are important. Brout (1954a) has made a quantum-mechanical calculation by the method of distorted waves, and obtains, $P_{20} = 3.04 \times 10^{-3}$ ($Z_{20} = 329$) and $P_{31} = 2.96 \times 10^{-3}$ ($Z_{31} = 338$). These values compare well with the experimental value, Z^* ($= Z\beta$) $= 350$, calculated from ultrasonic dispersion measurements on $n - H_2$ at 0°C by Stewart and Stewart (1952). Takayanagi (1957, 1959) has performed similar calculations for H_2 and D_2, which also give fair approximation to the experimental results, and finds that the transfer probability is much larger for HD, where single quantum transitions are allowed.

For heavier homonuclear diatomic molecules ($M > 20$), similar theoretical methods (Brout, 1954b) show that the mean transition probability may be taken as approximately equal to $\frac{1}{2}(d_0/r_0)^2$, where d_0 is the internuclear distance in the molecule, and r_0 the kinetic collision diameter. The reasons for the simplicity of this result are interesting. Temperature disappears from the relation because the increased spread of occupied rotational levels at higher temperatures causes a lowering of the probability, which cancels out the favorable effect of increased velocity of approach. Mass disappears because a larger moment of inertia brings the rotational levels closer and increases the probability, canceling out the adverse effect of the decreased velocity of approach of heavier molecules. The intermolecular potential is unimportant, as the collision time is always short with respect to the frequency of the rotation. The resulting values of $Z^* = 1/P$ are 17 for O_2 and 23 for N_2, which are higher than the most recent experimental values obtained by ultrasonic dispersion; $Z^* = 5.3$ for N_2 and 4.1 for O_2 (Greenspan, 1959). The latter are confirmed by shock wave measurements by Andersen and Hornig (1959) giving, for N_2, $Z^* = 5.5$. A similar low value, $Z^* = 7$, has recently been found from acoustic absorption measurements on HCl (Breazeale and Kneser, 1960). This is particularly surprising, as the HCl molecule has a much smaller moment of inertia than O_2 or N_2 combined with a similar over-all mass, and would be expected to show much less efficient transfer. Breazeale and Kneser (1960) suggest that the high *dipole moment* of the molecule may have a profound effect on the efficiency. There is independent evidence for an effect of dipole moment on the nature of the intermolecular collision process, from observations on gas kinetic transport properties (Schäfer, 1943; Craven and Lambert, 1951). The high efficiency of equilibration of rotational energy among HCl molecules is confirmed by shock tube measurements (Andersen and Hornig, 1959). The OH radical, which resembles HCl, is also shown by spectroscopic examination of detonation waves (Kistiakowsky and Tabbutt, 1959) to have a low rotational relaxation time, corresponding to $Z^* = 10$. Shock tube measurements on the heavier triatomic molecules N_2O and CO_2 give values of Z^* between 1 and 2 collisions (Andersen and Hornig, 1959), and it may be inferred that for the majority of polyatomic molecules rotational relaxation is, for all practical purposes, nonexistent.

Polyatomic *spherical top* molecules have been theoretically treated by Wang Chang and Uhlenbeck (1951), who find for homomolecular collisions the simple expression,

$$Z^* = \tfrac{3}{8}(1 + 2b)^2/b,$$

where $b = I/Ma^2$; I is the moment of inertia, M the reduced mass of the collision, and a the sum of the molecular radii. This gives for CH_4, $Z^* = 18$, which is in good agreement with the experimental values, 14-17, obtained ultrasonically by Kelly (1957). Heteromolecular collisions between a spherical top molecule and an inert gas are treated by Widom (1960), who finds a theoretical expression,

$$Z^* = \tfrac{3}{8}(1 + b^2)/b,$$

giving values of Z^* ranging from 2 for $(CCl_4 + Ar)$ to 22 for $(CH_4 + Ar)$. No experimental data are available for comparison.

4 Vibrational Relaxation in Pure Gases

4.1 Diatomic Molecules

The theoretical problem of vibrational relaxation for a diatomic molecule is less intractable than either vibrational relaxation in a polyatomic molecule, or rotational relaxation. In its lower energy states, the diatomic molecule can be regarded as a simple harmonic oscillator, and, in the majority of cases, $h\nu > kT$, so that only $0 \rightleftharpoons 1$ or $1 \rightleftharpoons 2$ transitions, which are of equal energy, come into consideration. Experimental investigation is unfortunately difficult, as most diatomic molecules have high vibrational frequencies which are appreciably active only at higher temperatures, where the more accurate ultrasonic techniques cannot be used. A physical picture of the problem may be expressed in terms of Ehrenfest's adiabatic principle: if a changing external force acts on a periodic motion, the process will be adiabatic if the change of force is small during a period of the motion, and nonadiabatic if the change is large during this time. The criterion for a nonadiabatic process, involving energy transfer, is thus a collision which is of short duration compared with the period of oscillation. The brevity of the collision depends on the velocity of approach of the molecules, proportional to $(T/M)^{1/2}$, and on the steepness of the intermolecular repulsion potential; the period of the oscillation is ν^{-1}. The probability of energy transfer is thus favored by low mass, increase of temperature, steep intermolecular repulsion potential, and low vibrational frequency. An early classical quantitative treatment based on these principles by Landau and Teller (1936) gave a striking approximation to the results of more recent theoretical treatments.

Modern quantum-mechanical theory treats the collision as a parallel stream of molecules, which may be regarded as a de Broglie wave, falling on a molecule regarded as stationary. Vibrational excitation corresponds to inelastic scattering of the wave. The problem may be handled quantitatively by the method of distorted waves, due to Jackson and Mott (1932). The treatment originated by Schwartz *et al.* (1952), usually referred to as the S.S.H. method, has been developed in various ways, which are discussed in detail by Herzfeld and Litovitz (1959). The transfer probability, P_{1-0}, is given by Schwartz and Herzfeld (1954) in the form:

$$P_{1-0} = 0.716(1 + C/T)^{-1} (r_c/r_0)^2 (\pi/3)^{1/3} (8\pi^3 \mu h\nu/\alpha^* h)^2 \times V_{1-0}^2 \chi^{1/2} \exp(-3\chi + h\nu/2kT + \epsilon/kT) \quad (8)$$

where V_{1-0} is the oscillatory "matrix element" for the $0 \rightarrow 1$ quantum jump, C is Sutherland's constant, r_c is the distance of closest approach, r_0 is the zero of the intermolecular potential energy curve, μ is the reduced mass of the colliding particles, α^* is the particular value of the intermolecular repulsion potential corresponding to an exponential expression, $V = V_0 \exp(-\alpha r) + \epsilon$, ϵ is the depth of the minimum in the potential energy curve, and

$$\chi = \left[\frac{2\pi^2\mu^2(h\nu)^2}{\alpha^* h^2 kT}\right]^{1/3}.$$

This expression enables, in principle, an *a priori* calculation of P_{10} and β to be made in terms of the mass, vibrational frequency, temperature, and intermolecular force field. P_{10} is very strongly dependent on the intermolecular repulsion parameter, α^*, and a critical factor in the calculation is the fitting of the exponential expression, $V = V_0 \exp(-\alpha r) + \epsilon$, to an actual intermolecular potential function, such as the

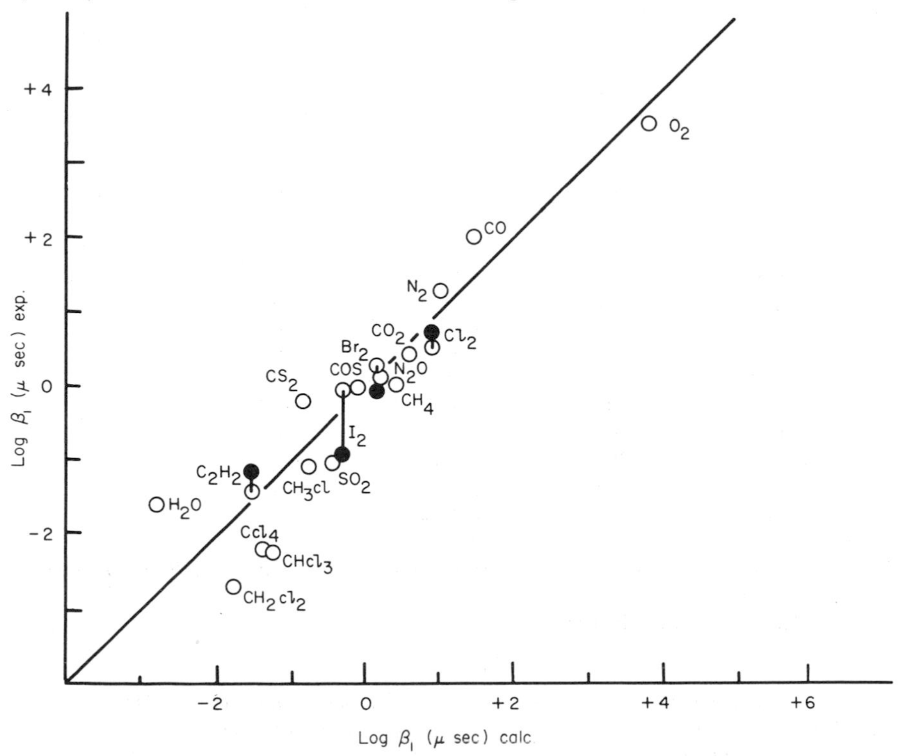

FIG. 1. Comparison of calculated and experimental relaxation times (Dickens and Ripamonti, 1961). This figure is reproduced by kind permission of the Council of the Faraday Society.

Lennard-Jones. This problem is discussed in detail by Herzfeld and Litovitz (1959), who conclude that the best method involves making a two-point fit between the exponential and Lennard-Jones curves. Dickens and Ripamonti (1961) have compared values of β calculated by this method with some of the more reliable experimental values available. These are plotted in Fig. 1, together with data for some polyatomic molecules, which are discussed in § 4.2, and the relevant data and sources are listed in Table I. It will be seen that for the diatomic

TABLE I

EXPERIMENTAL AND CALCULATED RELAXATION TIMES[a, b]

Molecule[c]	T(°K)	β_1 (μ sec) calc	β_1 (μ sec) exptl	References
O_2	288	5840	3180	(a)
N_2	3480	10	19	(b)
CO	2200	26.7	100	(c)
Cl_2	290	7.88	3.4	(d)
			5.0	(e)
Br_2	331	1.42	1.6	(d)
			0.81	(e)
I_2	453	0.495	0.85	(d)
			0.104	(e)
C_2H_2	298	0.0285	0.041	(f)
			0.067	(g)
N_2O	273	1.53	1.18	(h)
COS	288	0.827	0.902	(h)
CS_2	288	0.44	0.64	(h)
CO_2	300	3.80	2.53	(h)
H_2O	585	0.00158	0.03	(i)
SO_2*	373	0.36	0.089	(j)
CH_2Cl_2*	303	0.0158	0.00195	(k)
CH_4	303	2.70	1.06	(l)
CH_3Cl	303	0.147	0.078	(k)
$CHCl_3$	303	0.0510	0.0060	(k)
CCl_4	303	0.0420	0.0062	(k)

[a] This table is reproduced by kind permission of the Council of the Faraday Society.
[b] Dickens and Ripamonti (1961).
[c] *—Denotes shorter relaxation time.

REFERENCES

(a) Knötzel and Knötzel (1948).
(b) Blackman (1956).
(c) Windsor *et al.* (1959).
(d) Richardson (1959).
(e) Shields (1960).
(f) Lambert and Salter (1959).
(g) Edmonds and Lamb (1958b).
(h) Herzberg and Litovitz (1959).
(i) Huber and Kantrowitz (1947).
(j) Lambert and Salter (1957).
(k) Sette *et al.* (1955).
(l) Eucken and Aybar (1940).

molecules there is general agreement between theory and experiment (though this is far from true for many earlier computations, cf. Herzfeld and Litovitz, 1959). Dickens and Ripamonti (1961) also show that, using values of α listed in this way, a fair approximation to the correct linear dependence of $\log P_{10}$ and $T^{-1/3}$ is given, where experimental data are available over large ranges of temperature. They conclude that the S.S.H. treatment is satisfactory for diatomic molecules, *provided* an adequate fit is used for the intermolecular potential. The systematic errors which have been reported previously by Arnold and associates (1957) are due mainly to the use of a theoretical expression which omits part of the exponential term in (8).

The quantum-mechanical treatment has been extended to energy transfer probabilities for diatomic molecules in higher vibrational levels by Shuler and his collaborators (cf. Shuler and Zwanzig, 1960), who show that there is appreciable probability for jumps involving more than one quantum. This will be important in the high-temperature phenomena discussed in § 2.2 and § 2.3, rather than in ultrasonic measurements, and the experimental techniques are not yet sufficiently refined for comparing theory with experiment.

4.2 Polyatomic Molecules

The problem for polyatomic molecules is complicated by such molecules possessing more than one fundamental mode of vibration. The frequencies of these modes differ, so that there are a number of different possibilities for translation-vibration transitions. The possibilities for a molecule with two active vibrational modes, of frequency ν_1 and ν_2, are illustrated in Fig. 2. There are three possible transitions;

(a) Transfer of translational energy to $0 \rightarrow 1$ excitation of the mode ν_1, with relaxation time, β_1.

(*b*) Transfer of translational energy to $0 \rightarrow 1$ excitation of the mode ν_2, with relaxation time, β_2.

(c) The complex transfer of a quantum of vibrational energy from mode, ν_1, plus the necessary increment of translational energy, to give $0 \rightarrow 1$ excitation of mode, ν_2. Owing to the quantized nature of the process, this can occur only in collision, and will correspond to a relaxation time, β_{12}, which will be pressure dependent in the same way as β_1 and β_2.

For the vast majority of polyatomic molecules, which have been investigated experimentally, $\beta_2 \gg \beta_1 \gg \beta_{12}$, and a *single* relaxation process is observed, with a relaxation time corresponding to β_1; the energy then rapidly flows in collisions involving complex transfer to the second mode, ν_2 (and other modes, if present). Process (a) is thus the rate controlling step for a relaxation of the whole of the vibrational energy

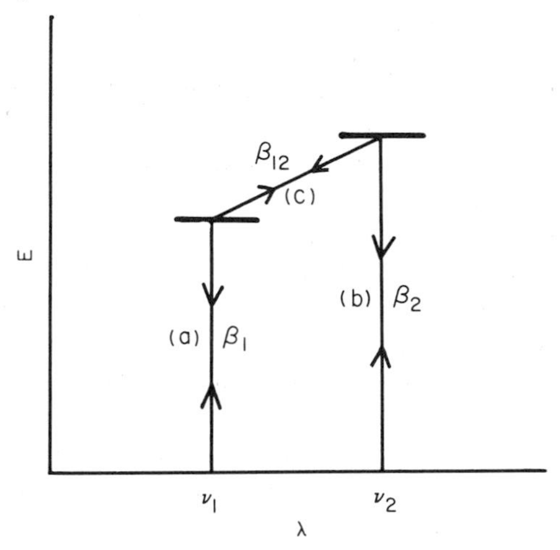

FIG. 2. Energy level diagram showing possible transitions for a molecule with two fundamental vibrational modes.

of the molecule. It may be shown that, if the over-all relaxation time is β, then $\beta_1 = (C_1/C_s)\,\beta$, where C_1 is the specific heat contribution due to mode 1 alone, and C_s the total vibrational specific heat.

For the few molecules where there is a large frequency difference between ν_1 and ν_2, the probability of the complex transfer can be much smaller, and the condition, $\beta_2 \gg \beta_{12} > \beta_1$, applies. This gives rise to a double relaxation process: the longer relaxation time, β_{12}, involves a vibrational specific heat, $(C_s - C_1)$; the shorter relaxation time, β_1, involves a vibrational specific heat, C_1. This has been observed for CH_2Cl_2 ($\nu_1 = 283$; $\nu_2 = 704$ cm^{-1}) by Sette *et al.* (1955), and for SO_2 ($\nu_1 = 525$; $\nu_2 = 1150$ cm^{-1}) by Lambert and Salter (1957). Both these gases, in fact, have three active fundamental modes: the energy levels for SO_2 ($\nu_3 = 1360$ cm^{-1}) are shown in Fig. 3. Detailed theoretical calculations by Dickens and Linnett (1957), using the S.S.H. method, show that the shorter observed relaxation time corresponds to the direct excitation of the first level of ν_1, and the longer relaxation time to the complex transition between the *second* level of ν_1 and the *first* level of ν_2. The complex transition from the first level of ν_2 to the first level of ν_3 has a much higher probability and is not rate determining. A similar theoretical picture was shown by Tanczos (1956) for CH_2Cl_2. It appears in general that the approximate criterion for a molecule to show double relaxation is the condition that, $\nu_2 > 2\nu_1$, a distribution of frequencies which is rare among polyatomic molecules generally.

The torsional oscillations, which occur in more complex molecules,

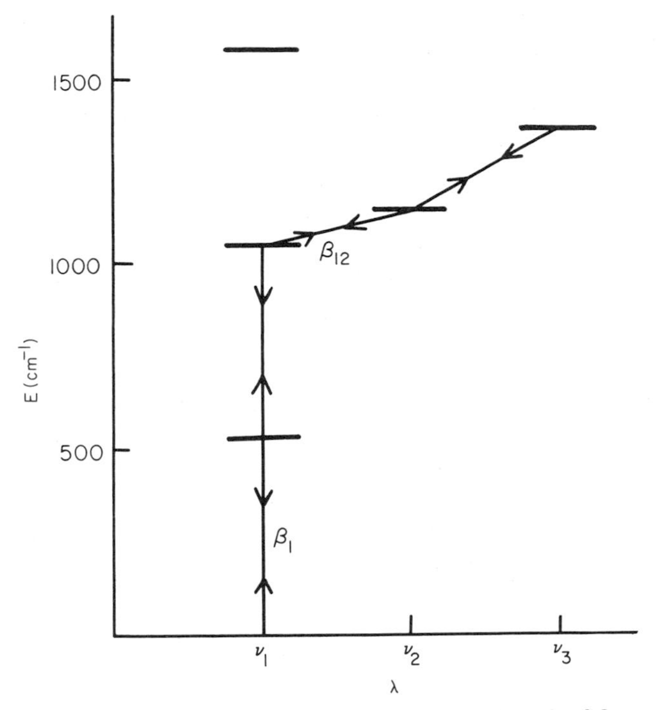

FIG. 3. Energy level diagram showing transitions for SO_2.

due to hindered internal rotation, behave similarly to other fundamental modes in respect to the overall molecular relaxation process (Lambert and Salter, 1959). In the few cases where there is a large gap between the torsional frequency and the other modes, as in C_2H_6, a double relaxation process is observed. Otherwise vibrational energy enters all modes via the torsional mode, giving rise to a single relaxation process involving the whole of the molecular vibrational energy, with a relaxation time characteristic of the torsional mode. Since torsional modes of complex molecules, such as higher hydrocarbons, usually have very low frequencies, (< 200 cm^{-1}), energy transfer between translation and vibration is very easy ($Z_{10} < 5$).

Values of β_1, measured by acoustic techniques, are now available for a fairly large range of polyatomic molecules which show the single relaxation process described above. Theoretical prediction of relaxation times may be made by the S.S.H. method, as extended by Tanczos (1956), but there is greatly increased difficulty in finding and fitting an adequate intermolecular potential for polyatomic molecules, especially when polar. Tanczos (1956) and Dickens and Linnett (1957) both used a Krieger potential for calculations on CH_2Cl_2 and SO_2, respectively, and both succeeded in predicting the double nature of the relaxation process and the approximate ratio between the two relaxation times, but the absolute calculated transition probabilities were too small by a factor of 5 to 10. A calculation for CH_4 by Cottrell and Ream (1955) showed similar discrepancies with experiment. The theoretical values for polyatomic molecules shown in Fig. 1 were all calculated by similar methods, and it will be seen that agreement with experiment is far less satisfactory, particularly for polar molecules and where short relaxation times are involved, than for the simple diatomic molecules.

Consideration of the available experimental data for polyatomic molecules at temperatures close to 30°C by Lambert and Salter (1959) shows that there is a striking and very simple empirical relationship between the frequency of the lowest fundamental mode, $\nu_{\min}$, and the value of Z_{10}: this is shown in Fig. 4. Molecules fall into two classes, differentiated by the presence or absence of hydrogen atoms; each class showing a linear relation between log Z_{10} and $\nu_{\min}$. The existence of this simple relation, which is entirely independent of mass or interaction potential, covering a wide range of molecules, suggests that the *kinetic energy* involved in the collision (rather than the *velocity* of approach) and the size of the vibrational quantum are the main factors determining the probability of transfer. This view was proposed by Fogg *et al.* (1953), based on the physical assumption that transfer only occurs when the two colliding molecules possess sufficient translational energy to penetrate deeply into one another's force fields. If the energy required is ϵ, and is proportional to the size of the vibrational quantum, $h\nu_{\min}$, the probability of transfer in a single collision is

$$P_{10} = \exp(-\epsilon/kT) = \exp(-Xh\nu_{\min}/kT)$$

and

$$Z_{10} = 1/P_{10} = \exp(Xh\nu_{\min}/kT) \quad (9)$$

which gives the required linear relation between log Z_{10} and $\nu_{\min}$, with values of $X \sim 2$ and 4 for the two classes shown in Fig. 4.

The effect of the presence of hydrogen atoms in the molecule in facilitating energy transfer was first noticed by Rossing and Legvold (1955) for a series of freon molecules; a qualitative explanation was suggested by Lambert and Salter (1959). In any vibration involving a

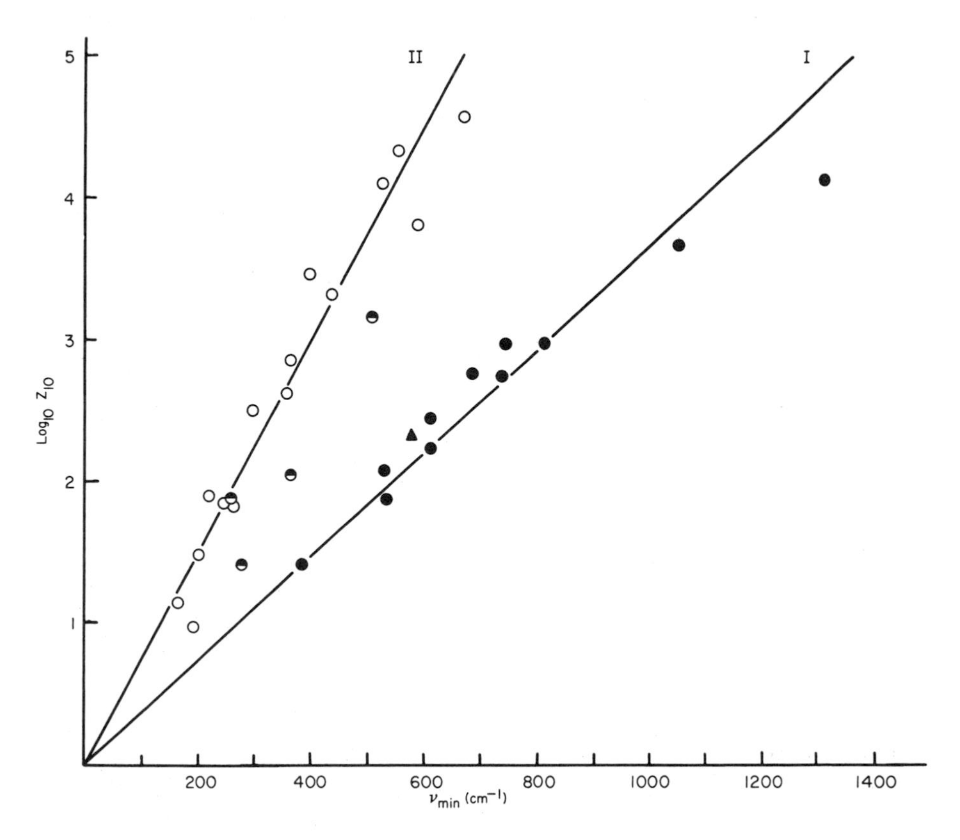

FIG. 4. Relation between Z_{10} and ν_{min} for molecules showing a single relaxation process (Lambert and Salter, 1959). Values listed in order from the bottom upwards (temp., 300°K). (Letters in parentheses refer to the references listed below.) ○, molecules containing no hydrogen atom; C_2F_4 (*a*); CF_2Br_2 (*b*); CF_2BrCl (*b*); CF_2Cl_2 (*b*); $CFCl_3$ (*b*); CCl_4 (*c*); CF_3Br (*b*); CF_3Cl (*b*); SF_6 (*d*); CF_4 (*a*); CS_2 (*e*); N_2O (*f*); COS (*g*); Cl_2 (*h*); CO_2 (*f*). ◒, molecules containing one hydrogen atom; $CHCl_2F$ (*b*); $CHCl_3$ (*c*); $CHClF_2$ (*b*); CHF_3 (*a*, *b*). ●, molecules containing two or more hydrogen atoms; CH_2ClF (*b*); CH_3I (*i*); CH_2F_2 (*a*); CH_3Br (*b*, *i*); C_2H_2; CH_3Cl (*b*); C_2H_4O (*e*); C_2H_4 (*j*); *cyclo*-C_3H_6 (*j*); CH_3F (*i*); CH_4 (*k*). ▲, deuterated molecule; CD_3Br. This figure is reproduced by kind permission of the Council of the Royal Society.

Key to references:

(a) Fogg and Lambert (unpublished results)
(b) Rossing and Legvold (1955)
(c) Sette *et al.* (1955)
(d) O'Connor (1954)
(c) Sette *et al.* (1955)
(e) Angona (1953)
(f) Eucken and Nümann (1937)
(g) Eucken and Aybar (1940)
(h) Eucken and Becker (1934)
(i) Hanks and Lambert (unpublished results)
(j) Corran *et al.* (1958)
(k) Cottrell and Martin (1957)

bond between a heavy atom, such as carbon, and a much lighter atom, such as hydrogen, the latter will execute far more extensive motions relative to the centre of mass than the former. This extensive motion will have a correspondingly greater effect on the intramolecular force field. Conversely, a collision in which the impact is received on a peripheral hydrogen atom might be expected to result in a greater perturbation of the molecular force field, than when the impact is received on a heavier atom. Since the ratio between the mass of a hydrogen atom and the total molecular weight is of a different order to the corresponding ratio for any other atomic species, it is to be expected that molecules containing hydrogen atoms should fall in a class of their own.

The temperature dependence of vibrational relaxation in polyatomic molecules can usually be investigated only over a limited range, owing to thermal instability at high temperatures. Amme and Legvold (1959) have made measurements on a number of freons at temperatures up to 400°C, and find fair agreement with the S.S.H. theory (8), particularly if a steeper intermolecular repulsion potential than the Lennard-Jones, 6-12, is used. Discrepancies reported by Arnold *et al.* (1957) for a number of gases, and by Corran *et al.* (1958) for ethylene *cyclo*propane and carbon tetrafluoride, are due mainly to the use of an inaccurate abbreviation of (8), (cf. § 4.1). The alternative relation, (9), predicts in all cases a temperature dependence several times greater than observed, and is not satisfactory in this respect. On the other hand, certain strongly polar molecules, such as SO_2, CH_3F, and CH_3Br, show a definite reversal of temperature dependence at lower temperatures, so that P_{10} is larger at 20°C than at 100°C (Corran *et al.*, 1958). This cannot be explained on the basis of either theory, and would accord with the view recently expressed by both Herzfeld and Litovitz (1959) and Dickens and Ripamonti (1961), that collisions between complex polyatomic molecules, especially when polar, are in general of much longer duration than collisions between simpler molecules, so that the physical basis of (8) no longer applies; and that energy transfer may occur via a "transition state" involving some form of intermolecular association complex. There is independent evidence from virial coefficient measurements for such association occurring in polar vapors (Lambert, 1953), and this is of course more marked at lower temperatures. The unexpectedly high transfer efficiency recently reported for gaseous NO (ν_{min} = 1878 cm^{-1}; Z_{10} = 2700 at 293°K) by Bauer *et al.* (1959) and Robben (1959) may be due to intermolecular association, for which, in this case, there is spectroscopic evidence (D'Or *et al.*, 1951). If such association is regarded as a chemical reaction, requiring an energy of activation, there is some justification for the exponential form of (9), but the physical basis of this equation is at present much too crude, and it is quite unclear how a theory of this kind will link with the S.S.H. treatment.

It is interesting that vibrational relaxation phenomena in *liquids*, where the molecules can be regarded as in continual close association, show the same general features as for the corresponding gases, e.g., double relaxation for CH_2Cl_2 and SO_2 (Andreae, 1957; Bass and Lamb, 1957), and energy transfer would appear to occur in binary collisions with the same transfer probability per collision as in the gas (Litovitz, 1957; Bass and Lamb, 1958). This is also the case with very highly compressed gaseous CO_2, at a temperature above the critical temperature, where the relaxation time remains inversely proportional to the density up to a pressure of 250 atm (Henderson and Peselnick, 1957).

5 Vibrational Relaxation in Gas Mixtures

In a mixture of two gases, A and B, three types of intermolecular collision are possible. If A is a relaxing gas and B a nonrelaxing gas, such as helium, or hydrogen at room temperature, there are two collision processes by which vibrational-translational energy transfer may occur:

$$(1)\quad A^* + A \rightarrow A + A \qquad (\text{vib} \rightarrow \text{trans})$$

$$(2)\quad A^* + B \rightarrow A + B \qquad (\text{vib} \rightarrow \text{trans}).$$

Since (1) and (2) will have different probabilities per collision, the result will be a composite relaxation time given by

$$\frac{1}{\beta} = \frac{x_A}{\beta_{AA}} + \frac{x_B}{\beta_{AB}} = \frac{1 - x_B}{\beta_{AA}} + \frac{x_B}{\beta_{AB}} \tag{10}$$

where x_A and x_B are the mole fractions of A and B present. β_{AB} will be related to the transfer probability per A/B collision ($P_{AB} = 1/Z_{AB}$) in the same manner as β_{AA} to P_{AA}. A number of ultrasonic measurements have been made on mixtures of a relaxing gas, A, with small quantities of a nonrelaxing gas, B. The general behavior of such mixtures is illustrated by the data for carbon dioxide and ethylene, which are shown in Table II. It is immediately clear that the transfer probabilities for heteromolecular collisions involving a single dispersing gas differ very widely indeed. Some of the effects, such as the high transfer efficiency shown by very light molecules, $H_2 > D_2 \sim He$, in collision with both CO_2 and C_2H_4, or the low efficiency shown for C_2H_4/N_2 collisions,

TABLE II

VALUES OF Z_{AB} FOR HETEROMOLECULAR COLLISIONS

Additive	Relaxing gas	
	CO_2 (293°K)	C_2H_4 (288°K)
Self	108,000[a]	1070[b]
H_2	500[a]	70[a]
He	1500[a]	—
D_2	1500[a]	550[b]
N_2	—	1400[b]
H_2O	130[a]	24[b]
C_3H_8	—	31[b]
n-C_4H_{10}	—	26[b]
n-C_5H_{12}	—	19[b]

[a] McCoubrey and McGrath (1957).
[b] Arnold *et al.* (1958).

can be semiquantitatively explained in terms of the S.S.H. theory as due to variations in mass and intermolecular potential. The more striking effects, such as the very high transfer probability in CO_2/H_2O and C_2H_4/hydrocarbon collisions, seem to require a different type of explanation, perhaps corresponding more with the associated intermolecular transition complex suggested for polyatomic molecules in (§ 4.2). The high efficiency of CO_2/H_2O collisions was attributed by Eucken and Franck (1933) to incipient chemical combination; a detailed quantum-mechanical treatment in terms of potential energy curves was proposed by Widom and Bauer (1953). Ubbelohde and his collaborators (cf. Arnold *et al.*, 1958) have attributed the efficiency of long chain hydrocarbons in activating the ethylene molecule to a "wrestling collision," in which the flexible hydrocarbon molecules undergo prolonged contact with the ethylene molecule. There is independent evidence from gas kinetic transport properties for a special type of collision process for flexible molecules (Lambert *et al.*, 1955). An important practical consequence of the profound effect which small traces of an impurity may have on the relaxation time of a gas is that scrupulous attention must be paid to the chemical purity of gases used in relaxation experiments. Many reported discrepancies between different observers are undoubtedly due to impure gases being used.

There are few experimental data about the temperature dependence of transfer probabilities in heteromolecular collisions. Eucken and

Nümann (1937) found that P_{AB} for CO_2/H_2O collisions increases with rising temperature in the lower temperature range, but begins to decrease consistently at higher temperatures. Küchler (1938) reports a barely significant change in P_{AB} for CO_2/H_2 over a range of 300°, but an increase of P_{AB} with rising temperature for CO_2/He. Eucken and Aybar (1940) found the normal increase of P_{AB} with rising temperature for CH_4/H_2, CH_4/CO_2, and COS/A mixtures. It thus appears that heteromolecular transfer probabilities show the same kind of anomalous temperature dependence as observed for homomolecular collisions between strongly polar gases (§ 4.2). This lends further support to the hypothesis that a specific intermolecular transition complex may play an important role in energy transfer.

A fresh factor, which may be important in heteromolecular collisions, is the direct transfer of vibrational energy between A and B. On theoretical grounds it would appear unlikely that there is any hindrance to collision transfer of a quantum of vibrational energy from one molecule to another in a homomolecular collision. This process cannot be experimentally followed by acoustic methods, which only measure a transfer between vibration and translation, but chemical spectroscopic evidence, discussed in § 2.3 confirms that equilibrium distribution of vibrational energy between molecules of a single species is very rapidly attained. The case for heteromolecular collisions is different. The possible transfer processes are now:

$$(1)\quad A^* + A \rightarrow A + A \qquad (\text{vib} \rightarrow \text{trans})$$

$$(2)\quad A^* + B \rightarrow A + B \qquad (\text{vib} \rightarrow \text{trans})$$

$$(3)\quad A^* + B \rightarrow B^* + A \qquad (\text{vib} \rightarrow \text{vib})$$

$$(4)\quad B^* + B \rightarrow B + B \qquad (\text{vib} \rightarrow \text{trans})$$

$$(5)\quad B^* + A \rightarrow B + A \qquad (\text{vib} \rightarrow \text{trans}).$$

If A is a relaxing molecule, such as ethylene, and B a complex hydrocarbon, which will have a very small relaxation time (cf. § 4.2), a rapid process (3) may be followed by the even more rapid process (4), and a greatly enhanced vibrational-translational transfer probability will result, with a value of β_{AB} characteristic of process (3). By analogy with the theory of complex transfer, discussed in § 4.2, easy vibration-vibration transfer [rapid process (3)] would be expected between molecules with near resonant fundamental vibration frequencies. Complex hydrocarbon molecules possess a wide range of vibrational frequencies, lying fairly close to one another, and are particularly suitable for fulfilling this criterion.

For a similar mixture where B also shows some degree of relaxation (though $\beta_{BB} \ll \beta_{AA}$) and process (4) is less rapid than process (3), the former will be the rate determining step for transfer in heteromolecular collisions. As the rate of (4) is proportional to x_B^2 (instead of to x_B), the over-all relaxation time will no longer follow (10), which predicts a linear variation of $1/\beta$ with x_B. Boudart (1955; cf. Herzfeld and Litovitz, 1959) has investigated by a shock tube method the variation of relaxation time for O_2 ($\nu = 1580$ cm^{-1}) with small additions of H_2O ($\nu_2 = 1595$), where this state of affairs would be expected, and finds a nonlinear dependence of $1/\beta$ on the mole fraction of H_2O present. For mixtures of O_2 with D_2O ($\nu_2 = 1178$ cm^{-1}), which has no near-resonant vibration frequency so that easy vibration-vibration transfer would not be expected, the normal linear dependence is found, suggesting that here only processes (1) and (2) come into play.

A mixture of two gases, both of which show relaxation which is experimentally measurable, should exhibit a double relaxation phenomenon (Herzfeld and Litovitz, 1959), corresponding to two distinct relaxation times, β_A and β_B, such that

$$\frac{1}{\beta_A} = \frac{x_A}{\beta_{AA}} + \frac{x_B}{\beta_{AB}}$$

and

$$\frac{1}{\beta_B} = \frac{x_B}{\beta_{BB}} + \frac{x_A}{\beta_{BA}}$$

where β_{AB} corresponds to the process

$$A^* + B \rightarrow A + B \qquad \text{(vib} \rightarrow \text{trans)}$$

and β_{BA} to the process

$$A + B^* \rightarrow A + B \qquad \text{(vib} \rightarrow \text{trans).}$$

Experimental measurements by Amme and Legvold (1957) on binary mixtures of freons show no conclusive evidence for or against this (Calvert and Amme, 1960). It is difficult to obtain experimental resolution of relaxation times differing by a factor of less than $\times$ 10, and further measurements on a variety of binary systems are needed to investigate this point.

REFERENCES

Amme, R., and Legvold, S. (1957) *J. chem. Phys*, **26**, 514.

Amme, R., and Legvold, S. (1959) *J. chem. Phys.* **30**, 163.

Andersen, W. H., and Hornig, D. F. (1959) *Molecular Phys.* **2**, 49.

Andreae, J. H. (1957) *Proc. Phys. Soc.* **B71**, 71.

Angona, F. C. (1953) *J. Acoust. Soc. Amer.* **25**, 1116.

Arnold, J. H., McCoubrey, J. C., and Ubbelohde, A. R. (1957) *Trans. Faraday Soc.* **53**, 738.

Arnold, J. H., McCoubrey, J. C., and Ubbelohde, A. R. (1958) *Proc. Roy. Soc.* **A248**, 445.

Bass, R., and Lamb, J. (1957) *Proc. Roy. Soc.* **A243**, 94.

Bass, R., and Lamb, J. (1958) *Proc. Roy. Soc.* **A247**, 168.

Bauer, H. J., Kneser, H. O., and Sittig, E. (1959) *J. chem. Phys.* **30**, 1119.

Bethe, H. A., and Teller, E. (1941) *Aberdeen Proving Ground Rep.* X, 117.

Blackman, V. H. (1956) *J. Fluid. Mech.* **1**, 61.

Boudart, M. J. (1955) Princeton University Tech. Note 7, Contract AF33(038)-23976.

Breazeale, M. A., and Kneser, H. O. (1960) *J. Acoust. Soc. Amer.* **32**, 885.

Brout, R. (1954a) *J. chem. Phys.* **22**, 934.

Brout, R. (1954b) *J. chem. Phys.* **22**, 1189.

Calvert, J. B., and Amme, R. (1960) *J. chem. Phys.* **33**, 1270.

Cashion, J. K., and Polanyi, J. C. (1960) *Proc. Roy. Soc.* **A258**, 529.

Chapman, S., and Cowling, T. G. (1939) "Mathematical Theory of Non-Uniform Gases." Cambridge Univ. Press, London and New York.

Clarke, J. K., and Ubbelohde, A. R. (1957) *J. chem. Soc.* p. 2050.

Clouston, J. G., Gaydon, A. G., and Glass, I. I. (1958) *Proc. Roy. Soc.* **A248**, 429.

Clouston, J. G., Gaydon, A. G., and Hurle, I. R. (1959) *Proc. Roy. Soc.* **A252**, 143.

Corran, P., Lambert, J. D., Salter, R., and Warburton, B. (1958) *Proc. Roy. Soc.* **A244**, 212.

Cottrell, T. L., and Martin, P. E. (1957) *Trans. Faraday Soc.* **53**, 1157.

Cottrell, T. L., and Ream, N. (1955) *Trans. Faraday Soc.* **51**, 159.

Craven, P. M., and Lambert, J. D. (1951) *Proc. Roy. Soc.* **A205**, 439.

Danby, C. J., Lambert, J. D., and Mitchell, C. M. (1957) *Proc. Roy. Soc.* **A239**, 365.

Dickens, P. G., and Linnett, J. W. (1957) *Proc. Roy. Soc.* **A243**, 84.

Dickens, P. G., and Ripamonti, A. (1961) *Trans. Faraday Soc.* **57**, 735.

D'Or, L., de Lattre, A., and Tarte, P. (1951) *J. chem. Phys.* **19**, 1064.

Edmonds, P. D., and Lamb, J. (1958a) *Proc. Phys. Soc.* **71**, 17.

Edmonds, P. D., and Lamb, J. (1958b) *Proc. Phys. Soc.* **72**, 940.

Eucken, A. (1913) *Phys. Z.* **14**, 324.

Eucken, A., and Aybar, S. (1940) *Z. phys. Chem.* **B46**, 195.

Eucken, A., and Becker, R. (1934) *Z. phys. Chem.* **B27**, 235.

Eucken, A., and Franck, J. (1933) *Z. phys. Chem.* **B20**, 460.

Eucken, A., and Nümann, E. (1937) *Z. phys. Chem.* **B36**, 163.

Fogg, P. G. T., Hanks, P. A., and Lambert, J. D. (1953) *Proc. Roy. Soc.* **A219**, 490.

Frey, H. M., and Kistiakowsky, G. B. (1957) *J. Amer. Chem. Soc.* **79**, 6373.

Garvin, D., Broida, H. P., and Kostowski, H. J. (1960) *J. chem. Phys.* **32**, 880.

Gaydon, A. G. (1948) "Spectroscopy and Combustion Theory." Chapman & Hall, London.

Green, E. F., and Hornig, D. F. (1953) *J. chem. Phys.* **21**, 617.

Greenspan, M. (1959) *J. Acoust. Soc. Amer.* **31**, 155.

Harrington, R. E., Rabinovitch, B. S., and Hoare, M. R. (1960) *J. chem. Phys.* **33**, 744.

Heaps, H. S., and Herzberg, G. (1952) *Z. Phys.* **133**, 48.
Henderson, M. C., and Pcsclnick, L. (1957) *J. Acoust. Soc. Amer.* **29**, 1074.
Herzfeld, K. F., and Litovitz, T. A. (1959) "Absorption and Dispersion of Ultrasonic Waves." Academic Press, New York.
Huber, P. W., and Kantrowitz, A. (1947) *J. chem. Phys.* **15**, 275.
Jackson, J. M., and Mott, N. F. (1932) *Proc. Roy. Soc.* **A137**, 703.
Jacox, M. E., and Bauer, S. H. (1957) *J. phys. Chem.* **61**, 833.
Kantrowitz, A. (1946) *J. chem. Phys.* **14**, 150.
Kelly, B. T. (1957) *J. Acoust. Soc. Amer.* **29**, 1005.
Kistiakowsky, G. B., and Tabbutt, F. D. (1959) *J. chem. Phys.* **30**, 577.
Knötzel, H., and Knötzel, L. (1948) *Ann. Phys.* (Leipzig) [6]**2**, 393.
Küchler, L. (1938) *Z. phys. Chem.* **B41**, 199.
Lambert, J. D. (1953) *Disc. Faraday Soc.* **15**, 226.
Lambert, J. D., and Salter, R. (1957) *Proc. Roy. Soc.* **A243**, 78.
Lambert, J. D., and Salter, R. (1959) *Proc. Roy. Soc.* **A253**, 277.
Lambert, J. D., Cotton, K. J., Pailthorpe, M. W., Robinson, A. M., Scrivins, J., Vale, W. R. F., and Young, R. M. (1955) *Proc. Roy. Soc.* **A231**, 280.
Landau, L. D., and Teller, E. (1936) *Phys. Z. Sowjetunion* **10**, 34.
Lipscomb, F. J., Norrish, R. G. W., and Thrush, B. A. (1956) *Proc. Roy. Soc.* **A233**, 455.
Litovitz, T. A. (1957) *J. chem. Phys.* **26**, 469.
Lukasik, S. J., and Young, J. E. (1957) *J. chem. Phys.* **27**, 1149.
McCoubrey, J. C., and McGrath, W. D. (1957) *Quart. Revs.* (*London*) **11**, 87.
O'Connor, C. L. (1954) *J. Acoust. Soc. Amer.* **26**, 361.
Penney, H. C., and Aroeste, H. (1955) *J. chem. Phys.* **23**, 1281.
Richardson, E. G. (1955) *Rev. mod. Phys.* **27**, 15.
Richardson, E. G. (1959) *J. Acoust. Soc. Amer.* **31**, 152.
Robben, F. (1959) *J. chem. Phys.* **31**, 420.
Rossing, T. D., and Legvold, S. (1955) *J. chem. Phys.* **23**, 1128.
Schäfer, K. (1943) *Z. phys. Chem.* **B53**, 149.
Schwartz, R. N., and Herzfeld, K. F. (1954) *J. chem. Phys.* **22**, 767.
Schwartz, R. N., Slawsky, Z. I., and Herzfeld, K. F. (1952) *J. chem. Phys.* **20**, 1591.
Sette, D., Busala, A., and Hubbard, J. C. (1955) *J. chem. Phys.* **23**, 787.
Shields, F. D. (1960) *J. Acoust. Soc. Amer.* **32**, 180.
Shuler, K. E., and Zwanzig, R. (1960) *J. chem. Phys.* **33**, 1778.
Slobodskaya, P. V. (1948) *Izv. Akad. Nauk SSSR, Ser. fiz.* **12**, 656.
Smiley, E. F., and Winkler, E. H. (1954) *J. chem. Phys.* **22**, 2018.
Stevens, B., and Boudart, M. J. (1957) *Ann. N. Y. Acad. Sci.* **67**, 570.
Stewart, J. L., and Stewart, E. S. (1952) *J. Acoust. Soc. Amer.* **24**, 194.
Takayanagi, K. (1957) *Proc. Phys. Soc.* **A70**, 348.
Takayanagi, K. (1959) *Sci. Rep. Saitama Univ.* **A3**, 65.
Tanczos, F. I. (1956) *J. chem. Phys.* **25**, 439.
Waelbroeck, F. G., and Zuckerbrodt, P. (1958) *J. chem. Phys.* **28**, 524.
Wang Chang, C. S., and Uhlenbeck, G. E. (1951) University of Michigan Report CM-681, Project N-Ord-7924.
Widom, B. (1960) *J. chem. Phys.* **32**, 913.
Widom, B., and Bauer, S. H. (1953) *J. Chem. Phys.* **21**, 1670.
Windsor, M. W., Davidson, N., and Taylor, R. (1957) *J. chem. Phys.* **27**, 315.
Windsor, M.W., Davidson, N., and Taylor, R. (1959) *Symposium on Combustion 7th Symposium London, 1958*, p. 80.

21.

Chemical Processes

J. C. Polanyi

1 *Categorization of Reaction Rates*

1.1 Dependence of Rates on Concentrations

It is customary to categorize reactions according to the dependence of the rate of reaction on the concentrations of reagents for the generalized reaction,

$$aA + bB \rightarrow mM + nN + \ldots; \tag{1}$$

the rate can be defined as the rate of decrease in concentration of one or another reagent (A, B, etc.), or the rate of production of one or another product (M, N, etc.). If the number of molecules of A participating in the reaction is different from the number of molecules of B, i.e., $a \neq b$, then the rate defined in terms of these two species will be different. Similarly, if $b \neq m$ or $m \neq n$, it is important to state the manner in which rate was defined. If we define rate in terms of reagent A, having concentration c_A at time t,

$$-\frac{dc_A}{dt} = k\, c_A^a\, c_B^b \ldots, \tag{2}$$

a is referred to as the *order* of reaction with respect to reagent A, b is the order with respect to B. The over-all *order of the reaction* is $a + b + \ldots.$

If a reaction takes place principally at a surface, the rate may be found to be independent of the concentration of reagents; this is referred to as a zero order reaction.

When an energized molecule, A, decomposes or rearranges, we have a reaction which is first order and *unimolecular*,

$$-\frac{dc_A}{dt} = k\, c_A. \tag{3}$$

Integration of (3) leads to

$$-\log c_A = kt + \text{const.} \tag{4}$$

Calling the initial concentration of A, a_0, at $t = 0$, the constant becomes $-\log a_0$, and (4) can be written

$$c_A = a_0\, e^{-kt}. \tag{5}$$

It is seen that the concentration of reagent decays exponentially with time, as might be expected from the fact that at any instant the rate of reaction is proportional to the concentration of reagent remaining.

If we write the concentration of product at time t as x, the instantaneous concentrations of reagent and product at any time will be as follows:

$$\begin{array}{cc} \mathrm{A} & \rightarrow \mathrm{M} \\ a_0 - x & x. \end{array} \tag{6}$$

Putting $a_0 - x$ in place of c_A in (5), we get

$$k_1 = \frac{1}{t} \log \frac{a_0}{a_0 - x}. \tag{7}$$

The constant k_1 is the *rate constant*; it is the reaction rate at unit concentration of each reagent [cf. (2)]. It is a function of temperature (see below) but not of concentration.

When the rate of reaction is proportional to the product of two concentrations, reaction is second order and *bimolecular*,

$$\begin{array}{ccccc} \mathrm{A} & + & \mathrm{B} & \rightarrow \mathrm{M} \\ a_0 - x & & b_0 - x & x. \end{array} \tag{8}$$

In the simplest case where initial concentrations are the same, the rate law is simply

$$-\frac{dc_A}{dt} = -\frac{d(a_0 - x)}{dt} = \frac{dx}{dt} = k_2(a_0 - x)^2; \tag{9}$$

integration leads to

$$k_2 = \frac{x}{a_0 t(a_0 - x)}. \tag{10}$$

For the more complex case of a bimolecular reaction in which initial reagent concentrations are not the same,

$$\frac{dx}{dt} = k_2(a_0 - x)(b_0 - x) \tag{11}$$

and integration gives

$$k_2 = \frac{1}{t(a_0 - b_0)} \log \frac{b_0(a_0 - x)}{a_0(b_0 - x)}. \tag{12}$$

For a *termolecular* reaction, in the most common type of example where two of the participating species are identical and consequently present in the same concentration,

$$\begin{matrix} A & + & 2B & \rightarrow M \\ a_0 - x & & b_0 - 2x & x, \end{matrix} \tag{13}$$

the rate law is

$$\frac{dx}{dt} = k_3(a_0 - x)(b_0 - 2x)^2 \tag{14}$$

which integrates to

$$k_3 = \frac{1}{t(2a_0 - b_0)^2} \left[\frac{(2a_0 - b_0)\, 2x}{b_0(b_0 - 2x)} + \log \frac{a_0(b_0 - 2x)}{b_0(a_0 - x)} \right]. \tag{15}$$

The units customarily used for the rate constants of reactions of the various orders referred to previously are listed in Table I.

The order of a reaction may be determined experimentally in a number of ways, of which two will be mentioned here. The most obvious method in the light of the foregoing discussion is to measure the concentration of product x at various times t, starting with known concentrations of reagents, a_0, b_0, ... These quantities can then be inserted into the various integrated rate equations, of which (7), (10), (12), and (15) are typical, to see which yields a rate constant, k, constant

TABLE I

UNITS OF RATE CONSTANTS (k) FOR REACTIONS OF VARIOUS ORDER (n)

Order, n: 0	1	2	3
$k_0 \quad \frac{\text{mole}}{\text{liter sec}}$	$k_1 \quad \text{sec}^{-1}$	$k_2 \quad \frac{\text{liter}}{\text{mole sec}}$	$k_3 \quad \frac{\text{liter}^2}{\text{mole}^2\ \text{sec}}$
$k_0 = \frac{\text{mole}}{\text{cc sec}} 10^3$	$k_1 = \text{sec}^{-1}$	$k_2 = \frac{\text{cc}}{\text{mole sec}} 10^{-3}$	$k_3 = \frac{\text{cc}^2}{\text{mole}^2\ \text{sec}} 10^{-6}$
$k_0 = \frac{\text{molecule}}{\text{cc sec}} \cdot 1.660 \cdot 10^{-21}$	$k_1 = \text{sec}^{-1}$	$k_2 = \frac{\text{cc}}{\text{molecule sec}} \cdot 6.025 \cdot 10^{20}$	$k_3 = \frac{\text{cc}^2}{\text{molecule}^2\ \text{sec}} \cdot 3.630 \cdot 10^{41}$

to within experimental error. This method has the disadvantage that a large number of trials may be required before a successful integrated rate equation is found—this is especially true if the system is complex (involving successive or competing reactions) and consequently has an over-all order which is fractional. In this case it may be preferable to use the method of half-lives.

The half-life of a reaction, τ, is the time required for the concentration of reagent to fall to one-half of the initial concentration. For a first order reaction (as for a radioactive decay—which is typical of a first order process) τ is independent of the concentration of starting material. On writing $x = a_0/2$ into (7),

$$k_1 = \frac{1}{\tau} \log \frac{a_0}{a_0 - (a_0/2)}; \tag{16}$$

therefore

$$\tau = \frac{1}{k_1} \log 2. \tag{17}$$

For a second order reaction τ is inversely proportional to the first power of the starting concentration, a_0. From (10)

$$k_2 = \frac{a_0/2}{a_0 \tau [a_0 - (a_0/2)]}; \tag{18}$$

therefore

$$\tau = \frac{1}{a_0 k_2}. \tag{19}$$

For reaction of nth order, assuming that all reagents are initially at concentration a_0,

$$\tau = \frac{2^{n-1} - 1}{(n-1)\, k_n a_0^{n-1}}; \tag{20}$$

that is to say,

$$\tau \propto \frac{1}{a_0^{n-1}}. \tag{21}$$

To solve (20) for the order of reaction, n, it is only necessary to determine half-lives for two (known) initial concentrations of reagents.

The importance of the order of reaction is that it provides a guide to the *molecularity* of the reaction, that is to say, the number of molecules that must collide in order for the reagents to pass over into products.

This obviously is a quantity of fundamental importance in any theoretical discussion of the molecular event that constitutes a chemical reaction.

However, it is not safe simply to equate order of reaction with molecularity. For example, if a reaction occurs in the presence of a large excess of one reagent (this reagent might be the solvent) the concentration of this reagent will not alter appreciably in the course of the reaction. Its effect on the rate can, therefore, easily be overlooked. The order ascribed to the reaction will then be n, despite the fact that the molecularity is $n + 1$. This does not raise any insuperable problems (in the case cited, the order of reaction can be determined, by way of a check, in another solvent), but the example illustrates the need for caution in equating order and molecularity.

A problem of a more fundamental nature is raised by those gas reactions which are found to be first order at high pressures and second order at low pressures. These reactions are unimolecular at high pressures. They are described as quasi-unimolecular reactions at low pressures, even though at these pressures a bimolecular energy-transfer process (which is not a chemical reaction) has become the rate-determining factor. This is discussed further in § 3.1.1.

1.2 Dependence of Rates on Temperature

It has been known for some time that increase in temperature generally increases the rate of chemical reaction (a very rough rule of thumb is that the rate doubles for every 10°C rise in temperature).

Arrhenius, in 1889, put forward the view that for a reaction to take place reagent molecules must become "activated," and that there exists an equilibrium between normal and activated molecules. This view is still regarded as being essentially correct. What is in doubt is the nature of the *activated complex*, and the extent to which it can properly be regarded as being in equilibrium with the reagents.

Van't Hoff proposed an expression for the temperature variation of an equilibrium constant

$$\frac{d \log K_c}{dT} = \frac{E}{RT^2} \tag{22}$$

where K_c is the equilibrium constant expressed in terms of concentrations and E is the difference in internal energy between the states in equilibrium (if K is at constant pressure, E is replaced by H; heat content).

Later Arrhenius proposed the use of the analogous relation

$$\frac{d \log k}{dT} = \frac{E_a}{RT^2} \tag{23}$$

where k is the rate constant in concentration units and E_a is the difference in internal energy between the activated molecules and the normal reagent molecules. If the model is correct and E_a is a true constant, (23) can be integrated to give

$$\log k = \text{const} - \frac{E_a}{RT}. \tag{24}$$

On writing const = log A,

$$k = \text{A} \cdot \exp(-E_a/RT). \tag{25}$$

E_a is referred to as the activation energy. Since it is expressed in the same units as RT, E_a/RT is dimensionless. A has the same units as the rate constant, k. For a first order reaction A therefore has the units $\sec^{-1}$ (Table I) and the common description *frequency factor* is appropriate. However, for reactions of higher order this description is confusing, and A has been termed the *temperature-independent factor*. This too is unfortunate since the quantity, on other theories than that of Arrhenius, is temperature dependent. A is today most often referred to as the *pre-exponential factor* or simply the *A-factor*. We shall make use of the last description.

After 70 years the Arrhenius equation retains its value as an empirical representation of the dependence of rate constants on temperature. To within experimental error, a plot of log k vs. $1/T$ is found to be linear. E_a can be obtained from the slope of this line, and A by an extrapolation to $1/T = 0$.

Nonetheless, there is good reason to suppose that with improved experimental techniques it will be found that both E_a and A, as defined by Arrhenius, are temperature dependent. The temperature dependence of A will be discussed in § 3.1. The temperature dependence of E_a follows from the established temperature dependence of E [the energy of reaction, (22)] in the van't Hoff isochore. The true expression for the dependence of equilibrium constant on temperature is more complex than was supposed in Arrhenius' time. Since the equilibrium constant is simply the ratio of two rate constants, it follows that the temperature dependence of the rate constants must be more complex than that given by the Arrhenius equation, which provides only for a temperature dependence of the van't Hoff type.

Moreover, if the Arrhenius model for reaction, in which normal and activated molecules are in equilibrium, is examined in detail from the point of view of statistical mechanics, the requirement that E_a shall be temperature dependent emerges. When E_a is expressed in terms of molecular energies, it is found that

$$E_a = N_0(\epsilon^{\ddagger} - \bar{\epsilon}) \tag{26}$$

where N_0 is Avogadro's number, $\epsilon^{\ddagger}$ is the average energy of the activated molecules (those with a total energy above some threshold energy, ϵ_a), and $\bar{\epsilon}$ the average energy of all molecules including the activated molecules (Tolman, 1927). $\epsilon^{\ddagger}$ and $\bar{\epsilon}$ are both functions of temperature, and it would be purely fortuitous if they showed the same temperature dependence. It follows that E_a, obtained from experiment by the application of Arrhenius' equation, will be temperature dependent.

1.3 Dependence of Rates on Kinematic Factors

Until recently almost all investigations in the field of reaction kinetics were concerned with the variation of reaction rate with changing concentrations of reagents (§ 1.1) and changing temperature (§ 1.2). Since both of these variables are statistical in nature, this had the result that theories of reaction rate could be tested only insofar as their predictions were statistical. It would be a great deal more satisfactory if the underlying molecular assumptions of the various theories, concerning the mechanics of molecular encounter and rearrangement, could be put to an experimental test.

At the present time two main avenues are being explored in the hope of obtaining direct experimental evidence concerning the detailed kinematics of chemical reaction. The first of these, the measurement of the rate of reaction (or "reaction cross section," to avoid a term involving concentration) in crossed molecular beams, involves a control of the energies of the colliding species. The second method is the measurement by spectroscopic means of the energies of freshly formed product molecules. By this means it should eventually be possible to obtain what might be called the fine structure of the reaction rate: rate constants for reaction into specific vibrational and rotational states of the product molecules. The second method can be made to yield information of the same type as the first, since, by the principle of microscopic reversibility, those energies that are most favored among the products of reaction would lead to correspondingly high rates of reaction, for reaction in the reverse direction.

Work along these lines has only begun in recent years and the results so far obtained are, on the whole, inconclusive. The methods are discussed in § 2.2. We shall not enter in detail into the question of the interpretation of the results obtained to date. It will be sufficient to point out that if it is found (as seems probable) that the reaction cross section for reacting species with molecular energy greater than the critical amount ϵ_a does not remain constant, then the activation energy, E_a, obtained by application of the Arrhenius equation to experimental data, will be a function of temperature; that is to say, the Arrhenius equation will fail.

This can be understood very simply in qualitative terms. Suppose that particles in the energy range ϵ_a to $\epsilon_a + \delta\epsilon_a$ have a different reaction cross section from those in the higher energy range ϵ_b to $\epsilon_b + \delta\epsilon_b$. As the temperature of an assembly of reagent molecules increases (supposing that a Boltzmann distribution is maintained despite depletion of the high-energy tail of the curve through consumption of reagents) the number of molecules with energy greater than ϵ_a, expressed as a fraction of the total number, will increase rapidly owing to the changing shape of the Boltzmann curve. This is the origin of the rapid change in reaction rate with temperature. However, the same changing shape of the Boltzmann curve has the result that of the reagent molecules with energy greater than ϵ_a the fraction having an energy in the range ϵ_b to $\epsilon_b + \delta\epsilon_b$ is increasing rapidly relative to the fraction having an energy in the range ϵ_a to $\epsilon_a + \delta\epsilon_a$. According to our hypothesis, the dependence of the rate on the temperature will be different in the temperature range where ϵ_a to $\epsilon_a + \delta\epsilon_a$ molecules account for the bulk of reaction than in the range where ϵ_b to $\epsilon_b + \delta\epsilon_b$ molecules do so. The coefficient E_a which describes the dependence of the rate constant on the temperature must therefore be a function of temperature.

A mathematical statement of this general type of effect is to be found in the work of Fowler (1936) and of Kassel (1930).

2 *Measurement of Reaction Rates in Gases*

2.1 Measurement of the Dependence of Rates on Concentration and Temperature

2.1.1 *Unimolecular Reactions*

Between the wars a large number of investigations were made of the gas-phase thermal decompositions of organic compounds. These reactions were thought to be unimolecular. The availability of experimental results stimulated a great deal of interest in the theory of unimolecular reactions (§ 3.1.1). Though the theories developed at that time are still thought to be substantially correct, the experimental data to which they were applied is now believed to be irrelevant since these decompositions were not unimolecular, but complex.

Nonetheless, there exist today a number of apparently bona fide unimolecular gas reactions. The most significant feature of these reactions is that their first order rate constants frequently exhibit a decrease at low pressures. The rate constant can be restored to the high-pressure value by the addition of any gas (not necessarily the reagent) at sufficient pressure. It is thought that all unimolecular reactions would exhibit the phenomenon of pressure-dependent rate constant if the pressure could be sufficiently reduced. Experimentally, the limit to further reduction of pressure is reached when the pressure of product material formed over a reasonable length of time becomes too small to measure with any accuracy.

Investigations of these reactions have been designed in the first place to establish that they are gas-phase, and unimolecular at sufficiently high pressure. Secondly, it is of interest to determine the shape of the rate-constant-versus-pressure curve, and thirdly to investigate the effect of changing gas composition on the shape of this curve.

It has been found that the simpler the decomposing molecule, that is to say, the smaller the effective number of oscillators, the higher is the pressure up to which quasi-unimolecular behavior (defined in § 1.1) can be observed. The simpler molecules are therefore greatly to be preferred in investigating this effect. In the case of the very simplest molecules (see the pyrolysis of O_3, below) more than an atmosphere pressure of most foreign gases would be required to take the reaction out of the quasi-unimolecular pressure range.

In view of the potential importance of quasi-unimolecular behavior in throwing light on the mechanism of chemical reaction, and as a tool in the study of energy transfer processes, a list is given of 10 gases whose unimolecular or quasi-unimolecular decomposition has been studied in recent years. The list is arranged in order of increasing complexity of the vibrations of the decomposing molecules (cf. § 3.1.1): O_3 (Benson and Axworthy, 1957); N_2O (Johnston, 1951); H_2O_2 (Giguère and Liu, 1957); N_2O_5 (Mills and Johnston, 1951; Johnston and Perrine, 1951); C_2H_6 (decomposing to $2CH_3$; Dodd and Steacie, 1954; Hoare and Walsh, 1957); C_2H_5Cl (Howlett, 1952); C_3H_6 *cyclo*propane (isomerization; Pritchard *et al.*, 1953); C_4H_8 *cyclo*butane (isomerization; Kern and Walters, 1952); $(CH_2O)_3$ trioxymethylene (to CH_2O; Burnett and Bell, 1938); CH_3NNCH_3 azomethane (to $C_2H_6 + N_2$; Ramsperger, 1927; Rice and Sickman, 1936).

The first two of these reactions, in their truly unimolecular pressure region, are of the types

$$A_3 \rightarrow A_2 + A \tag{27}$$

and

$$A_2B \rightarrow A_2 + B. \tag{28}$$

The first reaction, ozone decomposition, has been investigated by static pyrolysis (thermal decomposition) and pyrolysis in a flow system. The considerable body of experimental data has been extended and reviewed by Benson and Axworthy (1957).

Benson and Axworthy's work will be described as an illustration of the large amount of valuable information that can be obtained from a careful study of a moderately complex chemical system.

Benson and Axworthy found that all the experimental data, their own and that of earlier workers, could be accounted for in terms of the following mechanism [based on that suggested a number of years ago by Jahn (1906)]:

$$O_3 + M \underset{2}{\overset{1}{\rightleftharpoons}} O_2 + O + M \tag{29}$$

$$O + O_3 \xrightarrow{3} 2O_2. \tag{30}$$

This mechanism implies that in the accessible range of pressures the ozone decomposition is sensitive to the pressure of "M," which was O_3, O_2 or added N_2, CO_2, or He. The reaction is therefore to be regarded as quasi-unimolecular over the observed pressure range.

In reactions where an intermediate (atomic oxygen in the present system) is present in only small concentration compared with reagent and product concentrations [(O_3) and (O_2)], it is usual to assume that the rate of change of concentration of the intermediate with time is zero. This is referred to as the *stationary state hypothesis*. This will not, of course, be true at the very start of the reaction when the intermediate's concentration is building up to a stationary value (the *induction period*), nor would it be true in the final stage of reaction, if the reaction went to completion. The induction period can be significant if the reaction is being measured only over a short period immediately following the mixing of reagents. The longest induction periods are encountered in reactions where a long chain is present which regularly re-forms some intermediate (for detailed discussion see Semenov, 1939; Benson, 1952).

On making use of the stationary state hypothesis we can write

$$\frac{d(\mathrm{O})}{dt} = 0 = k_1(\mathrm{O_3})\,(\mathrm{M}) - k_2(\mathrm{O_2})\,(\mathrm{O})\,(\mathrm{M}) - k_3(\mathrm{O_3})\,(\mathrm{O}) \tag{31}$$

where the terms in parentheses represent the instantaneous concentrations (moles/liter) of the species indicated; the first term on the right-hand side of (31) gives the rate of formation of atomic oxygen, the following two terms give the rate of removal. From (31)

$$(\mathrm{O})_{\mathrm{SS}} = \frac{k_1(\mathrm{O_3})\,(\mathrm{M})}{k_2(\mathrm{O_2})\,(\mathrm{M}) + k_3(\mathrm{O_3})} \tag{32}$$

where $(\mathrm{O})_{\mathrm{SS}}$ is the stationary-state concentration of atomic oxygen. The rate of reaction of ozone is

$$-\frac{d(\mathrm{O_3})}{dt} = k_1(\mathrm{O_3})\,(\mathrm{M}) + k_3(\mathrm{O_3})\,(\mathrm{O}) - k_2(\mathrm{O_2})\,(\mathrm{O})\,(\mathrm{M}). \tag{33}$$

On putting (32) into (33)

$$-\frac{d(\mathrm{O_3})}{dt} = \frac{2k_1k_3(\mathrm{O_3})^2(\mathrm{M})}{k_2(\mathrm{O_2})\,(\mathrm{M}) + k_3(\mathrm{O_3})}. \tag{34}$$

On making the substitution

$$k_{\mathrm{S}} = \frac{d}{dt}\left(\frac{1}{(\mathrm{O_3})}\right) = \frac{1}{(\mathrm{O_3})^2}\left[-\frac{d(\mathrm{O_3})}{dt}\right], \tag{35}$$

(34) becomes

$$k_{\mathrm{S}}k_2(\mathrm{O_2})\,(\mathrm{M}) + k_{\mathrm{S}}k_3(\mathrm{O_3}) = 2k_3k\,(\mathrm{M}) \tag{36}$$

which on division by $2k_{\mathrm{S}}k_1k_3\,(\mathrm{O_3})$ gives

$$\frac{(\mathrm{M})}{k_{\mathrm{S}}(\mathrm{O_3})} = \frac{k_2}{2k_1k_3}\,\frac{(\mathrm{O_2})\,(\mathrm{M})}{(\mathrm{O_3})} + \frac{1}{2k_1}. \tag{37}$$

(M) must be treated as the weighted sum of the contributions of different gases to reactions 1 and 2 (29). Thus, if we express the efficiencies of all possible M's relative to $\mathrm{O_3}$, we can write

$$(\mathrm{M}) = (\mathrm{O_3}) + \sum_i a_i(\mathrm{M}_i) \tag{38}$$

where a_i represents the relative efficiency of M_i.

From (37) it can be seen that if $(M)/k_S\ (O_3)$ is plotted against $(O_2)(M)/(O_3)$, both ratios being experimentally determinable, a straight line should be obtained with slope $k_2/2\ k_1k_3$ and intercept $\frac{1}{2}\ k_1$. Benson and Axworthy measured these ratios in the presence of various foreign gases, in different reaction vessels, and over a range of temperatures (70°-120°C). Good linear plots were obtained in every case. The following expression for (M) gave the most consistent set of rate constants:

$$(M) = (O_3) + 0.44(O_2) + 0.41(N_2) + 1.06(CO_2) + 0.34(He). \qquad (39)$$

The efficiency of O_3 is here assumed to be unity. Figure 1 shows Benson

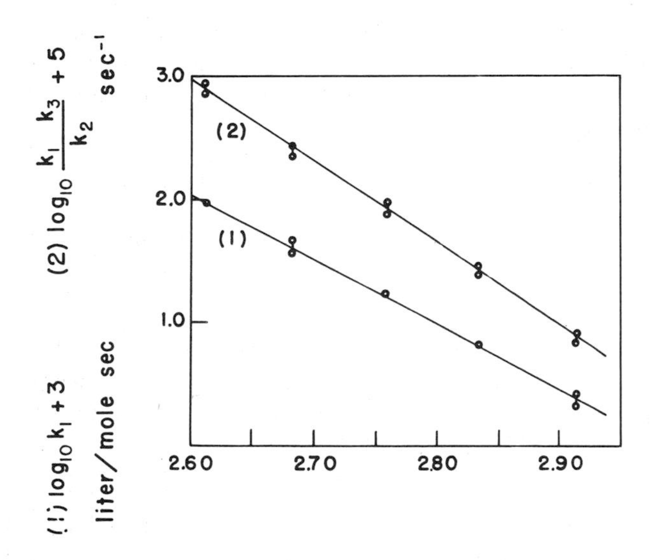

FIG. 1. Arrhenius plots for the pyrolysis of ozone (Benson and Axworthy, 1957).

and Axworthy's values for k_1k_3/k_2 and for k_1 on an Arrhenius plot ($\log_{10}k$ vs. $1/T$). The slope of $\log_{10}k_1$ vs. $1/T$ is $-E_1/2.303\ RT$. The slope of $\log_{10}k_1k_3/k_2$ is $-(E_1 - E_2 + E_3)/2.303RT$. The intercepts with $1/T = 0$ are $\log_{10}A_1$ and $\log_{10}A_1A_3/A_2$. From the figure the following two Arrhenius equations are obtained:

$$k_1 = (4.61 \pm 0.25) \times 10^{12} \exp(-24{,}000/RT) \text{ liters/mole sec} \qquad (40)$$

$$k_1k_3/k_2 = (2.28 \pm 0.16) \times 10^{15} \exp(-30{,}600/RT) \text{ sec}^{-1}. \qquad (41)$$

Equation (41) agrees well with the results of Garvin's (1954) pyrolysis of O_3 by a flow method.

The equilibrium constant for reactions 1 and 2, $K_{1.2}(= k_1/k_2)$, and the temperature dependence of the equilibrium constant, can be calculated from thermodynamic and spectroscopic data,

$$K_{1.2} = k_1/k_2 = 7.7 \times 10^4 \exp(-24{,}600/RT) \text{ moles/liter.} \tag{42}$$

The quantity $-$ 24,600 kcal/mole, is $E_1 - E_2$. This is a difference in internal energy, $\Delta E°$. It should be noted that heat of reaction is usually tabulated as $\Delta H°$, a change in heat content (since heat of reaction is conventionally measured at constant pressure whereas kinetic studies are made in reacting systems of constant volume). If there is a change in the number of moles (Δn), then $\Delta H°$ must be corrected; $\Delta H° - \Delta nRT = \Delta E°$. For reaction 1 (29) $\Delta H° = 25.2$, $\Delta E° = 24.6$ kcal/mole.

Subtraction of $\Delta E°$ of (42) from the over-all E_a of (41) gives E_3.

k_1 has been expressed in Arrhenius form in (40); k_2 and k_3 can now be similarly expressed:

$$k_2 = (6.00 \pm 0.33) \times 10^7 \exp(+600/RT) \text{ liter}^2/\text{mole}^2 \text{ sec} \tag{43}$$

$$k_3 = (2.96 \pm 0.21) \times 10^{10} \exp(-6000/RT) \text{ liter/mole sec.} \tag{44}$$

These results exhibit a number of interesting features:

(i) Reaction 1 has a high A-factor compared with the "normal" figure of 10^{11}. In terms of transition state theory this implies that entropy of activation is larger in a quasi-unimolecular than in a true bimolecular process (see Table II; and also § 3.1.1).

(ii) The activation energy of reaction 1 is less than the bond dissociation energy of the bond being broken: 24 as against 24.6 kcal; that is to say, the rate of decomposition of O_3 does not increase as rapidly with temperature as would be expected. Benson and Axworthy suggest that this is due to increasing depletion of the population of O_3 in the highest vibrational energy states, due to rapid decomposition from these states. If the equilibrium concentration in these states cannot be maintained, the rate of decomposition falls off (cf. § 2.2.3).

(iii) The activation energy of reaction 2 is negative. This is a general phenomenon for termolecular reactions (§ 2.1.3).

(iv) The oxygen molecules formed in reaction 3 have 93 kcal heat of reaction + 6 kcal activation energy, distributed among them. Nonetheless, it seems that they cannot bring about collisional decomposition of O_3, despite the fact that the $O_2 - O$ bond energy in ozone is only 24.6 kcal (see also Schumacher, 1960; Benson, 1960b); that is to say, a reaction O_2(hot) + $O_3 \rightarrow 2O_2 + O$, does not have to be included in the reaction scheme.

This would suggest that even though a molecule has high kinetic or vibrational energy, it cannot in general dissociate a collision-partner in a few collisions. This fits in with the current view (§ 3.2) that dissociation occurs as a result of multiple collisions, which cause a vibrator to climb its "vibrational ladder," one or two levels (or several levels, see Montroll and Shuler, 1958) at a time, in a random walk which can lead to dissociation. The discussion under (ii) above is also in accord with this multicollision theory of dissociation, for if a vibrator could pass readily from its ground vibrational state to a high state close to dissociation, it would not be possible to postulate a depletion of population in these highest vibrational states relative to lower ones.

(v) The relative efficiencies (39) of various gases in promoting dissociation are also their relative efficiencies as "third bodies," M, in the association reaction (reaction 2, the reverse of 1). These efficiencies conform to the rough generalization (see, for example, Trotman-Dickenson, 1955) that the efficiency of energy transfer of this type increases with increasing complexity of M, soon reaching a maximum efficiency beyond which no further increase occurs for further increase in complexity of M.

It has been remarked previously that the simpler the decomposing molecule the higher is the pressure at which true unimolecular kinetics is observed. O_3 is an extremely simple molecule. Benson and Axworthy found that they could not pursue their experiments to high enough pressures in order to observe true unimolecular kinetics. Johnston (1951) in his investigation of the pyrolysis of N_2O, the next most complex molecule on the list of ascending complexity, was able to study the reaction in the quasi-unimolecular and in the unimolecular pressure region.

2.1.2 *Bimolecular Reactions*

The most widely studied bimolecular reactions are exchange reactions (or "metathetical reactions"), of which the simplest type involves reaction between an atom and a diatomic molecule,

$$A + BC \rightarrow AB + C.$$

Of this class of exchange reactions the most important from a theoretical standpoint are those in which A, B, and C are H or its isotopes.

There are eight possible reactions of this type (leaving out of account tritium):

$$H + H_2 \xrightarrow{1} H_2 + H \tag{45}$$

$$D + D_2 \xrightarrow{2} D_2 + D \tag{46}$$

$$D + H_2 \xrightarrow{3} DH + H \tag{47}$$

$$H + D_2 \xrightarrow{4} HD + D \tag{48}$$

$$H + HD \xrightarrow{5} H_2 + D \tag{49}$$

$$D + DH \xrightarrow{6} D_2 + H \tag{50}$$

$$H + DH \xrightarrow{7} HD + H \tag{51}$$

$$D + HD \xrightarrow{8} DH + D. \tag{52}$$

The energy interactions between three H atoms have been calculated by approximate quantum-mechanical treatments, and activation energies for these 8 reactions have been obtained. *A*-factors have also been calculated for all these 8 reactions, by means of the "theory of absolute reaction rates." These calculations will be discussed in § 3.1.2. They are mentioned here in order to emphasize the importance of these 8 reactions. This is the only set of reactions to which the theory of absolute reaction rates can be applied with some degree of rigor.

The idea has gained currency that the rate constants of these 8 reactions are experimentally known. However, at the time of writing, rate constants have been measured only for reactions 1 and 2. With the exception of k_3/k_5 and k_4/k_6, which define calculable equilibrium constants, not even relative rate constants are known for the other 6 reactions. (Attempts have been made to measure these constants. However, to extract them from the experimental data, it was found necessary to combine the experimental quantities with theoretical speculation. It is these results which have been mistaken for experimental values.)

Reactions 1 and 2 can be followed by measuring the rate of thermal reconversion of para-hydrogen ($p - H_2$) into normal hydrogen containing 75% ortho-hydrogen ($o - H_2$) and 25% $p - H_2$. This proceeds according to the mechanism (Farkas, 1930, 1931; Farkas and Farkas, 1935)

$$\underset{\uparrow}{H} + p - \underset{\uparrow\downarrow}{H_2} \underset{k_1'}{\overset{k_1}{\rightleftharpoons}} o - \underset{\uparrow\uparrow}{H_2} + \underset{\downarrow}{H}. \tag{53}$$

The arrows denote orientation of nuclear spin. To obtain almost pure $p - H_2$ for this reaction, it was only necessary to cool hydrogen to 20°K (the boiling point of liquid hydrogen) in the presence of a catalyst such as charcoal; and equilibrium mixture was obtained containing 99.8% $p - H_2$. This was then heated, in the absence of catalyst, to a temperature in the region of 1000°K, where thermal reconversion would occur at a conveniently measurable rate. The o/p composition of hydrogen was measured (to better than 1%) by thermal conductivity. The concentration of atomic hydrogen was obtained from a knowledge of the equilibrium constant of the process

$$H_2 \rightleftharpoons 2H. \qquad (54)$$

The equilibrium constant for the reaction at high temperature was calculated by statistical mechanics, from the spectroscopically determined spacing of vibrational energy levels in H_2. The equilibrium constant is slightly different for $p - H_2$ and $o - H_2$, so the concentration of H changes slightly as the reaction proceeds.

Measurement of the rate of approach to equilibrium leads to a value for $k_1 + k_1'$. At equilibrium the forward rate $[k_1(H)(p - H_2)]$ equals the backward rate $[k_1'(H)(o - H_2)]$. The equilibrium constant is

$$K = \frac{(H)(o - H_2)}{(H)(p - H_2)} = \frac{k_1}{k_1'}. \qquad (55)$$

Since K is known, k_1/k_1' can be combined with $k_1 + k_1'$ to yield a value for k_1.

Van Meersche (1951) extended the measurements of Farkas on para-ortho hydrogen conversion, and confirmed the results of the earlier work.

However, the thermal reconversion technique cannot conveniently be used to obtain k_1 over a wide temperature range. To extend the temperature range two other techniques have been developed.

Geib and Harteck (1931) measured the rate of para-ortho hydrogen conversion in the neighborhood of room temperature (283°-373°K). Hydrogen coming from a discharge (Wood, 1922) contains a high concentration of atomic hydrogen. This concentration can be measured by means of a Wrede-Harteck gage (see Farkas and Melville, 1939). A small amount of $p - H_2$ was injected into the flowing H atoms, and the concentration of $p - H_2$ and H was remeasured after a known time interval (that is to say, a known distance downstream). The rate constant, k_1, could then be calculated. The chief uncertainty in this technique was in the measurement of the time interval just referred to:

it was not known how long the reagents spent in the reaction zone in a fully mixed condition. However, the rate constants were probably good to within a factor of 2 or 3.

These results received support from a second determination of k_1 at room temperature. Melville and Robb (1949) produced atomic hydrogen by mercury sensitized photolysis of hydrogen: light of 2537 A wavelength was shone into a mixture of H_2 + Hg vapor,

$$Hg\,(6^1S_0) + h\nu \rightarrow Hg^*\,(6^3P_1) \tag{56}$$

$$Hg^* + H_2 \rightarrow Hg + 2H. \tag{57}$$

The rate of production of H atoms could be calculated from the amount of 2537 A light absorbed. The reaction cell had a depth of only a few millimeters, bounded on one side by a 3-cm-diameter quartz window admitting the ultraviolet, and on the other by a layer of molybdenum oxide powder which it was believed removed all H atoms at their first collision. Assuming that all H atoms were formed just inside the quartz and diffused one-dimensionally to the oxide surface where they were destroyed, it was possible to calculate a mean lifetime for H, which combined with the rate of formation lead to a value for the stationary-state concentration. This value for the concentration of H should be unaltered in the presence of H_2 since reaction 1 (45) does not remove hydrogen atoms. The reaction vessel was filled with $p - H_2$ and the rate of conversion to $o - H_2$ in the presence of a known H atom concentration was measured. This yielded a value for k_1.

The results of Farkas and Van Meersche at the high temperature, and those of Geib and Harteck and also Melville and Robb at a low temperature taken together permit a log k_1 vs. $1/T$ plot to be made over a considerable temperature range. A correction should be applied to the results of Farkas and those of Van Meersche, as suggested by Boato *et al.* (1956; Cimino *et al.*, 1956). These workers point out that Farkas and Van Meersche's rate constants would have been uniformly lower by a factor of $\times$ 0.5 if they had taken pains to eliminate the diffusion of oxygen (which acts as a catalyst) through the quartz walls of the reaction vessel. The slope of the graph then yields $E_1 = 7.5 \pm 1$ kcal/mole; the *A*-factor is $10^{10.7}$ liter mole^{-1} sec^{-1}, which is "normal" (Table II).

Farkas and Farkas (1935), using the method of Farkas (1930, 1931), measured the rate of thermal reconversion of $o - D_2$ to $p - D_2$. Since the temperature range of observation was small (903°-981°K), and no experiments have been performed at lower temperature by other methods, the *A*-factor and activation energy are unreliable. However, it is probably significant that the value obtained for k_2 at 1000°K was 0.53 × that obtained previously by the same method for k_1. Van Meersche (1951) recalculated k_2 from Farkas and Farkas' data using better figures for the dissociation equilibrium of H_2, and found k_2/k_1 to be 0.35 at 1000°K.

As remarked previously, experimental values for k_3, k_4, k_5, k_6, k_7, and k_8 are widely quoted in the literature. These are attributed to Farkas and Farkas (1935), Van Meersche (1951), and Boato *et al.* (1956), all of whom have measured the rate of reaction of $H_2 + D_2$ to form HD, over a total temperature range 720°-1010°K. Unfortunately, the only quantity that can be derived from this type of experiment is $k_3k_4/(k_3 + k_4)$ and its temperature dependence (Farkas and Farkas, 1935; p. 143; Cimino *et al.*, 1960). If this expression could be made to yield k_3 and k_4 (which it cannot, at present), then k_5 and k_6 could be calculated since they relate to the reverse of reactions 3 and 4, and the required equilibrium constants are calculable. k_7 and k_8 could, however, never be obtained from these experimental data, since reactions 7 and 8 do not occur in this system to a significant extent.

To sum up: A_1 and E_1 are known; k_1 and k_2 are known at 1000°K; k_3, k_4, ..., k_8 are unknown, but it is known that $k_3/k_5 = 2.6$ and $k_4/k_6 = 1.5$ at 1000°K.

A-factors and activation energies have been measured for a large number of atomic exchange reactions (for tabulations see Steacie, 1954; Trotman-Dickenson, 1955; Semenov, 1958; Benson, 1960a). The greater part of the data comes from detailed studies of complex systems (an example, the pyrolysis of ozone, has been discussed previously). However, a number of techniques exist for more direct measurement of rates of atomic reactions. Two of these have just been outlined (Geib and Harteck, 1931; Melville and Robb, 1949). A few more will be indicated in the following paragraphs.

A number of years before the systematic investigation of atomic reactions began, the velocity of one such reaction was determined (independently by Herzfeld, 1919; Christiansen, 1919; Polanyi, 1920) from an analysis of the chain of events involved in the reaction between molecular hydrogen and molecular bromine (Bodenstein and Lind, 1906). The reaction was

$$Br + H_2 \rightarrow BrH + H. \tag{58}$$

The *A*-factor was found to be simply equal to the number of gas kinetic collisions $Br + H_2$. The activation energy was simply the heat of reaction. It followed that the activation energy for the reverse reaction was zero. This led to the view that atomic reactions proceeding in the exothermic direction had no "chemical inertia": reaction occurred at every collision. This simple view had to be modified in the light of evidence concerning the reactions of alkali metal atoms.

Three related techniques yielded a large body of data concerning the reactions of atoms of the alkali metals, with the result that far more rate constants of this type of gas reaction are known than of any other. The first of these techniques, the method of dilute flames (Beutler and Polanyi, 1925), and the second, the diffusion flame method (reviewed by Polanyi, 1932; Bawn, 1942; Warhurst, 1951), both measure rate of reaction by comparison with rate of diffusion. The alkali metal which was most extensively studied was sodium.

In the dilute flame method, sodium at 10^{-2} mm pressure (saturation pressure for about 300°C) was introduced from one end of a 1-meter-long, 3-cm-diameter tube, and halogen at the same pressure was introduced from the other end of the tube. Since the mean free path at these pressures is greater than the diameter of the tube the two gases interpenetrated by diffusion. The Na penetrating into X_2 (the halogen) was consumed and its partial pressure fell to zero (curve p_{Na} of Fig. 2). The same

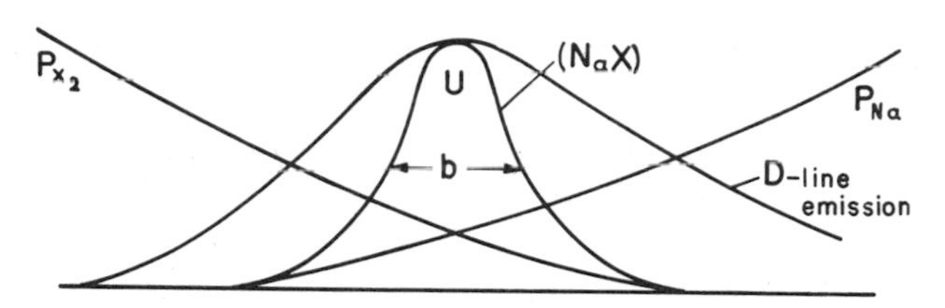

FIG. 2. The "dilute flame" method for measuring rates of atomic reactions having collision efficiencies in the range 1-10^{-1}.

thing happened for X_2 penetrating Na (curve p_{X_2} of Fig. 2). The relative rates of reaction were given by the curve (NaX), $\alpha P_{Na} P_{X_2}$.

Experimentally, the curve (NaX) comes from the rate of deposition of product NaX at various points along the tube, as evidenced by changing opacity of the tube. The slower the reaction the greater the interpenetration of reagents, i.e., the greater the half-width, b, of the (NaX) curve. It can be shown that the rate constant is given by,

$$k = \frac{27}{b^3} \frac{1}{2qUK_{Na}K_{X_2}} \tag{59}$$

where q is the cross section of the tube, U is the total amount of reaction, and K_{Na} and K_{X_2} are the low-pressure viscosities, or "diffusion resistances" of Na and X_2 in the tube employed (these were determined in separate experiments).

It was found that the reaction Na + X_2 occurred at more than every gas kinetic collision; that is to say, it appeared that the collision diameter for chemical reaction was over twice as great as that derived from the kinetic theory of gases (for a discussion see Polanyi, 1959).

In addition to the curve (NaX), Fig. 2 includes a curve labeled "*D*-line emission." The distribution of this emission along the tube showed it to be due to a secondary reaction. This will be discussed in § 2.2.2.

The method of dilute flames is inconvenient for reactions having a collision efficiency (ratio of gas kinetic collisions leading to reaction, to total number of collisions) $< 10^{-1}$, since b becomes too large. To overcome this difficulty a technique was developed by von Hartel and Polanyi (1930, Fig. 3) in which a small amount of alkali metal diffused

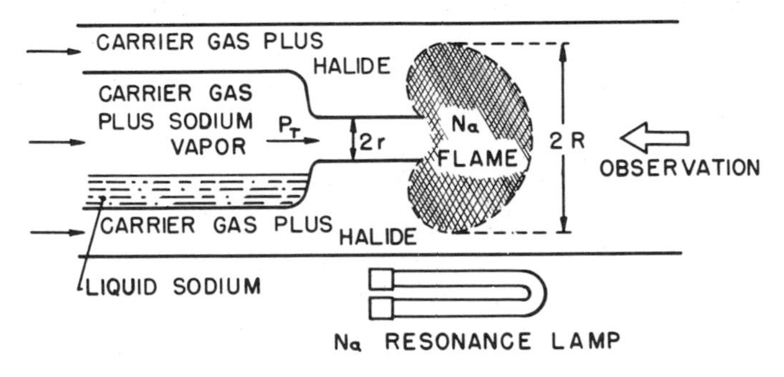

FIG. 3. The "diffusion flame" method for measuring rates of atomic reactions having collision efficiencies in the range 10^{-1}-10^{-5}.

from a nozzle (of radius $r = 1$-2 mm) into a large excess of reagent, RX. The distance that the sodium diffused into the reagent could be measured if the sodium was illuminated by a sodium resonance lamp. The lower the collision efficiency of the reaction, the greater was the radius of the "flame" (R). Suppose that at a distance R from the center of the nozzle the partial pressure of sodium had fallen from its initial value P_{T} (usually about 2×10^{-3} mm; the sodium vapor pressure at 250°C) to the limit of detectability P_0 (determined in a separate experiment as $\sim 5.10^{-6}$ mm). The halide being present in excess could be regarded as having a constant pressure, P_{RX}. The rate constant was then given by

$$k = \frac{(\log P_{\mathrm{T}}/P_0 - \log R/r)^2\, \delta}{(R - r)^2\, P_{\mathrm{RX}}} \tag{60}$$

where δ was the diffusion constant of sodium in the inert carrier gas (the principal content of the reaction zone).

The technique was found to be applicable to the determination of rate constants for reactions having collision efficiencies in the range 10^{-1}-10^{-5}. Owing to the difficulty in determining rate constants over a range of temperatures by this method, activation energies were calculated on the assumption that the *A*-factor had the "normal" value of $10^{11.7}$ liter mole^{-1} sec^{-1} (cf. Table II). This assumption was based on measurements of the temperature coefficient of *k* for the reactions Na + CH_3Cl and Na + CH_3Br (von Hartel and Polanyi, 1930) and Na + bromobenzene (Warhurst, 1939).

More recently, Cvetanovič and LeRoy (1952) in a detailed study of the system Na + C_2H_5Cl have shown that the *A*-factor is "normal." Nonetheless, there are good theoretical reasons for supposing that the *A*-factor will be 2 or 3 times lower when the halide under attack is a complex organic molecule than when it is a relatively simple one (such as CH_3Cl). In such cases small apparent differences in activation energy calculated by the isothermal comparison method outlined above may not be meaningful.

A third variant on the diffusion method avoids the considerable hydrodynamic difficulties of the diffusion flame method [for a discussion of these difficulties see Heller (1937) and Reed and Rabinovitch (1955)]. This third method is the "life period method" (Frommer and Polanyi, 1934). The method makes use of the fact that, for a steady state, atoms must enter and leave the reaction vessel at the same rate,

$$n = kNP_{\text{RX}} \tag{61}$$

where *n* is the number of atoms entering the reaction vessel per second, and *N* is the number of atoms present in the vessel under steady state conditions. The equation can be rewritten

$$k = (\tau P_{\text{RX}})^{-1}$$

where $\tau = N/n$, the mean lifetime of an Na atom in the reaction vessel. *N* is measured by absorption spectroscopy, *n* by observation of flows and pressures.

Results obtained by this method agree reasonably well with those obtained by the diffusion flame method (Haresnape *et al.*, 1940).

In recent years two new variants on the diffusion method have been developed: the temperature pattern method for reactions in which a resonance "flame" cannot be obtained (Garvin and Kistiakowsky, 1952; Smith and Kistiakowsky, 1959), and the product-emitter method for chemiluminescent reactions (Garvin *et al.*, 1960b).

An important "direct" technique for the investigation of reactions involving atomic hydrogen involves the production of hydrogen atoms at a tungsten filament in a flow system (Tollefson and LeRoy, 1948; Berlie and LeRoy, 1952). The H atoms at a partial pressure of $\sim 10^{-2}$ mm (total H_2 pressure 1-10 mm) are mixed with reagent. The gas then flows down a thermostated tube past a movable platinum-wire detector operated as an "isothermal calorimeter," and the loss of hydrogen atoms due to reaction is determined. As a check, a rate constant can also be calculated from the amount of product formed. LeRoy and his co-workers have investigated a number of reactions between hydrogen atoms and simple hydrocarbons in this way. Activation energies and A-factors can be determined experimentally by temperature variation.

Elias and Schiff (1960) have developed an analogous technique for the measurement of absolute rates of reactions between O atoms and hydrocarbons. The O atoms are formed in a microwave discharge instead of at a hot filament, and their concentration along the length of the thermostated tube is measured by photometric monitoring of their chemiluminescent reaction with a small amount of NO added as an indicator.

A considerable body of data is now available concerning relative rates of reaction of fluorine, chlorine, and bromine atoms with alkanes. The method is referred to as "competitive halogenation" (Pritchard *et al.*, 1955; Mercer and Pritchard, 1959; Fettis *et al.*, 1960). Halogen atoms are produced by photolysis of halogen in the presence of two reagents (for example methane and ethane); the rate of consumption of each reagent is measured.

Various simple generalizations arise from these "direct" studies, and also the "indirect" studies, of the exchange reactions between atoms and molecules [see p. 292 of Benson (1960a) for a recent partial tabulation]. In the first place, the A-factors appear to be $\sim 0.1 \times$ the value of 10^{11} liter mole^{-1} sec^{-1} predicted by collision theory (§ 3.1.2). In the second place, the activation energies for the exothermic exchange reactions are very much less than the energy of the bond to be broken. This discrepancy is surely connected with the fact that a new chemical bond is being formed at the same time as the old one is being broken; a theory of chemical reactions must take this into account. Thirdly, it appears that a parallelism exists between the heats of certain similar reactions, and their activation energies (Evans and Polanyi, 1938; see, however, Trotman-Dickenson, 1955, 1959; and § 3.1.2).

As will be seen in the following section, the basic theoretical problems involved in a fundamental understanding of the simpler atomic exchange reactions, are at the present time insoluble. Nonetheless, the experimental study of more complex systems has proceeded apace, with benefit to the semiempirical theories of reaction rate.

For example, a sizeable body of evidence is available concerning A-factors and activation energies for the metathetical reactions of polyatomic free radicals (a free radical is a molecule containing one or more unpaired electrons—all the atomic species referred to above, were free radicals). The largest body of data relates to reactions of the type

$$R_1 + R_2H \rightarrow R_1H + R_2 \tag{62}$$

where R_1 is the methyl radical (CH_3), ethyl radical (C_2H_5) or trifluoromethyl radical (CF_3). R_2 is any radical. Methyl radicals can most conveniently be formed by the photolysis of acetone, mercury dimethyl, or azomethane. Probably the most important series of reactions to be studied by this technique is the series of eight isotopic reactions, $CH_3 + H_2$, $CD_3 + D_2$, $CD_3 + H_2$, $CH_3 + D_2$, $CH_3 + HD$, $CD_3 + DH$, $CH_3 + DH$, $CD_3 + HD$ (Majury and Steacie, 1953; Whittle and Steacie, 1953). These correspond to the 8 possible isotopic reactions between atomic and molecular hydrogen [(45)-(52)]. This is the simplest extensive series of isotopic reactions for which individual rate constants have been measured. The results are therefore of importance in connection with theories of the effect of isotopic substitution on reaction rate (§ 3.1.2).

The foregoing leaves the false impression that all bimolecular reactions involve free radicals (monatomic or polyatomic). Though this is the field that has received the most intensive study in recent years, many reactions between stable molecules are also known. The simplest type,

$$A_2 + B_2 \rightarrow 2AB, \tag{63}$$

is exemplified by the reaction

$$H_2 + I_2 \rightarrow 2HI \tag{64}$$

which was extensively studied by Bodenstein in 1899. Kistiakowsky (1928) confirmed Bodenstein's results, and greatly extended the concentration range. He found the reaction to be homogeneous and bimolecular over the whole range (for recent discussions see Benson and Srinivasan, 1955; Magee, 1957; Sullivan, 1959).

2.1.3 *Termolecular Reactions*

Studies of reactions in this category have been largely concerned with cases in which the third body acts to remove energy evolved in an association process, such as

$$A + A + M \rightarrow A_2 + M \tag{65}$$

or

$$A + A_2 + M \rightarrow A_3 + M. \tag{66}$$

Examples of reaction (65) are the recombinations, in the presence of various third bodies, M, of hydrogen atoms (Steiner, 1935), oxygen atoms (Golden and Myerson, 1958; Morgan *et al.*, 1960), bromine atoms (Rabinowitch and Wood, 1936a,b; Strong *et al.*, 1957; Burns and Hornig, 1960), and iodine atoms (see below). An example of (66) is the combination of oxygen atoms with molecular oxygen to form ozone (§ 2.1.1).

The combination of iodine atoms has been the subject of intensive study in recent years owing to the development of two new techniques. In the original investigations by Rabinowitch and Wood, direct photometric measurement was made of the Br_2 stationary-state concentration in a system where the halogen was being continuously photolyzed; $Br_2 + h\nu \rightleftharpoons 2Br$. In the first of the new techniques, flash photolysis (Norrish and Porter, 1949; Porter, 1950; Norrish and Thrush, 1956), an intense flash of light, typically 2000 joules discharged in 10^{-4} sec, dissociated in the region of 50% of the halogen gas into atoms (10^{-6} sec discharge times are now feasible).

A second lamp, shining continuously through the reaction cell onto a phototube, gave a measure of the instantaneous concentration of the molecular species in the tube. The output from the phototube was displayed on an oscilloscope. Following a flash, a descending curve was observed on the oscilloscope. The curve could be analyzed to give a rate constant for the rate of combination of the halogen atoms to re-form the halogen (Christie *et al.*, 1953; Marshall and Davidson, 1953; Russell and Simons, 1953; Christie *et al.*, 1955; Strong *et al.*, 1957; Engleman and Davidson, 1960). A later version of this apparatus will be referred to in § 2.2.2. It is illustrated in Fig. 7.

The second of the new techniques for studying atom recombination involves thermal dissociation. Previously this had not been feasible owing to the difficulty in obtaining the very high temperatures required to bring about appreciable dissociation. However, if the halogen gas is heated by a shock wave, temperatures in the region 1000°-2000°K are readily obtained. Fig. 4 shows the main features of such an apparatus. When diaphragm, B, is ruptured, the high-pressure "driving" gas, A, causes a wave of compression to pass through the low-pressure $M + I_2$ mixture. The shock front passes the 1-mm light beam at D in about 1 μsec heating the gas M (argon for example) to 1000° to 2000°K in the time of a few collisions. The subsequent rate of dissociation of I_2, $M + I_2 \rightarrow M + I + I$, is indicated by the increasing signal from the phototube, E. The rate constant for the termolecular association can then bc calculated from the equilibrium constant.

Burns and Hornig (1960), in their study of the analogous bromine reaction, have combined flash photolysis (to bring about dissociation)

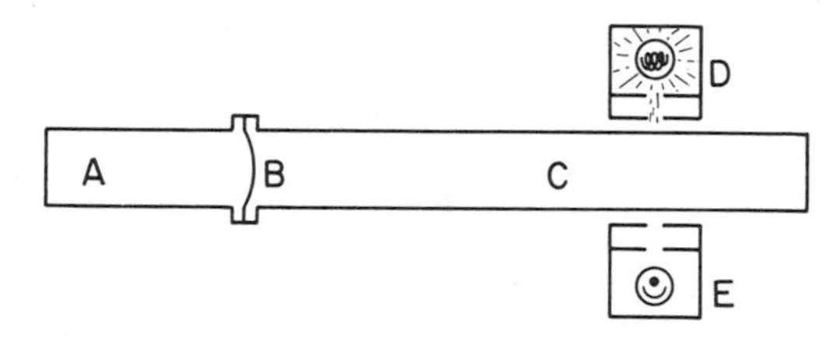

FIG. 4. Schematic diagram of a shock tube apparatus. A, hydrogen or helium "driving" gas; B, diaphragm; C, $I_2 + M$; D, light source; E, phototube linked to oscilloscope.

with the shock wave technique (to bring about heating of the dissociated gas). This permitted them to study recombination at elevated temperatures.

These investigations have led to a few generalizations (in decreasing order of certainty): the association reactions have negative activation energies (cf. § 2.1.1) the efficiencies of the simpler third bodies M tend to increase with complexity of M; third bodies with "affinity" for the combining atoms are more efficient [the affinity may be of the Van der Waals type as for cyclohexane (see Russell and Simons, 1953), or chemical as for NO (see Engleman and Davidson, 1960)]; the negative activation energy is greater the more efficient the third body, M (Bunker and Davidson, 1958; Porter and Smith, 1959; Engleman and Davidson, 1960).

2.2 Measurement of the Dependence of Rates on Kinematic Factors

The scope of this field of investigation has been outlined in § 1.3. The following paragraphs will give some indication of the work that is presently being undertaken. Few generalizations of a high degree of certainty have yet emerged.

2.2.1 *Unimolecular Reactions*

Trischka and co-workers (Luce and Trischka, 1953; Marple and Trischka, 1956) have studied radio-frequency spectra of alkali halides by the molecular-beam electric resonance method. Molecules reaching a hot tungsten-wire detector are in a completely specified state with respect to vibrational energy, angular momentum ($J = 1$) and space orientation of angular momentum ($M_J = 0$). Klemperer and Herschbach (1957) have analyzed the results of Marple and Trischka on LiF to show an apparent dependence of the rate of the gas phase dissociation of LiF on the vibrational state. Though fresh experimental work by Moran and Trischka (1961) has invalidated this conclusion for the particular case examined by Klemperer and Herschbach, work continues along these lines.

2.2.2 *Bimolecular Reactions*

As has been remarked previously (§ 1.3), experimental evidence concerning the detailed kinematics of bimolecular reactions, $A + BC \rightarrow AB + C$, is currently being sought by two different routes.

The first of these, the measurement of reaction cross section in crossed molecular beams, was pioneered by Taylor and Datz (1955) who studied the reaction $K + HBr \rightarrow KBr + H$. Greene *et al.* (1960), using a more refined apparatus with mechanical velocity selection on the

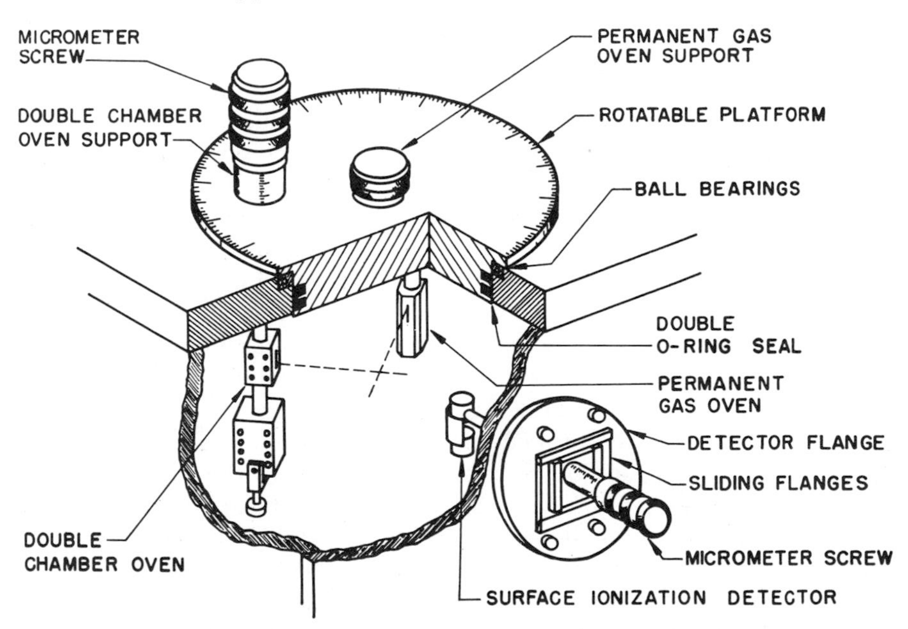

FIG. 5. Cut-away view of a "crossed molecular beam" apparatus. Broken lines emerging from the ovens indicate the paths of the beams. Ovens are mounted to permit measurement of in-plane and out-of-plane scattering. Entire scattering chamber cooled to liquid N_2 temperature (Kwei *et al.*, 1960).

K beam, and with HBr velocity-spread diminished by cooling, found that the cross section for the reaction at first increased with relative kinetic energy of the colliding species and then decreased. However, it now appears likely that the effect was spurious (Herschbach, 1960). It is mentioned here as an indication of the type of information being sought.

A crossed-beam apparatus is shown in Fig. 5. The apparatus is cooled to liquid nitrogen temperature. As a result, background scattering is extremely slight. The apparatus has been applied to the reaction $K + CH_3I$ and $K + C_2H_5I$ (Kwei *et al.*, 1960; Herschbach *et al.*, 1961) and also $Rb + CH_3I$ (Kinsey *et al.*, 1961). First results show that the final velocity vector tends to lie along the direction of the initial velocity vector. It also appears that about three-fourths of the heat of reaction goes into vibrational and rotational excitation of the products.

The second method involves the measurement of the energies of freshly formed product molecules.

The first indication that exothermic reactions, $A + BC \rightarrow AB + C$, result in the direct formation of $AB^\dagger$ (vibrationally excited AB in its ground electronic state) came from the analysis of the sodium + halogen chemiluminescence (§ 2.1.2). Detailed investigation of this system by M. Polanyi and associates (see Evans and Polanyi, 1939; Polanyi, 1959 for a review) led to the conclusion that the process $X + Na_2 \rightarrow NaX^\dagger + Na$ was occurring with high collision efficiency to produce $NaX^\dagger$ containing 50-90% of the heat of reaction as vibration in the newly formed bond. The presence of $NaX^\dagger$ was not shown directly but was inferred from the behavior of the chemiluminescence which gave every appearance of being due to $NaX^\dagger + Na \rightarrow NaX + Na^*$, followed by $Na^* \rightarrow Na + h\nu$.

Direct evidence for the presence of $AB^\dagger$ among the products of reaction $A + BC$, came from the observation of infrared chemiluminescence ($\lambda \approx 3\mu$) arising from the room temperature reaction $H + Cl_2 \rightarrow HCl^\dagger + Cl$ (Cashion and Polanyi, 1958). Subsequently, Cashion and Polanyi (1959b, 1960a; $H + Br_2$, 1960b) attempted to derive the distribution of freshly formed product molecules among vibrational states, from the observed distribution. However, the calculation of initial energy distribution from the stationary-state distribution which has been modified by collision, is doubtful. Experiments are currently being performed at such low pressures (10^{-2} mm total pressure, $\sim 10^{-4}$ mm partial pressure of $HCl^\dagger$) and high flow rates, that the initial distribution can be obtained more directly (Charters and Polanyi 1962). The new results differ significantly from those obtained from "high pressure" data. The product, $HCl^\dagger$, at the low pressure, is in a markedly non-Boltzmann distribution among vibrational levels, demonstrating that it is of chemical rather than thermal origin. A second difference is that, whereas earlier work suggested that the entire heat of the reaction could go into vibration, the lowpressure study indicates that no appreciable amount of $HCl^{\dagger}$ is formed directly with more than 65% of the heat of reaction present as vibration. Rate of

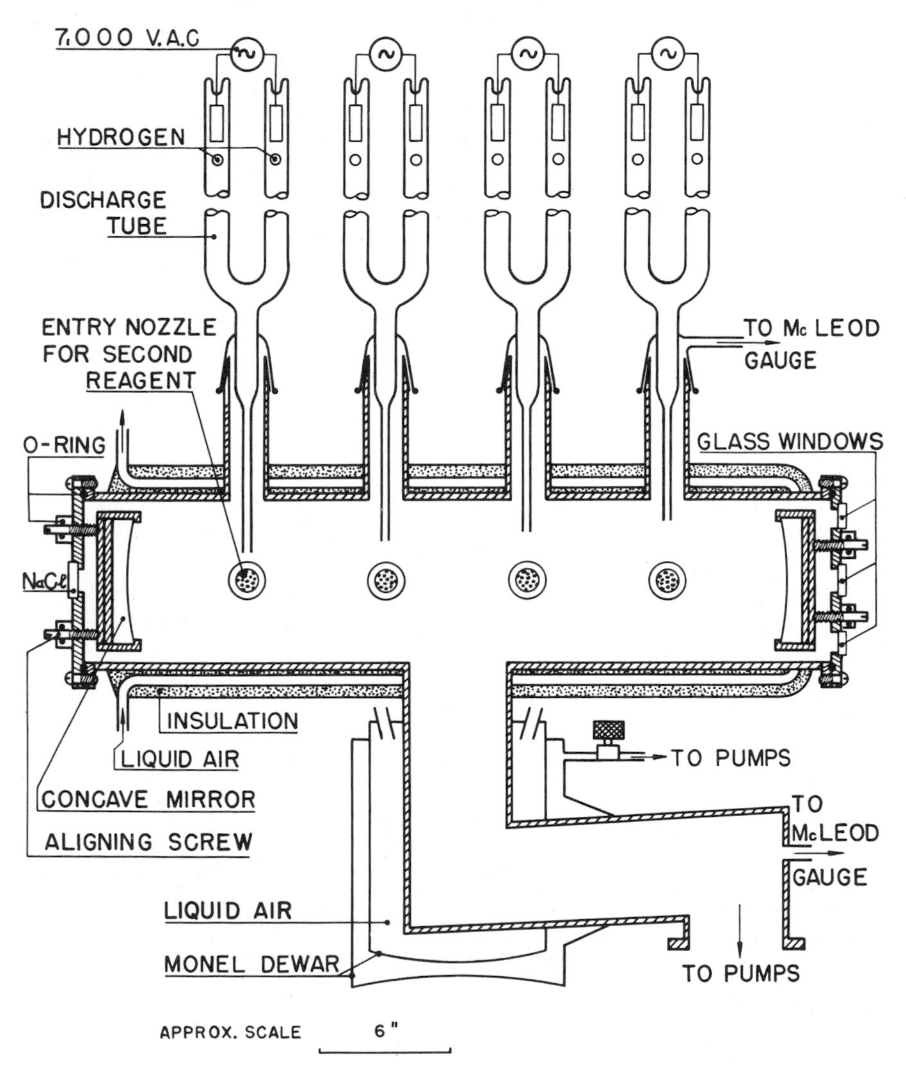

FIG. 6. Low-pressure "infrared chemiluminescence" apparatus. H atoms enter from discharge tubes, second reagent from side nozzles. Adjustable mirrors collect the emission, which passes through the NaCl window at the left into an infrared grating spectrometer (Charters and Polanyi, 1960).

reaction appears, in this case, to be greater into successively lower vibrational levels. The apparatus employed in the low-pressure work (Charters and Polanyi, 1960) is shown in Fig. 6.

The reverse process to that discussed above would be an endothermic reaction $AB^{\dagger} + C \rightarrow A + BC$. Diffusion of sodium into hydrogen coming from a discharge, results in bimolecular consumption of the sodium, with formation of NaH. However, the reaction $Na + H_2$ is too endothermic to proceed at an appreciable rate. Electronically excited H_2, it has been shown, cannot be present. The only remaining possibility appears to be $H_2^{\dagger} + Na \rightarrow H + NaH$; a reaction of the $AB^{\dagger} + C$ variety (Polanyi and Sadowski, 1962).

The first direct evidence of high vibrational energy among reaction products came from the study of a reaction of the type $A + BCD \rightarrow AB^{\dagger} + CD$. McKinley *et al.* (1955; Cawthorn and McKinley, 1956; Kraus, 1957; Garvin, 1959) observed a visible glow from the system $H + O_3$, which they showed to be due to $OH^{\dagger}$ among the products. Garvin *et al.* (1960a) extended these observations into the infrared, at pressures in the region 2-5 mm and attempted to calculate the original distribution of $OH^{\dagger}$ among vibrational levels. Cashion and Polanyi (1961) did the same for $HCl^{\dagger}$ formed from $H + NOCl$. Unfortunately, these calculations are subject to considerable uncertainty (see $H + Cl_2$ above). A detailed appraisal of the results must, therefore, await investigation at lower pressures.

Direct evidence for the formation of vibrationally excited product in reactions of the type under discussion, $A + BCD \rightarrow AB^{\dagger} + CD$, has

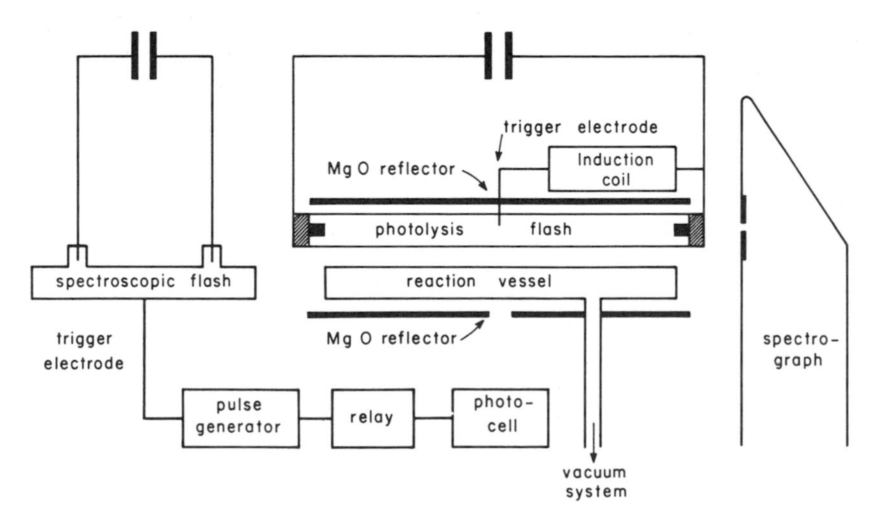

FIG. 7. Schematic diagram of a flash photolysis apparatus (Norrish and Thrush, 1956).

been obtained in a number of cases by the powerful technique of flash photolysis, followed by absorption spectroscopy (Fig. 7). The absorption spectrum involves electronic transitions; consequently, the method is not restricted, as is that described previously, to product molecules possessing a permanent dipole moment. Furthermore, the interval between reaction and observation can be made extremely short. In practice, the latter advantage tends to be offset by the fact that much higher pressures are employed. Systems that have been studied are $O + ClO_2$, $O + NO_2$ (Lipscomb *et al.*, 1956) and $O + O_3$ (McGrath and Norrish, 1957, 1960); producing $O_2^\dagger$ in each case. The systems $Br + O_3$ and $Cl + O_3$, which give rise to $BrO^\dagger$ and $ClO^\dagger$, have also been studied (McGrath and Norrish, 1957, 1958). The vibrational population distributions among the products, observed by this method, may be the initial distributions (Basco and Norrish, 1960). However, this has not yet been established, nor has a quantitative estimate of the populations been made.

All that can be said at the present time with any degree of confidence concerning the exothermic reactions A + BCD, is that these reactions too can result in the *direct* formation of product $AB^\dagger$ as a highly vibrating species. In the system H + NOCl (Cashion and Polanyi, 1961) the product is in a non-Boltzmann distribution among vibrational states; the excitation process must, therefore, be chemical since *thermal* excitation, subsequent to formation, could not account for this.

2.2.3 *Termolecular Reactions*

In § 2.1.1 evidence was adduced to show that dissociation takes place from high vibrational levels in a quasi-unimolecular process, such as $AB + M \rightarrow A + B + M$. It is a corollary that association, $A + B + M \rightarrow AB + M$, takes place *into* high vibrational levels of AB.

The only direct experimental evidence for this comes from observations (Cashion and Polanyi, 1960a, b) of infrared chemiluminescence arising from the systems H + HCl, H + DCl, D + HCl, and H + HBr. The results were consistent with an initial hydrogen abstraction reaction ($H + HX \rightarrow H_2 + X$) followed by a termolecular association reaction ($H + X + M \rightarrow HX^\dagger + M$) to form the highly vibrating halide responsible for the infrared emission. HCl or DCl acting as third body, M, was not excited to levels higher than $v = 1$.

When H atoms were diffused into NO (Cashion and Polanyi, 1959a), infrared emission was obtained from vibrationally excited HNO. The excited adduct is probably formed by a process of the type $A + BC + M \rightarrow ABC^\dagger + M$.

3 *Theory of Reaction Rates in Gases*

3.1 Theory of the Dependence of Rates on Concentration and Temperature

3.1.1 *Unimolecular Reactions*

(i) *Collision Theory.* Starting from the Arrhenius hypothesis (§ 1.2) that reaction is restricted to "activated" molecules, a theory of reaction rates must postulate a mechanism for the activation process. Trautz (1918), Lewis (1918), and Perrin (1919) independently suggested that, in thermal reactions, activation was produced by absorption of infrared radiation emitted by the vessel walls. One of the reasons for rejecting this theory was that it implied that a reaction proceeding in the reverse direction would lead to the emission of infrared radiation. There is now good evidence (§ 2.2.2) that exothermic reactions proceeding at low pressure to form infrared-active product molecules, do in fact emit infrared radiation. Under these special circumstances radiational activation could, therefore, make a contribution to the reverse reaction.

In general, however, activation will be collisional. Only collisional activation can account both for the unimolecular kinetics observed at high pressures, and the transition to bimolecular kinetics at lower pressure (Christiansen, 1921; Lindemann, 1922). Suppose that ABC is dissociating to give A + BC in a bath of molecules M:

$$\mathrm{M} + \mathrm{ABC} \underset{-2}{\overset{2}{\rightleftharpoons}} \mathrm{M} + \mathrm{ABC}^{\ddagger} \tag{67}$$

$$\mathrm{ABC}^{\ddagger} \xrightarrow{1} \mathrm{A} + \mathrm{BC} \tag{68}$$

where $\mathrm{ABC}^{\ddagger}$ is the activated molecule, present in small concentration. According to the stationary-state hypothesis (§ 2.1.1)

$$k_2\,(\mathrm{M})\,(\mathrm{ABC}) = k_{-2}\,(\mathrm{M})\,(\mathrm{ABC}^{\ddagger}) + k_1\,(\mathrm{ABC}^{\ddagger}) \tag{69}$$

Suppose that in the high-pressure region the rate of unimolecular decomposition of $\mathrm{ABC}^{\ddagger}$ is small compared to the rate of collisional deactivation; $k_{-2}\,(\mathrm{M})\,(\mathrm{ABC}^{\ddagger}) \gg k_1\,(\mathrm{ABC}^{\ddagger})$. Then to a very good approximation (69) becomes

$$k_2(\mathrm{M})\,(\mathrm{ABC}) = k_{-2}(\mathrm{M})\,(\mathrm{ABC}^{\ddagger}) \tag{70}$$

and

$$(\mathrm{ABC}^{\ddagger}) = K_{\mathrm{eq}}^{\ddagger} (\mathrm{ABC}),$$

but

$$- d(\mathrm{ABC}^{\ddagger})/dt = k_1(\mathrm{ABC}^{\ddagger}) \tag{71}$$

$$= k_1 \cdot K_{\mathrm{eq}}^{\ddagger}(\mathrm{ABC}). \tag{72}$$

The observed rate of dissociation will be first order. However, at sufficiently low pressure, $k_{-2}\,(\mathrm{M})\,(\mathrm{ABC}^{\ddagger}) \ll k_1\,(\mathrm{ABC}^{\ddagger})$ so that

$$k_2(\mathrm{M})\,(\mathrm{ABC}) = k_1(\mathrm{ABC}^{\ddagger}) \tag{73}$$

and

$$(\mathrm{ABC}^{\ddagger}) = k_2(\mathrm{M})\,(\mathrm{ABC})/k_1$$

substituting into (71)

$$- d(\mathrm{ABC}^{\ddagger})/dt = k_2(\mathrm{M})\,(\mathrm{ABC}). \tag{74}$$

The rate of dissociation at the low-pressure limit is second order.

This theory in its simple form fails to account for the fact that when complex molecules are being dissociated, first order kinetics can be observed down to relatively low pressures. This implies a very efficient activation process able to maintain the equilibrium concentration of activated molecules even at low pressures. Hinshelwood (1927) suggested that a complex molecule can be regarded as activated if it has a total energy greater than E distributed among v normal modes of vibration. The probability of this is given by,

$$\exp\left(- E/RT\right) \frac{(E/RT)^{v-1}}{(v - 1)!}\,. \tag{75}$$

When by chance, this energy eventually finds its way into one particular normal mode of vibration, the molecule dissociates. Kassel (1928, 1932) and Rice and Ramsperger (1927, 1928) have elaborated this picture to include the plausible notion that the rate with which energy passes into the critical vibration is greater, the greater the total energy in the activated molecule.

There exists a second model for the mechanism of unimolecular dissociation, based on a quite different assumption from that of Hinshelwood, Kassel, Rice, and Ramsperger. Whereas the HKRR theory is based on the idea that energy flows freely between normal modes of vibration, the second theory assumes that no such energy flow occurs. This second view has its origins in the work of Polanyi and Wigner (1928) who assumed that a molecule dissociated, not when a single bond gathered a large energy, but when it was greatly stretched as a result of "phase coincidence" between normal modes; that is to say, as a result of a fortuitous juxtaposition of normal modes, involving large stretching of the critical bond. Polanyi and Wigner were only able to apply the argument approximately to a simple case. Slater (1939, 1948, 1959) has succeeded in developing a precise and generalized theory.

Gill and Laidler (1958, 1959a, b, c) have compared the two theories, by employing each in the calculation of the decline of the first order rate constant with decreasing pressure. The cases examined are the majority of the reactions listed in order of increasing complexity in § 2.1. Gill and Laidler find that (with the exception of O_3 where Slater's theory succeeds) the Slater theory gives poor results for less complex molecules, good results for complex ones. They suggest that in small molecules dissociation implies more energy per normal mode, and hence increased likelihood of the flow of energy about the molecule; making the Slater-type argument untenable.

(ii) *Transition State Theory*. This theory, outlined in more detail in the following section, presumes an equilibrium between reagents and activated species. It is therefore inappropriate to the discussion of quasi-unimolecular behavior, which has its origins in a failure to maintain such an equilibrium. However, in the high-pressure region of true unimolecular reaction, we can write in terms of the transition state theory, for the reaction $ABC \rightarrow AB + C$,

$$k_1 = \frac{kT}{h} \frac{Q^*_{ABC}}{Q_{ABC}} \exp(-E_0/RT)$$

or

$$k_1 = \frac{kT}{h} \exp(\Delta S^*/R) \exp(-\Delta H^{\ddagger}/RT).$$

Here Q^*_{ABC} is the complete partition function for the activated molecule, with the exception of the degree of freedom along the coordinate of decomposition. ΔS^* is the entropy of activation with the exception of the entropy term for the degree of freedom corresponding to dissociation. kT/h has a value $\sim 10^{13}$ sec^{-1} at normal temperatures. In decompositions where the Arrhenius A-factor differs markedly from the normal experimental value of 10^{13}, this can often be understood in terms of a significant contribution from ΔS^*. ΔS^* will be significant only if there is a change in the number of low-frequency vibrations (in practice, this usually means bending modes) or rotations. Thus, the opening of a ring, as in the isomerization of cyclopropane, should be accompanied by a very high A-factor. The observed A-factor is $1.5 \cdot 10^{15}$ (Pritchard *et al.*, 1953). Conversely, the closing of a ring should be accompanied by a "low" A-factor, $\ll 10^{13}$ (For a review see Szwarc, 1950; Gowenlock, 1960).

3.1.2 *Bimolecular Reactions*

(i) *Collision Theory.* This theory (Lewis, 1918; Herzfeld, 1919; Polanyi, 1920) supposes that there are two requirements for bimolecular reaction. (a) The reagent molecules must approach to within a distance σ_{AB} of one another, where σ_{AB} is the mean of the gas kinetic collision diameters, σ_{AA} and σ_{BB}, in the pure gases. The rate of this binary collision process in a reaction A + BC is given by

$$Z'_{A,BC} = p(A)\,(BC)\,Z \tag{76}$$

$$= p(A)\,(BC)\,\sigma^2_{AB}\,(8\pi RT/\mu)^{1/2} \text{ liter mole}^{-1}\text{ sec}^{-1}. \tag{77}$$

where p is a symmetry factor which is unity for the present case, but would be 0.5 if the reagent molecules were identical; (A) and (BC) are concentrations in moles per liter; μ is the reduced mass $M_A M_{BC}/M_A + M_{BC}$, where M_A and M_{BC} are masses of 1 mole of A and BC; Z is the "collision number," that is to say, the collision frequency at unit concentration of unlike molecules.

(b) the second requirement is that the colliding species have a relative translational energy along the line of centers, in excess of E. Expression (75) gives the probability that a molecule will have energy greater than E in v vibrations, that is to say, in $2v$ "square terms" (oscillation involves two energy "square terms": kinetic and potential). Two degrees of freedom of translation, one belonging to each of two approaching molecules, also constitute two square terms. On writing $v = 1$ in (75) we have $\exp(-E/RT)$ for the probability that the colliding species have a relative translational energy along the line of centers in excess of E. [This result is correct, but the derivation is not rigorous. See Fowler and Guggenheim (1939) for a direct derivation.]

Combining factors (a) and (b) we have for the rate of bimolecular reaction, according to collision theory,

$$\text{Rate} = p(A)\,(BC)\,Z\exp(-E/RT) \tag{78}$$

and the rate constant,

$$k_2 = pZ\exp(-E/RT). \tag{79}$$

TABLE II

"NORMAL" A-FACTORS OBTAINED BY EXPERIMENT

Order, *n*:	1	2	3
A-factor	10^{13}-10^{14} sec^{-1}	$10^{10.5}$-$10^{11.5}$ $\frac{\text{liter}}{\text{mole sec}}$	10^{9}-10^{10} $\frac{\text{liter}^2}{\text{mole}^2 \text{ sec}}$

The Arrhenius A-factor is seen to be given by pZ. The A-factor should, therefore, increase slowly with temperature, $\alpha T^{1/2}$ [see (77)]. The values obtained for A at normal temperatures, using reasonable values for σ_{AB} and μ, are in the range $10^{10.5}$-$10^{11.5}$ liter mole^{-1} sec^{-1}. This accords well with the A-factors normally obtained by experiment (Table II). A-factors several powers of 10 lower than $10^{10.5}$ liter mole^{-1} sec^{-1} are, however, quite usual for more complex bimolecular reactions [see Steacie and Szwarc (1951) for a compilation, and also Table IV]. To take account of this, an empirical "steric factor," P, is frequently placed ahead of the collision number,

$$k = PpZ \exp(-E/RT). \tag{80}$$

P cannot be calculated by the collision theory. However, it can easily be shown (Laidler, 1950) that P should be related to the entropy change on going to the activated state. It is the great strength of the transition state theory, that follows, that it permits calculation of P along these lines.

(ii) *Transition State Theory.* Unlike the collision theory, this theory attempts to take into account all the internal motions of the reacting molecules in their initial state and in the activated state. The treatment involves two separate calculations. First, it is necessary to calculate a potential energy surface for the system, that is to say, the potential energies of the reacting species as a function of internuclear separation. It is supposed that the system passes across the energy barrier separating reagent configuration from product configuration at its lowest point. The region in configuration space corresponding to this low pass in the energy barrier is called the "transition state." A system in the transition state we will refer to as an "activated complex." In the second part of the calculation the concentration of activated complexes is calculated by statistical or thermodynamic argument, assuming an equilibrium between reactant molecules and activated complexes. By a simple device the activated complex concentration is related to the rate of passage across the barrier from reagents to products, that is to say, the rate of reaction (Pelzer and Wigner, 1932; Eyring, 1935; Evans and Polanyi, 1935).

The potential-energy surface for a reaction $A + BC \rightarrow AB + C$ can be plotted in three dimensions if it is supposed that the atoms are always in a straight line. London (1928, 1929) showed that the energy of a linear arrangement of three atoms is given by

$$E = A' + B' + C' + \{\tfrac{1}{2}[(\alpha - \beta)^2 + (\beta - \gamma)^2 + (\alpha - \gamma)^2]\}^{1/2} \tag{81}$$

where A′, B′, and C′ are "Coulombic" energies, and α, β, and γ are "exchange" energies, all of which are functions of internuclear separation. If the integrals A′, B′, C′, α, β, and γ were calculated, a poor estimate would be obtained of E, owing to the approximate nature of London's equation (Coolidge and James, 1934). To circumvent this difficulty, Eyring and Polanyi (1931) looked for a semiempirical solution. They argued that the sums, $A' + \alpha$, $B' + \beta$, and $C' + \gamma$, can be equated to the experimentally determined potential energies of the diatomic molecules A_2, B_2, and C_2, expressed in the form of Morse functions. This left an adjustable parameter—the ratio of Coulombic energy to exchange energy. On using a plausible fraction of $\sim 10\%$ for this ratio (based on the equations of Sugiura, 1927) Eyring and Polanyi obtained an activation energy of 15 kcal/mole for the reaction $H + H_2 \rightarrow H_2 + H$ (experimental value 7.5 ± 1, see § 2.1.2). Having regard to the method of calculation, this semiempirical activation energy has more qualitative than quantitative interest. Through its links with quantum mechanics the method of calculation was able for the first time to provide an explanation of the low activation energies of exchange reactions, in comparison with the energy of the bond that was broken (7.5 kcal as against 103 kcal, in the case under discussion).

More recently a number of purely quantum-mechanical calculations of the H_3 surface have been made by variational calculations (Hirschfelder *et al.*, 1936a; Ransil, 1957; Kimball and Trulio, 1958; Boys and Shavitt, 1959). Boys and Shavitt calculated, *a priori*, an activation energy of 15.4 kcal/mole. Their calculation employed six 1*s* orbitals, two of them, with different screening parameters, on each of the three hydrogen atoms, with the additional variation of an over-all scaling factor. It is probable that a very much more elaborate treatment would be required to improve significantly on their value for the activation energy.

The *statistical calculation of absolute rate* begins (in the formulation of Eyring, 1935) with the allocation to the activated complex of an arbitrary length δ in the reaction coordinate of the transition state region. (The reaction coordinate is defined by the path of minimal energy across the barrier.) The rate of reaction is then $c^{\ddagger}\bar{v}/\delta$, where $c^{\ddagger}$ is the concentration of activated complexes, and $\bar{v}$ is their mean velocity along the same coordinate to which δ refers. On substituting the Maxwell-Boltzmann expression for $\bar{v}$, and taking note of the fact that reaction rate $= k_r c_A c_{BC}$,

$$k_r = \frac{c^{\ddagger}}{c_A \cdot c_{BC}} \left(\frac{kT}{2\pi m^{\ddagger}}\right)^{1/2} \frac{1}{\delta} \tag{82}$$

where $m^{\ddagger}$ is the effective mass of the activated complex along the reaction coordinate. On the equilibrium hypothesis the concentration ratio in (82) is given by a ratio of partition functions. In these partition functions energy appears in exponential terms; by taking out $\exp(-E_0/RT)$, where E_0 is the difference in the zero level energy (per mole) between reagents and activated complex, all partition functions, Q, can be referred to the same zero of energy;

$$k_r = \frac{Q^{\ddagger}_{ABC}}{Q_A \cdot Q_{BC}} \exp(-E_0/RT) \left(\frac{kT}{2\pi m^{\ddagger}}\right)^{1/2} \frac{1}{\delta}. \tag{82a}$$

It is convenient to use instead of $Q^{\ddagger}_{ABC}$, which is the complete partition function for the activated complex, Q^{*}_{ABC} from which has been abstracted the partition function for the translational degree of freedom along the coordinate of decomposition; $(2\pi m^{\ddagger}\, kT)^{1/2}\, \delta/h$. Equation (82a) becomes

$$k = \frac{kT}{h} \frac{Q^{*}_{ABC}}{Q_A \cdot Q_{BC}} \exp(-E_0/RT). \tag{83}$$

kT/h has the dimensions of frequency and represents the frequency with which *any* activated complex crosses a barrier at temperature T ($\sim 6.10^{12}$ at 300°K; of the same order of magnitude as vibrational frequency). In some cases it is necessary to introduce a transmission coefficient, κ, and also a correction for quantum-mechanical tunneling, ahead of the terms on the right-hand side of (83) (Glasstone *et al.*, 1941).

The product $(kT/h)\,(Q^{*}_{ABC}/Q_A \cdot Q_{BC})$ corresponds to the Arrhenius A-factor of the reaction A + BC. The temperature dependence of this A-factor will depend on the spacing among vibrational levels (particularly the bending vibrations in $ABC^{\ddagger}$, since these are closely spaced) and rotational levels of reagents and activated complex. The temperature dependence for A + BC will range from $T^{-1/2}$ to $T^{1/2}$ (see Frost and Pearson, 1953).

Expression (83) is, of course, applicable to other reactions than the A + BC type, if the appropriate partition functions are employed.

The calculation of Q^{*}_{ABC} involves a knowledge of the force constants for the stretching and bending modes of vibration in the activated complex and a knowledge of the internuclear separations. These are taken from the potential energy surface. In the semiempirical calculations of Hirschfelder, Eyring, and Topley on the system $H + H_2$, the percentage of Coulombic energy was adjusted to give a surface with a barrier height in accord with the experimental activation energy. This method led to a highly asymmetric activated complex, ($r_{AB} \approx 2r_{BC}$). The purely quantum-mechanical calculations, on the other hand, invariably lead to a symmetrical activated complex ($r_{AB} = r_{BC}$). It is

TABLE III

Observed Rate Constants (1000°K) Compared with Rate Constants Calculated by the Collision Theory and by the Theory of Absolute Reaction Rates[a]

Reaction	Observed	Calculated		
		Collision theory	Absolute theory	
			Semi-empirical	Purely quantum mechanical
$H + H_2 \xrightarrow{1} H_2 + H$	$k_1 = 1.00$[b]	1.00[d]	1.00[e]	1.00[f]
$D + D_2 \xrightarrow{2} D_2 + D$	$k_2 = 0.52$[b]	0.71[d]	0.51[e]	0.48[f]
$D + H_2 \underset{5}{\overset{3}{\rightleftharpoons}} DH + H$	$k_3/k_5 = 2.4$[c]	1.75[d]	1.78[e]	2.5[f]
$H + D_2 \underset{6}{\overset{4}{\rightleftharpoons}} HD + D$	$k_4/k_6 = 1.35$[c]	2.4[d]	0.99[e]	1.43[f]

[a] All constants expressed relative $k_1 = 1$.
[b] Farkas and Farkas (1935).
[c] Boato *et al.* (1956)
[d] Van Meersche (1951).
[e] Hirschfelder *et al.* (1936b).
[f] Shavitt (1959).

TABLE IV

A-Factors (400°K) for Hydrogen Abstraction Reactions[a]

Reaction	A_{calc} $\frac{\text{liter}}{\text{mole sec}}$	A_{obs} $\frac{\text{liter}}{\text{mole sec}}$
$H + H_2$	$6.0 \cdot 10^{10}$[b]	$5.4 \cdot 10^{10}$[d]
$CH_3 + H_2$	$1.9 \cdot 10^{9}$[b]	$1.6 \cdot 10^{9}$[e]
$CF_3 + H_2$	$3.8 \cdot 10^{8}$[c]	$3.2 \cdot 10^{8}$[e,f]

[a] Calculated values from absolute rate theory.
[b] Polanyi (1955, 1956); see also Bywater and Roberts (1952).
[c] Ayscough and Polanyi (1956).
[d] Boato *et al.* (1956).
[e] Majury and Steacie (1953); Whittle and Steacie (1953).
[f] Pritchard *et al.* (1956).

surprising that when (83) is applied to either the symmetric or the asymmetric surface (Table III), a good correlation is obtained between experimental and theoretical rate constants in the isotopic series $H + H_2$, $D + D_2$, etc. (cf. the isotopic series $CH_3 + H_2$, $CD_3 + D_2$, etc.; Polanyi, 1955).

That the transition state theory represents a very real improvement over the collision theory can be seen from the results presented in Table IV (Ayscough and Polanyi, 1956). On using a collision diameter of 4.0*A* for CF_3 and 2.8*A* for H_2, the collision theory predicts an *A*-factor of 4.6 10^{11} at 400°K for the reaction $CF_3 + H_2 \rightarrow CF_3H + H$. The transition state theory (using a three-dimensional model for the activated complex) leads to an *A*-factor smaller by a factor of 10^{-3}—in good agreement with experiment.

For other studies on these lines, see Knox and Trotman-Dickenson (1956) and Herschbach *et al.* (1956).

The *thermodynamic formulation of absolute rate theory* has its origin in the equilibrium $A + BC \rightleftharpoons ABC^{\ddagger}$ [cf. (82)] for which we can write an equilibrium constant $K^{\ddagger} = \exp(-\Delta F^{\ddagger}/RT) = \exp(\Delta S^{\ddagger}/R) \exp(-\Delta H^{\ddagger}/RT)$. It can be shown that

$$k_r = \frac{kT}{h} \exp(\Delta S^*/R) \exp(-\Delta H^{\ddagger}/RT) \tag{84}$$

where ΔS^* differs from $\Delta S^{\ddagger}$, the entropy change on activation, by a small amount owing to the fact that ΔS^* lacks an entropy term for the degree of freedom corresponding to dissociation. kT/h once again represents the frequency of dissociation.

A variant on the semiempirical method of plotting potential energy surfaces, is the *energy profile* method (Ogg and Polanyi, 1935; Evans and Polanyi, 1938). This method expresses the total energy in terms of the strengths of the bonds being broken and formed, and also the repulsions between reagents and between products. A relation between heats of reactions and activation energies arises out of the treatment;

$$E = \alpha \Delta H + c.$$

α was calculated to be ≈ 0.3 (Butler and Polanyi, 1943). The validity of the relationship is still in doubt (see Trotman-Dickenson, 1955, 1959; Dodd, 1957).

3.1.3 *Termolecular Reactions*

(i) *Collision Theory*. Bodenstein (1922) suggested that the ratio of bimolecular to termolecular collisions should be the ratio of the duration of a binary collision to the time required to traverse a mean free path, which is roughly the ratio of molecular diameter to the length of a mean free path. At 1-atm pressure this is 10^{-3}. Tolman (1927) has given a more accurate method of calculation based on a hard-sphere model of the colliding atoms,

$$k_3 = pZ_3 \exp(-E/RT)$$

where Z_3, the termolecular collision number (collision rate at unit concentrations), is given by

$$Z_3 = 8\sqrt{2}\,\pi^{3/2}\,\sigma_{AB}^2\,\sigma_{BM}^2\,d\sqrt{kT}\left(\frac{1}{\sqrt{\mu_{AB}}} + \frac{1}{\sqrt{\mu_{BM}}}\right). \tag{86}$$

Here σ_{AB} and σ_{BM} are mean collision diameters [see (77)] in the termolecular reaction $A + B + M \rightarrow AB + M$; d is the distance that an atom can move while still remaining within interaction range of another (this determines the time interval during which A and B constitute a pair, and can be "successfully" approached by M); μ_{AB} and μ_{BM} are reduced masses.

(ii) *Transition State Theory*. We can apply a relation analogous to (83).

$$k_r = \frac{kT}{h}\frac{Q^*_{ABM}}{Q_A \cdot Q_B \cdot Q_M}\exp(-E_0/RT). \tag{87}$$

A consideration of the actual partition functions, when A, B, and M are atoms (Q_A, Q_B, and Q_M, all translational), leads to the result that (87) is identical, in this case, with the collision theory expression (Benson, 1960a).

More sophisticated statistical calculations of the rate of termolecular association reactions ($A + B + M \rightarrow AB + M$) have been made by a variational method devised by Wigner (1937). By use of "trial" potential-energy surfaces, a surface can be chosen which yields an upper limit for the reaction rate (Keck, 1958, 1960). As it turns out this agrees quite well with the experimental rate.

Recently a quantum-mechanical calculation of the rate of the recombination reaction $O + O + O_2$ has been successfully carried out (Bauer and Salkoff, 1960). The calculation, by a "perturbed-stationary-state" method (Salkoff and Bauer, 1958), was applied to a model in which a colinear collision was presumed to take place with resultant transfer of the newly formed O_2 into a vibrational state near its dissociation limit. The calculated rate constant agreed to within a factor of three with the observed value.

Two theories have been advanced to account for the decrease in the rates of association reactions at increased temperatures (described by a negative activation energy, $E_a \approx -1.5 \pm 1$ kcal/mole). The first theory (Christie *et al.*, 1953) suggests that the association goes by way of $A + M \rightarrow AM$ followed by $AM + B \rightarrow AB + M$. At increased temperature AM has a shorter life and consequently $AM + B$ becomes less probable. This argument has been offered only as a qualitative explanation.

The second view (Polanyi, 1959) is that $A + B + M$ leads to the formation of highly vibrating $AB^{\dagger}$ close to dissociation. This $AB^{\dagger}$ is for the most part stabilized at subsequent collisions. However, a certain fraction redissociates. The fraction redissociating is greater at higher temperatures. This argument leads to negative activation energies of the right order of magnitude.

3.2 Theory of the Dependence of Rates on Kinematic Factors

The theories outlined in the foregoing embody assumptions concerning the kinematics of reaction. Eliason and Hirschfelder (1959) have shown what restrictions must be applied to a generalized collision theory in order to make it equivalent to the transition state theory. An even more general collision approach, based on the quantum-statistical mechanical theory of irreversible processes, has been developed by Yamamoto (1960). One important limitation of the transition state theory is evident from an examination of (83); the partition functions in this expression are only relevant if translation, vibration, and rotation are thermally equilibrated in the reagents and in the activated complex.

It is a common kinematic assumption (see, however, Careri, 1953) that dissociation of a molecule AB will occur more rapidly from a highly vibrating state. Rates of thermal dissociation have been calculated on the basis of a model in which the molecule climbs a "vibrational ladder" by a random walk, to reach a particular level from which dissociation can occur (Montroll and Shuler, 1958; Shuler, 1959; Kim, 1958; Nikitin, 1958; Widom, 1958, 1959). By using simple models, calculations have been made of relative rates of dissociation from various vibrational levels in the neighborhood of dissociation (Pritchard, 1961). These calculations confirm, in an approximate but quantitative fashion, the general expectation that the collision efficiency for dissociation will increase rapidly as the level of vibrational excitation is increased toward the dissociation limit.

Since the reaction to which these calculations are applied is the "quasi-unimolecular" process $AB + M \rightarrow A + B + M$, the calculations throw some light on the termolecular association which is the reverse reaction. By the principle of microscopic reversibility if the rate of dissociation is greatest from the high vibrational levels of AB then the rate of association should be greatest into these levels. As already remarked, this may account for the "negative activation energies" of association reactions (§ 3.1.3). Moreover, the view that recombination forms highly vibrating product receives additional theoretical support from the fact that it forms the basis of a successful quantum-mechanical calculation of the rate of association (§ 3.1.3).

Various models have been proposed for the calculation of energy distribution among the products of exothermic exchange reactions, $A + BC$, (or, conversely the most favorable energy distribution among the reagents for the reverse, endothermic, reaction). Though the problem is essentially quantum mechanical, some insight can be obtained from classical arguments. If quantization of vibration is ignored it becomes possible to represent the exchanges of internal energy within a reacting system by the motion of a sliding point mass across a potential-energy surface with skewed coordinates (Eyring and Polanyi, 1931; Wall *et al.*, 1958). Alternatively, a simple valence bond description of the activated complex $ABC^{\ddagger}$, combined with evidence concerning the efficiency of transfer of vibrational energy at a collision, leads to the result that the major part of the heat of reaction will appear as vibration in the new molecule AB (Polanyi, 1959). This is in accord with the result obtained classically by Smith (1959).

The quantum theory of reaction rates has been developed formally by Golden (1949), Bauer and Wu (1953), Mazur and Rubin (1959), and Bauer (1958). The limitation on quantitative calculation lies in the inadequacy of the quantum-mechanical calculations of potential energy surfaces. Golden and Peiser (1949) have applied a quantum-mechanical formulation of the reaction-rate problem to reaction across a semi-empirical (Eyring-Polanyi) potential energy surface. The reaction studied was the endothermic process $Br + H_2 \rightarrow HBr + H$. It was found that 95% of the rate came from hydrogen molecules in the first excited vibrational state. The "rotational temperature" of the newly formed HBr was predicted to be roughly one-half the initial H_2 temperature. This, of course, implies that the exothermic reaction, $H + HBr$, should result in the formation of vibrationally and rotationally excited H_2. The result gains added significance from the fact that it is found to be insensitive to changes in the percentage Coulombic energy employed in constructing the semiempirical energy surface.

REFERENCES

Arrhenius, S. (1889) *Z. phys. Chem.* **4**, 226.
Ayscough, P. B., and Polanyi, J. C. (1956) *Trans. Faraday Soc.* **52**, 960.
Basco, N., and Norrish, R. G. W. (1960) *Canad. J. Chem.* **38**, 1769.
Bauer, E. (1958) New York University Inst. Math. Sci. Research Report CX-33.
Bauer, E., and Salkoff, M. (1960) *J. chem. Phys.* **33**, 1202.
Bauer, E., and Wu, T.-Y. (1953) *J. chem. Phys.* **21**, 726, 2072.
Bawn, C. E. H. (1942) *Ann. Rep. Chem. Soc.* **39**, 36.
Benson, S. W. (1952) *J. chem. Phys.* **20**, 1605.
Benson, S. W. (1960a) "The Foundations of Chemical Kinetics." McGraw-Hill, New York.
Benson, S. W., (1960b) *J. chem. Phys.* **33**, 939.
Benson, S. W., and Axworthy, A. E. (1957) *J. chem. Phys.* **26**, 1718.
Benson, S. W., and Srinivasan, R. (1955) *J. chem. Phys.* **23**, 200.
Berlie, M. R., and LeRoy, D. J. (1952) *J. chem. Phys.* **20**, 200.
Beutler, H., and Polanyi, M. (1925) *Naturwissenschaften* **13**, 711.
Boato, G., Careri, G., Cimino, A., Molinari, E., and Volpi, G. G. (1956) *J. chem. Phys.* **24**, 783.
Bodenstein, M. (1899) *Z. phys. Chem.* **29**, 295.
Bodenstein, M. (1922) *Z. phys. Chem.* **100**, 118.
Bodenstein, M., and Lind, S. C. (1906) *Z. phys. Chem.* **57**, 168.
Boys, S. F., and Shavitt, I. (1959) University of Wisconsin Naval Research Lab. Tech. Report WIS-AF-13.
Bunker, D. L., and Davidson, L. (1958) *J. Amer. Chem. Soc.* **80**, 5085.
Burnett, R. le G., and Bell, R. P. (1938) *Trans. Faraday Soc.* **34**, 420.
Burns, G., and Hornig, D. F. (1960) *Canad. J. Chem.* **38**, 1702.
Butler, E. T., and Polanyi, M. (1943) *Trans. Faraday Soc.* **39**, 19.
Bywater, S., and Roberts, R. (1952) *Canad. J. Chem.* **30**, 773.
Careri, G. (1953) *J. chem. Phys.* **21**, 749.
Cashion, J. K., and Polanyi, J. C. (1958) *J. chem. Phys.* **29**, 455.
Cashion, J. K., and Polanyi, J. C. (1959a) *J. chem. Phys.* **30**, 317.
Cashion, J. K., and Polanyi, J. C. (1959b) *J. chem. Phys.* **30**, 1097.
Cashion, J. K., and Polanyi, J. C. (1960a) *Proc. Roy. Soc.* **A258**, 529.
Cashion, J. K., and Polanyi, J. C. (1960b) *Proc. Roy. Soc.* **A258**, 570.
Cashion, J. K., and Polanyi, J. C. (1961) *J. chem. Phys.* **35**, 600.
Cawthorn, T. M., and McKinley, J. D. (1956) *J. chem. Phys.* **25**, 583.
Charters, P. E., and Polanyi, J. C. (1960) *Canad. J. Chem.* **38**, 1742.
Charters, P. E., and Polanyi, J. C. (1962) *Disc. Farad. Soc.* **33**.
Christie, M. I., Norrish, R. G. W., and Porter, G. (1953) *Proc. Roy. Soc.* **A216**, 152.
Christie, M. I., Harrison, A. J., Norrish, R. G. W., and Porter, G. (1955) *Proc. Roy. Soc.* **A231**, 446.
Christiansen, J. A. (1919) *K. Danske Vidensk. Selsk. fys. Medd.* **1**, 14.
Christiansen, J. A. (1921) Reaktion Kinetiske Studier, p. 50. Thesis, Copenhagen.
Cimino, A., Molinari, E., and Volpi, G. G. (1956) *Gazz. chim. ital.* **86**, 609.
Cimino, A., Molinari, E., and Volpi, G. G. (1960) *J. chem. Phys.* **33**, 616.
Coolidge, A. S., and James, H. M. (1934) *J. chem. Phys.* **2**, 811.
Cvetanovič, R. J., and LeRoy, D. J. (1952) *J. chem. Phys.* **20**, 1016; (1951) *Canad. J. Chem.* **29**, 597.
Dodd, R. E. (1957) *J. chem. Phys.* **26**, 1353.

Dodd, R. E., and Steacie, E. W. R. (1954) *Proc. Roy. Soc.* **A223**, 283.
Elias, L., and Schiff, H. I. (1960) *Canad. J. Chem.* **38**, 1657.
Eliason, M. A., and Hirschfelder, J. O. (1959) *J. chem. Phys.* **30**, 1426.
Engleman, R., Jr., and Davidson, N. R. (1960) *J. Amer. Chem. Soc.* **82**, 4770.
Evans, M. G., and Polanyi, M. (1935) *Trans. Faraday Soc.* **31**, 875.
Evans, M. G., and Polanyi, M. (1938) *Trans. Faraday Soc.* **34**, 11.
Evans, M. G., and Polanyi, M. (1939) *Trans. Faraday Soc.* **34**, 178.
Eyring, H. (1935) *J. chem. Phys.* **3**, 107.
Eyring, H., and Polanyi, M. (1931) *Z. phys. Chem.* **B12**, 279.
Farkas, A. (1930) *Z. Elektrochem.* **36**, 782.
Farkas, A. (1931) *Z. phys. Chem.* (*Leipzig*) **B10**, 419; **B14**, 371.
Farkas, A., and Farkas, L. (1935) *Proc. Roy. Soc.* **A152**, 124.
Farkas, A., and Melville, H. W. (1939) "Experimental Methods in Gas Reactions." Macmillan, New York.
Fettis, G. E., Knox, J. H., and Trotman-Dickenson, A. F. (1960) *Canad. J. Chem.* **38**, 1643.
Fowler, R. H. (1936) "Statistical Mechanics." Cambridge Univ. Press, London and New York.
Fowler, R. H., and Guggenheim, E. A. (1939) "Statistical Thermodynamics." Cambridge Univ. Press, London and New York.
Frommer, L., and Polanyi, M. (1934) *Trans. Faraday Soc.* **30**, 519.
Frost, A. A., and Pearson, R. G. (1953) "Kinetics and Mechanism." Wiley, New York.
Garvin, D. (1954) *J. Amer. Chem. Soc.* **76**, 1523.
Garvin, D. (1959) *J. Amer. Chem. Soc.* **81**, 3173.
Garvin, D., and Kistiakowsky, G. B. (1952) *J. chem. Phys.* **20**, 105.
Garvin, D., Broida, H. P., and Kostkowski, H. J. (1960a) *J. chem. Phys.* **32**, 880.
Garvin, D., Gwyn, P. P., and Moskowitz, J. W. (1960b) *Canad. J. Chem.* **38**, 1795.
Geib, K., and Harteck, P. (1931) *Z. phys. Chem. Bodenstein Festband*, p. 849.
Giguère, P. A., and Liu, I. D. (1957) *Canad. J. Chem.* **35**, 283.
Gill, E. K., and Laidler, K. J. (1958) *Canad. J. Chem.* **36**, 1570.
Gill, E. K., and Laidler, K. J. (1959a) *Trans. Faraday Soc.* **55**, 753.
Gill, E. K., and Laidler, K. J. (1959b) *Proc. Roy. Soc.* **A250**, 121.
Gill, E. K., and Laidler, K. J. (1959c) *Proc. Roy. Soc.* **A251**, 66.
Glasstone, S., Laidler, K. J., and Eyring, H. (1941) "The Theory of Rate Processes." McGraw-Hill, New York.
Golden, J. A., and Myerson, A. L. (1958) *J. chem. Phys.* **28**, 978.
Golden, S. (1949) *J. chem. Phys.* **17**, 620.
Golden, S., and Peiser, A. M. (1949) *J. chem. Phys.* **17**, 630.
Gowenlock, B. G. (1960) *Quart. Revs.* (*London*) **14**, 133.
Greene, E. F., Roberts, R. W., and Ross, J. (1960) *J. chem. Phys.* **32**, 940.
Haresnape, J. N., Stevels, J. M., and Warhurst, E. (1940) *Trans. Faraday Soc.* **36**, 465.
Heller, W. (1937) *Trans. Faraday Soc.* **33**, 1556.
Herschbach, D. R. (1960) *J. chem. Phys.* **33**, 1870.
Herschbach, D. R., Johnston, H. S., Pitzer, K. S., and Powell, R. E. (1956) *J. chem. Phys.* **25**, 736.
Herschbach, D. R., Kwei, G. H., and Norris, J. A. (1961) *J. chem. Phys.* **34**, 1842.
Herzfeld, K. F. (1919) *Z. Elektrochem.* **25**, 301.
Hinshelwood, C. N. (1927) *Proc. Roy. Soc.* **A113**, 230.
Hirschfelder, J. O., Eyring, H., and Rosen, N. (1936a) *J. chem. Phys.* **4**, 121.
Hirschfelder, J. O., Eyring, H., and Topley, B. (1936b) *J. chem. Phys.* **4**, 170.

Hoare, D. E., and Walsh, A. D. (1957) *Trans. Faraday Soc.* **53**, 1102.
Howlett, K. E. (1952) *J. chem. Soc.* p. 3695.
Jahn, S. (1906) *Z. anorg. Chem.* **48**, 260.
Johnston, H. S. (1951) *J. chem. Phys.* **19**, 663.
Johnston, H. S., and Perrine, R. L. (1951) *J. Amer. Chem Soc.* **73**, 4728.
Kassel, L. S. (1928) *J. phys. Chem.* **32**. 225.
Kassel, L. S. (1930) *Proc. Nat. Acad. Sci. U.S.A.* **16**, 358.
Kassel, L. S. (1932) "Kinetics of Homogeneous Gas Reactions," Chapt. 5. Reinhold, New York.
Keck, J. C. (1958) *J. chem. Phys.* **29**, 410.
Keck, J. C. (1960) *J. chem. Phys.* **32**, 1035.
Kern, F., and Walters, W. D. (1952) *Proc. Nat. Acad. Sci. U.S.A.* **43**, 937.
Kim, S. K. (1958) *J. chem. Phys.* **28**, 1057.
Kimball, G. E., and Trulio, J. G. (1958) *J. chem. Phys.* **28**, 493.
Kinsey, J. L., Kwei, G. H., and Herschbach, D. R. (1961) *Amer. Phys. Soc. Meeting Monterey, California.*
Kistiakowsky, G. B. (1928) *J. Amer. Chem. Soc.* **50**, 2315.
Klemperer, W., and Herschbach, D. R. (1957) *Proc. Nat. Acad. Sci. U.S.A.* **43**, 429.
Knox, J. H., and Trotman-Dickenson, A. F. (1956) *J. phys. Chem.* **60**, 1367.
Kraus, F. (1957) *Z. Naturforsch.* **12a**, 479.
Kwei, G. H., Norris, J. A., and Herschbach, D. R. (1960) *Bull. Amer. Phys. Soc.* [2] **5**, 503.
Laidler, K. J. (1950) "Chemical Kinetics." McGraw-Hill, New York.
Lewis, W. C. McC. (1918) *J. chem. Soc.* **113**, 471.
Lindemann, F. A. (1922) *Trans. Faraday Soc.* **17**, 598.
Lipscomb, F. J., Norrish, R. G. W., and Thrush, B. A. (1956) *Proc. Roy. Soc.* **A233**, 455.
London, F. (1928) "Probleme der modernen Physik" (Sommerfeld Festschrift), p. 104. S. Hirzel, Leipzig.
London, F. (1929) *Z. Elektrochem.* **35**, 552.
Luce, R. G., and Trischka, J. W. (1953) *J. chem. Phys.* **21**, 105.
McGrath, W. D., and Norrish, R. G. W. (1957) *Proc. Roy. Soc.* **A242**, 265.
McGrath, W. D., and Norrish, R. G. W. (1958) *Z. phys. Chem. (Frankfurt)* [N. S.] **15**, 245.
McGrath, W. D., and Norrish, R. G. W. (1960) *Proc. Roy. Soc.* **A254**, 317.
McKinley, J. D., Garvin, D., and Boudart, M. J. (1955) *J. chem. Phys.* **23**, 784.
Magee, E. M. (1957) *J. Amer. Chem. Soc.* **79**, 5375.
Majury, T. G., and Steacie, E. W. R. (1953) *Disc. Faraday Soc.* **14**, 45.
Marple, D. T. F., and Trischka, J. W. (1956) *Phys. Rev.* **103**, 597.
Marshall, R., and Davidson, N. R. (1953) *J. chem. Phys.* **21**, 659.
Mazur, J., and Rubin, R. J. (1959) *J. chem. Phys.* **31**, 1395.
Melville, H. W., and Robb, J. C. (1949) *Proc. Roy. Soc.* **A196**, 445.
Mercer, P. D., and Pritchard, H. O. (1959) *J. phys. Chem.* **63**, 1468.
Mills, R. L., and Johnston, H. S. (1951) *J. Amer. Chem. Soc.* **73**, 938.
Montroll, E. W., and Shuler, K. E. (1958) *Advances in Chem. Phys.* **1**, 361.
Moran, T. I., and Trischka, J. W. (1961) *J. chem. Phys.* **34**, 923.
Morgan, J. E., Elias, L., and Schiff, H. I. (1960) *J. chem. Phys.* **33**, 930.
Nikitin, E. E. (1958) *Dokl. Akad. Nauk. SSSR* **121**, 991.
Norrish, R. G. W., and Porter, G. (1949) *Nature* **164**, 658.
Norrish, R. G. W., and Thrush, B. A. (1956) *Quart. Revs. (London)* **10**, 149.
Ogg, R. A., and Polanyi, M. (1935) *Trans. Faraday Soc.* **31**, 604, 1375.

Pelzer, H., and Wigner, E. P. (1932) *Z. phys. Chem.* **B15**, 445.
Perrin, J. (1919) *Ann. Phys.* (*Paris*) **11**, 5.
Polanyi, J. C. (1955) *J. chem. Phys.* **23**, 1505.
Polanyi, J. C. (1956) *J. chem. Phys.* **24**, 493.
Polanyi, J. C. (1959) *J. chem. Phys.* **31**, 1338.
Polanyi, J. C., and Sadowski, C. M. (1962) *J. chem. Phys.* In press.
Polanyi, M. (1920) *Z. Elektrochem.* **26**, 49, 228.
Polanyi, M. (1932) "Atomic Reactions." Williams and Norgate, London.
Polanyi, M., and Wigner, E. P. (1928) *Z. phys. Chem.* **A139**, 439.
Porter, G. (1950) *Proc. Roy. Soc.* **A200**, 284.
Porter, G., and Smith, J. A. (1959) *Nature* **184**, 446.
Pritchard, G. O., Pritchard, H. O., Schiff, H. I., and Trotman-Dickenson, A. F. (1956) *Trans. Faraday Soc.* **52**, 849.
Pritchard, H. O. (1961) *J. phys. Chem.* **65**, 504.
Pritchard, H. O., Sowden, R. G., and Trotman-Dickenson, A. F. (1953) *Proc. Roy. Soc.* **A217**, 563.
Pritchard, H. O., Pyke, J. B., and Trotman-Dickenson, A. F. (1955) *J. Amer. Chem. Soc.* **77**, 2629.
Rabinowitch, E., and Wood, W. C. (1936a) *Trans. Faraday Soc.* **32**, 907.
Rabinowitch, E., and Wood, W. C. (1936b) *J. chem. Phys.* **4**, 497.
Ramsperger, H. C. (1927) *J. Amer. Chem. Soc.* **49**, 912, 1495.
Ransil, B. J. (1957) *J. chem. Phys.* **26**, 971.
Reed, J. F., and Rabinovitch, B. S. (1955) *J. phys. Chem.* **59**, 261.
Rice, O. K., and Ramsperger, H. C. (1927) *J. Amer. Chem. Soc.* **49**, 1617.
Rice, O. K., and Ramsperger, H. C. (1928) *J. Amer. Chem. Soc.* **50**, 617.
Rice, O. K., and Sickman, D. V. (1936) *J. chem. Phys.* **4**, 242.
Russell, K. E., and Simons, J. (1953) *Proc. Roy. Soc.* **A217**, 271.
Salkoff, M., and Bauer, E. (1958) *J. chem. Phys.* **29**, 26.
Schumacher, H. J. (1960) *J. chem. Phys.* **33**, 938.
Semenov, N. N. (1939) *J. chem. Phys.* **7**, 683.
Semenov, N. N. (1958) "Some Problems in Chemical Kinetics and Reactivity." Princeton Univ. Press, Princeton, New Jersey.
Shavitt, I. (1959) *J. chem. Phys.* **31**, 1359.
Shuler, K. E. (1959) *J. chem. Phys.* **31**, 1375.
Slater, N. B. (1939) *Proc. Cambridge Phil. Soc.* **35**, 36.
Slater, N. B. (1948) *Proc. Roy. Soc.* **A194**, 112.
Slater, N. B. (1959) "Theory of Unimolecular Reactions." Methuen, London.
Smith, F. T. (1959) *J. chem. Phys.* **31**, 1352.
Smith, F. T., and Kistiakovsky, G. B. (1959) *J. chem. Phys.* **31**, 621.
Steacie, E. W. R. (1954) "Atomic and Free Radical Reactions." Reinhold, New York.
Steacie, E. W. R., and Szwarc, M. (1951) *J. chem. Phys.* **19**, 1309.
Steiner, W. (1935) *Trans. Faraday Soc.* **31**, 623.
Strong, R. L., Chien, J. C. W., Graf, P. E., and Willard, J. E. (1957) *J. chem. Phys.* **26**, 1287.
Sugiura, Y. (1927) *Z. Phys.* **45**, 484.
Sullivan, J. H. (1959) *J. chem. Phys.* **30**, 1292.
Szwarc, M. (1950) *Chem. Revs.* **47**, 75.
Taylor, E. H., and Datz, S. (1955) *J. chem. Phys.* **23**, 1711.
Tollefson, E. L., and LeRoy, D. J. (1948) *J. chem. Phys.* **16**, 1057.
Tolman, R. C. (1927) "Statistical Mechanics," p. 245. Chemical Catalogue Co., New York.

Trautz, M. (1918) *Z. anorg. Chem.* **102**, 81.
Trotman-Dickenson, A. F. (1955) "Gas Kinetics." Butterworths, London.
Trotman-Dickenson, A. F. (1959) "Free Radicals." Methuen Monograph. Wiley, New York.
Van Meersche, M. (1951) *Bull. Soc. Chim. Belg.* **60**, 99.
von Hartel, H., and Polanyi, M. (1930) *Z. phys. Chem.* **B11**, 97.
Wall, F. T., Hiller, L. A., and Mazur, J. (1958) *J. chem. Phys.* **29**, 255.
Warhurst, E. (1939) *Trans. Faraday Soc.* **35**, 674.
Warhurst, E. (1951) *Quart. Revs.* (*London*) **5**, 44.
Whittle, E., and Steacie, E. W. R. (1953) *J. chem. Phys.* **21**, 993.
Widom, B. (1958) *J. chem. Phys.* **28**, 918.
Widom, B. (1959) *J. chem. Phys.* **31**, 1387.
Wigner, E. P. (1937) *J. chem. Phys.* **5**, 720.
Wood, R. W. (1922) *Proc. Roy. Soc.* **A102**, 1.
Yamamoto, T. (1960) *J. chem. Phys.* **33**, 281.

AUTHOR INDEX

Numbers in italic show the page on which the complete reference is listed.

C

D

E

F

G

H

I

J

K

Q

R

T

U

V

W

Y

Z

SUBJECT INDEX

Species of Atom or Molecule

D

F

H

O

P

S

X

SUBJECT INDEX

General

D

E

F

G

H

I

J

K

L

M

N

O

P

Q

R

S

T

U

V